Developments in Geomathematics 1

GEOMATHEMATICS

Developments in Geomathematics 1

GEOMATHEMATICS

Mathematical Background and Geo-Science Applications

by

F.P. AGTERBERG

Head, Geomathematics Section, Geological Survey of Canada, Ottawa, Ontario, Canada

Elsevier Scientific Publishing Company

Amsterdam London New York 1974

Elsevier Scientific Publishing Company
335 Jan van Galenstraat
P.O. Box 211, Amsterdam, The Netherlands

American Elsevier Publishing Company, Inc.
52 Vanderbilt Avenue
New York, New York 10017

Library of Congress Card Number: 72–97418

ISBN: 0–444–41091–0

With 128 illustrations and 50 tables

Printed in The Netherlands

To Codien

FOREWORD

With few exceptions, geologists have been slow to make use of mathematics and statistics. It is an intriguing paradox that mathematics the oldest of the sciences is, when applied to studies of the earth in the form of geomathematics, the youngest of the geosciences.

It has been my privilege to be closely involved with the developing interest in applications of mathematics to geology, first on Canada's national committee on data processing and subsequently on the international Cogeodata committee. The great expansion of geophysics and geochemistry in the post-war years generated huge volumes of numerical data for whose processing and evaluation, mathematics (including statistics), proved to be essential. The same three decades have seen rapid development of computers that can now be programmed to integrate and process both numerical and descriptive data. Thus the stage is set for geomathematics to play a major role in providing a secure statistical basis for interpretation by geologists.

Undoubtedly a majority of today's geologists remain unconvinced that mathematics can be effective in a science dependent largely upon inductive reasoning. They feel, as somebody remarked, that many people use statistics the way the traditional inebriate used a lamppost – more for support than illumination. On the contrary it is my belief that geological data when ordered, collated, integrated, evaluated and tested by mathematical procedures, provide a more complete and objective basis for geological interpretation, classification, testing of concepts, solution of problems and prediction of probabilities, than a basis limited to any one individual's experience and memory. Moreover, such a basis may be used as common ground by many geologists each of whom adds his own experience and knowledge to reach his own interpretation. Such common ground could well be a major factor in advancing geology as a science.

Means must be found to encourage geologists to test the usefulness of mathematics to geology; particularly those in universities who are responsible for the training of the new generation. To stimulate this interest a rapport between geologists, geomathematicians and computer scientists must be achieved. Probably the first step is to demonstrate by means of case histories how useful geomathematics has proved to be in pioneer studies. The second step requires a review of powerful new mathematical procedures and an indication of their potential usefulness to geology. Finally it is essential that such presentations be made by geologists using terms that other geologists comprehend.

Geomathematics is becoming indispensable to the earth sciences as the huge volume and wide variety of observations and measurements increase. In fact geomathematics offers the only means of assimilating, assessing and reducing to comprehensible form the rapidly increasing flow of data from geophysical, geochemical and satellite sources and merging it with field observations.

S.C. ROBINSON

PREFACE

This book is intended for advanced geology students, research workers and teachers with an interest in using mathematical techniques for problem-solving. Advanced concepts from petrology, economic geology, sedimentology and structural geology are used in later chapters without detailed explanations. Competence in calculus and statistics will be required to read the entire book.

I assumed that the mathematical background of potential readers varies considerably. Some may have taken mathematics courses as a student, later forgetting most of the concepts because these are not generally used by geologists. The first part of the book consists of reviews of basic mathematical methods and these may serve as a set of brief refresher courses. Elementary mathematical concepts are illustrated by simple geological examples.

Readers who need further help in elementary or more advanced mathematical subjects will find lists of selected books for further reading at the end of the book.

The degree to which mathematical explanations are included in later chapters varies according to the topic under discussion. Methods of matrix algebra and least squares are discussed extensively with numerical examples or geological applications in most chapters. Other subjects including the theory of generating functions, stochastic processes, and harmonic analysis are introduced briefly before they are applied to data. The serious student then should consult the mathematical textbooks for a more comprehensive treatment and proofs. The purpose of including many subjects is to make available to earth scientists a large variety of mathematical techniques.

The reader will soon discover that, apart from the introductory chapters, the text consists mainly of case histories, often of considerable geological complexity. To a large extent, these were derived from my own research during the past ten years with the Geological Survey of Canada. Care was taken to provide each case history with an introduction and geological interpretation of results. My hope is to reach the geologist who is in a position where he may, with the aid of computers, apply mathematical techniques to his data but who also wishes to know more about the background of these techniques and to make a practical choice between alternatives.

I am indebted to many individuals and organizations for support. Since 1968 when I was contacted by the publishers about writing this book, F.W.B. van Eysinga of Elsevier has provided guidance and encouragement.

Thanks are due to Y.O. Fortier, Director, Geological Survey of Canada, for encouragement and approval of publication.

A number of chapters are in part based on notes prepared for three courses: (1) statistics in geology, a one-semester course for undergraduates, University of Ottawa (Chapters 6 and first parts of Chapters 7–10 and 14); (2) ore estimation for mining geologists, Continued Education Program, Queen's University, 1969 (Chapters 9–11 and parts of 15); (3) advanced introduction, Kansas Geological Survey, University of Kansas,

1970 (Chapters 9–13). This enabled me to test parts of the material on audiences with different mathematical backgrounds. Several chapters were prepared during 1969–1970, when I was a visiting research scientist with the Kansas Geological Survey.

I am grateful to S.C. Robinson with whom I prepared a paper on mathematical problems in geology, presented in 1971 to the 38th Session of the International Statistical Institute (Agterberg and Robinson, 1972). Many of the ideas expressed in that paper were incorporated in Chapter 1.

Invaluable help during preparation of the manuscript was provided by C.F. Chung and A.G. Fabbri; for Chapters 2 and 4 by A.M. Kelly; and for Chapter 3 by J.C. Davis.

Data for the case history studies and help in interpreting results were originally provided by many, especially, I. Banerjee, W.R.A. Baragar, T.N. Irvine, I.D. MacGregor, C.H. Smith, C.H. Stockwell, and H.P. Trettin. For the last chapter, I have made extensive use of results obtained in collaboration with staff of the Geomathematics Section.

I am most indebted to those who provided comments which led to many improvements. J.C. Griffiths, R. Kretz and D.F. Merriam have gone through the entire manuscript; and D.P. Harris, D.G. Krige, W.C. Krumbein, H.S. Sichel, R.W. van Bemmelen, M.I. Watson and E.H.T. Whitten have read parts of it.

F.P. AGTERBERG

CONTENTS

CHAPTER 15. SPATIAL VARIABILITY OF MULTIVARIATE SYSTEMS

BIBILIOGRAPHY

Chapter 1

MATHEMATICAL MODELS IN GEOLOGY

1.1. Introduction

Geomathematics, in its broadest sense, includes all applications of mathematics to studies of the earth's crust. The first part of this book consists of reviews of calculus, matrix algebra, geometry and mathematical statistics. In later chapters, mathematical concepts and techniques will be applied to actual examples in selected case history studies. The emphasis will be on basic principles of general usage in geology and on data analysis in practice. However, geologists need a variety of geomathematical techniques in all of the following fields of activity:

(1) *Data acquisition and processing.* Systematic recording, ordering and comparison of data for different parameters; and methods for graphical display of results.

(2) *Data analysis.* Identification of trends, clusters and simple or multiple correlations for which geological explanations are needed.

(3) *Sampling.* Provision of a statistical basis for practical problems of data acquisition.

(4) *Hypothesis testing.* Verification of concepts or models of processes believed to explain the origin and provenance of specific phenomena (includes simulation models).

(5) *Quantitative prediction in applied geology.* Provision of solutions to specific problems such as computing the probability of occurrence of specific types of mineral deposits in a given region; probabilities of occurrence of volcanic eruptions and earthquakes.

In this chapter, we will discuss difficulties in the application of mathematics to geological problems in these fields. These difficulties stem from: (1) the nature of geological phenomena, e.g., paucity of exposure and restriction of observations to the record of past events; and (2) from the nature of traditional geological methods of research which are largely nonmathematical.

When mathematics is applied to geological problems, the parameters must be defined in a manner sufficiently rigorous to permit nontrivial derivations. The initial hurdle is to choose variables that are substantially meaningful. During the design of models, which will be used for deduction, it is important to keep in mind Chamberlin's (1899) warning: "The fascinating impressiveness of rigorous mathematical analysis, with its atmosphere of precision and elegance, should not blind us to the defects of the premises that condition the whole process. There is, perhaps, no beguilement more insidious and dangerous than an elaborate and elegant mathematical process built upon unfortified premises."

Geological models

Krumbein and Graybill (1965) have distinguished three types of models in geology: (1) scale-models; (2) conceptual models; and (3) mathematical models. Traditionally, geologists have been concerned with scale-models and conceptual models mainly.

Examples of scale-models are the geological map and cross-sections where the spatial variability of attributes is represented at a reduced scale for topographic surface and vertical planes, respectively. Geologic processes also can be represented by scale-models. A classic discussion of this subject was given by Hubbert (1937). Conceptual models are mental images of a phenomenon or process. Mathematical models involve use of equations consisting of variables and constants. They are statistical or deterministic depending on whether one or more random variables are used in the equation or system of equations to express uncertainty. Mathematical equations generally can be represented geometrically by curves or surfaces.

The three types of models listed are not mutually exclusive. Scale-models can be based on mathematical criteria and conceptual models may be partly or entirely quantitative.

Most mathematical models in geology have some important aspects of uncertainty and, for this reason, are statistical. The problem may consist of eliminating the random variations from data so that a deterministic expression is retained representing the relationship between averages for assemblages of attributes rather than between single features. Statistical components for the uncertainty provide a way of expressing a range of different extrapolations for single features, all of which are possible, but with different probabilities of occurrence. This method replaces that of extrapolating a phenomenon with absolute certainty.

Geology differs from physics, chemistry and other sciences in that the possibility of doing controlled experiments is more limited. The observations are restricted to a record of past events, making geology a historical science. Generally, the final product of many interrelated processes is exposed at the surface of the earth in an imperfect manner. These mainly physico-chemical processes seldom reached a state of equilibrium; most came to a halt before reaching equilibrium at one or more specific points of time.

Construction of geological maps and profiles

A field geologist is concerned with collecting numerous observations from those places where rocks are exposed at the surface. Observation is hampered by poor exposure. For most areas, 90% or more of the bedrock surface is covered by unconsolidated overburden restricting observation to available exposures, for example, along rivers. The existence of these exposures may be a function of the rock properties. In glaciated areas, for example, the only rock that may be seen may be hard knobs of pegmatite or granite, whereas the dominant softer rocks may never be exposed.

The observable facts must be correlated with one another, for example in a strati-

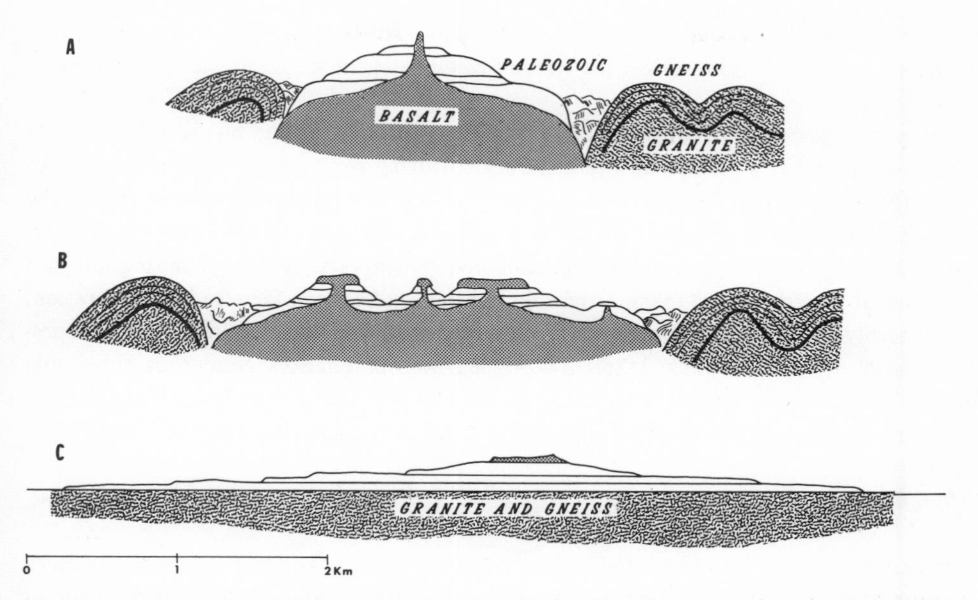

Fig.1. Schematic sections compiled by Nieuwenkamp (1968) indicating that a conceptual understanding of time-dependent geological processes is required for three-dimensional extrapolation of geological features observable at the surface. Sections A and B across Kinnehulle, Sweden, are after Von Buch (1842) illustrating his genetic interpretation. Basaltic caps on hills would have central pipes connecting them to primeval basalt. Von Buch had modified theory of Neptunism by assuming that basalt penetrated upwards in a viscous (nonvolcanic) state; and, locally, Paleozoic strata were changed into gneiss by metamorphic activity of primeval granite. Section C is after Westergård et al. (1943).

graphic column. Continuously, trends must be established and projected into the unknown. This is because rocks are three-dimensional media which can only be observed two-dimensionally. We can look at a rock formation but usually not within it. Areal features must be extrapolated across volumes where they supposedly remain applicable.

During the past century, geologists have acquired a capability of imagining three-dimensional configurations by conceptual methods. This skill was not obtained easily. In Fig.1, according to Nieuwenkamp (1968), an example is shown of a typical geological extrapolation problem. The tops of the hills consist of basaltic rocks; sedimentary rocks of Paleozoic age are exposed on the slopes; granites and gneisses in the valleys. The first two projections into depth for this area in Sweden were made by a prominent geologist (Von Buch, 1842) at an early date. We can assume that today most geologists would rapidly arrive at the third reconstruction (Fig.1C) by Westergård et al. (1943). At the time of Von Buch, it was not common knowledge that basaltic flows can form extensive plates within sedimentary sequences.

It is well known that geological maps can become obsolete after a number of years. This is not because the observable features in the rocks have changed but because of changing geological concepts. Harrison (1963) has given an example of geological maps for the same region but made at different times. The striking discrepancies, in that situation, to a large extent reflect the state of collective geological knowledge at different points in time.

Many geologists regard mapping as a creative art. From scattered bits of evidence, one must piece together a picture at a reduced scale which covers at least most of the surface of bedrock in an area. Usually, this cannot be done without an understanding of the geological processes that were operative in the region. A large amount of interpretation is involved. Many situations can only be evaluated by experts. Although most geologists agree that it is desirable to make a rigorous distinction between facts and interpretation, this is hardly possible in practice, partly because results for larger regions must be represented at scales of 1:25,000, 1:250,000, or less for compilation. Numerous observable features cannot be represented in the scale-models and the many changes within geological environments from place to place make it difficult to develop consistent rules.

Different types of earth-science data

Recent progress in several fields has enhanced the possibility of amassing large amounts of data, many of which are obtained secondarily from rock samples collected in the field and afterwards processed in the laboratory. This presents the problem of choosing samples for further analysis. Not only must bias from secondary processes such as weathering be eliminated, but a specimen must be representative of a larger entity. Results based on a rock specimen are restricted to the collection site if another specimen taken at a distance of several centimeters, meters or hundreds of meters, would produce a different result in the laboratory.

As a rule, sampling should be done by experts but these may be biased in their views on the field of "influence" to be assigned to a single measurement. Geomathematics can contribute to the solution of problems of this type. It should, however, be realized that the complexity of most geological environments makes it inadvisable to collect samples without a thorough understanding of the local geology based on scale-models and conceptual interpretations by experts.

Geologists, geophysicists and geochemists produce different types of maps for the same region. Geophysical measurements are mainly indirect. They include gravity and seismic data, variations in the earth's magnetic field, electrical conductivity and others. Geochemists may do element determinations from chips of rock in situ but also from samples of water or mud. The latter approach gives indirect measurements which are weighted averages of possibly large parts of a geological environment.

As many as eight different maps may be needed for a single area to represent the various types of earth-science data. Each map has a different explanation and considerable knowledge is required to be able to read all maps in order to obtain a comprehensive picture for a single area.

Apart from the diversity in terms and concepts that entered geology with the sub-disciplines, geology, as a qualitative science, has obtained a large and complex terminology. Unfortunately, there exists a wide disparity in the usage of terms. S.C. Robinson has remarked that one geologist's "granite" is another's "quartz monzonite" (cf. Agterberg and Robinson, 1972). Professional geologists may be able to stay abreast of changing concepts and terminology. A consensus or, at least, understanding of each other's approach can be reached during joint excursions and discussions. However, as a science, geology has become increasingly more difficult to understand by nongeologists.

1.2. Scientific methods in geology

Van Bemmelen (1961) has stated that the personality of the geologist is of essential importance in the way he analyzes the past. Geologists can be distinguished as two types: one type considers geology as a creative art; and the other regards it as an exact science. According to Wright (1958), some scientific investigators aim for classification of their objects of study wishing to force the rigid discipline of certain schemes upon them. Others are more receptive of new concepts and less rigorous in schematizing the objects. Ostwald (1910) distinguished between "classicists" and "romanticists". Wegmann (1958) has pointed out that, among the early geologists, Werner was a classicist and Hutton a romanticist.

At the end of the 18th century, Werner postulated the pan-sedimentary theory of "Neptunism" according to which all rocks including granites were deposited in a primeval ocean without later changes. Nieuwenkamp (1968) has demonstrated that this theory was related to the philosophical concepts of Schelling and Hegel. The latter supported Werner's view when it was criticized by other early geologists who assumed processes of change during geological time. Hegel (cf. Nieuwenkamp, 1968) replied by comparing the structure of the earth with that of a house. One looks at the complete house only and it is trivial that the basement was built first and the roof last. Initially, Werner's conceptual model provided an appropriate and adequate classification system, although Von Buch, who was a follower of Werner, rapidly ran into problems during his attempts to apply the model to explain the occurrence of rocks in different parts of Europe. This was illustrated in Fig.1 where the primeval granite was thought to be active later changing sediments into gneisses, while the primeval basalt became a source for volcanoes.

Werner's contemporary and opponent Hutton postulated "Plutonism" instead of Neptunism. He was a "romanticist" (Wegmann, 1958) proposing a model of continuously recurrent cycling of matter in the earth. His theory provided not a scheme for rigid classification but it opened the mind to possible occurrences in the course of time of a great variety of interrelated plutonic and tectonic processes.

Geology has known an exceptionally large amount of controversy and polemics closely related to the personality and experience of the participants. For example, Neptunism began in areas where sedimentary rocks were situated on top of granites and gneisses whereas Plutonism originated with Hutton in Scotland where tectonism with granitic intrusives is more apparent.

Griffiths (1970) has pointed out that the lively debates of the more recent past about the origin of "granite" are now more lucidly posed as "how much granite is formed this way or how much that". He also argued that the new question would be unresolvable without a more quantitative (meta) language for expression.

How much mathematics?

According to Van Bemmelen (1961), Ostwald's classicist type of scientist may consider physics and chemistry as the only true natural sciences. Any relationship should be measurable and expressed mathematically. The classicist is inclined to compare geologists with "barbaric stone-age" scientists struggling toward the remote but ultimate goal of total quantification.

On the other hand, most geologists hesitate to go too far in the direction of what they regard as "robot-geology". Some postulate that geology is "an inexact science and an art" (Link, 1954); others state that originally geology was essentially descriptive but by the middle of the 20th century, it had developed into an exact physical science "making liberal use of chemistry, physics and mathematics" (Leet and Judson, 1954).

The question of whether geology should be more quantitative than it has been was considered continuously by geologists in the past. Fisher (1954) has suggested that geology with Lyell (1833) was evolving as a more quantitative science. Rapidly, the opposition to this development grew to the extent that Lyell's elaborate tables and statistical arguments (60 pages long) for his subdivision of the Tertiary were omitted from later editions of his *Principles of Geology*.

Most geologists agree that Chamberlin's (1897) scientific method of "multiple working hypotheses" is ideal for geology. A multiple approach is needed because of the great complexity of geological processes. Unfortunately, the sheer complexity and volume of observations entailed by this method make it impractical. Instead, geologists became accustomed to synthesize from a few broadly relevant observations in order to develop a preferred hypothesis. Alternatives may have been considered but the end product usually reflects a preferred method of approach.

Some geologists have become weary of far-fetched speculations and restrict themselves to a limited number of facts that can be established with absolute certainty. A review of geological problems of a wider scope such as mountain-building or metallogenesis makes it clear that a consensus of opinion among experts is rarely, if ever, achieved, particularly if date of publication also is considered. A limited number of facts are beyond doubt but the discrepancies between conceptual models tend to widen with the broadness of their scope (also see Fig.114 to be discussed later).

Griffiths (1962) has stressed the analogy between Chamberlin's method and Fisher's

(1960) description of statistical analysis. By using different statistical models and formal inference, it is feasible to test different hypotheses for the same problem provided that the geological facts can be expressed adequately in numerical form. A larger-scale quantification of geological data at first could to a large extent consist of digitizing geological maps, sections and other scale-models prepared by experts. These results are not necessarily an end product of geological research but they can be used for computations based on multivariate statistical models.

The quantification of geology has been stimulated by awareness of the fact that the other sciences, including the social sciences – economics, sociology and psychology –, have benefitted from an increased use of mathematics. The advent of the computer in the 1950's has had its impact on quantitative geology. On the other hand, this development has sometimes resulted in a ready acceptance of mathematical techniques of established applicability in other sciences where data can be collected in a more systematic manner and experiments can be designed with more freedom in the choice of control.

Most existing applications of mathematics to earth-science data are based on quantitative measurements. In reality, the number of quantitative data that can be collected for a geological entity usually is limited. Many data are of the presence-absence type. For example, at a given place sandstone is present and not shale; and at another place, shale instead of sandstone.

If many discontinuities or abrupt changes in state occur at a given place, the information may be restricted to the statement that one particular rock type is predominant. On a number of geological maps, the only information recorded systematically is the age of rock formations as based on a time-stratigraphic column. Remarks on the other attributes of the rocks may be given verbally in explanatory comments.

For lack of precise data, the scope of geomathematical problems could not be nearly as broad as that of the conceptual models in qualitative geology where data of many different types and places are integrated in the mind of the geologist. Van Bemmelen (1972) has stated that, during a lifetime, a geologist memorizes numerous facts and concepts constituting a background from which he can draw relevant notions while practicing geology. When confronted with a new situation, the geologist can remember analogous situations wherein certain factors were more clearly expressed. Only part of this experience can be communicated to others. It would be useful if this accumulation of interconnected facts could be stored in mechanical brains so that it would be available to all.

Electronic storage and retrieval of geoscience data

The need to have data available in standard form has been felt strongly and expressed by several authors notably S.C. Robinson (1969) and Hubaux (1972). In a number of places, computer oriented files were developed containing many facts of the same types, to be retrieved selectively (Laffitte, 1969; Dixon, 1970).

Examples of coding forms to record geological data were given by many authors including S.C. Robinson (1970) and Bostick (1970). A bibliography of 336 publications dealing with computer-based storage and retrieval systems for geoscience data has been prepared by Hruška and Burk (1971). A list of references of interest on data processing in the geological sciences has been compiled by Merriam (1971).

Three practical approaches to the standardization of field data applied in Canada, in order that systematic ordering by computer can be employed, are representative of early attempts to bring consistency and an objective basis to recording ground truth in the field. They are Operation Pioneer in Manitoba (Haugh et al., 1967), the Coast Range Mountains Project in British Columbia (Hutchison and Roddick, 1972) and the Grenville Mapping Program in Quebec (Wynne-Edwards et al., 1970). All used the same standard coding form at each station. The terrains consisted mainly of igneous and metamorphic rock. In the Quebec project, detail of information was sufficient to allow automatic naming of the rock by computer. In the other two, an agreed set of definitions for specific rock names was used. In the projects by Hutchison and Roddick and by Wynne-Edwards et al., automatic plotters were used to represent stations, and attributes measured at these stations, on maps.

Similar projects on sedimentary strata in Greenland (Alexander-Marrack et al., 1970) and Spitzbergen (Piper et al., 1970) have used standardized records for sedimentary sections. The Scandinavian countries are jointly using GEOMAP, a system for data storage and processing in geological mapping (Berner et al., 1971).

A new method of coordinating geoscience information would consist of developing national and international computer-based networks of data files (Burk, 1968). Index files for geological data can have elaborate systems for cross-referencing (McGee, 1969).

Computer technology is progressing rapidly. Digital computers now are accessible directly in many places or through a terminal connected by telephone line to a central data-processing facility. Computers can have a core memory in which more than a hundred thousand single facts can be manipulated simultaneously by means of fast algorithms for multivariate mathematical techniques. In addition, they have a virtually unlimited peripheral memory consisting of magnetic tapes and discs where millions or billions of facts can be stored for selective retrieval and printout or manipulation in the core memory.

Geology could benefit more from this possibility provided that suitable frameworks can be found for storing the facts. Ideally, millions or billions of data for a given region are stored according to their location in three-dimensional space or in relation to a network of cells of equal volume. Such a comprehensive data base could be used both for constructing conventional geological maps and sections, and for calculations, by which the facts and variables are correlated with one another. The major problem in endeavours of this type is to find methods to weight the individual facts. For example, a strike and dip measurement for a layer at a given place has no significance, unless there is considerable information about the surrounding geological environment. Interesting proposals to record individual facts as part of a more complex entity were made by Briggs and Briggs (1969) and Loudon (1971).

Use of computer by geologists

A problem for the geologist is that, in order to use the computer, his repertory of terms and concepts must again be enlarged, this time by elements from mathematics and computer science. Traditionally, mathematics has been low on the list of priorities of subjects to be taught to geology students (cf. Harbaugh, 1965). Geologists using the computer must rely heavily on people more familiar with these other languages. This can result in misunderstanding. A geologist should not expect the expert at the computing center to suggest a solution to geological problems such as how to sample a rock. He may help in preparing computer programs in the language COBOL or FORTRAN for storage and retrieval problems, or in FORTRAN or ALGOL for mathematical computing.

For the solution of specific problems involving many basic operations, use can be made of efficient algorithms available in scientific subroutine packages. The basic operations are logical statements of Boolean algebra and arithmetic calculations such as addition, multiplication and powering. Although all basic operations are simple, they are performed at such speed that elaborate numerical problems of matrix algebra, integration and harmonic analysis can be solved rapidly, paving the way for multivariate statistical analysis of geological data.

It is noted that the availability of these mathematical data-processing techniques offers a temptation to the mathematically inclined geologist to do extensive computing from limited data. This may create an impression of robot-geology which is met with scepticism by other geologists.

Induction and deduction

The arrangement of observations into a pattern of relationships is a mental process of "induction". It involves postulating a hypothesis or theory that is in agreement with the facts. A working hypothesis initially based on intuition gains in functional validity if it leads to "deductions" that can be verified. Logical deductions from a hypothesis provide predictions for facts that have not been observed. The validity of a hypothesis is tested on the basis of new facts not considered when it was framed. New facts may lead to modification of the hypothesis or theory. The result is a recurrent cycle of inductions and deductions. Van Bemmelen (1961) has referred to this approach as the prognosis-diagnosis method of research.

This method is commonly used in the conceptual models of qualitative geology. A problem is that different scientists may assign a different amount of relevance to basic observations when exploring them by means of a theory. This can lead to considerable differences between conceptual models.

The same holds true for mathematical models which may differ from one another and where the data can be weighted differently. As mentioned above, the scope of these models could not be as broad as that for the large-scale conceptual models of qualitative geology. It may be possible in the future to enlarge the scope of geomathematical methods by designing increasingly complex, multivariate methods where spatial location, setting within the environment, and age of the features are being considered. These methods would be applied to data bases containing millions or billions of individual facts. It will be more important than ever to check such models for validity. Using a computer, it is relatively easy to make multiple applications of a complex method. The prognosis–diagnosis method of science provides one method of testing. A method is not only applied to the entire data set but also to a number of subsets. Predictions based on the subsets are evaluated for validity on sets of control data not used for developing the prediction equations.

If done systematically, the procedure becomes comparable with the Jackknife method (Tukey, 1970; Mosteller, 1971; Gray and Schucany, 1972). Krige (1966a) has developed methods where equations are developed from limited data and later checked against "reality" which is based on many more data. He also has made systematic attempts at applying different types of mathematical models to the same sets of data and comparing the results with one another. The latter approach constitutes a direct application of Chamberlin's method of multiple working hypotheses, which is the second principal method to be used for testing complex multivariate methods.

In Chapter 15, we will discuss some more complex methods applied to many data of different types. Geological and geophysical data for a larger region will be combined into a single mathematical model. It will be seen, however, that the information systematically available for larger regions is limited in comparison with the detail available for selected, smaller subareas. In order to test the validity of the resulting predictions, subsets of data are deliberately omitted from the data base.

To offset the drawback of mathematical models that they are narrower in scope, we mention two advantages as compared with conceptual models: (1) the validity of the approach can be tested quantitatively; and (2) many different factors for many places can be considered simultaneously by using multivariate methods. The computer-based methods are less flexible than the human mind. However, this rigor can be an advantage.

The DELPHI-method

Recent methods where it is attempted to reach a consensus of opinions by experts include the so-called DELPHI-method as reviewed by Dalkey and Rourke (1971). Different experts are asked to make predictions for a rather specific problem. The results are compared with one another and mean or median is taken.

Harris et al. (1971) have applied the DELPHI-method to the geological problem of mineral potential evaluation. Twenty different exploration geologists were asked to make a quantitative guess of mineral potential (hidden resources) for different commodities, in a number of cells of an area in the Yukon and northern British Columbia. A basic requirement was that the experts have a thorough knowledge of the area of study and entertain no contact with each other when making the predictions. Barry and Freyman (1970) have pointed out that results can be obtained within a relatively short time (several months) by means of this method of subjective probabilities. Discrepancies in opinion between individuals can be assessed systematically by the DELPHI-method and to some extent eliminated. Of course, it is not feasible to correct for bias in the collective opinion of all experts and it is possible that one expert's opinion is better than the average of several. Nevertheless, the method of subjective probabilities clearly indicates those situations where a consensus could be reached. Recently, Harris has modified the original approach by letting the experts also choose shapes and sizes for subareas where they assume mineral potential (Harris and Euresty, 1973).

Geological time

Geological hypotheses are based on processes taking place today (principle of actualism), and on experiments performed in the laboratory. However, geological evolution is a continuing process which is directed along the time-axis. Not only may the complexity of a geological environment increase with its age but processes that took place a long time ago (e.g., two or three billion years) may have differed from those of the more recent past. It is difficult to relate an observed effect to a specific cause in all but the youngest rocks.

A great length of time was required for most geological processes to come to a halt, commonly of the order of 10^8 years. It may be difficult or even impossible to simulate these processes in the laboratory. There are, of course, catastrophic events such as earthquakes, volcanic eruptions, and landslides. These are important features but they comprise a small minority of all processes and have had only a minor effect on the present constitution of the earth's crust.

Direct measurement in the course of geological time is not possible. Instead of this, geologists work with indirect systems of analogies. For example, at the earth's surface one may find, next to one another, different types of geological environment reflecting different stages of the same evolutionary sequence.

In small areas, for example, at outcrop-scale, the situation may be clear-cut and local extrapolations and interpolations can be accomplished. However, the real challenge comes with the attempt to fit many bits of evidence together into a single, comprehensive process-model where time is a factor and with a wider range of applicability in other areas and in depth.

1.3. Use of geomathematics

An annotated bibliography by Vistelius (1969) contains abstracts of 712 selected publications where mathematics is used for problem-solving in various subfields of geology. More than 3000 papers in geomathematics and related fields were listed in the first three volumes of GEOCOM Bulletin (1968–70).

A fairly large number of geomathematical papers were published in the proceedings of conferences many of which were oriented toward the mineral industry. Most of these are listed in the November, 1970, issue of GEOCOM Bulletin.

The American Geological Institute has published a *Bibliography of Statistical Applications in Geology* (Howard, 1968). In 1968, the International Association of Mathematical Geology was formed and has sponsored a journal since 1969.

An annotated bibliography of textbooks and monographs on mathematics in geology is given at the end of the book.

A trend in several subfields of geology from qualitative description and interpretation, via quantitative measurement, to mathematical analysis was documented by Griffiths (1970).

Krumbein (1969) has classified computer methods in geology by type of method and subject matter. His paper concluded a set of fourteen review papers by different authors covering usage in the subfields of geology (Merriam, 1969). Krumbein's block diagram, where time with two-year intervals forms a third dimension, is shown in Fig.2A. One horizontal slice through the block diagram shows time of entry of computer applications into stratigraphy (Fig.2B). The vertical slice (Fig.2C) is for factor analysis. The dashed lines indicate precomputer applications of the methods shown.

Stratigraphy is not entirely representative of what happened in the other fields. However, the diagram illustrates that since the 1960's computers have been used for most methods. Some methods had precomputer applications in geology; others – such as factor analysis, discriminant analysis and simulation – entered geology after their usefulness had been established outside geoscience. For example, factor analysis has had many precomputer applications in psychology; it became computerized in this field and then was adopted by several other disciplines including geology.

Because, in this book, we will cover a wide range of geological subjects and mathematical methods, it is useful to represent the contents approximately in terms of Krumbein's block diagram. A relatively high proportion of the examples comes from petrology (Chapters 5 and 13), structure (Chapter 14), economic geology (Chapter 10 and elsewhere), sedimentology (Chapters 10 and 11) and stratigraphy (Chapter 12). In addition, we include examples from geophysics and geochemistry (Chapter 7). Several methods would fall under regression analysis (Chapters 8 and 15), trend-surface analysis (Chapter 9), time-series analysis (Chapter 10–13), and Markov process-models (Chapters 12 and 13). Not all methods can be treated in detail in this book but attention will be given to analogies between methods. For example, the definition of communalities (see sec-

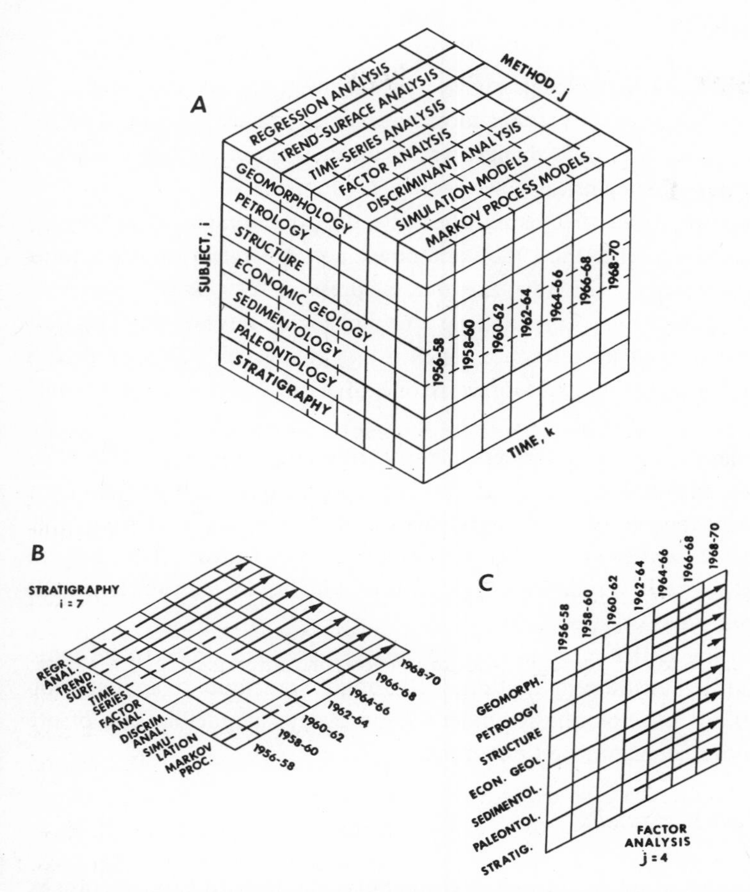

Fig.2. Use of digital computers for problem solving in various fields of classical geology after Krumbein (1969). Horizontal slice (B) and vertical slice (C) through block diagram (A) are shown for example. Broken lines are for precomputer geomathematics.

tion 5.1) in factor analysis is analogous to that of signal and noise components in models for filtering space series. Discriminant functions can arise as a special case of multiple regression analysis (section 15.3).

Only limited attention is given to the promising methods of computer simulation. A book on this subject was recently published by Harbaugh and Bonham-Carter (1970). We are mainly concerned with data analysis. By means of simulation models, entire geological processes in three-dimensional space and with time as a parameter may be modeled. An interesting example is the digital model of evaporite sedimentation in a basin by Briggs and Pollack (1967).

Place of geomathematics in the framework of the earth sciences

Griffiths (1970) has stated that changes in the paradigms (patterns of thought) of geology are needed if we are to benefit from a more widespread application of geomathematics. He goes on to predict that such changes will be brought about only by successful demonstration that the new paradigms can solve geological problems.

The geosciences have improved their applications by the development of geophysics since the 1930's, geochemistry since the 1950's and now a new synergism may be leading to geomathematics. According to Griffiths, the greatest pressure for change arises from areas of application of geoscience. For example, the impact of physics was first felt strongly in the search for petroleum resources which is an applied field. Geochemistry's constribution to searching for natural resources is one of its main sources for growing strength.

Geophysics is now developing in part independently from applied geology. One of its recent contributions to the understanding of the earth is the discovery of new facts supporting horizontal movement of continental masses. It is unlikely that geomathematics will develop into a separate discipline such as geophysics. Instead, it is likely to remain closer to geology itself. Probably its main role will be to contribute to the integration of the various types of geoscience data.

It is worth emphasizing Griffiths' point that geomathematics, in order to be more widely accepted, should demonstrate its usefulness. It should contribute to applied geology, e.g., by new methods of exploring for hidden mineral deposits, delineation of ore reserves and evaluation of regional mineral potential.

Evaluation of geomathematical results

The approach to problem-solving using the combination of automatic recorders, mathematics, computers and instruments for display, is sometimes called "black-box technology". Of course, the results can be compared with those obtained from the more conventional intuitive mode of reasoning from observation to conclusions. Suppose that the outcomes match for a specific problem; according to Griffiths (1970) not much is gained although possibly time and expertise is saved by using the automatic system. In fact, geomathematics would duplicate existing methodology, which is not a sufficient reason to justify its application.

Intuitive reasoning makes use of many facts but most are of the presence–absence type. The simplest type of information consists of yes–no data for two mutually exclusive features using a "nominal" scale. In many situations, more than two classifications of this type can be made. It may be possible to rank the data by employing an ordinal scale with "more than" or "less than" as the essential criteria. These methods of coding and classification are mainly qualitative. They may result in complex, hierarchical classification schemes by which each object may be uniquely characterized by a number of overlapping binary codes. However, in order that full use is made of black-box technology, it is required that the original information should be as complete as possible and numbers should be assigned to attributes where this is feasible. The logic of conceptual models is close to the paths of thinking described by Boole (1854) in his *Laws of Thought.* A large number of yes–no decisions is made requiring binary data for operation.

Computers are well suited to cope with problems of Boolean algebra. In fact, the smallest entity for storage in the memory is one binary digit (bit) which can be zero or one. Other numbers are formed as strings of bits and can be manipulated by arithmetic operations. In geology, the majority of basic facts are of a binary nature (presence–absence data) but combinations for larger volumes of rock can give sums or averages which are fully quantitative. An example is petrographic modal analysis.

Individual counts for minerals in a thin section provide binary data but combining many individual counts gives volume percentages of minerals in a rock. Likewise, instead of considering that granite occurs in a region, we can determine how much granite occurs.

In the next section, we will discuss several examples of Boolean algebra applied to geological problems where the procedures come fairly close to a simulation of conceptual modelling although computers were used for solution. The input may consist of whether a specific feature is present within a larger region, whereas questions of how much of the feature is present and in what way are hardly considered.

The first attempts to do a geomathematical analysis for larger regions were hampered because more quantitative information is needed than for conceptual models. However, suppose that we succeed in collecting an extensive base of quantitative facts for a larger region, and subject it to black-box technology. This is feasible to some extent, because geological maps are widely available and they can be digitized. Several other types of data, in particular geophysical measurements, which are systematic, are also widely available.

The results of a geomathematical analysis then remain to be interpreted. The only criteria available for comparison are those based on conceptional models. This imposes another type of restriction because we may well be inclined to accept only those results which also were predicted by the more conventional methods.

The problems inherent in doing measurements on geological objects, dictate not only that ground truth be established by experts, but also that mathematical results be further interpreted by geologists employing qualitative concepts. As geologists come to realize the major contribution that mathematical correlation and statistical appraisal of observable or measurable variables make to their ability to solve geological problems, it is predictable that they will define additional variables and take steps to see that relevant observations are made in a systematic manner. The end result should be a combined effort by concept-motivated geologists and geomathematicians. Changes in geological methodology and philosophy are desirable in order to apply mathematics. In a partnership, the change in paradigms to which Griffiths (1970) refers should come about as an evolutionary process.

Three examples of the mathematical treatment of presence–absence data

Perhaps the earliest large-scale attempt on record to quantify a geologic problem dates back to Lyell as early as 1833. The objects to be classified were three series of strata to which a fourth, the "present", was added for comparison. The elements in Lyell's matrix denote presence or absence of 7816 species of recent and fossil shells. By counting, for each series, the number of fossil shells of species living today and recomputing the resulting numbers to percentage values, Lyell (1833) established his well-known subdivision of the Tertiary Period into Pliocene, Miocene and Eocene. Later, Paleocene, Oligocene and Pleistocene were added, thus providing the break-down into epochs of the Cenozoic Era in the geological time table using names that reflect the magnitude of the percentage values.

More recently, Botbol (1971) coded presence–absence of 366 features (rock types, minerals, dips of fractures) in thirty mining districts of the U.S.A. Copper, lead and zinc districts were treated separately. Botbol counted the number of times a feature occurs in different districts. To this was added the number of times the feature occurs together with any other feature as a pair. The result is the so-called typicality of the feature. For example, the mineral chalcopyrite has the highest typicality for copper districts because it is the principal copper mineral and occurs in all of them. This would be known without calculating the typicality but other results suggest new relationships and any new district can be compared with the norm consisting of all typicalities. The resulting index reflects the similarity between the new district and the set of districts used for calculating the norm.

A third example consists of a method developed by Vyshemirskiy et al. (1971). They classified 21 giant oil pools from the Arabian, North African, Russian, West Siberian and Turan platforms on the basis of 179 features. Every giant pool is characterized by a row of yes–no data for properties of sequences of strata within, underneath, or above the oil-bearing suite and parameters for the structural setting. If the data for the features are regarded as a set of separate columns, we can imagine forming all possible combinations of columns. By counting, we then determine the number K of those combinations of columns which have at least two equal rows if only one column is deleted. The number of times that a specific feature i is represented by a column in one of these combinations can be written as K_i. The ratio $P_i = (K-K_i)/K$ is called the identifying informative weight of the i-th feature. The informative weight T_j for a given giant oil pool is obtained by adding the P_i's for all nonzero entries in the row (j) for that pool.

By this method, Vyshemirskiy et al. (1971) have found that the ratio P_i tends to be large for geological factors also believed to play an important role in the genesis of giant oil pools when intuitive reasoning is used.

Secondly, the informative weight T_j for a pool tends to be large if its resources are larger than average. Hence, the largest giants are characterized by a relatively large proportion of those features which all giant pools tend to have in common. The potential resources of an undrilled area could be estimated by evaluating its informative weight T_j and comparing this number with those for the giant pools used for the description.

The previous three examples have in common that presence–absence data are used for classification. Lyell made use of the fact that the number of species of shells changed during geological time according to what is now thought to be a stochastic birth-and-death process.

In the other two examples, the attempt was made to characterize objects in terms of associated features. It seems that combinations of geological features normally provide more information than single features. It was pointed out, however, that these classifications are limited by the fact that original information of a feature was restricted to its occurrence or absence within rather large regions. It is likely that, in other sciences, the questions such as how much is present, and where, would be asked immediately. In geoscience, these additional questions may be difficult to answer. It is physically laborious to collect the information when many different regions are involved. Again, the primary information may never have been reported in a systematic manner and it may be a matter of subjective judgement to assign a number to a feature if it is present.

Another type of inconsistency occurs in the distribution of information; a considerable amount of exploration work (e.g., drilling) is done in places where the presence of economically important objects, such as giant oil pools and mining areas, has been indicated. This is reflected in the amount of data available about the objects to be classified. The geological features coexisting with these objects may also occur in areas where no discoveries have been made but this exploration information is much scarcer.

1.4. Mathematical techniques in use

A number of mathematical techniques that can be used for analyzing geological data are shown in Table I. Most will be discussed in more detail in later chapters. The present discussion of Table I is restricted to remarks on the classification that was used.

Dimension-free methods differ from dimensional methods in that parameters of space and time are not contained in the equations applied to the data. Dimension-free methods are subdivided further by number of populations and number of variables. These models are appropriate if spatial variability and change during geological time can be deleted as continuous variables. Differences between geological entities of different location or age can be considered in these models by defining two or more populations whose parameters can represent these differences.

Dimensional methods are distinguished firstly on the basis of whether use is made of the powerful assumption of stationarity. A domain is stationary or homogeneous if measurements from any of its parts are representative for the entire domain. In stationary situations, the values of attributes usually are expressed in terms of measurements made at regular intervals in time or space. Relationships of this type hold true elsewhere as long

TABLE I: SOME TECHNIQUES IN USE FOR THE MATHEMATICAL ANALYSIS OF GEOLOGICAL DATA

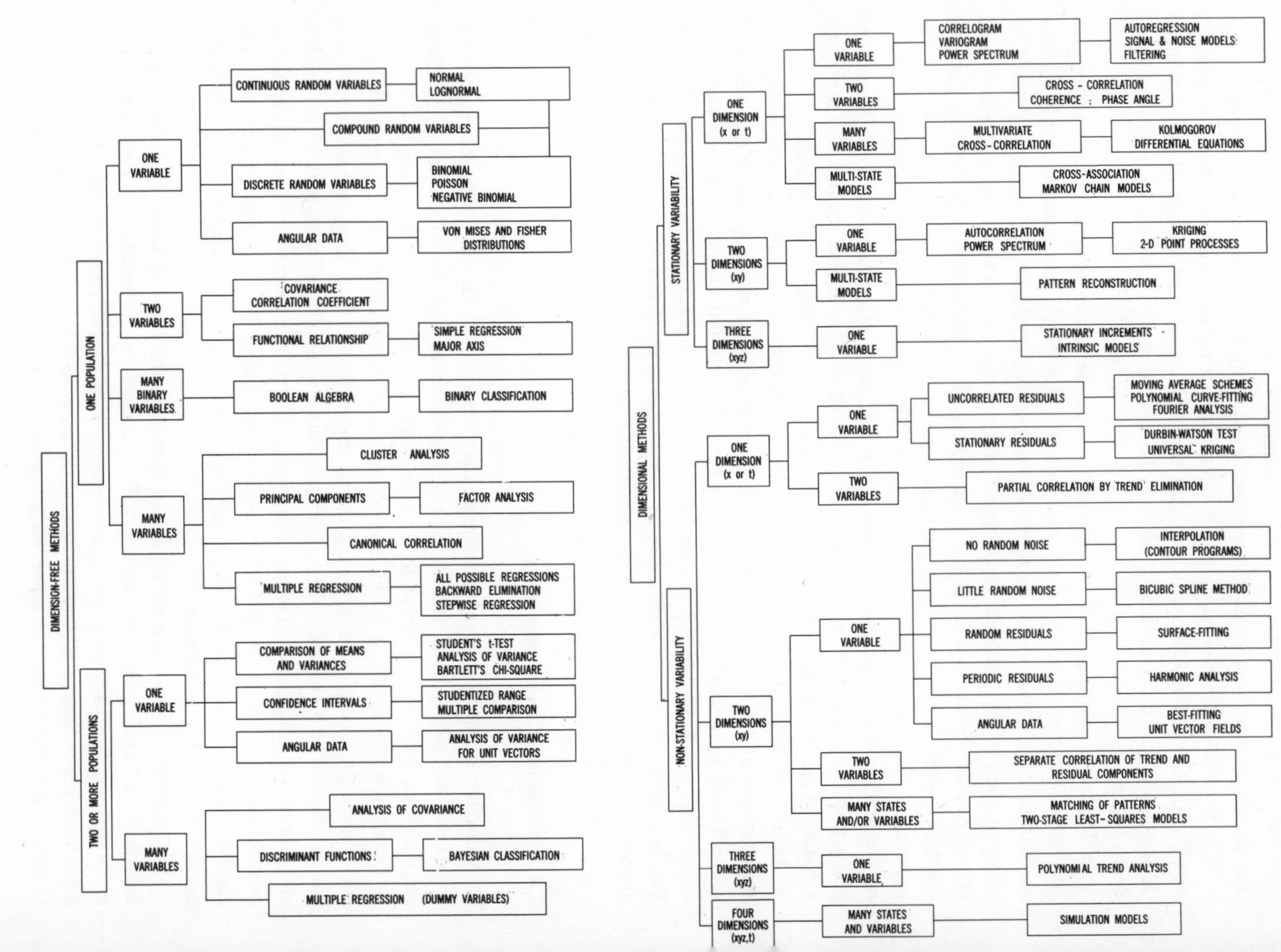

as the condition of stationarity is satisfied. For most existing techniques, the observation points are equally spaced along lines or on regular grids. In some applications, however, the sampling interval itself is a random variable. A simple example of a stationary model is that the average attitude of a layer in a given area can be approximated by a plane passing through three measurement points on the layer in that area.

In several dimensional methods, the dimensions of time and distance are interchangeable and the forward property of time is not considered. For example, filters to remove small-scale irregular fluctuations when applied to geological time or space series may have bilateral symmetry (see sections 10.2 and 11.8). On the other hand, time differs from distance in situations where the forward direction of the time-axis is relevant. Examples of this are models for Markov chains with cyclical components (section 12.7) and cross-correlation (section 10.6) of time-dependent variables subject to differences in phase with respect to one another. Forward equations, where future events are expressed in terms of past events, are more commonly used in situations of this type, but backward equations may give comparable results when time is reversed (sections 13.8 and 13.9).

The mathematical apparatus of dimensional methods is closely related to that of methods of time-series analysis well known in statistical theory of communications and several other branches of science. A difference is that, in geology, we are usually not interested in predicting the future from past events. Instead of this, the object may be to obtain a time-dependent expression for an entire process (stochastic-process model), or to predict a value at a specific point in space from known data at other points ("kriging", see Chapter 10).

Many data in geology are of the presence–absence type. This may result in sets of binary variables for statistical analysis or in distinct states with discontinuous boundaries in the dimensional space if physical extent in space or time is considered. As mentioned before, the combination of many one–zero data can give quantitative numbers (e.g., volume percent estimates in petrographic modal analysis).

Directional features are important in geology, particularly for problems in structural geology involving folds and faults. The resulting measurements on planes or lines commonly represent directed or undirected lines in three-dimensional space. They are referred to as angular data or unit vectors in Table I.

Computer algorithms are widely available for most statistical methods listed in Table I. A good starting point for building a library of computer techniques is to select usable algorithms from the scientific subroutine packages commonly available at computing centers.

For the introduction of computing techniques into geology, an important role is played by several series of geological computer programs including the Special Distribution Publications of the Geological Survey of Kansas from 1963 to 1966 (see Merriam, 1966), the one initiated in 1966 by the Geology Department of Northwestern University, Evanston, Ill. (see, e.g., Krumbein, 1966), and the Computer Contribution Series of the Geological Survey of Kansas (1966–70). The latter series contains over 50 detailed programs mainly in FORTRAN that are widely used. During the past few years, the frequency of publication of computer programs by geologists has declined significantly. One reason for this may be that many geologists now have learned how to use the computer. They are able to solve numerical problems by using the algorithms available in the literature and in extensive subroutine libraries or can do their own programming. For most specific applications, fairly extensive modifications of existing algorithms are required. The many different types of computing facilities in existence demand other modifications. In order to make use of time-sharing by remote control on a large computing system, familiarity with a highly specific Job Control Language (JCL) may be required. In some instances, the preparation of JCL-programs, that must precede the FORTRAN or COBOL programs, requires as much programming and attention as the latter.

Davis (1973), has prepared a textbook on computing in geology. Basic principles of FORTRAN programming with reference to geological data processing were discussed by Smith (1966). The SAFRAS-system developed by Sutterlin (1971) exemplifies a system of elaborate subroutines for the building of geological data files, editing, and selective retrieval according to a flexible format. More comprehensive systems would enable the geologist to manipulate his data simultaneously for editing, selective retrievals and mathematical analysis. Systems to do both retrieving and computing have been developed in FORTRAN by Chayes (1972), and in APL (A Programming Language) by McIntyre (1969). Advances have been made in automatic contouring by plotters and other methods of graphical data display; for an extensive review with many references, see Tomlinson (1972).

1.5. Use of empirical functions

It can be argued that all geological processes are deterministic. However, it is usually not possible to use purely deterministic expressions in the mathematical equations because of uncertainties or unknown causes. In many geological problems, the spatial variability of measurable features can be divided into regional variations of a systematic nature (loosely called "trends") and more local, unpredictable, fluctuations (residuals from the trend). The trends may have been generated by broad-scale deterministic processes. For example, average grain size of sand particles increases towards a beach. However, larger and smaller particles may coexist everywhere in the sampled area in different proportions.

The systematic variations called "trends" are, in part, determined by the density of sampling points. If more measurements are done locally, a residual from a more regional trend can become the trend for the more local survey area. Semantically, the expression "trend" has disadvantages because of the subjective manner in which it may be defined and interpreted. Matheron (1970a) preferred to use the word "drift" in geostatistical applications.

It is appropriate to use deterministic functions for "trends" when it is kept in mind that the results do not necessarily describe the effect of deterministic processes. We have a choice of using empirical functions such as polynomials (see later) or functions corresponding to curves or surfaces that are theoretical predictions for geological phenomena. Sometimes a mixture of these two procedures can be applied. For example, Vistelius and Janovskaja (1967) have pointed out that the partial differential equations for diffusion processes underlying the concentration of some chemical elements in some rocks, usually have exponential functions as a solution. For trend fitting they suggested the use of exponential functions with polynomials placed in the exponent. In that situation, one may not know the exact shape of the trends to be fitted to the data, but by using a specific class of functions a satisfactory fit is obtained more readily.

Any continuous function for a geometrical shape can be expanded into an infinite series (e.g., polynomial or Fourier series) and a restricted number of terms of the infinite series can provide an adequate approximation for all possible geometrical configurations. Theoretically, even shapes with discontinuities (e.g., breaks related to a fault or contact) can be represented by a truncated series. In practice, however, this may not be feasible because of the large number of terms and data points that would be required. The fitting of trends should be restricted to geological entities with features that are subject to gradational change without sudden breaks.

An example of curve-fitting is shown in Fig.3. The original data represent concentration of the element strontium (in parts per million) determined for equally spaced rock

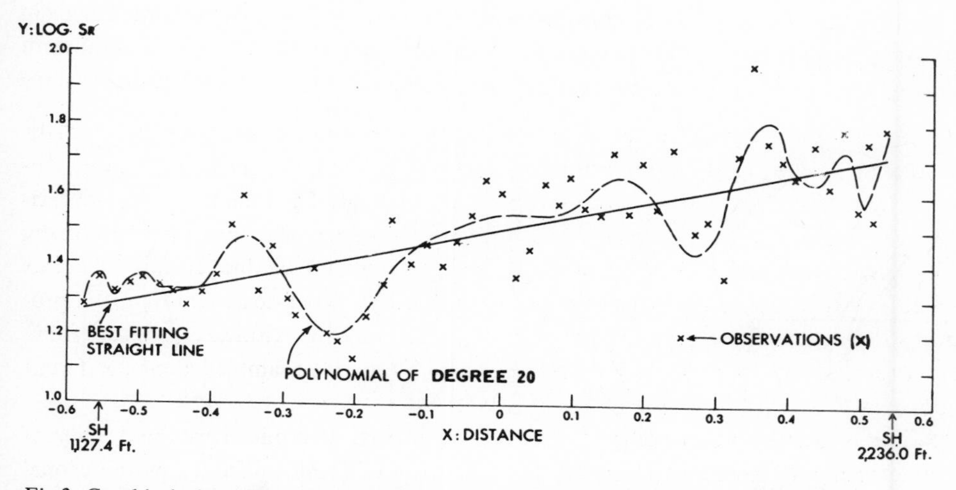

Fig.3. Graphical plot of 58 strontium determinations along a drillhole through dunite layer of Muskox intrusion, Canada. Two results for a continuous "trend" obtained by polynomial curve-fitting. (After Agterberg, 1967a.)

samples along a drillhole through a dunite–serpentinite layer of the Muskox ultramafic intrusion, District of Mackenzie (data from Findlay and Smith, 1965).

The logarithm of Sr was plotted against distance and polynomials of degrees 1 and 20 were fitted to the logarithmically transformed data by the method of least squares.

The "trend" is rather well approximated by a straight line in Fig.3. The logarithm of Sr content decreases gradationally in the upward direction of the borehole. It can be shown that this implies that untransformed Sr content decreases exponentially in that direction.

The data points in Fig.3 deviate from the trend. These deviations are called "residuals". They are irregular but not randomly distributed about the best-fitting straight line. In order to improve the closeness of fit, a polynomial of a higher degree could be fitted. It can be shown by statistical inference that the fit of the polynomial of degree 20 in Fig.3. is "significantly" better. It was based on solving an equation with 21 unknown coefficients whereas only two coefficients were needed to obtain the straight line.

It is questionable that the fluctuations in the second curve are meaningful. A polynomial curve of degree 10 or 30 shows other patterns and, for lack of data, we cannot say which fit is best.

In most practical situations, it is best to restrict the fitting of curves or surfaces to shapes represented by relatively few coefficients. As an indicator of the "trend", we would prefer the straight line in Fig.3. In problems of trend analysis, it is usually not warranted to fit functions with many terms which are more flexible in shape.

The spatial variability in this example is rather irregular. Most practical situations we will be dealing with are of this type. The problem therefore is statistical instead of deterministic.

Another type of problem is illustrated in Fig.4. according to Leopold and Langbein (1966). A meandering stream is represented on the map in Fig.4A. For a number of points at the center of the stream, Leopold and Langbein measured both distance along

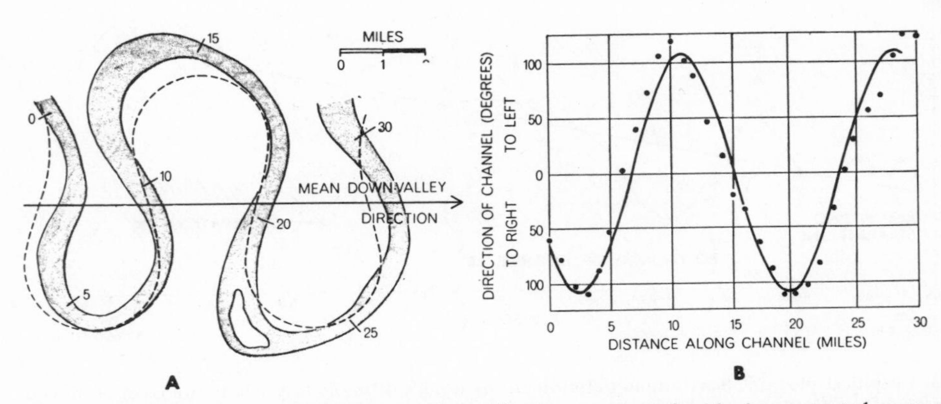

Fig.4. Meander in Mississippi River near Greenville, Mo., compared with sine-generated curve representing theoretical geometrical shape. From *River Meanders* by L.B. Leopold and W.B. Langbein. Copyright 1966, by Scientific American, Inc.

the channel and angle of deviation between channel and mean down-valley direction. These points are represented in Fig.4B. A continuous line could be passed through these points. We could draw this curve by hand or use an automatic interpolation technique such as the one based on spline functions (Ahlberg et al., 1967; Bhattacharyya and Holroyd, 1971). This would provide an adequate representation of the natural phenomenon which is continuous itself. Leopold and Langbein have fitted the sine-curve shown as a wave in Fig.4B. It provides a close approximation for the solution of an elliptic integral which, in turn, is the theoretical shape of a meandering stream in a homogeneous medium.

The approach in the second example differs from that adopted for Fig.3. Leopold and Langbein were able to develop a theory for meandering streams resulting in a theoretical equation for the shape. They have tested this theory against nature and the discrepancy between river and dotted line in Fig.4A represents the extent by which the natural phenomenon satisfies the purely theoretical shape. Other geomorphological examples of geometrical forms in nature were given by Cole and King (1968). One field of geology where important advancements have been made in mathematically based theory, leading to predictions for geometrical forms in nature, is the theory of deformation of rocks by folding (see, e.g., Ramberg, 1967; Dieterich, 1970). Theoretical predictions of the coexistence of minerals in igneous and metamorphic rocks as based on thermodynamical considerations including work by Kretz (1970) fall in this category. The theory of heat flow applied to cooling of igneous intrusions provides another example of theoretical curves derived for homogeneous media used in a predictive model (see, e.g., Irvine, 1970).

In quantitative geological models such as the one of Fig.4, the predicted shapes are compared with natural phenomena in order to test the theory. In geological data analysis, relatively little use has been made of theoretical results of this type for several reasons. Suppose that field data are collected in a systematic manner, as in the example of Fig.3, and that trends are to be established by statistical methods. Occasionally, it may be possible to develop a theoretical equation based on a model for the underlying geological process. In general, this equation will be of a more complicated form than a polynomial equation. The result may be that the problem of fitting the theoretical expression to data becomes difficult from a statistical point of view. For example, a so-called nonlinear model may have to be applied as discussed by James (1967). Nonlinear models may not be solved without a fair initial guess at the outcome.

In addition to the mathematical difficulties related to nonlinear models, it is only rarely that a predicted shape is exactly satisfied in nature and can be used as a trend. For example, in Fig.4, the meandering stream did not exactly satisfy the theoretical shape (dotted line) for homogeneous media. In most practical problems, the objective is to describe a natural phenomenon as closely as possible and not its idealized form related to a theoretical model.

The result of these practical difficulties is that, in most situations, we are able to use empirical functions only. The purpose of the mathematical techniques then is to establish trends, clusters or other patterns for which geological explanations are needed.

The drawback of any empirical approach is that it may not be substantially meaningful. Conceivably, more information could be extracted from data if the statistical model is based on a quantitative conceptual model for the underlying geological process (see Krumbein, 1970).

At present, most results of geomathematics are based on the use of empirical functions. Relatively few classes of functions are used and the emphasis is on selecting the best variables to be included in relatively simple equations. For example, most functions used in problems of multivariate analysis are simply linear combinations of variables of the type $y = a_1 x_1 + a_2 x_2 + \ldots$ where the variable y is related to the variables $x_1, x_2, \ldots$ and where $a_1, a_2, \ldots$ are constants. It is unlikely that the underlying geological processes satisfy equations of such a simple structure. In geological situations, however, the deviations from almost any type of model may rapidly become enormous. Under those circumstances, it is good practice to choose and test the simplest possible forms for extrapolation and prediction. Also, optimum methods for measurement procedures are more readily developed for simpler models.

Nevertheless, it should be attempted continuously to develop methods which are also substantially meaningful and where concepts for the time-dependent processes are more directly considered.

Chapter 2

REVIEW OF CALCULUS

2.1. Introduction

The purpose of this chapter is to recapitulate a number of basic concepts from calculus. The treatment of subjects selected for discussion is not comprehensive. In order to learn calculus, the reader should consult one or more out of the many mathematical textbooks in existence. Some of these are listed in section 2.1 of the list for Selected Reading at the end of the book.

The more experienced reader can omit reading this elementary chapter. Apart from providing an introduction to the geologist with little background in mathematics, it can be used: (1) as a refresher course, and (2) as an explanation of terms and notation to be used in later chapters.

A large portion of the theoretical material is introduced by relating it to examples. Three subjects treated in detail are: (1) the graphical representation and properties of elementary curves; (2) the propagation of measurements errors; and (3) simple methods of differentiation and integration to solve geological problems. This includes the calculation of average values from contour maps and the reconstruction of patterns from directional attributes such as the attitudes of folded layers and paleocurrent indicators.

2.2. Classical geometry

Geometry deals with points, lines, surfaces and solids. These concepts are useful to the geologist for representing phenomena which can be observed in the real world. Two-dimensional geometry was developed by Euclid. The fundamental theorem of Euclidean geometry is the Pythagorean theorem for the right-angled triangle represented in Fig. 5A. If the sides of this triangle have lengths a, b, and c, then:

$$a^2 + b^2 = c^2$$

This type of geometry can be extended to three-dimensional space (see Chapter 4). It can also be extended to n-dimensional space with n orthogonal axes which is useful for the study of multivariate systems.

Analytic geometry is the mathematics which consists of the application of algebra to geometrical problems. It was developed by Descartes. Points, lines, surfaces, and solids are studied with reference to a coordinate system. The axes of this so-called Cartesian coor-

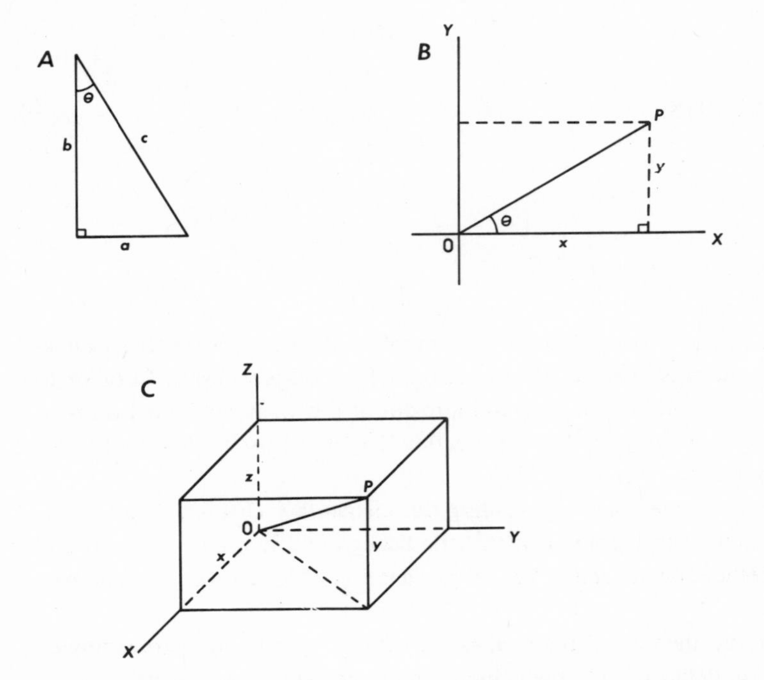

Fig.5. A. Right-angled triangle for which $a^2 + b^2 = c^2$. B. Representation of a point P in plane analytic geometry. C. Ditto, using three-dimensional Cartesian coordinate system.

dinate system are at right angles to one another. For example, in plane analytic geometry, there are two axes: the horizontal X-axis and the vertical Y-axis.

A point P in the plane can be described by its coordinates x and y as shown in Fig. 5B. The point of intersection between X- and Y-axis is the origin O. According to the Pythagorean theorem, OP, which is the distance between O and P, satisfies:

$$OP^2 = x^2 + y^2$$

The distances x and y are referred to as the abscissa and the ordinate of the point P.

A three-dimensional Cartesian coordinate system is shown in Fig. 5C. A Z-axis has been added and it can be readily shown that:

$$OP^2 = x^2 + y^2 + z^2$$

In n-dimensional space, there are n axes. Any one axis is perpendicular to the $(n - 1)$ others. Of course, this situation can not be imagined or represented by means of a diagram. However, the concept is useful as will be seen later. If the n axes are labelled $X_1, X_2, \ldots, X_n$, then:

$$OP^2 = x_1^2 + x_2^2 + \ldots + x_n^2$$

or:

$$OP^2 = \sum_{i=1}^{n} x_i^2$$

where the sigma sign represents the sum with i going from 1 to n.

Let us briefly review the trigonometric functions which are important in analytic geometry. In Fig. 5A, θ represents the angles between sides b and c of the triangle. The angle between sides a and b is 90°. Some definitions are:

$$\text{sine of } \theta = \sin\theta = a/c$$

$$\text{cosine of } \theta = \cos\theta = b/c$$

$$\text{tangent of } \theta = \tan\theta = a/b$$

Also:

$$\theta = \arcsin(a/c) = \arccos(b/c) = \arctan(a/b)$$

This means that θ is the angle (subtending an arc on a circle) for which the sine is equal to the ratio a/c, etc. The angle θ can be represented by using radians or by using degrees. The relationship between radians and degrees is:

$$1 \text{ radian} = 360°/2\pi = 57.3°$$

or:

$$1° = 2\pi/360 = 0.01745 \text{ radians}$$

From the definitions, it follows that:

$$a = c\sin\theta;\; b = c\cos\theta$$

The Pythagorean theorem then becomes:

$$c^2 = a^2 + b^2 = c^2(\sin^2\theta + \cos^2\theta)$$

and:

$$\sin^2\theta + \cos^2\theta = 1$$

Suppose that P_1 and P_2 are two points in the plane with coordinates $P_1 = (x_1, y_1)$ and

$P_2 = (x_2, y_2)$. We can translate the coordinate system by moving the origin to a point O' as follows. If the coordinates of P_1 and P_2 with respect to the new system are $P_1 = (x_1', y_1')$ and $P_2 = (x_2', y_2')$ with:

$$x_1' = 0 \qquad y_1' = y_1 - y_2$$

$$x_2' = x_2 - x_1 \qquad y_2' = 0$$

then the Pythagorean theorem gives:

$$P_1P_2^2 = (x_2 - x_1)^2 + (y_1 - y_2)^2$$

The value of P_1P_2 which is the distance between the points P_1 and P_2 is the square root of this expression or:

$$P_1P_2 = \sqrt{(x_1 - x_2)^2 + (y_1 - y_2)^2}$$

Another way of writing this is:

$$P_1P_2 = \{(x_1 - x_2)^2 + (y_1 - y_2)^2\}^{\frac{1}{2}}$$

Note that the values of x_1 and x_2 have been interchanged. This may be done because $(x_2 - x_1)^2 = (x_1 - x_2)^2$. The reader can verify this application of the Pythagorean theorem by the construction of a diagram.

In plane analytic geometry, the point $P = (x, y)$ can have negative values for x or y. The trigonometric functions can assume negative values also. The reader should be familiar with these possibilities and also with elementary trigonometric identities such as $\sin(\pi/2 - x) = \cos x$ which will be used in the text without proof.

2.3. Variables and functions

A symbol representing any unspecified number of a given set of numbers is called a variable. The numbers of the set are the values of the variable; the entire set of numbers represents the range of the variable.

In the real number system, we can distinguish between: (1) the integers . . . , – 2, – 1, 0, 1, 2, . . . , which arise from counting and do not have a decimal point; and (2) all other real numbers with a decimal point. Complex numbers are of the form $a + bi$ with $i = \sqrt{-1}$. The absolute value of a real number a is positive or zero, and is $-a$ when a is negative; it is written as $|a|$.

If two variables are related in such a manner that to each value of one variable in a given range, there corresponds one value (or more than one value) of another variable, the other variable is called a function of the first variable. The first variable is termed the independent variable and the other (the function) is the dependent variable.

The equation $y = f(x)$ denotes that y is a function of x. The function $f(x)$ can be represented as a curve in plane analytic geometry. $f(a)$ represents the value of $f(x)$ when x has the value a. For example, when $y = f(x) = a + bx + cx^2$, and in a given situation, $a = 1$, $b = 2$, and $c = 3$, then $f(2) = 17$. Setting $x = 2$ determines a point on the curve for y with abscissa 2 and ordinate 17. This point also can be written as (2,17).

Continuity of functions

A function $f(x)$ is continuous for a value $x = a$ if both $\lim_{x \to a} f(x)$ and $f(a)$ exist with:

$$\lim_{x \to a} f(x) = f(a)$$

This means that in the limit, $f(x)$ assumes the value $f(a)$ as x approaches a. The values of x as used in the expression $\lim_{x \to a} f(x)$ can be larger as well as smaller than a.

Examples of functions

Straight line

If c is a constant, $y = f(x) = c$ means that y is equal to c for all values of x. The graphical representation of y is a horizontal line. The expression $y = f(x) = x$ means that, for all values of x, y is equal to x. This equation is represented by a straight line that passes through the origin with (0, 0). The angle of slope is 45°. One can construct functions x^2, $x^3, \ldots, x^n$ where n is a positive integer. If $c_0, c_1, c_2, \ldots, c_n$ denote arbitrary constants:

$$f(x) = c_0 + c_1 x + c_2 x^2 + \ldots + c_n x^n$$

denotes a function that is called a polynomial in x of degree n. It can be said that $f(x)$ is a linear combination of the functions $1, x, x^2, \ldots x^n$. A linear polynomial is of the first degree, e.g., $f(x) = 1 + 2x$. Polynomials of the second, third, fourth, fifth and sixth degree are called quadratic, cubic, quartic, quintic, and sextic, respectively.

The linear polynomial $y = a + bx$ can be represented as a straight line with intercept a and slope b (Fig.6A). If ϕ is the angle of slope, then $b = \tan \phi$. This is equivalent to writing $\phi = \arctan b$ with $-\pi/2 \leqslant \phi \leqslant \pi/2$.

Sine curve

The sine function:

$$y = a \sin(\omega x + \phi) \qquad [2.1]$$

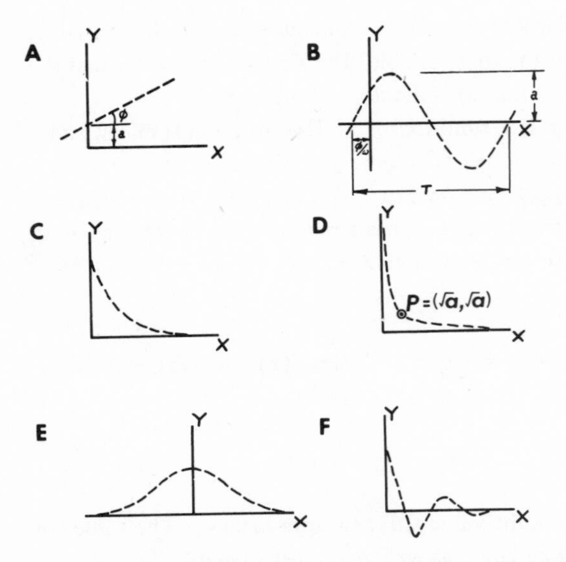

Fig.6. A. Straight line with equation $y = a + bx$ where a is intercept and b = arctan ϕ represents slope. B. Sine-curve with amplitude a, period T and phase ϕ. C. Exponential curve beginning on Y-axis and with X-axis as asymptote. D. Hyperbola $xy = a$ with two asymptotes. E. Gaussian curve. F. Damped cosine-curve.

is shown as a sine curve in Fig.6B. The constant a represents the amplitude or maximum height of the curve; ω is the angular frequency or number of waves over the distance 2π or 360° along the X-axis; the angle ϕ is the phase. The ratio $-\phi/\omega$ is the abscissa of the point on the X-axis where the positive half of the wave has its beginning. The constant $T = 2\pi/\omega$ is called the wavelength or period. It represents the distance from any point on the wave to the corresponding point on the next wave. The constant $f = 1/T$ is called the frequency; it is the number of waves per unit of x.

For a given sine wave, a, f and ϕ are constants. One or more of these constants can be replaced by variables, e.g., if the amplitude is itself a function of x, the maximum height of the new curve may increase or decrease along the X-axis. The constants a and ϕ can be replaced by new constants p and q with $p = a \cos \phi$ and $q = a \sin \phi$, and [2.1] becomes:

$$y = p \sin \omega x + q \cos \omega x \qquad [2.2]$$

The relationships between the constants a and ϕ in [2.1] and p and q in [2.2] also can be written as:

$$a^2 = p^2 + q^2; \quad \phi = \arctan (q/p)$$

In [2.2], $y = f(x)$ is a linear combination of a sine function and a cosine function. Suppose that for [2.1] we write $v = \sin u$ with $u = \omega x + \phi$ and $v = y/a$. Graphically, this means that we are changing the scales along the coordinate axes, whereas the origin is shifted along the horizontal axis so that the phase becomes zero. If, along the new U-axis, the origin is given another shift equal to $\pi/2$ or 90° towards the right, the result is a curve with equation $v' = \sin(u + \pi/2)$. By applying elementary trigonometric considerations, it may be shown that:

$$v' = \sin(\pi/2 - u) = \cos u$$

Consequently, the sine function $\sin \omega x$ and the cosine function $\cos \omega x$ in [2.2] are identical except for a shift in phase equivalent to 90°.

Exponential

The exponential function has equation:

$$y = ce^{-ax} \qquad [2.3]$$

The exponential curve is shown in Fig.6C for positive constants a and c, and positive values of x only. The constant c is the ordinate for the point of intersection with the Y-axis; a determines the rate of decrease in y, e.g., y halves for each increase of ${}^{e}\log(2/a)$ in x. Here ${}^{e}\log$ represents the natural logarithm (base e). When x in [2.3] is replaced by $|x|$, the absolute value of x, negative values of x yield the same values of y as positive values of x with symmetry about the Y-axis.

Hyperbola

The function:

$$y = a/x \qquad [2.4]$$

provides an example of a hyperbola. For large values of x the curve approaches the X-axis, and for large values of y the Y-axis. Straight lines to which a curve approaches are called asymptotes. This hyperbola is shown in Fig. 6D for positive constant a and positive values of x. The point called P in this diagram has both ordinate and abscissa equal to $\sqrt{a}$.

Gaussian curve

The probability curve has equation:

$$y = ce^{-ax^2} \quad (a \text{ and } c > 0) \qquad [2.5]$$

It is represented in Fig.6E. The maximum occurs at the point $(0, c)$; there are two inflection points with abscissa $x = \pm 1/\sqrt{2a}$. An inflection point is defined as a point where a curve changes from concave upward to concave downward or vice versa.

Damped cosine curve

The damped cosine function has equation:

$$y = ce^{-ax} \cos \omega x \qquad [2.6]$$

It is shown in Fig. 6F for positive constants a and c and $x > 0$. It can be considered as a cosine curve with angular frequency ω and phase zero whose amplitude decreases in the positive X-direction. The maxima fall on a curve with [2.3].

2.4. Transformations

Eq. [2.1] reduces to a simpler form by changing the coordinate system. An operation of this type is called a transformation.

If $y = f(x)$, the variables x and y can be transformed by defining new variables u and v, which are functions of x and y, respectively. If $v = h(y)$ is plotted against $u = g(x)$, the result is another representation of $y = f(x)$. In general, (1) the origin with $x = 0$ and $y = 0$ is moved to a new position where $u = 0$ and $v = 0$; and (2) the scales along the X-axis and Y-axis are changed so that they become linear scales for the U-axis and V-axis. When the transformation consists of moving the origin without a change in scales, it is called a translation.

A more general transformation is to define new functions u and v which are both a function of x and a function of y. This implies that, in addition to the changes mentioned above, the U- and V-axes may appear as curves in the rectangular coordinate system for the variables x and y, and vice versa.

Suppose that u and v are both linear combinations of x and y, with:

$$u = ax + by; v = cx + dy$$

This is an example of a linear transformation. If the angle between the U- and V-axes is equal to 90°, this transformation is called a rotation. Examples of rotations and other linear transformations will be given later (Chapter 4). Calculations for this type of transformation are greatly simplified by the use of matrix algebra (Chapter 3).

Straight line

If the variable x is transformed into $u = p + qx$, the equation $y = a + bx$ becomes:

$$y = g(u) = (a - pb/q) + (b/q)u$$

$y = g(u)$ also represents a straight line but its intercept and slope differ from those of $y = f(x)$. If $p = a$, and $q = b$, it follows that $y = g(u) = u$ with intercept zero and slope equal to one.

The transformation $u = p + qx$ is linear in x. By subjecting both x and y to linear trans-

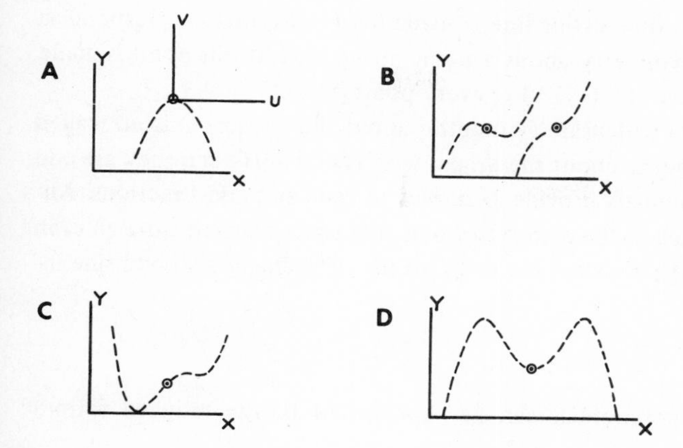

Fig.7. Polynomial curves with origin (O) of translated (*UV*) coordinate system. A. Parabola. B. Cubic curves with and without two extrema. C. Quartic curve. D. Symmetrical quartic (without linear term).

formations of this type, it is possible to simplify the form of polynomials such as the quadratic, cubic and quartic functions.

Parabola

If:

$$u = x + b/2c; \qquad v = y - a + b^2/4c$$

the quadratic polynomial $y = a + bx + cx^2$ becomes $v = cu^2$. This operation is a translation (see Fig. 7A). A further transformation $v' = v/c$ would yield $v' = u^2$ and all quadratic polynomials can be reduced to this form.

The equation $v = cu^2$ represents a parabola. It is symmetrical with respect to the V-axis; the parabola shown in Fig. 7A has c equal to a negative constant, indicating that the curve is concave upwards. The absolute value of c determines the rate at which the curve falls off.

Symmetry

The functions x^2, x^4, x^6, . . . are even powers of x; x, x^3, x^5, . . . are odd powers. Polynomials which are linear combinations of even powers of x, are symmetric with respect to a line (Y-axis). If the polynomial contains odd powers of x only, it is symmetric with respect to a point (the origin when the constant term is zero). A polynomial which contains both even and odd powers of x can be divided into two parts, one for the even terms and the other for the uneven terms. The curve for the first part is symmetric with respect to a line, and that for the second part is symmetric with respect to a point.

Symmetry about a line means that if this line is made the V-axis, there corresponds a $(-u, v)$ to every point (u, v). Symmetry about a point means that, if this point is made the origin, there corresponds a point $(-u, -v)$ to every point (u, v).

Functions with $f(x) = f(-x)$ which are symmetric about the vertical axis are called even functions. If they are symmetric about the origin, with $f(x) = -f(-x)$, they are odd functions. The low-degree polynomials provide examples of even and odd functions. Another example is y in [2.2] which is the sum of an odd function ($p \sin \omega x$) and an even function ($q \cos \omega x$). A shift of the Y-axis equal to $\frac{1}{4}T$ to the right changes an odd sine into an even cosine curve.

Cubic polynomial

By a translation any cubic polynomial can be represented in the reduced form :

$$v = au + bu^3 \qquad [2.7]$$

The curve then becomes symmetric with respect to the new origin, which falls at its point of inflection; a represents the slope at the origin. Two cubic curves are shown in Fig.7B.

Quartic polynomial

A quartic polynomial can be represented by:

$$v = au + bu^2 + cu^4 \qquad [2.8]$$

The cubic term has been eliminated by translation. The average abscissa of the three extrema occurs at the new origin and a represents the slope at this point (see Fig.7C). A quartic function can have three extrema (maxima or minima). In general, the maximum possible number of extrema in a polynomial curve is one less than its degree. The curve of degree 20 in Fig. 3 shows 14 extrema.

After translation, the linear and cubic polynomial become odd functions, and the quadratic an even function. The quartic polynomial can not be changed into an odd or even function by translation. Eq. [2.8] consists of a linear part which is odd and the even part:

$$v' = bu^2 + cu^4 \qquad [2.9]$$

This is the equation of a symmetrical quartic. The central extremum now occurs at the point where $v' = u = 0$ (see Fig. 7D).

Polar coordinates

A point $P = (x, y)$ in the rectangular coordinate system also can be located by its polar coordinates r and θ. In Fig. 5B, the distance OP is equal to $r = \sqrt{x^2 + y^2}$ and θ is the angle between OP and the X-axis as measured in the counter-clockwise direction. Hence, $\theta = \arctan(y/x)$.

A circle has the property that all its points are at the same distance from the center. When the center falls at the origin, the equation $x^2 + y^2 = r_1^2$ represents a circle with radius r_1. If polar coordinates are used, this equation reduces to the form $r = r_1$.

A more complicated example is as follows. Suppose that a curve satisfies the equation:

$$r = ce^{-a\theta}$$

This is [2.3] with x and y replaced by θ and r. If r and θ are polar coordinates, the corresponding curve in the XY-system is the logarithmic spiral. It resembles the winding curve described by certain ammonites.

The shape of meanders

An example of a geometrical form in nature are river meanders as discussed in section 1.5. Leopold and Langbein (1966) proposed that the theoretical curve for an idealized meander satisfies the expression:

$$\theta = a \sin (\omega u + \phi)$$

which is equivalent to [2.1]. In the XY-system, u represents distance as measured along the channel of the meandering river; θ is the angle between the channel and the average down-valley direction. A representation of a theoretical meander in the XY-system is given by the dashed line in Fig.4A. The advantage of using transformed variables here is that it is relatively easy to compare the observed data with a sine curve while checking for departures between the mathematical model and reality.

2.5. Other concepts of plane analytic geometry

Points on a straight line always satisfy the general equation of the first degree:

$$px + qy + r = 0 \qquad [2.10]$$

where p, q and r are constants. This expression is equivalent to $y = a + bx$ as used before. Division by $\sqrt{p^2 + q^2}$ gives the general equation:

$$\frac{px}{\sqrt{p^2 + q^2}} + \frac{qy}{\sqrt{p^2 + q^2}} + \frac{r}{\sqrt{p^2 + q^2}} = 0$$

Because p, q, and $\sqrt{p^2 + q^2}$ can be seen as the sides of a right-angled triangle, we may define:

$$\cos \alpha = \frac{p}{\sqrt{p^2 + q^2}} \; ; \quad \sin \alpha = \frac{q}{\sqrt{p^2 + q^2}}$$

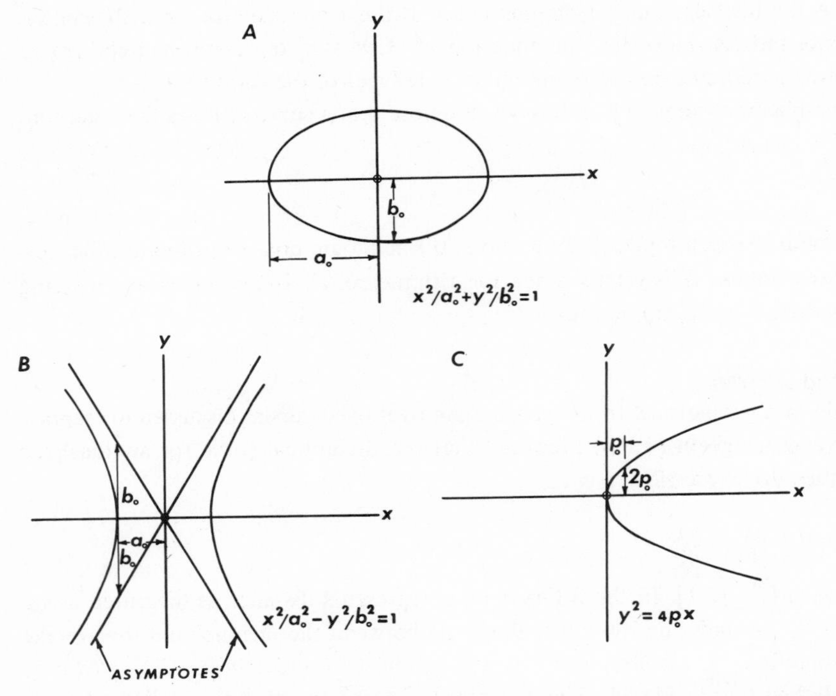

Fig.8. Three possible shapes of curves arising from the general equation of the second degree. A. Ellipse. B. Hyperbola. C. Parabola. Origin was translated and coordinate system was rotated to obtain simplified equations with coefficients that can be readily interpreted.

The general equation then can be written in its so-called normal form:

$$x \cos \alpha + y \sin \alpha = d_0 \qquad [2.11]$$

where:

$$d_0 = \frac{-r}{\sqrt{p^2 + q^2}}$$

From comparing this result with the expression $y = a + bx$, it follows that $\tan \alpha = -1/b$. The constant d_0 represents the perpendicular distance from the origin to the line.

The general equation of the second degree in x and y is:

$$ax^2 + bxy + cy^2 + dx + ey + f = 0 \qquad [2.12]$$

When $b = c = 0$, this expression reduces to y as quadratic polynomial in x. If $a=c$ and $b = 0$, the curve satisfying the equation is a circle which is a special form of the ellipse.

The three main types of curves that satisfy [2.12] are shown in Fig.8. The curve is an ellipse if in [2.12] $b^2 > 4ac$, or a hyperbola if $b^2 < 4ac$. In the special case $b^2 = 4ac$, the curve is a parabola whose axis of symmetry is not necessarily parallel to the Y-axis as when $b = c = 0$.

It will be shown later (Chapter 4) that [2.12] can always be reduced to the form $pu^2 + qv^2 = 1$ by employing a translation and a rotation. The sign of the product pq then indicates the type of the curve. It is positive for an ellipse, negative for a hyperbola and zero for a parabola. A third method of writing these equations is given in Fig.8.

2.6. Differential calculus

Differential calculus is concerned with finding the rate of change of a given function. The first derivative of $y = f(x)$ with respect to x is defined by:

$$f'(x) = \mathrm{d}y/\mathrm{d}x = \lim_{\Delta x \to 0} \frac{\Delta y}{\Delta x} \qquad [2.13]$$

The first derivative at $x = a$ or $[f'(x)]_{x=a} = f'(a)$ represents the slope of the tangent to the curve at the point with $x = a$. If ϕ is the angle of slope of $y = f(x)$, then:

$$\frac{\mathrm{d}y}{\mathrm{d}x} = \tan \phi \qquad [2.14]$$

This concept is illustrated in Fig.9A.

A slightly different form of the definition of the first derivative is:

$$f'(x) = \lim_{\Delta x \to 0} \frac{f(x + \Delta x) - f(x)}{\Delta x} \qquad [2.15]$$

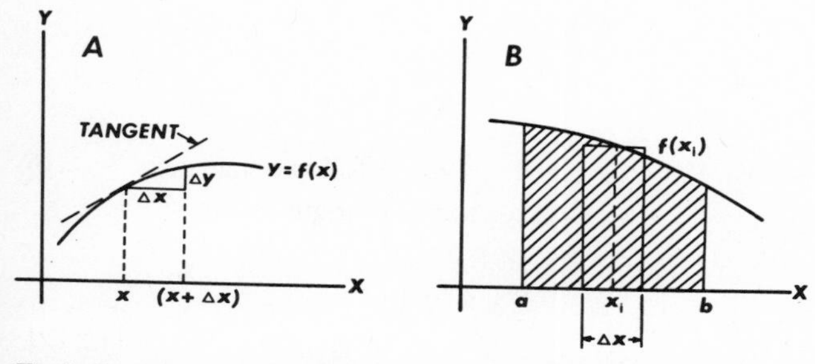

Fig.9. Graphical representation of (A) first derivative (limiting form of ratio $\Delta Y/\Delta X$) at point with abscissa (x), and (B) definite integral representing area under curve between $x = a$ and $x = b$.

Note that the limit cannot exist if the slope changes abruptly at the value of x being considered : $f'(x)$ would then be discontinuous. Ordinarily, only functions with continuous derivatives are differentiated.

The so-called differential $\mathrm{d}y$ of the function $y = f(x)$ satisfies the equation:

$$\mathrm{d}y = f'(x)\,\mathrm{d}x$$

Some important derivatives which occur frequently are presented below without proof.

If c is a constant, $y = c$ represents a horizontal line with slope zero and $\mathrm{d}c/\mathrm{d}x = 0$. A straight line through the origin that dips 45° has equation $y = x$ and derivative $\mathrm{d}y/\mathrm{d}x = 1$.

Other derivatives are:

$$\frac{\mathrm{d}(cx)}{\mathrm{d}x} = c \qquad \frac{\mathrm{d}(x^p)}{\mathrm{d}x} = px^{p-1}$$

$$\frac{\mathrm{d}(\sin x)}{\mathrm{d}x} = \cos x \qquad \frac{\mathrm{d}(\cos x)}{\mathrm{d}x} = -\sin x$$

$$\frac{\mathrm{d}(\mathrm{e}^x)}{\mathrm{d}x} = \mathrm{e}^x \qquad \frac{\mathrm{d}(\ln x)}{\mathrm{d}x} = 1/x$$

When u is a function of x and $f(u)$ a function of u, then:

$$\frac{\mathrm{d}f(u)}{\mathrm{d}x} = \frac{\mathrm{d}f}{\mathrm{d}u}\,\frac{\mathrm{d}u}{\mathrm{d}x}$$

When u and v are functions of x:

$$\frac{\mathrm{d}(u+v)}{\mathrm{d}x} = \frac{\mathrm{d}u}{\mathrm{d}x} + \frac{\mathrm{d}v}{\mathrm{d}x}$$

$$\frac{\mathrm{d}(uv)}{\mathrm{d}x} = u\frac{\mathrm{d}v}{\mathrm{d}x} + v\frac{\mathrm{d}u}{\mathrm{d}x}$$

$$\frac{\mathrm{d}(u/v)}{\mathrm{d}x} = \frac{v\dfrac{\mathrm{d}u}{\mathrm{d}x} - u\dfrac{\mathrm{d}v}{\mathrm{d}x}}{v^2}$$

Higher derivatives

These are obtained by differentiation of successive derivatives. For example, the second derivative of $f(x)$ satisfies :

$$f''(x) = \frac{df'(x)}{dx}$$

Partial derivatives

If y is a function of a number of independent variables $x_1, x_2, \ldots, x_n$ or:

$$y = f(x_1, x_2, \ldots, x_n)$$

then $\partial y/\partial x_i$ ($i = 1, 2, \ldots, n$) is the partial derivative of y with respect to x_i. It means that y is differentiated with respect to x_i whereby all other independent variables x_k ($k \neq i$) are kept constant.

The so-called total differential of y is dy with :

$$dy = \frac{\partial y}{\partial x_1} dx_1 + \frac{\partial y}{\partial x_2} dx_2 + \ldots + \frac{\partial y}{\partial x_n} dx_n$$

If u is a function of x and y, the second partial derivatives of u are:

$$\frac{\partial}{\partial x}\left(\frac{\partial u}{\partial x}\right) = \frac{\partial^2 u}{\partial x^2}$$

$$\frac{\partial}{\partial y}\left(\frac{\partial u}{\partial y}\right) = \frac{\partial^2 u}{\partial y^2}$$

$$\frac{\partial}{\partial x}\left(\frac{\partial u}{\partial y}\right) = \frac{\partial}{\partial y}\left(\frac{\partial u}{\partial x}\right) = \frac{\partial^2 u}{\partial x \partial y}$$

The third relationship holds true when the first and second partial derivatives are continuous. In this case:

$$\frac{\partial^2 u}{\partial x \partial y} = \frac{\partial^2 u}{\partial y \partial x}$$

and the order of differentiation is immaterial. Similar notation is used for higher partial derivatives.

Extrema and inflection points

A function $f(x)$ has a maximum or a minimum at $x = a$, when $f'(a) = 0$. This follows directly from the definition of the first derivative: At an extremum, the slope or rate of change of a curve is equal to zero. The second derivative $f''(x)$ will not be zero at an extremum. It represents the change in slope and is negative around a maximum and positive for a minimum.

A function $f(x)$ has an inflection point at $x = a$ if the second derivative $f''(a) = 0$. At such a point the curve changes from concave upward to concave downward or vice versa.

Example

The probability curve has $f(x) = ce^{-ax^2}$. Putting $u = -ax^2$ gives $f'(x) = ce^u\,du/dx = -2acx\,e^{-ax^2}$. Further, $f''(x) = 2ac\,(2ax^2 - 1)\,e^{-ax^2}$. For the extremum $f'(x) = 0$ or $x = 0$. At this point $f''(0)$ is negative which shows that the extremum is a maximum.

The two inflection points satisfy $f''(x) = 0$ or $2ax^2 - 1 = 0$. This gives $x^2 = 1/2a$ with solutions $x_1 = 1/\sqrt{2a}$ and $x_2 = -1/\sqrt{2a}$.

Other examples of differentiation will be given in section 2.8 which deals with the propagation of errors. An introduction to infinite series is presented first.

2.7. Infinite series

An infinite power series is defined as:

$$f(x) = \sum_{n=0}^{\infty} c_n x^n = c_0 + c_1 x + \ldots + c_n x^n + \ldots$$

A power series in powers of $(x - a)$ instead of x is:

$$f(x) = \sum_{n=0}^{\infty} c_n (x-a)^n = c_0 + c_1(x-a) + \ldots + c_n(x-a)^n + \ldots \qquad [2.16]$$

This is the Taylor series expansion of $f(x)$ about $x = a$ if the coefficients c_n satisfy:

$$c_0 = f(a)\ ;\ c_1 = \frac{f'(a)}{1!}\ ;\quad c_2 = \frac{f''(a)}{2!}\ ;\ \ldots\ ;\ c_n = \frac{f^{(n)}(a)}{n!}$$

Here $f^{(n)}(a)$ represents the n-th derivative of $f(x)$ with respect to x at point $x = a$ and $n!$ means factorial n or:

$$n! = n(n-1)(n-2)\ldots 3\cdot 2\cdot 1$$

The expressions for the coefficients c_n are derived as follows. We have:

$$f(x) = c_0 + c_1(x-a) + c_2(x-a)^2 + c_3(x-a)^3 + \ldots$$

$$f'(x) = c_1 + 2c_2(x - a) + 3c_3(x - a)^2 + \dots$$

$$f''(x) = 2c_2 + 3 \cdot 2c_3(x - a) + \dots$$

$$f^{(n)}(x) = n(n - 1)(n - 2) \dots 3 \cdot 2c_n + \dots$$

If, in the successive derivatives, x is set equal to a, all terms on the right-hand side of these expressions disappear except the first one. A Taylor series therefore can be written as:

$$f(x) = f(a) + \frac{f'(a)}{1!}(x - a) + \frac{f''(a)}{2!}(x - a)^2 + \dots \qquad [2.17]$$

In the case when $a = 0$, the Taylor series of $f(x)$ becomes:

$$f(x) = f(0) + \frac{f'(0)}{1!}x + \frac{f''(0)}{2!}x^2 + \dots \qquad [2.18]$$

This is called the Maclaurin series.

Example

The exponential function $f(x) = e^x$ has:

$$f(x) = f'(x) = f''(x) = \dots = f^{(n)}(x) = \dots = e^x$$

and $f^{(n)}(0) = 1$ for any value of n. Consequently, its Maclaurin expansion is:

$$e^x = 1 + x + \frac{x^2}{2!} + \frac{x^3}{3!} + \dots + \frac{x^n}{n!} + \dots \qquad [2.19]$$

The reader may verify himself that:

$$\left.\begin{aligned} \sin x &= x - \frac{x^3}{3!} + \frac{x^5}{5!} - \frac{x^7}{7!} + \dots \\ \cos x &= 1 - \frac{x^2}{2!} + \frac{x^4}{4!} - \frac{x^6}{6!} + \dots \end{aligned}\right| \qquad [2.20]$$

Note that the infinite series for $\sin x$ is an odd function and that for $\cos x$ an even function.

Taylor series of a function of many variables

Suppose that $f(x_1, x_2, \ldots, x_n)$ represents a function of n variables x_i ($i = 1, 2, \ldots, n$). The Taylor expansion about a general point $(a_1, a_2, \ldots, a_n)$ satisfies:

$$f(x_1, x_2, \ldots, x_n) = f(a_1, a_2, \ldots, a_n) + \sum_{i=1}^{n} \left[\frac{\partial f}{\partial x_i}\right]_{x_i=a_i} (x_i - a_i)$$

$$+ \tfrac{1}{2}\sum_{i} \left[\frac{\partial^2 f}{\partial x_i^2}\right]_{x_i=a_i} (x_i - a_i)^2 + \sum_{\substack{i,j \\ i \neq j}} \left[\frac{\partial^2 f}{\partial x_i \partial x_j}\right]_{\substack{x_i = a_i \\ x_j = a_j}} (x_i - a_i)(x_j - a_j)$$

$$+ \text{higher order terms.}$$

If terms higher than linear are neglected, the result also can be written as:

$$f(x_1, x_2, \ldots, x_n) = f(a_1, a_2, \ldots, a_n)$$

$$+ \left[\frac{\partial f}{\partial x_1}\right]_{x_1=a_1} (x_1 - a_1) + \left[\frac{\partial f}{\partial x_2}\right]_{x_2=a_2} (x_2 - a_2) + \ldots \qquad [2.21]$$

The corresponding Maclaurin expansion is obtained by setting $a_1 = a_2 = \ldots = a_n = 0$.

2.8. Propagation of error

Suppose that two variables y and x are related by $y = f(x)$; one or more values of x can be measured and the corresponding values of y calculated by using the functional relationship. The error in x is known to be equal to Δx. The problem is to determine Δy which is the resulting error in y.

From the definition of the derivative (see Fig.9A), it follows that:

$$\Delta y \approx \frac{dy}{dx} \Delta x \qquad [2.22]$$

if Δy and Δx are both small. This is the rule for the propagation of error or error rule.

Propagation of rounding-off errors

Suppose that $y = x^2$ and that the observed value of x is $x_0 = 3$ whereas the true value of x is $x_t = 2.5$. The error in x_0 can be defined as:

$$\Delta x_0 = x_0 - x_t = 0.5$$

Let us first calculate the "true" error in y to be denoted as Δy_t. From $x_t = 2.5$, it follows that $y_t = 6.25$. On the other hand, $y_0 = x_0^2 = 9$.
Hence:

$$\Delta y_t = y_0 - y_t = 9 - 6.25 = 2.75$$

If the error rule is applied, it follows from $dy/dx = 2x$, that $\Delta y \approx 2x\Delta x$. This leads to $\Delta y_0 = 2x_0\Delta x_0 = 2 \times 3 \times 0.5 = 3$. It is concluded that there is a discrepancy between $\Delta y_0 = 3$ and $\Delta y_t = 2.75$. This difference is equal to $3 - 2.75 = 0.25$ and illustrates the approximate nature of the error rule.

Suppose now that $y = x^2$ and $x_0 = 3$, but that $x_t = 3.01$. The reader may verify that the error rule yields $\Delta y_0 = -0.06$ which differs from Δy_t by only 0.0001.

These two examples illustrate that Δx and Δy should be small as compared to x and y before the error rule is applicable. It can be applied to solve problems such as how many digits should be retained in a number that is derived from another number by some calculation if this other number has a round-off error.

Until now, it has been assumed that both the observed and true value of x are known. In other applications, one may only know the observed value of x whereas the information on the error in x is of a more general nature. For example, the standard error of x may be known. Then the problem consists of calculating the propagation of the standard error. This concept will be briefly explained.

Standard error

The standard error of x may be obtainable by the repetition of measurements for a single value of x. It is calculated in the same manner as the standard deviation, which is a more general concept from the theory of statistics (see Chapter 6). Estimates for the standard error and the standard deviation will both be indicated as $s(x)$.

If an unknown value of x is measured n times, the observed values can be written as:

$$x_1, x_2, x_3, \ldots , x_n$$

Usually, the best estimate of the unknown value of x is provided by the mean or average:

$$\bar{x} = (x_1 + x_2 + \ldots + x_n)/n$$

$$\bar{x} = (1/n) \sum_{i=1}^{n} x_i \qquad [2.23]$$

Approximate errors in individual observations are obtained by subtracting $\bar{x}$ from the data which yields the deviations:

$$x_1 - \bar{x},\ x_2 - \bar{x},\ x_3 - \bar{x},\ \ldots\ ,\ x_n - \bar{x}$$

The standard error then is calculated as:

$$s(x) = \sqrt{\frac{\sum_i (x_i - \bar{x})^2}{n-1}} \qquad [2.24]$$

The square of the standard error is called the variance $s^2(x)$ which is approximately equal to the average square deviation from the mean $\bar{x}$. The sum of the squared deviations $\Sigma_i(x_i - \bar{x})^2$ is divided by $(n-1)$ instead of by n to account for the fact that $\bar{x}$ is an estimate of a true mean. Therefore, $\bar{x}$ is subject to an error which decreases when n increases (also see Chapter 6). This error of the mean implies that use of n instead of $(n-1)$ in [2.24] results in slight underestimation of the standard error.

It is frequently assumed that the errors have a Gaussian probability distribution. The probability then is small that a single error will exceed two or three times the standard error at either side of the mean. An error larger than two times the standard error would occur with a frequency of once per twenty measurements.

An error of more than three times the standard error has a frequency of less than one out of a hundred. Contrary to Δx_0 in the previous examples, which can be negative, the standard error $s(x)$ is always positive.

If $s(y)$ and $s(x)$ are small as compared to y and x which, as before, are related by $y = f(x)$, then:

$$s(y) \approx \frac{dy}{dx}\, s(x)$$

Error in X-ray determinations for olivine

Jambor and Smith (1964) in their table II have listed a number of X-ray measurements for olivines used in a determinative curve. The measurements in this example from mineralogy are for the 2θ angle of the 174-reflection by diffractometer. This angle is reported in degrees. Eighteen out of 26 measurements were done by different workers. From this, Agterberg (1964c) calculated that the standard error $s(2\theta)$ amounts to 0.0453°. Bragg's equation:

$$2d_{174} \sin\theta = 1.93579$$

can be used to obtain values for the cell edge d_{174} of olivine. A problem consists of determining, in a rapid way, $s(y)$ from the known value of $s(2\theta)$ where y represents the cell edge d_{174}.

First, we derive the relationship:

$$s(\theta) = \tfrac{1}{2} s(2\theta) = 0.0277^\circ \quad 0.0227^\circ$$

If θ is referred to as x, Bragg's equation can be written as:

$$y = 1.93579/(2 \sin x) = c/\sin x$$

Further:

$$\frac{dy}{dx} = c \frac{d(1/\sin x)}{dx} = -c \frac{\cos x}{\sin^2 x}$$

Hence:

$$s(y) \approx c \left[\frac{\cos x}{\sin^2 x}\right] s(x)$$

The minus sign was omitted because it is immaterial if standard errors are being used.

All measurements are on values of 2θ which are close to 140°. Hence $x \approx 70°$ which yields:

$$\left[\frac{\cos x}{\sin^2 x}\right] \approx 0.387$$

Consequently:

$$s(y) \approx 0.96789 \times 0.387 \times 0.0227^\circ = 0.00850^\circ = 1.48 \cdot 10^{-4} \text{ Å}..$$

Discussion

The purpose of this example was to illustrate the use of the error rule. A more accurate estimate of $s(y)$ can be obtained by application of Bragg's equation to all original data of 2θ and then estimating $s(y)$ as $s(x)$ was estimated before. This more accurate method gives $s(y) = 1.51 \cdot 10^{-4}$ Å. The advantage of using the error rule is that it is quicker but, in this example, it is used at the expense of committing an error equal to $0.03 \cdot 10^{-4}$ Å or about 2%.

2.9. Application of Taylor series

In the beginning of section 2.8, use was made of x_t and x_0. If these quantities are substituted for x and a in [2.17], the result is:

$$f(x_t) = f(x_0) + f'(x_0)(x_t - x_0) + \frac{f''(x_0)}{2!}(x_t - x_0)^2 + \ldots$$

When by our definition, $\Delta x_0 = x_0 - x_t$, and $\Delta y_t = y_0 - y_t = f(x_0) - f(x_t)$, it follows that

$$\Delta y_t = f'(x_0)\,\Delta x_0 - \frac{f''(x_0)\,\Delta x_0^2}{2!} + \ldots \qquad [2.25]$$

For the example $y = x^2$; $f''(x_0) = 2$ and all higher derivatives are equal to zero.

The first term in [2.25] gives 3 as before. It is not surprising that the second term $-f''(x_0)\,\Delta x_0^2/2 = -0.25$ is exactly equal to the discrepancy between Δy_0 and Δy_t as calculated previously.

The error rule is equivalent to expanding a function as a Taylor series and then neglecting all terms with powers higher than one. If $y = f(x_1, x_2, \ldots, x_n)$ represents a function of n variables $x_1, x_2, \ldots, x_n$, then the error Δy in y satisfies:

$$\Delta y \approx \frac{\partial f}{\partial x_1}\Delta x_1 + \frac{\partial f}{\partial x_2}\Delta x_2 + \ldots\ \ldots + \frac{\partial f}{\partial x_n}\Delta x_n \qquad [2.26]$$

This result can be derived from the definition of the total differential or, alternatively, from the Taylor expansion shown in [2.21].

Squaring both sides of [2.26] and writing f_i' instead of $\partial f/\partial x_i$ yields:

$$\Delta y^2 \approx (f_1'\,\Delta x_1)^2 + (f_2'\,\Delta x_2)^2 + \ldots + (f_n'\,\Delta x_n)^2 + 2f_1' f_2'\,\Delta x_1\,\Delta x_2 + 2f_1' f_3'\,\Delta x_1\,\Delta x_3 + \ldots \qquad [2.27]$$

Now let all errors be replaced by standard errors. This is equivalent to allowing the errors $\Delta x_i (i = 1, 2, \ldots, n)$ to take on all possible values within their respective ranges of variation. The result is:

$$s^2(y) = \left(\frac{\partial f}{\partial x_1}\right)^2 s^2(x_1) + \left(\frac{\partial f}{\partial x_2}\right)^2 s^2(x_2) + \ldots + \left(\frac{\partial f}{\partial x_n}\right)^2 s^2(x_n) \qquad [2.28]$$

The square of the standard error is the variance and [2.28] is known as the propagation of variance. The standard error $s(y)$ is obtained by taking the square root on both sides of [2.28]. Note that the cross-product terms $2f_1'f_2'\Delta x_1\Delta x_2$, etc., which occur in [2.27] were omitted in [2.28]. This can only be done when the errors in $x_1, x_2, \ldots, x_n$ cancel out during the summation of their cross-products. By going from [2.27] to [2.28], each term of [2.27] is replaced by its average value and the average value for the product of uncorrelated errors $\Delta x_i \Delta x_j$ $(i \neq j)$ is equal to zero.

Error in the arithmetic average

The average of n observations is defined as $y = \bar{x} = (1/n)(x_1 + x_2 + \ldots + x_n)$. Suppose that the measurements are all subject to the same standard error $s(x)$. The propagation of variance is:

$$s^2(y) = (1/n)\, s^2(x) \qquad [2.29]$$

Because the mean $\bar{x}$ is a linear combination of the observations, all higher derivatives such as $\partial^2 y/\partial x_i \partial x_j$ are zero, and [2.29] is free of errors due to neglecting terms in the Taylor expansion. Eq. [2.29] represents one of the more important theorems of statistical theory.

2.10. Integral calculus

Integral calculus is concerned with the inverse problem to differential calculus, that is of finding a function when its rate of change is given.

If $F(x)$ is a function whose first derivative is $f(x)$ so that $F'(x) = f(x)$, then $F(x)$ is called an integral of $f(x)$. The most general integral of $f(x)$ can be denoted as $\int f(x)\, dx$ and is of the form $F(x) + C$ where C is an arbitrary constant. The symbol $\int$ is a stylized S representing "Sum". Because C is arbitrary, $\int f(x)\, dx$ is called an indefinite intregal.

Example

When $f(x) = a + bx + cx^2$, $\int f(x)\, dx = \int (a + bx + cx^2)\, dx = ax + bx^2/2 + cx^3/3 + C$, because $d/dx \int f(x)\, dx = f(x)$. During differentiation the constant C is eliminated, because $dC/dx = 0$ regardless of the value of C.

The definite integral

The definite integral arises when the integration is performed between defined limits. For example, the function $y = f(x)$ is graphically represented in Fig. 9B. The definite integral:

$$\int_a^b f(x)\,\mathrm{d}x$$

is equal to the area under the curve between the limits $x = a$ and $x = b$. It can be defined by the limiting process:

$$\int_a^b f(x)\,\mathrm{d}x = \lim_{\substack{n \to \infty \\ \Delta x \to 0}} \sum_{i=1}^{n} f(x_i)\,\Delta x \qquad [2.30]$$

where x_i denotes a sequence of values along the X-axis with $i = 1, 2, \ldots, n$ so that the value $(b - a)$ is divided into n equal parts Δx.

Proof

From elementary geometry, it follows that the area under the curve between limits a and b is:

$$\text{Area} \approx f(x_1)\,\Delta x + f(x_2)\,\Delta x + \ldots + f(x_n)\,\Delta x = \sum_{i=1}^{n} f(x_i)\,\Delta x$$

This expression provides an approximation for the definite integral. The difference between this area and the definite integral can be made as small as we wish by decreasing Δx and increasing $n (n \to \infty$ and $\Delta x \to 0)$.

The relationship between definite and indefinite integrals is:

$$\int_a^b f(x)\,\mathrm{d}x = \left[\int f(x)\,\mathrm{d}x\right]_a^b = [F(x)]_a^b = F(b) - F(a) \qquad [2.31]$$

Example

A parabola with equation $f(x) = -x^2 + 6x - 5$ has $F(x) = -x^3/3 + 3x^2 - 5x + C$. Its curve intersects the X-axis at the points $x = 1$ and $x = 5$. The area enclosed by the curve and the X-axis therefore satisfies:

$$\int_1^5 f(x)\,\mathrm{d}x = F(5) - F(1) = 10\tfrac{2}{3}$$

The mean value or average of a function $f(x)$ between the limits a and b is defined as $1/(b - a) \int_a^b f(x)\,dx$. In the example, the mean of the parabola between the points $x = 1$ and $x = 5$ amounts to $2\frac{2}{3}$.

A continuous function $f(x)$ without discontinuous changes in slope always has a first derivative with respect to x which can be calculated readily. However, the indefinite integral of $f(x)$ does not necessarily exist as a simple analytic expression. For example, a simple expression for the indefinite integral of $f(x) = ce^{-ax^2}$ (probability curve) can not be found, although definite integrals $\int_{x_1}^{x_2} f(x)\,dx$ $(x_2 > x_1)$ can be calculated for any limits x_1 and x_2 by using truncated series expansions.

The purpose of the following examples is to familiarize the reader with the concepts of graphical integration and differentiation. The geologist is frequently faced with situations where it may not be advisable to represent observed curves by equations (e.g., folded strata in a cross-section). In situations, where the prime purpose is representation of reality, replacing the observed curves by analytic expressions might introduce discrepancies which can be avoided by employing graphical methods.

Reconstruction of an isoclinal fold by graphical integration

A number of attitudes of strata suggesting an anticline were observed along a line across the topographic surface. The observation points are labelled 0, 1, 2, . . . , n ($n = 4$ in Fig. 10). The objective is to reconstruct a complete pattern for this fold in vertical cross-section.

It will be assumed that the isoclines (lines of equal dip) in the profile are parallel to a line through O which coincides with the Y-axis in Fig. 10. The projections of points 0, 1, 2, . . . , n on the corresponding X-axis through O are called $x_0, x_1, x_2, \ldots, x_n$, or, in abbreviated notation, x_i $(i = 0, 1, 2, \ldots, n)$.

Let α_i be the angles of dip with respect to the X-axis, then the curve for a function $f(x) = \tan\alpha$ can be constructed by using the $(n + 1)$ known values α_i. The points P_i in Fig.10 have ordinates equal to $\tan\alpha_i$.

Integration of $f(x)$, which is a smooth curve passing through the P_i, will give a function y with $(dy/dx) = f(x)$ and:

$$y = \int f(x)\,dx = F(x) + C$$

As before, C represents an arbitrary constant.

In Fig.10, use is made of the method of graphical integration. The curve $f(x)$ is replaced by a staircase function $f^*(x)$ with the property:

$$\int_{x_i}^{x_{i+1}} f(x)\,dx = \int_{x_i}^{x_{i+1}} f^*(x)\,dx$$

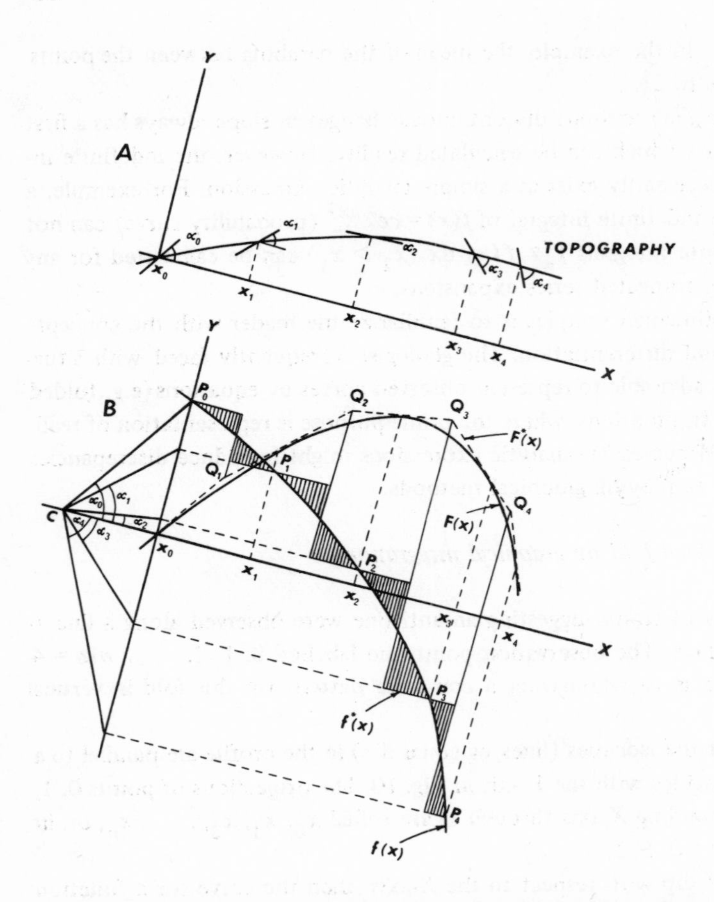

Fig.10. Reconstruction of isoclinal fold by graphical integration as done by Trooster (1950). A. Five dips along topographic surface are given; (*XY*)-system is constructed with *Y* parallel to isoclines. B. Curve *F*(*x*) has the known dips at intersection points with isoclines through observation points. (Modified after Van Landewijk, 1954.)

where x_i and x_{i+1} are abscissas of two successive points P_i and P_{i+1} $(i = 0, 1, \ldots, n-1)$. The function $f^*(x)$ is readily integrated yielding $F^*(x) + C$. The constant C is specified by letting $F^*(x)$ pass through O. Note that $F^*(x)$ consists of a succession of straight line segments between points Q_i and Q_{i+1} and with dips equal to α_i. $F(x)$ is a smooth curve that coincides with $F^*(x)$ in points with abscissas x_i.

Double and triple integrals

If $f(x, y)$ is a function of two independent variables x and y, it can be represented by using a three-dimensional Cartesian coordinate system with the values of $f(x, y)$ plotted along the vertical Z-axis. Any contour map provides a graphical representation in the XY-plane of a function of two variables x and y. The expression:

$$\iint f(x, y)\, dx\, dy$$

represents a two-dimensional integral (or double integral). Similarly, a volume integral, which is a three-dimensional integral, may be denoted by:

$$\iiint f(x, y, z)\, dx\, dy\, dz$$

The expression:

$$\int_{x_1}^{x_2} \int_{y_1}^{y_2} f(x, y)\, dx\, dy$$

means that $f(x, y)$ is integrated for a rectangle in the map bounded by the coordinates x_1 and x_2 in the X-direction and by y_1 and y_2 in the Y-direction. In a three-dimensional coordinate system, it represents the volume under the surface described by the function over the rectangle. If the area over which the function is integrated is written as A we can use the abbreviated notation $\int f dA$. The average value of a function of position f over a plane area A is defined as $(1/A) \int f dA$. Similarly, the mean value throughout a volume V is $(1/V) \int f\, dV$ where the integral is taken throughout the volume.

Hypsometric curves

Geophysicists have been interested in studying the surface of the earth by using hypsometric curves. A hypsometric curve for an area is a plot of the percentage of this area above a certain height level against height. A set of data defining the hypsometric curve of the earth is shown in Table II. These numbers are based on information by Kossinna (1933).

The elevation with respect to average sea level is shown in column 1 of Table II. The symbols ∞ and $-\infty$ correspond to the highest and lowest point on earth, respectively. The area of the earth's surface above a given level is shown in column 2 in millions of square kilometers. The corresponding percentage values are shown in column 3. The percent area between successive levels is given in column 4. These numbers can be plotted against height (see Fig. 11). The resulting distribution curve can be considered as the first derivative with respect to height of the hypsometric curve. (Some authors use the term "hypsometric curve" for this derivative.) The derivative is comparable to the frequency curves of statistical theory to be discussed in Chapter 6. It shows two distinct peaks that correspond to the most common elevation in oceans and on continents re-

TABLE II
Data for hypsometric curves for surface of entire earth (columns 2 and 4) and continents (5 and 6)

1 Level (km)	*2* Area above level (10^6 km^2)	*3* Area (%)	*4* Area between levels (%)	*5* Area (%)	*6* Area between levels (%)
∞	0.0	0.0		0.0	
			0.1		0.3
5	0.5	0.1		0.3	
			0.4		1.5
4	2.7	0.5		1.8	
			1.1		3.9
3	8.5	1.6		5.7	
			2.3		7.6
2	19.7	3.9		13.3	
			4.5		15.3
1	42.3	8.3		28.6	
			20.7		71.4
0	148.1	29.0		100.0	
			8.5		
–1	191.8	37.6			
			3.0		
–2	207.0	40.6			
			4.8		
–3	231.4	45.4			
			13.9		
–4	302.2	59.3			
			23.3		
–5	421.3	82.6			
			16.4		
–6	505.0	99.0			
			1.0		
–∞	510.0	100.0			

spectively. Columns 5 and 6 are comparable to columns 3 and 4. They are related to a hypsometric curve for the continents only which is also shown in Fig.11.

A problem that can be solved by methods of integration is to calculate the average elevation on the different continents. The formula to be used can be written as:

$$\bar{h} = \frac{1}{A} \int h \, dA$$

where h represents height. The total area of continents is equal to $A = 148.1 \cdot 10^6$ km^2. Clearly, the expression $\int h \, dA$ can be interpreted as the total volume of mass on earth above sea level. The advantage of working with percent values as in column 5 of Table II is that the constant A is reduced to unity so that:

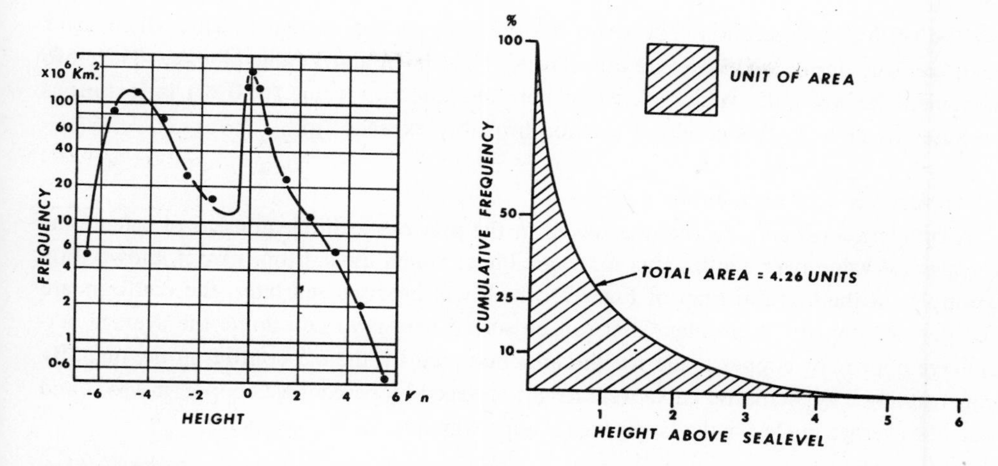

Fig.11. Derivative of hypsometric curve for earth's surface (after Scheidegger, 1963); and hypsometric curve for continents. By measuring the total area under hypsometric curve, the average height can be determined.

$$\bar{h} = \int h \mathrm{d}A'$$

where A' expresses area as a percentage.

A rapid method of calculating the average would consist of using the frequency values in the last column of Table II. For example, 7.6% of the area on the continents has an elevation between 2 and 3 km. It can be assumed that the average elevation for this subarea is 2.5 km. By following the procedure of taking midpoints for average height in all subareas defined by successive levels, a weighted average $\bar{h}$ is obtained with:

$$\bar{h} = 0.714 \times 0.5 + 0.153 \times 1.5 + 0.076 \times 2.5 + 0.039 \times 3.5 + 0.015 \times 4.5 + 0.003 \times 5.5 = 0.997 \text{ km}$$

Unfortunately, this procedure seems to be rather inaccurate mainly because the midpoint between levels 0.0 and 1.0 km overestimates the average height for this subarea. Heiskanen and Vening Meinesz (1958) report that the mean elevation of the different continents is approximately 850 m.

A rapid and relatively precise method for obtaining $\bar{h}$ from a hypsometric curve is to measure the area under the curve as shown in Fig. 11. This can be done by using a planimeter or by counting squares on graph paper. For the hypsometric curve of continents, the unit distance along the vertical axis is five times that for height which is plotted along the horizontal axis. The total area under the curve between limits $h = 0$ and $h = \infty$ is approximately 4.26 as large as the unit of area. This value must be divided by five to account for the vertical exaggeration. The result is 0.852 km for the average height, which, after rounding off, duplicates the value reported by Heiskanen and Vening Meinesz. The close agreement between the value calculated here and the true value (850 m) is fortuitous because the error in this graphical method probably exceeds 2m.

Average for area on contour map

A problem analogous to the one solved in the previous example consists of calculating the average value over a given area for a variable of which the contour map is known. For example, in the contour map of Fig. 47, which will be discussed later, the contours are for percent copper. A problem that can be solved here is to determine the average percentage copper or copper grade for the area bounded by the 0.5 contour in the map. By constructing a hypsometric curve with height replaced by percentage copper, it was found that the average grade for the larger area is approximately 0.99% copper.

2.11. Differential equations

A differential equation is an equation that contains derivatives (or differentials) of one or more unknown functions. If it involves derivatives of an unknown function of a single independent variable, it is called an ordinary differential equation. If the differential equation contains partial derivatives of an unknown function of two or more independent variables, it is called a partial differential equation. Two examples of differential equations are:

$$(1)\ \frac{dy}{dx} = 3x\ ; \qquad (2)\ \frac{\partial^2 u}{\partial x^2} + \frac{\partial^2 u}{\partial y^2} = 0$$

The order of a differential equation is the order of the highest derivative in the equation. Thus, eq. 1 is an ordinary differential equation of the first order, and eq. 2. is a partial differential equation of the second order.

The general solution of a differential equation of the n-th order is defined as a solution that contains n independent arbitrary constants. On the other hand, a particular solution is a solution that can be obtained from the general solution by assigning particular values to the arbitrary constants. These particular constants are usually derived from conditions that apply to the situation for which the differential equation was developed. Suppose that the differential equation is of the form:

$$\frac{d^2y}{dx^2} - k^2 y = 0 \qquad [2.32]$$

The general solution then is:

$$y = c_1 \mathrm{e}^{kx} + c_2 \mathrm{e}^{-kx} \qquad [2.33]$$

where c_1 and c_2 are arbitrary constants. This answer is verified readily by taking the second derivative of y with respect to x.

The degree of a differential equation is equal to the degree of the derivative of the highest order that occurs in it. For example, the equation:

$$\left(\frac{\mathrm{d}y}{\mathrm{d}x}\right)^2 = k^2 - y^2$$

is of the first order but of the second degree. A differential equation of the first order and first degree can be written in the form:

$$\frac{\mathrm{d}y}{\mathrm{d}x} = f(x, y) \qquad [2.34]$$

where $f(x, y)$ is a function of x and y. Alternatively, it can be presented in its differential form:

$$U\mathrm{d}x + V\mathrm{d}y = 0 \qquad [2.35]$$

where U and V are functions of x and y. It is not possible to solve every differential equation of the first order and first degree by using elementary methods. Usually, however, a solution can be obtained by employing methods of numerical analysis. Our discussion will be restricted to a few fairly simple examples.

If in [2.35] U is a function of x and V a function of y only, the equation can be solved by integration. It is then said that the variables are separable. The solution:

$$\int U\mathrm{d}x + \int V\mathrm{d}y = 0$$

will contain one arbitrary constant.

Shape of layers deposited in a basin

Suppose that: (1) clastic sediments are transported by currents into a basin; (2) the concentration of sediment held in suspension in the water is proportional to the velocity of the current; and (3) the velocity decreases in the current direction due to uniform friction along the bottom. These three assumptions specify a simple model from which we can solve the form of the thickness profile for a layer of sediment deposited on the bottom in the course of time.

If the X-axis is taken in the current direction, the decrease in velocity v, because of friction along the bottom, can be written as:

$$\frac{\mathrm{d}v}{\mathrm{d}x} = -av \qquad [2.36]$$

where a is a friction constant. The general solution of this differential equation is:

$$v = C\mathrm{e}^{-ax} \qquad [2.37]$$

where C is the arbitrary constant. This result is verified readily by taking the first derivative of v with respect to x. Suppose that the velocity is known at a point along the X-axis near the shoreline. This point then may be taken as the origin with $x = 0$. The velocity at the origin is $v = C$ or $v = v_0$. Hence [2.37] can be written as:

$$v = v_0\,\mathrm{e}^{-ax} \qquad [2.38]$$

which represents the particular solution of the differential equation $\mathrm{d}v/\mathrm{d}x = -av$.

If it is assumed that this exponential profile for the velocity is independent of time, then the amount of sediment $\mathrm{d}S$ deposited from the current during the time interval $\mathrm{d}t$ is, according to the second assumption made above, proportional to change in velocity in the current direction. This can be written as:

$$\frac{\partial S}{\partial t} = -b\,\frac{\mathrm{d}v}{\mathrm{d}x}$$

where b is a positive constant. It follows that:

$$\frac{\partial S}{\partial t} = abv = abv_0\,\mathrm{e}^{-ax}$$

The solution of this new equation is:

$$S = abv_0\,\mathrm{e}^{-ax}\,t + C_2$$

where C_2 is a arbitrary constant. It is logical to assume that the amount of sediment deposited is $S = 0$ when $t = 0$. It follows that $C_2 = 0$. If a new constant $abv_0 = c$ is defined, the final solution becomes:

$$S = c\mathrm{e}^{-ax}\,t \qquad [2.39]$$

This means that, under the conditions of the model, the thickness of deposited sediments falls off exponentially in the current direction and, at any point along the X-axis, increases linearly with time. If the process would be interrupted suddenly at a given time, the result would be a single layer with a thickness that wedges out exponentially towards the basin.

Reconstruction of a delta in the Bjorne Formation, Arctic Archipelago

The Bjorne Formation is a predominantly sandy unit of Early Triassic age. It was deposited at the margin of the Sverdrup Islands of the Canadian Arctic Archipelago. On northwestern Melville Island, the Bjorne Formation consists of three separate members which can be distinguished, mainly on the basis of clay content which is lowest in the upper member, C. The total thickness of the Bjorne Formation does not exceed 500 ft. on Melville Island. The formation forms a prograding fan-shaped delta. The paleocurrent directions are indicated by such features as the dip azimuths of planar foresets and axes of spoon-shaped throughs. The average current direction for 43 localities of Member C is shown in Fig.13A. These localities occur in a narrow belt where the sandstone member is exposed at the surface. The azimuth of the paleocurrents changes along the belt. This is also shown in Fig.12 where the 43 observed azimuths have been plotted against distance along a line that approximately coincides with the belt.

The variation pattern shows many irregularities but is characterized by a linear trend (systematic regional variation of constant rate of change). In the (U, V) plane for the coordinates (see Fig.13), this trend can be represented by the expression:

$$x = 292.133 + 27.859u + 19.932v \text{ (degrees)}$$

This is the equation of a plane. The variation of this plane along the belt gave the straight line called "trend" in Fig.12. By using a plane instead of a line for the trend, our model becomes more flexible in the horizontal plane. The present equation was derived by using the method of trend-surface analysis to be discussed later. The reader should consult Fig.40 for other reconstructions based on this model fitted to the data of Fig.13A.

In general, the regional paleocurrent direction x (in degrees) can be expressed as a function $F(u, v)$ of rectangular coordinates u and v with:

$$x = F(u, v) \qquad [2.40]$$

At each point on the map (U, V-plane), x represents the tangent of a curve for the paleocurrent trend that passes through that point.

The following definitions should be kept in mind to understand subsequent derivations: (1) The U-and V-axis in Fig.13 point toward the south and east, respectively; and (2) the azimuth x is $0°$ for the north-direction (= negative U-axis) and increases clockwise to $360°$. It then can be seen that:

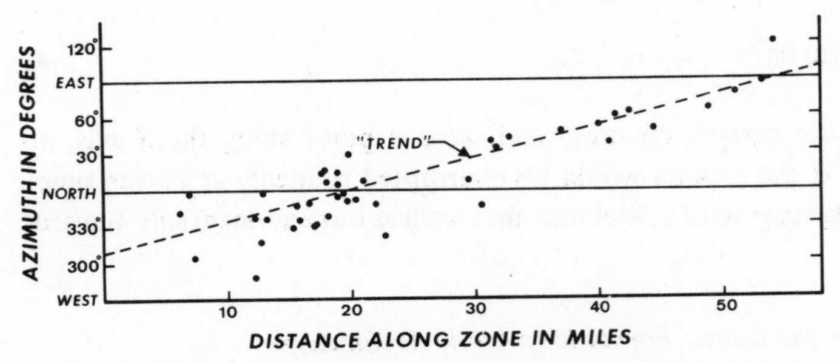

Fig.12. Variation of preferred paleocurrent direction along straight line coinciding with surface outcrop of Member C, Bjorne Formation (Triassic, Melville Island, Canada); systematic change in azimuth represented by trend line.

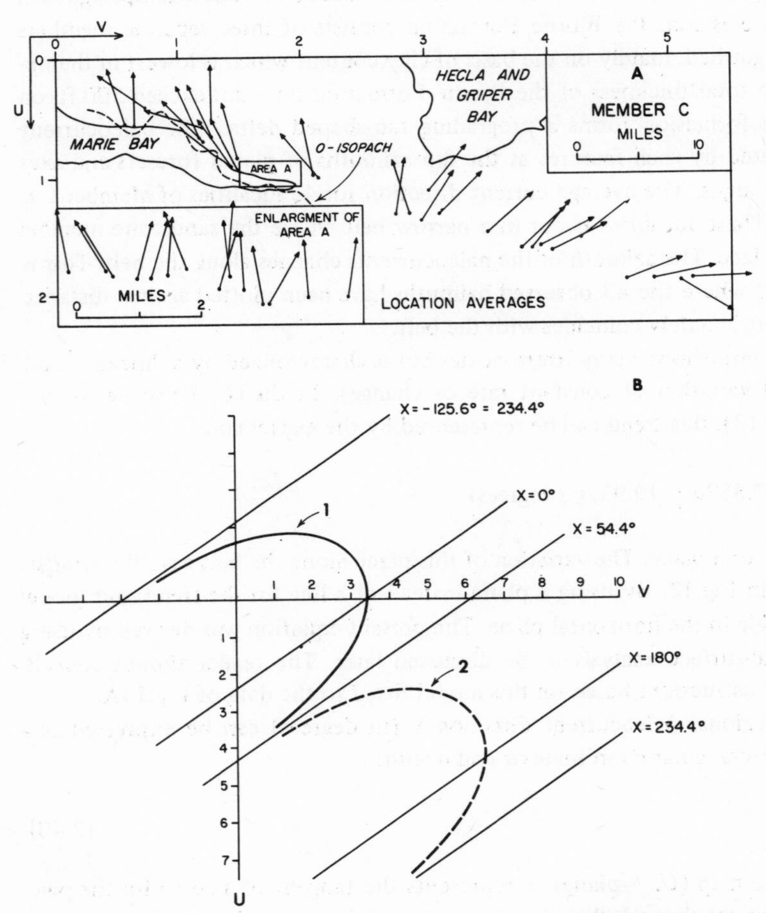

Fig.13. Reconstruction of approximate preferred paleocurrent direction and shape of paleodelta from preferred current directions, Member C, Bjorne Formation, Melville Island (after Agterberg, Hills and Trettin, 1967). A. Preferred directions of paleocurrent indicators at measurement stations; 0-isopach applies to lower part of Member C only. B. Graphical representation of solution of differential equation for paleocurrent trends. Inferred approximate direction of paleoriver near its mouth is N54.4°E. For comparison, see Fig.41.

$$\frac{dv}{du} = -\tan x \qquad [2.41]$$

In fact, this expression is equivalent to the definition of the first derivative as formulated in [2.14].

Continuous curves for the paleocurrent direction can be solved from [2.40] and [2.41] for all points in the map area. This may be accomplished by solving the differential equation:

$$\frac{dv}{du} = -\tan F(u, v)$$

When $F(u, v)$ is a linear polynomial in two dimensions, then:

$$x = F(u, v) = a_0 + a_1 u + a_2 v \qquad [2.42]$$

Differentiation with respect to u gives:

$$\frac{dx}{du} = a_1 + a_2 \frac{dv}{du}$$

With [2.41] this becomes:

$$\frac{dx}{du} = a_1 - a_2 \tan x \qquad [2.43]$$

This is an ordinary differential equation of the first order. The variables x and u can be separated by employing the differential form:

$$du = \frac{dx}{a_1 - a_2 \tan x}$$

Integration of both sides gives:

$$\int du = \int \frac{dx}{a_1 - a_2 \tan x}$$

or:

$$u = C + \frac{1}{a_1^2 + a_2^2} [a_1 x - a_2 \ln (a_1 \cos x - a_2 \sin x)] \text{ if } \tan x < a_1/a_2$$

and:

$$u = C + \frac{1}{a_1^2 + a_2^2} [a_1 x - a_2 \ln(-a_1 \cos x + a_2 \sin x)] \qquad \text{if } \tan x > a_1/a_2 \qquad [2.44]$$

This result can be obtained by consulting tables of integrals. It is readily verified by differentiation with respect to x. In the example, $a_0 = 292.133$; $a_1 = 27.859$; and $a_2 = 19.932$.

The result applies when x is expressed in degrees. When $x \geqslant 360°$, the quantity 360° may be subtracted from x, because azimuths are periodic with period 360°. Hence, we can also write $a_0 = -61.867$. In terms of radians:

$$x = -1.1845 + 0.4862u + 0.3479v \text{ (radians)}$$

By using this form, [2.44] becomes:

(A) $u = C + 1.3603x - 0.9733 \ln(0.4862 \cos x - 0.3479 \sin x)$

if $234°.4 < x \leqslant 360°$; $0° \leqslant x < 54°.4$

and:

(B) $u = C + 1.3603x - 0.9733 \ln(-0.4862 \cos x + 0.3479 \sin x)$

if $54°.4 < x < 234°.4$ [2.45]

The constant C is arbitrary. A value can be assigned to it by inserting specific values for u and x into [2.45]. For example, if $u = 0$ and $x = 0$, $C = -0.7018$.

The value $C = -0.7018$ was used to calculate a set of values for u forming a sequence of values for x. By using [2.42], the corresponding values of v then are readily found.

The paleocurrent trend is constructed by plotting the values of v against those for u in the map area. The resulting curves (1) and (2) for $C = -0.7018$ are shown in Fig. 13B. If the value for C is changed, the curves (1) and (2) become displaced in the $x = 54°.4$ direction. In this way, a set of curves is created that represent the paleocurrent direction for all points in the area.

Suppose that the paleocurrents were flowing in directions perpendicular to the average topographic contours at the time of sedimentation of the sand. If curve (1) of Fig. 13B is moved in the 144°.4 direction over a distance that corresponds to 90° in x, the result represents a set of directions which are perpendicular to the paleocurrent trends. Four of these curves which may approximate the shape of the delta are shown in Fig. 41C. These contours, which are labelled a, b, c, and d, satisfy [2.45] for different values of C and with x replaced by $(x + 90°)$ or $(x - 90°)$. Definite values cannot be assigned to these contours because x represents a direction and not a vector with both direction and magnitude.

Discussion

It can be argued that [2.45] and Fig.13B constitute a complicated choice for approximating the shape of a delta. However, this representation is the direct result of the simple hypothesis that, on the average, the current direction changes linearly along any line perpendicular to the axis of the delta. This axis points in the N54°E direction for the example.

The degree of fit of this model can be tested by using methods of statistical inference (see section 9.3). These methods indicated that the assumption of a more complex shape (e.g., by using a quadratic function for the gradual change in azimuth) is not warranted by the limited amount of data available in this situation, an inference also suggested by the simplified representation shown in Fig.12.

All flowlines plotted in Fig.41A converge to a single line-shaped source. They have the same asymptote in common which suggests the location of a river. Locally, there occur departures from the regional pattern constructed here as can be seen from the original data in Fig.12 and 13A.

An independent method for locating the source of the sand consists of mapping the grade of the largest clasts contained in the sandstone at a given place. Four grades of largest clasts could be mapped in the area of study. They are : (1) pebbles, granules, coarse sand; (2) cobbles, pebbles; (3) boulders, cobbles; and (4) boulders (max. 23 inches). The size of the clasts is larger where the velocity of the currents was higher. Approximate grade contours for classes 1 – 4 are shown in Fig.41C for comparison. This pattern corresponds to the contours constructed for the delta.

2.12. Complex numbers

In a graphical representation, the parabola with equation:

$$y = x^2 - 1$$

intersects the X-axis in two points, when $x_1 = 1$ or $x_2 = -1$. The values x_1 and x_2 are the roots of the polynomial $y = P(x) = x^2 - 1$. However, the parabola with equation $y = x^2 + 1$ does not intersect the X-axis because y cannot be smaller than 1. An attempt to solve the equation $x^2 + 1 = 0$ or $x^2 = -1$ results in the introduction of complex numbers of the form $a + bi$ where a and b are real numbers and i the imaginary unit with the property $i^2 = -1$. Thus, the solution of $x^2 = -1$ yields $x_1 = i$ and $x_2 = -i$.

The algebraic equation $ax^2 + bx + c = 0$ has the general solution:

$$x_{1,2} = \frac{-b \pm \sqrt{b^2 - 4ac}}{2a}$$

These two roots will form a complex pair of the type $p \pm qi$ if $b^2 < 4ac$.

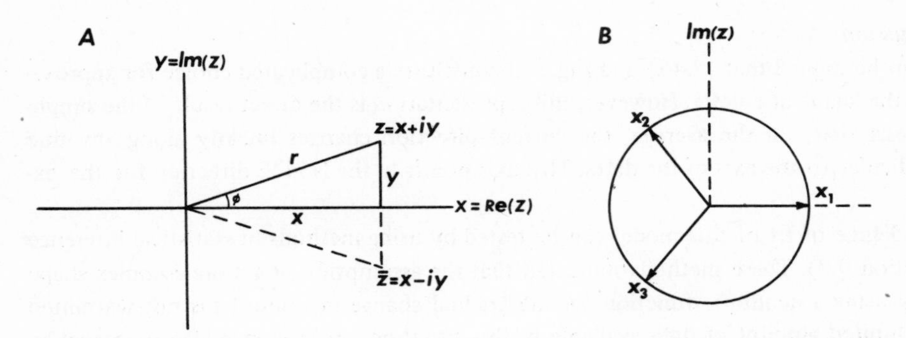

Fig.14. A. Representation of complex numbers in Z-plane. B. Diagram for three solutions of the third root of unity $x = \sqrt[3]{1}$.

In general, the n-th degree polynomial with equation:

$$P(x) = c_0 + c_1x + c_2x^2 + \ldots \ldots + c_nx^n$$

has n roots $x_k (k = 1, 2, \ldots, n)$ and can be written in the form:

$$P(x) = c_n(x - x_1)(x - x_2) \ldots \ldots (x - x_n)$$

If x_k is a real number it is also a point on the X-axis for the curve that represents $P(x)$. A number of the roots can be complex. If one of these roots is $a + ib$, its complex conjugate $a - ib$ also is a root.

Complex numbers can be identified with points in a plane (Z-plane) with a rectangular coordinate system consisting of the X-axis for the real part and the Y-axis for the imaginary part (see Fig.14A). The complex number $z = x + iy$ then corresponds to the point P with coordinates (x, y). The real numbers have $y = 0$ and fall on the X-axis. Purely imaginary numbers lie on the Y-axis. If $z = x + iy$, $\bar{z} = x - iy$ is called the complex conjugate of z. By the Pythagorean theorem:

$$|z| = r = \sqrt{x^2 + y^2}$$

and r is the absolute value or modulus of z. The angle $\phi = \arctan(y/x)$ which is positively measured counterclockwise from the X-axis is called the argument of z.

Summarizing, if $z = x + iy$, we have:

$x = \text{Re}\ (z)$ = real part of z

$y = \text{Im}\ (z)$ = imaginary part of z

$\phi = \arg z =$ argument of z

$r = |z| =$ absolute value or modulus of z

$\bar{z} = x - iy =$ conjugate of z .

The rules for addition and multiplication are:

$$(x_1 + iy_1) + (x_2 + iy_2) = x_1 + x_2 + i(y_1 + y_2)$$

$$(x_1 + iy_1)(x_2 + iy_2) = x_1x_2 - y_1y_2 + i(x_1y_2 + x_2y_1)$$

Likewise, for division:

$$z_1/z_2 = (x_1 + iy_1)/(x_2 + iy_2) = \frac{(x_1 + iy_1)(x_2 - iy_2)}{(x_2 + iy_2)(x_2 - iy_2)}$$

$$= \frac{x_1x_2 + y_1y_2 + i(x_2y_1 - x_1y_2)}{x_2^2 + y_2^2}$$

These rules are consistent with the basic definition $i = \sqrt{-1}$ so that $i^2 = -1$. When the polar coordinates r and ϕ are used:

$$z = x + iy = r\cos\phi + ir\sin\phi = r(\cos\phi + i\sin\phi)$$

The expression $r(\cos\phi + i\sin\phi)$ is the polar form of z. The rule of multiplication now becomes:

$$z_1z_2 = r_1(\cos\phi_1 + i\sin\phi_1)\cdot r_2(\cos\phi_2 + i\sin\phi_2)$$

$$= r_1r_2(\cos\phi_1\cos\phi_2 - \sin\phi_1\sin\phi_2) + i(\sin\phi_1\cos\phi_2 + \cos\phi_1\sin\phi_2)$$

$$= r_1r_2\cos(\phi_1 + \phi_2) + i\sin(\phi_1 + \phi_2)$$

The n-th power of z therefore satisfies:

$$z^n = [r(\cos\phi + i\sin\phi)]^n = r^n(\cos n\phi + i\sin n\phi)$$

The n-th root of z is derived as follows: if $z = z_1^n$ with $z_1 = r_1(\cos\phi_1 + i\sin\phi_1)$ then:

$$r_1^n(\cos n\phi_1 + i \sin n\,\phi_1) = r(\cos\phi + i\sin\phi)$$

Hence, $r_1 = \sqrt[n]{r}$ and $n\phi_1 = \phi \pm 2\pi k$ ($k = 1, 2, \ldots$). Only n of the resulting complex numbers are different from each other:

$$\sqrt[n]{z} = z^{1/n} = \sqrt[n]{r}\cos(\phi/n + 2\pi k/n) + i\sin(\phi/n + 2\pi k/n)$$

with $k = 0, 1, 2, \ldots (n-1)$. The n-th root of unity satisfies:

$$\sqrt[n]{1} = \cos(2\pi k/n) + i\sin(2\pi k/n); \quad k = 0, 1, \ldots, (n-1).$$

Example

The equation $x^3 = 1$ or $x = \sqrt[3]{1}$ has three roots $x_1 = 1$, if $k = 0$; $x_2 = \cos 2\pi/3 + i \sin 2\pi/3$, when $k = 1$, and $x_3 = \cos 4\pi/3 + i\sin 4\pi/3$, when $k = 2$. A schematic representation of the third root of unity is given in Fig.14B.

The exponential function

e^z for complex z is defined as:

$$e^z = e^{x+iy} = e^x(\cos y + i\sin y).$$

It follows that:

$$e^{i\phi} = \cos\phi + i\sin\phi$$

$$z = r(\cos\phi + i\sin\phi) = re^{i\phi} \quad [2.46]$$

The n-th root of unity now can be written as:

$$\sqrt[n]{1} = e^{i2\pi k/n} (k = 0, 1, \ldots, n-1) \quad [2.47]$$

Also:

$$\cos\phi = \tfrac{1}{2}(e^{i\phi} + e^{-i\phi}); \quad \sin\phi = \frac{1}{2i}(e^{i\phi} - e^{-i\phi}) \quad [2.48]$$

Chapter 3

ELEMENTARY MATRICES

3.1. Introduction

A matrix is a rectangular array of numbers which are arranged in rows and columns as in a double entry table and which obey certain rules of addition and multiplication. Matrices are important to geologists. For example, if p chemical analyses are made on n specimens, and the data for the specimens are entered as successive rows in a table, the results is a matrix consisting of $(n \times p)$ values.

Because of missing observations, the information stored in a geological data base may not occur as rectangular arrays. The following example illustrates how matrices can be extracted from geological information. It should be kept in mind that matrices usually are formed for a specific purpose which involves further derivations based on the operations discussed in this chapter.

Some mineralogical data for the Mount Albert ultramafic intrusion in Quebec are shown in Table III. These data are not usable as a matrix because the variables y_1, y_2, y_3, and y_4 could not be determined for a number of the samples. Several matrices can be extracted from this table which is part of a larger data base.

TABLE III
Some mineralogical data for the Mount Albert Peridotite intrusion, Quebec

Sample no.	u	v	y_1	y_2	y_3	y_4
SDM 6	30,320	18,330	88.9	87.5	8.150	2.81
SDM 7c	27,680	18,840				2.61
SDM 8	27,980	19,020	91.2	87.5	8.150	2.75
SDM 9	27,970	19,370				2.66
SDM 12	28,740	18,890	90.0	88.3		2.69
SDM 15	29,720	19,720	92.1			

u = east–west geographic coordinate (feet)
v = north–south geographic coordinate (feet)
y_1 = olivine composition (% forsterite)
y_2 = orthopyroxene composition (% enstatite)
y_3 = chrome spinel cell edge (Å)
y_4 = rock density (g/cm^3)

The (4 X 2) matrix:

$$\begin{bmatrix} 30{,}320 & 18{,}330 \\ 27{,}980 & 19{,}020 \\ 28{,}740 & 18{,}890 \\ 29{,}720 & 19{,}720 \end{bmatrix}$$

contains the location data for the (4 X 1) matrix:

$$\begin{bmatrix} 88.9 \\ 91.2 \\ 90.0 \\ 92.1 \end{bmatrix}$$

which consists of the data available for the composition of olivine.

The (3 X 2) matrix:

$$\begin{bmatrix} 88.9 & 87.5 \\ 91.2 & 87.5 \\ 90.0 & 88.3 \end{bmatrix}$$

contains the composition data for co-existing mineral pairs of olivine and orthopyroxene.

The (2 X 4) matrix:

$$\begin{bmatrix} 88.9 & 87.5 & 8.150 & 2.81 \\ 91.2 & 87.5 & 8.150 & 2.75 \end{bmatrix}$$

consists of data on samples for which all four variables could be measured.

3.2. Basic definitions

If there is only one row or one column of values, a matrix reduces to a row vector or a column vector, respectively. An example of a row vector is $[0\ 1\ -3\ 0]$. This vector contains four elements. As a column vector, these elements, in the same order, are denoted as:

$$\begin{bmatrix} 0 \\ 1 \\ -3 \\ 0 \end{bmatrix}$$

We can say that the (2 X 4) matrix:

$$\begin{bmatrix} 0 & 1 & -3 & 0 \\ 2 & 5 & 0 & 4 \end{bmatrix}$$

consists of 2 row vectors, or alternately, 4 column vectors.

If the number of rows n is equal to the number of columns p, a matrix is a square matrix, e.g. :

$$\mathbf{A} = \begin{bmatrix} 1 & 2 & 0 \\ 3 & 0 & -1 \\ 0 & 5 & 2 \end{bmatrix}$$

The matrix **A** has the elements 1, 0, and 2 along its main diagonal. A diagonal matrix has zero elements except along its main diagonal, e.g.:

$$\begin{bmatrix} 1 & 0 & 0 \\ 0 & 2 & 0 \\ 0 & 0 & 3 \end{bmatrix}$$

A symmetrical matrix has symmetry in elements with respect to the diagonal, for example :

$$\begin{bmatrix} 1 & 4 & 123 \\ 4 & 2 & -7 \\ 123 & -7 & 3 \end{bmatrix}$$

Addition

Two ($n \times p$) matrices are added to one another by addition of all individual elements.

Example:

$$\overset{\mathbf{A}}{\begin{bmatrix} 1 & 2 & 3 \\ -1 & 5 & 0 \\ 0 & 0 & 2 \end{bmatrix}} + \overset{\mathbf{B}}{\begin{bmatrix} 1 & 4 & 0 \\ 0 & 0 & 4 \\ 6 & 0 & 0 \end{bmatrix}} = \overset{\mathbf{C}}{\begin{bmatrix} 2 & 6 & 3 \\ -1 & 5 & 4 \\ 6 & 0 & 2 \end{bmatrix}}$$

When the three matrices of this example are written as **A**, **B**, and **C**, the addition can be written as $\mathbf{A} + \mathbf{B} = \mathbf{C}$. The subtraction of two matrices is performed by subtracting individual elements in the same position.

The following relationships are in agreement with the rule for addition (and subtraction) of matrices. When $\mathbf{A} + \mathbf{B} = \mathbf{C}$, then:

$$\mathbf{B} + \mathbf{A} = \mathbf{C}$$

$\mathbf{C} - \mathbf{B} = \mathbf{A}$

$\mathbf{C} - \mathbf{A} = \mathbf{B}$

$-\mathbf{C} + \mathbf{B} = -\mathbf{A}$

The elements of a matrix may be indicated by subscripts, e.g. a_{ij}. In the example, $a_{11} = 1$, $b_{23} = 4$, and $c_{31} = 6$.

Transposition

The transpose of a matrix $\mathbf{A}$ is a matrix $\mathbf{A}'$ whose columns are the same as the rows of $\mathbf{A}$, or $a_{ij} = a'_{ji}$.

Example:

$$\mathbf{A} = \begin{bmatrix} 1 & 2 & 0 \\ 3 & 0 & -1 \\ 0 & 5 & -2 \end{bmatrix} \qquad \mathbf{A}' = \begin{bmatrix} 1 & 3 & 0 \\ 2 & 0 & 5 \\ 0 & -1 & -2 \end{bmatrix}$$

In general, $(\mathbf{A}')' = \mathbf{A}$. If $\mathbf{A} + \mathbf{B} = \mathbf{C}$, it follows that:

$\mathbf{A}' + \mathbf{B}' = \mathbf{C}'$ and $(\mathbf{A} + \mathbf{B})' = \mathbf{C}'$

Multiplication

A matrix is multiplied by a scalar by multiplication of all its elements by this constant.

Example:

$$\mathbf{A} = \begin{bmatrix} 1 & 2 & 0 \\ 3 & 0 & -1 \\ 0 & 5 & -2 \end{bmatrix} \quad 2\mathbf{A} = \begin{bmatrix} 2 & 4 & 0 \\ 6 & 0 & -2 \\ 0 & 10 & -4 \end{bmatrix} \quad -\mathbf{A} = \begin{bmatrix} -1 & -2 & 0 \\ -3 & 0 & 1 \\ 0 & -5 & 2 \end{bmatrix}$$

Multiplication of two matrices is accomplished by forming a new matrix $\mathbf{C}$ with elements:

$$c_{ij} = \sum_{k=1}^{t} a_{ik} b_{kj} \qquad [3.1]$$

where t represents the number of columns of $\mathbf{A}$ and the number of rows of $\mathbf{B}$. If $\mathbf{A}$ is an $(n \times p)$ matrix and $\mathbf{B}$ a $(m \times q)$ matrix, the product matrix $\mathbf{C}$ can only be formed if $p = m = t$, i.e., when the matrices are conformable for multiplication.

Example:

$$\mathbf{A} = [0 \quad 1] \quad \mathbf{B} = \begin{bmatrix} 0 & 2 & -1 \\ 5 & 0 & 0 \end{bmatrix} \quad \mathbf{C} = \mathbf{AB} = [5 \quad 0 \quad 0]$$

A practical method of multiplying two matrices by hand is to draw the following cross with four quadrants:

II	I
III	IV

If the matrices **A** and **B** are placed in quadrants II and IV, the product **C** can be formed in quadrant I by multiplying the individual elements in rows of **A** with the corresponding elements in columns of **B** and adding the products.

Example:

		2	3	0	15	4	− 2
A	**C**	0	1	4	5	4	− 8
		1	0	0	0	2	− 1
	=						
					0	2	− 1
	B				5	0	0
					0	1	− 2

The element c_{11} with value 15 has row [2 3 0] to the left of it, and column $\begin{bmatrix}0\\5\\0\end{bmatrix}$ below it, with $2 \times 0 + 3 \times 5 + 0 \times 0 = 15$. Likewise, $c_{23} = 0 \times (-1) +$ $+ 1 \times 0 + 4 \times (-2) = -8$.

In general, **AB** ≠ **BA**, or if **A** is postmultiplied by **B**, the product is not equal to that of **A** premultiplied by **B**.

A unit matrix or identity matrix is a diagonal matrix with ones along the main diago - nal, e.g. :

$$\mathbf{I} = \begin{bmatrix} 1 & 0 & 0 \\ 0 & 1 & 0 \\ 0 & 0 & 1 \end{bmatrix}$$

The following relationships between three matrices **A, B,** and **C** are in agreement with the multiplication rule:

$$\mathbf{AI} = \mathbf{IA} = \mathbf{A}$$

$$(\mathbf{AB})\mathbf{C} = \mathbf{A}(\mathbf{BC})$$

$$(\mathbf{A} + \mathbf{B})\mathbf{C} = \mathbf{AC} + \mathbf{BC}$$

$$\mathbf{A}(\mathbf{B} + \mathbf{C}) = \mathbf{AB} + \mathbf{AC}$$

$$(\mathbf{AB})' = \mathbf{B}'\mathbf{A}'$$

3.3. Determinants

Determinants are useful in the solution of linear equations and for the process of matrix inversion. In the situation of small matrices, the calculations for the procedures discussed in this section can be done by hand. If the matrices are large, computing rapidly becomes time-consuming. The methods discussed in this section have been programmed for digital computer and, at most computer installations, it is possible to call as subroutines, the algorithms which are available for these procedures.

Consider the two simultaneous linear equations:

$$a_{11}x_1 + a_{12}x_2 = b_1$$

$$a_{21}x_1 + a_{22}x_2 = b_2$$

These two equations are linear in the two unknowns x_1 and x_2 which can be solved from the constants a_{11}, a_{12}, a_{21}, a_{22}, b_1, and b_2.

In matrix form, the equations are:

$$\begin{bmatrix} a_{11} & a_{12} \\ a_{21} & a_{22} \end{bmatrix} \begin{bmatrix} x_1 \\ x_2 \end{bmatrix} = \begin{bmatrix} b_1 \\ b_2 \end{bmatrix}$$

as can be readily verified by applying the rules discussed in the previous section. The following solutions x_1 and x_2 can be verified by substituting them into the original equations:

$$x_1 = \frac{b_1 a_{22} - a_{12} b_2}{a_{11} a_{22} - a_{12} a_{21}}$$

$$x_2 = \frac{a_{11} b_2 - b_1 a_{21}}{a_{11} a_{22} - a_{12} a_{21}} \qquad [3.2]$$

These expressions are called the determinantal equations for the unknows x_1 and x_2. They can be written in the form:

$$x_1 = \begin{vmatrix} b_1 & a_{12} \\ b_2 & a_{22} \end{vmatrix} \Bigg/ \begin{vmatrix} a_{11} & a_{12} \\ a_{21} & a_{22} \end{vmatrix}$$

$$x_2 = \begin{vmatrix} a_{11} & b_1 \\ a_{21} & b_2 \end{vmatrix} \Bigg/ \begin{vmatrix} a_{11} & a_{12} \\ a_{21} & a_{22} \end{vmatrix}$$

where the vertical bars denote the determinants of a matrix in each case. For example :

$$|\mathbf{A}| = \begin{vmatrix} a_{11} & a_{12} \\ a_{21} & a_{22} \end{vmatrix} = a_{11}a_{22} - a_{12}a_{21} \qquad [3.3]$$

The determinant of a (2×2) matrix is thus obtained by subtracting the product of the elements which are not along the diagonal from the product of the diagonal elements.

The determinant of an $(n \times n)$ matrix may be defined by the following induction on n. The method of induction is frequently used in mathematics and consists of: (1) defining an expression for a relatively simple situation, and (2) defining the steps by which successively more complicated situations can be derived from the simple situation. The following definition of determinants is an example of this method. Suppose that:

$$|\mathbf{A}| = \begin{vmatrix} a_{11} & a_{12} & a_{13} & \cdots & a_{1n} \\ a_{21} & a_{22} & a_{23} & \cdots & a_{2n} \\ a_{31} & a_{32} & a_{33} & \cdots & a_{3n} \\ \vdots & \vdots & \vdots & & \vdots \\ a_{n1} & a_{n2} & a_{n3} & \cdots & a_{nn} \end{vmatrix}$$

To each element a_{ij}, there corresponds an $(n - 1) \times (n - 1)$ matrix consisting of the elements of $\mathbf{A}$ except those in row i and column j. The determinant of this matrix is called the minor of a_{ij}. For example, when $n = 3$, the minors of the elements a_{11}, and a_{12}, and a_{13} are:

$$\begin{vmatrix} a_{22} & a_{23} \\ a_{32} & a_{33} \end{vmatrix}, \quad \begin{vmatrix} a_{21} & a_{23} \\ a_{31} & a_{33} \end{vmatrix}, \text{ and } \begin{vmatrix} a_{21} & a_{22} \\ a_{31} & a_{32} \end{vmatrix}$$

respectively. In general, if the minor corresponding to element a_{ij} is written as c_{ij}, the determinant of the $(n \times n)$ matrix **A** satisfies:

$$|\mathbf{A}| = \sum_{j=1}^{n} (-1)^{i+j} a_{ij}c_{ij} \qquad [3.4]$$

The value of $(-1)^{i+j}$ is equal to 1 when $(i+j)$ is even, and equal to -1 when $(i+j)$ is odd. The value of the determinant is independent of the choice of the value of i; for example, we can take $i = 1$. In turn, the minors c_{ij} can be developed as linear combinations of the minors for matrices of a smaller size. When the original matrix **A** has size $(n \times n)$, the process can be continued until all minors are determinants of (2×2) matrices which are readily calculated. For example, when the determinant of a (3×3) matrix is expressed in terms of the minors of the elements in its first row, the result is:

$$|\mathbf{A}| = \begin{vmatrix} a_{11} & a_{12} & a_{13} \\ a_{21} & a_{22} & a_{23} \\ a_{31} & a_{32} & a_{33} \end{vmatrix} = a_{11} \begin{vmatrix} a_{22} & a_{23} \\ a_{32} & a_{33} \end{vmatrix} - a_{12} \begin{vmatrix} a_{21} & a_{23} \\ a_{31} & a_{33} \end{vmatrix} + a_{13} \begin{vmatrix} a_{21} & a_{22} \\ a_{31} & a_{32} \end{vmatrix}$$

From [3.4], it follows that:

$$|\mathbf{A}| = a_{11}a_{22}a_{33} - a_{11}a_{23}a_{32} - a_{12}a_{21}a_{33} + a_{12}a_{23}a_{31} + a_{13}a_{21}a_{32} - a_{13}a_{22}a_{31}$$

The same result is obtained by making i equal to 2 or 3 in [3.4].

If the determinant of a scalar is defined as having the same value as this scalar, it can readily be verified that the application of [3.4] to a (2×2) matrix results in [3.3]. The minors of a (2×2) matrix then are scalars. For example, the minors of the first row of

$$\begin{bmatrix} a_{11} & a_{12} \\ a_{21} & a_{22} \end{bmatrix}$$

are a_{22} for the element a_{11}, and a_{21} for a_{12}. It follows immediately that:

$$\begin{vmatrix} a_{11} & a_{12} \\ a_{21} & a_{22} \end{vmatrix} = a_{11}a_{22} - a_{12}a_{21}$$

Our usage of determinants will be limited and not all of their properties will be discussed in detail here.

A property of determinants is that, when **B** is a matrix obtained from **A** by interchanging any pair of adjacent rows, or columns, the relationship between the determinants is :

$$|\mathbf{B}| = -|\mathbf{A}|$$

For interchanging rows, this equality follows from [3.4] which is valid for any value of i. For example, if **B** is obtained by interchanging the first and second rows of **A**, **B** as evaluated for $i = 1$, is equal to **A** for $i = 2$, except for opposite signs for all individual terms in the two evaluations. The reader may verify himself, e.g., for a (3 × 3) matrix, that the sign of a determinant changes if two of its columns are interchanged. The following equation amplifies [3.4] :

$$\sum_{j=1}^{n} (-1)^{i+j} a_{ij} c_{kj} = 0 \quad \text{if} \quad k \neq i \qquad [3.5]$$

In this situation, the elements in the i-th row of **A** are multiplied by the minors for the elements of the k-th row. Eq. [3.5] holds true in general. We will verify it for the (3 × 3) matrix **A**.

The minors of the second row of **A** are:

$$c_{21} = \begin{vmatrix} a_{12} & a_{13} \\ a_{32} & a_{33} \end{vmatrix}, \; c_{22} = \begin{vmatrix} a_{11} & a_{13} \\ a_{31} & a_{33} \end{vmatrix} \text{ and } c_{23} = \begin{vmatrix} a_{11} & a_{12} \\ a_{31} & a_{32} \end{vmatrix}$$

It follows that, for $i = 1$ and $k = 2$, [3.5] becomes:

$$-a_{11}(a_{12}a_{33} - a_{13}a_{32}) + a_{12}(a_{11}a_{33} - a_{13}a_{31}) - a_{13}(a_{11}a_{32} - a_{12}a_{21}) = 0$$

A more rapid method for obtaining the determinant of an $(n \times n)$ matrix will be given in section 3.6.

3.4. The inverse matrix

The inverse or reciprocal matrix of **A** is written as $\mathbf{A}^{-1}$ and satisfies:

$$\mathbf{A}\mathbf{A}^{-1} = \mathbf{I} \qquad [3.6]$$

Example:

36	16	4	1	0	0
15	9	3	0	1	0
6	4	2	0	0	1
			1/8	−1/3	1/4
			−1/4	1	−1
			1/8	−1	7/4

It can be readily verified that for the (2×2) matrix $\mathbf{A} = \begin{bmatrix} a & b \\ c & d \end{bmatrix}$:

$$\mathbf{A}^{-1} = \frac{1}{(ad - bc)} \begin{bmatrix} d & -b \\ -c & a \end{bmatrix} \qquad [3.7]$$

Example:

$$\begin{bmatrix} 2 & 2 \\ 1 & 4 \end{bmatrix}^{-1} = (1/6) \begin{bmatrix} 4 & -2 \\ -1 & 2 \end{bmatrix} = \begin{bmatrix} 2/3 & -1/3 \\ -1/6 & 1/3 \end{bmatrix}$$

A general method for obtaining the inverse of a matrix will now be discussed; more rapid methods will be presented in section 3.6. The so-called cofactor a_{ij}^{c} of the element a_{ij} of a matrix $\mathbf{A}$ is equal to $(-1)^{i+j}$ times the minor of a_{ij}.

Because each of the elements a_{ij} of $\mathbf{A}$ has its cofactor a_{ij}^{c}, a cofactor matrix $\mathbf{A}^{c}$ can be defined. It can be shown that:

$$\mathbf{A}(\mathbf{A}^{c})' = |\mathbf{A}|\mathbf{I} \qquad [3.8]$$

where $\mathbf{I}$ is the $(n \times n)$ identity matrix. On both sides of the equality sign, there occurs a $(n \times n)$ matrix. If the individual elements of this matrix are compared to one another, the resulting equation is identical to [3.4] for the diagonal elements and to [3.5] for the remaining elements. If both sides of this equation are divided by $|\mathbf{A}|$, and premultiplied by $\mathbf{A}^{-1}$, it follows that:

$$\mathbf{A}^{-1} = \frac{(\mathbf{A}^{c})'}{|\mathbf{A}|} \qquad [3.9]$$

Example. If:

$$\mathbf{A} = \begin{bmatrix} 2 & 2 \\ 1 & 4 \end{bmatrix}, \text{ it follows that } \mathbf{A}^c = \begin{bmatrix} 4 & -1 \\ -2 & 2 \end{bmatrix} \text{ and } (\mathbf{A}^c)' = \begin{bmatrix} 4 & -2 \\ -1 & 2 \end{bmatrix}$$

$$|\mathbf{A}| = 2 \times 4 - 2 \times 1 = 6$$

Hence:

$$\mathbf{A}^{-1} = (1/|\mathbf{A}|) \cdot (\mathbf{A}^c)' = (1/6) \begin{bmatrix} 4 & -2 \\ -1 & 2 \end{bmatrix}$$

The following relationships between matrices follow from the definition of the inverse:

$$\mathbf{A}\mathbf{A}^{-1} = \mathbf{A}^{-1}\mathbf{A} = \mathbf{I}$$

$$(\mathbf{A}^{-1})^{-1} = \mathbf{A}$$

$$(\mathbf{AB})^{-1} = \mathbf{B}^{-1}\mathbf{A}^{-1}$$

$$(\mathbf{A}^{-1})' = (\mathbf{A}')^{-1}$$

When $\mathbf{AB} = \mathbf{C}$, $\mathbf{B} = \mathbf{A}^{-1}\mathbf{C}$

We have now discussed all the fundamental laws of matrix algebra. The three basic operations are:

Addition (and subtraction)
Multiplication (and inversion)
Transposition.

The third operation has no equivalent in ordinary algebra. Another difference is that the so-called commutative law of multiplication is valid in ordinary algebra ($ab = ba$), but generally fails in matrix algebra ($\mathbf{AB} \neq \mathbf{BA}$).

3.5. Linear equations

The solution of linear equations is an example of the use of matrices. A system of n linear equations with n unknowns x_i can be written as:

$$
\begin{array}{l}
a_{11}x_1 + a_{12}x_2 + \ldots + a_{1n}x_n = c_1 \\
a_{21}x_1 + a_{22}x_2 + \ldots + a_{2n}x_n = c_2 \\
\ldots\ldots\ldots\ldots\ldots\ldots\ldots\ldots \\
\ldots\ldots\ldots\ldots\ldots\ldots\ldots\ldots \\
a_{n1}x_1 + a_{n2}x_2 + \ldots + a_{nn}x_n = c_n
\end{array}
$$

When matrix notation is used, this system can be written as:

$$\mathbf{AX} = \mathbf{C} \qquad [3.10]$$

where **A** is a $(n \times n)$ matrix consisting of the known elements a_{ij}, **X** is a column vector for the unknowns x_i, and **C** is a column vector for the known elements c_i. When the relationship is represented by the method of quadrants, **A**, **X**, and **C** should be placed in quadrants II, IV and I, as follows:

$$
\begin{array}{cccc|c}
a_{11} & a_{12} & \ldots & a_{1n} & c_1 \\
a_{21} & a_{22} & \ldots & a_{2n} & c_2 \\
. & . & \ldots & . & . \\
. & . & \ldots & . & . \\
a_{n1} & a_{n2} & \ldots & a_{nn} & c_n \\
\hline
 & & & & x_1 \\
 & & & & x_2 \\
 & & & & . \\
 & & & & . \\
 & & & & x_n
\end{array}
$$

The solution of [3.10] is:

$$\mathbf{X} = \mathbf{A}^{-1}\mathbf{C} \qquad [3.11]$$

Example. The three linear equations:

$$36x_1 + 16x_2 + 4x_3 = 8$$

$$15x_1 + 9x_2 + 3x_3 = 3$$

$$6x_1 + 4x_2 + 2x_3 = 4$$

can be represented by the matrix equation:

$$\begin{bmatrix} 36 & 16 & 4 \\ 15 & 9 & 3 \\ 6 & 4 & 2 \end{bmatrix} \begin{bmatrix} x_1 \\ x_2 \\ x_3 \end{bmatrix} = \begin{bmatrix} 8 \\ 3 \\ 4 \end{bmatrix}$$

By using the inverse matrix shown in the first example of section 3.4, [3.11] yields:

$$\begin{bmatrix} x_1 \\ x_2 \\ x_3 \end{bmatrix} = \begin{bmatrix} 1/8 & -1/3 & 1/4 \\ -1/4 & 1 & -1 \\ 1/8 & -1 & 7/4 \end{bmatrix} \begin{bmatrix} 8 \\ 3 \\ 4 \end{bmatrix} = \begin{bmatrix} 1 \\ -3 \\ 5 \end{bmatrix}$$

The solution for the three coefficients is $x_1 = 1$, $x_2 = -3$, and $x_3 = 5$. From [3.9] and [3.11], it follows that the solution of the linear equations also can be written as:

$$\mathbf{X} = \frac{(\mathbf{A}^{\mathrm{c}})'}{|\mathbf{A}|}\mathbf{C} \qquad [3.12]$$

The reader should verify that the unknown coefficients x_i (with $i = 1, 2, \ldots, n$) satisfy:

$$x_i = \frac{c_1 a_{1j}^{\mathrm{c}} + c_2 a_{2i}^{\mathrm{c}} + \ldots + c_n a_{ni}^{\mathrm{c}}}{|\mathbf{A}|} \qquad [3.13]$$

From [3.4] it follows that the numerator of this expression can be considered as the determinant of matrix **A** in which the elements $a_{1i}, a_{2i}, \ldots$ have been replaced by the elements $c_1, c_2, \ldots$. In the case $n = 2$, this results in the determinantal equations [3.2].

3.6. Method of pivotal condensation

This method is illustrated in Table IV for the numbers used in the example of section 3.5. The method can be used to calculate: (1) the determinant of a matrix; (2) the unknown coefficients for a system of linear equations; (3) the inverse matrix. In the case of small matrices and vectors, e.g. $n = 3$, the calculations may be done by hand. If n is larger

TABLE IV
Pivotal condensation

	1	*2*	*3*	*4*	*5*	*6*	*7*	*Sum Check* *8*
1	36	16	4	8	1	0	0	65
2	15	9	3	3	0	1	0	31
3	6	4	2	4	0	0	1	17
10 P.R.	1	4/9	1/9	2/9	1/36	.	.	1.8056
11		7/3	4/3	−1/3	−5/12	1	.	3.9167
12		4/3	4/3	8/3	−1/6	.	1	6.1667
20 P.R.		1	4/7	−1/7	−5/28	3/7	.	1.6786
21			4/7	20/7	1/14	−4/7	1	3.9286
3′			1	5	1/8	−1	7/4	6.8750
2′		1	.	−3	−1/4	1	−1	−2.2500
1′	1	.	.	1	1/8	−1/3	1/4	2.0139

P.R. = pivotal row

than 4, the process of doing the calculations by hand or desk calculator may become too time-consuming and can be programmed for digital computers advantageously.

Discussion of procedure followed in Table IV

(1) The first three values in columns *1* – *3*, and *5* – *7*, are the same as quadrants II and I in the example at the beginning of section 3.4. The first three values in column *4* are the constants c_i of the example of section 3.5. Column 8 is the Sum Check column with values equal to the sum of those in the previous seven columns. The values in each column are subjected to the same procedure. If a dot occurs, the value need not be computed.

(2) Row *10* is called a pivotal row. All elements in row *1* have been divided by the first element of row *1*, e.g., the second element of row *10* is equal to 16/36 = 4/9.

(3) Row *11* is obtained by multiplication of row *10* by the first element of row *2*, and subtracting the result from row *2*, e.g., the first value in row *11* is equal to *9* – *15* ×(4/9) = 7/3.

(4) Row *12* is obtained in the same manner as row *11* by using row *3* instead of row *2*, e.g., the first value in row *12* is equal to *4* – *6* × (4/9) = 4/3.

(5) Row *20* is the second pivotal row. All elements of row *11* have been divided by the first element of row *11*, e.g., the second value in row *20* is equal to (4/3) : (7/3) = 4/7.

(6) Row *21* is obtained by multiplication of row *20* by the first value of row *12* and subtracting the result from row *12*, e.g., its first value is (4/3) – (4/3) × (4/7) = 4/7.

(7) Row *3′*, is row *21* divided by the first value in row *21*, e.g., (20/7) : (4/7) = 5; (1/14) : (4/7) = 1/8. The values in columns *5*–*7* of row *3′* constitute the bottom row of

the inverse matrix (see section 3.4). The value in column *4* is equal to x_3.

(8) The value of x_2 and the second row of the inverse matrix occur in row *2'*, which is obtained by multiplying the value in row *3'*, by the second value in row *20* (= 4/7) and subtracting the result from row *20*, e.g., (– 5/28) – (1/8) × (4/7) = –1/4.

(9) The value of x_1 and the first row of the inverse matrix (row *1'*) are obtained by multiplying the values in rows *3'* and *2'* which occur in the same column by the third and second values in row *10*, respectively; the result is then subtracted from row *10*, e.g., (1/36) – (1/9) × (1/8) – (4/9) × (– 1/4) = 1/8.

(10) The determinant of the matrix **A** is equal to the product of the first values in rows *1*, *11*, and *21*, or $|\mathbf{A}| = 36 \times (7/3) \times (4/7) = 48$. This result can be obtained by using columns *1–3* and rows *1–21* only.

(11) The coefficients of the linear equations of section 3.5 occur (in the reverse order) in column *4*, rows *1'*, *2'* and *3'*. This solution can be obtained without solving the inverse matrix (columns *5–7*).

(12) The final results are readily verified. In more complicated situations, when the results are rounded off during the calculations, the Sum Check column (*8*) can be used to check if sufficient precision has been retained in the values. The values in column *8* are calculated in the same manner as those for the other columns. Any newly derived row has the property that its values (for columns *1–7*) add to that in column *8*. If discrepancies occur, the calculations should be repeated using greater precision.

The method of pivotal condensation is also known as the abbreviated Doolittle method. It is a rapid, numerical evaluation of [3,9] which can be used in programs for digital computers. For a more detailed discussion, see Conté (1965) or Wilkinson (1965).

3.7. Homogeneous equations; singularity

A set of linear equations is homogeneous if all of the constants on the right-hand sides of the equality signs are equal to zero :

$$a_{11}x_1 + a_{12}x_2 + \ldots + a_{1n}x_n = 0$$

$$a_{21}x_1 + a_{22}x_2 + \ldots + a_{2n}x_n = 0$$

. . . .

. . . .

$$a_{n1}x_1 + a_{n2}x_2 + \ldots + a_{nn}x_n = 0$$

In matrix notation, this becomes:

$$\mathbf{AX} = \mathbf{O} \qquad [3.14]$$

where **O** represents a column vector consisting of zeros. An obvious solution of these equations is:

$$\mathbf{X} = \mathbf{O}$$

which is called the trivial solution.

If the solution of the unknowns x_i is written by using [3.12], it follows that:

$$x_i = \frac{0}{|\mathbf{A}|} \quad \text{or} \quad \mathbf{X} = \frac{\mathbf{O}}{|\mathbf{A}|}$$

This indicates, that a vector **X** with nonzero elements can only exist if $|\mathbf{A}|$ is equal to zero.

Example. The two linear equations:

$$3x_1 + 5x_2 = 0$$

$$9x_1 + 15x_2 = 0$$

or:

$$\begin{bmatrix} 3 & 5 \\ 9 & 15 \end{bmatrix} \begin{bmatrix} x_1 \\ x_2 \end{bmatrix} = \begin{bmatrix} 0 \\ 0 \end{bmatrix}$$

have:

$$\begin{vmatrix} 3 & 5 \\ 9 & 15 \end{vmatrix} = 3 \times 15 - 5 \times 9 = 0$$

and the determinant is equal to zero.

Both equations lead to the following relationship between x_1 and x_2:

$$x_1 = -(5/3)\, x_2$$

If an arbitrary value is assigned to x_2 for example $x_2 = 6$, it follows that $x_1 = -10$. The vector $\begin{bmatrix} -10 \\ 6 \end{bmatrix}$ is therefore a solution of the two homogeneous equations in this example.

The vector **X** can be expressed as:

$$\mathbf{X} = k \begin{bmatrix} -5 \\ 3 \end{bmatrix}$$

where k can have any value (including zero).

A matrix whose determinant is zero is called singular. From [3.9] it follows that a singular matrix cannot be inverted. Its inverse matrix does not exist.

If the method of pivotal condensation is applied to a singular matrix, the nonexistence of the inverse would manifest itself by the appearance of a pivotal element which is equal to zero. Unless stated otherwise, it will always be assumed here that the inverse matrices in the equations exist.

3.8. Partitioned matrices

An $(m \times n)$ matrix $\mathbf{A}$ can be partitioned as:

$$\mathbf{A} = \begin{bmatrix} \mathbf{A}_{11} & \mathbf{A}_{12} \\ \mathbf{A}_{21} & \mathbf{A}_{22} \end{bmatrix}$$

Example:

$$\mathbf{A} = \begin{bmatrix} 1 & 3 & 2 & 4 \\ 4 & 0 & 6 & 8 \\ 9 & 3 & 1 & 5 \end{bmatrix}$$

and:

$$\mathbf{A}_{11} = \begin{bmatrix} 1 & 3 & 2 \\ 4 & 0 & 6 \end{bmatrix} \qquad \mathbf{A}_{12} = \begin{bmatrix} 4 \\ 8 \end{bmatrix}$$

$$\mathbf{A}_{21} = \begin{bmatrix} 9 & 3 & 1 \end{bmatrix} \qquad \mathbf{A}_{22} = \begin{bmatrix} 5 \end{bmatrix}$$

If matrix $\mathbf{B}$ is partitioned in the same manner as $\mathbf{A}$, the product $\mathbf{C}$ satisfies:

$$\mathbf{C} = \begin{bmatrix} \mathbf{C}_{11} & \mathbf{C}_{12} \\ \mathbf{C}_{21} & \mathbf{C}_{22} \end{bmatrix} = \begin{bmatrix} \mathbf{A}_{11}\mathbf{B}_{11} + \mathbf{A}_{12}\mathbf{B}_{21} & \mathbf{A}_{11}\mathbf{B}_{12} + \mathbf{A}_{12}\mathbf{B}_{12} \\ \mathbf{A}_{21}\mathbf{B}_{11} + \mathbf{A}_{22}\mathbf{B}_{21} & \mathbf{A}_{21}\mathbf{B}_{12} + \mathbf{A}_{22}\mathbf{B}_{22} \end{bmatrix} \qquad [3.15]$$

This result follows directly from the multiplication rule for matrices.

The inverse of the partitioned matrix $\mathbf{A} = \begin{bmatrix} \mathbf{B} & \mathbf{C} \\ \mathbf{D} & \mathbf{E} \end{bmatrix}$ is:

$$\mathbf{A}^{-1} = \begin{bmatrix} \mathbf{B}^{-1}(\mathbf{I} + \mathbf{C}\mathbf{F}^{-1}\mathbf{D}\mathbf{B}^{-1}) & -\mathbf{B}^{-1}\mathbf{C}\mathbf{F}^{-1} \\ -\mathbf{F}^{-1}\mathbf{D}\mathbf{B}^{-1} & \mathbf{F}^{-1} \end{bmatrix} \qquad [3.16]$$

where **I** is an identity matrix and $\mathbf{F} = \mathbf{E} - \mathbf{D}\mathbf{B}^{-1}\mathbf{C}$.

The proof of [3.16] may consist of calculating the matrix $\mathbf{C} = \mathbf{A}\mathbf{A}^{-1}$ as above. The result should be $\mathbf{C}_{11} = \mathbf{I}$, $\mathbf{C}_{12} = \mathbf{N}$, $\mathbf{C}_{21} = \mathbf{N}$, and $\mathbf{C}_{22} = \mathbf{I}$, where **I** are identity matrices and **N** are null matrices that consist of zeroes only.

This result is readily obtained, for example:

$$\mathbf{C}_{11} = \mathbf{B}\mathbf{B}^{-1}(\mathbf{I} + \mathbf{C}\mathbf{F}^{-1}\mathbf{D}\mathbf{B}^{-1}) - \mathbf{C}\mathbf{F}^{-1}\mathbf{D}\mathbf{B}^{-1} = \mathbf{I}$$

3.9. Canonical form, eigenvalues and eigenvectors

The problems discussed in the following sections are important in several areas of geomathematics. A number of problems which involve matrices are simplified if the matrix is presented in its canonical form which is determined by the eigenvalues and eigenvectors.

If **A** is an $(n \times n)$ square matrix and **X** an unknown column vector the product $\mathbf{Y} = \mathbf{A}\mathbf{X}$ can be considered as a vector whose elements are linear combinations of all unknowns x_i $(i = 1, 2, \ldots, n)$. The so-called eigenvalue problem arises if the linear substitution $\mathbf{Y} = \mathbf{A}\mathbf{X}$ is subjected to the condition $\mathbf{Y} = \lambda\mathbf{X}$, where λ is a scalar quantity. If this condition is to be satisfied, the vector **X** can be solved from the set of linear equations:

$$\mathbf{A}\mathbf{X} = \lambda\mathbf{X} \qquad [3.17]$$

or:

$$(\mathbf{A} - \lambda\mathbf{I})\,\mathbf{X} = \mathbf{0} \qquad [3.18]$$

This is a set of homogeneous equations which, in detailed form, are:

$$\left.\begin{array}{l} (a_{11} - \lambda)x_1 + a_{12}x_2 + \ldots + a_{1n}x_n = 0 \\ a_{21}x_1 + (a_{22} - \lambda)x_2 + \ldots + a_{2n}x_n = 0 \\ \ldots\ldots\ldots\ldots\ldots\ldots\ldots\ldots\ldots\ldots \\ a_{n1}x_1 + a_{n2}x_2 + \ldots + (a_{nn} - \lambda)x_n = 0 \end{array}\right| \qquad [3.19]$$

The nontrivial solutions of **X** only will exist if the determinant of the coefficient matrix is zero (see section 3.7) or:

$$|\mathbf{A} - \lambda \mathbf{I}| = \begin{vmatrix} a_{11} - \lambda & a_{12} & \dots a_{1n} \\ a_{21} & a_{22} - \lambda & \dots a_{2n} \\ \dots\dots\dots & \dots\dots & \dots \\ \dots\dots\dots & \dots\dots & \dots \\ a_{n1} & a_{n2} & \dots a_{nn} - \lambda \end{vmatrix} = 0 \qquad [3.20]$$

This equation is called the characteristic equation for **A.** Values of λ which satisfy this equation are called eigenvalues and the corresponding solutions for **X** are called eigenvectors.

If the determinant in [3.20] is expanded, the result is a polynomial expression in terms of λ. The highest power term is λ^n and the characteristic equation has the general form:

$$\lambda^n + p_{n-1} \lambda^{n-1} + \dots + p_0 = 0 \qquad [3.21]$$

For example, the characteristic equation of

$$\mathbf{A} = \begin{bmatrix} a_{11} & a_{11} \\ a_{21} & a_{22} \end{bmatrix} \text{ is:}$$

$$\begin{vmatrix} a_{11} - \lambda & a_{12} \\ a_{21} & a_{22} - \lambda \end{vmatrix} = 0$$

or: $(a_{11} - \lambda)(a_{22} - \lambda) - a_{12}a_{21} = 0$

$$\lambda^2 - (a_{11} + a_{22})\lambda + (a_{11}a_{22} - a_{12}a_{21}) = 0$$

In simplified notation, this becomes:

$$\lambda^2 + p_1\lambda + p_0 = 0$$

This equation has two roots:

$$\lambda_{1,2} = \frac{-p_1 \pm \sqrt{p_1^2 - 4p_0}}{2}$$

These roots will be real if $p_1^2 > 4p_0$. They are equal to one another if $p_1^2 = 4p_0$ or form a pair of complex values when $p_1^2 < 4p_0$.

Example. The eigenvalues of $\begin{bmatrix} 5 & 2 \\ -6 & -3 \end{bmatrix}$ satisfy:

$$\lambda^2 - 2\lambda - 3 = 0$$

with solution $\lambda_1 = 3$ and $\lambda_2 = -1$.

If one or more of the eigenvalues of a matrix are zero, it follows from [3.20] that the determinant of the matrix is zero, or that the matrix is singular. The rank of a matrix is equal to its number of nonzero eigenvalues.

The theory of matrix algebra also is concerned with the situation when two or more nonzero eigenvalues are equal to one another (e.g. $p_1^2 = 4p_0$ in the preceding example). This situation is not likely to occur in problems of mathematical geology and will not be treated in detail.

The eigenvectors $\mathbf{X}_i$ of a matrix $\mathbf{A}$ satisfy [3.17] and can be obtained by substituting the values of $\lambda_i (i = 1, 2, \ldots, n)$ into this equation.

In the example, one of the eigenvalues is $\lambda_1 = 3$. Hence:

$$(\mathbf{A} - \lambda_1 \mathbf{I})\mathbf{X}_1 = \begin{bmatrix} 5-3 & 2 \\ -6 & -3-3 \end{bmatrix} \begin{bmatrix} x_{11} \\ x_{12} \end{bmatrix} = \begin{bmatrix} 0 \\ 0 \end{bmatrix}$$

The homogeneous equations to be solved are:

$$2x_{11} + 2x_{12} = 0 \quad \text{and} \quad -6x_{11} - 6x_{12} = 0$$

When x_{11} is arbitrarily put equal to 1, it follows that $x_{12} = -1$, consequently:

$$\mathbf{X}_1 = k_1 \begin{bmatrix} 1 \\ -1 \end{bmatrix}$$

where k_1 is an arbitrary constant. For $\lambda_2 = -1$, we obtain:

$$\mathbf{X}_2 = k_2 \begin{bmatrix} 1 \\ -3 \end{bmatrix}$$

The eigenvectors of a matrix form the columns of the so-called modal matrix $\mathbf{V}$. By using a partitioned form, $\mathbf{V}$ can be written as:

$$\mathbf{V} = [\mathbf{X}_1 \; \mathbf{X}_2 \ldots \mathbf{X}_n]$$

Because of the arbitrary constants k_i, the columns of **V** are not uniquely determined. The first element of each eigenvector can be made equal to unity. The modal matrix for the example then is:

$$\mathbf{V} = \begin{bmatrix} 1 & 1 \\ -1 & -3 \end{bmatrix}$$

Another method of writing the modal matrix is to normalize the eigenvectors. An eigenvector is called normalized when the squares of its elements add to unity. In that case, the modal matrix will be written as $\mathbf{V}_n$. In the example :

$$\mathbf{V}_n = \begin{bmatrix} \frac{1}{2}\sqrt{2} & \frac{1}{10}\sqrt{10} \\ -\frac{1}{2}\sqrt{2} & -\frac{3}{10}\sqrt{10} \end{bmatrix}$$

The normalized vectors are uniquely determined except for the scalar factor ± 1.

Canonical form of a matrix

Each of the n eigenvectors of the matrix **A** satisfies an equation:

$$\mathbf{AX}_i = \lambda_1 \cdot \mathbf{X}_i \qquad [3.22]$$

If a diagonal matrix $\mathbf{\Lambda}$ is defined as:

$$\mathbf{\Lambda} = \begin{bmatrix} \lambda_1 & 0 \ldots & 0 \\ 0 & \lambda_2 \ldots & 0 \\ \ldots & \ldots & \ldots \\ \ldots & \ldots & \ldots \\ 0 & 0 \ldots & \lambda_n \end{bmatrix}$$

the n equations [3.22] can be written as a single matrix equation:

$$\mathbf{AV} = \mathbf{V\Lambda} \qquad [3.23]$$

where **V** is the modal matrix.

Postmultiplication of both sides of [3.23] by the inverse of **V** yields:

$$A = \mathbf{V\Lambda V}^{-1} \qquad [3.24]$$

which is called the canonical form of **A**.

Premultiplication of both sides of [3.23] by $\mathbf{V}^{-1}$ gives:

$$\mathbf{V}^{-1}\mathbf{A}\mathbf{V} = \mathbf{\Lambda} \qquad [3.25]$$

This equation represents the so-called diagonalization of **A**, which is the transformation of **A** into a diagonal matrix.

Example. For:

$$\mathbf{A} = \begin{bmatrix} 5 & 2 \\ -6 & -3 \end{bmatrix}, \qquad \mathbf{\Lambda} = \begin{bmatrix} 3 & 0 \\ 0 & -1 \end{bmatrix} \quad \text{and} \quad \mathbf{V} = \begin{bmatrix} 1 & 1 \\ -1 & -3 \end{bmatrix}$$

it follows that:

$$\mathbf{V}^{-1} = \begin{bmatrix} \frac{3}{2} & \frac{1}{2} \\ -\frac{1}{2} & -\frac{1}{2} \end{bmatrix}$$

Hence:

$$\mathbf{A} = \begin{bmatrix} 1 & 1 \\ -1 & -3 \end{bmatrix} \begin{bmatrix} 3 & 0 \\ 0 & -1 \end{bmatrix} \begin{bmatrix} \frac{3}{2} & \frac{1}{2} \\ -\frac{1}{2} & -\frac{1}{2} \end{bmatrix} = \begin{bmatrix} 5 & 2 \\ -6 & -3 \end{bmatrix}$$

Symmetric matrices

From $\mathbf{AV} = \mathbf{V\Lambda}$, it follows that $(\mathbf{AV})' = (\mathbf{V\Lambda})'$, or:

$$\mathbf{V}'\mathbf{A}' = \mathbf{\Lambda}'\mathbf{V}' = \mathbf{\Lambda}\mathbf{V}' \qquad [3.26]$$

because $\mathbf{\Lambda}' = \mathbf{\Lambda}$. Premultiplication of [3.26] by $(\mathbf{V}')^{-1}$ gives:

$$\mathbf{A}' = (\mathbf{V}')^{-1}\,\mathbf{\Lambda}\mathbf{V}' \qquad [3.27]$$

It follows that a matrix and its transpose have the same $\mathbf{\Lambda}$ matrix and consequently the same eigenvalues. If **A** is symmetrical with $\mathbf{A} = \mathbf{A}'$, then:

$$\mathbf{A} = \mathbf{V\Lambda V}' = (\mathbf{V}')^{-1}\mathbf{\Lambda}\mathbf{V}' \qquad [3.28]$$

Suppose $\mathbf{P} = \mathbf{V}'\mathbf{V}$, then $\mathbf{P}^{-1} = \mathbf{V}^{-1}(\mathbf{V}')^{-1}$. Premultiplication of **A** by $\mathbf{V}^{-1}$ and post-multiplication by $(\mathbf{V}')^{-1}$ yields:

$$\mathbf{\Lambda} = \mathbf{P}^{-1}\mathbf{\Lambda}\mathbf{P}$$

or:

$$\mathbf{P\Lambda} = \mathbf{\Lambda P} \qquad [3.29]$$

Since $\mathbf{\Lambda}$ is a diagonal matrix, [3.29] results in:

$$p_{ij}\lambda_j = \lambda_i p_{ij} \qquad [3.30]$$

This relationship only can be satisfied when $p_{ij} = 0$ if $i \neq j$. It follows that **P** is a diagonal matrix. Consequently, $\mathbf{V'V} = \mathbf{P}$ can be rewritten as:

$$(\mathbf{VQ})' (\mathbf{VQ}) = \mathbf{I}$$

where **Q** is a diagonal matrix with elements $q_{ii} = (p_{ii})^{-\frac{1}{2}}$. From the definition of the inverse matrix, we have:

$$(\mathbf{VQ})' = (\mathbf{VQ})^{-1}$$

The sum of squares of elements is equal to unity for all columns and rows of **VQ**. The matrix **VQ** therefore is the modal matrix consisting of normalized eigenvectors discussed above, or:

$$\mathbf{VQ} = \mathbf{V}_n \qquad [3.31]$$

The matrix $\mathbf{V}_n$ for a symmetric matrix has the property that its inverse is equal to its transpose. In general, a matrix **A** which satisfies $\mathbf{A}' = \mathbf{A}^{-1}$ is called an orthogonal matrix (see section 3.12).

A general, nonsymmetrical matrix can have eigenvalues that are complex numbers. However, a symmetric matrix has real eigenvalues only. For a proof of this property, see Albert (1949, p.97).

The sum of the diagonal elements of a matrix **A** is called the trace of **A** and is denoted as tr (**A**). It can be shown that:

$$\text{tr}(\mathbf{A}) = \sum_{i=1}^{n} a_{ii} = \sum_{i=1}^{n} \lambda_i \qquad [3.32]$$

It can also be proved that the determinant of a matrix is equal to the product of its eigenvalues or:

$$|\mathbf{A}| = \lambda_1 \lambda_2 \ldots \lambda_n \qquad [3.33]$$

These properties for the sum and product of the eigenvalues can be useful for the checking of eigenvalues calculated by some numerical method. The reader may verify these properties for the examples presented in this chapter.

3.10. Powers of a matrix

If $\mathbf{A}$ is a square matrix, the product $\mathbf{AA}$ can be written as $\mathbf{A}^2$. Likewise $\mathbf{AAA} = \mathbf{A}^2\mathbf{A} = \mathbf{A}^3$. In general, $\mathbf{A}^s$ represents $\mathbf{A}$ to the power s. When $\mathbf{A}$ is written in its canonical form:

$$\mathbf{A} = \mathbf{V}\mathbf{\Lambda}\mathbf{V}^{-1}$$

it follows that:

$$\begin{aligned} \mathbf{A}^2 &= \mathbf{AA} = (\mathbf{V\Lambda V}^{-1})(\mathbf{V\Lambda V}^{-1}) \\ &= (\mathbf{V\Lambda})(\mathbf{V}^{-1}\mathbf{V})\,\mathbf{\Lambda V}^{-1} \\ &= (\mathbf{V\Lambda})\mathbf{I}(\mathbf{\Lambda V}^{-1}) \\ &= \mathbf{V}(\mathbf{\Lambda\Lambda})\mathbf{V}^{-1} \\ &= \mathbf{V\Lambda}^2\mathbf{V}^{-1} \end{aligned}$$

Likewise:

$$\mathbf{A}^s = \mathbf{V\Lambda}^s\mathbf{V}^{-1} \qquad [3.34]$$

Fractional powers

If $\mathbf{A}^s = \mathbf{B}$, we may write $\mathbf{A} = \mathbf{B}^{1/s}$ and refer to $\mathbf{A}$ as the s-th root of $\mathbf{B}$. For example, the square root of a matrix $\mathbf{A}$ is written as $\mathbf{A}^{\frac{1}{2}}$. When the canonical form of $\mathbf{A}$ is employed :

$$\mathbf{A}^{\frac{1}{2}} = \mathbf{V\Lambda}^{\frac{1}{2}}\mathbf{V}^{-1} \qquad [3.35]$$

Negative powers

These are defined as $\mathbf{A}^{-s} = (\mathbf{A}^{-1})^s$ where $\mathbf{A}^{-1}$ is the inverse of $\mathbf{A}$. The canonical form of $\mathbf{A}^{-1}$ is:

$$\mathbf{A}^{-1} = \mathbf{V\Lambda}^{-1}\mathbf{V}^{-1} \qquad [3.36]$$

Proof:

$$\mathbf{AA}^{-1} = (\mathbf{V\Lambda V}^{-1})(\mathbf{V\Lambda}^{-1}\mathbf{V}^{-1}) = \mathbf{V}(\mathbf{\Lambda\Lambda}^{-1})\,\mathbf{V}^{-1} = \mathbf{VV}^{-1} = \mathbf{I}$$

Example. The canonical form of

$$\mathbf{A} = \begin{bmatrix} 5 & 2 \\ -6 & -3 \end{bmatrix}$$

has been derived in the example of section 3.9. Application of [3.36] gives:

$$\mathbf{A}^{-1} = \begin{bmatrix} 1 & 1 \\ -1 & -3 \end{bmatrix} \begin{bmatrix} -\frac{1}{3} & 0 \\ 0 & -1 \end{bmatrix} \begin{bmatrix} \frac{3}{2} & \frac{1}{2} \\ -\frac{1}{2} & -\frac{1}{2} \end{bmatrix} = \frac{1}{3} \begin{bmatrix} 3 & 2 \\ -6 & -5 \end{bmatrix}$$

3.11. Spectral components of a matrix

The canonical form of **A** can also be written as:

$$\mathbf{A} = \mathbf{V \Lambda T} \qquad [3.37]$$

with $\mathbf{T} = \mathbf{V}^{-1}$.

The rows of **T** are row vectors which will be denoted as $\mathbf{T}_i'$ $(i = 1, 2, \ldots, n)$ so that:

$$\mathbf{T} = \begin{bmatrix} \mathbf{T}_1' \\ \mathbf{T}_2' \\ \cdot\cdot \\ \cdot\cdot \\ \mathbf{T}_n' \end{bmatrix}$$

It will be recalled that $\mathbf{V} = [\mathbf{V}_1 \ \mathbf{V}_2 \ldots \ \mathbf{V}_n]$.

If **A** is a (3 X 3) matrix, the canonical form can be written in full as:

$$\mathbf{A} = [\mathbf{V}_1 \ \mathbf{V}_2 \ \mathbf{V}_3] \begin{bmatrix} \lambda_1 & 0 & 0 \\ 0 & \lambda_2 & 0 \\ 0 & 0 & \lambda_3 \end{bmatrix} \begin{bmatrix} \mathbf{T}_1' \\ \mathbf{T}_2' \\ \mathbf{T}_3' \end{bmatrix}$$

or:

$$\mathbf{A} = \begin{bmatrix} v_{11} & v_{21} & v_{31} \\ v_{12} & v_{22} & v_{32} \\ v_{13} & v_{23} & v_{33} \end{bmatrix} \begin{bmatrix} \lambda_1 & 0 & 0 \\ 0 & \lambda_2 & 0 \\ 0 & 0 & \lambda_3 \end{bmatrix} \begin{bmatrix} t_{11} & t_{12} & t_{13} \\ t_{21} & t_{22} & t_{23} \\ t_{31} & t_{32} & t_{33} \end{bmatrix} \qquad [3.38]$$

Then we can write:

$$\mathbf{A} = \lambda_1 \begin{bmatrix} v_{11} \\ v_{12} \\ v_{13} \end{bmatrix} [t_{11} \ t_{12} \ t_{13}] + \lambda_2 \begin{bmatrix} v_{21} \\ v_{22} \\ v_{23} \end{bmatrix} [t_{21} \ t_{22} \ t_{23}] + \lambda_3 \begin{bmatrix} v_{31} \\ v_{32} \\ v_{33} \end{bmatrix} [t_{31} \ t_{32} \ t_{33}]$$

or:

$$\mathbf{A} = \lambda_1 \mathbf{V}_1 \mathbf{T}_1' + \lambda_2 \mathbf{V}_2 \mathbf{T}_2' + \lambda_3 \mathbf{V}_3 \mathbf{T}_3'$$

In general, for an $(n \times n)$ matrix:

$$\mathbf{A} = \sum_{i=1}^{n} \lambda_i \mathbf{V}_i \mathbf{T}_i' \qquad [3.39]$$

or:

$$\mathbf{A} = \sum_{i=1}^{n} \lambda_i \mathbf{Z}_0(\lambda_i)$$

with:

$$\mathbf{Z}_0(\lambda_i) = \mathbf{V}_i \mathbf{T}_i' \qquad [3.40]$$

The matrices of $\mathbf{Z}_0(\lambda_i)$ are the spectral components of **A**.
From [3.34] and [3.40], it follows that:

$$\mathbf{A}^s = \sum_{i=1}^{n} \lambda_i^s \mathbf{Z}_0(\lambda_i) \qquad [3.41]$$

The spectral components are not affected by the power s. The matrix $\mathbf{A}^s$ is the sum of n terms, each of which is proportional to the s-th power of one of the eigenvalues. When the n eigenvalues are real and distinct, one value λ_1 generally will be larger than the others. The value of λ_1^s increases more rapidly than λ_i^s if $\lambda_1 > \lambda_i$ $(i = 2, \ldots, n)$ and $\mathbf{A}^s$ approaches $\lambda_1^s \, \mathbf{Z}_0(\lambda_1)$ if s increases.

The dominant eigenvalue λ_1 and the corresponding component $\mathbf{Z}_1 = \mathbf{Z}_0(\lambda_1)$ can be solved by raising the matrix **A** to a sufficiently high power. This principle is at the basis of most iterative methods and computer algorithms for the solution of the eigenvalues of a matrix. The other eigenvalues λ_i and components $\mathbf{Z}_i$ can be found in the same manner.

Example

$$\mathbf{A} = \begin{bmatrix} 5 & 2 \\ -6 & -3 \end{bmatrix}$$

$$\mathbf{A}^2 = \begin{bmatrix} 13 & 4 \\ -12 & -3 \end{bmatrix}$$

$$\mathbf{A}^4 = \begin{bmatrix} 121 & 40 \\ -120 & -39 \end{bmatrix}$$

$$\mathbf{A}^8 = \begin{bmatrix} 9841 & 3280 \\ -9840 & -3279 \end{bmatrix}$$

$$\mathbf{A}^9 = \begin{bmatrix} 29525 & 9842 \\ -29526 & -9843 \end{bmatrix}$$

In this example, the powering was speeded up by successively computing:

$$\mathbf{A}^2, \quad \mathbf{A}^4 = \mathbf{A}^2 \cdot \mathbf{A}^2, \quad \mathbf{A}^8 = \mathbf{A}^4 \cdot \mathbf{A}^4, \quad \mathbf{A}^9 = \mathbf{A}^8 \cdot \mathbf{A}$$

The ratios of corresponding elements in $\mathbf{A}^9$ and $\mathbf{A}^8$ are:

$$\begin{matrix} 3.0002 & 3.0006 \\ 3.0006 & 3.0019 \end{matrix}$$

Each of these ratios approximates the value 3 which is the largest of the two eigenvalues of this matrix.

If 3 is accepted as an estimate of the dominant eigenvalue in this case, the first spectral component can be estimated from:

$$\mathbf{A}^9/3^9 = \begin{bmatrix} 1.50002 & 0.50002 \\ -1.50008 & -0.50008 \end{bmatrix}$$

and: $$\mathbf{Z}_1 = \begin{bmatrix} 1.5 & 0.5 \\ -1.5 & -0.5 \end{bmatrix}$$

The first row of this matrix can be taken for $\mathbf{T}_1'$ with:

$$\mathbf{T}_1' = [1.5 \quad 0.5]$$

In that case, $\mathbf{V}_1 = \begin{bmatrix} 1 \\ -1 \end{bmatrix}$ and $\lambda_1 \mathbf{Z}_1 = \begin{bmatrix} 4.5 & 1.5 \\ -4.5 & -1.5 \end{bmatrix}$

When λ_1 and $\mathbf{Z}_1$ are known, the product $\lambda_1 \mathbf{Z}_1$ can be subtracted from $\mathbf{A}$. If this difference is raised to a high power, it will approach $\lambda_2^s \mathbf{Z}_2$. Thus the real eigenvalues can be found in their order of decreasing absolute value.

This method will be generalized in Chapter 12 so that it will also account for the case of conjugate pairs of complex eigenvalues. It can be further illustrated by using the preceding example.

When $\lambda_1 \mathbf{Z}_1$ is subtracted from $\mathbf{A}$, this results directly in:

$$\lambda_2 \mathbf{Z}_2 = \begin{bmatrix} 5 & 2 \\ -6 & -3 \end{bmatrix} - \begin{bmatrix} 4.5 & 1.5 \\ -4.5 & -1.5 \end{bmatrix} = \begin{bmatrix} 0.5 & 0.5 \\ -1.5 & -1.5 \end{bmatrix}$$

If this matrix is squared, the result is:

$$\begin{bmatrix} -0.5 & -0.5 \\ 1.5 & 1.5 \end{bmatrix} = \begin{bmatrix} 0.5 & 0.5 \\ -1.5 & -1.5 \end{bmatrix}^2$$

from which it follows that $\lambda_2 = -1$ and:

$$\mathbf{Z}_2 = \begin{bmatrix} -0.5 & -0.5 \\ 1.5 & 1.5 \end{bmatrix} \quad \text{with } \mathbf{V}_2 = \begin{bmatrix} 1 \\ -3 \end{bmatrix} \quad \text{and } \mathbf{T}_2' = [-0.5 \quad -0.5] .$$

In general, it is not necessary to power the entire matrix for finding the dominant eigenvalue. A more rapid computational procedure will be applied in section 12.4.

3.12. Miscellaneous concepts

This section contains a number of concepts which are frequently used in matrix algebra. Some will be applied in later chapters. A further development of some aspects of matrix algebra will be presented in Chapter 12, which is oriented to general, asymmetrical matrices which may have complex eigenvalues and eigenvectors.

Some properties of the matrices discussed below are given without proof. The interested reader may consult textbooks on matrix algebra for more details, e.g., Frazer et al. (1960) and Wilkinson (1965).

Orthogonal matrix

A square matrix $\mathbf{A}$ is orthogonal if its transpose is its inverse or:

$$\mathbf{A}'\mathbf{A} = \mathbf{I} \qquad [3.42]$$

As discussed in section 3.9, the normalized modal matrix $\mathbf{V}_n$ for a symmetric matrix is orthogonal.

Another property is that if $\mathbf{A}$ is orthogonal, its determinant $|\mathbf{A}|$ satisfies:

$$|\mathbf{A}| = \pm 1 \qquad [3.43]$$

Commuting matrices

If **A** and **B** are two square matrices, then, generally, $\mathbf{AB} \neq \mathbf{BA}$. However, if **A** and **B** have the same eigenvectors so that $\mathbf{A} = \mathbf{V}\mathbf{\Lambda}_1\mathbf{V}^{-1}$ and $\mathbf{B} = \mathbf{V}\mathbf{\Lambda}_2\mathbf{V}^{-1}$, they also commute with:

$$\mathbf{AB} = \mathbf{BA}$$

In general, the opposite also holds true. If **A** and **B** have the property of commutation, they also have the same eigenvectors. However, this property is subject to a condition as discussed by Wilkinson (1965, p.52). If the eigenvalues of **A** and **B** are distinct their eigenvectors will be identical. The identity matrix commutes with an arbitrary matrix, but its eigenvalues are not distinct. This is an example for which the property does not hold.

Symmetric idempotent matrix

If **A** is symmetric with $\mathbf{A}' = \mathbf{A}$ and, in addition to this has the property $\mathbf{AA} = \mathbf{A}$, it is called symmetric idempotent. The identity matrix provides an example.

Companion matrix of a polynomial

It is readily verified that the polynomial equation:

$$\lambda^n + p_1\lambda^{n-1} + \ldots\ldots + p_n = 0$$

is the characteristic equation of the matrix:

$$\mathbf{A} = \begin{bmatrix} 0 & 1 & 0 & \ldots & 0 \\ 0 & 0 & 1 & \ldots & 0 \\ \cdots & \cdots & \cdots & \cdots & \cdots \\ \cdots & \cdots & \cdots & \cdots & \cdots \\ \cdots & \cdots & \cdots & \cdots & \cdots \\ 0 & 0 & 0 & \ldots & 1 \\ -p_n & -p_{n-1} & -p_{n-2} & \ldots & -p_1 \end{bmatrix} \qquad [3.44]$$

The matrix **A** then is called the companion matrix of the polynomial. It follows that practical methods to obtain the eigenvalues of a matrix also can be used to solve for the roots of a polynomial equation.

Bilinear and quadratic forms

If $\mathbf{A}$ is an $(n \times n)$ symmetric matrix, $\mathbf{X}$ an $(n \times 1)$ column vector, and $\mathbf{Y}$ an $(1 \times n)$ row vector, the product $\mathbf{YAX}$ is called a bilinear form. If $\mathbf{Y} = \mathbf{X}'$, the bilinear form becomes a quadratic form. From the multiplication rule for matrices it follows that a quadratic form satisfies:

$$\mathbf{X}'\mathbf{A}\mathbf{X} = \sum_{i=1}^{n} \sum_{j=1}^{n} a_{ij} x_i x_j \qquad [3.45]$$

Differential calculus for matrices

The elements of a matrix $\mathbf{A}$ may be functions of a variable, say t. In that case, we can define $\mathrm{d}\mathbf{A}/\mathrm{d}t = \mathbf{DA}$, where $\mathbf{D}$ is a diagonal matrix with the differentiation operator $\mathrm{d}/\mathrm{d}t$ on the diagonal line.

Example. If $\mathbf{A}$ is a (2×2) matrix:

$$\mathbf{DA} = \begin{bmatrix} \mathrm{d}/\mathrm{d}t & 0 \\ 0 & \mathrm{d}/\mathrm{d}t \end{bmatrix} \begin{bmatrix} a_{11} & a_{12} \\ a_{21} & a_{22} \end{bmatrix} = \begin{bmatrix} \mathrm{d}a_{11}/\mathrm{d}t & \mathrm{d}a_{12}/\mathrm{d}t \\ \mathrm{d}a_{21}/\mathrm{d}t & \mathrm{d}a_{22}/\mathrm{d}t \end{bmatrix} \qquad [3.46]$$

Some immediate deductions are:

$$\mathrm{d}(\mathbf{A} + \mathbf{B})/\mathrm{d}t = \mathrm{d}\mathbf{A}/\mathrm{d}t + \mathrm{d}\mathbf{B}/\mathrm{d}t$$

and:

$$\mathrm{d}(\mathbf{AB})/\mathrm{d}t = \mathbf{A}\mathrm{d}\mathbf{B}/\mathrm{d}t + (\mathrm{d}\mathbf{A}/\mathrm{d}t)\mathbf{B}$$

or:

$$\mathbf{D}(\mathbf{A} + \mathbf{B}) = \mathbf{DA} + \mathbf{DB}$$

and:

$$\mathbf{D}(\mathbf{AB}) = \mathbf{ADB} + \mathbf{D}(\mathbf{A})\mathbf{B}$$

In partial differentiation, the matrix $\mathbf{D}$ has elements $\partial/\partial t$ along its diagonal.

If $\mathbf{X}$ and $\mathbf{A}$ are column vectors, $y = \mathbf{X}'\mathbf{A}$ represents a scalar. The partial derivative $\partial y/\partial \mathbf{X}$ then denotes the operation:

$$\frac{\partial y}{\partial \mathbf{X}} = \begin{bmatrix} \partial y/\partial x_1 \\ \cdots\cdots \\ \cdots\cdots \\ \cdots\cdots \\ \partial y/\partial x_n \end{bmatrix} \qquad [3.47]$$

It follows that:

$$\partial y/\partial \mathbf{X} = \partial(\mathbf{X}'\mathbf{A})/\partial \mathbf{X} = \mathbf{A} \qquad [3.48]$$

Likewise, if **A** represents an $(n \times n)$ matrix, the quadratic form $\mathbf{X}'\mathbf{A}\mathbf{X}$ satisfies:

$$\partial(\mathbf{X}'\mathbf{A}\mathbf{X})/\partial \mathbf{X} = 2\mathbf{A}\mathbf{X} \qquad [3.49]$$

Definite matrices

If **A** is a square symmetric matrix, it is said to be positive definite if the quadratic form $\mathbf{X}'\mathbf{A}\mathbf{X}$ is positive for every nonzero $(n \times 1)$ vector **X**. A property of a positive definite matrix **A** is that all its eigenvalues are positive.

A nonnegative definite (or positive semidefinite) matrix has a quadratic form that is nonnegative for every **X**. If **P** is an arbitrary matrix, then $\mathbf{P}'\mathbf{P}$ always is nonnegative definite.

If **A** is nonnegative definite, but not positive definite, its smallest eigenvalue is zero and **A** is singular.

Hermitian matrices

If the elements of a matrix **A** are complex numbers, the matrix $\mathbf{A}^*$ with the corresponding conjugate complex elements is called the conjugate of **A**. A matrix **A** is Hermitian if $\mathbf{A}^* = \mathbf{A}'$.

Chapter 4

GEOMETRY

4.1. Introduction

This chapter is more advanced than the introduction to geometry given in section 2.2 and other sections of Chapter 2. Most observable geological attributes are related to geometrical forms in nature lying in three-dimensional (3D-) space. Many can be approximated by points, lines, surfaces or solids, justifying a more detailed review of principles of classical solid analytic geometry.

The geometrical interpretation of vectors and matrices and operations such as diagonalization (Chapter 3) is emphasized in this chapter, mainly for situations in 2D- and 3D-space. The properties of quadric surfaces are reviewed in order to prepare the reader for methods of trend analysis (Chapter 9). Most methods of optiminization, such as the method of least squares, are based on the partial differentiation of a quadratic form with respect to unknown coefficients. This form is represented by an ellipsoid in 3D-space which is a special type of quadric surface.

Several methods of fitting lines to data to represent functional relationships in bivariate situations are discussed in detail. Most methods of this chapter are extended in later chapters for situations in higher-dimensional space.

4.2. Geometrical interpretation of vectors and matrices

A point P in three-dimensional space is uniquely determined by its three coordinates x, y, and z with respect to a rectangular, Cartesian coordinate system consisting of X-, Y- and Z-axis (see Fig. 5C). An equation for representing the point is $P = (x, y, z)$.

Suppose that as in section 3.1, p variables $X_1, X_2, \ldots, X_p$ were measured for individual rock specimens. The result then can be represented by the formula $P = (x_1, x_2, \ldots, x_p)$ where x_i, $i = 1, 2, \ldots\, p$, are values assumed by the variables X_i for a rock specimen. The point P now lies in p-space and is determined by p coordinates for p mutually perpendicular axes. This is a direct extension of the geometry for two- or three-dimensional space.

The point P can also be considered as the end point of a vector OP in p-space which has its beginning in the origin O of the coordinate system. If the p measurements for a single rock specimen are recorded as an $(1 \times p)$ row vector $\mathbf{X} = [x_1\ x_2\ \ldots\ x_p]$, this can be interpreted as a vector OP in p-space. Consequently, an $(n \times p)$ matrix that contains the values for n different rock specimens can be interpreted either as a cluster of n points in p-space, or as an assemblage of vectors OP_k with k going from 1 to n.

The $(n \times p)$ data matrix, alternately, can be interpreted as p points or vectors in n-space. In that case, the n data for each column are considered separately. The geometrical interpretation of a column vector is then analogous to that discussed in more detail for the row vector. In the sequel, we will represent an $(n \times p)$ data matrix both in n-space or in p-space depending on the mathematical technique that is applied. It is important to keep in mind that for an $(n \times p)$ data matrix in a multivariate situation, the rows indicate observation points or stations whereas the columns refer to different variables that have been measured. This convention, for example, is adhered to in most textbooks on mathematical statistics. However, some workers prefer $(p \times n)$ row matrices instead of $(n \times p)$ column matrices.

Suppose that n measurements have been made for a single variable. We then have an $(n \times 1)$ column vector **X** or OP whose length can be denoted as $|OP|$. By extension of the Pythagorean theorem:

$$|OP| = \left(\sum_{k=1}^{p} x_k^2\right)^{\frac{1}{2}} \qquad [4.1]$$

The vector OP determines a line passing through the origin. However, all points on this line would uniquely determine the line. Therefore, a line may be characterized by considering only a single point P_1 on it for which:

$$|OP_1| = 1 \qquad [4.2]$$

OP_1 is the unit vector corresponding to OP. It has been stated in section 3.9 that a vector can be normalized by dividing all of its elements by the square root of their sum of squares. This is the same as dividing by $|OP|$. Geometrically, the normalization of a vector is equivalent to moving the point P to the point P_1 which also is on the line OP but at unit distance from the origin O.

In general, the sum of two vectors

$$\mathbf{A} = \begin{bmatrix} a_1 \\ a_2 \\ \cdot \\ \cdot \\ \cdot \\ a_n \end{bmatrix} \quad \text{and } \mathbf{B} = \begin{bmatrix} b_1 \\ b_2 \\ \cdot \\ \cdot \\ \cdot \\ b_n \end{bmatrix} \quad \text{satisfies } \mathbf{A} + \mathbf{B} = \begin{bmatrix} a_1 + b_1 \\ a_2 + b_2 \\ \cdot \\ \cdot \\ \cdot \\ a_n + b_n \end{bmatrix}$$

Let this sum be written as **C**. A schematic representation of **A** + **B** = **C** is shown in Fig. 15A. It is not necessary to have a single fixed origin for all vectors in the geometrical re-

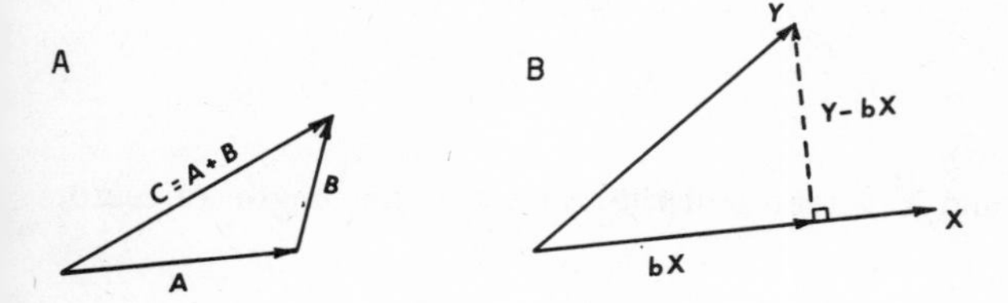

Fig.15. A. Addition of vectors A and B giving the sum vector $C = A + B$. B. Perpendicular projection of vector Y onto X giving vector bX.

presentation. The vector **B** in Fig. 15**A** has its point of beginning at the endpoint of **A**. If $\mathbf{A} + \mathbf{B} = \mathbf{C}$, we also have $\mathbf{B} = \mathbf{C} - \mathbf{A}$ and $\mathbf{A} = \mathbf{C} - \mathbf{B}$. For example, if **C** and **A** are given, **B** is obtained by subtracting the vector **A** from **C** so that the relation $\mathbf{A} + \mathbf{B} = \mathbf{C}$ is satisfied.

Orthogonal projection of a vector onto another vector

The purpose of the following example is to illustrate that relatively complicated relationships between vectors and matrices can have a simple geometrical interpretation.

Suppose that n measurements were made on two variables X and Y which are linearly related by the function $Y = cX$ where c is a constant. However, if the observed data are plotted in a planar diagram, the points (x_k, y_k), $k = 1, 2, \ldots, n$, do not exactly fall on a straight line because of measurements errors in Y.

Let the data be represented as two column vectors:

$$\mathbf{X} = \begin{bmatrix} x_1 \\ x_2 \\ \cdot \\ \cdot \\ \cdot \\ x_n \end{bmatrix} \quad \text{and } \mathbf{Y} = \begin{bmatrix} y_1 \\ y_2 \\ \cdot \\ \cdot \\ \cdot \\ y_n \end{bmatrix}$$

Suppose that $b\mathbf{X}$, where b is a scalar, represents the perpendicular (orthogonal) projection of **Y** onto **X** in n-dimensional space. The vectors **X**, **Y**, $b\mathbf{X}$, and also the difference $\mathbf{Y} - b\mathbf{X}$ are schematically shown in Fig. 15B.

By definition, the inner product of two vectors is a scalar which is obtained by writing one of the vectors as a row vector and then postmultiplying by the other vector which is written as a column vector. If **A** and **B** represent two $(n \times 1)$ column vectors, their inner product is $\mathbf{A}'\mathbf{B}$ or $\mathbf{B}'\mathbf{A}$. The inner product of a vector multiplied by itself yields the sum of squares of the elements, or:

$$\mathbf{X'X} = \sum_{k=1}^{n} x_k^2$$

From [4.1], it follows that **X′X** and **Y′Y** represent the squares of the lengths of vectors **X** and **Y** or:

$$|\mathbf{X}|^2 = \mathbf{X'X} \quad \text{and} \quad |\mathbf{Y}|^2 = \mathbf{Y'Y}$$

If the Pythagorean theorem is applied to the situation shown in Fig. 15B, the result is:

$$|\mathbf{Y} - b\mathbf{X}|^2 = |\mathbf{Y}|^2 - |b\mathbf{X}|^2 \tag{4.3}$$

In matrix notation, this becomes:

$$(\mathbf{Y} - b\mathbf{X})'(\mathbf{Y} - b\mathbf{X}) = \mathbf{Y'Y} - (b\mathbf{X})'(b\mathbf{X})$$

or:

$$\mathbf{Y'Y} - b\mathbf{Y'X} - b\mathbf{X'Y} + b^2\mathbf{X'X} = \mathbf{Y'Y} - b^2\mathbf{X'X}$$

and:

$$2b\mathbf{X'X} = \mathbf{X'Y} + \mathbf{Y'X}$$

Keeping in mind that $\mathbf{Y'X} = \mathbf{X'Y}$, it follows readily that:

$$b\,\mathbf{X'X} = \mathbf{X'Y}$$

or:

$$b = (\mathbf{X'X})^{-1}\,\mathbf{X'Y} \tag{4.4}$$

In this result, the scalar b is expressed in terms of the original vectors **X** and **Y**. If we had defined **X** and **Y** as row vectors instead of column vectors, the result would be:

$$b = (\mathbf{XX'})^{-1}\,\mathbf{YX'}$$

When the angle between **X** and **Y** in Fig.15B is written as θ: $\cos^2\theta = (b\mathbf{X})'(b\mathbf{X}) / (\mathbf{Y'Y})$. After some manipulation, it follows from [4.4] that:

$$\cos\theta = \frac{\mathbf{X'Y}}{[(\mathbf{X'X})\,(\mathbf{Y'Y})]^{\frac{1}{2}}} \tag{4.5}$$

The inner product, therefore, is proportional to the angle between two vectors.

It will be shown in section 4.6 that b in [4.4] provides an estimate of c in $Y=cX$ for the linear relationship between the two variables X and Y with Y subject to error and whose measurements were given by the $(n \times 1)$ vectors **X** and **Y**.

Suppose that in the example, **X** is not an $(n \times 1)$ column vector but an $(n \times p)$ column matrix. Then, the $(n \times 1)$ column vector **Y** can be "projected" onto the space defined by the p vectors $\mathbf{X}_1, \mathbf{X}_2, \ldots, \mathbf{X}_p$ which are the columns of the matrix **X**.

The perpendicular projection of **Y** onto this space is according to a vector **XB** where **B** is a column vector consisting of p coefficients. This vector satisfies $\mathbf{B} = (\mathbf{X}'\mathbf{X})^{-1}\ \mathbf{X}'\mathbf{Y}$ which resembles [4.4]. In section 4.3, we will prove this equation for **B** in the situation that $n = 3$ and $p = 2$.

First, the subject of lines and planes in space is discussed with reference to a (XYZ)-coordinate system. It should be kept in mind that most results can be extended readily into multidimensional space.

4.3. Lines and planes in three-dimensional space

The point $P = (x, y, z)$ determines a vector OP. Suppose that α, β, and γ are the angles of this vector with X-, Y-, and Z-axes. It follows that:

$$x = |OP| \cos \alpha \; ; \qquad y = |OP| \cos \beta \; ; \qquad z = |OP| \cos \gamma \qquad [4.6]$$

When P_1 is a point on OP with $|OP_1| = 1$, we can define $P_1 = (\alpha, \beta, \gamma)$ with:

$$\lambda = \cos \alpha \; ; \qquad \mu = \cos \beta \; ; \qquad \nu = \cos \gamma \qquad [4.7]$$

Also, from [4.6] :

$$\lambda = x/|OP| \; ; \qquad \mu = y/|OP| \; ; \qquad \nu = z/|OP| \qquad [4.8]$$

It is verified readily that:

$$\cos^2 \alpha + \cos^2 \beta + \cos^2 \gamma = 1 \qquad [4.9]$$

The numbers $\lambda = \cos \alpha$, $\mu = \cos \beta$, and $\nu = \cos \gamma$ are called the direction cosines of OP. Any line passing through the origin and an arbitrary point is fully determined by its direction cosines, whose sum of squares is equal to one.

The general equation of a plane can be presented by the equation:

$$z = a + bx + cy \qquad [4.10]$$

where a, b, and c are constants. This plane is graphically represented as ABC in Fig. 16A.

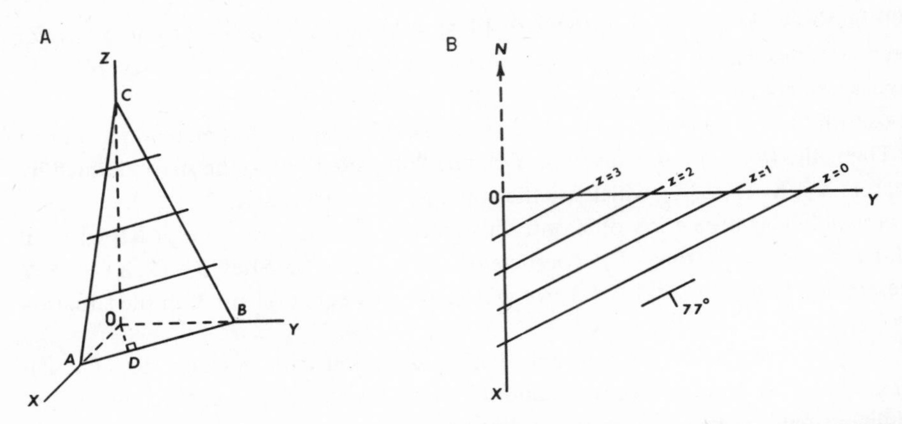

Fig.16. A. Representation of a plane in three-dimensional Cartesian coordinate system. B. Contours in the XY-plane of the plane with equation $z = 4 + x + 2y$, striking N63°E and dipping 77°.

The coefficient a is the intercept or the distance from the origin to the point where the plane intersects the Z-axis. The coefficients b and c define the slopes of the lines CA and CB.

The equation of a plane can also be written in the form:

$$\lambda x + \mu y + \nu z = |OP| \qquad \text{with} \qquad \lambda^2 + \mu^2 + \nu^2 = 1 \qquad [4.11]$$

This is called the equation of a plane in normal form. It is related to the general equation [4.10] by:

$$\lambda = -b/R, \quad \mu = -c/R, \quad \nu = 1/R \text{ and } |OP| = a/R$$

where:

$$R = (b^2 + c^2 + 1)^{\frac{1}{2}}.$$

From [4.8] and [4.11], it follows that $|OP|$ represents the distance from the origin to a point P with coordinates $\lambda|OP|$, $\mu|OP|$, and $\nu|OP|$. The point P lies in the plane because its coordinates satisfy [4.11] and λ, μ, ν are the direction cosines of the line OP.

Suppose that the coordinates of a point P in space are known. We can calculate λ, μ and ν and $|OP|$ from these coordinates. The equation $\lambda x + \mu y + \nu z = |OP|$ then represents a plane passing through the point P which is perpendicular to the line OP. The line OP is called the normal of this plane.

Distance from a point to a plane

If the point is $P_1 = (x_1, y_1, z_1)$ and the plane $\lambda x + \mu y + \nu z = p$, then the distance d from P_1 to the plane satisfies:

$$d = \lambda x_1 + \mu y_1 + \nu z_1 - p \qquad [4.12]$$

Proof. The equation $\lambda x + \mu y + \nu z = p + d$ where d is a variable distance, defines a family of planes which are parallel to the original plane. P_1 must lie in one of these planes. Its coordinates satisfy the equation which specifies the unknown distance d.

Strike and dip

In structural geology, a plane is characterized by its strike and dip. The strike and dip of the plane ABC in Fig. 16A would be as follows. Suppose that the north direction points in the negative X-direction. The strike δ then may be defined as the angle between the line AB and the negative X-axis. By elementary geometrical manipulations, it is derived that:

$$\tan \delta = OB/OA = b/c = \mu/\lambda \qquad [4.13]$$

The dip of the plane is equal to the angle between the lines OD and CD in Fig. 16A. This angle is equal to γ. Hence, strike and dip of a plane are readily derived from the general equation.

Contours of a plane

Suppose that the plane satisfies:

$$z = 4 - 4x - 2y$$

From [4.13], it follows that the strike δ is:

$$\delta = \arctan (b/c)$$

$$= \arctan 2 = 63^\circ.4$$

The dip γ is obtained by normalizing the vector $[-b \ -c \ 1] = [4 \ 2 \ 1]$ which results in:

$$[\lambda \ \ \mu \ \ \nu] = [0.873 \ \ 0.436 \ \ 0.218]$$

Hence:

$$\cos \gamma (= \nu) = 0.218$$

$$\gamma = \arccos 0.218 = 77^\circ.4$$

When δ and γ are calculated from the coefficients in the general equation, the relationship between plane, coordinate system and north direction should be considered carefully. The equations presented here apply to the specific situation of the example only. In general, the dip γ can be defined as subject to the condition $0° < \gamma < 90°$. The strike δ then may be measured clockwise from the north direction to a direction 90° less than the direction of dip of the plane.

The plane with equation $z = 4 - 4x - 2y$ can be represented in mapform just as a topographic surface is represented by means of a contour map. The result is shown in Fig. 16B. The individual contours are equidistant straight lines. For example, the contour with $z = 3$ satisfies the equation:

$$3 = 4 - 4x - 2y, \text{ or } y = 0.5 - 2x$$

All contours point in the direction of strike with $\delta = 63°.4$. The distance between contours is controlled by the contour interval, the dip γ, and the scale of the map.

Angle between two lines

Suppose that $P_1 = (x_1, y_1, z_1)$ and $P_2 = (x_2, y_2, z_2)$ are two points and that the lines OP_1 and OP_2 have direction cosines $(\lambda_1, \mu_1, \nu_1)$ and $(\lambda_2, \mu_2, \nu_2)$, respectively.

Let P'_1 and P'_2 be points on OP_1 and OP_2 which are at unit distance from the origin. The angle θ between OP_1 and OP_2 is the same as that between OP'_1 and OP'_2. It satisfies:

$$\cos\theta = \lambda_1\lambda_2 + \mu_1\mu_2 + \nu_1\nu_2 \qquad [4.14]$$

The proof can be based on [4.5]. From [4.1] and [4.6], it follows that [4.14] also can be written as:

$$\cos\theta = (x_1x_2 + y_1y_2 + z_1z_2)\,(x_1^2 + y_1^2 + z_1^2)^{-\frac{1}{2}}\,(x_2^2 + y_2^2 + z_2^2)^{-\frac{1}{2}} \qquad [4.15]$$

which is [4.5] repeated in a different form.

Orthogonal projection of a vector onto a plane

Let $P_1 = (x_1, x_1, z_1)$, $P_2 = (x_2, y_2, z_2)$ and $P_3 = (x_3, y_3, z_3)$ be three distinct points. The points P_1 and P_2, together with the origin O, define a plane onto which P_3 can be projected. Let Q be the foot of the perpendicular from P_3 to the plane. Because of orthogonality, the line P_3Q, whose direction cosines are denoted as λ, μ, and ν, satisfies:

$$\left.\begin{aligned} \lambda x_1 + \mu y_1 + \nu z_1 &= 0 \\ \lambda x_2 + \mu y_2 + \nu z_2 &= 0 \end{aligned}\right| \qquad [4.16]$$

Since $\lambda^2 + \mu^2 + \nu^2 = 1$, the line is fully determined. From [4.12], we know that the distance d from P_3 to the plane satisfies:

$$d = \lambda x_3 + \mu y_3 + \nu z_3 \tag{4.17}$$

The length $|OQ|$ according to the Pythagorean theorem satisfies:

$$|OQ|^2 = |OP_3|^2 - d^2 \tag{4.18}$$

$|OQ|$ also is the shortest distance between O and P_3Q.

Matrix algebra can be employed to express the vector OQ in terms of the vectors OP_1, OP_2 and OP_3. Because of an interpretation to be given below, these three vectors will be written as $\mathbf{X}_1$, $\mathbf{X}_2$, and $\mathbf{Y}$, respectively.

From elementary geometrical considerations it follows that any vector can be expressed as a linear combination of two known vectors. Therefore, if the projection of $\mathbf{Y}$ onto the plane is written as $\hat{\mathbf{Y}}$, we may write:

$$\hat{\mathbf{Y}} = a\mathbf{X}_1 + b\mathbf{X}_2 \tag{4.19}$$

where a and b are two unknown scalars. Using vector notation, an equivalent expression for [4.18] is:

$$\hat{\mathbf{Y}} = \mathbf{Y} - \mathbf{E} \tag{4.20}$$

where $\mathbf{E}$ denotes the vector P_3Q. Consequently,

$$a\mathbf{X}_1 + b\mathbf{X}_2 = \mathbf{Y} - \mathbf{E}\,.$$

We know that $\mathbf{X}_1$ and $\mathbf{X}_2$ are both orthogonal to $\mathbf{E}$ or $\mathbf{X}_1'\mathbf{E} = \mathbf{X}_2'\mathbf{E} = 0$. Hence:

$$\mathbf{X}_1'(a\mathbf{X}_1 + b\mathbf{X}_2) = \mathbf{X}_1'\mathbf{Y}\,; \quad \mathbf{X}_2'(a\mathbf{X}_1 + b\mathbf{X}_2) = \mathbf{X}_2'\mathbf{Y}$$

This is equivalent to writing:

$$\begin{bmatrix} \mathbf{X}_1'\mathbf{X}_1 & \mathbf{X}_1'\mathbf{X}_2 \\ \mathbf{X}_2'\mathbf{X}_1 & \mathbf{X}_2'\mathbf{X}_2 \end{bmatrix} \begin{bmatrix} a \\ b \end{bmatrix} = \begin{bmatrix} \mathbf{X}_1'\mathbf{Y} \\ \mathbf{X}_2'\mathbf{Y} \end{bmatrix}$$

Obviously, the unknown coefficients satisfy:

$$\begin{bmatrix} a \\ b \end{bmatrix} = \begin{bmatrix} \mathbf{X}_1'\mathbf{X}_1 & \mathbf{X}_1'\mathbf{X}_2 \\ \mathbf{X}_2'\mathbf{X}_1 & \mathbf{X}_2'\mathbf{X}_2 \end{bmatrix}^{-1} \begin{bmatrix} \mathbf{X}_1'\mathbf{Y} \\ \mathbf{X}_2'\mathbf{Y} \end{bmatrix} \qquad [4.21]$$

Notation can be simplified further by defining a (3×2) matrix $\mathbf{X}$ with:

$$\mathbf{X} = [\mathbf{X}_1 \ \mathbf{X}_2]$$

From results obtained in section 3.8, it follows that:

$$\mathbf{X}'\mathbf{X} = \begin{bmatrix} \mathbf{X}_1'\mathbf{X}_1 & \mathbf{X}_1'\mathbf{X}_2 \\ \mathbf{X}_2'\mathbf{X}_1 & \mathbf{X}_2'\mathbf{X}_2 \end{bmatrix} \quad \text{and} \quad \mathbf{X}'\mathbf{Y} = \begin{bmatrix} \mathbf{X}_1'\mathbf{Y} \\ \mathbf{X}_2'\mathbf{Y} \end{bmatrix}$$

If the vector $\begin{bmatrix} a \\ b \end{bmatrix}$ is denoted as $\mathbf{B}$, [4.21] becomes:

$$\mathbf{B} = (\mathbf{X}'\mathbf{X})^{-1} \ \mathbf{X}'\mathbf{Y} \qquad [4.22]$$

This result was obtained for b in [4.4] for the projection in n-space of observations for a dependent variable Y onto a vector described by simultaneous observations on an independent variable X. We have now shown that this result holds true for two independent variables when $n = 3$. It can be generalized to a situation of n observations on one dependent variable and p independent variables X_i, $i = 1, 2, \ldots, p$. In this general case, $\mathbf{X}$ represents an $(n \times p)$ matrix.

Equation of a line in symmetric form

When $P_1 = (x_1, y_1, z_1)$ and $P_2 = (x_2, y_2, z_2)$ are two distinct points in 3D-space, the origin of the coordinate system can be translated to P_1. In terms of the new coordinates x', y', and z', the equations for P_1 and P_2 satisfy:

$$P_1' = (0, 0, 0) \; ; \quad P_2' = (x_2 - x_1, y_2 - y_1, z_2 - z_1)$$

Suppose that $P = (x, y, z)$ represents an arbitrary point. After translation, this becomes $P' = (x - x_1, y - y_1, z - z_1)$. The point P lies on the line P_1P_2 if PP_1 has the same direction cosines as P_1P_2. This is true if (and only if):

$$x - x_1 = t\,(x_2 - x_1) \; ; \qquad y - y_1 = t(y_2 - y_1) \; ; \qquad z - z_1 = t\,(z_2 - z_1) \qquad [4.23]$$

where $t = |PP_1|/|P_1P_2|$. If λ, μ, and ν represent the direction cosines of P_1P_2, the ratio $\lambda/\mu/\nu$ is equal to:

$$\lambda/\mu/\nu = (x_2 - x_1)/(y_2 - y_1)/(z_2 - z_1) \qquad [4.24]$$

By elimination of the parameter t from [4.23] and using [4.24], one obtains what is called the equation of the line P_1P_2 in symmetric form:

$$\frac{x - x_1}{\lambda} = \frac{y - y_1}{\mu} = \frac{z - z_1}{\nu} \qquad [4.25]$$

A point $P = (x, y, z)$ only lies on the line P_1P_2 if its coordinates satisfy [4.25].

Distance between a point and a line

Let l represent a line with direction cosines λ, μ, and ν. l passes through a point $P_1 = (x_1, y_1, z_1)$ and satisfies [4.25]. Further, let P_2 be a point with $P_2 = (x_2, y_2, z_2)$ that does not necessarily lie on l.

The problem to be treated consists of finding the distance between P_2 and l. If P is the foot of the perpendicular from P_2 to l, then the required distance is equal to $|PP_2|$.

When θ represents the angle between P_1P_2 and l, then:

$$|PP_2| = |P_1P_2| \sin\theta \qquad [4.26]$$

Consequently:

$$|PP_2|^2 = |P_1P_2|^2 \sin^2\theta = |P_1P_2|^2 (1 - \cos^2\theta)$$

$$= [(x_2 - x_1)^2 + (y_2 - y_1)^2 + (z_2 - z_1)^2] - |P_1P_2|^2 \cos^2\theta \qquad [4.27]$$

From [4.14], it follows that:

$$\cos\theta = \frac{1}{|P_1P_2|} [\lambda(x_2 - x_1) + \mu(y_2 - y_1) + \nu(z_2 - z_1)]$$

Hence:

$$|P_1P_2|^2 \cos^2\theta = [\lambda(x_2 - x_1) + \mu(y_2 - y_1) + \nu(z_2 - z_1)]^2 \qquad [4.28]$$

The required solution as derived from [4.27] and [4.28] is:

$$|PP_2|^2 = (x_2 - x_1)^2 + (y_2 - y_1)^2 + (z_2 - z_1)^2$$

$$- [\lambda(x_2 - x_1) + \mu(y_2 - y_1) + \nu(z_2 - z_1)]^2 \quad [4.29]$$

This result can be generalized to the situation of a point and a line in p-space where p may be any positive integer value.

4.4. Quadric surfaces

Our more detailed discussion of curved surfaces will be restricted to the class of quadric surfaces. The general equation of a quadric surface in three-dimensional space is:

$$ax^2 + 2bxy + 2cxz + dy^2 + 2eyz + fz^2 + gx + hy + iz + j = 0 \quad [4.30]$$

which contains ten different coefficients. If z is made equal to a constant, [4.30] reduces to the general equation of the second degree in the plane ([2.12] of section 3.5). This equation always can be reduced to a form:

$$px^2 + qy^2 + r = 0 \quad [4.31]$$

which, depending on the signs of the coefficients, represents one out of three basic forms: ellipse, hyperbola, or parabola. The subject of reducing equations of the second degree to a simpler form is discussed in the next section with reference to [4.30]. It will be shown that by using matrix algebra, [4.30] usually can be reduced to the form:

$$px^2 + qy^2 + rz^2 + s = 0 \quad [4.32]$$

In general this quadric form represents only one out of three types of surfaces: ellipsoid, hyperboloid of one sheet, or hyperboloid of two sheets. In addition to this, we will consider the special case when one of the coefficients p, q, or r is equal to zero, say $r = 0$. In that case, [4.30] does not reduce to [4.32] but to a form:

$$px^2 + qy^2 + rz = 0 \quad [4.33]$$

This form, which is linear in z, has two basic representations: elliptic paraboloid or hyperbolic paraboloid.

Ellipsoid

If p, q, and r in [4.31] have the same sign (positive or negative), the quadric is an

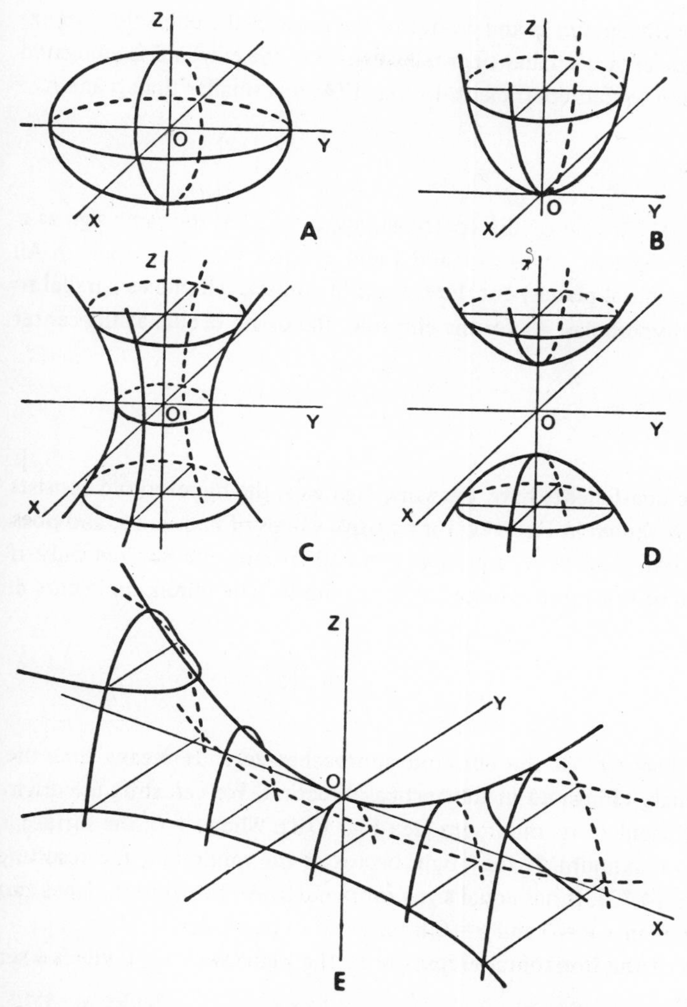

Fig.17. Five possible shapes of quadric surfaces in three-dimensional space. A. Ellipsoid. B. Elliptic paraboloid. C. Hyperboloid of one sheet. D. Hyperboloid of two sheets. E. Hyperbolic paraboloid.

ellipsoid (Fig. 17A). The constant s must be of the opposite sign since x^2, y^2, and z^2 always are positive numbers. If $p = q = r$, the solid enclosed by the surface is a sphere. The point where the X-axis intersects an ellipsoid can be found by setting y and z equal to zero. This point has coordinates $(-s/p, 0, 0)$. Likewise, $(0, -s/q, 0)$ and $(0, 0, -s/r)$ represent intersections with positive Y- and Z-axis, respectively.

The origin O in Fig.17A occurs at the so-called center of the ellipsoid. If an ellipsoid intersects a plane, the resulting curve of intersection is always an ellipse. Intersection with the horizontal XY-plane gives the ellipse $px^2 + qy^2 + s = 0$, because the equation of the plane then is $z = 0$. The coefficients p, q and r control the form of the ellipsoid. For example, if one of these coefficients is smaller than the other two, the ellipsoid is elongated in the direction corresponding to that coefficient. In Fig. 17A, q is smaller than p and r.

Hyperboloid of one sheet

This quadric satisfies [4.31] if one of the coefficients p, q, or r has the same sign as s. It is shown in Fig. 17C for negative values of r and s and positive values of p and q. All planes parallel to $z = 0$ (horizontal planes) cut the surface in ellipses. All planes parallel to $x = 0$, or to $y = 0$, cut it in hyperbolas. As for the ellipsoid, the origin occurs at the center of this quadric.

Hyperboloid of two sheets

If two of the first three coefficients have the same sign as s, the hyperboloid consists of two separate sheets. It is shown in Fig. 17D for negative values of p, q and s, and positive value of r. A horizontal plane with equation $z = c$ then cuts the surface only if $c > \sqrt{-s/r}$ and the section of this intersecting plane is an ellipse. The minimum occurs at the point with $z = \sqrt{-s/r}$.

Elliptic paraboloid

Suppose that the coefficient r for the ellipsoid approaches zero. It means that the ellipsoid becomes increasingly elongated in the vertical direction. We can shift the origin from the center of the ellipsoid to its minimum (see Fig. 17B). When $r = 0$, the surface is a paraboloid which has no maximum. If the origin occurs at the minimum, the resulting elliptic paraboloid satisfies [4.33]. It has equal signs for p and q. All horizontal planes cut this surface in ellipses. The planes $x = 0$ and $y = 0$ intersect it in parabolas.

The vertical projection of the horizontal ellipses onto the plane with $z = 0$ yields a set of contours for this surface.

Hyperbolic paraboloid

This quadric also satisfies [4.33] but has opposite signs for the coefficients p and q. It is shown in Fig. 17E for positive p and negative q and r. The hyperbolic paraboloid is comparable to the elliptic paraboloid except that the horizontal planes cut the surface in hyperbolas instead of ellipses.

4.5. Centers and axes of quadrics

Suppose that $f(x, y)$ represents a quadric function in two variables x and y, with:

$$f(x, y) = ax^2 + 2bxy + cy^2 + dx + ey + f \qquad [4.34]$$

For any specific value of $f(x, y)$, this reduces to the general equation of the second degree in the plane. The function itself can be represented as a surface $z = f(x, y)$ in three-dimensional space. Its equation always can be reduced to [4.31] by carrying out a translation and a rotation. The surface, therefore, is either an elliptic or a hyperbolic paraboloid. In both Fig. 17B and E, the origin of the coordinate system used to describe these quadrics occurs at the extremum. Suppose that the coordinates in the XY-plane of this minimum are (x_o, y_o). A translation of the origin to the vertical projection of the minimum in the XY-plane can then be performed by the linear transformation $u = x - x_o$, $v = y - y_o$. In terms of the new coordinates u and v, [4.34] becomes:

$$z = a(u + x_o)^2 + 2b(u + x_o)(v + y_o) + c(v + y_o)^2 + d(u + x_o) + e(v + y_o) + f \qquad [4.35]$$

or:

$$z = au^2 + 2buv + cv^2 + (2ax_o + 2by_o + d)u + (2bx_o + 2cy_o + e)v + (ax_o^2 + 2bx_oy_o + cy_o^2 + dx_o + ey_o + f) \qquad [4.36]$$

The coordinates x and y can be solved by setting the terms in u and v equal to zero. The result are the linear equations:

$$\left.\begin{aligned} 2ax_o + 2by_o + d &= 0 \\ 2bx_o + 2cy_o + e &= 0 \end{aligned}\right| \qquad [4.37]$$

with solution:

$$x_o = \frac{cd - be}{2(b^2 - ac)}\,; \qquad y_o = \frac{bd - ae}{2(b^2 - ac)}$$

By writing, $z_o = ax_o^2 + 2bx_oy_o + cy_o^2 + dx_o + ey_o + f$, [4.35] becomes:

$$z - z_o = au^2 + 2buv + cv^2 \qquad [4.38]$$

This constitutes a simplification with respect to the original form. The origin of the UV-system now falls on a vertical line through the extremum of the surface. One could move the origin along this line to the extremum by defining a W-axis with $w = z - z_o$. However, in general, we will keep the Z-axis untransformed. The value assumed by $z = f(x, y)$ at its extremum is equal to z_o. The same simplification can be achieved by using the method of partial differentiation. From the original equation [4.34], we obtain:

$$\frac{\partial f}{\partial x} = 2ax + 2by + d \, ; \qquad \frac{\partial f}{\partial y} = 2bx + 2cy + e \qquad [4.39]$$

Setting these partial derivatives equal to zero yields [4.37]. Working with partial derivatives simplifies the procedure if there are more than two variables.

At this point, the geometrical interpretation of partial differentiation may be visualized as follows. The expression $\partial f/\partial x$ represents the rate of change in f when y is kept constant. We are considering the first derivative of the curve of intersection between the surface $z = f(x, y)$ and a plane perpendicular to the Y-axis. By inspection of Fig. 17B and E, it becomes clear that the plane $y = 0$ intersects the surface according to a curve which has its minimum at $\partial f/\partial x = 0$. In the plane $x = 0$, $\partial f/\partial y = 0$ occurs at a maximum in Fig. 17E. The fact that the origin coincides with the extremum is immaterial in this respect.

The expression $z - z_o = au^2 + 2buv + cv^2$ can be reduced further by a rotation around the Z-axis. The expression $au^2 + 2buv + cv^2$ is a quadratic form and in matrix notation [4.38] can be written as:

$$z - z_o = [u \quad v] \begin{bmatrix} a & b \\ b & c \end{bmatrix} \begin{bmatrix} u \\ v \end{bmatrix} \qquad [4.40]$$

The matrix $\mathbf{A} = \begin{bmatrix} a & b \\ b & c \end{bmatrix}$ can be written in its canonical form:

$$\mathbf{A} = \mathbf{V}\Lambda\mathbf{V}'$$

or, if the eigenvectors are normalized, as:

$$\mathbf{A} = \mathbf{V}_n\Lambda\mathbf{V}_n'$$

If $\mathbf{V}_n = \begin{bmatrix} v_{11} & v_{21} \\ v_{12} & v_{22} \end{bmatrix}$, we have $v_{11}^2 + v_{12}^2 = 1$; $v_{21}^2 + v_{22}^2 = 1$.

Hence, $|v_{11}| \leqslant 1$, and we can write $v_{11} = \cos\varphi$ and $v_{12} = \sin\varphi$. Since $\mathbf{V}_n$ is orthogonal, $\mathbf{V}_n^{-1} = \mathbf{V}_n'$, or:

$$\begin{bmatrix} v_{22} & -v_{21} \\ -v_{12} & v_{11} \end{bmatrix} = \begin{bmatrix} v_{11} & v_{12} \\ v_{21} & v_{22} \end{bmatrix}$$

or $v_{22} = v_{11}$ and $v_{21} = -v_{12}$.

Collecting the results, we can write:

$$z - z_o = [u \quad v] \begin{bmatrix} \cos\varphi & -\sin\varphi \\ \sin\varphi & \cos\varphi \end{bmatrix} \begin{bmatrix} \lambda_1 & 0 \\ 0 & \lambda_2 \end{bmatrix} \begin{bmatrix} \cos\varphi & \sin\varphi \\ -\sin\varphi & \cos\varphi \end{bmatrix} \begin{bmatrix} u \\ v \end{bmatrix} \qquad [4.41]$$

New coordinates u' and v' can be defined with:

$$\begin{bmatrix} u' \\ v' \end{bmatrix} = \begin{bmatrix} \cos\varphi & \sin\varphi \\ -\sin\varphi & \cos\varphi \end{bmatrix} \begin{bmatrix} u \\ v \end{bmatrix} \qquad [4.42]$$

or:

$$u' = u\cos\varphi + v\sin\varphi$$

$$v' = -u\sin\varphi + v\cos\varphi$$

Hence:

$$z - z_o = [u' \quad v'] \begin{bmatrix} \lambda_1 & 0 \\ 0 & \lambda_2 \end{bmatrix} \begin{bmatrix} u' \\ v' \end{bmatrix} \qquad [4.43]$$

or:

$$z - z_o = \lambda_1 u'^2 + \lambda_2 v'^2 \qquad [4.44]$$

Geometrically, φ represents the angle by which the U- and V-axes are rotated in order to obtain the U'- and V'-axes.

It should be kept in mind that the matrix $\begin{bmatrix} \lambda_2 & 0 \\ 0 & \lambda_1 \end{bmatrix}$ represents a second diagonal form of **A**. A change in the order of the eigenvalues in the diagonal matrix **Λ** is equivalent to an interchange of columns in the matrix $\mathbf{V}_n$. Furthermore, it has been stated in section 3.9 that the signs of the elements of any eigenvector can be changed. The geometrical interpretation of these possibilities is that a diagonalization of the matrix **A** uniquely determines two lines coinciding with the U'- and V'-axis. Any direction along these two lines can be defined as the positive U'-axis. Then there remain two possibilities for defining the positive V'-axis. The transformation truly represents a rotation, only if the V'-axis is chosen at 90° from the U'-axis in the anticlockwise direction.

The coefficients in the reduced form $z - z_o = \lambda_1 u'^2 + \lambda_2 v'^2$ are the eigenvalues of the matrix **A**.

A similar procedure can be followed in a three-dimensional situation. For example, suppose that:

$$f(x, y, z) = ax^2 + 2bxy + 2cxz + dy^2 + 2eyz + fz^2 + gx + hy + iz + j \qquad [4.45]$$

For a specific value of f this yields [4.30]. Let us take the ellipsoid (Fig. 17A) for example. Eq. [4.45] represents a so-called hypersurface whose contours in 3D-space are ellipsoids corresponding to different values of $f(x, y, z)$. The projection in 3D-space of the extremum falls at the center which coincides with the origin. The coordinates (x_o, y_o, z_o) of the center for [4.45] can be found by partial differentiation of $f(x, y, z)$ with respect to x, y, and z, and setting the results equal to zero. The value assumed by f at the center is either a minimum or a maximum.

The results of the translation can be written as:

$$f(x, y, z) - f(x_o, y_o, z_o) = au^2 + 2buv + 2cuw + dv^2 + 2evw + fw^2$$

$$= [u \;\; v \;\; w] \begin{bmatrix} a & b & c \\ b & d & e \\ c & e & f \end{bmatrix} \begin{bmatrix} u \\ v \\ w \end{bmatrix} \qquad [4.46]$$

If λ_1, λ_2, and λ_3 represent the eigenvalues of the matrix of the coefficients, a rotation analogous to that in the two-dimensional case gives:

$$f(x, y, z) - f(x_o, y_o, z_o) = \lambda_1 u'^2 + \lambda_2 v'^2 + \lambda_3 w'^2 \qquad [4.47]$$

For a further discussion of this result, the reader is referred to a practical example in section 9.3.

4.6. Method of least squares

This method has numerous applications and will be discussed in detail with reference to the practical example of Fig. 3. The expression "method of least squares" is commonly reserved for the following problem and its p-dimensional extensions. A set of data for two variables x and y is plotted in a diagram. The problem is to fit a straight line with equation $y = a + bx$ to the data whereby the sum of squares of the vertical distances from the points to the line is a minimum. The observations are represented as points $P_i = (x_i, y_i)$ with i going from 1 to n. If the vertical projection of P_i on the line is indicated as Q_i, the distance P_iQ_i represents the residual or deviation from the line. It also is written as e_i. The basic model is:

$$y_i = (a + bx_i) + e_i \qquad [4.48]$$

Every observed value y_i is assumed to be equal to a theoretical value on the line $(a + bx_i)$ plus a residual e_i. The coefficients a and b can be calculated by minimizing the sum of squares of the residuals $\Sigma_n e_i^2$. From [4.48], it follows that:

$$\Sigma\, e_i^2 = \Sigma\, (y_i - a - bx_i)^2 \qquad [4.49]$$

Dividing by n and putting $F(a, b) = \Sigma\, e_i^2 / n$ gives:

$$F(a, b) = (1/n)\, \Sigma\, (y_i - a - bx_i)^2$$

$$= (1/n)\, \Sigma\, (b^2 x_i^2 - 2bx_i y_i + y_i^2 + 2abx_i - 2ay_i + a^2)$$

$$= b^2\, \overline{x^2} - 2b\overline{xy} + \overline{y^2} + 2ab\bar{x} - 2a\bar{y} + a^2 \qquad [4.50]$$

where the bars denote values obtained by averaging x_i^2, $x_i y_i$, y_i^2, x_i and y_i over the n values.

Strontium in layer 5 of the Muskox intrusion

In Fig. 3, a point with $x = 0$ was selected at a location approximately halfway the sampled core and 1000 ft. was adopted as the unity of distance along the X-axis. For y, we used the scale shown along the Y-axis in this diagram. In order to obtain $F(a, b)$ for the 58 data, the following average values were calculated:

$$2\bar{x} = 0.02579\,; \quad 2\bar{y} = 2.9806\,; \quad \overline{x^2} = 0.10909\,; \quad 2\overline{xy} = 0.0476\,; \quad \overline{y^2} = 2.2546$$

It follows that [4.50] becomes:

TABLE V
Variation of $F(a, b)$ with respect to the coefficients a and b

b \ a	1.1	1.2	1.3	1.4	1.5	1.6	1.7	1.8	1.9
0.0	0.185	0.117	0.069	0.041	0.033	0.045	0.077	0.129	0.201
0.1	0.179	0.111	0.062	0.034	0.026	0.037	0.069	0.121	0.192
0.2	0.175	0.106	0.057	0.029	0.020	0.032	0.063	0.115	0.186
0.3	0.174	0.104	0.055	0.026	0.017	0.028	0.059	0.111	0.182
0.4	0.173	0.103	0.054	0.025	0.016	0.027	0.058	0.109	0.180
0.5	0.175	0.105	0.056	0.027	0.017	0.028	0.059	0.109	0.180
0.6	0.179	0.110	0.060	0.030	0.021	0.031	0.061	0.112	0.182
0.7	0.186	0.116	0.066	0.036	0.026	0.036	0.067	0.117	0.187
0.8	0.194	0.124	0.074	0.044	0.034	0.044	0.074	0.124	0.194

$$F(a, b) = a^2 + 0.02579\, ab + 0.10909\, b^2 - 2.9806\, a - 0.0476\, b + 2.2546$$

This function was evaluated for different values of a and b. The results are shown in Table V with a ranging from 1.1 to 1.9 and b from 0.0 to 0.8.

Every value in Table V represents the average sum of squares of residuals from a line with equation $y = a + bx$. The smallest value in Table V is $F(a, b) = 0.016$ when $a = 1.5$ and $b = 0.4$. An approximate expression for the best-fitting line therefore is $y = 1.5 + 0.4x$.

If the data for a problem are given, $F(a, b)$ is fully determined except for the two unknowns a and b. These are the variables in [4.50]. Table V is an example of a map of $F(a, b)$ in so-called coefficient space. Looking at maps for coefficient space can be useful, in particular if the coefficients a and b are subject to a constraint of the type $a = f(b)$. Then, $a = f(b)$ could be represented as a curve in the coefficient space and the constrained least-squares estimates would be determined by locating the minimum value of $F(a, b)$ along this curve. As a hypothetical example, which is not founded on reality, suppose that we know beforehand that $b = 0.5$. From Table V it then follows that the minimum of $F(a, b)$ along the line $b = 0.5$ is equal to 0.17 with $a = 1.5$. Hence, the method of least squares would give the result $y = 1.5 + 0.5x$ (given $b = 0.5$).

In general, the method of mapping the coefficient space is laborious and imprecise. In the present situation, a more precise solution can be rapidly obtained as follows. We know from the results in section 4.4, that $F(a, b)$ in [4.50] represents an elliptic paraboloid with a single minimum. The coordinates of this minimum can be solved by putting:

$$\frac{\partial F}{\partial a} = 0 \quad \text{and} \quad \frac{\partial F}{\partial b} = 0$$

For the example, this gives the linear equations:

$$\begin{aligned} a + 0.0129\, b &= 1.4903 \\ 0.0129\, a + 0.1091\, b &= 0.0238 \end{aligned}$$

Consequently, the best-fitting line satisfies the equation $y = 1.495 + 0.3949x$. This line was graphically represented in Fig. 3. The minimum value of $F(a, b)$ is equal to 0.01663.

A close inspection of the numbers in Table V reveals that there are two axes passing through the minimum at $a = 1.495$ and $b = 0.3949$. The values along these axes are relatively small. By applying the method of rotation discussed in section 4.5, we find that the matrix

$$\mathbf{A} = \begin{bmatrix} 1 & \overline{x} \\ \overline{x} & \overline{x^2} \end{bmatrix}$$

has eigenvalues $\lambda_1 = 1.0009$ and $\lambda_2 = 0.10890$. The angle of rotation φ is equal to 0°.83 which is less than one degree. If the point with $x = 0$ along the X-axis be moved initially to the point with $x = \overline{x}$, then the matrix $\mathbf{A}$ would be in diagonal form without further rotation. In that situation, $\lambda_1 = 1, \lambda_2 = \overline{x^2}$, and $\varphi = 0°$.

Best-fitting straight line (bivariate least squares)

We now will formalize the results obtained for the practical example, starting from the general expression for $F(a, b)$ in [4.50]. An evaluation of the equations $\partial F/\partial a = 0$ and $\partial F/\partial b = 0$ gives:

$$a + b\overline{x} = \overline{y} \qquad \text{and} \qquad a\overline{x} + b\overline{x^2} = \overline{xy} \qquad [4.51]$$

Hence:

$$b = \frac{\overline{xy} - \overline{x} \cdot \overline{y}}{\overline{x^2} - (\overline{x})^2} \qquad [4.52]$$

The coefficient a follows from $a = \overline{y} - b\overline{x}$. If both sides of [4.51] are multiplied by n, this gives the so-called Gaussian normal equations:

$$\left.\begin{aligned} an + b\,\Sigma\,x &= \Sigma\,y \\ a\,\Sigma\,x + b\,\Sigma\,x^2 &= \Sigma\,xy \end{aligned}\right| \qquad [4.53]$$

or, in matrix form:

$$\begin{bmatrix} n & \Sigma x \\ \Sigma x & \Sigma x^2 \end{bmatrix} \begin{bmatrix} a \\ b \end{bmatrix} = \begin{bmatrix} \Sigma y \\ \Sigma xy \end{bmatrix}$$

with solution:

$$\begin{bmatrix} a \\ b \end{bmatrix} = \begin{bmatrix} n & \Sigma x \\ \Sigma x & \Sigma x^2 \end{bmatrix}^{-1} \begin{bmatrix} \Sigma y \\ \Sigma xy \end{bmatrix}$$

$$= \frac{1}{n \, \Sigma x^2 - (\Sigma x)^2} \begin{bmatrix} \Sigma x^2 & -\Sigma x \\ -\Sigma x & n \end{bmatrix} \begin{bmatrix} \Sigma y \\ \Sigma xy \end{bmatrix} \qquad [4.54]$$

The reader may verify himself that this solution is identical to that given in [4.52].

The equation of the straight line is $y = a + bx$. If a is replaced by $a = \bar{y} - b\bar{x}$, we obtain:

$$y - \bar{y} = b(x - \bar{x}) \qquad [4.55]$$

Consequently, if both variables are corrected for their means, there is only a single unknown coefficient (b) to be found. From $a = \bar{y} - b\bar{x}$, it follows that the point with coordinates $(\bar{x}, \bar{y})$ must fall on the best-fitting line.

If we write $x' = x - \bar{x}$ and $y' = y - \bar{y}$, we have $\Sigma x' = \Sigma y' = 0$. Eq. [4.53] then reduces to $b \, \Sigma x'^2 = \Sigma x' y'$ with solution:

$$b = \Sigma x' y' \, / \, \Sigma x'^2 \qquad [4.56]$$

This result is identical to [4.52]. If the n deviations $x'_i = x_i - \bar{x}$ are written as a column vector $\mathbf{X}_0$ and the y'_i as $\mathbf{Y}_0$, then [4.56] becomes:

$$b = (\mathbf{X}'_0 \, \mathbf{X}_0)^{-1} \, \mathbf{X}'_0 \, \mathbf{Y}_0 \qquad [4.57]$$

This is equivalent to [4.4]. It follows that the determination of a best-fitting line in a bivariate situation has the geometrical interpretation of an orthogonal projection of the vector $\mathbf{Y}_0$ onto the vector $\mathbf{X}_0$ in n-space.

Finally, it is useful to consider one other approach to the problem of calculating the best-fitting straight line. Suppose that $\mathbf{J}_n$ represents a column vector consisting of n ones and that the vectors $\mathbf{X}$ and $\mathbf{Y}$ are column vectors for the original observations on the variables x and y. We then have:

$$\mathbf{J}_n = \begin{bmatrix} 1 \\ 1 \\ \cdot \\ \cdot \\ \cdot \\ 1 \end{bmatrix} \qquad \mathbf{X} = \begin{bmatrix} x_1 \\ x_2 \\ \cdot \\ \cdot \\ \cdot \\ x_n \end{bmatrix} \qquad \mathbf{Y} = \begin{bmatrix} y_1 \\ y_2 \\ \cdot \\ \cdot \\ \cdot \\ y_n \end{bmatrix}$$

It follows that $J_n'J_n = n$, $X'J_n = J_n'X = \Sigma x$ and $J_n'Y = \Sigma y$. Consequently, [4.53] can be written as:

$$\begin{bmatrix} a \\ b \end{bmatrix} = \begin{bmatrix} J_n' J_n & J_n' X \\ X' J_n & X' X \end{bmatrix}^{-1} \begin{bmatrix} J_n' Y \\ X' Y \end{bmatrix} \qquad [4.58]$$

This is identical to [4.21] in section 4.3 if $J_n = X_1$ and $X = X_2$. In n-space, the least-squares problem consists of the orthogonal projection of **Y** onto a plane passing through the vectors J_n and **X**. The geometrical interpretation of J_n is a vector with length n and direction cosines all equal to $1/\sqrt{n}$.

By moving the origin from the point (0, 0) to the point $(\bar{x}, \bar{y})$, the vectors **X** and **Y** can be changed into X_0 and X_0 (see above). If a vector J_n is defined for this modified situation, we have from $\Sigma y' = 0$ that $J_n' Y_0 = 0$. Hence, Y_0 is orthogonal to J_n. It follows that a perpendicular projection of Y_0 onto the plane through X_0 and J_n is equivalent to a direct projection of Y_0 onto X_0.

4.7. Major axis

A problem that can be solved by using [4.29] is the construction of the so-called major axis in a bivariate situation.

Suppose that observations have been made on two variables both of which are subject to measurement errors. An artificial example is shown in Fig. 18. As in section 4.6, it is assumed that the variables x and y were measured repeatedly under different circumstances, and are related by the linear equation $y = a + bx$ where a and b are coefficients. The observed values do not lie on this line because of measurement errors. A best-fitting

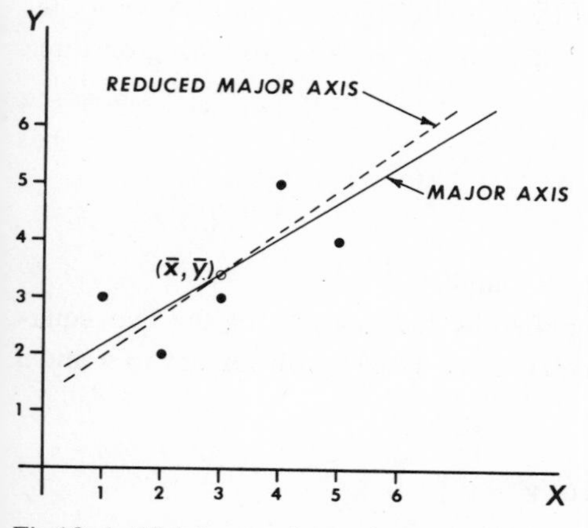

Fig.18. Artificial example; major axis and reduced major axis for hypothetical values (1,3), (2,2), (3,3), (4,5) and (5,4).

line $y = a + bx$ can now be calculated from the data by minimizing the sum of squares of the distances from the points to the line. A line with this property is called the major axis. The least-squares criterion for the major axis differs from that used in the previous section where a straight line was determined by minimizing the sum of squares of distances parallel to one of the axes of the coordinate system.

The general equation of a line l is:

$$\frac{x-u}{\lambda} = \frac{y-v}{\mu} \qquad [4.59]$$

where $P = (u, v)$ is an arbitrary point through which l passes. As before, λ and μ are the direction cosines of l. When α represents the angle of slope of l, then $\lambda = \cos\alpha$ and $\mu = \sin\alpha$. If there are n observations, these can be seen as a set of n points $P_i = (x_i, y_i)$ with i going from 1 to n. The perpendicular projections of the P_i onto l form another set of points to be denoted as the Q_i.

From [4.29], it follows that for each distance P_iQ_i:

$$P_iQ_i^2 = (x_i - u)^2 + (y_i - v)^2 - [\cos\alpha\,(x_i - u) + \sin\alpha\,(y_i - v)]^2$$

The sum of the n squared distances, therefore, is:

$$S = \sum_{i=1}^{n} P_iQ_i^2 = \sum_{i=1}^{n} \{(x_i - u)^2 + (y_i - v)^2 - [\cos\alpha\,(x_i - u) + \sin\alpha\,(y_i - v)]^2\} \qquad [4.60]$$

S is a function of the three unknowns u, v, and α. It means that for each set of distinct values for u, v, and α, S assumes a specific value. Of all possible values for S, one has the property that it is smaller than any of the others. This leads to the major axis for which S is a minimum. The minimum can be found by partial differentiation of S with respect to u, v, and α, and setting the three partial derivatives equal to zero. This gives three equations:

$$\frac{\partial S}{\partial u} = 0\,; \qquad \frac{\partial S}{\partial v} = 0\,; \qquad \frac{\partial S}{\partial \alpha} = 0$$

from which the three unknowns u, v, and α can be found.

The solution of the problem at hand is simplified by first considering the two equations $\partial S/\partial u = 0$ and $\partial S/\partial v = 0$. Partial differentiation of [4.60] with respect to u and v gives:

$$-\Sigma\,(x_i - u) + \Sigma\cos\alpha\,[\cos\alpha\,(x_i - u) + \sin\alpha\,(y_i - v)] = 0$$

$$-\Sigma\,(y_i - v) + \Sigma\sin\alpha\,[\cos\alpha\,(x_i - u) + \sin\alpha\,(y_i - v)] = 0$$

Elimination of the second term from these two equations gives:

$$\frac{\Sigma (x_i - u)}{\cos \alpha} = \frac{\Sigma (y_i - v)}{\sin \alpha} \qquad \text{or:} \qquad \frac{\Sigma x_i - nu}{\cos \alpha} = \frac{\Sigma y_i - nv}{\sin \alpha}$$

By introducing the arithmetic means $\bar{x} = (1/n)\, \Sigma x_i$ and $\bar{y} = (1/n)\, \Sigma y_i$, we can write:

$$\frac{\bar{x} - u}{\lambda} = \frac{\bar{y} - v}{\mu} \qquad [4.61]$$

This result only holds true if the point $(\bar{x}, \bar{y})$ lies on the line. It follows that the major axis must pass through the point whose coordinates are equal to the average values for x and y.

An important simplification is obtained by replacing the original equation [4.59] by the expression:

$$\frac{x - \bar{x}}{\lambda} = \frac{y - \bar{y}}{\mu} \qquad [4.62]$$

In this way, the two unknowns u and v have been eliminated. By using deviations from the mean:

$$x_i' = x_i - \bar{x} \; ; \qquad y_i' = y_i - \bar{y}$$

[4.60] reduces to:

$$S = \sum_{i=1}^{n} (x_i'^2 + y_i'^2) - \sum_{i=1}^{n} (x_i' \cos \alpha + y_i' \sin \alpha)^2 \qquad [4.63]$$

which is a function of α only.

The minimum can be found from the equation $dS/d\alpha = 0$ or:

$$-2\, \Sigma (x' \cos \alpha + y' \sin \alpha)(-x' \sin \alpha + y' \cos \alpha) = 0$$

Hence:

$$\Sigma \{(x'y' \cos^2 \alpha - x'y' \sin^2 \alpha) - (x'^2 \sin \alpha \cos \alpha - y'^2 \sin \alpha \cos \alpha)\} = 0$$

$$\cos 2\alpha\, \Sigma x'y' - \tfrac{1}{2} \sin 2\alpha\, \Sigma (x'^2 - y'^2) = 0$$

$$\tan 2\alpha = \frac{2\, \Sigma x'y'}{\Sigma (x'^2 - y'^2)} \qquad [4.64]$$

Artificial example

The data plotted in Fig. 18 are (1,3), (2,2), (3,3), (4,5), and (5,4). The average values are $\bar{x} = 3$ and $\bar{y} = 3.4$. Hence, the deviations x' $(= x - \bar{x})$ and $y' = (y - \bar{y})$ satisfy:

x'	y'
– 2	– 0.4
– 1	– 1.4
0	– 0.4
1	1.6
2	0.6

It follows that $\Sigma\, x'y' = (-2) \times (-0.4) + (-1) \times (-1.4) + 0 \times (-0.4) + 1 \times 1.6 + 2 \times 0.6 = 5.0$. Also, $\Sigma\, x'^2 = 10.0$ and $\Sigma\, y'^2 = 5.2$. Consequently, [4.64] becomes:

$$\tan 2\alpha = \frac{2 \times 5.0}{10 - 5.2} = 2.0833$$

$$2\alpha = \arctan 2.0833 = 64^\circ \text{ or } -116^\circ$$

A second solution $2\alpha = -116^\circ$ was added because we also have $\tan(64^\circ - 180^\circ) = 2.0833$.

The result can be written in the form $\alpha_1 = 32^\circ$ and $\alpha_2 = -58^\circ$. This solution is graphically represented in Fig. 18. The angle of dip of the major axis is given by $\alpha_1 = 32^\circ$. It corresponds to the required minimum value for S. In general, the equation $dS/d\alpha = 0$ has two solutions for α both of which correspond to an extremum of S. One of these extrema is a minimum and the other is a maximum. In the example, $\alpha_2 = -58^\circ$ gives the maximum value of S. Geometrically, it corresponds to a line passing through the point $(\bar{x}, \bar{y})$ that is perpendicular to the major axis. It represents the line through $(\bar{x}, \bar{y})$ with the poorest fit.

Reduced major axis

The major axis has an important drawback in that it is not independent of scale (unit distances along X- and Y-axes). For this reason, the data for x and y may be "standardized" before the major axis is constructed. The result is known as the "reduced major axis".

The procedure is applied to transformed variables z_1 and z_2 which are related to x and y by:

$$z_1 = x'/s(x)\,; \qquad z_2 = y'/s(y) \tag{4.65}$$

where $x' = x - \bar{x}$, and $y' = y - \bar{y}$; $s(x)$ and $s(y)$ are the standard deviations as calculated by [2.24] from the data for x and y.

It is found that the standarized data have the property $\Sigma\, z_1^2 = \Sigma\, z_2^2 = n - 1$. Eq. [4.64], therefore, reduces to $\tan 2\alpha = \infty$. Hence, $2\alpha = 90°$, and $\alpha = 45°$.

In a (Z_1Z_2)-diagram, the resulting axis is simply a straight line through the origin with equation $z_1 = z_2$. With [4.65], this gives:

$$y' = x'\, s(y)/s(x) \tag{4.66}$$

This line can be plotted in the original (XY)-diagram. It is the reduced major axis which also passes through the point $(\bar{x}, \bar{y})$. The angle of dip α^* now satisfies:

$$\alpha^* = \arctan\,[s(y)/s(x)] \tag{4.67}$$

Eq. [4.67] also can be written as:

$$\alpha^* = \arctan\,[\Sigma y'^2/\Sigma x'^2]^{\frac{1}{2}}$$

If this result is applied to the data of the previous example, it follows that:

$$\alpha^* = \arctan \left[\frac{5.2}{10.0}\right]^{\frac{1}{2}} = \arctan 0.72 = 36°$$

The angle of slope of the reduced major axis therefore differs by 4 ° from the result $\alpha_1 = 32°$ obtained for the major axis (see Fig. 18). For other properties of the reduced major axis, the reader is referred to Miller and Kahn (1962).

It will be seen in the next chapter that the construction of a major axis (or a reduced major axis) is an application to a bivariate situation of a more general method called component analysis.

Chapter 5

FACTOR ANALYSIS

5.1. Introduction

In section 1.4, factor analysis was taken as an example of a technique initially developed outside geology and afterwards adopted and modified by geologists for analysis of geo-science problems (cf. Griffiths, 1966c). The original methods have been reviewed by Kaiser (1958), Harman (1960) and Cattell (1965). Some of these methods are applied in this chapter. However, our main concern will be to obtain substantially meaningful expressions for linear relationships between variables, the original measurements of which are subject to various sources of variability. The scatter in the original data taken for example is large and geological relationships are obscured initially.

Depending on the mathematical model that is selected, we may find solutions that differ significantly from one another. By refining the specifications of the model, earlier results, which are biased, can be improved. Factor analysis as applied in this chapter principally is an attempt to divide the scatter for each variable into a noise component which is unpredictable and a component for the interrelationship between the variables in the system. The noise component may be due to various sources of variability. Harman (1960) and Kendall (1965), systematically, recognized two sources of variability for individual variables called specific factors and error terms, respectively.

In order to evaluate different best-fitting lines for the same relationship between a pair of variables, these results are compared to one another and to a solution that will be derived later by trend analysis (Chapter 9).

The methodology to be employed here, on the basis of what was discussed in the previous three chapters, is less elaborate than the methods that will be at our disposal later. Methods of matrix algebra and geometry are applied without complicating the situation mathematically by considering the spatial location of the observation points. Methods of mathematical statistics are not used in this chapter. This is not such a disadvantage as it may seem at first, because statistical parameters such as the standard error of the slope of a best-fitting line only may be calculated if the underlying statistical model is applicable.

Geomathematics is concerned with three interrelated aspects of a problem: (1) quantification and measurement of the geologic variables; (2) formulation of geologic theory in mathematical terms; and (3) data analysis or the testing of multiple working hypotheses. In this chapter, these three aspects will be discussed for a practical example (Mount Albert Peridotite intrusion).

If the location points of the measurements are not directly considered, methods of factor analysis seem to yield the best estimates for the example. However, there are many methods by which factors can be calculated and, in particular if the number of variables is large, we may not be able to know which procedure will lead to the optimum result. Not only may it be difficult to identify all possible sources of variability, but, in addition to this, the relationships between the variables may not be linear.

Measurements for geologic variables generally are done at locations that lie in three-dimensional space. If the variables are subject to systematic spatial variations, it is usually fruitful to consider the spatial coordinates of the observation points as variables in the mathematical model. This may result in the elimination of error-terms and in an adequate representation of relationships between variables that are not linearly related as in models for factor analysis.

Correlation matrix

Suppose that we have measurements on p variables for each of n specimens. As was done in the previous two chapters, these data can be represented by the $(n \times p)$ matrix $\mathbf{X}$. We assume that the number of observations is larger than the number of variables so that $\mathbf{X}$ is a column matrix. The symmetrical matrix $\mathbf{C} = \mathbf{X}'\mathbf{X}/(n-1)$ is called the variance–covariance matrix of $\mathbf{X}$. The elements along the main diagonal of $\mathbf{C}$ are the variances. The off-diagonal elements are the covariances, one for each pair of variables. The matrix $\mathbf{X}'\mathbf{X}$ is divided by $(n-1)$ instead of n to obtain unbiased estimates as discussed in section 2.8 for the variance. The subject of bias will be considered in more detail in Chapter 6 and 8.

In section 4.7, we introduced the method of standardizing the data for a variable (see [4.65]). If all data in $\mathbf{X}$ are standardized, we obtain the new matrix $\mathbf{Z}$ with variance–co-variance matrix $\mathbf{R} = (\mathbf{Z}'\mathbf{Z})/(n-1)$. $\mathbf{R}$ is the correlation matrix. Its diagonal elements are all equal to one. The off-diagonal elements are the correlation coefficients with:

$$r_{ij} = [\sum_n (x_i - \bar{x}_i)(x_j - \bar{x}_j)]\ [\sum_n (x_i - \bar{x}_i)^2 \sum_n (x_j - \bar{x}_j)^2]^{-\frac{1}{2}} \qquad [5.1]$$

where x_i and x_j denote two of the variables measured for the n observation points. (Note that $r_{ii} = 1$ and $r_{ij} = r_{ji}$.)

From the geometric concepts discussed in the previous chapter (cf. [4.5] and [4.15]) it follows that a correlation coefficient can be interpreted as the cosine of the angle between two vectors $(\mathbf{X}_i - \bar{\mathbf{X}}_i)$ and $(\mathbf{X}_j - \bar{\mathbf{X}}_j)$ in n-space. This constitutes an extension from 3D-space to an n-dimensional situation. The subtraction of the vectors $\bar{\mathbf{X}}_i$ and $\bar{\mathbf{X}}_j$ indicates that the n observations for a variable were corrected for the mean of that variable. If $r_{ij} = 1$, the two vectors $(\mathbf{X}_i - \bar{\mathbf{X}}_i)$ and $(\mathbf{X}_j - \bar{\mathbf{X}}_j)$ coincide. We have $r_{ij} = 0$ if the two vectors are orthogonal, and $r_{ij} = -1$ if they coincide but point in opposite directions.

The method called factor analysis in the $\mathbf{R}$-mode is based on the correlation matrix $\mathbf{R}$. In order to eliminate noise components (e.g., due to measurement errors) from the measured variables, the elements equal to one along the diagonal of $\mathbf{R}$ may be replaced by communalities h_{ii}^2 which are less than one (see later). In most geological applications, this is not done and the total variation in the multivariate system, $T = \Sigma h_{ii}^2$, which is equal to the sum of the communalities, amounts to p, representing the total number of variables. If **R** with $T = p$ is written as the sum of the p spectral components in its canonical form, each of these may be interpreted as corresponding to a new variable (called "principal component") whose contribution to T is given by its eigenvalue. The original variables contribute equally to T but some of the principal components may contribute more than one and the contribution of others may be close to zero. The latter may be deleted from the system with the result that the p variables are replaced by fewer than p principal components.

The principal components may be standardized as was done for the original data. In order to interpret the results, it may be necessary to rotate the principal components, e.g., by using the varimax method (Kaiser, 1958).

If communalities instead of ones are used, **R** is replaced by **R***, and the preceding method, as applied to **R***, is called factor analysis instead of component analysis. However, in some situations, the principal components can be interpreted as factors and it is not possible to make a clear-cut distinction between the two methods (Kendall, 1965).

Cameron (1968) has applied factor analysis to geochemical data. He has found that factors which explain relatively little of the total variation in the system may yet be important for interpretation of the results. Other applications of factor analysis in geochemistry include those by Spencer (1966) and Garrett and Nichol (1969).

Imbrie and Purdy (1962), Imbrie (1963) and Imbrie and Van Andel (1964) have pioneered factor analysis in geology. It became clear to them that useful information on geological environments could be extracted by doing a so-called **Q**-mode factor analysis. The initial matrix then is not **R** but $\mathbf{Q} = [n/p(n-1)]\ \mathbf{ZZ}'$. As in the expression for the $(p \times p)$ matrix **R**, **Z** denotes the $(n \times p)$ matrix of standardized data but **Q** is an $(n \times n)$ matrix. Imbrie (1963) introduced so-called vector analysis in the **Q**-mode consisting of computing **Q** from data not corrected for their means. The measure of similarity between two variables then is the cosine of the angle between two vectors in so-called sample space where the observations are points whose coordinates are given by the columns of **Z**. **R** and **Q** have equal rank and are related because $\mathbf{Z}'\mathbf{Z}$ and $\mathbf{ZZ}'$ have p eigenvalues in common.

Methods of multivariate analysis are not only discussed in this Chapter but also in Chapters 13 and 15. Two additional comments are made before the method will be applied. In Chapter 10, we will deal with a problem in the statistical analysis of space series which is similar to that of determining communalities. In that situation, the noise component applies to unsystematic, local variations that cannot be correlated from place to place, whereas the systematic component is for gradational changes between observation points.

It was pointed out that a limitation of factor analysis may be that the location of the observation points in 3D-space is not directly considered. Another limitation may occur if the data do not form a single cluster in p- or n-dimensional space. A method for locating clusters of variables which are relatively strongly interrelated is cluster analysis as discussed by Sokal and Sneath (1963). An interesting geological application of this method was made by Merriam (1970). A detailed review of cluster and factor analysis, with emphasis on geometrical interpretation, has also been written by Isnard et al. (1972).

5.2. Practical example: Mount Albert Peridotite intrusion

The Mount Albert intrusion is the largest ultramafic mass (approx. 17 sq. miles or 44 km^2) in the Gaspé (Quebec) portion of the so-called Appalachian ultramafic belt. It is probably $530 \cdot 10^6$ year old (Lowdon et al., 1963). For an outline of the mass see Fig. 43 in Chapter 9. The intrusion was mapped and sampled in 1959 by C.H. Smith and I.D. MacGregor of the Geological Survey of Canada, who made available to the author for statistical analysis a set of largely unpublished analytical data for their samples (cf. Agterberg, 1964b).

The petrography of the body is relatively simple. Prior to serpentinization, it consisted of from 80 to 90% olivine, the remainder being primarily orthopyroxene with some chrome spinel (up to 1%) and diopsidic clinopyroxene. An attempt was made to collect specimens from intersection points of a rectangular grid with approximately 1000 ft. (300 m) spacing. The actual number of specimens that could be selected was decreased because of overburden. The number of mineralogical determinations was further reduced by serpentine alteration.

The following four variables were determined for as many collected specimens as possible: (1) cell edge d_{174} of olivine; (2) index of refraction N_z of orthopyroxene; (3) unit cell dimension of chrome spinel; and (4) specific gravity of the whole rock.

The first two mineralogical parameters were converted to percentage magnesium in olivine and orthopyroxene, respectively, and the results are reported as mol. percent forsterite (Mg-olivine) and enstatite (Mg-orthopyroxene).

A few data for the variables were shown in Table III as an example of how to form matrices from observed data. The chrome spinel cell edge is reported in Ångströms and rock density in g/cm^3. Some statistics for all available measurements are shown in Table VI. The variables are written as y_i instead of x_i because, in Chapter 9, they will be treated as functions of variables for location.

TABLE VI
Some statistics for mineralogical variation in Mount Albert intrusion

Variables	*Number of data*	*Mean*	*Standard deviation*	*Range*	*Approx. precision of measurements*
y_1	167	90.19	0.98	88–94	±2
y_2	174	89.20	1.77	86–94	±1
y_3	189	8.2209	0.0577	8.145–8.320	±0.005
y_4	359	2.725	0.159	2.5–3.3	±0.01

TABLE VII

Correlation matrix for mineralogical data, Mount Albert intrusion. Number of pairs of values on same specimen given in brackets

	y_2	y_3	y_4
y_1	0.3159 (99)	0.4106 (108)	−0.3102 (121)
y_2		0.6824 (141)	−0.3833 (151)
y_3			−0.3623 (156)

Ditto for 70 fitted values, $\hat{y}_i$ (see text)

	$\hat{y}_2$	$\hat{y}_3$	$\hat{y}_4$
$\hat{y}_1$	0.8663	0.8237	−0.2144
$\hat{y}_2$		0.9125	−0.3193
$\hat{y}_3$			−0.0240

The correlation coefficients satisfy [5.1].

Besides number of data, mean and standard deviation, Table VI contains information on the range from smallest to largest observation and an estimate of the precision of individual measurements. The determination error for y_1 (olivine) is relatively large. It could be calculated statistically as a 95-% confidence interval from replicate determinations. This aspect will be discussed in more detail. The other measures of precision, e.g., ± 0.01 for specific gravity, indicate that individual data such as those reported previously in Table III probably do not deviate from true values by more than this measure. Thus, a density value reported as 2.69 approximates a true value so that the latter will almost certainly not be greater than 2.70 or smaller than 2.68.

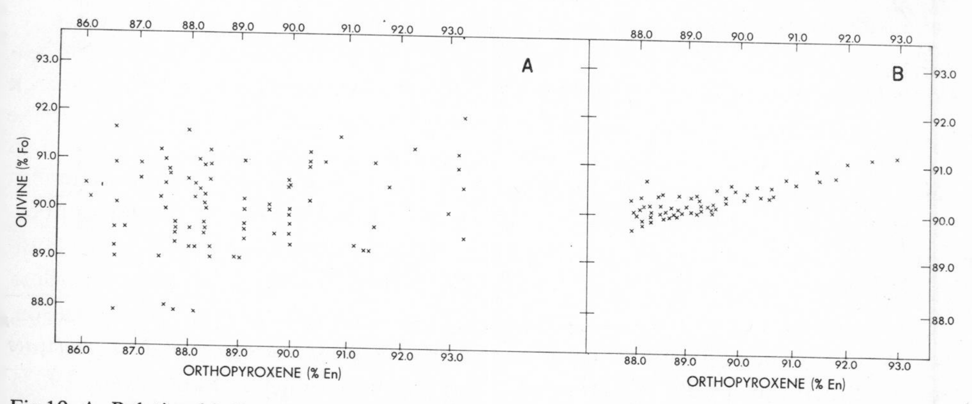

Fig.19. A. Relationship between measured percent forsterite in olivines, and enstatite in orthopyroxenes for coexisting mineral pairs in Mount Albert Peridotite intrusion, Gaspé, Canada. Note effect of round-off errors in enstatite determinations. B. Ditto, for values on a regular grid across quadratic trend surfaces for these variables. An approximate linear relationship is suggested. (From Agterberg, 1964b.)

The correlation matrix for the four variables is presented in Table VII with, in brackets, the number of pairs of values on which an individual correlation coefficient is based.

The first two variables y_1 and y_2 are correlated $r_{12} = 0.3159$. This value is based on 99 pairs of coexisting olivines and orthopyroxenes. The original values for this situation are represented in Fig. 19A. This scattergram does not suggest a definite relationship between the two variables. For comparison, another scattergram is shown in Fig. 19B. This concerns 70 values $\hat{y}_1$ and $\hat{y}_2$ as read from a regular grid superimposed on quadratic trend surfaces for the two variables $\hat{y}_1$ and $\hat{y}_2$ shown in Fig. 42. The details of the method by which the trend values $\hat{y}_1$ and $\hat{y}_2$ were derived will be discussed later (section 9.6). Correlation coefficients for the fitted values are shown in the second part of Table VII. The plot of Fig. 19B has 0.8663. It suggests a linear relationship with $\hat{y}_1$ increasing from approximately 90 to 91 when $\hat{y}_2$ increases from 88 to 93. The absolute value of this correlation coefficient is relatively large (close to unity) and the scattergram suggests a more or less linear relationship between the two variables. If data for two variables plot on a single straight line, $r_{ij} = 1$. The absolute value can never be greater than one because r_{ij} is the cosine of an angle in n-space.

The correlation coefficients for fitted values suggest that the three variables $\hat{y}_1$, $\hat{y}_2$ and $\hat{y}_3$ are linearly related. However, the correlation coefficients with respect to $\hat{y}_4$ are not larger than those in the first part of Table VII. By various methods, we will extract linear relationships from these data using methods already discussed. It will be argued later (Chapter 9) that the relationships between the variables are probably best expressed by correlating the fitted trend values for the $\hat{y}_i$. This result is anticipated here for the purpose of evaluating the accuracy of various lines to be fitted to the observed data for the y_i.

In the following sections, we will discuss successively: (1) quantification and measurement of the four variables; (2) some concepts from theoretical petrology; and (3) results obtained by data analysis.

Measurement of mineralogical variables

Olivine

Mg-content of the 167 olivines was determined by using the d_{174} method also referred to in section 2.8. The crystals used for analysis were either obtained from thin sections or picked from crushed concentrates. Precision of measurements is as low as ± 2% due to measurement error and local zoning. Because the difference between largest and smallest observation for the entire intrusion is only 6%, it was decided to determine the precision of the method in detail by repeating measurements for six specimens. The data resulting from this experiment by J.L. Jambor and R.N. Delabio (personal communication, 1963) in the **X**-ray Laboratory of the Geological Survey of Canada in Ottawa are shown in Table VIII. The range is as large as 2.5 for specimen No. SDM 650. A so-called pooled standard error s_m can be calculated from the five variances reported in Table VIII by taking the

TABLE VIII
Replicate olivine determinations on six specimens

Sample no.	*SDM 126*	*SDM 246*	*SDM 612*	*SDM 650*	*No number*
%Fo (chemical)		91	89		
%Fo (X-ray)	90.2	89.2	89.0	88.0	89.0
	90.6	89.7	89.5	89.0	89.0
	91.0	90.6	90.3	89.0	89.2
	91.2	91.0	91.3	90.5	89.5
	90.2	90.0	89.0		
	90.2	90.0	89.5		
	90.2	91.0	89.7		
	90.7	91.0	90.2		
	91.2	91.6	90.5		
Mean	90.61	90.46	89.89	89.13	89.18
Variance (s^2_m)	0.191	0.598	0.556	1.063	0.056

square root of the weighted average variance (cf. section 6.10). This gives s_m = 0.69. Since the standard deviation of 167 values for specimens from the whole mass is 0.98 (Table VI), the variations between 88 and 94% forsterite could be interpreted as reflecting experimental errors only. The conclusion then would be that the mass is homogeneous in olivine composition. By a detailed statistical analysis, however, the measurement error can be almost eliminated and we will see that there are some slight systematic variations in olivine composition across the mass.

For the three remaining variables, the measurement error is considerably smaller than the variations between sampling points and a more detailed study of the precision can be omitted.

Orthopyroxene

This mineral was measured optically for 184 specimens by determining the index of refraction N_z. The results were converted to percentage enstatite (Mg-content) or orthopyroxene by using the determinative curve of Hess (1952). This curve was chosen because it fitted several check measurements on orthopyroxene specimens. It is not absolutely certain that the Hess curve can be applied here. The percentage enstatite values were rounded off which accounts for the clustering of data along the horizontal axis in Fig. 19A.

Chrome spinel

The measured unit cell dimension for 189 chrome spinels ranges from 8.145 Å to 8.320 Å. MacGregor and Smith (1963) have established determinative curves for this variable in the Mount Albert intrusion. The observed range indicates a considerable variation in the chemical composition of the chrome spinels. The unit cell dimension increases linearly with increasing chromium and iron content, and decreases with increasing aluminum and magnesium content of chrome spinel from Mount Albert. In fact, an increase of

Cr^{3+} at the expense of Al^{3+} in chrome spinel results in a greater unit cell but substitution of Mg^{2+} for Fe^{2+} would yield a smaller cell edge. These substitutions compensated one another with the Cr/Al one having a dominant effect (MacGregor and Smith, 1963).

A positive correlation between y_1, y_2, and y_3 as suggested by the correlation coefficients indicates that an increase in Mg-content for olivine and orthopyroxene would be accompanied by a decrease in Mg-content for chrome spinel and vice versa. On thermodynamic grounds, one may say that, at constant temperature, an increase in Mg-content of olivine or orthopyroxene should be accompanied by an increase in Mg-content of spinel. An increase–decrease relationship could develop in a temperature gradient.

Rock density

The 359 specific gravity determinations were made on unweathered specimens over 100 g in weight as taken from outcrops. The values indicate the degree of serpentinization of the rock. They range from 2.5 to 3.3 with the lowest values corresponding to serpentinite and the largest values to nearly unaltered peridotite. The relationship between volume percent serpentine and rock density is approximately a straight line.

5.3. Some thermodynamical considerations

In geomathematical problems, one usually asks if there are theoretical reasons to assume that certain variables are related to one another and if theoretical geology can provide guidelines for statistical models to be fitted to the data. Generally, the geologist can observe and reconstruct only part of a very complicated process involving many variables that has taken place in the course of time under changing circumstances.

In practice, the resulting mathematical model may then consist of fitting to the data a few linear terms out of a Taylor expansion ([2.21]) for the mainly unknown geological process. Quantitative geologic considerations are indispensable, first for selection of the variables to be measured, for designing the mathematical model, and later for interpretation and evaluation of the results. The following remarks are mainly restricted to the thermodynamics of coexisting olivine and orthopyroxene. The behavior of the chemically more complicated mineral chrome spinel in this system will be summarized.

Both olivine and pyroxene are silicates that contain a variable amount of magnesium (Mg^{2+}) and iron (Fe^{2+}). Pure magnesium olivine that does not contain iron is forsterite. The other end member of the olivine series with iron instead of magnesium is called fayalite. Chemical expressions for olivine are $(Mg, Fe)_2\ SiO_4$ or $(Mg_{x_o}\ Fe_{1-x_o})_2\ SiO_4$ where x_o denotes the fraction of Mg atoms in olivine. The variable x_o then is equivalent to our variable y_1 which is measured in percent. When $x_o = 1$ (or $y_1 = 100$), the formula becomes $Mg_2\ SiO_4$ for forsterite.

Orthopyroxene has chemical composition $(Mg_{x_p}\ Fe_{1-x_p})\ SiO_3$. When $x_p = 1$ ($y_2 = 100$), we have the end member enstatite; when $x_p = 0$, the pyroxene is called ferrosilite.

In a system where olivine and pyroxene coexist, a chemical process may take place in that some Fe^{2+} ions go from olivine to pyroxene or vice versa. Simultaneously, the same quantity of Mg^{2+} ions goes in the opposite direction. This reaction can be written as:

$$\begin{array}{ll} (Mg'_{x_o}\, Fe'_{1-x_o})_2\, SiO_4 + 2\; (Mg'_{x_p}\, Fe'_{1-x_p})\, SiO_3 \rightleftharpoons & \\ \text{olivine}' \qquad\qquad\qquad \text{orthopyroxene}' & \\ (Mg''_{x_o}\, Fe''_{1-x_o})_2\, SiO_4 + 2\; (Mg''_{x_p}\, Fe''_{1-x_p})\, SiO_3 & [5.2] \\ \text{olivine}'' \qquad\qquad\qquad \text{orthopyroxene}'' & \end{array}$$

The minerals at the left hand side of this reaction are indicated as olivine′ and orthopyroxene′. This implies that in a given situation (constant temperature, pressure and other variables), $x_o = x'_o$ and $x_p = x'_p$. If conditions are changed we obtain olivine″ with $x_o = x''_o$ and pyroxene″ with $x_p = x''_p$. If the system is in equilibrium, the reaction is reversible. Eq. [5.2] may be simplified to $Mg(SiO_4)_{\frac{1}{2}} + FeSiO_3 = Fe(SiO_4)_{\frac{1}{2}} + MgSiO_3$ [5.3]

The functional relationship between x_o and x_p was studied by several authors including Ramberg and DeVore (1951). It depends on several factors the most important one being temperature. When other factors such as pressure are kept constant, the equilibrium constant K satisfies:

$$K = e^{-\Delta G/RT} \qquad [5.4]$$

where ΔG is the change in Gibbs free energy for the exchange reaction in simplified form ([5.3]); R is the gas constant, and T absolute temperature.

If both minerals behave as ideal solutions:

$$K = \frac{x_p}{1 - x_p} \cdot \frac{1 - x_o}{x_o} \qquad [5.5]$$

Ramberg and DeVore (1951) found that in natural rocks, x_p is generally smaller than x_o when $x_o > 0.8$ (or $y_1 > 80\%$) whereas the reverse holds true when $x_o < 0.8$. According to their calculations based on [5.5], percent enstatite in orthopyroxene would always be greater than percent forsterite in coexisting olivine. For a recent review of this problem, see Grover and Orville (1969). Olivine does not behave exactly as an ideal solution at its forsterite end. Experimental data by Nafziger and Muan (1967) suggest that $x_o = x_p$ for $x_o > 0.8$.

For Mount Albert, the average value for x_p is $\bar{x}_p = 0.89$, and $\bar{x}_o = 0.90$ (Table VI). The averages are nearly equal to one another which would be in agreement with the above-mentioned findings for other natural rocks and the experimental data.

The theoretical relationship between x_o and x_p for a constant value of K can be approximated as follows. From [5.4] :

$$x_o = \frac{x_p}{K + (1 - K)x_p} \qquad [5.6]$$

The midpoint of the range of observed x_p-values is 0.90. Setting $K = 1.15$ as was done by Irvine (1965) for Mount Albert, we obtain from [5.6] that $x_o = 0.89$ from $x_p = 0.90$ and $K = 1.15$. This result also is close to the observation data. The value $x_p = 0.90$ can be used for a Taylor expansion of x_o in terms of x_p which according to [2.17] gives:

$$x_o = [x_o]_{x_p=0.9} + \left[\frac{K}{\{K + (1 - K)x_p\}^2}\right]_{x_p=0.9} \cdot (x_p - 0.9) \qquad [5.7]$$

It is readily verified that the higher-order terms of the Taylor expansion can be safely neglected for the observational range $0.85 < x_p < 0.95$, and that:

$$x_o = -\,0.1179 + 1.1163x_p \qquad [5.8]$$

This can be represented as a straight line. If x_o is plotted along the vertical axis, this line dips 48° towards the origin. As mentioned before, the experimental results gave $x_o = x_p$ with 45° dip. A positive correlation between the variables x_o and x_p (or y_1 and y_2) therefore is predicted. However, we will see that a best-fitting line computed statistically from the data dips 16°, as already suggested by the pattern of Fig. 19B. Clearly, olivine in the mass varies less than orthopyroxene and this result departs from what is predicted thermodynamically for a system in equilibrium. Mueller (1961) has assumed that discrepancies between results predicted by the thermodynamical models and nature is caused by disequilibrium conditions in nature. The problem of disequilibrium also has been discussed by Kretz (1961, 1963).

The Mount Albert mass intruded at a high temperature with a well-developed contact-metamorphic aureole. The sequence of crystallization of the minerals in the body was that olivine formed first. Later, orthopyroxene and chrome spinel crystallized at about the same time, in part from an interstitial liquid. The attitudes of certain layered structures in the mass indicate that the magma was largely solid (crystal mush) while piercing through the surrounding rocks according to a mechanism that can be compared to salt-dome formation. At the same time and later, the mass was serpentinized by a more or less separate process.

MacGregor and Smith (1963) have suggested that the availability of water during the final phase of crystallization caused the serpentinization. This may have resulted in a higher partial oxygen pressure which, in turn, caused the coexistence of more Mg-rich silicates (olivine and orthopyroxene) and more ferric iron-(and chromium-)rich spinel.

Irvine (1965, 1967) has made a study of chrome spinel as a petrological indicator with examples that include the Mount Albert intrusion. The chemical formula of a spinel mineral is $A^{2+} B_2^{3+} O_4$ where A can be Mg or Fe^{2+}, and B can be Cr, Al, or Fe^{3+}. For example, magnetite Fe_3O_4 which occurs in many natural rocks is one of the many possible end members of chrome spinel. Under the assumption that coexisting olivine, orthopyroxene and chrome spinel form ideal solutions and are in equilibrium under temperature and pressure conditions during crystallization, Irvine constructed equipotential surfaces predicting that Cr/Al ratio of spinel and Mg/Fe ratio of olivine or orthopyroxene are positively correlated. This explains the positive correlation between cell edge (y_3) and Mg-content in olivine or orthopyroxene.

Another factor that may control the composition (and cell edge) of chrome spinel is partial oxygen pressure. An increase in this factor should induce a reaction which produces magnetite at the expense of ferrous iron in both olivine and orthopyroxene and could be written :

$$\underset{\text{(in olivine)}}{Fe_2\,SiO_4} + \underset{\text{(in pyroxene)}}{Fe\,SiO_3} + O_2 = Fe_3O_4 \text{ (not balanced)} \qquad [5.9]$$

An increase in partial oxygen pressure would favor the formation of iron-rich chrome spinel. At the same time, it would lead to olivine and pyroxene enriched in Mg and to a change in the olivine/pyroxene volume ratio.

The effect of this reaction may not be noticeable in olivine and orthopyroxene because these minerals are more abundant than chrome spinel which is present in minute amounts only.

In the next section, the attempt will be made to calculate by statistical methods the relationships between the variables y_1–y_4. The theoretical considerations suggest that a positive correlation may exist between the variables y_1, y_2, and y_3, and possibly a negative correlation between the three variables ($y_1 - y_3$) and y_4. However, part of the theory is based on the assumption of perfect chemical equilibrium and we have already seen that this condition was probably not fulfilled in Mount Albert. The statistical problems are aggravated by relatively large measurement errors in particular for olivine. The following analysis is for example mainly. Fewer data of higher accuracy and precision would give results more rapidly in this situation. The advantage of using rapid methods of limited precision and accuracy was that a large data base became available for the study of spatial variation patterns (see Chapter 9).

5.4. Statistical relationship between the mineralogical variables

The relationship between the variables y_1 and y_2 will be considered first. There are 99 pairs of coexisting olivine and pyroxene measurements with r_{12} = 0.32 (see Table VII). We can calculate the reduced major axis by employing the method discussed in section 4.7. For the 99 pairs, $\bar{y}_1 = 90.06$, $s_1 = 0.849$; $\bar{y}_2 = 89.03$, $s_2 = 1.799$.

Hence, $s_1/s_2 = 0.472$. It is noted that these values for means and standard deviations differ slightly from those reported in Table VI which are based on more data. The reduced major axis passes through the point with $y_1 = 90.06$ and $y_2 = 89.03$. Its dip amounts to arctan $(s_1/s_2) = 25\frac{1}{2}°$. It would dip to the left in Fig. 19.

When using the method of least squares (see section 4.6), we have the choice of projecting the observations along lines which are either parallel to the Y_1-axis or to the Y_2-axis. Both best-fitting lines pass through the point with coordinates $y_1 = 90.06$ and y_2 89.03. The first one has slope $b_{12} = r_{12} \cdot s_1/s_2$ and would dip 8½° toward the origin in Fig.19. The second line (direction of projection parallel to Y_2-axis) has slope $b_{21} = r_{12} \cdot s_2/s_1$ and dips 56°.

These expressions for the slope in terms of the correlation coefficient and the standard deviations can be verified readily from [5.1] and the original definitions of parameters given in Chapter 4. Since $r_{12} \leqslant 1$, the slope $b_{12} = r_{12} \cdot s_1/s_2$ always is equal to or less than that for the reduced major axis with $b_{12} = s_1/s_2$. In fact, we have:

$$b_{12} < b < b_{21} \qquad [5.10]$$

The second part of this inequality follows from interchanging the axes in the coordinate system used for the scattergram.

We now will include the variable y_3 in the system by applying component analysis. This method is an extension to a p-dimensional situation ($p \geqslant 3$) of the methods of fitting a major axis or a reduced major axis. Component analysis may be applied to the variance–covariance matrix **C** or to the correlation matrix **R**. In a bivariate situation, the result is the major axis if **C** is used and the reduced major axis for **R**. These two methods do not give the same answer (cf. Fig. 17), a problem that will be further discussed in Chapter 13 for multivariate situations.

For a geometrical interpretation and formal derivations of the method of component analysis, the reader is referred to one of the books on multivariate analysis in the list for selected reading at the end of the book.

Briefly, by writing **R** in its canonical form $\mathbf{R} = \mathbf{V}_n \Lambda \mathbf{V}'_n$, it follows that $\mathbf{Z} = \mathbf{V}_n(\mathbf{V}'_n \mathbf{Z})$, and we can define the matrix:

$$\Xi = \mathbf{V}'_n \mathbf{Z} \qquad [5.11]$$

The p columns ξ_i of Ξ are the principal components. Of course, we also have:

$$\mathbf{Z} = \mathbf{V}_n \Xi \qquad [5.12]$$

The total variation (T) in the system satisfies $T = p$. The first normalized eigenvector $\mathbf{V}_{n1}$, which corresponds to the largest eigenvalue λ_1, consists of the direction cosines for an axis fitted to the n data points in p-space. The sum of squares of the perpendicular distances from the n points to this axis is a minimum. If S represents this sum of squares divided by n, we have (cf. Kendall, 1965):

$$S = T - \lambda_1 \qquad [5.13]$$

Hence, the magnitude of λ_1 is a measure for the closeness of fit of the axis. Another result is that λ_1 represents the mean square of the n elements of the first principal component ξ_1.

Similar results hold true for the other principal components. If the n data may be regarded as an ellipsoidal cluster in p-space, the eigenvectors of **R** represent the orthogonal principal axes of this cluster. $\mathbf{V}_{n1}$ is for the longest axis, $\mathbf{V}_{n2}$ for the next longest one, etc. In the situation that $p = 3$, a perfect fit of the first axis would imply $\lambda_1 = 3$ and $\lambda_2 = \lambda_3 = 0$. All points would occur on a single straight line. If a (3×3) matrix **R** has one eigenvalue equal to zero ($\lambda_3 = 0$), it means that the n points lie in a plane with $\mathbf{V}_{n3}$ giving the normal of this plane. All values of ξ_3 are then equal to zero.

By taking the appropriate values from Table VII, we have for the example:

$$\mathbf{R} = \begin{bmatrix} 1.0000 & 0.3159 & 0.4106 \\ 0.3159 & 1.0000 & 0.6824 \\ 0.4106 & 0.6824 & 1.0000 \end{bmatrix}$$

From **R**, the matrices $\mathbf{V}_n$ and Λ must be derived to obtain the canonical form $\mathbf{R} = \mathbf{V}_n \Lambda \mathbf{V}'_n$. We could do this by calculating the three cubic roots of the characteristic equation $|\mathbf{R} - \lambda \mathbf{I}| = 0$ and then using the methods discussed in section 3.9. A more rapid method, which can be used on the desk calculator, will be presented later (section 12.4). Moreover, practically all computer centers have subroutines for solving this numerical problem. These subroutines generally are based on a method to be outlined in section 12.6. The solution for the problem at hand is:

$$\mathbf{V}_n = \begin{bmatrix} 0.47 & 0.87 & 0.13 \\ 0.61 & -0.43 & 0.66 \\ 0.64 & -0.23 & -0.74 \end{bmatrix} \qquad \Lambda = \begin{bmatrix} 1.96 & 0 & 0 \\ 0 & 0.73 & 0 \\ 0 & 0 & 0.31 \end{bmatrix}$$

The three eigenvalues are 1.96, 0.73 and 0.31. They add to 3 which also represents the number of variables in the system. The largest root $\lambda_1 = 1.96$ is the variance of the first principal component ξ_1 with:

$$\xi_{1k} = 0.47\, z_{1k} + 0.61\, z_{2k} + 0.64\, z_{3k}$$

where z_1, z_2, and z_3 are standardized y_1, y_2, and y_3. The subscript k denotes observation number. Consequently, the best-fitting line in 3D-space (standardized variables) has direction cosines $\lambda = 0.47$, $\mu = 0.61$, and $\nu = 0.64$.

In order to compare this result with the best-fitting lines obtained in the beginning of this section, we must go through several steps.

First, the line may be projected onto the (Z_1Z_2)-plane. The direction cosines of the projected line are:

$$\lambda' = \lambda/\sqrt{\lambda^2 + \mu^2}\ ; \qquad \mu' = \mu/\sqrt{\lambda^2 + \mu^2}$$

or, in numerical values, $\lambda' = 0.614$ and $\mu' = 0.790$. Because $\lambda'^2 + \mu'^2 = 1$, one of these values is redundant. If φ' represents the angle between projected line and standardized orthopyroxene axis, then $\cos\varphi' = \mu' = 0.790$. Hence, $\varphi' = 38°$. This angle of slope in the (Z_1Z_2)-plane can be compared to that of the reduced major axis which dips 45° in this plane. The angle φ' can be converted to φ_1' in the (Y_1Y_2)-plane by changing the scales along the coordinate axes as was done for the reduced major axis. From $\cos\varphi' = 0.790$, it follows that $\tan\varphi' = 0.777$. Further:

$$\tan\varphi_1' = (s_1/s_2)\tan\varphi' = 0.367\ , \text{ and } \quad \varphi_1' = 20°$$

In rapid succession, we have obtained four different statistical estimates of the angle of dip for the line expressing the linear relationship between y_1 and y_2. The four best-fitting lines are based on four different mathematical models. They all pass through the point with coordinates $y_1 = 90.06$, $y_2 = 89.03$ but the dips toward the origin are:

Reduced major axis: $25\frac{1}{2}°$

Least squares (y_1 on y_2): $8\frac{1}{2}°$

Least squares (y_2 on y_1): 56°

Projection of first principal component: 20°

Chemical equilibrium (see section 5.3): 45° – 48°

The large discrepancies between individual results exist because the data for y_1 and y_2 are poorly correlated (see Fig. 19A).

On the contrary, the data plotted in Fig. 19B are relatively strongly correlated with $r_{12} = 0.8663$. Some comparable estimates for this situation are:

Reduced major axis: 16°

Least squares ($\hat{y}_1$ on $\hat{y}_2$): $13\frac{1}{2}°$

Least squares ($\hat{y}_2$ on $\hat{y}_1$): $17\frac{1}{2}°$

These three estimates are close together and we can accept 16° as a fair estimate of the angle of slope. It will be seen later (section 9.4), that this also represents the angle of slope to be expected for the linear relationship between the measurements on the original variables y_1 and y_2. This enables us to evaluate results obtained for Fig. 19A.

Closest to 16° is the angle of slope for the projected first principal component (20°). This illustrates the principle that, if variables individually are poorly correlated, it may yet be possible to estimate the relationship between them by applying multivariate analysis. By considering a larger number of variables simultaneously, the sources of variability that obscure the relationships between individual variables can often be eliminated to a considerable extent. This particular result can be improved further by applying factor analysis (see Section 5.5).

The next best result is provided by the least-squares model (y_1 on y_2) where the observations are projected along lines parallel to the olivine-axis. This model is equivalent to assuming that all the errors occur in the olivine determinations, rather than in the orthopyroxene determinations. If seen in the light of the precision data listed in Table VI, this model would seem to be more reasonable than the third one which gave a 56° dip.

Finally, it is of interest to apply component analysis to the four variables $y_1 - y_4$. The correlation matrix is:

$$\mathbf{R} = \begin{bmatrix} 1.00 & 0.32 & 0.41 & -0.31 \\ 0.32 & 1.00 & 0.68 & -0.38 \\ 0.41 & 0.68 & 1.00 & -0.36 \\ -0.31 & -0.38 & -0.36 & 1.00 \end{bmatrix}$$

This yields the eigenvalues $\lambda_1 = 2.26$; $\lambda_2 = 0.74$; $\lambda_3 = 0.70$; and $\lambda_4 = 0.31$. The coefficients of the first eigenvector are (0.43, 0.55, 0.57, – 0.44). The corresponding first principal component accounts for 2.26/4 or 55% of the total variation in the system. The remaining principal components account for 19, 17 and 8% respectively. It therefore is reasonable to assume that there is a single linear relationship or "factor" in the system. The signs of the coefficients of the first eigenvector suggest that the fourth variable (rock density) is negatively correlated to the first three variables ($y_1 - y_3$) which are positively correlated. However, it will be seen later that the behavior of y_4 with respect to the other variables is rather complicated and can hardly be approximated by the relatively simple linear mathematical model that underlies component analysis.

It may be possible to refine the results of component analysis by a more detailed consideration of the sources of variability for individual variables. These methods (factor analysis) will be introduced in the next section.

5.5. Factor analysis

So far, the analysis has been restricted to component analysis and calculation of the principal components. The first principal component was used to approximate the relations in chemical composition between three minerals in the Mount Albert intrusion. The results for this situation also can be written as follows:

$$\xi_{1k} = 0.47\, z_{1k} + 0.61\, z_{2k} + 0.64\, z_{3k}$$

$$\xi_{2k} = 0.87\, z_{1k} - 0.43\, z_{2k} - 0.23\, z_{3k}$$

$$\xi_{3k} = 0.13\, z_{1k} + 0.66\, z_{2k} - 0.74 z_{3k}$$

Alternatively, we can write:

$$z_{1k} = 0.47\, \xi_{1k} + 0.87\, \xi_{2k} + 0.13\, \xi_{3k}$$

$$z_{2k} = 0.61\, \xi_{1k} - 0.43\, \xi_{2k} + 0.66\, \xi_{3k}$$

$$z_{3k} = 0.64\, \xi_{1k} - 0.23\, \xi_{2k} - 0.74\, \xi_{3k}$$

These equations follow from writing [5.11] and [5.12] in full.

Factor analysis consists of extracting a number of linear relationships (factors) from the data. It assumes that the variables that can be measured may be largely determined by a small number of relevant factors; errors in each variable that may influence the results are being considered. A sharp distinction between component analysis and factor analysis does not exist in practice. In fact, component analysis can be seen as a special case of factor analysis.

Thus, in the situation of Mount Albert, we can make the assumption that the first principal component (ξ_1) has physico-chemical significance, whereas the other two components (ξ_2 and ξ_3) reflect local variability and measurement errors.

It is common practice to define factors f_i in such a manner that they have unit variance. For example, the factor f_1 is related to ξ_1 by:

$$f_1 = \xi_1/\sqrt{\lambda_1}$$

or:

$$f_{1k} = 0.175\, \xi_{1k} = 0.34\, z_{1k} + 0.44\, z_{2k} + 0.46\, z_{3k}$$

Since the values ξ_{1k} have variance equal to $\lambda_1 = 1.96$, the values f_{1k} will have variance equal to one, as required. The values f_{1k} as calculated for all possible values of k ($k = 1, 2, \ldots, n$) are sometimes called the factor scores.

Suppose that we define a factor f_i for each ξ_i. Then, the f_i are the columns of an $(n \times p)$ matrix $\mathbf{F}$ with:

$$\mathbf{F} = \Lambda^{-\frac{1}{2}}\, \Xi \qquad [5.14]$$

If a $(p \times p)$ matrix is defined as:

$$\mathbf{A} = \mathbf{V}_n \, \Lambda^{\frac{1}{2}} \qquad [5.15]$$

then:

$$\mathbf{AF} = \mathbf{V}_n \, \Lambda^{\frac{1}{2}} \, \Lambda^{-\frac{1}{2}} \, \Xi = \mathbf{V}_n \, \Xi = \mathbf{Z}$$

The standardized data therefore satisfy:

$$\mathbf{Z} = \mathbf{AF} \qquad [5.16]$$

Each variable z_i is expressed in terms of p factors with:

$$z_{ik} = \sum_{j=1}^{p} a_{ij} f_{jk} \qquad [5.17]$$

Therefore, if three factors are defined for the example:

$$f_{1k} = 0.34 \, z_{1k} + 0.44 \, z_{2k} + 0.46 \, z_{3k}$$

$$f_{2k} = 1.02 \, z_{1k} - 0.51 \, z_{2k} - 0.27 \, z_{3k}$$

$$f_{3k} = 0.24 \, z_{1k} + 1.20 \, z_{2k} - 1.32 \, z_{3k}$$

and:

$$z_{1k} = 0.66 \, f_{1k} + 0.75 \, f_{2k} + 0.07 \, f_{3k}.$$

$$z_{2k} = 0.85 \, f_{1k} - 0.37 \, f_{2k} + 0.37 \, f_{3k}$$

$$z_{3k} = 0.89 \, f_{1k} - 0.20 \, f_{2k} - 0.41 \, f_{3k}$$

f it is assumed that f_2 and f_3 reflect local variability such as measurement errors, then imply:

$$z_{1k} = 0.66 \, f_{1k} + e_{1k} \, ; \qquad z_{2k} = 0.85 \, f_{1k} + e_{2k} \, ; \qquad z_{3k} = 0.89 \, f_{1k} + e_{3k}$$

'his expression can be seen as a simplified version of the more general factor model:

$$z_i = \sum_{j=1}^{q} a_{ij} f_j + b_i \epsilon_i \qquad [5.18]$$

where $i = 1, 2, \ldots, p$. The subscript k was dropped in [5.18] and the quantity $q < p$ denotes the number of factors recognized in the system. In the situation of a single factor $q = 1$.

In the factor model of [5.18], the f_j are the so-called common factors which occur in more than one variable z_i. The ϵ_i are error terms which occur in one variable (z_i) only.

The f and ϵ are regarded as having unit variance. They are assumed to be mutually independent. Since the z_i also have unit variance, it is shown readily that:

$$\sum_{j=1}^{q} a_{ij}^2 + b_i^2 = 1 \qquad [5.19]$$

for each value of i. The quantity $\Sigma\, a_{ij}^2$ is called "communality". Usually it is written as h_i^2

Our treatment of the factor model is abbreviated. There is an extensive nomenclature of which we are mentioning the most commonly used terms only. For a full review, more extensive proofs and alternative methods, the reader is referred to Kendall (1965) or Harman (1960). Our prime purpose of introducing factor analysis is that it may constitute a significant refinement of component analysis.

Suppose that a variance–covariance matrix **C** with elements c_{ij} is determined for the variables z_i. Then:

$$c_{ij} = \sum_{m=1}^{q} a_{im}\, a_{jm} + b_i b_j \text{ ave } (\epsilon_i, \epsilon_j) \qquad [5.20]$$

where "ave" denotes average cross-product. Because of the independence and unit variance assumptions, this further reduces to:

$$\left.\begin{aligned} c_{ij} &= \Sigma\, a_{im}\, a_{jm} \quad (i \neq j) \\ 1 &= \Sigma\, a_{im}^2 + b_i^2 \quad (i = j) \end{aligned}\right| \qquad [5.21]$$

The second expression duplicates [5.19]. We can eliminate the effect of the error term from each variable and define the reduced variables:

$$z_i' = z_i - b_i\, \epsilon_1 = \Sigma\, a_{im}\, f_m \qquad [5.22]$$

The variance–covariance matrix for the reduced variables z_i' will be the same as that for the z_i except in the diagonal elements where:

$$c_{ii}' = 1 - b_i^2 = \Sigma\, a_{im}^2 = h_i^2 \qquad [5.23]$$

The difference between factor analysis and component analysis is that instead of using the data z_{ik} in p-space, we operate on reduced data z'_{ik} that also lie in p-space. It means that the correlation matrix **R** is replaced by a matrix:

$$\mathbf{R}^* = \begin{bmatrix} h_1^2 & r_{12} & \cdots & r_{1p} \\ r_{21} & h_2^2 & \cdots & r_{2p} \\ \cdot & \cdot & \cdots & \cdot \\ r_{p1} & r_{p2} & \cdots & h_p^2 \end{bmatrix} \qquad [5.24]$$

When the diagonal elements (equal to one in **R**) have been replaced by the communalities h_i^2 in **R***, the method of extracting factors from **R*** is the same as that used for treating **R**. A number of different methods have been developed to estimate communalities in a multivariate situation (cf. Harman, 1960).

5.6. Application of factor analysis

With regard to the chemical composition of the minerals in Mount Albert, we may argue as follows. The variance of y_1 as estimated from 169 olivines is $s_1^2 = 0.952$. In section 5.2, it was derived that the variance due to measurement errors for olivine is 0.474. Thus, for the variable y_1, half of its variance for specimens from the entire intrusion can be explained as caused by measurement errors. If this source of variability is eliminated from the standardized data $\mathbf{Z}_1$, the result is a communality $h_1^2 = 0.5$. No exact data are available for the measurement errors of the variables y_2 and y_3. However, it can be assumed safely that less than 10% of their variances for the entire intrusion is caused by measurement errors. This is mainly because the chemical composition of orthopyroxenes and chrome spinels changes more strongly across the intrusion than that for olivine.

For this reason, let us assume:

$$\mathbf{R}_1^* = \begin{bmatrix} 0.5 & 0.3159 & 0.4106 \\ 0.3159 & 1.0 & 0.6824 \\ 0.4106 & 0.6824 & 1.0 \end{bmatrix}$$

The largest eigenvalue of $\mathbf{R}_1^*$ is 1.875. It would account for 1.875/2.5 or 75% of the total ariation which now is of the form $T = (n-1)^{-1}[\Sigma z_1'^2 + \Sigma z_2^2 + \Sigma z_3^2] = 2.5$. This is a omewhat larger percent value than that obtained for **R** which amounts to 1.96/3 or 65% see section 5.4). The first eigenvector of **R*** provides the direction cosines $\lambda = 0.35$, $\mu =$ .65, and $\nu = 0.67$. Projection of the best-fitting line onto the $(Z_1'Z_2)$-plane and conversion o the (Y_1Y_2)-plane yields a line that dips 14½° in the negative Y_2-direction (orthopyroxene-axis). This is close to the angle of 16° which, in the previous section, was accepted as our best available estimate of the dip.

By applying a simple method of factor analysis, we were able to improve our estimate of the linear relationship between olivine and orthopyroxene composition data. However, one possibly important source of variability has not been considered yet.

The measurement error for the olivine determinations, which was eliminated by considering the communality, mainly accounts for two sources of "noise". These are: (1) errors due to apparatus and measuring; and (2) zoning in individual olivine crystals. In a statistical analysis based on olivines from different localities, the results are likely to be influenced by a third source of noise consisting of irregular variations between olivines from different localities. Similar noise components for variations between localities may exist for the orthopyroxene and chrome spinel determinations. These additional sources of variation should be considered in the factor analysis if they are specific for each variable and cannot be correlated from variable to variable.

The direction cosines for the best-fitting line in $(Z_1Z_2Z_3)$-space are modified as follows by assuming $h_1^2 = 0.5$.

	λ	μ	ν
R (component analysis)	0.47	0.61	0.64
$\mathbf{R}_1^*$(factor analysis)	0.35	0.65	0.67

These results for the relationship between the three variables can be compared with those obtained by component analysis applied to the correlation matrix $\mathbf{R}_T$ for the fitted trend values. Then:

$$\mathbf{R}_T = \begin{bmatrix} 1.00 & 0.87 & 0.82 \\ 0.87 & 1.00 & 0.91 \\ 0.82 & 0.91 & 1.00 \end{bmatrix}$$

The matrices $\mathbf{V}_n$ and Λ for $\mathbf{R}_T$ become:

$$\mathbf{V}_n = \begin{bmatrix} 0.57 & 0.79 & 0.24 \\ 0.59 & -0.18 & -0.79 \\ 0.58 & -0.59 & -0.56 \end{bmatrix} \quad \text{and} \quad \Lambda = \begin{bmatrix} 2.73 & 0 & 0 \\ 0 & 0.19 & 0 \\ 0 & 0 & 0.08 \end{bmatrix}$$

The largest eigenvalue $\lambda_1 = 2.73$ accounts for 2.73/3 or 91% of the total variation in the system. It means that the data are close to a single line in $(Z_1Z_2Z_3)$-space with direction cosines:

$$\lambda_T = 0.57 \qquad \mu_T = 0.59 \qquad \nu_T = 0.58$$

This line is close to the line with direction cosines $(1/\sqrt{3}, 1/\sqrt{3}, 1/\sqrt{3})$ or (0.577, 0.577, 0.577) which makes equal angles with the three axes. Projection of the line with equal direction coefficients onto the (Z_1Z_2)-, (Z_1Z_3)- or (Z_2Z_3)-plane yields the reduced major axis for the corresponding bivariate relationships. In the previous section, we have made use of the reduced major axis in the (Z_1Z_2)-plane that after conversion to the (Y_1Y_2)-system dips 16° with respect to the orthopyroxene axis.

If we wish to compare the results for $\mathbf{R}_T$ with those for $\mathbf{R}$ or $\mathbf{R}_1^*$, we must consider the fact that the fitted trend values have standard deviations that differ from those for the original data. They are shown as $s(\hat{y}_i)$ in the following table:

	$s(y_i)$	$s(\hat{y}_i)$
Olivine	0.849	0.334
Orthopyroxene	1.799	1.213
Chrome spinel	0.0557	0.0281

If the line with $(\lambda_T, \mu_T, \nu_T)$ is to be plotted in the space of $\mathbf{R}$ and $\mathbf{R}_1^*$, the coefficients should be modified. For example, $\lambda_T = 0.5676$ first is multiplied by $s(\hat{y}_1) = 0.334$ and then divided by $s(y_1) = 0.849$ yielding 0.2234. The corresponding values for the other two variables are 0.3962 and 0.2910. The three numbers are normalized to give the three modified direction consines (0.41, 0.73, 0.54).

If this set of values is compared to the (λ, μ, ν) for $\mathbf{R}$ and $\mathbf{R}_1^*$ we note a considerable discrepancy. The similarity is restricted to the ratio (λ/μ) for the olivine–orthopyroxene relationship. It indicates that further modifications should be made in the communalities of $\mathbf{R}_1^*$, particularly in that for the chrome spinel cell edge.

Let us suppose that:

$$\mathbf{R}_2^* = \begin{bmatrix} 0.5 & 0.32 & 0.41 \\ 0.32 & 1.0 & 0.68 \\ 0.41 & 0.68 & x \end{bmatrix} \qquad [5.25]$$

where x represents the unknown communality for y_3. A simple criterion by which x can be solved is to subject $\mathbf{R}_2^*$ to the condition that one of its eigenvalues is equal to zero. Because $\mathbf{R}_2^*$ is symmetric and therefore nonnegative definite, a zero eigenvalue also is the smallest eigenvalue λ_3.

The geometrical interpretation of this condition is that we are defining a new reduced variable z_3' to replace z_3 whose values lie in a single plane with the previously used values of z_1' and z_2. This plane that contains all data z_{1k}', z_{2k}, and z_{3k}' has an equation that can be calculated from the coefficients of the eigenvector corresponding to the zero eigenvalue.

The calculation of x will be done in detail. The characteristic equation of $\mathbf{R}_2^*$ is:

$$|\mathbf{R}_2^* - \lambda \mathbf{I}| = 0$$

or:

$$\begin{vmatrix} (0.5-\lambda) & 0.32 & 0.41 \\ 0.32 & (1-\lambda) & 0.68 \\ 0.41 & 0.68 & (x-\lambda) \end{vmatrix} = 0 \qquad [5.26]$$

Expansion of the determinant gives:

$$(0.5-\lambda)\begin{vmatrix} (1-\lambda) & 0.68 \\ 0.68 & (x-\lambda) \end{vmatrix} - 0.32\begin{vmatrix} 0.32 & 0.68 \\ 0.41 & (x-\lambda) \end{vmatrix} + 0.41\begin{vmatrix} 0.32 & (1-\lambda) \\ 0.41 & 0.68 \end{vmatrix} =$$

and:

$$(0.5-\lambda)(1-\lambda)(x-\lambda) - 0.68^2(0.5-\lambda) - 0.32^2(x-\lambda)$$

$$+ 0.32 \times 0.68 \times 0.41 + 0.41 \times 0.32 \times 0.68 - 0.41^2(1-\lambda) = 0 \qquad [5.27]$$

This is a cubic equation of the type $\lambda^3 + a\lambda^2 + b\lambda + c = 0$, or:

$$(\lambda - \lambda_1)(\lambda - \lambda_2)(\lambda - \lambda_3) = 0$$

It can be readily seen that the condition $\lambda_3 = 0$ implies that $c = 0$. The constant c consists of all terms of [5.27] that do not contain λ. Extraction of these terms and setting the result equal to zero gives:

$$0.5\,x - 0.2375 - 0.0998\,x + 0.1770 - 0.1686 = 0$$

The constants in this equation are reported in greater precision than those in [5.27] which were rounded off to two significant digits. It follows that $x = 0.5608$. Substitution of the value for x into [5.27] leads to the cubic equation:

$$\lambda^3 - 2.0608\,\lambda^2 + 0.6071\,\lambda = 0 \qquad [5.28]$$

Two roots λ_1 and λ_2 can now be solved from the quadratic equation:

$$\lambda^2 - 2.0608\,\lambda + 0.6071 = 0$$

The final result is:

$$\lambda_1 = 1.7047\,; \qquad \lambda_2 = 0.3561\,; \qquad \lambda_3 = 0$$

The first eigenvector $\mathbf{V}_1$ can be solved by developing the equation:

$$\mathbf{R}_2^* \mathbf{V}_1 = \lambda_1 \mathbf{V}_1$$

If, initially, v_{11} is set equal to one, this gives:

$$0.5 + 0.3159\, v_{12} + 0.4106\, v_{13} = 1.7047$$

$$0.3159 + v_{12} + 0.6824\, v_{13} = 1.7047\, v_{12}$$

The third linear equation is not reported. It is redundant in that it is a linear combination of the first two equations. Solution for v_{12} and v_{13} and normalization give:

$$\mathbf{V}_1 = \begin{bmatrix} 1 \\ 1.8850 \\ 1.4837 \end{bmatrix} \quad \text{and} \quad \mathbf{V}_{n1} = \begin{bmatrix} 0.3848 \\ 0.7253 \\ 0.5709 \end{bmatrix}$$

The vector $\mathbf{V}_{n1}$ determines the best fitting line for $\mathbf{R}_2^*$ ($x = 0.5608$). In the following table its direction cosines are compared to previous results.

	λ	μ	ν
$\mathbf{R}_1^*$	0.35	0.65	0.67
$\mathbf{R}_2^*$	0.39	0.73	0.57
$\mathbf{R}_T$	0.41	0.73	0.54

The first eigenvector of $\mathbf{R}_2^* = \begin{bmatrix} 0.5 & 0.32 & 0.41 \\ 0.32 & 1.0 & 0.68 \\ 0.41 & 0.68 & 0.56 \end{bmatrix}$

shows closer correspondence to that of $\mathbf{R}_T$. The first eigenvalue of $\mathbf{R}_2^*$ accounts for 1.7047/2.0608 or 83% of total variation. This percentage is less than the 91% for $\mathbf{R}_T$ but greater than the 75% obtained for $\mathbf{R}_1^*$.

Finally, we can solve the coefficients of the third eigenvector $\mathbf{V}_3$ that corresponds to the zero eigenvalue. If $v_{31} = 1$, v_{32} and v_{33} satisfy:

$$0.5 + 0.3159\, v_{32} + 0.4106\, v_{33} = 0$$

$$0.3159 + v_{32} + 0.6824\, v_{33} = 0$$

and:

$$\mathbf{V}_3 = \begin{bmatrix} 1 \\ 1.085 \\ -2.052 \end{bmatrix}; \quad \mathbf{V}_{n3} = \begin{bmatrix} 0.40 \\ 0.43 \\ -0.81 \end{bmatrix}$$

In the $(Z_1\ Z_2\ Z_3)$-coordinate system, all values for the variables z_1', z_2, and z_3' fall in a plane with equation $\xi_3 = 0$, or:

$$0.40\, z_1 + 0.43\, z_2 - 0.81\, z_3 = 0$$

If the Z_3-axis points in the vertical direction, this plane dips $36\frac{1}{2}°$ toward the origin. It represents the condition that we have imposed on the reduced variables. Of course, the data z_{1k}, z_{2k} and z_{3k} will scatter around this plane.

Chapter 6

PROBABILITY AND STATISTICS

6.1. Introduction

Probability calculus and statistical inference have contributed significantly to the solution of many geologic problems. In fact, statistical considerations form part of almost every application of mathematics in geology.

In the earth sciences, the object of study usually is an aggregate of many smaller objects, which can be studied individually, but, in some situations, only the properties of the aggregate may be meaningful. For example, a sample of sand consists of many grains each of which has many measurable properties. However, only statistical parameters for the entire sand sample such as an average grain size are of interest to the geologist. Another example is provided by the average properties of larger masses of rocks. A geophysicist may measure the velocity of seismic waves in various layers of the earth's crust. The resulting numbers are manipulated by using differential equations. These numbers, however, represent averages for assemblages of rocks. A stratigraphic sequence consisting of interbedded shales and sandstone will provide a single average velocity value. This number is useful but only within the framework for the mathematical model formulated by the seismologist. On a more local scale, the average is representative for neither sandstone nor shale.

Averages

Historically, the first average on record was taken by William Borough in 1581 for a set of compass readings (cf. Eisenhart, 1963). The procedure of averaging numbers was regarded with suspicion for a long period of time. In 1755, T. Simpson published a paper entitled *On the advantage of taking the mean of a number of observations in practical astronomy.* In his introduction, Simpson (1755) said: "It is well-known that the method practiced by astronomers to diminish the errors arising from the imperfections of instrument and of the organs of sense by taking the mean of several observations has not been so generally received but that some persons of note have publicly maintained that one single observation, taken with due care, was as much to be relied on, as the mean of a great number."

A word of caution remains in order. By repeating measurements and performing mathematical manipulations, one may be able to improve the precision of results. However, it is not possible to improve the accuracy in this manner.

Accuracy and precision

The term "accuracy" usually is reserved for situations that involve systematic discrepancies between measurements and true values caused by instrumental error. An accurate result has negligible bias although its precision may be poor (large standard error). If the true values are known, the systematic errors for a given method can be eliminated. An example is the chemical analysis of rocks.

During the past decade, rapid methods such as the one based on X-ray fluorescence have become available for the measuring of element-concentration values in rocks. The rapid methods generally have limited accuracy and precision in comparison with the classical methods of chemical analysis. The problem of restricted accuracy is solved by measuring a set of standards for which the true values are known with relatively great accuracy and precision. The measured values can be plotted in the vertical direction (Y-axis) of a scattergram against the "true" values (X-axis). If measurements for a given range are unbiased, the points in the diagram should scatter around the line with equation $y = x$.

Suppose that a line with equation $y = bx$ is fitted to the data by the method of least squares (y on x). The slope b of the best-fitting line then satisfies:

$$b = \frac{\Sigma x_i y_i}{\Sigma x_i^2}$$

Accurate data should have $b \approx 1$. If bias exists it can be eliminated by correcting for the difference between unity slope and least-squares estimate since:

$$\text{bias} = 1 - b = 1 - \Sigma xy/\Sigma x^2$$

The bias factor is a statistical parameter and subject to a random error that is related to the precision of the method and the number of data used. In fitting a line with equation $y = bx$, we are applying the method of least squares but with the constraint $a = 0$. The best-fitting line is forced through the origin with $x = y = 0$ which is usually a reasonable assumption for applications of this type. Sometimes it is better to fit the more conventional straight line $y = a + bx$ for which:

$$b = \Sigma x'y'/\Sigma x'^2$$

where $x' = x - \bar{x}$ and $y' = y - \bar{y}$. Both a and b then can be used to improve the accuracy of the method for a limited range of values. In other situations again, it may be wise to fit $y = a + bx + cx^2$, or it may be better to adopt a different type of least-squares model. Suppose that the measurement error is proportional to x. Then the best-fitting line through the origin has slope equal to $b = (1/n)\, \Sigma\, (y/x)$. On the other hand, the optimum (least squares) solution is $b = (\Sigma y)/(\Sigma x)$ if the square of the measurement error increases linearly with x. Models of this type were discussed by Deming (1943).

In this chapter, and later, it generally will be assumed that the numbers to which statistical techniques are applied are free of systematic errors and that there is no bias in the results because of poor accuracy of the instrument. By using statistical methods, we may be able to improve precision of results by mathematical manipulation but if initially all numbers are systematically too large or too small, this error would be retained in all further results.

Dispersion of values about the average

The individual numbers deviate from their average, This phenomenon can be graphically represented by means of a histogram (see, e.g., Fig. 21). The concept of a frequency distribution is an abstraction of the histogram for observed values. It may be useful to assume that the histogram is the result of what a limited set of data tells us about the underlying frequency distribution for all values that possibly could have been measured. The assemblage of all possible values is called the population. In geologic applications, the population usually may be considered as infinitely large for practical purposes.

We can make a hypothesis about the nature of the frequency distribution for the population. If the sample, or set of measured values, is small, the histogram will be consistent with different models assumed for the population. In many practical applications, the assumption is made that the frequency-distribution curve for the population is "normal". Its equation is:

$$y = c\ \mathrm{e}^{-ax^2}$$

where a and c are constants. A graphical representation of the "normal", "Gaussian" or "probability" curve was shown in Fig. 6E.

The next section contains a brief explanation of set theory. Other subjects to be treated are probability calculus and statistical inference as based on the normal frequency distribution. The reader without a background in set theory and elementary statistics is referred to the books listed in sections 2.4 and 2.5 of the list for Selected Reading. All books of section 2.5 of this list contain statistical tables for the tests to be discussed.

6.2. Set theory

We can think of a number of points in a space as a set. All points in the space form the universal set U. Suppose that A, B, and C denote three sets of points which are part of U. This situation is graphically represented in Fig. 20. The following definitions can be made for the relations among A, B, C, and U.

If a set is contained in another set, it is called a subset. The symbol $\subset$ is used to indicate subsets. Hence, $A \subset U$, $B \subset U$, and $C \subset U$. The complement of a set consists of all points in U that are not in the set. The superscript "c" is used to denote complements. In particular,

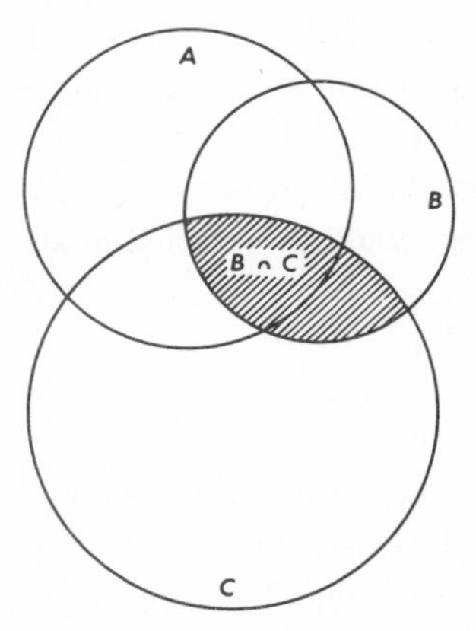

Fig.20. Venn diagram representing sets A, B, C, and the intersection of sets B and C.

$U^c = 0$ because there are no points that do not belong to the universal set.

Points that belong to two sets form the intersection of two sets. The expression $A \cap B$ is for points that are both part of A and part of B. The union of two sets A and B is written as $A \cup B$. It represents points that belong to either A or B. The expression $A-B$ denotes points in A not contained in B.

The following geological example illustrates several manipulations of the definitions for relations among sets.

Applications of set theory

Suppose that a large area is divided into several hundreds of square subareas or cells. The cells form the universal set U. Rock types in the area were mapped and a number of the cells are known to contain ultramafic intrusions. These cells form the set $A \subset U$. Only part of the cells in the universal set have been explored in greater detail to see if nickel deposits are present. Cells that contain nickel deposits belong to set B. The well-explored cells form set C. A geologic evaluation of what is known for the area has indicated that nickel deposits tend to be associated with ultramafic intrusions.

Nickel deposits have only been found in cells that belong to set C. Hence, our information with regard to set B is restricted to a subset $B_m = B \cap C$ and we lack data on $B \cap C^c$.

The reader may verify on the basis of the definitions and the diagram (Fig. 20) that our information can be classified as follows:

$$A \cap B_m = A \cap B \cap C$$

$$A^c \cap B_m = A^c \cap B \cap C$$

$$A \cap B_m^c = A \cap (B \cap C)^c = A \cap (B^c \cup C^c)$$

$$A^c \cap B_m^c = A^c \cap (B \cap C)^c = A^c \cap (B^c \cup C^c)$$

This information can be used to make predictions about the sets $A \cap B \cap C^c$ and $A^c \cap B \cap C^c$. Suppose that $B \subset A$. This would imply $A^c \cap B = 0$ and further exploration would be restricted to the set $A \cap C^c$. In an ideal situation, $A = B$.

Several useful extensions of the classification theory are suggested by this example. It may be useful: (1) to assign probabilistic measures to the sets, and (2) to define "conditional" probabilities for indicating the chances that a particular cell will belong to one set when it is known that it belongs to another set. These additional concepts will be formulated below and also in Chapter 8.

Set theory is important for various reasons. The previous example can be considered as a formulation of the logic applied in exploration geology. The type of algebra that is used is Boolean algebra. A diagram such as Fig. 20 is called a Venn diagram. Set theory provides the foundation of probability calculus. Another application is that the logic is compatible with what can be done by using digital computers. For example, in a version of the computer language FORTRAN (for FORmula TRANslation), the expression .OR. is equivalent to union and the word .AND. to intersection. These expressions can be used in so-called conditional IF statements. For example, a value may be tested as follows. If it belongs to the set $A \cap (B \cup C)$, it is subjected to one kind of manipulation. If it is part of $A \cap (B \cup C)^c$, it receives other treatment. Boolean algebra is used for classification purposes and to retrieve selected pieces of information stored in large geologic data banks.

Sample space and events

If we perform a geologic measurement, the result can be considered as having originated from one point in a sample space. The points in the sample space account for all possible measurements. The sample space can be seen as a universal set. Aggregates of sample points, which are equivalent to sets, are called events. For example, if we take a single measurement, the specific event A may or may not occur. More than a single event can occur if a single measurement is made. We can assign probabilities to all events in the sample space. This determines the chances of a given event occurring when a measurement is made.

In probability theory, the sample space and the events are considered to be known. In practice, the success of an experiment that consists of taking a number of measurements, usually depends strongly on how accurately the sample space and its events have been defined.

Sample and population

A population is the assemblage of all possible measurements for which a sample space and the events contained in it have been defined. The definition of a sample space and events can be considered as laying the framework for a population.

A sample is a group of measurements obtained from the population. The purpose of mathematical statistics is to obtain information on the population by drawing one or more samples from it.

In some geologic applications, the points of the sample space that may be represented in a sample constitute another event. In that situation, the sample is not representative for the population. This has resulted in the concepts of sampled population and target population (cf. Cochran et al., 1954; Szameitat and Schäffer, 1963; Krumbein and Graybill, 1965).

Sampled population versus target population

In a number of situations, a distinction should be made between the "target" population which is the purpose of investigation and the "sampled" population. Three possibilities for lack of coverage between target population and sampled population are:

(1) Measurements belonging to the target population are not included in the sampled population. For example, because of differential weathering, certain rocks with attributes belonging to the target population are less accessible for measurement and they may be underrepresented or absent in the sampled population.

(2) Measurements belonging to the target population are contained in the sampled population several times. This situation can arise in geologic sampling when the observation points occur in clusters and groups of approximately equal values are measured in areas that are relatively small.

(3) Measurements in the sampled population do not belong to the target population. For each population, one has to define a frame of measuring which is equivalent to defining the sample space. When there exist two or more populations with overlapping frames, it may not be possible to establish if a given value belongs to the target population or to another population with an overlapping frame.

6.3. Calculus of probabilities

The probability of an event is an abstraction of the idea of the relative frequency by which this event occurs in a sequence of trials or measurements during a given experiment.

An area overlain by a network of cells with different colours can be used as a reference model. Suppose that the area contains N equally large cells of two types (black and white) and that the value 1 is assigned to every black cell and 0 to every white cell. If a single cell is selected at random, e.g., by an undirected bombing process, the probability that it is black is equal to the relative frequency of black cells in the area. If there are N_b black cells, the relative frequency amounts to $p = N_b/N$.

The value of a cell drawn at random is either one or zero. This uncertainty can be indicated by using a random variable to be written as X. The probability that $X = 1$ is written as:

$$P(X = 1) = p$$

or:

$$P(X = 1) = N_b/N$$

The probability that the cell is white, or that $X = 0$, satisfies:

$$P(X = 0) = q$$

Obviously, $p + q = 1$, and we can write:

$$P(X = 0) = 1 - p$$

The experiment of selecting one cell whose value is either one or zero is called a Bernoulli trial.

In general, a random variable X can assume different values x_k in such a manner that each value x_k has a given probability $P(X = x_k)$. A random variable differs from an ordinary variable because of the uncertainty and this difference is brought out in notation by using X instead of x.

The sample space for the Bernoulli trial is relatively simple. It consists of two sample points with values one and zero. There are only two events (one and zero) which are mutually exclusive. The probabilities for the events are p and q, respectively, with $p + q = 1$.

Minerals in an amphibolite

A given amphibolite consists of dark minerals (predominantly hornblende) and a light mineral (plagioclase). The probability that a certain type of mineral occurs at an arbitrary point in a thin section is equal to volume percentage of this mineral in the rock. If dark (D) minerals are coded as one, and the light (L) mineral as zero, then:

$$P(X = 1) = P(\mathrm{D})\ ; \quad P(X = 0) = P(\mathrm{L})$$

where $P(\mathrm{D})$ and $P(\mathrm{L})$ represent volume percentage values with $P(\mathrm{D}) + P(\mathrm{L}) = 1$.

Addition of probabilities

The probability of an event occurring in one of several possible ways is calculated as the sum of the probabilities of the occurrence for the several possible ways.

Suppose that the amphibolite consists of (1) hornblende, (2) plagioclase, (3) epidote, (4) chlorite, (5) sphene, and (6) iron ore with percentage values equal to $P(i); i = 1, 2, \ldots, 6$, respectively. Then:

$$P(X = 1) = P(\mathrm{D}) = P(1) + P(3) + P(4) + P(5) + P(6)$$

$$P(X = 0) = P(\mathrm{L}) = P(2)$$

During the addition of probabilities, care should be taken to account for possible overlap of events in the outcome of an experiment. The probability $P(\mathrm{S})$ that a mineral is a silicate is equal to:

$$P(\mathrm{S}) = P(1) + P(2) + P(3) + P(4)$$

If $P(\mathrm{S} \cup \mathrm{D})$ denotes the probability that a mineral is either a silicate or a dark mineral then $P(\mathrm{S} \cup \mathrm{D}) \leqslant P(\mathrm{S}) + P(\mathrm{D})$. In fact, $P(\mathrm{S} \cup \mathrm{D}) = P(\mathrm{S}) + P(\mathrm{D}) - P(\mathrm{S} \cap \mathrm{D})$ where $P(\mathrm{S} \cap \mathrm{D})$ denotes the probability of a mineral that is both a silicate and a dark mineral. It is readily verified that $P(\mathrm{S} \cup \mathrm{D}) = 1$ and $P(\mathrm{S} \cap \mathrm{D}) = P(1) + P(3) + P(4)$.

Multiplication of probabilities

Probabilities are to be multiplied if the probability of two or more events in succession is considered. An important qualification of the multiplication law is that the events must be mutually independent.

Suppose that one cell is selected at random from the mosaic of black and white cells. As before, we have:

$$P(X_1 = 1) = p\,; \quad P(X_1 = 0) = 1 - p$$

The subscript 1 is assigned to X to indicate that this selection will be followed by others. If a new cell is selected at random, we can denote the probabilities for the second event as:

$$P(X_2 = 1) = p\,; \quad P(X_2 = 0) = 1 - p$$

There are three possible outcomes for the total experiment that consists of two independent Bernoulli trials: two black cells, one black cell and one white cell, and two white cells. The sample space for the total experiment can be defined in two different ways. We can say that it consists of four points (1,1), (1,0), (0,1), and (0,0). If the order of events is ignored, the points (1,0) and (0,1) are equivalent and we have only three points in the sample space with the following probabilities:

	Sample point	*Probability*
Two black cells	(1,1)	p^2
One black cell and one white cell	(1,0)	$2pq$
Two white cells	(0,0)	q^2

A random variable $X = X_1 + X_2$ can be defined for the experiment consisting of two draws. When both cells are one, the probability that X is equal to 2 is p^2. Hence:

$$P(X = 2) = p^2$$

Likewise:

$$P(X = 1) = 2pq$$

$$P(X = 0) = q^2$$

The addition of the random variables X_1 and X_2 is equivalent to a multiplication of the probabilities for the events associated with X_1 and X_2 separately.

In general, if X and Y denote two different random variables, then the multiplication law can be written as:

$$P(X = x_j \text{ and } Y = y_k) = P(X = x_j)\,P(Y = y_k) \qquad [6.1]$$

Another way of writing $P(X = x_j$ and $Y = y_k)$ is $P(x_j, y_k)$ or simply p_{jk}. It denotes the joint probability that $X = x_j$ and $Y = y_k$ where x_j and y_k are values that can be assumed by X and Y.

The multiplication law holds true only if X and Y are mutually independent random variables. In fact, [6.1] can be used as a definition of statistical or stochastic independence. In this chapter, emphasis is on mutually independent variables. Stochastic dependence, where the multiplication law does not hold true will be treated in Chapter 8.

6.4. The binomial distribution

Suppose that the experiment of selecting one cell at random from the mosaic is repeated n times. The probability that all cells are black is p^n or:

$$P\left(\sum_{i=1}^{n} X_i = n\right) = p^n$$

For example, if $p = \frac{1}{2}$, the probability of ten black cells in succession is $(\frac{1}{2})^{10} = 1/1024$.

For n Bernoulli trials, there are $(n + 1)$ possible outcomes: n, $(n - 1)$, . . . , 2, 1, 0. The $(n + 1)$ probabilities for these outcomes are given by the successive terms of the series obtained by expanding the expression $(p + q)^n$. For $n = 2$, this series is $p^2 + 2pq + q^2$. Hence, as in the previous section:

$$P(X_1 + X_2 = 2) = p^2$$

$$P(X_1 + X_2 = 1) = 2pq$$

$$P(X_1 + X_2 = 0) = q^2$$

Since $p + q = 1$, $(p + q)^n = 1$ for any value of n, and the sum of all probabilities for ΣX_i is always equal to one.

The $(n + 1)$ probabilities, together, form the so-called binomial distribution. The probability that there are exactly k black cells in n trials satisfies:

$$P\left(\sum_{i=1}^{n} X_i = k\right) = \binom{n}{k} p^k q^{n-k} \qquad [6.2]$$

where $\binom{n}{k}$ is a binomial coefficient.

The binomial coefficient $\binom{n}{k}$ can be obtained from the triangle of Pascal:

n											
1					1		1				
2				1		2		1			
3			1		3		3		1		
4		1		4		6		4		1	
5	1		5		10		10		5		1

In this triangle, the value $\binom{n}{k}$ is the k-th value in the n-th row if k is counted from the lef with $k = 0$ for the first value that is always equal to one.

Each value in Pascal's triangle is the sum of two adjacent values in the row above it. I can be shown that:

$$\binom{n}{k} = \frac{n!}{k!\,(n-k)!} \qquad [6.3]$$

where $n!$ denotes factorial n with $n! = n(n-1)(n-2)\ldots 3 \cdot 2 \cdot 1$.

Boreholes in a vein-type mineral deposit

A mineral deposit consists of many ore-bearing veins distributed in a barren host roc A large number of boreholes was drilled into the deposit during a development program The percentage of holes that intersected one or more veins amounts to 90%. This implie that 10% of the holes did not intersect any veins. Later, five new holes were drilled ne the margin of the deposit. It was found that four of the new holes did not intersect an veins. We can evaluate this situation as follows:

The probability that a single new borehole intersects ore is assumed to be $p = 0.9$ c the basis of past experience for many holes. Hence, $q = 0.1$. The probability that onl one out of five new holes intersects ore therefore satisfies:

$$P\left(\sum_{i=1}^{5} X_i = 1\right) = \binom{5}{1} \times 0.9 \times 0.1^4$$

From [6.3], we have $\binom{5}{1} = 5$. The probability becomes $5 \times 0.9 \times 0.0001 = 0.00045$. It means that the situation near the margin of the deposit of ore in only one of five holes is anomalous if considered in the light of past experience for the entire deposit. This event would happen once per two thousand trials. It is judged to be unlikely to happen, although it would not be impossible. With little uncertainty, it may be assumed that the frequency of ore veins in the ground is less than 90% in the area of the five new holes.

If $p = 0.9$ we can calculate all probabilities for the six possible outcomes. They are given in Table IX.

TABLE IX
Binomial distribution for $p = 0.9$ and $n = 5$

k	*Probability*
0	0.00001
1	0.00045
2	0.00810
3	0.07290
4	0.32805
5	0.59049
Sum =	1.00000

For example, the probability that a set of five new holes will all intersect veins is 59%. The addition law can be used to answer such questions as: What is the probability that four or five out of five holes will intersect ore? The answer is 0.32805 + 0.59049 = 0.91854 or 92%.

6.5. Discrete and continuous frequency distributions

An example of a theoretical discrete frequency distribution was shown in Table IX. The term "discrete" indicates that the random variable which can be written as K assumes specific values only (integers in the example), each of which has its own probability of occurring during an experiment.

The theoretical frequency distribution, like the random variable, is an abstraction. If we do a practical experiment and then list the observed relative frequencies they are bound to deviate from the theoretical frequencies. If the model is applicable, the observed frequencies will deviate at random from the theoretical frequencies predicted by the model.

The data of Table IX also can be reported by using cumulative frequencies. The result is the cumulative discrete frequency distribution shown in Table X. A continuous

TABLE X
Cumulative frequencies for Table IX

k	*Cumulative probability*
0	0.00001
0 or 1	0.00046
0 or 1 or 2	0.00856
$\leqslant 3$	0.08146
$\leqslant 4$	0.40941
$\leqslant 5$	1.00000

random variable X can be defined on the basis of its cumulative frequency distribution $F(x)$ with:

$$P(X \leqslant x) = F(x) \qquad [6.4]$$

The sample space for X is continuous and X has the property that it can assume any value within a given range. In many situations, this range is the so-called real line which extends from $-\infty$ to $+\infty$.

The continuous random variable X can assume any value less than x with probability $F(x)$. Eq. [6.4] is an ordinary function of the ordinary variable x. It has two important properties: (1) it can not decrease for increasing values of x; (2) its smallest and largest value are equal to zero and one, respectively. $F(x)$ can be considered as the definite integral:

$$F(x) = \int_{-\infty}^{x} f(x)\, \mathrm{d}x \qquad [6.5]$$

of the so-called frequency-distribution function $f(x)$.

Some basic definitions of integral calculus (section 2.10) are applicable to this situation. We can write:

$$f(x) = \frac{\mathrm{d}F}{\mathrm{d}x} = \frac{F(x + \mathrm{d}x) - F(x)}{\mathrm{d}x}$$

From [6.4], it follows that:

$$P(X \leqslant x + \mathrm{d}x) = F(x + \mathrm{d}x)$$

$$P(X \leqslant x + \mathrm{d}x) - P(X \leqslant x) = F(x + \mathrm{d}x) - F(x)$$

and, finally:

$$P(x < X \leqslant x + \mathrm{d}x) = f(x)\, \mathrm{d}x \qquad [6.6]$$

Consequently, the expression $f(x)\,dx = dF(x)$ denotes the probability that X belongs to the interval bounded by x and $(x + dx)$.

6.6. The normal distribution

The general form of the normal or Gaussian distribution was given by [2.5]. In mathematical statistics, this usually is written as:

$$f(x) = \frac{1}{\sigma\sqrt{2\pi}} \exp\left\{-\tfrac{1}{2}\left(\frac{x-\mu}{\sigma}\right)^2\right\} \qquad [6.7]$$

where μ and σ represent mean and standard deviation, respectively. It follows that the cumulative normal frequency distribution has equation:

$$F(x) = (1/\sigma\sqrt{2\pi}) \int_{-\infty}^{x} \exp\left\{-\tfrac{1}{2}\left(\frac{x-\mu}{\sigma}\right)^2\right\} dx \qquad [6.8]$$

These equations are simplified by defining the standardized normal random variable $Z = (X - \mu)/\sigma$ which has a mean equal to zero and a standard deviation equal to one. $f(z)$ and $F(z)$ can be written as $\phi(z)$ and $\Phi(z)$. Then:

$$\phi(z) = \frac{1}{\sqrt{2\pi}} e^{-\frac{1}{2}z^2} \qquad [6.9]$$

$$\Phi(z) = (1/\sqrt{2\pi}) \int_{-\infty}^{z} e^{-\frac{1}{2}z^2} dz \qquad [6.10]$$

$\phi(z)$ and $\Phi(z)$ are useful because every normal frequency distribution reduces to this form. Graphically, the standardization of a normal distribution $f(x)$ consists of moving the origin to $x = \mu$ and changing the scale along the X-axis so that a distance of one standard deviation (σ) becomes equal to unity along the Z-axis.

Some values of $\Phi(z)$ are shown in Table XI.

From results derived in section 2.6, it follows that the inflection points of $\phi(z)$ fall at $z = 1$ and $z = -1$. Extensive tables for $\Phi(z)$ and $\phi(z)$ occur in most handbooks with statistical and mathematical tables. Because of symmetry with respect to the central peak, $\phi(z) = \phi(-z)$. This corresponds to $\Phi(z) = 1 - \Phi(-z)$.

If a set of observed values x_i all are representations of the same normally distributed random variable X, their histogram will resemble the normal curve. Since $\Phi(0) = 0.5$, approximately 50% of deviations from the mean will be positive and the other 50% negative. A value from a normal distribution has probability $\Phi(1) - \Phi(-1) = 0.841 - 0.159 = 0.682$ or 68% of deviating less than one standard deviation for the mean. This is because:

TABLE XI
Selected values of $\Phi(z)$

z	$\Phi(z)$
$-\infty$	0.000
-3	0.001
-2	0.023
-1	0.159
0	0.500
1	0.841
2	0.977
3	0.999
∞	1.000

$$P(Z \leqslant 1) = \Phi(1)\ ; \quad P(Z \leqslant -1) = \Phi(-1) \qquad [6.11]$$

Hence:

$$P(|Z| \leqslant 1) = \Phi(1) - \Phi(-1) = 0.682$$

and:

$$P\left(\left|\frac{X-\mu}{\sigma}\right| < 1\right) = 0.682$$

or:

$$P(\mu - \sigma < X < \mu + \sigma) = 0.682$$

Likewise:

$$P(\mu - 2\sigma < X < \mu + 2\sigma) = 0.977 - 0.023 = 0.954$$

and:

$$P(\mu - 3\sigma < X < \mu + 3\sigma) = 0.997$$

These are the probabilities that underlie the statements on errors made in section 2.8.

A useful procedure is to screen a set of values x_i for possibly anomalous data or outliers. This method consists of estimating mean and standard deviation and then standardizing the data by forming $z_i = (x_i - \bar{x})/s(x)$. Values z_i whose absolute value is greater than 3 may be anomalous since, on the average, they would occur with a relative frequency of only three per thousand.

Some other values of $\Phi(z)$ that are widely used are $\Phi(1.64) = 0.95$ and $\Phi(1.96) = 0.975$, from which it follows that:

$$P(|Z| < 1.96) = 0.95 \text{ or } P(|Z| > 1.96) = 0.05 \qquad [6.12]$$

and:

$$P(Z < 1.64) = \Phi(1.64) = 0.95$$

or:

$$P(Z > 1.64) = 0.05$$

6.7. Significance test and confidence interval

In the last part of the previous section, we have applied the concept of testing a hypothesis by checking that all observations come from the same normal population. We will now formalize this concept for the so-called z-test of significance. This provides a prototype for all other tests of significance. The z-test is more general than the procedure of screening data which forms a special case ($n = 1$). If n numbers are drawn from a normal population with mean μ with variance σ^2, then the sample mean $\bar{X}$ is normally distributed about μ with variance σ^2/n and standard deviation $\sigma/\sqrt{n}$. This statement is in agreement with [2.29]. It follows that:

$$Z = \frac{\bar{X} - \mu}{\sigma/\sqrt{n}}$$

is normally distributed about 0 with standard deviation 1.

Suppose that we have some outside information suggesting that $\mu = \mu_0$, where μ_0 is a constant. We can frame the *test- or null-hypothesis* H_0 that $\bar{x}$ comes from a normal distribution with mean μ_0 and standard deviation $\sigma/\sqrt{n}$. Momentarily, we assume that σ is given or can be estimated with sufficient precision. The test hypothesis $\mu = \mu_0$ is rejected if the statistical test suggests that $\mu \neq \mu_0$. It would be reasonable to reject H_0 if the quantity

$$|z| = \left|\frac{\bar{x} - \mu_0}{\sigma/\sqrt{n}}\right| \qquad [6.13]$$

is large, and to accept it when $|z|$ is small. If the hypothesis is rejected when $|z| > 1.96$, it will be rejected in 5% of the cases when it is true and accepted in 95% of such cases. The value $|z| = 1.96$ is called the *significance limit* for the test.

If $|(\bar{x} - \mu_0)/(\sigma/\sqrt{n})| < 1.96$ is chosen as the criterion for accepting the test hypothesis $\mu = \mu_0$, the interval for the sample mean:

$$\mu_0 - 1.96\,\sigma/\sqrt{n} < \bar{x} < \mu_0 + 1.96\,\sigma/\sqrt{n} \qquad [6.14]$$

is called the *region of acceptance.* If the hypothesis is true, we have:

$$P\left(\left|\frac{\bar{x} - \mu_0}{\sigma/\sqrt{n}}\right| < 1.96\right) = 0.95 \quad \text{and} \quad P\left(\left|\frac{\bar{x} - \mu_0}{\sigma/\sqrt{n}}\right| > 1.96\right) = 0.05$$

Suppose that μ_0 can not be specified. This does not prevent us from determining the 95% *confidence interval* of $\bar{x}$ with:

$$P(\bar{x} - 1.96\,\sigma/\sqrt{n} < \mu < \bar{x} + 1.96\,\sigma/\sqrt{n}) = 0.95 \qquad [6.15]$$

The first step in constructing a significance test is to choose a *level of significance* α. For example, in the z-test we used $\alpha = 0.05$. To some extent, the choice of α is arbitrary. In some problems of economic decision-making, it is possible to calculate an optimum value for α by using methods developed by Wald (1947). However, when α has been selected, the power of the test can be calculated.

Power of a significance test

Two types of erroneous assumption can be made by basing a conclusion on the result of a significance test. They are called Type 1 error (a) and Type 2 error (b), respectively.

(a) Rejection of the test hypothesis H_0 when it is true. The probability of this happening is given by the level of significance α. We have chosen $\alpha = 0.05$.

(b) Acceptance of H_0 when it is false. This probability is denoted as $\beta = 1 - \pi$ where π is the power of the test. Toleration of a large risk of rejecting a true H_0 results in a relatively large power for the test. On the other hand, a large value of α implies that we run a large risk of accepting a wrong hypothesis.

One- and two-tailed tests

The z-test as used so far is a two-tailed test. Suppose that additional outside information allows us to take $\mu < \mu_0$ or $\mu > \mu_0$ as the null-hypothesis instead of $\mu = \mu_0$. This results in one-tailed tests because the null-hypothesis would automatically be accepted when $\bar{x} < \mu_0$ or $\bar{x} > \mu_0$, respectively. For $\mu < \mu_0$, the significance limit is $z = 1.64$ since $\Phi(1.64) = 95\%$; for $\mu > \mu_0$ we can use $z = -1.64$.

6.8. Method of moments

Suppose that a continuous random variable X has a frequency distribution $f(x)$. We wish to determine statistical parameters such as mean and variance for continuous frequency distribution functions. This can be done by the concept of "expected value" or "expectation" written as E.

The r-th moment of X is defined as:

$$\mu_r' = E(X^r) = \int_{-\infty}^{\infty} x^r f(x)\,\mathrm{d}x \qquad [6.16]$$

The definite integral is determined for all possible values of x. For convenience, we will delete the limits $-\infty$ and ∞. The zeroth moment $\mu'_0 = 1$ represents total area $F(\infty) = 1$ in [6.5] and the first moment μ'_1 represents the mean μ with:

$$\mu = \mu'_1 = EX = \int x f(x)\, dx \qquad [6.17]$$

Although [6.16] defines moments for a continuous random variable, it is also applicable to discrete random variables. In that case, we can use sums instead of integrals. Eq. [6.17] becomes $\mu = EX = \Sigma\, xP(X = x)$. For example, the expected value for a Bernoulli trial is $EX = 0 \times \frac{1}{2} + 1 \times \frac{1}{2} = \frac{1}{2}$.

Moments about the mean

Rather than operating on the moments μ'_r for a random variable, it is easier to use so-called moments about the mean, μ_r, if r is equal to 2 or larger:

$$\mu_r = E[(X - \mu)^r] = \int (x - \mu)^r f(x)\, dx \qquad [6.18]$$

The first moment about the mean μ_1 is equal to zero. The second moment about the mean μ_2 is the variance σ^2 with:

$$\sigma^2 = \mu_2 = E(X - \mu)^2 = \int (x - \mu)^2 f(x)\, dx \qquad [6.19]$$

This can be expressed in terms of $\mu = \mu'_1$ and μ'_2 by rewriting [6.19] as follows:

$$\sigma^2 = \int (x^2 - 2x\mu + \mu^2) f(x)\, dx$$

$$= \int x^2 f(x)\, dx - 2\mu \int x f(x)\, dx + \mu^2 \int f(x)\, dx$$

$$= \mu'_2 - 2\mu \cdot \mu + \mu^2, \text{ or } \sigma^2 = \mu'_2 - \mu^2 \qquad [6.20]$$

The positive square root of the variance is the standard deviation σ. Both variance σ^2 and standard deviation σ provide measures of the dispersion of values about their mean μ. If [6.17] and [6.19] are applied to [6.7], the results are the identities $E(X) = \mu$ and $\sigma^2(X) = \sigma^2$ for the normal frequency distribution with [6.7]. The calculation of moments is facilitated by the use of moment generating functions to be introduced in the next chapter.

As before, summation signs can be used for discrete random variables. The variance of the Bernoulli trial satisfies $\sigma^2 = \Sigma\, (x - \mu)^2 P(x)$ or $\sigma^2 = (1 - \frac{1}{2})^2 \times \frac{1}{2} + (-\frac{1}{2})^2 \times \frac{1}{2} = \frac{1}{4}$.

By utilizing properties of mean and variance to be discussed in section 6.9 it is shown readily that, for the binomial distribution, $\mu = np$ and $\sigma^2 = npq$. This result is used in the following example.

Petrographic modal analysis

Suppose that 12% by volume of a rock consists of a given mineral "A". The method of point-counting is applied to a thin section for this rock. Suppose that 100 points are counted. From $p = 0.12$ and $n = 100$, it follows that $q = 0.88$ and:

$$\mu = np = 12$$

$$\sigma = \sqrt{npq} = \sqrt{12 \times 0.88} = 3.25$$

If the experiment of counting 100 points would be repeated many times, the number of times (K) that the mineral "A" is counted would describe a binomial distribution with mean 12 and standard deviation 3.25. According to the so-called central limit theorem, a binomial distribution approaches a normal distribution if n increases. Hence, we can say for a single experiment that:

$$P(\mu - 1.96\sigma < K < \mu + 1.96\sigma) = 0.95$$

or:

$$P(5.6 < K < 18.4) = 95\%$$

The resulting value of K is between 5.6 and 18.4 with a probability of 95%.

This precision can be increased by counting more points. For example, if $n = 1000$ then $\mu = 120$ and $\sigma = 10.3$ and:

$$P(99.8 < K < 140.2) = 95\% \qquad [6.21]$$

This is sometimes written as $k = 120 \pm 20.2$. In order to compare the experiment of counting 1000 points to that for 100 points, the result must be divided by 10, giving $k' = 12 \pm 2.02$. This number represents the estimate of volume percent for the mineral "A". It is $3.25/(10.3/10) = 3.2$ times as precise as the first estimate for 100 points only. It is noted that this increase in precision is in agreement with [2.29] which in terms of population variances reads $\sigma^2(\bar{x}) = \sigma^2(x)/n$. Since there are ten times as many counted values in the second experiment, the result is $\sqrt{10}$ or 3.2 times more precise.

The method of modal analysis has been discussed in more detail by Chayes (1956).

Systems of frequency-distribution functions

Several systems have been developed for continuous functions $f(x)$ that may fit histograms for observed data. Systems as developed by Pearson (1895), Kapteyn (1903), Burr (1942) or Johnson (1949), are sufficiently flexible to cover many of the frequency curves that have been observed for geological data.

A comprehensive treatment of these systems can be found in Kendall and Stuart (1958).

The Pearson-curve system (also see Elderton and Johnson, 1969) is based on the so-called hypergeometric distribution which arises as follows.

Suppose that we randomly designate n cells on a mosaic that contains N cells (black and white). These n cells can be used with and without reuse. The use of one cell ($n = 1$) is a Bernoulli trial. If this experiment is repeated n times, whereby the cell is reused each time, the number of black cells (k) among the n cells satisfies the binomial distribution. However, if the cells are not reused on the mosaic, and a cell can only be selected once, the theory has to be modified. The sampling of n cells without reuse is the same as designating n cells at once.

A note on randomness is in order here. We must always make a careful distinction between randomness in nature and random sampling. When the mosaic is a regular grid where the black cells occur at random, any method of sampling will produce a random sample. On the other hand, if the black cells form a regular pattern in the mosaic, then only random sampling (designating cells at random) will produce a random sample.

It can be shown that the number of black cells (k) per n cells now satisfies the discontinuous distribution:

$$P\left(\sum_{i=0}^{n} X_i = k\right) = \frac{1}{N^n}\binom{n}{k}(Np)^k\,(Nq)^{n-k} \qquad [6.22]$$

where $k = 0, 1, 2, \ldots, n; Np \geqslant n \leqslant Nq$. This form is called the hypergeometric distribution. There is a strong analogy with [6.2] for the binomial form. In fact, if N approaches infinity, [6.22] approaches the binomial form. This implies that designating n cells on the mosaic with reuse is identical to using n cells without reuse from an infinitely large mosaic. Of course, the ratio of black to white cells then must be the same in finite and infinite population.

It can be argued that, in geology, we usually are sampling from infinite populations. For example, the size of a specimen that is chemically analyzed for the elements contained in it generally has a size that is small when compared to the entire volume of the rock unit from which it was taken. Hazen (1967) has suggested that many frequency distributions for element concentration values from mineral deposits where the ore occurs in distinct grains are binomial.

The more abundant minerals in a rock may be binomially distributed (Krumbein, 1955; Chayes, 1956; Griffiths, 1960). On the other hand, there are some situations in the earth sciences where a model based on a finite instead of an infinite population is justified. Becker (1964) has expanded the theory of the hypergeometric distribution and applied it to mineral deposits, mainly to broken ore sampling.

Skewness and kurtosis

These two quantities may be written as γ_1 (skewness) and γ_2 (kurtosis). They are based on moments about the mean with:

$$\gamma_1 = \mu_3 \mu_2^{-3/2} \; ; \qquad \gamma_2 = \mu_4 \mu_2^{-2}$$

For a normally distributed random variable, the skewness is equal to zero and the kurtosis equal to 3. This property can be used for a normality test consisting of estimating the moments about the mean for a sample by:

$$m_r = (1/n) \sum (x - \bar{x})^r \qquad [6.23]$$

with $r = 2$, 3, and 4. This equation is an application of [6.18] to the n values of a sample. As will be discussed in the next section, the m_r are not unbiased estimates of the μ_r in that they converge to other values if r increases. However, for this normality test we can use the biased estimates m_r. We calculate skewness: $\sqrt{b_1} = m_3/m_2^{\;3/2}$ and kurtosis: $b_2 = m_4/m_2^2$.

If one or both of the values $\sqrt{b_1}$ and b_2 are significantly different from 0 and 3, respectively, the hypothesis of normality is rejected. Tables with the confidence limits for this test for levels of significance $\alpha = 0.05$ and $\alpha = 0.01$ (one-tailed test) are given by Pearson and Hartley (1958).

Harris (1965) has used skewness and kurtosis as follows. His objective was to normalize a number of geological variables because of subsequent use of statistical tests based on the assumption of normality. For this reason, each variable was subjected to a large number of transformations. Skewness and kurtosis of transformed data were calculated in each situation. Transformed values yielding the absolute values for skewness and kurtosis closest to 0 and 3, respectively, were accepted by Harris as coming from approximately normal populations and used in further calculations.

6.9. Properties of expectation (E) and variance (σ^2)

If X is a random variable and c a constant, it follows from [6.17] that for the expectation:

$$E(X + c) = EX + c \qquad [6.24]$$

$$E(cX) = cEX \qquad [6.25]$$

When X and Y are two random variables, their joint frequency distribution can be defined as a two-dimensional function $f(x, y)$. If [6.17] is extended to this bivariate situation, then:

$$E(X + Y) = \iint (x + y) f(x, y) \, \mathrm{d}x \, \mathrm{d}y \qquad [6.26]$$

It will be shown that:

$$E(X + Y) = EX + EY \qquad [6.27]$$

regardless of the properties of $f(x, y)$ which may involve statistical dependence of X and Y.

In [6.26], the infinitesimal quantity $f(x, y)\,dx\,dy$ expresses the probability that $(X + Y)$ assumes value $(x + y)$ in an area $dxdy$. Suppose that $f(x, y)$ is integrated over all possible values of y for a strip of width dx in the X-direction. This gives:

$$\left[\int_{-\infty}^{\infty} f(x, y)\,dy\right]dx = f(x)\,dx$$

which denotes the probability that $X = x$ for interval dx or $P(x < X < x + dx)$. If X is considered by itself, the value assumed by Y is immaterial. Eq. [6.17] now can be written as:

$$E(X) = \int\int x\,f(x + y)\,dxdy$$

Since $E(Y)$ satisfies a similar expression, [6.26] becomes:

$$E(X + Y) = EX + EY \qquad [6.27, \text{repeated}]$$

It follows immediately that, if $X_1, X_2, \ldots, X_n$ are n random variables, we have:

$$E(\Sigma X_i) = \Sigma EX_i \qquad [6.28]$$

A corresponding general theorem for the expectation of products does not exist. If X and Y are two random variables, then:

$$E(XY) = \int\int xy\,f(x, y)\,dxdy \qquad [6.29]$$

Let the double integral be approximated by a double sum, then:

$$E(XY) = \sum_j \sum_k x_j y_k\,P(x_j, y_k) \qquad [6.30]$$

In section 6.3, mutual independence of two random variables was defined as:

$$P(x_j, y_k) = P(x_j)\,P(y_k) \qquad [6.31]$$

Substituting [6.31] into [6.30] gives:

$$E(XY) = \sum_j \sum_k [x_j P(x_j)]\,[y_k P(y_k)]$$

Hence:

$$E(XY) = (EX)\,(EY) \qquad [6.32]$$

but for independent random variables only.

Properties of the variance

From [6.19], it follows that:

$$\sigma^2(X + c) = \sigma^2(X) \qquad [6.33]$$

$$\sigma^2(cX) = c^2\sigma^2(X) \qquad [6.34]$$

The variance of the sum of two independent random variables X and Y satisfies:

$$\sigma^2(X + Y) = \sigma^2(X) + \sigma^2(Y) \qquad [6.35]$$

This can be proved from eq. [2.19] and the properties of the expectation.

Estimation of parameters from the sample

The mean $\mu = E(X)$ and standard deviation $\sigma = \sqrt{\sigma^2(X)}$ are parameters with specific numerical values for a given population. Statistical theory is concerned with methods of estimating parameters such as μ and σ from samples of observed data.

The sample estimate of a parameter is called unbiased if its expectation is equal to the value of this parameter for the population.

By using [6.24] through [6.35], we will prove several well-known formulas for sample means and variances.

Sample mean

The sample mean $\bar{x} = (1/n)(x_1 + x_2 + \ldots + x_n)$ is an unbiased estimate of the population mean $\mu = E\,\bar{X}$. The proof is simple: If $EX_1 = EX_2 = \ldots = EX$, $E(\bar{X}) = nEX/n = EX$, or:

$$E(\bar{X}) = \mu \qquad [6.36]$$

Variance of sample mean

By extending [6.34] we have:

$$\sigma^2(\bar{X}) = (1/n^2)\sigma^2\,(X_1 + X_2 + \ldots + X_n)$$

If $X_1, X_2, \ldots, X_n$ are independent random variables from the same population with variance $\sigma^2(X)$, then:

$$\sigma^2(\bar{X}) = (1/n^2)n\;\sigma^2(X)$$

or:

$$\sigma^2(\bar{X}) = \sigma^2(X)/n \qquad [6.37]$$

This result was derived in section 2.9 by using the rule for the propagation of variance.

Sample variance

From [6.19] we know:

$$\sigma^2(X) = E(X - EX)^2$$

If EX is replaced by the mean $\bar{x}$ for a sample of n data, the sample variance could be estimated by:

$$s^2(x - \bar{x}) = (1/n)\sum_{i=1}^{n}(x_i - \bar{x})^2 \qquad [6.38]$$

This gives an estimate of $\sigma^2(X - \bar{X})$. It will be shown that $s^2(x - \bar{x})$ provides a biased estimate of $\sigma^2(X)$.

We derive:

$$\begin{aligned}\sigma^2(X) &= \sigma^2(X - EX)\\ &= \sigma^2(X - \bar{X} + \bar{X} - EX)\\ &= \sigma^2(X - \bar{X}) + \sigma^2(\bar{X} - EX)\\ &= \sigma^2(X - \bar{X}) + \sigma^2(\bar{X})\\ &= \sigma^2(X - \bar{X}) + (1/n)\sigma^2(X)\end{aligned}$$

From this, we obtain:

$$\sigma^2(X) = [n/(n-1)]\;\sigma^2(X - \bar{X}) \qquad [6.39]$$

In terms of sample statistics, this can be written as:

$$s^2(x) = [n/(n-1)]\, s^2(x-\bar{x}) \tag{6.40}$$

The unbiased estimate $s^2(x)$ satisfies:

$$E[s^2(x)] = \sigma^2(X) \tag{6.41}$$

The factor $n/(n - 1)$ is also known as Bessel's correction. It results in [2.24], which was used before, or:

$$s^2(x) = \frac{\sum_{i=1}^{n} (x_i - \bar{x})^2}{n-1} \tag{6.42}$$

Duplicate measurements

A set of duplicate X-ray measurements for 18 specimens is shown in Table XII. This experiment also has been discussed in section 2.8. The original measurements were for angle 2θ which has been converted to d_{174} in Table XII by using Bragg's equation. The

TABLE XII
Duplicate X-ray measurements for olivines by two observers. (From Jambor and Smith, 1964, table II.)

k	$2\theta_{174}$ *Fe-K*α_1 *radiation*		d_{174}		$\lvert\Delta_k\rvert$
	(1) Jambor	*(2) Delabio*	*(1)*	*(2)*	
1	143.62°	143.61°	1.01880	1.01883	0.00003
2	142.87	142.91	1.02102	1.02090	0.00012
3	142.86	142.81	1.02105	1.02120	0.00015
4	142.77	142.84	1.02132	1.02111	0.00021
5	142.86	142.80	1.02105	1.02123	0.00018
6	142.70	142.70	–	–	0.0
7	142.01	142.03	1.02363	1.02368	0.00005
8	141.89	141.90	1.02400	1.02397	0.00003
9	140.95	140.90	1.02694	1.02710	0.00016
10	140.40	140.51	1.02871	1.02835	0.00036
11	140.00	139.96	1.03001	1.03014	0.00013
12	139.85	140.01	1.03050	1.02997	0.00053
13	140.00	140.04	1.03001	1.02988	0.00013
14	138.64	138.64	–	–	0.0
15	136.77	136.85	1.04110	1.04081	0.00029
16	136.81	136.86	1.04096	1.04078	0.00018
17	136.38	136.44	1.04253	1.04230	0.00023
18	136.53	136.60	1.04197	1.04171	0.00026

problem to be solved is the estimation of the measurement error. Suppose that the sample variance for the k-th pair of measurements is written as s_k^2.

From [6.42], it follows that:

$$s_k^2 = \frac{(\frac{1}{2}\Delta_k)^2 + (\frac{1}{2}\Delta_k)^2}{2-1} = \frac{1}{2}\Delta_k^2 \qquad [6.43]$$

where Δ_k denotes the difference between the two values.

The overall variance s^2 can be obtained by averaging the 18 sample variances s_k^2, or $s^2 = 2.29 \cdot 10^{-8}$, since:

$$s^2 = (1/m) \sum_{k=1}^{m} s_k^2$$

Suppose that Bessel's correction had not been applied. In that situation, the result would have been a biased estimate of the variance being exactly twice as small as our result $s^2 = 2.29 \cdot 10^{-8}$. This illustrates that the factor $n/(n-1)$ should not be neglected in some situations. The reader may verify this result for himself.

6.10. Continuous frequency distributions derived from the normal distribution

Most methods of statistical inference are based on the assumption that the underlying populations have normal frequency distributions. Another method of approach is distribution-free statistics where no assumptions are made on the nature of the distribution of the population.

Normality of a population should never be taken for granted. If a histogram for the sample indicates that the population is not normal, it may be possible to achieve approximate normality by applying a transformation. For example, if the histogram is positively skew in that it has a long tail extending toward high values, it can be useful to replace all values by their logarithms.

χ^2-distribution

If the mean $\mu = E(X)$ of a normal distribution is known, the sample variance satisfies:

$$s^2 = (1/n) \sum_{i=1}^{n} (x_i - \mu)^2 = (\sigma^2/n) \sum \left(\frac{x_i - \mu}{\sigma}\right)^2$$

$$s^2 = (\sigma^2/n) \sum_{i=1}^{n} z_i^2 \qquad [6.44]$$

where $z_i = (x_i - \mu)/\sigma$ is standardized x_i. The random variable $\chi^2(f)$ is defined as:

$$\chi^2(f) = \sum_{i=1}^{f} Z_i^2 \qquad [6.45]$$

The chi-square distribution for f degrees of freedom is the distribution of the sum of squares of f standardized normal variables Z_i.

A comparison of [6.44] with [6.45] shows that the sample variance satisfies:

$$s^2 = \frac{\sigma^2 \chi^2(f)}{f} \qquad [6.46]$$

with $f = n$ if the mean μ is known. If the mean is not known, we have [6.42] which becomes:

$$s^2 = [\sigma^2/(n-1)] \sum \left(\frac{x_i - \bar{x}}{\sigma}\right)^2$$

or.

$$s^2 = [\sigma^2/(n-1)]\ \chi^2(n-1)$$

This also is a representation of [6.46] with $f = (n-1)$.

Since σ^2 is a constant, the sample variance s^2 is distributed as χ^2 with $f = (n-1)$ degrees of freedom.

Addition theorem for χ^2

If $f = p + q$, then:

$$\chi^2(f) = \chi^2(p) + \chi^2(q) \qquad [6.47]$$

This is called the addition theorem for χ^2. It follows directly from the definition of χ^2 in [6.45].

The expectation $E\chi^2(1)$ of χ^2 with one degree of freedom satisfies $E\chi^2(1) = E(Z^2)$. A standard random variable has $\sigma^2(Z) = 1$. It follows that:

$$E\chi^2(1) = 1 \qquad [6.48]$$

because:

$$\sigma^2(Z) = E(Z^2) - (EZ)^2 = E(Z^2)$$

From [6.28] then follows:

$$E\chi^2(f) = f \tag{6.49}$$

Taking the expectation of both sides of [6.46] gives [6.41] or:

$$E(s^2) = \sigma^2$$

A set of sample variances can be pooled by averaging them. If they are based on m samples of unequal size, a weighted average variance s^2 may be calculated as:

$$s^2 = \frac{\sum_{k=1}^{m} f_k s_k^2}{\sum_{k=1}^{m} f_k} \tag{6.50}$$

where $f_k (k = 1, 2, \ldots, m)$ represents number of degrees of freedom in the k-th sample and s_k^2 the corresponding sample variance.

Eq. [6.50] is based on the addition theorem for χ^2 since:

$$s^2 \sum_{k=1}^{m} f_k = \sum_{k=1}^{m} f_k s_k^2$$

also can be written as:

$$\sigma^2 \chi^2(\Sigma f_k) = \Sigma\, \sigma^2 \chi^2(f_k)$$

which reduces to [6.47] in extended form or:

$$\chi^2(\Sigma f_k) = \chi^2(f_1) + \chi^2(f_2) + \ldots + \chi^2(f_m) \tag{6.51}$$

Example. If [6.50] is used to calculate a pooled variance s_m^2 for the data shown in Table VIII, the result is:

$$s_m^2 = (1/30)\,(8 \times 0.191 + 8 \times 0.598 + 8 \times 0.556 + 3 \times 1.063 + 3 \times 0.056) = 0.4706$$

By taking the square root it follows that for the measurement error $s_m = 0.69$.

F-distribution

Suppose that the variances are calculated for two samples from a normal distribution with sizes n_1 and n_2, respectively. We have:

$$s_1^2 = \frac{\sigma^2 \chi_1^2}{f_1}; \quad s_2^2 = \frac{\sigma^2 \chi_2^2}{f_2}$$

where $f_1 = n_1 - 1$ and $f_2 = n_2 - 1$.

The ratio F of these two variances satisfies:

$$F = s_1^2/s_2^2 = \frac{\chi_1^2/f_1}{\chi_2^2/f_2} \quad 0 \leqslant F < \infty \qquad [6.52]$$

The ratio F does not contain the population variance σ^2. It depends only on the numbers of degrees of freedom f_1 and f_2. This distribution can be used to test two variances for equality.

Another application is analysis of variance where means from different samples are tested for equality. Suppose that m samples from the same normal population contain n data per sample. The sample means then satisfy:

$$\sigma^2(\bar{X}) = \sigma^2(X)/n$$

By calculating a variance $s^2(\bar{x})$ from the m means and multiplying this estimate by n, we obtain an estimate of the variance $\sigma^2(X)$. This so-called variance between groups is written as:

$$s_1^2 = ns^2(\bar{x}) \qquad [6.53]$$

By application of [6.50] we obtain a second estimate of the variance:

$$s_2^2 = (1/m) \sum_{k=1}^{m} s_k^2 \qquad [6.54]$$

This is the variance within groups. The ratio s_1^2/s_2^2 is distributed according to $F(m-1, mn-m)$ or F with $(m-1)$ and $(mn-m)$ degrees of freedom.

The first estimate of variance has $(m-1)$ degrees of freedom because it is based on a set of m values $\bar{x}_k$. If the population means for the different samples are not equal to one another, s_1^2 is likely to be too large. The result is an one-tailed F-test. If the F-ratio as calculated for an experiment of this type is significantly larger than one, the test hypothesis that the samples all originate from the same population is rejected.

t-distribution

Since the mean of n data from a normal distribution with mean μ and variance σ^2 i normally distributed with mean μ and variance σ^2/n, we can write:

$$Z = \frac{\bar{X} - \mu}{\sigma/\sqrt{n}} \quad [6.55]$$

which is normally distributed with zero mean and unit variance. This expression has been used as the foundation of the z-test of significance in section 6.7. However, usually σ^2 is unknown and must be replaced by the sample variance s^2. This leads to:

$$t = \frac{\bar{X} - \mu}{s/\sqrt{n}} \quad [6.56]$$

From [6.55] and [6.56], it follows that $t = Z\,\sigma/s$, or:

$$t = Z/\sqrt{\chi^2/f} \quad [6.57]$$

Consequently, the random variable is independent of μ and σ. It depends only on the number of degrees of freedom f.

If both sides of [6.57] are squared, it follows from the definitions of χ^2 and F that $t^2(f) = F(1, f)$, or:

$$t(f) = \pm\sqrt{F(1, f)} \quad [6.58]$$

Like the z-distribution, t-distributions are symmetric about the zero point.

The use of the t-distribution instead of the z-distribution is preferable if the number of data is small. It is also known as Student's t-distribution.

6.11. Age determinations for structural provinces in the Canadian shield

In the final sections of this chapter, we will apply several methods discussed in the previous ten sections. As a practical example, we will discuss the ages of rocks in the Canadian shield. The classification and terminology for this example are according to Stockwell (1964).

Histograms for potassium/argon ages as determined for minerals in rocks from the structural provinces are shown in Fig. 21. Our considerations will be restricted mainly to the Superior, Churchill, and Grenville provinces. Most data are for these three provinces where the structural deformation mainly took place during the Kenoran, Hudsonian and Grenvillian orogenies, respectively.

Stockwell (1964) has argued as follows. The average or mean age can be calculated for ages belonging to any one of the major peaks in Fig. 21 which are for the larger provinces. For example, the Churchill province has 141 ages in the range 1450–2010 millions of

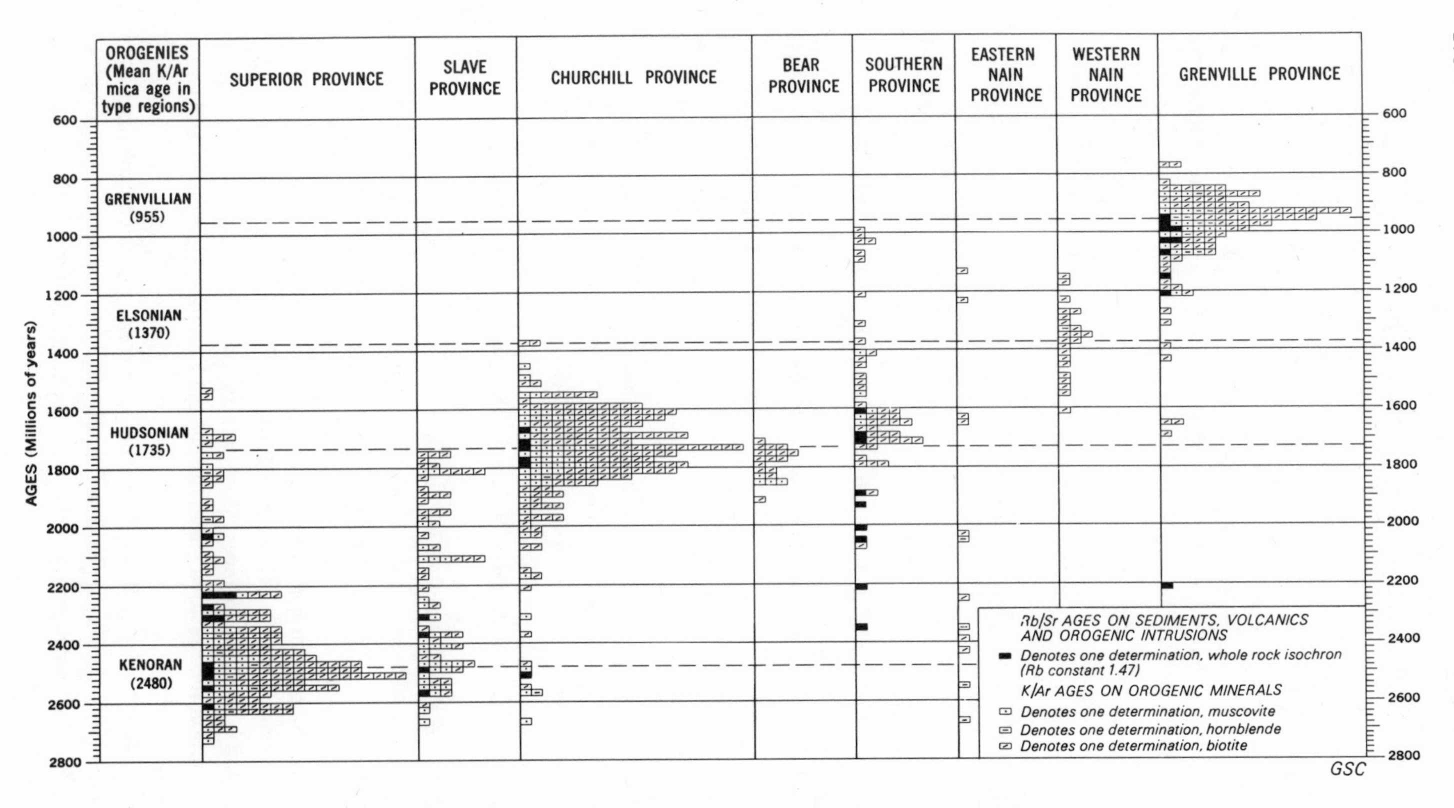

Fig.21. Histogram of K/Ar and Rb/Sr dates from structural provinces and orogenies in the Canadian shield. (From Douglas, 1970; fig. IV-3 by C.H. Stockwell.)

years. The mean age (M) amounts to 1735 millions of years (m.y.). The standard deviation is 95 m.y. Stockwell assumed that the Hudsonian orogeny, by approximation, took place between MM = 1735 – 95 = 1640 m.y. and MP = 1735 + 95 = 1830 m.y. age. The expressions MM and MP are convenient abbreviations for Mean Minus (one standard deviation) and Mean Plus (one standard deviation), respectively. These limits include about two thirds of all ages for the Churchill province.

Stockwell has argued that isotopic dates are subject to various sources of variation. Usually, the age changes with the analytical error, the material dated, the method used, and with changing geologic conditions in the course of time that may have altered the normal ratio of decay products.

For statistical analysis, the data were confined to one type of material (micas) and one age method (potassium/argon method). This mainly eliminates two sources of variability. Of course, one must distinguish between random and systematic variation. By confining the data in this manner, systematic discrepancies between measurements and true ages can be introduced. The relative age will not be affected by a bias of this type if the bias is important.

The remaining sources of variability are analytical error, time span of the orogeny, and geological conditions affecting the normal ratio of decay products. The latter factor would consist of an anomalous loss or gain in argon.

The analytical error is probably random and normally distributed. It reflects the joint influence of several separate factors as discussed by Wanless et al. (1965). It differs from determination to determination. For example, in the 1964 age report of the Geological Survey of Canada (Wanless et al., 1965) most determinations on micas from rocks from the Churchill province have 95% confidence limits of ± 60 or ± 70 m.y. The precision of geochronological methods decreases with time.

Our statistical analysis will include many pre-1964 ages and we will accept for all ages the following confidence limits suggested for pre-1964 ages by Wanless.

Age	*95-% conf. limits*
500 m.y.	± 35 m.y.
1000	± 60
2500	± 125

From these numbers, we can derive approximately standard deviations s_m for orogenies. For example, the Hudsonian orogeny has M = 1735 m.y. Interpolation between ± 60 for 1000 m.y. and ± 125 for 2500 m.y. gives ± 87 as 95% confidence limits for 1735 m.y. This value must then be divided by 2 to give $s_m \approx 44$ m.y. This final estimate is crude of necessity but can be used for the exploratory model in the next section. Consequently, one source of variability (analytical error) could be determined in some detail, and, to some extent, be eliminated from the statistical results. However, it is not feasible to distinguish from each other the other sources of variability such as time span of the orogeny and anomalous changes in argon content of micas.

The data to be used for statistical analysis have been published in age reports through 1968. Ages for the period 1964–68 were added by Dr. C.H. Stockwell to the ages used for statistical analysis in 1964. Our statistical analysis is based on this larger collection of data (C.H. Stockwell, personal communication, 1970). A distinction is made between muscovite and biotite ages and determinations on other micaceous minerals and hornblende are not considered.

In recent years, many geologists have adopted the opinion that the structural provinces in the Canadian shield, which give the different age determinations, were caused by a remobilization of older material with a simultaneous overprinting of earlier radiometric ages (cf. Wynne-Edwards and Hason, 1970). In fact, the K/Ar time clock may have been set to zero when temperature and pressure conditions exceeded a critical point at which all argon escaped from the micas without simultaneous deformation. This has replaced the older opinion that the Canadian shield has grown outwards from its Kenoran core by accretion.

An inspection of Fig.21 indicates that there occur some older rocks in the Grenville province which is a well-defined structural unit. The histogram is positively skewed in that it has a relatively long tail towards older dates. Likewise, the histogram for the Superior province is negatively skew with a longer tail pointing towards younger ages. The anomalous ages, which do not fall in the main peaks for these two orogenies, mainly come from geographic locations restricted to relatively narrow zones along the boundaries between provinces. For example, near the so-called Grenville front at Chibougamau, Quebec, it was found that muscovite yields discrepantly young ages for at least 2.5 miles (4 km) from the front into the Superior province (M = 2480 m.y.). Stockwell's interpretation of this phenomenon was that it is caused by a loss in argon due to the Grenville orogeny (M = 960 m.y.). A strongly anomalous result was found in a single specimen from this zone where muscovite gave 1630 m.y., biotite 3300 m.y. and whole rock 1673 m.y. This would indicate that the biotite was enriched in argon whereas the muscovite lost argon.

It is possible to distinguish in the Grenville province zones which, in addition to having been involved in the Grenville orogeny, were involved in older orogenies. Stratigraphic formations can be correlated across the Grenville front from older into younger rocks and vice versa (Stockwell, 1964). It seems that the Canadian shield existed as a continental plate from a date preceding the Kenoran orogeny (M = 2480 m.y.).

These considerations indicate that it is difficult to decide for an age in a given province that falls outside the main age peak for this province, whether it should be included in the sample from which statistical parameters are calculated. For ages until 1968, the following results were obtained:

Province	*Range (m.y.)*	*No. data*	*M (m.y.)*	*S.D. (m.y.)*	s_m *(m.y.)*	s_r *(m.y.)*
Grenville	770 – 1210	94	959	90	27	86
Churchill	1490 – 2070	204	1736	116	44	107
Superior	2190 – 2730	159	2477	118	61	101

In this table, *M* represents mean, *S.D.* is standard deviation, s_m is approximate analytical error, and $s_r = \sqrt{s^2 - s_m^2}$ is the approximate standard deviation for a province after elimination of the analytical error (cf. [6.35]).

The age peaks suggest a cyclical recurrence pattern of orogenies. Several explanations for this phenomenon have been suggested in recent literature (e.g. Runcorn, 1965). Zwart (1969) has proposed that a distinction can be made between two types of orogeny on earth. The orogenies represented in the three largest provinces of the Canadian shield would be of the so-called Hercyno-type. The mean ages can be listed as follows:

Orogeny	*Age (m.y.)*	*Difference (m.y.)*
Hercynian	300	
Grenvillian	959	659
Hudsonian	1736	777
Kenoran	2477	741
Average difference		729

A cyclicity with an average period of some 730 m.y. is indicated.

The other type of orogeny distinguished by Zwart is the Alpino-type whose peak periods would occur more recently and alternate with those of the last two Hercyno-type orogenies.

Alpino-type orogenic belts are not present on the Canadian shield. However, Stockwell (1964) has proposed the Elsonian orogeny on the basis of ages from the western Nain province in Labrador. In 1968, 22 age determinations from the western Nain province indicated a mean age of 1372 m.y. with standard deviation equal to 120 m.y. The Elsonian orogeny (see Fig. 21) would fit in with the cyclicity of the younger Alpino-type orogenies.

The periodicity (or quasi-periodicity) of orogenic activity is well-known to geologists. Umbgrove (1947) has used the expression "pulse of the earth" for this phenomenon.

Van Bemmelen (1954) has pointed out that orogenesis or mountain building seems to be a rhythmic process. A period of orogeny involves the rapid release of considerable energy that had accumulated at greater depths during longer periods of time preceding the orogenic event.

With regard to more recent crustal deformations, Van Bemmelen (1954) presented convincing evidence that orogenic phases of the Sunda Mountain and Banda systems in Indonesia occurred rhythmically. Seven or eight such phases can be distinguished for the past 250 m.y. at an average interval of about 30 m.y. During the time span, the main centers of deformation moved in steps across the surface of the earth.

6.12. A simple mathematical model to estimate time span

Some of the age peaks of Fig. 21 show resemblance to the Gaussian curve. This justifies the usage of statistical methods based on the normality assumption as will be done in the next section.

However, in view of what was discussed in section 6.11, we must be careful in our definition of the random variables to be used. For example, the Grenville has $M = 959$ m.y. and a standard deviation equal to 90 m.y. We might be tempted to evaluate the precision of this mean by using the formula:

$$s(\bar{x}) = s(x)/\sqrt{n}$$

which would give $s(\bar{x}) = 90/\sqrt{94} = 9.3$ m.y. Approximate 95% confidence limits on $M = 959$ m.y. would satisfy $\pm$ 1.96 $s(\bar{x})$ or $\pm$ 18 m.y. However, this procedure applies only if the orogeny took place within a short time interval and simultaneously in the entire province.

It is more reasonable to assume that the orogeny took place during a longer span of time and affected rocks in different parts of the province at different times.

It seems to be difficult to frame a mathematical model for the orogenies for lack of sufficient data. We will work out a simple model.

The Grenville province is considerably smaller in area than the other two provinces. Its standard deviation s_r also is smaller. This fact is illustrated in the following table where the unit of area is the Grenville province. Of course, more precise data are available for areas of provinces but not all parts of the provinces were sampled. We do not really know how representative single ages are for larger domains. The values, therefore, are approximate only.

	s_r *(m.y.)*	*x (= area)*
Grenville	86	1
Churchill	107	3
Superior	101	2.5

There are indications that smaller areas within a province tend to have standard deviations smaller than those for the entire province.

Let us assume that, in addition to analytical error whose effect has been eliminated by approximation, there are two main sources of variation for the ages in a province : (1) time span of orogeny; and (2) irregular, unpredictable changes in argon content of micas. If these factors are independent, we have:

$$\sigma_r^2 = \sigma_t^2 + \sigma_a^2 \qquad [6.59]$$

where σ_t^2 denotes variance related to time span and σ_a^2 variance for anomalous argon content.

It is further assumed that the pattern of orogenic activity for a large area during time can be approximated by a rectangular wave. The base of this wave is as wide as the total time span for the orogeny. If such a form is sampled at random, the result is a random variable with so-called uniform frequency distribution.

When t denotes the variable time, the uniform distribution satisfies:

$$\left.\begin{aligned} f(t) &= \frac{1}{2a}, \text{ if } t_1 < t < t_2 \\ f(t) &= 0, \text{ otherwise} \end{aligned}\right| \qquad [6.60]$$

In this expression, t_1 and t_2 represent beginning and end of the orogeny, and $2a = t_2 - t_1$ is total time span.

The mean age of the orogeny occurs at $M = (t_1 + t_2)/2$. We may choose $t = 0$ at M. Then the variance can be calculated by using [6.19], or:

$$\sigma_t^2 = E(t^2) = \int_{-a}^{+a} t^2 f(t)\,dt = \frac{1}{2a}\int_{-a}^{+a} t^2\,dt = \frac{1}{2a}[t^3/3]_{-a}^{+a} = a^2/3$$

Hence:

$$\sigma_t^2 = a^2/3 \;;\; \sigma_t = a/\sqrt{3} \qquad [6.61]$$

We assume further that the larger the area (x) of a province, the larger the time span of the orogeny that has been sampled. When a is taken to be proportional to x, we can write:

$$\sigma_t^2 = a^2/3 = B_1 x^2 \qquad [6.62]$$

where B_1 is an unknown constant. With $\sigma_a^2 = A_1$, [6.59] becomes:

$$\sigma_r^2 = B_1 x^2 + A_1 \qquad [6.63]$$

This expression contains two variables x and σ_r^2, and two unknown coefficients A_1 and B_1. The latter can be solved from the three pairs of values for s_r and x which are given. Writing $y = s_r^2 \cdot 10^{-4}$, the data can be tabulated in the form:

	y	x^2
Grenville	0.740	1
Churchill	1.145	9
Superior	1.020	6.25

The method of least squares may be used to solve for A_1 and B_1. The equation for the model can be written as:

$$y = A_1 + B_1 x^2 + \epsilon_1 \qquad [6.64]$$

where ϵ_1 is the residual term. If $\hat{y}_1 = y - \epsilon_1$, we have:

$$\hat{y}_1 = A_1 + B_1 x^2 \qquad [6.65]$$

By applying the method of section 4.6, whereby x is replaced by x^2, we obtain:

$$\begin{bmatrix} A_1 \\ B_1 \end{bmatrix} = \begin{bmatrix} n & \Sigma x^2 \\ \Sigma x^2 & \Sigma x^4 \end{bmatrix}^{-1} \begin{bmatrix} \Sigma y \\ \Sigma x^2 y \end{bmatrix}$$

or, using numbers:

$$\begin{bmatrix} A_1 \\ B_1 \end{bmatrix} = \begin{bmatrix} 3 & 61.25 \\ 16.25 & 121.0625 \end{bmatrix}^{-1} \begin{bmatrix} 2.9046 \\ 17.4193 \end{bmatrix} = \begin{bmatrix} 0.6917 \\ 0.0510 \end{bmatrix}$$

Consequently, [6.65] becomes $\hat{y}_1 = 0.6917 + 0.0510\,x^2$.

The calculated values $\hat{y}_1$ in comparison to the original values y are:

	y	$\hat{y}_1$	ϵ_1
Grenville	0.740	0.743	– 0.003
Churchill	1.145	1.151	– 0.006
Superior	1.020	1.011	0.009

The residuals are small and can be neglected. By eliminating the auxiliary variable y the results for the orogenies are:

	s_r (m.y.)	s_t (m.y.)	$2a$ (m.y.)
Grenvillian	86	23	78
Hudsonian	107	68	235
Kenoran	101	56	196

In this table, s_t is the estimate for σ_t used in [6.59] and [6.62]. A time span ($2a$) of about 200 m.y. is indicated for the Kenoran and Hudsonian orogenies.

The Grenvillian estimate is small because this province has a smaller area than the other two provinces and area was weighted heavily in the mathematical model. Suppose that material of Grenville age exists outside the boundaries of the present Grenville province. This additional area, according to our model would give ages that differ somewhat from those now available for the Grenville. This, is turn, would increase the standard deviations s_r and s_t for the province.

The value of s_a which estimates σ_a of [6.59] is equal to 83 m.y. for all three provinces. It would represent the standard deviation for variations in argon content of micas that are not systematically related to the orogeny.

Clearly, the specifications for this model are tentative. The problem here is not so much lack of age determinations but difficulties encountered in laying the framework for the populations to be used for problem-solving. Nevertheless, the construction of simple mathematical models in situations of uncertainty can be of help in the consideration of alternative solutions to a problem.

6.13. Statistical inference of age determinations

The data available in 1968 for the Grenville province are shown in Table XIII. They are mainly for micas from granitic rocks. A distinction was made between biotite and muscovite ages. The measurements were grouped in classes which are 20 m.y. wide. Midpoints of these classes are given in the table. For example, there are six ages of 890 m.y. meaning that six measured dates were between 880 and 900 m.y.

If the central values of classes are called x_i, and n_i represents number of ages per class, the sample mean $\bar{x} = \Sigma\, x/n$ also can be calculated by the expression:

$$\bar{x} = \frac{\Sigma\, n_i x_i}{\Sigma\, n_i} \qquad [6.66]$$

TABLE XIII
Potassium/argon ages, Grenville province

Biotite		*Muscovite*	
age class (m.y)	*no. ages*	*age class (m.y)*	*no. ages*
770	2		
830	1		
850	6		
870	5	870	3
890	6		
910	7	910	1
930	12	930	3
950	11		
970	5	970	2
990	2	990	4
1010	3	1010	2
1030	3		
1050	4	1050	1
1070	1	1070	1
1090	2		
1110	1		
1130	1		
1170	1		
1190	2		
1210	1	1210	1
	$n_1 = 76$		$n_2 = 18$

The average biotite age for 76 single data is 954 m.y. and that for muscovite is 976 m.y. The hypothesis that these two means are equal, and may be combined with one another to give a single mean, can be tested by a t-test. This test is based on the assumption that the variance of the population for biotite ages is equal to that for muscovite ages. The hypothesis of equality of variance will be tested first by using an F-test.

Both t- and F-test are based on the assumption that the underlying populations are normal. The normality assumption can also be tested by using a χ^2-test (see later).

Practical calculation of variances

The equation for the sample variance is:

$$s^2(x) = [1/(n-1)] \; \Sigma \, (x - \bar{x})^2$$

The calculation of the sum $\Sigma(x - \bar{x})^2$ can be facilitated by using the following simplification:

$$\Sigma (x - \bar{x})^2 = \Sigma (x^2 - 2x\bar{x} + \bar{x}^2)$$

$$= \Sigma x^2 - 2\bar{x}\,\Sigma x + \Sigma \bar{x}^2$$

$$= \Sigma x^2 - 2n\bar{x}\,\bar{x} + n\bar{x}^2$$

$$= \Sigma x^2 - n\bar{x}^2$$

onsequently:

$$s^2(x) = \frac{1}{n-1} [\Sigma x^2 - (\Sigma x)^2/n] \qquad [6.67]$$

or example, from the biotite ages, we have:

$\Sigma x = 72540$; $\Sigma x^2 = 69862800$

q. [6.67] gives:

$$s_1^2(x) = (1/75)\,[69862800 - 72540^2/76] = 8337$$

onsequently:

$s_1^2(x) = 8337$; $s_1(x) = 91.3$

he subscript 1 refers to biotite ages. For muscovite (subscript 2) the result is:

$s_2^2(x) = 6861$; $s_2(x) = 82.8$

-test for comparing two variances

The F-test for equality of variance consists of calculating the ratio:

$$\hat{F}(75,17) = s_1^2/s_2^2 = \frac{8337}{6861} = 1.22$$

f two population variances are equal to one another, the $\hat{F}$-ratio should be close to one. or level of significance $\alpha = 0.05$, we must look up the 97.5-% fractile of the cumulative -distribution in statistical tables. For 75 and 17 degrees of freedom, $F_{0.975} = 2.36$, and e test hypothesis is accepted. The 97.5-% fractile is used because this test is two-tailed. e also could have formed the ratio:

$$\hat{F}(17,75) = s_2^2/s_1^2 = 0.82$$

This quantity must be larger than:

$$F_{0.025} = 1/F_{0.975} = 0.42$$

for the test hypothesis to be accepted.

Because the population variances for biotite and muscovite ages are probably equal to one another, we may form the pooled variance by using [6.50]. This results in:

$$s^2 = \frac{75 \times 8337 + 17 \times 6861}{92} = 8064$$

Hence:

$$s^2(x) = 8064 \; ; \qquad s(x) = 89.8$$

t-test for comparing two means

The t-test is used for the hypothesis that the two population means are equal. It consists of calculating the quantity:

$$\hat{t}(f_1 + f_2) = \frac{|\bar{x}_1 - \bar{x}_2|}{s(x)\sqrt{1/n_1 + 1/n_2}} \qquad [6.68]$$

This gives:

$$\hat{t}(92) = \frac{975 - 954}{89.8\sqrt{1/76 + 1/18}} = 0.89$$

The t-test also is two-tailed. For $\alpha = 0.05$, we find in statistical tables that:

$$t_{0.975}(92) = 1.98$$

Because this value exceeds $\hat{t}(92) = 0.89$, the test hypothesis is accepted. It is noted that for a t with 92 degrees of freedom, there is little difference between t-test and z-test (see section 6.7). The level of significance for the two-tailed z-test is 1.96 regardless of the number of degrees of freedom.

A brief explanation for the derivation of [6.68] is as follows. In [6.56], t has been defined for the difference $(\bar{x} - \mu)$ where μ is the population mean. In the present application, there are two sample means $\bar{x}_1$ and $\bar{x}_2$ and we are interested in the distribution of their difference:

$$(\bar{x}_1 - \bar{x}_2) = (\bar{x}_1 - \mu) - (\bar{x}_2 - \mu)$$

The quantities $(\bar{x}_1 - \mu)$ and $(\bar{x}_2 - \mu)$ are from normal distributions with variances equal to σ^2/n_1 and σ^2/n_2, respectively. Their difference, therefore, has variance $\sigma^2/n_1 + \sigma^2/n_2$ and standard deviation $\sqrt{\sigma^2/n_1 + \sigma^2/n_2}$, which represents the denominator in [6.68].

Inasmuch as both test hypotheses (equality of means and variances) have been accepted, we can form the overall mean $\bar{x} = \Sigma\, x/n$ or:

$$\bar{x} = \frac{n_1 \bar{x}_1 + n_2 \bar{x}_2}{n_1 + n_2} = 958.5$$

χ^2-test for normality

Suppose that a class with limits z_1 and z_2 is defined for a normal distribution in standard form. Then:

$$p = P(z_1 < Z < z_2) = \Phi(z_2) - \Phi(z_1) \qquad [6.69]$$

represents the probability that the random variable Z will fall in the class.

The probability that k out of n values will fall in the class satisfies the binomial form of [6.2] with mean $\mu = np$ and standard deviation:

$$\sigma = \sqrt{np(1-p)} \approx \sqrt{np}$$

The binomial distribution, under certain circumstances to be defined below, can be approximated by a normal distribution. In that situation, the standardized binomial variable:

$$\frac{K - \mu}{\sigma} = \frac{K - np}{\sqrt{np}}$$

is approximately normally distributed with mean 0 and standard deviation 1. It follows that the quantity:

$$\frac{(K - np)^2}{np}$$

s, by approximation, distributed as χ^2 with a single degree of freedom.

Suppose that we form m classes labelled i with theoretical probabilities equal to p_i. Then:

$$\sum_{i=1}^{m} \frac{(K_i - np_i)^2}{np_i} \approx \chi^2(m) \qquad [6.70]$$

This sum is approximately distributed as χ^2 with m degrees of freedom.

In practice, one proceeds as follows. First, class limits are set and the number of observations per class is counted. The result is the so-called observed frequencies f_o. Theoretical frequencies f_t are calculated by [6.69] and multiplying the resulting relative frequencies p_i by the total number of observations (n). According to [6.70] :

$$\chi^2 \approx \Sigma \frac{(f_o - f_t)^2}{f_t} \qquad [6.71]$$

Two important points must be considered here.

(1) The approximation is valid only if $f_t > 5$. Care must be taken that all classes are sufficiently wide to allow theoretical frequencies that are larger than 5 (see Cramér, 1947). This rule should also be considered in other applications of the χ^2-test.

(2) The sum of the observed frequencies f_o is equal to n. It can be shown (Cramér, 1947) that for this reason one degree of freedom must be subtracted from the number of classes m. More degrees of freedom are lost if the parameters for the frequency-distribution function used for obtaining the f_t are estimated from the same data as were used for obtaining the f_o. In general, the number of degrees of freedom $(m - 1)$ is to be reduced further by the number of parameters estimated for the distribution function.

For example, in order to fit a normal distribution function we must estimate two parameters (mean and variance). Consequently, the χ^2-test of normality has $(m - 3)$ degrees of freedom.

The biotite ages listed in Table XIII have been tested for normality. It is assumed that the parameters of the theoretical normal distribution are equal to those estimated from the 76 data, or $\mu = 954$ and $\sigma = 91$. The calculations are shown in Table XIV.

In total, there are seven classes each of which is 40 m.y. wide. The values z_i refer to the standardized upper class limits. For example:

$$z_2 = \frac{900 - 954}{91} = -0.60$$

The $\Phi(z_i)$ are looked up in statistical tables. They represent cumulative frequencies for the normal distribution. The relative theoretical frequencies are obtained by subtracting $\Phi(z_i - 1)$ for the lower class limit from $\Phi(z_i)$ for the upper class limit. The f_t are absolute frequencies obtained after multiplication by $n = 76$.

The value of $\hat{\chi}^2$ is computed by adding the values Δ^2/f_i in the last column with Δ representing the difference between observed frequencies f_o and theoretical frequencies f_t. We obtain $\hat{\chi}^2 = 9.1$.

The number of degrees of freedom is equal to the number of classes (7) minus three. From statistical tables we find for $\alpha = 0.05$ that:

$$\chi^2_{0.95}(4) = 9.49$$

TABLE XIV

χ^2-test for normality, biotite ages, Grenville province

Class i	*Limits*	f_o	z_i	$\Phi(z_i)$	$\Phi(z_i) - \Phi(z_{i-1})$	f_t	$\Delta = \lvert f_o - f_t \rvert$	Δ^2	Δ^2/f_t
1	<860	9	−1.04	0.149	0.149	11.3	2.3	5.29	0.47
2	860–900	11	−0.60	0.274	0.125	9.5	1.5	2.25	0.24
3	900–940	19	−0.16	0.436	0.162	12.3	6.7	44.89	3.65
4	940–980	16	0.28	0.610	0.174	13.2	2.8	7.84	0.59
5	980–1020	5	0.72	0.764	0.154	11.7	6.7	44.89	3.84
6	1020–1060	7	1.16	0.877	0.113	8.6	1.6	2.56	0.30
7	>1060	9	∞	1.000	0.123	9.3	0.3	0.09	0.01
									Sum = 9.1

The χ^2-test is one-tailed, because the hypothesis of normality always is accepted if χ^2 is small.

If $\chi^2_{0.95}(4)$ is the criterion, we would accept the test hypothesis. However, this approach may be too rigid in that the computed value $\hat{\chi}^2 = 9.1$ is only slightly smaller than the 95-% confidence limit. For comparison, we have looked up the value for $\alpha = 0.10$, which is:

$$\chi^2_{0.90}(4) = 7.78$$

In fact, it can be calculated that:

$$\chi^2_{0.941}(4) = 9.1$$

It means that a normal distribution would give a χ^2 equal to or larger than 9.1 in only 5.9% of events if this particular experiment would be repeated a large number of times for the theoretical normal distribution. Our conclusion is that a departure from normality may exists although it probably would not affect our previous results obtained by *F*- and *t*-test.

Positive skewness of biotite ages

The statistical analysis of the biotite ages can be continued. The data in Table XIV suggest that if a departure from normality exists it would be caused by the relatively long tail towards higher values. This would imply a positive skewness of the frequency distribution. In order to test this hypothesis, we calculate $\sqrt{b_1}$ for skewness (see section 6.8) from the sample moments by using [6.23]. The latter are: $m_2 = 8227$ and $m_3 = 620747$. Hence:

$$\sqrt{b_1} = m_3/m_2^{3/2} = 0.832$$

The estimates m_2, m_3 and $\sqrt{b_1}$ are biased. For example, we know that the unbiased estimate of variance is 8337 instead of m_2 = 8227. The Pearson-Hartley tables (1958) contain the following values on significance levels for skewness as based on $\sqrt{b_1}$:

Size of sample	*95 %*	*99 %*
70	0.459	0.673
80	0.432	0.631

By linear interpolation, the 95- and 99-% confidence limits for 76 values are determined as 0.44 and 0.65. These values are less than $\sqrt{b_1}$ = 0.83 and the hypothesis that skewness is not significant is rejected.

Summary of results for age data

The ages from the Canadian shield were used to exemplify several methods of statistical inference. We also may draw several conclusions of geologic interest. Biotites and muscovites from the Grenville province gave the same mean age and variance. Locally, in the boundary zone between the Superior and Grenville provinces (Grenville front), muscovites gave a younger age than biotites from the same specimen suggesting that muscovite may loose its argon before biotite during Precambrian "orogenic" events. However, a systematic discrepancy of this type does not persist throughout the Grenville province.

The Grenville biotite dates are approximately normally distributed but a positive skewness of the distribution seems to be statistically significant (also see Fig. 21). The "oldest" rocks occur relatively close to the Grenville front suggesting that overprinting of new ages on older rocks close to the older Superior was not as complete as elsewhere in the Grenville. A negatively skew histogram for the Superior in Fig. 21 suggests that the reverse may hold true for this province whose "younger" rocks occur relatively close to the Grenville front.

In section 6.12, we supposed that all ages from a single province do not satisfy the model of one random variable with the same mean and variance everywhere. If the theory of random variables is used, it is better to consider the measured age as the sum of two random variables, one giving the time that the K/Ar clock was set to zero at a given place and the other for a less regular variability related to analytical imprecision and anomalous loss of argon content. The topic of a frequency distribution for the sum of two independent random variables is discussed in next chapter.

Chapter 7

FREQUENCY DISTRIBUTIONS AND FUNCTIONS OF INDEPENDENT RANDOM VARIABLES

7.1. Introduction

Several methods of treating frequency histograms for geological data will be reviewed in more detail than in the previous chapter. Special attention is given to positively skew frequency distributions in geology. The concept that a frequency distribution can be the end product of a sequence of events ordered in time is illustrated by a computer simulation experiment.

From section 7.6 to the end, this chapter deals with more advanced concepts including the subject of probability generating functions. Several discrete distributions including the Poisson, negative binomial, and geometric distributions will be discussed in this context. An introduction is presented to the advanced problem of bodies of random size that are randomly distributed in space. Oil pools occurring in a sedimentary basin and orebodies on a crystalline shield, are examples of this.

For some statistical problems in geology, the classical method of moments may not give the best estimators of the parameters of the population. The last two sections are devoted to a discussion of desirable properties of estimators and include several applications of the method of maximum likelihood.

7.2. Functions of random variables

On the basis of the definitions of frequency functions for continuous random variables (see section 6.5), we may argue as follows.

Let X represent a random variable with cumulative distribution function $F_2(x)$. Suppose that we know the distribution $F_1[h(x)]$ where $h(x)$ is a function of x. When there is a one-to-one correspondence between x and $h(x)$, we must have:

$$F_2(x) = F_1[h(x)] \qquad [7.1]$$

Differentiation of both sides with regard to x gives:

$$\frac{\mathrm{d}F_2(x)}{\mathrm{d}x} = \frac{\mathrm{d}F_1[h(x)]}{\mathrm{d}h(x)} \cdot \frac{\mathrm{d}h(x)}{\mathrm{d}x} \qquad [7.2]$$

Writing $y = h(x)$, we obtain for the frequency functions $f_2(x)$ and $f_1(y)$:

$$f_2(x) = \frac{\mathrm{d}y}{\mathrm{d}x} f_1(y) \qquad [7.3]$$

Eq. [7.3] has several important applications.

Diameters and volumes of spherical grains

Assume that we know the frequency-distribution function $f_1(v)$ of the volumes v for a population of spherical grains. The frequency-distribution function $f_2(x)$ for the diameters x of the spheres now can be found as follows. The relationship between volume and diameter for any grain satisfies:

$$v = \tfrac{1}{6}\pi x^3$$

Hence:

$$\frac{\mathrm{d}v}{\mathrm{d}x} = \tfrac{1}{2}\pi x^2$$

and

$$f_2(x) = \tfrac{1}{2}\pi x^2 f_1(v) \qquad [7.4]$$

Lognormal distribution

A random variable X has lognormal frequency distribution $f_2(x)$ if the logarithms of its values are normally distributed.

If ${}^{e}\log X$ has the normal frequency distribution $f_1(\log x)$, the lognormal distribution $f_2(x)$ satisfies:

$$f_2(x) = \frac{\mathrm{d}(\log x)}{\mathrm{d}x} f_1(\log x)$$

or: $f_2(x) = \dfrac{1}{x} f_1(\log x)$ [7.5]

According to [6.7]:

$$f_1(\log x) = \frac{1}{\sigma\sqrt{2\pi}} \exp\left[-\frac{1}{2}\left(\frac{\log x - \mu}{\sigma}\right)^2\right] \qquad [7.6]$$

where μ and σ are mean and standard deviation of the normally distributed random variable $Y = \log X$. Consequently:

$$f_2(x) = \frac{1}{x\sigma\sqrt{2\pi}} \exp\left[-\frac{1}{2}\left(\frac{\log x - \mu}{\sigma}\right)^2\right] \quad [7.7]$$

where $x > 0$.

Sum of two random variables

A more general problem is to determine the frequency-distribution function of a random variable which is a function of two or more other random variables with known distributions.

This subject is part of the theory of probability calculus. It has been discussed by Wilks (1962), Gnedenko (1963), and, at a more advanced level, by Loève (1963). We will restrict our discussion to the sum of random variables. In general, the approach that provides the more rapid solution to this problem is through the method of generating functions (see section 7.6). However, in some situations, elementary probability calculus gives a rapid solution. We will illustrate this by applying a simple method to solve the theoretical frequency distribution for age determinations used in section 6.12.

Suppose that X_1 and X_2 are two random variables. We are interested in their sum $Y = X_1 + X_2$ with frequency-distribution function $f(y)$. As was done in section 6.9, the joint probability $P(X_1 = x_1$ and $X_2 = x_2)$ can be represented as $f(x_1, x_2)\, dx_1\, dx_2$ for infinitesimal area $dx_1\, dx_2$. We have:

$$P(Y < y) = F(y) = \iint_{x_1 + x_2 < y} f(x_1, x_2)\, dx_1\, dx_2 \quad [7.8]$$

The joint probability function of X_1 and X_2 is integrated over the area where $x_1 + x_2 < y$. This provides the total probability that Y is less than $y = x_1 + x_2$.

The situation is shown graphically in Fig. 22, where $f(x_1, x_2)$ is represented in the form

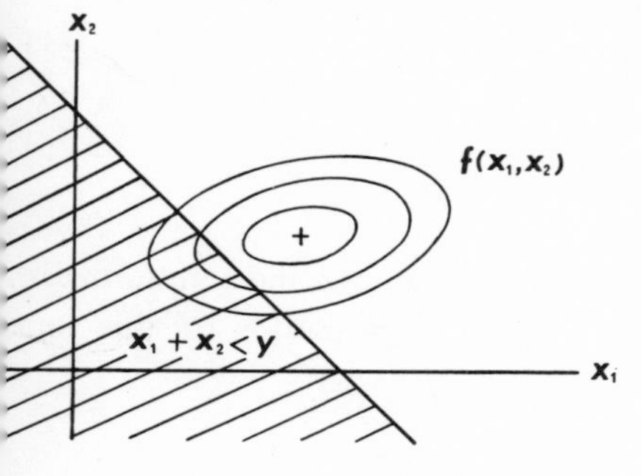

Fig.22. Graphical representation of cumulative frequency $F(y)$ of a random variable $Y = X_1 + X_2$ given the joint frequency distribution $f(x_1, x_2)$; $F(y)$ is equal to the volume under $f(x_1, x_2)$ bounded by the line $y = x_1 + x_2$.

of contours. Integration is done for the shaded area where $x_1 + x_2 < y$. In fact, this method is the same as integrating $f(x_1, x_2)$ first between limits $-\infty$ and $(y - x_1)$ in the X_2-direction, and then between $-\infty$ and $+\infty$ in the X_1-direction. Hence:

$$F(y) = \int_{-\infty}^{\infty} \left[\int_{-\infty}^{y-x_1} f(x_1, x_2)\, \mathrm{d}x_2 \right] \mathrm{d}x_1$$

For independent variables X_1 and X_2:

$$f(x_1, x_2) = f_1(x_1) \cdot f_2(x_2)$$

Writing $z = x_1 + x_2$ or $x_2 = z - x_1$, we have:

$$F(y) = \int \mathrm{d}x_1 \int_{-\infty}^{y-x_1} f_1(x_1) \cdot f_2(x_2)\, \mathrm{d}x_2$$

$$= \int \mathrm{d}x_1 \int_{-\infty}^{y} f_1(x_1) \cdot f_2(z - x_1)\, \mathrm{d}z$$

$$= \int_{-\infty}^{y} \mathrm{d}z \left[\int f_1(x_1) f_2(z - x_1)\, \mathrm{d}x_1 \right]$$

This allows us to write:

$$F(y) = \int f_1(x)\, F_2(y - x)\, \mathrm{d}x \qquad [7.9]$$

or:

$$F(y) = \int F_2(y - x)\, \mathrm{d}F_1(x) \qquad [7.10]$$

where $\mathrm{d}F_1(x) = f_1(x)\, \mathrm{d}x$. Differentiation of both sides gives:

$$f(y) = \int f_2(y - x)\, f_1(x)\, \mathrm{d}x \qquad [7.11]$$

Sum of a uniform and a normal distribution

In section 6.12, we calculated the variance of the sum of a normal and a uniform distribution. By using [7.10] or [7.11], the complete frequency distribution for this sum can be developed.

Suppose that the uniform distribution is defined as:

$$f(x) \begin{cases} = 0 \text{ if } x \leqslant a \text{ or } x > b \\ \\ = \dfrac{1}{b - a} \text{ if } a < x \leqslant b \end{cases}$$

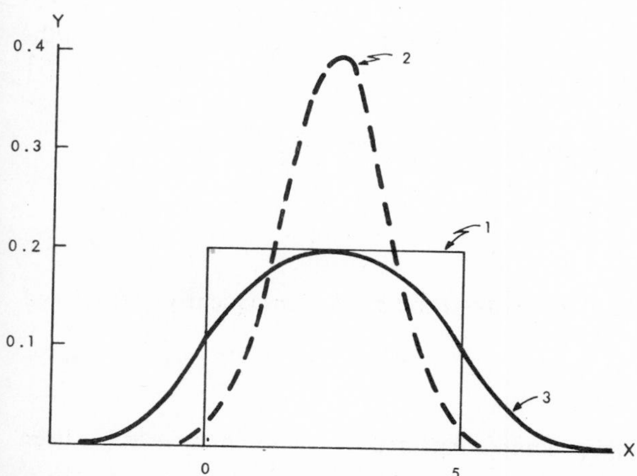

Fig.23. Frequency curve (*3*) for the sum of a rectangular distribution (*1*) and a Gaussian distribution (*2*). This situation arises by random sampling of a space-variable subject to a linear trend changing from 0 to 5, and with residuals that are normally distributed with zero mean and unity variance. The three curves have areas equal to one.

This distribution is shown graphically in Fig. 23 for $a = 0$ and $b = 5$. Suppose further that the normal distribution has the standard form $\varphi(x)$ which is also shown in Fig. 23.

If $g(x)$ represents the frequency-distribution function for the sum, [7.11] becomes:

$$g(x) = \int \varphi(x-y) f(y)\,\mathrm{d}y \tag{7.12}$$

or alternately:

$$g(x) = \int f(x-y)\,\varphi(y)\,\mathrm{d}y \tag{7.13}$$

One possible interpretation of these expressions is that we obtain g by "smoothing" f by means of φ. We may say that a value $g(x_1)$ for a point with given abscissa x_1 is obtained by: (1) centering the function φ so that its mean falls at x_1; and (2) calculating a weighted average of f whereby the weights decrease from x_1 in both positive and negative X-direction. The process of weighting or "filtering" one function by means of another function is also called convolution. The functions may be interchanged since both [7.12] and [7.13] hold true. This example provides a prototype for other convolutions to be applied later.

The uniform distribution is nonzero between $x = a$ and $x = b$ only. Therefore, [7.12] can be rewritten as:

$$g(x) = \frac{1}{b-a} \int_a^b \varphi(x-y)\,\mathrm{d}y \tag{7.14}$$

By putting $z = -y$, we obtain:

$$g(x) = \frac{1}{b-a} \int_{-b}^{-a} \varphi(x+z)\,\mathrm{d}z$$

or:

$$g(x) = \frac{1}{b-a} [\Phi(x-a) - \Phi(x-b)] \qquad [7.15]$$

By using the derivation of [6.61], mean and variance for $g(x)$ satisfy:

$$\mu = \frac{a+b}{2}; \qquad \sigma^2 = 1 + \frac{(b-a)^2}{12}$$

For the artificial example shown in Fig. 23, $a = 0$ and $b = 5$. Consequently, $\mu = 2.5$ and $\sigma^2 = 3.083$. Further:

$$g(x) = \tfrac{1}{5} [\Phi(x) - \Phi(x-5)]$$

Single values of $g(x)$ can be calculated and connected by a smooth curve as has been done in Fig. 23. For example, if $x = 2$:

$$\begin{aligned} g(x) &= \tfrac{1}{5} [\Phi(2) - \Phi(-3)] \\ &= \tfrac{1}{5} [0.9773 - 0.0014] \\ &= \tfrac{1}{5} \times 0.9759 = 0.1952 \end{aligned}$$

The three curves shown in Fig. 23 each have total area equal to one. The curve for $g(x)$ resembles a Gaussian curve in that it would be difficult to distinguish it from a Gaussian curve on the basis of a sample of n values unless n is large.

In the example, the base of the uniform frequency distribution $f(x)$ is five times the standard deviation of the normal curve $\varphi(x)$. In fact, we have used this standard deviation as unit of distance along the X-axis. The shape of $g(x)$ will change if the ratio (base of f/standard deviation of φ) is altered. If this ratio is decreased, $g(x)$ approaches the normal form. On the other hand, if it is increased, $g(x)$ begins to develop a flat top. Its shape then approaches that of $f(x)$.

Recently, Sichel (1972) has made an interesting application in geology of the theory of sums of random variables to the densities of occurrence and sizes of diamonds on ancient beaches in Southwest Africa. He assumed that, at anyone place, the density distribution conforms to the Poisson model (see later) and the size distribution to the lognormal model. However, by sampling larger entities, we obtain distributions which are the sum of the local distribution and a mixing distribution. Sichel used Pearson-type distributions for the mixing distributions.

7.3. Normal probability paper

The normal curve occurs frequently in nature. It is useful to compare a histogram of observed frequencies with a theoretical Gaussian curve. This is commonly done by using normal probability paper.

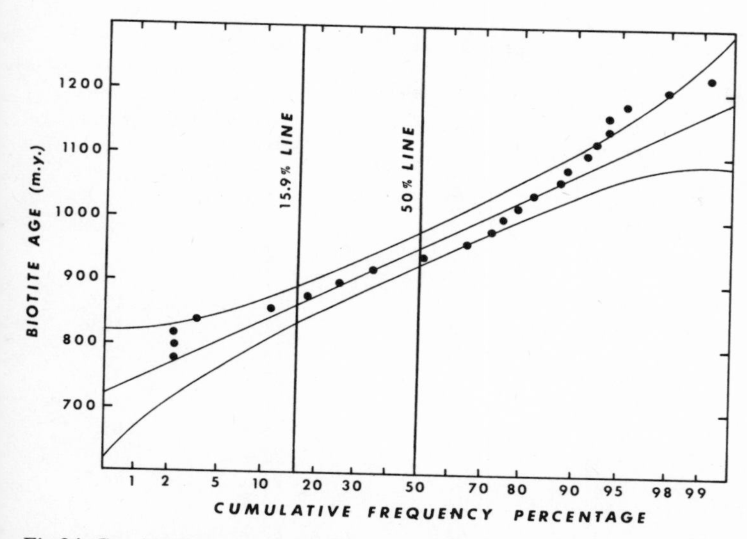

Fig.24. Graphical test of normality for 76 biotite ages from Grenville province, Canadian shield. Observed cumulative frequencies (solid dots) are not all contained within 95-% confidence belt on the theoretical normal curve (straight line), indicating departure from normality which is strongest in the upper tail.

In Fig.24, the scale along the vertical axis is as on ordinary graph paper but the horizontal scale has been changed in such a manner that the S-shaped curve for the cumulative normal distribution plots as a straight line. A normal curve always becomes a straight line on probability paper.

Fig.24 shows three types of plot for the 76 biotite ages listed in Table XIII; (1) original data (points); (2) theoretical normal curve (straight line); (3) a 95-% confidence belt on the theoretical normal curve. These three plots have been constructed as follows:

Plotting of observed frequencies

This procedure is outlined in Table XV. One may plot the upper class limits for the ages along the horizontal axis and the relative cumulative frequency along the vertical axis. Of course, then it is not possible to plot the value which is for class 1200–1220 m.y. A last class usually has cumulative frequency percent one hundred (100%) and this value does not occur on the probability scale. A slight refinement consists of plotting the data against so-called plotting percentages by using the equation:

$$\text{Plotting percentage} = \frac{3 \times \text{cumulative frequency} - 1}{(3n + 1)/100} \qquad [7.16]$$

where n represents number of data (Tukey, 1962; Koch and Link, 1971).

TABLE XV
Preparation of data to be plotted on probability paper; for biotite ages of Table XIII

Class center (in m.y)	*Frequency*	*Upper class limit (in m.y.)*	*Cumulative frequency*	*Cumulative frequency percentage*	*Plotting percentage*
770	2	780	2	2.6	2.2
790	0	800	2	2.6	2.2
810	0	820	2	2.6	2.2
830	1	840	3	4.0	3.5
850	6	860	9	11.8	11.4
870	5	880	14	18.4	17.9
890	6	900	20	26.3	25.8
910	7	920	27	35.5	34.9
930	12	940	39	51.3	50.7
950	11	960	50	65.8	65.1
970	5	980	55	72.4	71.6
990	2	1000	57	75.0	74.2
1010	3	1020	60	78.9	78.2
1030	3	1040	63	82.9	82.1
1050	4	1060	67	88.2	87.3
1070	1	1080	68	89.5	88.7
1090	2	1100	70	92.1	91.3
1110	1	1120	71	93.4	92.6
1130	1	1140	72	94.7	93.9
1150	0	1160	72	94.7	93.9
1170	1	1180	73	96.1	95.2
1190	2	1200	75	98.7	97.8
1210	1	1220	76	100.0	99.1

Theoretical normal curve

In section 6.13, it was calculated that, for the biotite ages, $\bar{x} = 954.5$ and $s(x) = 91.3$. This determines a theoretical normal distribution which is plotted by passing a straight line through (1) the mean $\bar{x}$ that theoretically has cumulative frequency equal to 50%, and (2) the point with abscissa $\bar{x} - s(x) = 863.2$ and a value of 15.9% along the probability scale.

95-% confidence belt

It may be helpful to test whether the observed frequencies do indeed belong to the theoretical normal distribution. An approximate test which has been applied by Vistelius (1960) is as follows.

If z_t represents the standardized value for a point on the straight line for the theoretical normal distribution and z an observed value that may range at random about the line, then the 95-% confidence belt satisfies:

$$z_t \pm 1.96\,\sigma(z) \qquad [7.17]$$

where $\sigma(z)$ is the square root of a variance determined by the expression:

$$\sigma^2(z) = (1/n)\,\frac{\Phi(z_t)\cdot\Phi(-z_t)}{\varphi^2(z_t)} \qquad [7.18]$$

$\Phi(z_t)$ and $\varphi(z_t)$ are as in [6.9] and [6.10].

This approximate belt is shown for $n = 76$ in Fig.24. It is relatively wide for the two tails of the frequency distribution.

If the data come from a normal population, a 95-% confidence belt, on the average, contains about 95-% of the plotted observations. Therefore, a single value outside the belt would not lead to rejection of the normality assumption. However, if the number of data is small, it is desirable to choose a level of significance larger than $\alpha = 0.05$. This test has been discussed by Hald (1952) and Vistelius (1960). We will prove [7.18] by using several results obtained in previous chapters.

Derivation of [7.18]

From [6.33] and [6.34], it follows that:

$$\sigma^2(a + bX) = b^2\sigma^2(X) \qquad [7.19]$$

where a and b are constants. Further, if $f(x)$ is a function of x that is approximately linear in the range of variation of x, then, according to Taylor's formula [2.17]:

$$f(x) \approx f(c) + (x - c)\,f'(c) \qquad [7.20]$$

where c is a constant.

From [7.19] and [7.20] it follows that:

$$\sigma^2\{f(X)\} \approx \{f'(c)\}^2\,\sigma^2(X) \qquad [7.21]$$

This result is applied as follows to the present situation:

Suppose that K out of n observed values are less than z in a situation corresponding to specific theoretical value z_t on the straight line for the normal plot. Then K has binomial distribution and variance:

$$\sigma^2(K) = np(1 - p) \qquad [7.22]$$

where $p = \Phi(z_t)$.

A relative cumulative frequency H can be defined as $H = K/n$. From [7.22]:

$$\sigma^2(H) = \frac{p(1-p)}{n} = \frac{\Phi(z_t)\cdot\Phi(-z_t)}{n} \qquad [7.23]$$

Application of [7.21] gives:

$$\sigma^2(z) \approx \sigma^2(H)\left(\frac{dz_t}{dp}\right)^2 \qquad [7.24]$$

with $H = K/n \approx p$. By using [7.23] and the relationship:

$$\frac{d\Phi(z_t)}{dz_t} = \frac{dp}{dz_t} = \varphi(z_t)$$

Eq. [7.24] gives:

$$\sigma^2(z) \approx (1/n)\,\frac{\Phi(z_t)\cdot\Phi(-z_t)}{\varphi^2(z_t)} \qquad [7.18, \text{repeated}]$$

and the proof is completed.

The quantity $\sqrt{n}\cdot\sigma(z)$ does not depend on the number of observations n. Some values for it are given in Table XVI.

The standard deviation $\sigma(z)$ is obtained by dividing the values in the last column of Table XVI by $\sqrt{n}$.

TABLE XVI
Calculation of $\sqrt{n}\cdot\sigma(z)$

z_t	$\Phi(z_t)$	$\Phi(-z_t)$	$\varphi(z_t)$	$\sqrt{n}\cdot\sigma(z)$
0.0	0.5	0.5	0.3989	1.253
0.5	0.6915	0.3085	0.3521	1.312
1.0	0.8413	0.1587	0.2420	1.510
1.5	0.9332	0.0668	0.1295	1.928
2.0	0.97725	0.02275	0.0540	2.762
2.5	0.99379	0.00621	0.0175	4.481

Example for biotite ages

Because there are 76 biotite ages, $\sqrt{n}\cdot\sigma(z)$, becomes $8.7178 \times \sigma(z)$ and we must divide by 8.7178 to obtain $\sigma(z)$. For example, if $z_t = 0$, then:

$$\sigma(z) = 1.253/8.7178 = 0.1437$$

The 95-% confidence belt satisfies:

$$z_t \pm 1.96\,\sigma(z)$$

or, if $z_t = 0$, 0.0 ± 0.2817. The relationship between z_t and x_t is:

$$z_t = \frac{x_t - \bar{x}}{s(x)} = \frac{x_t - 954.5}{91.3}$$

In order to plot the 95-% confidence belt in Fig.24, where the vertical scale is for x_t and not z_t, we must undo the transformation. Therefore, 0.0 ± 0.2817 becomes 954.5 ± 0.2817 X 91.3 or 954.5 ± 25.7. The 50-% line in Fig.24 intersects the 95-% confidence belt at values 954.5 + 25.7 = 980.2 and 954.5 – 25.7 = 929.1. The remainder of the belt can be constructed by using the same method. The belt is symmetric with respect to $z_t = 0.0$ or $x_t = 954.5$.

In Fig.24, seven points fall on or outside the confidence belt. This indicates that the test hypothesis of normality must be rejected. It already has been discussed that the departure in normality is caused by a relatively long tail of higher ages or significant positive skewness (see sections 6.11 – 13).

'.4. Other methods of frequency-distribution analysis

In this section, several other methods of analyzing frequency distributions will be discussed. A word of caution is in order here. A frequency histogram is readily compiled for any set of data. However, only rarely does the histogram by itself provide useful information regarding the underlying geological process. To begin a statistical analysis by doing a preliminary analysis on the histogram can provide useful guidelines for further analysis.

Histograms for data from Washington's (1917) compilation

In 1917, Washington published chemical analyses for all types of rocks on earth. From this compilation, Richardson and Sneesby (1922) have extracted data to make histograms. An example is shown in Fig.25 for SiO_2 in igneous rocks. It shows two peaks, one for the most frequent rock type which has basaltic composition (52.5% SiO_2) and the other for

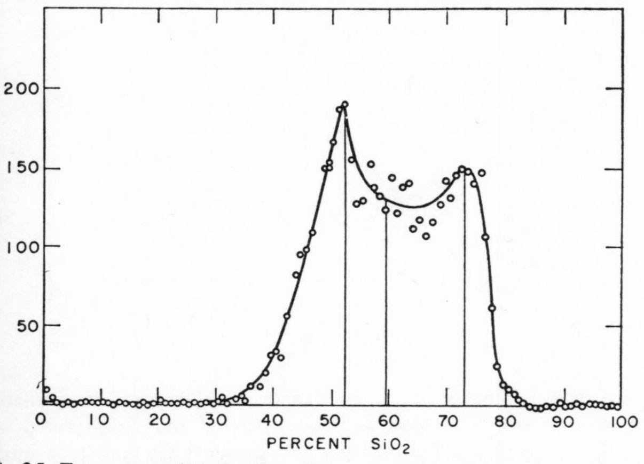

Fig.25. Frequency-distribution diagram of silica (SiO_2) in igneous rocks. Peaks correspond to basaltic and granitic rocks, respectively. (After Richardson and Sneesby, 1922.)

the next most frequent rock type with granitic composition (73.0 % SiO_2), Washington's "average" rock type has percentage silica equal to 59.1. It is less frequent than the other two types and has granodioritic composition.

A histogram such as the one shown in Fig.25 clearly is a mixture of many frequency distributions each of which may have been generated by a different geochemical process. However, the mixture itself may be considered as generated by a rock-making process which is broader in scope than the individual processes for its parts. Barth (1962) has stated that any theory on the origin of rocks must recognize and explain statistical facts such as the one represented in Fig.25.

Although Washington's data refer to relatively old analyses, they have remained a principal source for constructing histograms. For example, Ahrens (1964) has based a study on these data using probability paper.

An example of Ahrens' statistical analysis is shown in Fig.26 with plots of data on SiO_2 for 618 specimens of basalt and dolerite. Observed frequencies are shown as dots in Fig.26A. A smooth curve might be fitted to these dots. The data also were plotted on probability paper (Fig.26B). Rather than falling on a single straight line, they seem to follow a broken line with pieces called x and y in the diagram. Ahrens proposed to fit two straight lines in situations of this type. If this new fit is regarded as a single distribution function $F(x)$, the corresponding frequency distribution $f(x)$ can be plotted. The result is shown in Fig.26A. Incidentally, this result illustrates a feature of the method of differentiation (section 2.6). Because $\mathrm{d}F(x)/\mathrm{d}x = f(x)$, $f(x)$ is the first derivative of $F(x)$. Although $F(x)$ in Fig.26B is continuous, its first derivative (slope) is discontinuous. This causes the gap in the frequency curve of Fig.26A.

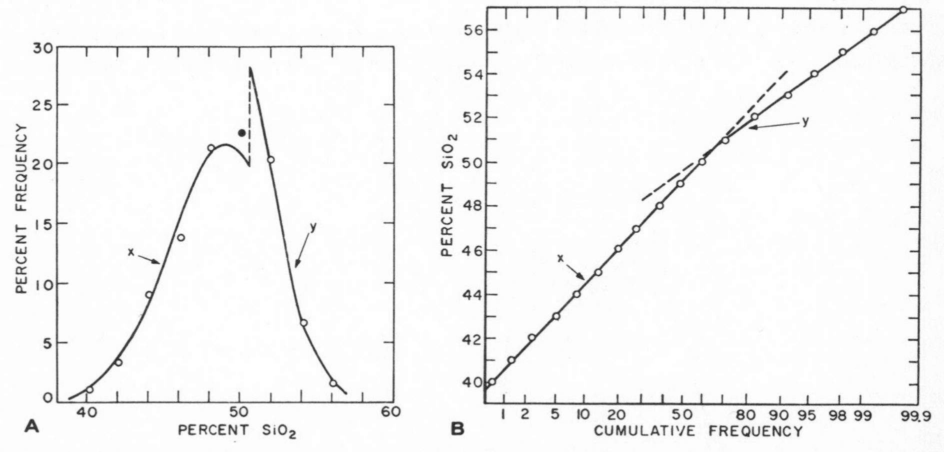

Fig.26. Frequency distribution of SiO_2 in 400 basalts and 218 dolerites. A. Observed frequencies indicated by circles; broken curve represents two truncated normal curves (x and y) obtained by fitting separate straight lines for x and y to points in B; and then converting these lines to normal curves in A. (From Ahrens, 1964.)

Mixture of two normal populations

Ahrens' method consists of fitting two normal curves which are truncated at the same point (50.7% SiO_2 in Fig.26). This method can be useful for descriptive purposes. It may be risky to assign petrological significance to the means and standard deviations of these truncated distributions.

Another approach taken to a situation, where the plot on probability paper shows a bend, is to consider the possibility of a mixture of two populations. A mixture of two random variables should not be confused with the sum of two random variables which was discussed in section 7.2.

An example of a mixture of two populations is shown for an artificial situation in Fig.27 that will also be discussed in section 7.5. The lines A and B in Fig.27 can be regarded as two normal populations if an arithmetic scale is used for value. The curve *C* would result if the sample population contained equal proportions from both normal distributions. This concept of mixtures of populations was introduced into geology by Doeglas (1946, fig.8).

If an observed distribution function resembles *C*, and the two underlying populations

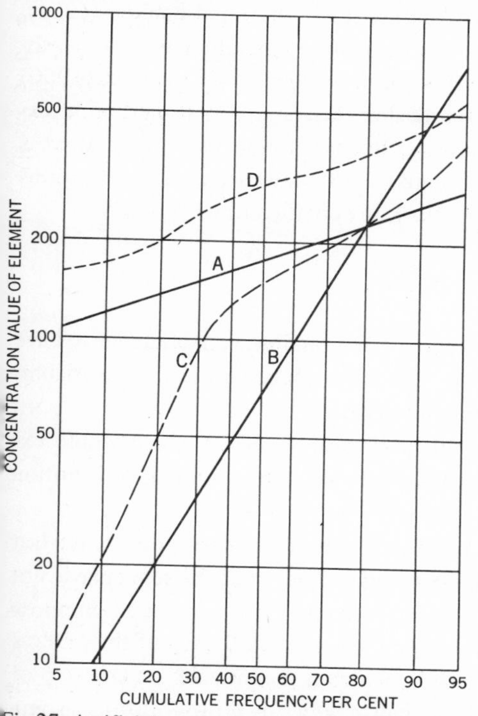

Fig.27. Artificial example. Plot on logarithmic probability paper of curve *C* representing heterogeneous population, and consisting of values from *A* (50%) and *B* (50%). The curve *D* was obtained by adding constant to values of *C* before plotting.

cannot be distinguished otherwise, one may try to reconstruct curves such as A and B from C, which is given. A satisfactory solution may be difficult to obtain in this situation. One statistical difficulty is that five parameters must be estimated simultaneously. These are the two means, two standard deviations, and also the proportion of each population in the mixture. A graphical method for this problem has been given by Hald (1952).

Another aspect of this problem is that, when other criteria to distinguish between the assumed two populations are missing, one usually also could speculate that more than two populations are involved in the mixture.

The lognormal distribution

Many geologic frequency distributions are distinctly not normal. Positive skewness occurs more frequently than negative skewness. A variety of geologic histograms can be described fairly well by using the lognormal distribution. Some examples are: (1) small-particle statistics; grain size (Krumbein, 1936); (2) thickness of sedimentary layers of different lithologies (Kolmogorov, 1951); (3) ore assays; element-concentration values in mineral deposits (Krige, 1951; Agterberg, 1961b); (4) size of oil and gas fields (Kaufman, 1963); (5) trace-element concentration values in rocks (Ahrens, 1953; Shaw and Bankier, 1954).

Our discussion of the lognormal distribution will follow two separate lines. Two methods of interpretation of skew frequency curves have been recognized by Vistelius (1960) in a fundamental study with reference to geochemical processes. The first method of explaining lognormality is based on the theory of proportionate effect. A second method is based on the theory that one may be sampling a mixture of many separate distributions which are mutually interrelated in that they have the same coefficient of variation.

Theory of proportionate effect

According to the so-called central-limit theorem, a given random variable tends to have a normal distribution if it is the resultant sum of many independent random variables each of which has a relatively small effect on the outcome.

Kapteyn (1903) has formulated the theory of proportionate effect which would explain that, under certain circumstances, the logarithms of values show normal distribution rather than the values themselves which then satisfy [7.7].

It is convenient to look at each individual value in a lognormal distribution as the last value of a sequence ordered through time. However, this ordering of the sequence is not an essential feature of the process (cf. Aitchison and Brown, 1957, chapter 3). Suppose that the random variable is initially equal to a constant x_0. At the i-th step of the process it is X_i and, eventually, after n steps, it becomes X_n which is measurable. At any step of the process, the variable is subject to a random change which is a proportion of some function $g(X_{i-1})$ of the value X_{i-1}. This leads to the expression:

$$X_i - X_{i-1} = \epsilon_i \, g(X_{i-1}) \qquad [7.25]$$

where $\epsilon_i (i = 1, 2, \ldots, n)$ denotes a sequence of independent random variables. We will return to [7.25] later. For the moment, let us look at the special cases $g(X_{i-1}) = 1$ and $g(X_{i-1}) = X_{i-1}$.

If g remains constant, we would have:

$$X_n = \sum_{i=1}^{n} (X_i - X_{i-1}) = \Sigma \epsilon_i \qquad [7.26]$$

Because of the central-limit theorem $\Sigma \epsilon_i$ and X_n assume normal form for large n.

Secondly, if $g(X_{i-1}) = X_{i-1}$, we have:

$$X_i - X_{i-1} = \epsilon_i X_{i-1} \qquad [7.27]$$

or: $$\frac{X_i - X_{i-1}}{X_{i-1}} = \epsilon_i$$

and: $$\sum_n \frac{X_i - X_{i-1}}{X_{i-1}} = \Sigma \, \epsilon_i \qquad [7.28]$$

Again, $\Sigma \epsilon_i$ will approach normal form. By making the individual steps $\Delta X_i = X_i - X_{i-1}$ small, we can write:

$$\sum_n \frac{X_i - X_{i-1}}{X_{i-1}} \approx \int_{X_0}^{X_n} \frac{\mathrm{d}X}{X} = \log X_n - \log X_0 \qquad [7.29]$$

Consequently:

$$\log X_n = \log X_0 + \Sigma \epsilon_i \qquad [7.30]$$

It follows that $\log X_n$ is asymptotically normally distributed. This treatment of the random process is rather heuristic. Kolmogorov (1941, 1951) has obtained the same result in a rigorous manner by using the theory of stochastic processes.

We will return to this method of approach in the next section. First, we will investigate the second manner by which skew frequency distributions may be explained.

Theory of a constant coefficient of variation

Vistelius (1960, p.11) argued as follows. Suppose that a geological process, at anyone time, yields a different population of values. If a given part of the crust of the earth is sampled, we obtain a set of values which may reflect various stages of the same geological process. In some places, the process may have developed further than in other places. The sampled population then consists of a mixture of many individual populations which are, however, interrelated by a single process.

Suppose that for each of these populations, the standard deviation σ is proportional to the mean μ. This is equivalent to assuming that the so-called coefficient of variation $\gamma = \sigma/\mu$ is constant. In that situation, the sampled population can easily have a positively skew frequency distribution because the subpopulations with relatively large means and standard deviations will generate a long tail of large values.

The frequency function $f(x)$ satisfies:

$$f(x) = \sum w_i f_i(x) \qquad [7.31]$$

where each $f_i(x)$ is the distribution representing the process at stage i; w_i is a weighting factor for that stage. It represents the proportion by which the subpopulation $f_i(x)$ occurs in the sampled population.

In order to test this theory, Vistelius (1960) has compiled averages and standard deviations for phosphorus (as weight percent P_2O_5) in granitic rock from various areas. He has found that $\bar{x}_i$ and s_i are positively correlated with a correlation coefficient $r = 0.56$. Individual distributions $f_i(x)$ are approximately normal but the joint distribution $f(x)$ with all values lumped together is positively skew.

Copper in the Muskox layered intrusion

We have tested the preceding theory for the element copper in various layers of the Muskox layered ultramafic gabbroic intrusion. This intrusion is located in the District of Mackenzie, northern Canada. It is about 1175 m.y. old. The layers were formed by crystallization differentiation of basaltic magma (Smith, 1962; Smith and Kapp, 1963). Individual layers are approximately homogeneous with regard to the major rock-forming minerals. Layers that are rich in olivine, sich as dunite and peridotite, tend to occur near the

TABLE XVII
Rock types of Muskox intrusion

Rocks containing olivine	*Without olivine*
Dunite (DN)	Orthopyroxenite (OPX)
Peridotite (PD)	Websterite (WB)
Feldspatic peridotite (FPD)	Melanogabbro
Troctolitic peridotite (TRPD)	Gabbro (GB)
Picrite (PC)	Granophyre-bearing gabbro (GR→GB)
Olivine clinopyroxenite (CPX)	
Picritic websterite (PCWB)	Granophyric gabbro (GRGB)
Olivine gabbro (OGB)	Mafic granophyre (MGR)
Bronzite gabbro (BGB)	Granophyre (GR)
Anorthositic gabbro (ANGB)	

bottom of the sequence of layers, whereas gabbroic layers occur closer to the top. The sequence is capped by an acidic layer of granophyric composition.

A number of rock types have been distinguished by Smith et al. (1963). In brackets we give abbreviations in Table XVII to be used later in Fig.29.

Fig.28A shows the frequency distribution of 116 copper values from specimens taken at the surface on a series of gabbros with increasing granophyre content. These gabbros are situated between a clinopyroxenite layer (at the bottom) and a mafic granophyre layer (at the top).

A logarithmic scale or ratio scale was used in Fig.28A for plotting the trace element copper in parts per million (p.p.m.). A straight line for the logarithms of values indicates approximate lognormality in Fig.28A. However, by using the classification given in Table XVII, the 116 specimens can be subdivided into three groups: GB, GR-GB, and GRGB. The resulting separate sets of copper values were plotted separately in Fig.28B.

The results prove that the frequency distribution shown in Fig.28A actually can be represented as a mixture of three separate populations with different properties. Average copper content increases with granophyre content of gabbro in this sequence.

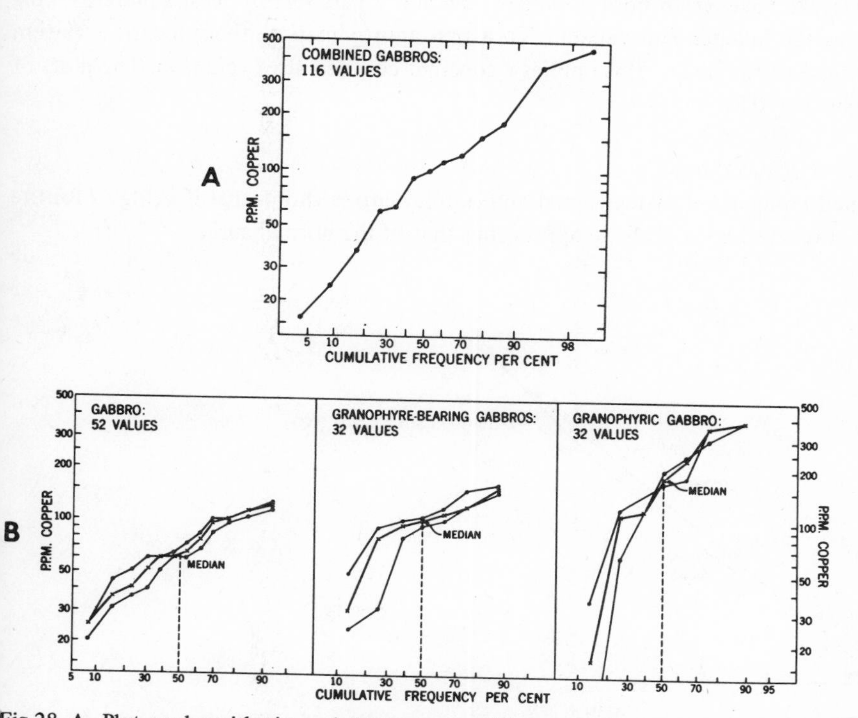

Fig.28. A. Plot on logarithmic probability paper of histogram for 116 copper determinations from gabbros in Muskox intrusion. B. Separate plots for different types of gabbro indicating that curve in A actually represents a mixture of several populations. Precision is indicated by area between curves for two subsamples, each consisting of 50% of the data.

Fractile graphical analysis

In order to test the stability of the separate frequency curves which are for relatively small samples, we used a simple procedure for small samples suggested by Mahalanobis (1960). Each sample of data is randomly divided into two subsamples. For example, if there are 52 values as for gabbro, we form two subsamples each consisting of 26 values. These subsamples are also plotted on the graph (Fig.28B). The area between the two frequency curves for the subsamples represents the so-called "error-area" of the curve for a combined sample. In fact, Mahalanobis (1960) has developed an approximate χ^2-test based on the error-areas to test for the separation between populations. In the present situation, the copper distribution for granophyric gabbro differs significantly from those for the other two rock types (GB and GR-GB).

Copper in other rock types

The Muskox intrusion has been sampled in detail both at the surface and by drilling. The average $\bar{x}$ and standard derivation s for copper in 17 different rock types have been plotted in Fig.29. There is positive correlation between these two statistical parameters ($r = 0.9002$). We have fitted both a straight line and a parabola by least squares, taking the means as the independent variable. To a first approximation, the standard deviation is proportional to the mean. This implies a constant coefficient of variation. The plots of Fig.29 suggest $c = 0.8$.

Coefficient of variation

A detailed comparison of lognormal and normal curves shows that if $\sigma(\log x)$ for the lognormal curve decreases, its form approaches that of the normal curve.

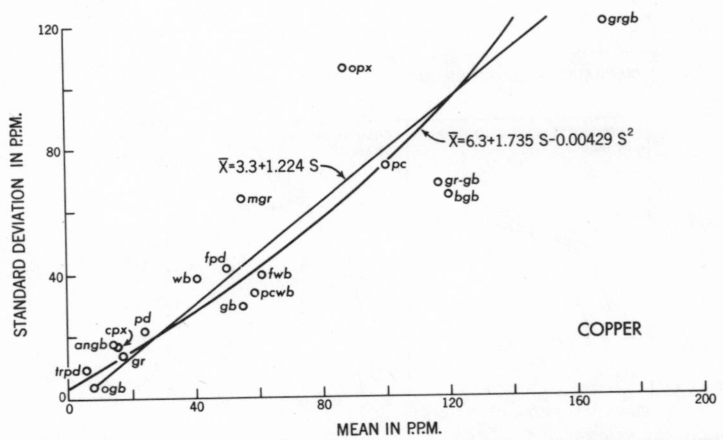

Fig.29. Standard deviation against mean for copper in different rock types of Muskox intrusion; two regression curves (linear and quadratic fit) are also shown. Diagram indicates that, to a first approximation, standard deviation is proportional to mean.

From [7.21], we derive:

$$\sigma^2(\log X) \approx \left[\frac{\mathrm{d}\,(\log x)}{\mathrm{d}x}\right]^2_{x=x_0} \sigma^2(X)$$

where x_0 is a constant value for x used in the Taylor approximation. We can choose $x_0 = \mu\,(X)$ where $\mu\,(X)$ represents the mean of X.

From $\mathrm{d}(\log x)/\mathrm{d}x = 1/x$, it follows that:

$$\sigma(\log X) \approx \sigma(X)/\mu(X) = \gamma \qquad [7.32]$$

Hence, the coefficient of variation γ is approximately equal to the standard deviation of the logarithmically transformed data. The approximation [7.32] is valid for $\gamma < 0.5$. From these considerations it follows that γ, or its estimate from the sample c, can be used as a criterion to discriminate between normality and lognormality. Hald (1952) has pointed out that if $c < 1/3$, a lognormal curve cannot be distinguished from a normal curve. In general, a sample for which $c > 0.5$ is distinctly nonnormal, but not necessarily lognormal. Because $c \approx 0.8$ for the copper distributions for individual rock types in the Muskox intrusion, nonnormality is indicated.

Composite frequency distribution

When the copper values x_{ik} for rock type i are divided by their rock type mean $\bar{x}_i$, we obtain a frequency distribution with mean equal to one. All values $x'_{ik} = x_{ik}/\bar{x}_i$ can be combined into a single distribution for 17 rock types with mean equal to one. The standard deviation of this composite frequency distribution for 622 copper values is 0.84 and, by approximation, represents the pooled coefficient of variation.

It is noted that the 17 frequency curves for separate rock types do not become exactly identical by this transformation. Bartlett's χ^2-test (see Hald, 1952) for equality of variances gave $\hat{\chi}^2 = 403.6$ before and $\hat{\chi}^2 = 90.3$ after the transformation for twelve degrees of freedom. Because the 95-% fractile of χ^2 (12 d.f.) is 26.2, statistical homogeneity of variance has not been achieved.

The frequency curve of the transformed copper values x' was studied by Agterberg (1965b). It is neither normal nor lognormal but can be fitted by a truncated three-parameter lognormal model as shown in Fig.30. The values x' were transformed into values $y' = {}^{e}\log(1 + x')$. A method and tables by Hald (1949) can be used to fit a normal distribution truncated at a known point (equal to zero in Fig.30).

The degree of fit for the curve shown in Fig.30 was tested by a χ^2-test (Kullback, 1959) similar to the test of section 6.11 but which allows smaller numbers (n_j) of observations in the classes (j). By approximation:

$$\hat{\chi}^2 \approx 2 \sum_j n_j \,{}^{e}\log\,(p_j/\hat{p}_j) = 18.44 \qquad [7.33]$$

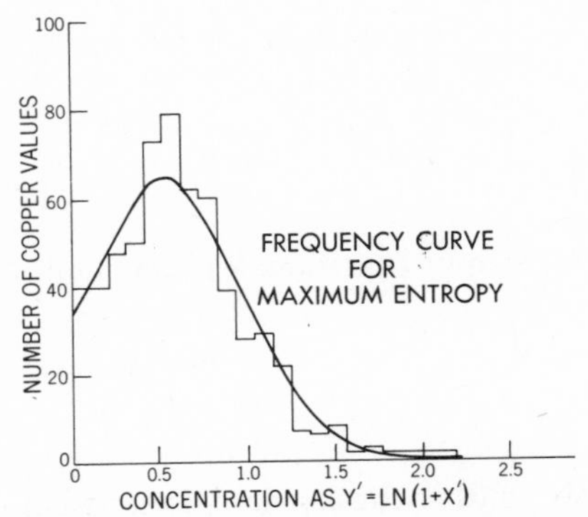

Fig.30. Composite histogram of copper values from Muskox intrusion. Original values x_{ik} from different rock types (i) were made comparable with one another by replacing them by $x'_{ik} = x_{ik}/\bar{x}_i$ where $\bar{x}_i$ represents rock-type mean; horizontal scale is for ${}^e\log(1 + x')$. The best-fitting curve is truncated normal (maximum-entropy solution), indicating truncated lognormality for original data.

where p_j and $\hat{p}_j$ are observed and theoretical relative frequencies, respectively. The classes coincided with the first 16 columns of Fig.30 and with all columns at the upper tail. Assuming (17 – 4 =) 13 degrees of freedom for $\hat{\chi}^2 = 18.44$ leads to acceptance of the model since the 95-% confidence limit then is 22.36.

When in [7.27], the right-hand side $(\epsilon_i X_{i-1})$ is replaced by $(\epsilon_i + \epsilon_i X_{i-1})$, [7.30] becomes $\log(X_n + 1) = \log X_0 + \Sigma e_i$. This approach would give a normal distribution for $\log(X_n + 1)$. Instead of this, we have a truncated normal distribution in Fig.30. The random process ϵ_i leading to $\Sigma\epsilon_i$ after n steps has to be modified when it is attempted to explain Fig.30 by this type of stochastic process-model.

Maximum entropy criterion

It is interesting to review some results of information theory in this context. For other geological applications of concepts from this field, see Dowds (1969) and McCammon (1970). The concept of entropy is used in thermodynamics for evaluating the amount of order or disorder in spatial configurations of attributes. For example, the molecules of an ideal gas can occur anywhere in a confined space and their spatial configuration is completely random at any time. The entropy of the system then is at a maximum (complete disorder). Shannon (1948) applied the concept of entropy to ordinary frequency distributions with n classes (i). His entropy:

$$S = -H \sum_{i=1}^{n} p_i \, {}^e\log p_i \qquad [7.34]$$

where H is an arbitrary constant can be maximized subject to various constraints. We always have $\Sigma_{i=1}^{n} p_i = 1$ for the sum of the relative frequencies in all classes. Suppose that another constraint is:

$$\sum_{i=1}^{n} p_i (x_i - \mu)^2 = \sigma^2$$

Shannon (1948, pp. 628 – 631) found the limiting form of the distribution curve for p_i by maximizing S subject to this constraint that the variance of the final distribution must be equal to the predetermined value σ^2. Partial differentiation of S with respect to the p_i and employing Lagrangian multipliers for the constraints, yields the maximum entropy curve which has the familiar Gaussian shape. Suppose that instead of a predetermined variance, we have a predetermined mean μ; then, the maximum entropy curve is exponential. When both mean and variance are predetermined for nonnegative x, the result becomes in the limit:

$$p(x) = \exp\left[-(A_0 + A_1 x + A_2 x^2)\right]$$

with $x > 0$. A_0, A_1 and A_2 are the Lagrangian multipliers. This result is due to Tribus (1962). Setting $c = -A_0 + A_1^2/4A_2^2$ we can write:

$$p(x) = c \exp\left[-(x + A_1/2A_2)^2\right] \qquad [7.35]$$

which is a Gaussian curve truncated at the origin with its peak at $x = -A_1/2A_2$. This maximum falls at $y' = 0.624$ in Fig.30 for copper.

Other trace elements

The transformation $x' = x/\bar{x}$ was applied to data for several other trace elements in the Muskox layered intrusion (Agterberg, 1965b). Pooled coefficients of variations were 0.84 (copper); 0.48 (zinc), 0.57 (chromium), 0.48 (scandium), 0.38 (nickel), 0.40 (cobalt), 0.80 (barium), 0.64 (strontium), and 0.46 (vanadium). With the exception of copper and barium showing strong truncation effects, all trace elements assumed approximately normal distributions after the transformation.

A full explanation of these results cannot be given without a better understanding of the spatial distribution patterns for crystals and the underlying physico-chemical processes. We know that copper occurs mainly in chalcopyrite crystals disseminated through the rocks. The concentration of these crystals (and to a lesser extent their size) varies according to complex spatial patterns of clusters. Within the clusters, the individual crystals may satisfy the binomial model, but the density of the clusters also changes randomly from place to place.

7.5. A computer simulation experiment

Methods of computer simulation as applied to geology have been discussed by Harbaugh and Bonham-Carter (1970). We use the technique here to illustrate some methods to generate frequency distributions by sequential processes such as that given by [7.25]. For convenience, it is assumed that the independent random variable ϵ, with values ϵ_i which control the individual steps of the process, is normally distributed with zero mean and variance $\sigma^2(\epsilon)$.

The random normal numbers used as the ϵ_i were generated by computer. This can be done by means of a so-called Monte Carlo technique (cf. Harbaugh and Bonham-Carter, 1970).

Three experiments were performed by using the same set of 6000 generated normal random numbers ϵ_i which was divided into 60 groups each consisting of 100 values. The situation which is simulated differs somewhat from that discussed in the context of [7.25]. In order to focus our thoughts, we might think of a geochemical process with the following characteristics. Initially, at time zero for the experiments, a chemical element A is evenly distributed throughout a geologic environment. This implies that a chemical analysis for any specimen would yield the same element concentration value. For convenience, this initial value is set equal to one.

Due to a metasomatic process, the element A becomes mobile and its ions are able to move through the environment where they may be captured in places that are relatively favorable. However, during this process, the system remains closed. The ions of A cannot leave the environment and the average concentration of A for the environment does not change in the course of time but remains equal to one.

Experiment A

When X_i is a random variable representing the concentration for an arbitrary place in the environment after step i of the process, and ϵ_i is the random increase (or decrease) in concentration from step $(i-1)$ to step i, we have:

$$X_1 = 1 + \epsilon_i$$

$$X_i = X_{i-1} + \epsilon_i \quad (i = 2, 3, \ldots, 100)$$

When i goes from 1 to 100, a single value X_i for a given place changes gradually from one to a value smaller or larger than one. In fact, because the values of ϵ_1 are randomly positive or negative, a single value X_i can change its sign several times during the process. A single run consisted of 100 steps.

In total 60 runs were performed so that the behavior of 60 values X_i can be studied for different values of i. If $\sigma(\epsilon) = 0.03$, the result is a curve of approximately normal shape that widens with increasing value of i. It is shown in Fig.31A as a frequency polygon for $i = 25$, 50, and 100, respectively. The standard deviation of X_i increases proportional to the square root of the number of steps.

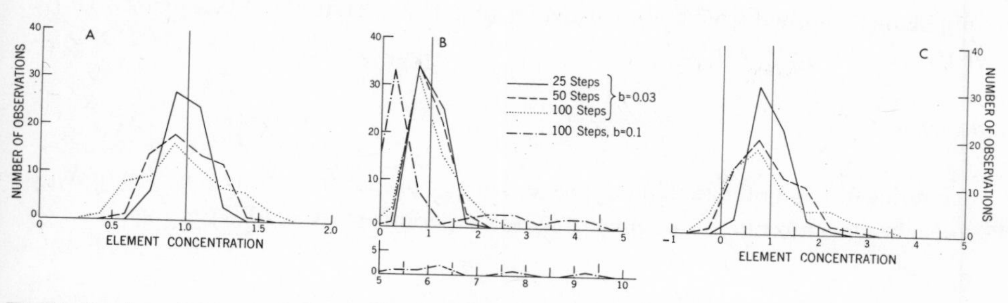

Fig.31. Results of three computer simulations based on random numbers. Experiment A (see text) results in Gaussian curves (A); experiment B in lognormal curves (B); and experiment C in three-parameter lognormal curves (C). Negative values in (C) are physically not possible but may be set equal to zero.

It is noted that the pattern of Fig.31A also may be regarded as an approximation for the diffusion equation of physical chemistry with:

$$p(t, x) = \frac{1}{\sqrt{2\pi Dt}} \exp\left[-\tfrac{1}{2}(x - ct)^2/Dt\right]$$

If particles in Brownian motion all begin their random walk at a point source, $p(t, x)$ is the probability that a single particle occurs at distance x after time t. At any given time, $p(t, x)$ describes a Gaussian curve with $\mu = ct$ and $\sigma^2 = Dt$. It also represents the density of particles at location x after time t. The constant c describes the "drift" or a linear change in location of the average; $c = 0$ in Fig.31. Scheidegger and Langbein (1966) used this model to represent the effect of erosion on mountain ranges.

Experiment B

The specifications for this process are:

$$X_1 = 1 + \epsilon_1$$

$$X_i = X_{i-1} + \epsilon_i \cdot X_{i-1}$$

Except for the first step, this process is identical to that described by [7.27] through [7.30]. It illustrates the law of proportionate effect. The frequency polygons are approximately lognormal in shape. This process differs from the one discussed previously in that random increments have zero mean. The mean remains equal to one, but the logarithmic variance increases in proportion to the number of steps.

If $\sigma(\epsilon) = 0.03$, the shape of the frequency polygon remains almost normal between steps $i = 25$ and 100 as is shown in Fig.31B. However, when $\sigma(\epsilon) = 0.1$, the result is a positively skew frequency distribution after 100 steps. This discrepancy is in agreement with previous result discussed in the context of [7.32]. If $\sigma(\epsilon) = 0.03$, $\gamma \approx \sigma(\log X) = 0.3$ after 100 steps and for such a small coefficient of variation γ, the lognormal distribution can not be distinguished from a normal distribution. However, if $\sigma(\epsilon) = 0.1$, $\sigma(\log X) = 1.0$ after 100 steps which is markedly skew.

Experiment C

According to [7.25], the random change ϵ_i can be a proportion of any function of X_{i-1}. We have chosen a mixture of the preceding two experiments (A and B), or:

$$X_1 = 1 + \epsilon_1$$

$$X_i = X_{i-1} + (X_{i-1} + 1)\,\epsilon_i$$

The results are again shown for steps $i = 25$, 50, and 100. (Fig.31C). The theoretical frequency curves in this situation are three-parameter lognormal. For this experiment, the third parameter is equal to one. In general, if a random variable X has the three-parameter lognormal frequency distribution, it means that $\log(X + \alpha)$ is normally distributed where α is a positive or negative constant.

If the results of experiment C are considered in the light of the physico-chemical situation that is being simulated, the result would not be feasible, because at the low-value end of the frequency curve we have obtained negative element-concentration values. In order to avoid this difficulty, we could subject the simulation experiment to an additional constraint, namely that the element-concentration value cannot become negative and remains at zero upon reaching this boundary. The result would be a discontinuity in the frequency distribution near zero where a relatively sharp peak would accumulate.

In the theory of diffusion processes, the zero position would be called an absorbing boundary. Otherwise, the result remains three-parameter lognormal after this modification.

Gold values in South African mines

The result of experiment C provides one possible explanation for the distribution of gold in the relatively thin conglomerate beds (reefs) of the Witwatersrand goldfields in South Africa. Until 1960, the general opinion was that these gold-concentration values satisfy a lognormal model. However, Krige (1960) has shown that a significant refinement in ore-evaluation methods is obtained by assuming that the underlying frequency distribution is three-parameter lognormal instead of lognormal (see Fig.32).

A remark on measures is in order here. The unit for measuring gold in South Africa was inch-dwt. value or inch pennyweight. The metric equivalent of a pennyweight is 64.8 mg. An inch is 2.54 cm. Hence, an inch-dwt. is equivalent to 0.165 cm g (centimeter-gram) and this measure is used for South African gold values (since 1971). The inch-dwt. value represents the amount of gold present in a column with a base of one square inch and perpendicular to the reef;

The frequency curve for 1000 values from the Merriespruit Mine is also shown in Fig.34A. A smooth curve was drawn by hand. There is a distinct departure from a straight

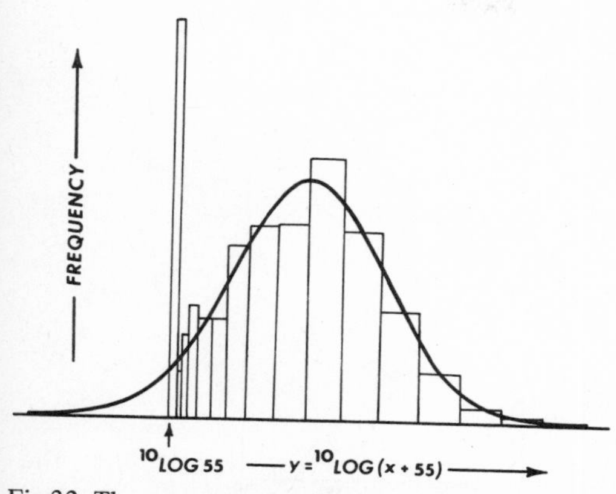

Fig.32. Three-parameter lognormal model for gold values in Merriespruit Mine, Witwatersrand goldfield, South Africa. (After Sichel, 1961, fig.2; original data from Krige, 1960.) The cumulative frequency curve is normal for positive values (cf. Fig.34A), because of an additional peak at point with $x = 0$. Note that this situation differs from that in Fig.30, although the transformation is of the same type. For positive values of x, the situation is equivalent to that in Fig.31C.

line which would indicate lognormality. However, if a value $\alpha = 55$ is added, it appears that $^{10}\log(x + 55)$ is normally distributed except for a discrepancy at the value $^{10}\log 55$ for $x = 0$. The best-fitting normal curve is shown in Fig. 32.

A graphical method for reconstructing the generating process

Our starting point for this discussion is [7.25] which can be written in the form:

$$dX_i = g(X_i) \cdot \epsilon_i \qquad [7.36]$$

where dX_i represents the difference $X_i - X_{i-1}$ in infinitesimal form.

Several systems of curve-fitting are in existence for relating an observed distribution to a normal form. An example of this is Johnson's (1949) system which makes use of the relationship given in [7.36]. An excellent review of several systems has been given by Kendall and Stuart (1958). The so-called Johnson S_B-system results in lognormal distribution for the variable $X/(1 + X)$. The function $g(X_i)$ in [7.36] for this system is $g(X_i) = X_i(1 - X_i)$. It has been used by Jizba (1959) and Rogers and Adams (1963) for geochemical processes.

It may be possible to obtain information on the function $g(X_i)$ by a graphical analysis based on the normal probability plot. This method is illustrated in Fig.33 for a lognormal situation.

A lognormal curve was plotted on normal probability paper (arithmetic scale) in Fig.33A. We then use an auxiliary variable z which is plotted along the horizontal axis and has the same scale as x along the vertical axis. The variable x is a function of z or $x = G(z)$. The slope of $G(z)$ is measured in a number of points giving the angle φ. For example, if $x = z = 20$ in Fig.33A, $\varphi = \varphi_1$. We have:

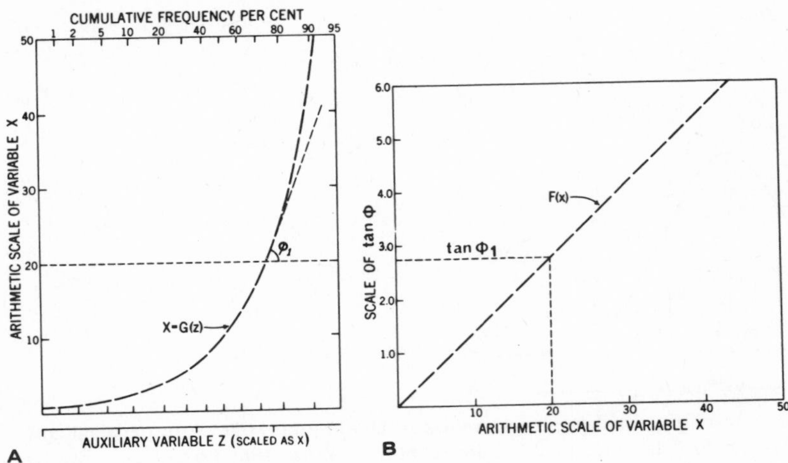

Fig.33. Graphical construction of $F(x)$ which is proportional to the tangent of the slope of $x = G(z)$ representing theoretical lognormal distribution plotted on arithmetic probability paper. (From Agterberg, 1964a.)

$$\frac{dG(z)}{dz} = \tan\varphi \qquad [7.37]$$

If, in a new diagram (Fig. 33B), $\tan\varphi$ is plotted against x, we obtain a function $F(x)$ which represents $g(X_i)$ of [7.36] except for a multiplicative constant that remains unknown.

This procedure is based on the formula:

$$Z = \int \frac{dX_i}{g(X_i)} = \Sigma\epsilon_i \qquad [7.38]$$

which is the general solution of [7.36]. It is assumed that the random variable Z is normally distributed.

The function $g(X_i)$ derived in Fig.33B for the lognormal curve of Fig.33A is simply a straight line through the origin. This result is in accordance with the theory of proportionate effect. In order to illustrate the graphical method, it has been applied in two other situations. The original curves are shown in Fig.34A on logarithmic probability paper. They are for the 1000 gold values from Merriespruit discussed before, and for a set of 84 copper concentration values for a sulphide deposit that surrounds the Muskox intrusion as a rim. The results are shown in Fig.34B.

The generating function $g(X_i)$ for Merriespruit is according to a straight line whose prolongation would intersect the X-axis at a point with negative abscissa $\alpha \approx 55$. The second situation (Muskox sulphides) gives a function $g(X_i)$ that resembles a broken line. It would suggest a rather abrupt change in the generating process for copper concentra-

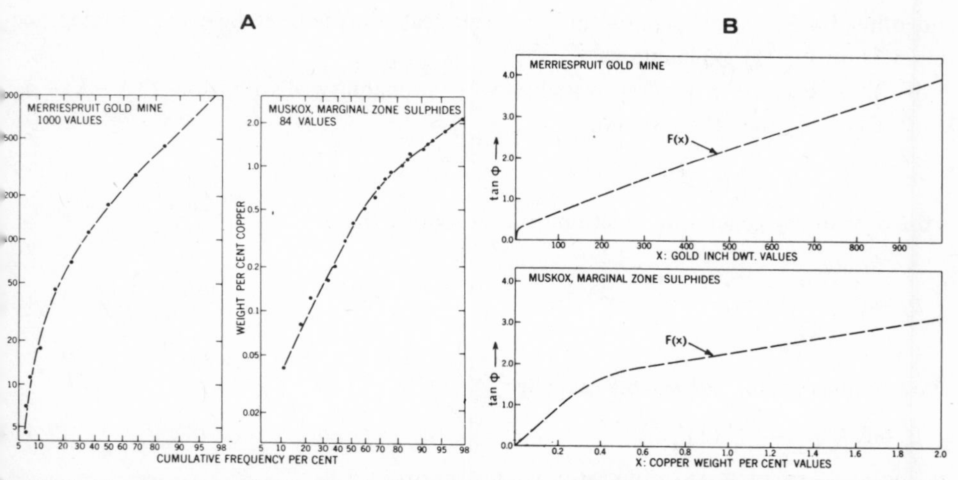

'ig.34. A. Two frequency distributions which depart from the lognormal model (logarithmic probabil- ·y paper). Example of 1000 values from Merriespruit gold mine is after Krige (1960). B. Method of 'ig.33 gave two types of $F(x)$, that for the Merriespruit Mine is approximately a straight line which 'ould intersect X-axis at a point with abscissa of about -55 (value used in Fig.32 for the same set of ata). Function $F(x)$ for copper values would suggest a change in the generating process after 0.5% opper concentration. $F(x)$ is proportional to $g(X_i)$ in [7.36].

on after the value $X_i \approx 0.5\%$. The influence of X_i on dX_i may have decreased with in- reasing X_i.

The usefulness of a graphical method of this type is limited, in particular because ran- om fluctuations in the original frequency histograms can not be considered carefully. owever, the method is rapid and offers suggestions with respect to the physico-chemical rocesses that may underly a frequency distribution.

In general, the outcome of a frequency-distribution analysis depends strongly on the ethod of approach taken to the problem. For example, curve C in Fig.27 shows some semblance to the two curves of Fig.34A. However, it was formed artificially as a mix- ıre of two lognormal distributions (curves A and B). By adding a constant $\alpha = 150$ to all alues of C and plotting the result on the same graph, we obtain curve D, which is not far om being a straight line. This indicates that C can be approximated by a three-parameter gnormal distribution with $\alpha = 150$ which is truncated at its lower end, because element- oncentration values cannot be negative.

.6. Probability generating functions

In the second part of this chapter, we discuss several examples of results for geological requency distributions obtained by applying more advanced methods. For reasons of pace, the theoretical introduction is kept brief. The reader is referred to Feller (1968) and other books on probability and mathematical statistics (see Selected Reading, section 5.3) for further explanations.

If X is a discrete random variable with probability distribution $P(X = k) = p_k$; $k = 0, 1, 2, \ldots$, then the function:

$$g(s) = p_0 + p_1 s + p_2 s^2 + \ldots \tag{7.39}$$

is the probability generating function of X. It follows that:

$$p_k = \frac{1}{k!}\left[\frac{d^k}{ds^k} g(s)\right]_{s=0} = \frac{1}{k!} g^{(k)}(0) \tag{7.40}$$

The r-th moment (cf. section 6.8) satisfies:

$$\mu'_r = \Sigma\, k^r p_k = g^{(r)}(1) \tag{7.41}$$

For the mean and variance (cf. Feller, 1968, p.360):

$$E(X) = g'(1); \qquad \sigma^2(X) = g''(1) + g'(1) - [g'(1)]^2 \tag{7.42}$$

Moment generating functions

If X is a continuous random variable, its moment generating function $m(u)$ satisfies:

$$m(u) = E(e^{uX}) = \int_{-\infty}^{\infty} e^{ux} f(x)\, dx \tag{7.43}$$

Now:

$$\mu'_r = \int x^r f(x)\, dx = m^{(r)}(0) \tag{7.44}$$

and:

$$E(X) = m'(0); \qquad \sigma^2(X) = m''(0) - [m'(0)]^2 \tag{7.45}$$

Moments of the lognormal distribution

The moment generating function of the normal distribution is:

$$m(u) = \exp[\mu u + \sigma^2 u^2/2] \tag{7.46}$$

If [7.7] for the lognormal distribution is substituted into [7.44] we find:

$$\mu'_r = \int x^r \phi({}^{e}\log x)/x\, dx$$

By writing $y = {}^{e}\log x$, this becomes:

$$\mu_r' = \int e^{ry} \phi(y)\, dy \qquad [7.47]$$

Consequently, the moments of the lognormal distribution constitute the moment generating function of the normal distribution, or:

$$\mu_r' = \exp\,[\mu r + \sigma^2 r^2/2] \qquad [7.48]$$

where $\mu = E({}^{e}\log X)$ and $\sigma^2 = \sigma^2({}^{e}\log X)$. The mean and variance of the lognormal are also written as α and β^2. Then:

$$\alpha = e^{\mu+\sigma^2/2}\,; \qquad \beta^2 = e^{2\mu+2\sigma^2} - e^{2\mu+\sigma^2} \qquad [7.49]$$

Characteristic functions

The characteristic function of X is defined as:

$$g(u) = E(e^{iuX}) = \int_{-\infty}^{\infty} e^{iux} f(x)\, dx \qquad [7.50]$$

This is also called the complex Fourier transform of $f(x)$. There exists an inverse relationship:

$$f(x) = (1/2\pi) \int_{-\infty}^{\infty} e^{-iux} g(u)\, du \qquad [7.51]$$

For the moments we now have:

$$\mu_r' = E(X^r) = i^{-r} g^{(r)}(0) \qquad [7.52]$$

The characteristic function of the normal distribution is:

$$g(u) = \exp\,[i\mu u - \sigma^2 u^2/2] \qquad [7.53]$$

Characteristic functions have a wider field of application than moment generating functions.

Sum of random variables

Suppose that X and Y are two independent discrete random variables with probability distributions:

$$P(X = k) = p_k\,; \qquad P(Y = k) = q_k$$

The distribution of their sum $Z = X + Y$ satisfies:

$$P(Z = k) = r_k = p_0 q_k + p_1 q_{k-1} + \ldots + p_k q_0 \quad [7.54]$$

The sequence r_k is called the convolution of the sequences p_k and q_k.

If $g_x(s), g_y(s)$ and $g_z(s)$ are the probability generating functions of X, Y and Z, respectively, it follows from [7.54] that:

$$g_z(s) = g_x(s) \cdot g_y(s) \quad [7.55]$$

If X, Y, and $Z = X + Y$ are continuous random variables, their characteristic functions satisfy:

$$g_z(u) = g_x(u) \cdot g_y(u) \quad [7.56]$$

7.7. Examples of discrete random variables

Binomial distribution

If X is a Bernoulli variable with $P(X = 0) = 1 - p, P(X = 1) = p$, its generating function is:

$$g(s) = (1 - p) + ps$$

The binomial distribution results from n Bernoulli trials. Repeated application of [7.55] therefore gives the generating function of the binomial with:

$$g(s) = [(1 - p) + ps]^n \quad [7.57]$$

Application of [7.42] leads to:

$$E(X) = np; \qquad \sigma^2(X) = np(1 - p) \quad [7.58]$$

Poisson distribution

If points occur at random in a plane, their frequency per equal-area cell satisfies the Poisson distribution. One method of deriving the Poisson distribution is to define a constant $\lambda = np$ for the binomial distribution and letting n and p approach infinity and zero respectively, whereas λ is kept constant. X satisfies the Poisson distribution if:

$$P(X = k) = e^{-\lambda} \lambda^k / k! \quad [7.59]$$

with: $g(s) = e^{-\lambda(1-s)}$ [7.60]

and: $E(X) = \sigma^2(X) = \lambda$. [7.61]

The mean and variance of the Poisson distribution are equal to one another.

Negative binomial distribution

Let us start a succession of Bernoulli trials by randomly selecting black or white cells from a mosaic. We may calculate the probability of how many times the trial must be repeated before $X = \Sigma_{i=1}^{m} X_i = r$ where r is a fixed positive integer number.

In this notation, m represents the number of experiments it takes to obtain r. We can write $m = k + r$, where k is another integer number.

The probability of a one (for a black cell) at the m-th trial is p. This probability must be multiplied by the probability that there were exactly k zeros (for white cells) during the preceding $m - 1$ $(= k + r - 1)$ experiments. This latter probability is binomial with:

$$\binom{r+k-1}{k} p^{r-1}(1-p)^k$$

The product:

$$p_k = \binom{r+k-1}{k} p^r q^k \quad [7.62]$$

is called the negative binomial distribution. Its generating function is:

$$g(s) = \left(\frac{p}{1-qs}\right)^r \quad [7.63]$$

By using eq. [7.42], we obtain:

$$E(X) = \mu = rq/p; \qquad \sigma^2(X) = rq/p^2 \quad [7.64]$$

where $q = 1 - p$.

For the negative binomial we always have $\sigma^2 > \mu$. On the other hand, for the ordinary binomial, $\sigma^2 < \mu$. The Poisson has $\sigma^2 = \mu$. A calculation of mean and variance for a set of discrete geologic data often provides a guideline as to which one of the three distributions should be fitted (cf. Ondrick and Griffiths, 1969).

Geometric distribution

If $r = 1$, the negative binomial distribution reduces to the so-called geometric distribution with:

$$p_k = pq^k; \qquad g(s) = \frac{p}{1-qs} \quad [7.65]$$

and: $\mu = q/p; \qquad \sigma^2 = q/p^2$ [7.66]

Observed distributions for thickness of lithologic units

Krumbein and Dacey (1969) have found that the thicknesses of lithologic units in the Oficina formation, eastern Venezuela, can be described by geometric distributions. Their data consisted of a series of letters A, B, C, and D, obtained by Scherer (1968) who has coded the rock types: (A) sandstone; (B) shale; (C) siltstone; and (D) lignite, in a well core at 2-ft. (61 cm) intervals.

For each rock type, the frequencies for sequences of consecutive letters can be determined and plotted in a diagram. For example, one counts how many times the sequences A, AA, AAA, . . . occur in the total series. The results are shown in Fig.35. Krumbein and Dacey (1969) have fitted geometric distributions to these data and tested the degree of fit by means of the χ^2-test. The theoretical distribution for shale that satisfies [7.65] is shown at the left-hand side of Fig.35. The good fit of the geometric distribution indicates that the deposition of these rock types was controlled by the following stochastic process.

Suppose that for a lithology, say sandstone, $p(A)$ denotes the probability that at a distance of 2 ft., another lithology (not sandstone) will occur. Obviously, $q(A) = 1 - p(A)$ then represents the probability that the same rock type (sandstone) will occur.

If for every lithology p(and q) remained constant during deposition of the entire series, then the probability that a sequence for any lithology is k letters long, satisfies the negative binomial form. We have $r = 1$, because each sequence is terminated at the point where it is replaced by another lithology. The result is a geometric distribution for each lithology.

It is possible to divide the probability p for every lithology into three parts, one for each of the three other lithologies. All probabilities can be arranged in the following matrix:

$$\begin{bmatrix} q(\mathrm{A}) & p(\mathrm{AB}) & p(\mathrm{AC}) & p(\mathrm{AD}) \\ p(\mathrm{BA}) & q(\mathrm{B}) & p(\mathrm{BC}) & p(\mathrm{BD}) \\ p(\mathrm{CA}) & p(\mathrm{CB}) & q(\mathrm{C}) & p(\mathrm{CD}) \\ p(\mathrm{DA}) & p(\mathrm{DB}) & p(\mathrm{DC}) & q(\mathrm{D}) \end{bmatrix}$$

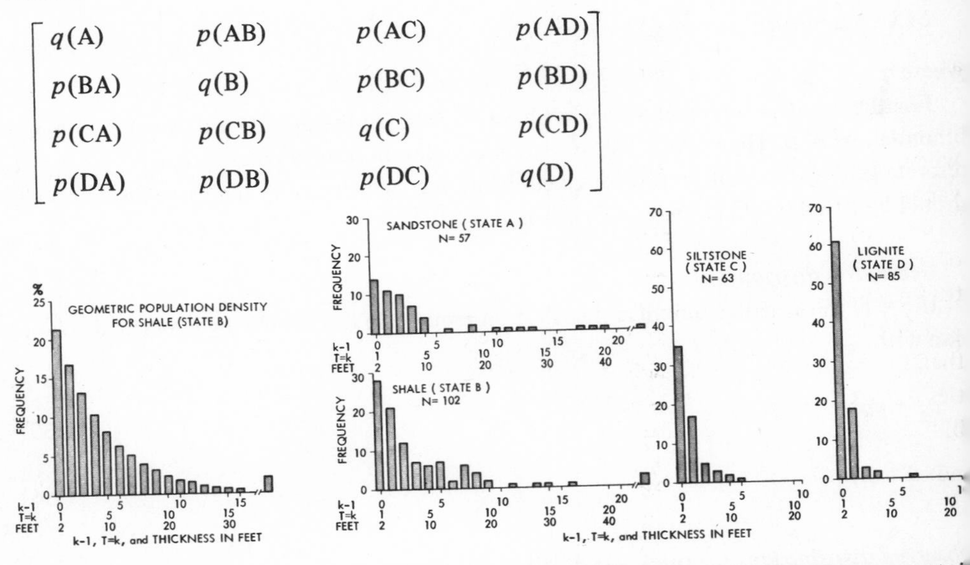

Fig.35. Theoretical frequency distribution for shale and distributions for all lithological components in Oficina Formation, eastern Venezuela. All are geometric distributions by close approximation. (From Krumbein and Dacey, 1969.)

where, e.g., $p(\mathrm{AB})$ denotes the probability of A being followed by B. This is called a transition matrix. The series of letters A, B, C, and D is called a Markov chain of the first order if the probabilities in the transition matrix remain constant. For our discussion here, we have considered only two probabilities p and q for each lithology. For example, sandstone (A) has:

$$p(\mathrm{A}) = p(\mathrm{AB}) + p(\mathrm{AC}) + p(\mathrm{AD})$$

$$q(\mathrm{A}) = 1 - p(\mathrm{A})$$

These two probabilities determined the geometric distribution for sandstone.

Krumbein and Dacey (1969) have calculated the full transition matrix as:

	A	B	C	D
A	0.787	0.071	0.075	0.067
B	0.048	0.788	0.061	0.103
C	0.105	0.316	0.430	0.149
D	0.182	0.388	0.132	0.298

The subject of Markov chains will be discussed in more detail in Chapters 11 and 12.

7.8. Compound random variables

Suppose that a number of random variables X_i have the same frequency distribution and generating function $g_x(s)$. We are interested in the sum Y with:

$$Y = X_1 + X_2 + \ldots + X_K \qquad [7.67]$$

where the number of terms K is a discrete random variable that is independent of the variables X_i.

For example, the X_i may represent sizes of ore deposits and K the number of deposits occurring in a cell of unit area. Then, Y represents the sum of the sizes of K deposits, or total amount of ore per cell.

Let us temporarily assume that the X_i are discrete random variables. The probability that the sum Y is equal to the integer number j then is equal to the sum of the probabilities $p_1 \cdot P(X_1 = j)$, $p_2 \cdot P(X_1 + X_2 = j), \ldots$, where the $p_1, p_2, \ldots$, represent the probabilities that K is equal to 1, 2, ... Consequently:

$$P(Y = j) = \sum_{i=0}^{\infty} p_i \cdot P(X_1 + X_2 + \ldots + X_i = j) \qquad [7.68]$$

This summation may start at $i = 0$, for which $P(X_0 = j)$ and $P(Y = j)$ are both equal to zero.

If $g_y(s)$ and $g_k(s)$ represent the generating functions for Y and K:

$$g_y(s) = \Sigma\, p_i\, \{g_x(s)\}^i$$

When this expression is compared to [7.39], it follows that we can write:

$$g_y(s) = g_k\, \{g_x(s)\} \qquad [7.69]$$

This expression is called a compound generating function.

The best-known application of this theory is the so-called compound Poisson distribution which arises when K is a Poisson variable. According to [7.60], [7.69] becomes:

$$g_y(s) = \exp\, [-\lambda + \lambda g_x(s)] \qquad [7.70]$$

An interesting possibility is that X also is a Poisson variable. This leads to the double Poisson process. If X satisfies the so-called logarithmic distribution Y becomes negative binomial (see Feller, 1968, p.291).

If X and K are both discrete random variables [7.41] can be used to derive the moments of Y from $g_y(s)$.

If X is a continuous random variable, whereas K is discrete, we obtain for the characteristic function of Y:

$$g_y(u) = g_k\, [g_x(u)] \qquad [7.71]$$

and [7.52] is used for the moments of Y. In particular, we may derive for the mean and variance:

$$E(Y) = E(K)\, E(X); \qquad \sigma^2(Y) = E(K)\, \sigma^2(X) + \sigma^2(K)\, E^2(X) \qquad [7.72]$$

Suppose now that X is normally distributed with:

$$g_x(u) = \exp\, [i\mu u - \sigma^2 u^2/2]$$

Then, if K is kept Poisson:

$$g_y(u) = \exp\, [-\lambda + \lambda(i\mu u - \sigma^2 u^2/2)] \qquad [7.73]$$

Eq. [7.72] gives:

$$E(Y) = \lambda\mu; \qquad \sigma^2(Y) = \lambda(\sigma^2 + \mu^2) \qquad [7.74]$$

If X has the lognormal distribution instead of the normal distribution, we obtain:

$$E(Y) = \lambda\alpha; \qquad \sigma^2(Y) = \lambda(\alpha^2 + \beta^2) \qquad [7.75]$$

where $\alpha = E(X)$ and $\beta^2 = \sigma^2(X)$. By using $\mu = E({}^{e}\log X)$ and $\sigma^2 = \sigma^2({}^{e}\log X)$, we have according to [7.49]:

$$E(Y) = \lambda\, e^{\mu + \frac{1}{2}\sigma^2}; \qquad \sigma^2(Y) = \lambda\, e^{2\mu + 2\sigma^2} \qquad [7.76]$$

These are basic results for an exploration strategy problem as formulated by Allais (1957) that will be discussed in the next section.

If the negative binomial is used for K instead of the Poisson distribution, whereas X is kept lognormal, we obtain for [7.71]:

$$g_y(u) = \left[\frac{p}{1 - q g_x(u)}\right]^r \qquad [7.77]$$

where p, q, and r are as in [7.63].

For negative binomial K and lognormal X, we also have:

$$E(Y) = (rq/p)\, e^{\mu+\sigma^2/2}$$

$$\sigma^2(Y) = (rq/p)\, e^{2\mu+\sigma^2}(e^{\sigma^2} + q/p) \qquad [7.78]$$

This result was derived by Uhler and Bradley (1970) and applies to a problem of exploration strategy conceived by Griffiths (1966b).

7.9. A problem of exploration strategy

One of the objectives of applied geology is to design methods for locating regions that are likely to contain hidden mineral deposits. Exploration is expensive mainly because large deposits are rare events and it is increasingly difficult to locate them. Several methods have been developed for the evaluation of regional mineral potential thus providing a starting point for the design of exploration strategies. Some of these techniques will be reviewed in the remainder of this chapter and later (see Chapter 15). For other, more complete reviews, see Griffiths and Singer (1971b) and Singer (1972).

McKelvey (1972) suggested that reserves for nonfuel minerals are roughly equal to their crustal abundance in regions which are sufficiently large. Early studies using similar, broadly based assumptions were performed by Nolan (1950) and Blondel and Ventura (1954). The latter authors analyzed annual values of mineral production per country on a world-wide basis. Later work on cumulative total production has resulted in the concept of unit regional value as a basis for predicting mineral resources of large regions (Griffiths and Singer, 1971a). According to Griffiths and Singer, the average value of mineral endowment per square mile is largely independent of geology and depends mainly on intensity of development. For example, the average value per square mile in the 48 conterminous states of the U.S.A. is about $250,000; 1967-values for Canada and Australia were $18,000 and $9,500, respectively, indicating a large potential for these countries where less exploration and development was done than in the U.S.A.

Allais (1957) has made a study to define an optimum strategy for exploring the Sahara Desert for mineral deposits. In this study, he made the following two assumptions:

(1) The sizes of mineral deposits can be measured by their total dollar value at some time and these values show a lognormal distribution if all commodities are lumped together.

(2) The places where the mineral deposits occur are randomly distributed across a large area.

The appropriate model then would be the Poisson distribution for the number of deposits occurring in a cell of equal area. Later studies by Griffiths (1966b) have indicated that mineral deposits over a large area usually do not occur independently of each other but tend to cluster. The negative binomial may then be used instead of the Poisson.

This approach has been considered by Uhler and Bradley (1970) for determining the economic prospects of petroleum exploration over large regions. They have taken the Province of Alberta in Canada for example.

The size of 314 oil deposits in Alberta is not far from being lognormal as is indicated by the following tabulation which is based on logarithmically transformed data.

Statistic	*Value*	*Confidence limits* (95%)	(99%)
Skewness	0.3989	±0.2300	±0.3290
Kurtosis	2.6908	±0.41	±0.54

As shown in this table, Uhler and Bradley have applied the normality test discussed in section 6.13. A departure from normality (or lognormality for original data) is indicated because the skewness is larger than permitted for a normal population. In order to test which spatial distribution should be used for occurrence of oil deposits, Uhler and Bradley (1970) divided the Alberta sedimentary basin into 8811 cells, each measuring 5 miles (8 km) on a side, and counted the number of deposits per cell. The results are shown in Table XVIII.

The results of Table XVIII show that the negative binomial provides a better fit than the Poisson. This is also borne out by a χ^2-test according to which the negative binomial model is acceptable.

We may now consider the problem of how much oil occurs in a grid area of given size. The total amount of oil for a larger area consisting of a larger number of grid areas can be calculated readily if the frequency distribution is given for smaller cells. Eq. [7.78], in addition to the mean, provides an estimate of the variance.

Uhler and Bradley (1970) applied the model by: (1) placing a random grid with unit areas of size 10 × 10 miles over the entire sedimentary basin of Alberta, and (2) selecting random samples consisting of 200 of such unit areas. Each of the random samples covers about 9% of total area and is fairly representative for the sedimentary basin.

The model may not be applicable if the sample is not representative for the entire area. For example, Uhler and Bradley (1970) pointed out that a single reservoir (Cardium Reservoir, Pembina Field) accounts for a substantial portion of the total original reserves in Alberta. In that situation, a set of adjacent grid areas of 10 × 10 miles, which is representative for a geographically distinct portion of the basin only, may yield a mean that is greater or smaller than accounted for by the statistical model. A general explanation for a

TABLE XVIII
Fits of the Poisson and negative binomial distributions to the spatial distribution of oil deposits for 5 × 5 miles grid areas, Alberta. (After Uhler and Bradley, 1970, table 1.)

Deposits	*Observed frequency*	*Poisson frequency*	*Negative binomial*
0	8586	8508.5	8584.3
1	176	303.0	176.8
2	35	5.4	39.1
3	13	0.1	11.3
4	6	0.0	3.6
5	1	0.0	1.2
6 or more	0	0.0	0.7

phenomenon similar to this one is that the random variables X_i and K of [7.67] are not independent but positively correlated.

Prediction of total value of mineral deposits in the Canadian shield

A second, more detailed example is given to provide another application of the model by predicting the mineral potential for a large area that has been explored only slightly. We used data from published sources and point out several problems associated with prediction on the basis of the compound distribution model.

Slichter (1960) compiled information on the number of valuable mines per unit area, for 185 units of 1000 sq. miles each in Ontario. The rocks for the entire area of size 185,000 sq. miles belong to different structural provinces and geological environments in the Canadian shield. In total, the area constitutes approximately 9% of the shield. Slichter initially tested the Poisson model for these data but found that this does not provide a fit that is satisfactory. He next fitted the so-called exponential distribution. Later, Griffiths (1966b) has shown that the negative binomial model can be applied to Slichter's data.

Suppose that the "control" area in Ontario is representative for the entire shield. Because there occur 147 mines in Slichter's area, the total number of mines for the entire shield would be approximately $(100/9) \times 147 \approx 1600$ mines. Of course, this estimate will be too low if not all mines in the control area have been discovered. On the other hand, it will be too large, if the control area is richer in mines than the remainder of the Precambrian shield.

The problem of predicting the total value of all orebodies in the Canadian shield was discussed by De Geoffroy and Wu (1970) who argued as follows. The Canadian shield occupies an area of about 2,146,000 sq. miles. Nearly 90% of commercial mineral deposits known to date occurs in volcanic belts and Lower Proterozoic sedimentary belts. These relatively favorable environments, which alternate with other rock types, occupy approximately 341,220 sq. miles or 15.9% of the total area. An area of 50,000 sq. miles

(Timmins–Kirkland Lake area, Ontario; Noranda–Val d'Or area, Quebec) for the favorable environments was treated in detail by De Geoffroy and Wu with the following results:

Size of area: 50,000 sq. miles

Total number of deposits: 254 orebodies

Average number of deposits per 10 × 10 sq. miles: $\bar{k} = 254/500 = 0.508$

Variance: $s^2(k) = 2.016$

Average value of deposits: $\bar{x} = 60.5 \cdot 10^6$ dollars (based on 1968-metal prices)

Logarithmic mean (base e): $\overline{\log x} = 2.858$ (unit of x is 10^6 dollars)

Variance: $s^2(\log x) = 3.103$

These statistics can be used for a preliminary prediction of number of deposits and their value in the 341,220 – 50,000 = 291,220 sq. miles of territory that, for the larger part, has not been explored in the same detail as the 50,000 sq. miles of favorable environment in the control area. According to the statistics compiled by De Geoffroy and Wu (1970), 0.508 × 341,220/100 = 1733 mines are predicted to exist on the shield. This number is fairly close to that based on Slichter's data. However, 129 of 254 orebodies of De Geoffroy and Wu fall in Slichter's area, and the control areas underlying the two estimated values overlap in part. Also, for the second estimate, deposits outside the more favorable environments were not considered.

De Geoffroy and Wu (1970) fitted the negative binomial to the number of orebodies per unit area with satisfactory results. They also fitted a normal distribution to the logarithms of values in dollars for mines. A χ^2-test as the one discussed in section 6.13 gave $\hat{\chi}^2 = 12.51$ for 6 degrees of freedom (9 classes). This value may be compared to $P\{\chi^2(6) < 12.5\} = 94.8\%$.

This result indicates that the lognormal model provides a degree of fit that is only moderately good. In fact, the histogram for logarithmically transformed data shows a positive skewness that is probably meaningful. We are now facing a practical estimation problem that also will be discussed in the next section. If lognormality is accepted, the risk is taken of the model yielding biased estimates because of a departure from lognormality. The present model [7.78] is fairly sensitive to departures from lognormality. This argument also applies to the previous example of oil pools in Alberta.

According to [7.49], the expected value of a lognormally distributed variable satisfies:

$$\alpha = \exp(\mu + \sigma^2/2) \qquad [7.79]$$

Setting $\mu = \overline{\log x} = 2.858$ and $\sigma^2/2 = \frac{1}{2}s^2(\log x) = 1.551$, we would obtain the estimate $\hat{\alpha}' = 82.18 \cdot 10^6$ dollars. The bias of this estimate will be discussed in section 7.10. An unbiased estimate can be obtained by replacing σ^2 by $0.985 \times s^2 = 3.056$ instead of $s^2 = 3.103$. This results in $\hat{\alpha} = 80.34 \cdot 10^6$ dollars. The numbers $\hat{\alpha}'$ and $\hat{\alpha}$ exceed the arithmetic average $\bar{x} = 60.5 \cdot 10^6$ dollars. It is noted that we may not use the antilog $\exp(\mu)$ as an estimator for average value. This would give $17.42 \cdot 10^6$ dollars which is definitely too small.

Inasmuch as the means $\bar{x} = 60.5 \cdot 10^6$ and $\overline{\log x} = 2.858$ are based on 254 data, they both are fairly good estimators. We may accept the ratio 60.5/17.42 = 3.473 as an alternative estimator for the factor $\exp(\sigma^2/2)$. This gives $\hat{\sigma}^2$ (log X) = 2.490 instead of s^2(log x) = 3.10 (see above). A reason for replacing the previous estimate of logarithmic variance is our suspicion that the distribution departs from the lognormal model. The previous estimates of variance may be too large because of positive skewness.

The number of orebodies in favorable environments outside the control area is estimated at (219,220/100) × 0.508 = 1478. With $\hat{E}(\bar{X}) = 60.5 \cdot 10^6$, their total value would amount to $1478 \times 60.5 \cdot 10^6 = 89.4 \cdot 10^9$ dollars.

The variable Y introduced in [7.67] would represent the value per unit area of 10 × 10 miles size with:

$$\hat{E}(Y) = 0.508 \times 60.5 \cdot 10^6 = 30.73 \cdot 10^6 \text{ dollars} \qquad [7.80]$$

The variance of the variable Y can be calculated by [7.78].

If the negative binomial model is satisfied, then:

$$E(K) = rq/p \approx \bar{k} = 0.508$$
$$\sigma^2(K) = rq/p^2 \approx s^2(k) = 2.016 \qquad [7.81]$$

With $p + q = 1$, this gives the estimated values $p' = 0.2520$; $q' = 0.7480$; $r' = 0.1711$. If $\hat{\mu} = 2.858$ and $\hat{\sigma}^2 = 2.490$, [7.78] yields:

$$\hat{E}(Y) = 30.73 \cdot 10^6 \text{ dollars}$$
$$\hat{\sigma}^2(Y) = 27922 \cdot 10^{12}; \qquad \hat{\sigma}(Y) = 167 \cdot 10^6 \text{ dollars} \qquad [7.82]$$

Our coefficients are such that the expression for $\hat{E}(Y)$ duplicates [7.80]. Until now, an analytical expression for the random variable Y has not been developed. However, because Y is based on both the negative binomial and lognormal distribution, its frequency distribution will be positively skew and multimodal. Because of the central-limit theorem, a variable $Z = \Sigma_{i=1}^{n} Y_i$ converges to normal form when n increases. For example, if $n = 400$, we have $\hat{E}(Z) = 12.30 \cdot 10^9$ and $\hat{\sigma}(Z) = 3.34 \cdot 10^9$ dollars. The coefficient of variation $\hat{\gamma}(Z) = 0.27$ is quite small. In that situation, it is permissible to determine a confidence interval $\pm 1.96\, \hat{\sigma}(Z)$ for $\hat{E}(Z)$. It would mean that for a sample consisting of 400 single 10 × 10 miles units and total size equal to 40,000 sq. miles, the predicted total value of ore is $12.3 \cdot 10^9$ dollars with 95-% confidence limits equal to $\pm 6.5 \cdot 10^9$ dollars. It should be kept in mind that the results are for deposits which could be located in the control area. Only the uppermost part of the earth's crust can be scrutinized for the occurrence of orebodies. Moreover, the concept of "ore" is mainly determined by economic considerations. Thus, a mineral deposit which would be mined elsewhere, remains inaccessible if it occurs in a more remote area.

Critical evaluation

A predicted value such as $12.3 \pm 6.5 \cdot 10^9$ dollars for an area of 40,000 sq. miles of relatively unexplored territory can be questioned because of three reasons.

(1) Several rather rigid assumptions have been made regarding the values of the parameters μ, σ, p, q, and r that occur in [7.78]. For example, if we accept the values $\hat{\mu} = 2.858$ and $\hat{\sigma}^2 = 3.103$ as computed from the data, our prediction is altered to $16.7 \pm 11.5 \cdot 10^9$ dollars. The subject of desirable properties of estimators will be discussed in more detail in the next two sections.

(2) The model is based on the assumption that we are dealing with independent random variables X_i and K. This assumption may be not valid. A model that admits the possibility of positive correlation between X_i and K would not result in [7.78]. As in the example for oil deposits in the Alberta sedimentary basin, the 400 squares then should constitute a random sample that is representative for all 2912 unit squares outside the control area. If this condition is satisfied, the prediction may yet be valid by approximation. However, usually one will be more interested in doing a prediction for a geographically distinct area.

(3) A more important objection can be made from a geological point of view. A lithological environment such as the volcanic belts in the Canadian shield may contain orebodies in some parts but be barren in others. A more precise investigation as to what constitutes a "favorable" environment improves the predictions to a considerable extent (cf. Chapter 15).

Early in this section, it was mentioned that Griffiths and Singer (1971a) have used unit regional value as a measure of potential for large regions. Their average value per sq. mile based on total production (until 1967) from 1,088,028 sq. miles of Precambrian in five Canadian provinces (Quebec, Ontario, Manitoba, Saskatchewan and Alberta) is \$ 29,176 which is small as compared to the value of about \$250,000 for the conterminous U.S.A. It is interesting to compare Griffiths and Singer's value for Precambrian rocks with that based on De Geoffroy and Wu's data. According to [7.82], the value per sq. mile of area underlain by Archean volcanic belts and Lower Proterozoic rocks is \$307,300. These rocks constitute about 15.9% of total area of the Canadian shield. Hence, the value per sq. mile of Precambrian would amount to \$48,800. Both this value and that of \$29,176, which is based on a larger region, are far below the most conservative expected unit regional value which is \$150,000 per sq. mile according to Griffiths and Singer (1971).

7.10. Desirable properties of estimators

Suppose that $\hat{\theta}$ represents a parameter of a population, θ, as estimated from data. The estimate of a parameter such as mean or variance also is called estimator.

Two desirable properties of the estimator $\hat{\theta}$ are: (1) it should be unbiased or $E\hat{\theta} = \theta$; and (2) it should have minimum variance or $E(\hat{\theta} - E\hat{\theta})^2 \leqslant E(\hat{\theta}' - E\hat{\theta}')^2$ where $\hat{\theta}'$ represents any other possible estimator. The estimator $\hat{\theta}$ is called minimum variance unbiased or best unbiased if both properties 1 and 2 are satisfied.

Until now, we have been mainly concerned with the first property (unbiasedness). For example, the sample mean $\bar{x} = \Sigma x/n$ and variance $s^2(x) = \Sigma(x - \bar{x})^2/(n - 1)$ always provide unbiased estimates of population mean and variance. We have used the method of moments to obtain unbiased estimators. Another method of estimation, which as yet has not been discussed in a statistical context, is the method of least squares. If the residuals behave as an independent random variable, this method as applied in section 4.6 provides minimum variance unbiased estimates for the coefficients.

In some situations, estimators obtained by the classical method of moments have the undesirable property of being imprecise, whereas more precise estimators could be obtained from the same data by using another method of estimation. For this reason, we include a brief discussion of another method of estimation which differs from that based on the theory of moments. One of the more important procedures for estimation is based on the principle of maximum likelihood.

The likelihood method was invented by Fisher (1922) who subsequently showed that, in this way, estimators can be obtained which usually are superior to those obtained by the method of moments. For this reason, it might be surprising that little use has been made of the likelihood method in geological situations. The reason for this is probably that in order to apply this method one must have some a-priori knowledge regarding the population which is being sampled. If this knowledge is not available beforehand it may be risky to use the likelihood method because the postulated model may not be realistic. Of course, the model can be tested a posteriori by using a χ^2-test. However, this may not provide conclusive evidence that the model is satisfactory, if the sample is small.

It may not be useful to apply the refinements offered by the likelihood method if these are negligible as compared to those which could be obtained by using theoretical geologic considerations to lay a better framework for the population that is being sampled.

The maximum likelihood method has been developed as an alternative to so-called Bayesian statistics named for Bayes who made an early application in 1763. In Bayesian statistics, sampling is regarded as a dynamic process. At each stage, one should use all information available at that time, postulate hypothetical distributions for the parameters and incorporate newly obtained data in order to obtain improved estimators. In principle, this is the endeavour of all scientific effort. However, it may be too difficult to quantify large amounts of a-priori information unless subjective notions are permitted to enter into the model. Nevertheless, there has recently occurred a revival of Bayesian statistics amongst statisticians and this could lead to a drastic change in approach to estimation problems including those in geology. At this time, it is more difficult to apply methods of Bayesian statistics to geologic problems than to use the likelihood method.

In many applications, it is best to calculate the arithmetic mean and variance in the usual way. In some instances, it may be preferable to use maximum-likelihood estimators.

7.11. Maximum likelihood and other methods of estimation

The principle of maximum likelihood is as follows. Suppose that X is a random variable with frequency distribution $f(x)$ that contains several parameters $\theta_1, \theta_2, \ldots$. Suppose further, that $f(x)$ is known beforehand except for the parameters which are to be estimated from a sample consisting of n values x_i assumed by X.

The probability that a value x_i will fall within a narrow interval Δx is approximately $f(x_i)\,\Delta x$. The probability that n values x_i will fall where they do is $f(x_i) \cdot f(x_2) \ldots f(x_n)\,\Delta x^n$ or $\Delta x^n \Pi_n f(x_i)$. As long as Δx is sufficiently narrow, the choice of it is immaterial and we can consider the product $\Pi_n f(x_i)$ only. This is the likelihood function. For given values of $\theta_1, \theta_2, \ldots$, the likelihood function will assume a specific value. If this value is relatively large, it may be assumed that our choice of the parameters is a good one. The likelihood method consists of maximizing the function $\Pi f(x_i; \theta_j)$ or its logarithm $\Sigma \log f(x_i; \theta_j)$ with respect to the θ_j. If we are able to calculate the maximum-likelihood estimators for a given frequency distribution function, it also is most probable that the data originate from a population with parameters equal to our estimators.

Maximum-likelihood estimators always have the minimum variance property. However, they may be biased.

Example of normal distribution

For n data from a normal distribution with mean μ and standard deviation σ, we have to maximize:

$$\prod_{i=1}^{n} \frac{1}{\sqrt{2\pi}\,\sigma} \exp\left[-\frac{1}{2\sigma^2}(x_i - \mu)^2\right]$$

If the logarithm (base e) of this expression is written as L,

$$L = -(n/2)\log 2\pi - (n/2)\log \sigma^2 - \frac{1}{2\sigma^2}\,\Sigma(x_i - \mu)^2 \qquad [7.83]$$

To localize the maximum, we calculate:

$$\frac{\partial L}{\partial \mu} = (1/\sigma^2)\,\Sigma(x_i - \mu)\,; \qquad \frac{\partial L}{\partial \sigma^2} = -n/2\sigma^2 + (1/2\sigma^4)\,\Sigma\,(x_i - \mu)^2 \qquad [7.84]$$

If these two partial derivatives are put equal to zero, we find the estimators:

$$\hat{\mu} = (1/n)\,\Sigma x_i = \bar{x}\,; \qquad \hat{\sigma}^2 = (1/n)\,\Sigma\,(x_i - \bar{x})^2$$

The estimator $\hat{\mu}$ is unbiased, but $\hat{\sigma}^2$ is biased by the factor $n/(n-1)$ as is known from the method of moments (see section 6.9).

The lognormal distribution

If the likelihood method is applied to logarithmically transformed data for this distribution, we obtain the same results as in the previous example. However, when we wish to estimate the mean α and standard deviation β for the distribution, the method turns out to be complicated from a mathematical point of view. For a method which is equivalent to the likelihood method, reference is made to an original publication by Finney (1941) and work by Sichel (1952, 1966). Our main concern here is how the resulting estimators for the lognormal distribution can be used. They have the optimum property of being minimum variance unbiased.

It has been discussed (see [7.79]) that $\hat{\alpha}'' = \exp(\overline{\log x})$, which represents the antilog of the mean log, may severely underestimate α. This estimate must be multiplied by a factor $\Psi_n(s^2/2)$ for obtaining the best unbiased estimator $\hat{\alpha}$ or:

$$\hat{\alpha} = e^{\overline{\log x}} \cdot \Psi_n(s^2/2) \qquad [7.85]$$

The function $\Psi_n(s^2/2)$ depends on number of observations n and logarithmic variance $s^2 = s^2(\log x)$. Setting $t = s^2/2$, we have:

$$\Psi_n(t) = 1 + \frac{n-1}{n}\, t + \frac{(n-1)^3}{n^2(n+1)}\, \frac{t^2}{2!} + \frac{(n-1)^5}{n^3(n+1)(n+3)}\, \frac{t^3}{3!} + \ldots \qquad [7.86]$$

Further, if $\hat{\beta}^2$ represents the best unbiased estimator of β^2, then:

$$\hat{\beta}^2 = e^{2\overline{\log x}} \left\{ \Psi_n(2s^2) - \Psi_n\left(\frac{n-2}{n-1}\, s^2\right) \right\} \qquad [7.87]$$

The series defining $\Psi_n(t)$ converges only slowly, but its values for variable n and t are calculated readily by digital computer.

Aitchison and Brown (1957) have published tables for $\Psi_n(t)$ where t goes from 0 to 2. Sichel's (1966) tables are for t up to 3. The latter paper also contains tables for confidence

ABLE XIX
bbreviated table to estimate correction factor $\psi_n(t)/\psi_\infty(t)$ for lognormal distributions

n	20	40	60	80	100	200	300	400	500	1000
1	0.9150	0.9542	0.9686	0.9761	0.9807	0.9902	0.9934	0.9950	0.9960	0.9980
2	0.7848	0.8754	0.9121	0.9321	0.9446	0.9712	0.9805	0.9853	0.9882	0.9941
3	0.6398	0.7764	0.8374	0.8722	0.8947	0.9440	0.9618	0.9710	0.9767	0.9882
4	0.5005	0.6685	0.7515	0.8011	0.8342	0.9094	0.9377	0.9525	0.9616	0.9804
5	0.3783	0.5607	0.6605	0.7232	0.7664	0.8687	0.9087	0.9300	0.9432	0.9708
6	0.2775	0.4593	0.5693	0.6424	0.6943	0.8230	0.8755	0.9040	0.9218	0.9595
8	0.1387	0.2898	0.4018	0.4849	0.5483	0.7215	0.7990	0.8428	0.8710	0.9319
0	0.0639	0.1703	0.2667	0.3469	0.4129	0.6139	0.7136	0.7725	0.8114	0.8984
5	0.0070	0.0352	0.0766	0.1231	0.1698	0.3638	0.4918	0.5786	0.6406	0.7938
0	0.0006	0.0055	0.0170	0.0345	0.0562	0.1850	0.3014	0.3940	0.4669	0.6709

intervals on the estimated means. Thöni (1969) has published similar tables, for $\Psi_n(t)$ with t going from 0 to 2.65.

In some problems of mathematical geology, the frequency distributions are extremely skew and t exceeds 2.65 or 3. Examples of these include the sizes of oil pools and ore-bodies. For this reason, we have calculated more values of $\Psi_n(t)$, which are reproduced in Table XIX.

These values were calculated by using the criterion applied by Thöni (1969). Eq. [7.87] was developed term by term and enough terms were included to make the last term smaller than 0.000008.

In order to facilitate interpolation we have tabulated $\Psi_n(t)/\Psi_\infty(t)$ rather than $\Psi_n(t)$. Because of [7.85]:

$$\Psi_\infty(t) = \exp(s^2/2) \qquad [7.88]$$

An application of the method is given in the following example.

Recoverable oil reserves (1965) in Leduc Formation, western Canada

McCrossan (1969) has compiled data on the probable ultimate recoverable oil reserves (1965) of the Leduc reef pools of western Canada. His histogram for the observed frequency distribution is given here in Table XX. The 52 data satisfy a lognormal distribution as shown by McCrossan (1969).

The largest pool has reserves equal to 817,000,000 barrels. By using the method given in section 6.13, it follows that Ave($^{10}\log x$) = 6.519 and $s_1^2(^{10}\log x)$ = 1.338.

The data have been combined in classes. For this application, it will be important to obtain estimators of $\overline{\log x}$ and $s^2(\log x)$ which are as accurate as possible. For this reason, we applied Sheppard's correction to the sample variance. This method is discussed by Kendall and Stuart (1958). Our estimate for the mean of grouped data cannot be improved. For the variance, Sheppard's correction amounts to $h^2/12$ where h represents class width. Because $h = 0.5$ in Table XX, $h^2/12 = 0.021$. This number must be sub-

TABLE XX
Leduc oil in barrels. (From McCrossan, 1969.)

^{10}*log (no. of barrels)*	*No. of pools*
4.0 – 4.5	2
4.5 – 5.0	3
5.0 – 5.5	7
5.5 – 6.0	5
6.0 – 6.5	7
6.5 – 7.0	11
7.0 – 7.5	7
7.5 – 8.0	3
8.0 – 8.5	5
8.5 – 9.0	2

tracted from the initial estimate s_1^2. Hence, $s^2({}^{10}\log x) = 1.338 - 0.021 = 1.317$; and $s({}^{10}\log x) = 1.148$.

Until now, our data analysis is satisfactory in that we can apply results for the normal distribution to the logarithmically transformed data. For example, 50% of the pools has reserves less than $10^{6.5} = 33$ million barrels. This is information that can be useful in predicting what to expect if a new pool is discovered. We may attempt to continue the analysis and estimate $E(X)$ which represents the amount of recoverable oil expected to be present in one pool. The arithmetic mean for original data amounts to $53 \cdot 10^6$ barrels. It provides an unbiased estimate of $E(X)$. Our estimators for logarithmic mean and variance were calculated from logarithms of data to the base 10 (cf. Table XX). Estimators based on natural logarithms (base e) are obtained by multiplication with ${}^{e}\log 10 = 2.3026$. The results are $\overline{\log x} = 15.011$; $s(\log x) = 2.642$. Further, $s^2 = s^2(\log x) = 6.983$; $t = s^2/2 = 3.491$.

By interpolation, we estimate a value of 0.77 for the correction factor in Table XIX, because $n = 52$ and $t = 3.49$.

A biased estimate of α is provided by:

$$\hat{\alpha}' = \exp\,(\overline{\log x} + s^2/2) = 108.5 \cdot 10^6$$

The best unbiased estimator of $\hat{\alpha}$ satisfies:

$$\hat{\alpha} = \hat{\alpha}' \cdot \Psi_n(t)/\Psi_\infty(t)$$

$$= 0.77 \times 108.5 \cdot 10^6 = 83.5 \cdot 10^6 \text{ barrels.}$$

This estimator is larger than the arithmetic mean $\bar{x} = 53 \cdot 10^6$ barrels.

Efficiency

The relative efficiency of one estimator with respect to another estimator is equal to the ratio of the variances for these estimators. If the lognormal model is accepted, $\hat{\alpha}$ is more efficient than any other estimator such as $\bar{x}$, because $\hat{\alpha}$ has minimum variance. Finney (1941) has developed equations for the variance of $\bar{x}$ and its relative efficiency as based on the variance of $\hat{\alpha}$. His method is approximate only. It was also discussed by Aitchison and Brown (1957). We only report results as obtained for the example and the reader is referred to the literature for details. Koch and Link (1971) discussed this subject in detail. The efficiency of $\bar{x}$ with respect to $\hat{\alpha}$ amounts to about 0.029. It means that the variance for $\hat{\alpha}$ is 3% of that for $\bar{x}$. Consequently, the precision of $\hat{\alpha}$, which is proportional to its standard deviation, is about six times greater than the precision of $\bar{x}$. Nevertheless, the standard deviation for $\hat{\alpha}$ is by approximation equal to $65 \cdot 10^6$, which is of the order of $\hat{\alpha}$ itself.

A drawback of using $\hat{\alpha}$ is that we are accepting lognormality of the data. In practice, $\bar{x}$ may be more reliable than $\hat{\alpha}$, because, if there is a departure from lognormality, $\hat{\alpha}$ may be biased because of this, whereas $\bar{x}$ remains unbiased.

If $\sigma^2({}^{e}\log x) = 1$, the efficiency of $\bar{x}$ with respect to $\hat{\alpha}$ amounts to 0.87. It means that for lognormally distributed data, $\bar{x}$ is nearly as precise as $\hat{\alpha}$ in that situation. On the other hand, the efficiency of $s^2(x)$ with respect to $\widehat{\beta^2}$ as calculated by means of [7.87] is only 0.16 if $\sigma^2({}^e\log x) = 1$. This illustrates that $s^2(x)$ usually is a poor estimator if the frequency distribution has a long tail at its high value end.

Mineral potential of Canadian shield, uncertainty of estimate

In section 7.9, it was shown that for 254 orebodies, $\hat{\alpha} = 80.34 \cdot 10^6$ versus $\bar{x} = 60.5 \cdot 10^6$ dollars. We accepted $\bar{x}$ rather than $\hat{\alpha}$. During the derivation of [7.82] from [7.78], it has been tacitly assumed than $s^2(x)$ can be estimated as:

$$\hat{\beta}'^2 = e^{2\overline{\log x} + s^2}(e^{s^2} - 1) = 40516 \cdot 10^{12}$$

By using [7.87], $\hat{\beta}'^2$ can be replaced by the best unbiased estimator $\widehat{\beta^2}$. We may investigate how [7.82] is affected by this replacement.

If $\overline{\log x} = 2.858$ and $s^2(\log x) = 2.490$, [7.87] gives:

$$\widehat{\beta^2} = 303.69\ [\Psi_n(4.980) - \Psi_n(2.500)] \cdot 10^{12}$$

If $n = 254$, interpolation based on Table XIX gives the correction factors 0.891 if $t = 4.980$ and 0.965 if $t = 2.500$. These numbers must be multiplied by exp (4.980) = 145.5 and exp (2.500) = 12.2, respectively, giving:

$$\Psi_{254}(4.980) = 129.6; \qquad \Psi_{254}(2.500) = 11.8$$

Consequently, $\widehat{\beta^2} = 35775 \cdot 10^{12}$.

In order to see how this result affects [7.82], we write [7.78] as:

$$\sigma^2(Y) = \frac{rq}{p}\beta^2 + \frac{rq}{p^2}\alpha^2 \qquad [7.89]$$

If β^2 is estimated by $\widehat{\beta^2}$ instead of by $\hat{\beta}'^2$, whereas the other parameters are left unchanged, we obtain:

$$\hat{E}(Y) = 30.73 \cdot 10^6$$

$$\hat{\sigma}^2(Y) = 25546 \cdot 10^{12}; \qquad \hat{\sigma}(Y) = 160 \cdot 10^6$$

The new standard deviation of Y differs only slightly from the previous estimate $\hat{\sigma}(Y) = 167 \cdot 10^6$ dollars and it seems that little is gained by refining the estimate of variance.

Application of the negative binomial model

Bliss and Fisher (1953) have developed a method for estimating maximum likelihood estimators for the negative binomial distribution. Ondrick and Griffiths (1969) have programmed this method in FORTRAN. The arithmetic mean also is the maximum-likelihood estimator of the population mean for a negative binomial distribution.

De Geoffroy and Wu (1970) have calculated the maximum-likelihood estimators for p, q, and r as used in the example of section 7.8. Our previous, less efficient estimators were $p' = 0.252$, $q' = 0.748$, $r' = 0.171$. The likelihood estimators are $p = 0.216$, $q = 0.784$, $r = 0.140$. This appears to influence the variance in [7.82] to some extent. However, this effect is small in comparison with that resulting from minor variations in the initial choice of $s^2\,({}^{e}\log x)$.

Gold occurrences in Ontario

In order to evaluate the usefulness of the negative binomial model for the spatial distribution of mineral deposits, we have carried out two experiments, the results of which are shown in Table XXI.

A grid with cells measuring 8 miles on a side was superimposed on the Timmins–Kirkland Lake area, Ontario, Canada, which is rich in gold mines and smaller gold deposits. For each cell, we (1) counted the number of known gold occurrences; and (2) checked from a geologic map if volcanics are exposed at the surface. This was done for 140 cells from an area of 8900 sq. miles (appr. 22,000 km^2) which is rectangular in shape and contains 572 known gold deposits. 116 of the 140 cells contain volcanic rocks. Only a single cell out of those without volcanics contains gold occurrences, 6 in total.

TABLE XXI

Observed frequencies (x_i) and negative binomial frequencies ($\hat{x}_i$) for two experiments; gold occurrences in Ontario

k_i	*Exp. 1*		*Exp. 2*		k_i	*Exp. 1*		*Exp. 2*	
	x_i	$\hat{x}_i$	x_i	$\hat{x}_i$		x_i	$\hat{x}_i$	x_i	$\hat{x}_i$
0	62	58.6	39	36.2	15	1	1.3	1	1.4
1	8	16.2	8	14.0	16	1	1.2	1	1.2
2	10	9.9	10	9.3	17	3	1.1	3	1.1
3	9	7.2	9	7.0	18	1	1.0	1	1.0
4	3	5.6	3	5.6	19	3	0.9	3	0.9
5	4	4.6	4	4.7	20	2	0.8	2	0.9
6	9	3.8	8	4.0	22	3	0.7	3	0.7
7	1	3.3	1	3.4	24	1	0.6	1	0.6
8	3	2.8	3	3.0	25	1	0.5	1	0.6
9	2	2.5	2	2.6	28	1	0.4	1	0.4
0	1	2.2	1	2.3	30	1	0.4	1	0.4
1	2	1.9	2	2.1	31	1	0.3	1	0.3
2	2	1.7	2	1.9	39	1	0.2	1	0.2
3	1	1.6	1	1.7	45	1	0.1	1	0.1
4	2	1.4	2	1.5					

	$\bar{x}$	s'^2	r'	r	p	q	s^2	$\hat{\chi}^2(7)$
xp. 1	5.51	73.8	0.445	0.291	0.050	0.950	110.3	10.95
xp. 2	6.60	81.9	0.597	0.411	0.059	0.941	112.8	8.02

By using the computer program by Ondrick and Griffiths (1969), a negative binomial distribution function was fitted to the two sets of data, for 140 and 166 cells, respectively. The observed frequencies for the two sets are identical except when $k = 0$ and $k = 6$. As shown in Table XXI, a good fit was obtained in both cases. By grouping the data in ten classes, each of which has a theoretical frequency larger than 5, a χ^2-test can be applied (see section 6.13). This gives $\hat{\chi}^2(7) = 10.95$ for the experiment with the 140 cells, and $\hat{\chi}^2(7) = 8.02$ for that with 116 cells. The 95-% confidence limit of χ^2 with 7 degrees of freedom amounts to 14.07. Consequently, the theoretical model can be accepted in both situations.

An interpretation of this result is that the negative binomial distribution is rather flexible in shape, in particular with regard to the number of values observed for the zero-class. If we include a number of cells without deposits, the parameters of the negative binomial are altered, but the model itself continues to provide a good fit. In the example, the 23 empty cells differ from the others in that either the bedrock is covered by a relatively thick layer of Pleistocene material or they contain other types of rock such as granitic material where gold deposits have not been found.

The following remarks are to explain the statistics presented in Table XXI. Ondrick and Griffiths (1969) initially estimate r as:

$$r' = \bar{x}^2/(s^2 - \bar{x}) \qquad [7.90]$$

where $\bar{x}$ and s^2 represent sample mean and variance. Eq. [7.90] can be derived from [7.64] if $E(X)$ and $\sigma^2(X)$ are replaced by $\bar{x}$ and s^2, respectively. The maximum likelihood estimator r is obtained by using an iterative procedure of Bliss and Fisher (1953). Since $\bar{x}$ also is the likelihood mean, p, q $(=1-p)$ and the new variance can be calculated from r and $\bar{x}$.

The theoretical frequencies are calculated by using a recurrence formula. From [7.62] it may be derived that:

$$P(X = 0) = p^r$$

$$P(X = k+1) = \frac{r+k}{k+1} \cdot q \cdot P(X = k) \qquad [7.91]$$

where $k = 0, 1, 2, \ldots$. If these probabilities are multiplied by the total number of cells we obtain the theoretical frequencies.

Chapter 8

STATISTICAL DEPENDENCE; MULTIPLE REGRESSION

8.1. Introduction

We begin this chapter by demonstrating the concept of random variables which are statistically dependent upon one another. This leads to statistical definitions of the covariance, the correlation coefficient and to the method of linear-regression analysis in a bivariate situation. A more general model, where a single variable is correlated to many others, is known as the general linear model. Matrix algebra can be used to describe the more important properties of this model.

If a number of random variables all have normal distributions and are mutually interrelated, they form a so-called multivariate normal distribution. Anyone of the variables can be made the dependent variable and its expected value can be expressed as a linear combination of the other variables by multiple regression. However, the general linear model can also be applied when the independent variables are not random variables, but functions of ordinary variables. Trend analysis (see Chapter 9), where the independent variables are functions of distance or time as measured along one or more coordinate axes, provides an example. The methods of multiple regression and trend analysis can both be based on the method of least squares which was discussed in Chapter 4.

A problem that can be solved by using the methods discussed in this chapter is how several dependent variables are related to a number of independent variables. It can be shown (Anderson, 1958, chapter 8) that, for this situation, every dependent variable may be regressed on the independent variables without considering the other dependent variables. The maximum likelihood estimator for the set of linear relationships arising from this situation is not influenced by a possible linear relationship between the dependent variables.

Another problem to be discussed is as follows: If a dependent variable is to be related to a large number of independent variables, we may wish to select those independent variables which are most important in that they, individually or combined, have the strongest correlation with the dependent variable. Methods of sequential multiple regression can be used to eliminate both irrelevant and redundant independent variables from the system.

Finally, we may wish to discriminate between linear relationships between variables in two or more different geological environments or geographically distinct areas. This can be done by using dummy variables to be discussed in a final section of this chapter.

8.2. Statistical dependence

The concept of statistical dependence can be illustrated by means of the following example.

Suppose that an oil company has drilled a sedimentary basin in n different locations. Oil has been found in $p_y \times n$ locations and the relative frequency of producing wells is equal to p_y. A geologist wishes to test the efficiency of a prediction device, which he has developed, by applying it to geologic information available before the drilling program was started. From this initial information, it is predicted that $p_x \times n$ holes will produce oil.

The prediction can be evaluated by means of the following contingency table.

Oil discovered	*Oil predicted*		*Sum*
	No	*Yes*	
Yes	p_{01}	p_{11}	p_y
No	p_{00}	p_{10}	$1-p_y$
Sum	$1-p_x$	p_x	1

The relative frequencies satisfy the following relationships: $p_{01} + p_{11} = p_y$; $p_{00} + p_{10} = 1-p_y$; $p_{01} + p_{00} = 1-p_x$; $p_{11} + p_{10} = p_x$.
For simplicity, we write the results as a matrix:

$$\mathbf{A} = \begin{bmatrix} p_{01} & p_{11} \\ p_{00} & p_{10} \end{bmatrix}$$

Two random variables X and Y may be defined such that X assumes value one if oil predicted, and zero if this is not so. Then, $P(X=1) = p_x$; $P(X=0) = 1-p_x$. When represents the true values, $P(Y=1) = p_y$; $P(Y=0) = 1-p_y$. If the two variables are independent, we would have:

$$P(X=1 \text{ and } Y=1) = p_x p_y = p_{11}$$

This would determine the other probabilities in the matrix **A** as follows:

$$p_{01} = p_y - p_{11} = p_y(1-p_x)$$

$$p_{10} = p_x - p_{11} = p_x(1-p_y)$$

$$p_{00} = 1 - p_x - p_{01} = (1-p_x)(1-p_y)$$

Consequently:

$$\mathbf{A} = \begin{bmatrix} p_y(1-p_x) & p_x p_y \\ (1-p_x)(1-p_y) & p_x(1-p_y) \end{bmatrix}$$

For further illustration, let $p_x = 0.2$ and $p_y = 0.3$. It would mean that the device predicts that 20 out of every 100 holes will produce oil. In reality, 30 per 100 holes are producers. We obtain:

$$\mathbf{A} = \begin{bmatrix} 0.24 & 0.06 \\ 0.56 & 0.14 \end{bmatrix}$$

The ratio p_{11}/p_{10} is equal to p_{01}/p_{00}.
Consequently, the percentage of producers for which no oil was predicted is equal to the percentage of producers for which oil was predicted correctly. It can then be concluded that the prediction device has no greater efficiency than a procedure consisting of the selection of 20 % of the locations at random and blindly guessing that oil will be found in these places.

A more efficient device has a matrix A which reflects a dependency of the random variables X and Y. In an ideal situation:

$$\mathbf{A} = \begin{bmatrix} 0 & p_y \\ 1-p_y & 0 \end{bmatrix} = \begin{bmatrix} 0 & 0.30 \\ 0.70 & 0 \end{bmatrix}$$

A property of the ideal device is that $p_x = p_y$. Incidentally, this illustrates that a prediction method can be evaluated in two ways. If it succeeds in predicting p_y, but is otherwise inefficient, because the variables X and Y are independent, it yet has merit because the mineral potential of the sedimentary basin is predicted correctly. For example, if $p_x = p_y = 0.30$, independent X and Y give:

$$\mathbf{A} = \begin{bmatrix} 0.21 & 0.09 \\ 0.49 & 0.21 \end{bmatrix}$$

The device may then not be useful for predicting the outcome for a single well. Nevertheless, it is useful if the objective is to predict the mineral potential of the entire basin.

8.3. Covariance, correlation coefficient and conditional probability

The expectation of the product of two random variables X and Y was defined in [6.29] as:

$$E(XY) = \int\int x\,y\,f(x, y)\,dx\,dy \qquad [8.1]$$

In the discrete case, this expression reduces to:

$$E(XY) = \Sigma x_j y_k P(x_j, y_k) \qquad [8.2]$$

For the elements of the matrix A, we have:

$P(X = 1, Y = 1) = p_{11}$

$P(X = 1, Y = 0) = p_{10}$

$P(X = 0, Y = 1) = p_{01}$

$P(X = 0, Y = 0) = p_{00}$

If $\mathbf{A} = \begin{bmatrix} 0.24 & 0.06 \\ 0.56 & 0.14 \end{bmatrix}$, then:

$$E(XY) = 1 \times 1 \times 0.06 + 1 \times 0 \times 0.14 + 0 \times 1 \times 0.24 + 0 \times 0 \times 0.56 = 0.06$$

In general, $E(XY) = p_{11}$ for the matrix **A**.

Covariance

The covariance of X and Y is defined as:

$$\sigma(X, Y) = E(X - \mu_x)(Y - \mu_y)$$
$$= E(XY) - \mu_x EY - \mu_y EX + \mu_x\mu_y$$

or: $\sigma(X, Y) = E(XY) - \mu_x\mu_y$ [8.3]

The covariance is the expectation of the product of the two random variables from which the product of the means is subtracted. This relationship applies to both continuous and discrete random variables.

For the example, $E(XY) = 0.06; \mu_x = p_x = 0.2; \mu_y = p_y = 0.3; \mu_x \mu_y = 0.06$.. Hence:

$$\sigma(XY) = 0.06 - 0.06 = 0$$

The covariance of two independent random variables is equal to zero.

On the other hand, suppose that:

$$\mathbf{A} = \begin{bmatrix} 0.12 & 0.18 \\ 0.68 & 0.02 \end{bmatrix}$$

Then, $\mu_x = 0.30; \mu_y = 0.20; \mu_x\mu_y = 0.06; E(XY) = 0.18$; and $\sigma(X, Y) = 0.18 - 0.06 = 0.12$.

Correlation coefficient

The covariance of two standardized random variables is the correlation coefficient

$$\rho_{xy} = \rho(X, Y) = \frac{\sigma(X, Y)}{\sigma(X) \cdot \sigma(Y)}$$ [8.4

'e recall that $\sigma(X)$ is the positive square root of $\sigma^2(X)$ with $\sigma^2(X) = E(X^2) - \mu_x^2$ and $(X^2) = \Sigma\, x_j^2 P(x_j)$. For the example, $E(X^2) = \mu_x$.

Thus, if $\mathbf{A} = \begin{bmatrix} 0.24 & 0.06 \\ 0.56 & 0.14 \end{bmatrix}$, we obtain:

$$\mu_x = 0.2\,; \quad \sigma_x^2 = 0.20 - 0.04 = 0.16\,; \quad \mu_y = 0.3\,; \quad \sigma_y^2 = 0.30 - 0.09 = 0.21$$

ence, $\rho_{xy} = 0 \times (0.16 \times 0.21)^{-\frac{1}{2}} = 0$

n the other hand, if $\mathbf{A} = \begin{bmatrix} 0.12 & 0.18 \\ 0.68 & 0.02 \end{bmatrix}$, then:

$$\rho_{xy} = \frac{0.12}{\sqrt{0.16 \times 0.21}} = \frac{0.12}{0.1833} = 0.655$$

he variables X and Y are positively correlated. An advantage of expressing the degree of sociation between two variables by using the correlation coefficient ρ_{xy} is that stan- ard statistical techniques are available for evaluating ρ_{xy} if X and Y are both normally istributed.

onditional probability

The conditional probability $P(Y = 1 | x = 1)$ represents the probability that $Y = 1$ vhere oil is found) when it is given that $x = 1$ (where oil is predicted). We consider only e column of the matrix $\mathbf{A}$ for which $x = 1$. It follows that:

$$P(Y = 1 | x = 1) = p_{11} / p_x$$

: $P(Y = 1 | x = 1) = P(X = 1,\ Y = 1) / P(X = 1)$

ikewise:

$$P(Y = 0 | x = 1) = P(X = 1,\ Y = 0) / P(X = 1)$$

nese two expressions are special cases of the more general equation:

$$P(Y = y | X = x) = P(X = x,\ Y = y) / P(X = x) \qquad [8.5]$$

nis also can be written as:

$$P(y | x) = P(x, y) / P(x)$$

$$P(x, y) = P(y | x) \cdot P(x) \qquad [8.6]$$

X and Y are independent, this reduces to the familiar form $P(x, y) = P(x) \cdot P(y)$.

Some other definitions for **A** are of interest. $P(X = 0, Y = 1) = p_{01}$ represents the Type 1 error. It denotes the probability of not predicting a well that produces. $P(X = 1, Y = 0) = p_{10}$ is the Type 2 error. It consists of predicting a no, where yes should have been the answer. The ratio p_{11}/p_y is sometimes called the success ratio. It represents the proportion of estimated yesses later proven to be correct.

8.4. The bivariate normal distribution

The bivariate normal distribution is a generalization of the normal distribution for a single variate. Its density has the form:

$$f(x, y) = \frac{1}{2\pi\sigma_x\sigma_y\sqrt{1-\rho^2}} \exp\left\{-\frac{1}{2(1-\rho^2)}\left[\left(\frac{x-\mu_x}{\sigma_x}\right)^2 - 2\rho\frac{x-\mu_x}{\sigma_x}\frac{y-\mu_y}{\sigma_y} + \left(\frac{y-\mu_y}{\sigma_y}\right)^2\right]\right\} \quad [8.7]$$

If the two random variables are independent, $\rho = 0$; the result is a two-dimensional normal distribution as shown graphically in Fig. 36A for $\mu_x = \mu_y = 0$ and $\sigma_x = \sigma_y = \sigma$. Suppose that $\rho = 0$, but that the variances of the two variables differ. The contour map of a surface of this type is shown in Fig. 36B.

If [8.4] is applied to [8.7], it appears that $\rho_{xy} = \rho$. Eq. [8.7] is simplified by using standardized variables with:

$$z_1 = \frac{x - \mu_x}{\sigma_x}; \quad z_2 = \frac{y - \mu_y}{\sigma_y}$$

We can write:

$$\varphi(z_1, z_2) = \frac{1}{2\pi\sqrt{1-\rho^2}} \exp\left[-\frac{1}{2(1-\rho^2)}(z_1^2 - 2\rho z_1 z_2 + z_2^2)\right] \quad [8.8]$$

If $f(x, y)$ and $\varphi(z_1, z_2)$ are equal to a constant, we have:

$$z_1^2 - 2\rho z_1 z_2 + z_2^2 = \text{constant} \quad [8.9]$$

This is the equation of an ellipse, whose shape is determined by ρ.
If $\rho = 0$, the density contours are circular. If $\rho = 1$, the ellipse reduces to a straight line.

On the other hand, if x or y are set equal to a constant, [8.7] reduces to the equation for a normal curve. Any vertical profile of the surface $f = f(x, y)$ provides a Gaussian curve.

According to results obtained in section 4.4, [8.9] always can be reduced to:

$$z_1'^2 + z_2'^2 = \text{constant} \quad [8.10]$$

by a rotation of the $(Z_1 Z_2)$-coordinate system about a vertical axis. The transformed variables Z_1' and Z_2' have a correlation coefficient that is equal to zero. Likewise, X and Y can be transformed into the independent normal variables X' and Y' by a rotation of the

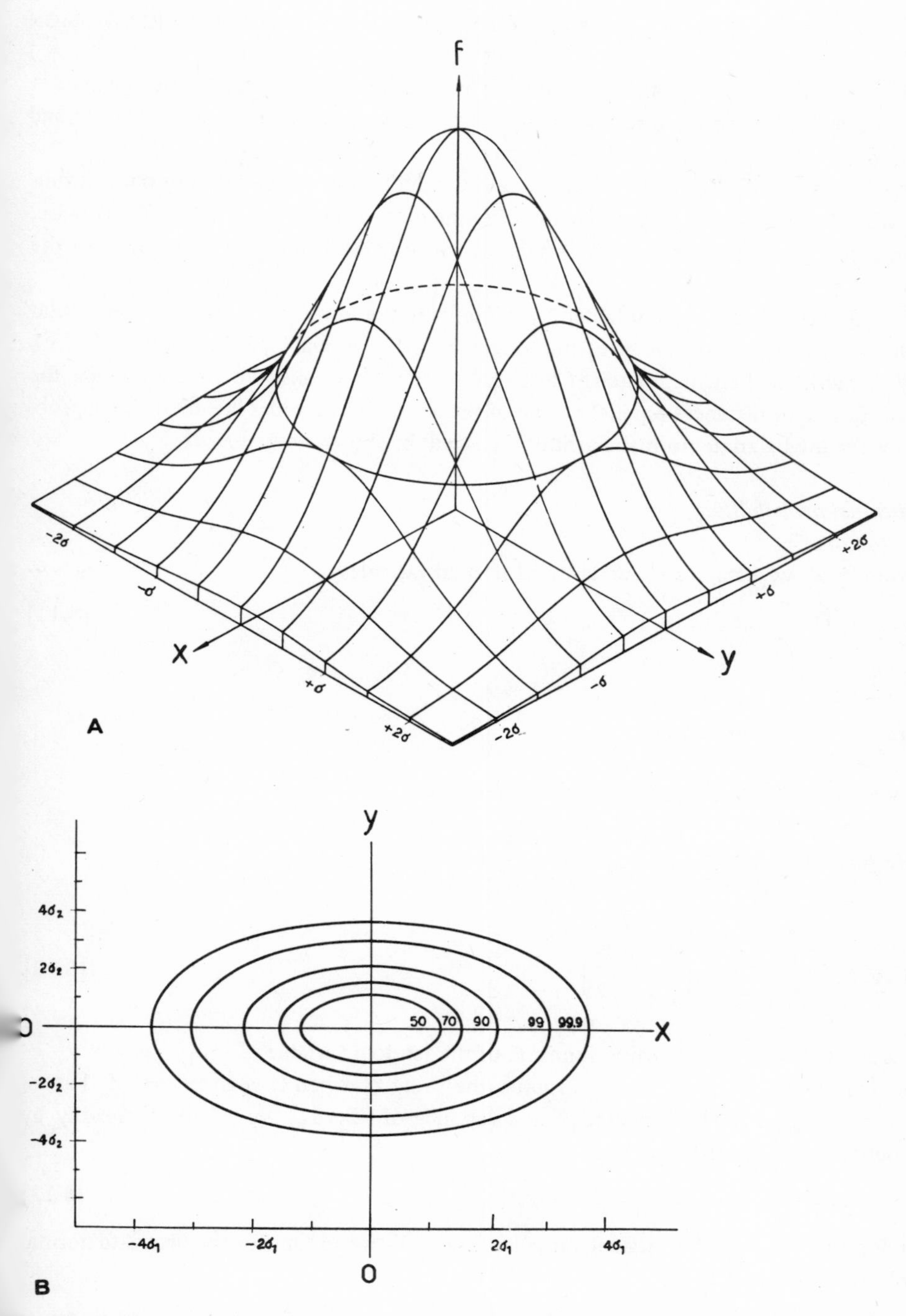

ig.36. A. Distribution surface $f = f(x, y)$ for isotropic two-dimensional normal distribution with $_x = \mu_y = 0$, $\sigma_x = \sigma_y = \sigma$. B. Contour ellipses for an anisotropic (2D-) normal distribution with correla- on coefficient 0; numbers indicate probabilities (in percent) that a bivariate random variable assumes value within an ellipse. (After Hald, 1952.)

(XY)-coordinate system about the vertical axis. The long axis of the elliptic contours is called the major axis of the bivariate normal distribution. If standardized variables Z_1 and Z_2 are used, we obtain the so-called reduced major axis.

Suppose that we have a sample of n pairs of values from a bivariate normal distribution. We can estimate the major axis by using the method given in section 4.7. Likewise, we may construct the reduced major axis, which in general does not coincide with the major axis.

Thus, a best-fitting line as obtained by minimizing the sum of squares of perpendicular distances from points to line gives the long axis of the probability distribution $f(x, y)$. Another method of fitting a straight line (see section 4.6) consisted of minimizing the sum of squares of distances parallel to one of the axes. This provides the conditional probabilities for the bivariate normal distribution as will be shown now.

Conditional probabilities

From [8.6], we obtain for standardized normal variables:

$$\varphi(z_2 \mid z_1) = \varphi(z_1, z_2)/\varphi(z_1) \qquad [8.11]$$

Because:

$$\varphi(z_1) = \frac{1}{\sqrt{2\pi}} \exp(-z_1^2/2)$$

eq. [8.8] and [8.11] give, after some manipulation:

$$\varphi(z_2 \mid z_1) = \frac{1}{\sqrt{2\pi}\sqrt{1-\rho^2}} \exp\left[\frac{-(z_2 - \rho z_1)^2}{2(1-\rho^2)}\right]$$

$$\varphi(z_2 \mid z_1) = \frac{1}{\sqrt{1-\rho^2}} \varphi\left(\frac{z_2 - \rho z_1}{\sqrt{1-\rho^2}}\right) \qquad [8.12]$$

This is the frequency-distribution function of the random variable $(Z_2 \mid z_1)$. It denotes the probability that Z_2 assumes the value z_2 when it is given that Z_1 has assumed the value z_1. The expectation and variance of $(Z_2 \mid z_1)$ are calculated readily by the method of moments, yielding:

$$E(Z_2 \mid z_1) = \rho z_1 ; \quad \sigma^2(Z_2 \mid z_1) = 1 - \rho^2 \qquad [8.13]$$

Transforming back to the original variables X and Y, we obtain for the bivariate normal distribution of X and Y:

$$E(Y \mid x) = \mu_y + \rho \frac{\sigma_y}{\sigma_x}(x - \mu_x); \quad \sigma^2(Y \mid x) = \sigma_y^2(1 - \rho^2) \qquad [8.14]$$

Eq. [8.14] shows that the expectation of Y, conditional upon X, is a linear function of the values assumed by the "independent" variable X. We can also write:

$$E(Y|x) = \beta_0 + \beta_1 x \tag{8.15}$$

where $\beta_0 = \mu_y - \rho \dfrac{\sigma_y}{\sigma_x} \mu_x;\quad \beta_1 = \rho \dfrac{\sigma_y}{\sigma_x}$

Estimation of correlation coefficients from data

The correlation coefficient ρ for X and Y can be estimated by using [5.1] or by:

$$r_{xy} = \frac{s(x, y)}{s(x) \cdot s(y)} \tag{8.16}$$

where the sample covariance satisfies $s(x, y) = \Sigma (x - \bar{x})(y - \bar{y})/(n - 1)$.
Most statistical textbooks and tables contain a chart or table for the frequency distribution of r on the assumption that X and Y are normally distributed. One also can use the fact that the auxiliary variable:

$$u = \tfrac{1}{2}\, {}^{e}\log \frac{1 + r}{1 - r} \tag{8.17}$$

has normal distribution with $\mu_u \approx \frac{1}{2} \log(1 + \rho)/(1 - \rho) + \rho/2(n - 1)$ and $\sigma_u^2 \approx 1/(n - 3)$.

Prediction of mineral potential in British Columbia

Kelly and Sheriff (1969) have laid a grid on top of the province of British Columbia with square cells measuring 20 × 20 miles (400 sq. miles or approx. 1000 km^2). Of the resulting 818 squares 211 appear to contain a rock type which was listed as "Upper Jurassic and Cretaceous intrusives of mainly acidic types" in the legend of the Geologic Map of British Columbia (G.S.C. Map 932A, 2nd ed.). On the other hand, 88 out of 818 cells contain orebodies whose combined value exceeds one million dollars.

Let this geological variable be written as X and total dollar value per cell as Y. We can ask the question that there exists an association between X and Y which could be useful for predicting mineral potential.

The 818 cells are divided into four groups as follows:

$(x = 0, y = 0)$ 567 cells
$(x = 1, y = 0)$ 163 cells
$(x = 1, y = 1)$ 48 cells
$(x = 0, y = 1)$ 40 cells

These absolute frequencies can be divided by 818 to give relative frequencies which are arranged in matrix form as:

$$\mathbf{A} = \begin{bmatrix} 0.049 & 0.059 \\ 0.693 & 0.199 \end{bmatrix}$$

with $p_x = 0.258$ and $p_y = 0.108$.

We can assume that the elements of **A** represent probabilities, e.g., the probability that a single cell, selected at random, has both $X = 1$ and $Y = 1$ is equal to $p_{11} = 0.059$ or 6%.

According to [8.5], the conditional probability $P(Y = 1 | x = 1)$ is equal to $p_{11}/p_x = 0.227$ which exceeds the nonconditional probability $P(Y = 1) = p_y = 0.108$. A positive association between X and Y is suggested. However, we may ask whether this correlation is statistically significant. We calculate the correlation coefficient as:

$$r_{xy} = \frac{p_{11} - p_x p_y}{\sqrt{(p_x - p_x^2)(p_y - p_y^2)}} = \frac{0.059 - 0.028}{\sqrt{0.191 \times 0.096}} = 0.229$$

If a succession of single cells are selected at random, X and Y both have the binomial distribution. A binomial distribution may be approximated by a normal distribution if the number of positive answers exceeds 10. Consequently, we could test r_{xy} by comparing it to correlation coefficients for the bivariate normal distribution. For a two-tailed test, we find in statistical tables $P(r_{xy} < 0.068) = 0.95$ and $P(r_{xy} < 0.090) = 0.99$. For a one-tailed test, the 95- and 99-% levels of significance are somewhat less. The conclusion would be that X and Y are indeed positively correlated because 0.229 exceeds 0.068 or 0.090. However, great care should be taken in evaluating a result of this type. The statistical model is valid only if both X and Y are randomly distributed over the 818 cells. If the cells with $x = 1$ or $y = 1$ show any systematic features in their spatial distribution, then the model is not satisfied.

A rapid inspection of the geologic map shows us that the pattern of cells with Jurassic and Cretaceous granites is far from being random. In fact, if a cell on the map has $x = 1$, there is a relatively large probability that a number of the adjoining squares also will have $x = 1$, because granite bodies usually fall in a number of contiguous cells. This phenomenon provides an example of so-called autocorrelation to be discussed in Chapter 10.

It is instructive to calculate the expected value of Y when x is given. According to [8.15]:

$$E(Y|x) = p_y + r_{xy} \sqrt{\frac{p_y - p_y^2}{p_x - p_x^2}} (x - p_x)$$

$$= 0.108 + 0.229 \sqrt{\frac{0.096}{0.191}} (x - 0.258)$$

$$= 0.066 + 0.162x$$

Since X is either one or zero, we have:

$$E(Y|x=0)=0.066; \quad E(Y|x=1)=0.228$$

This would mean that the expected gain is about three times larger for cells containing Upper Jurassic and Cretaceous intrusives. In this example, the two variables can assume the values one or zero only.

We can say that the probability of having $Y=1$ when $x=0$ is 6.6%, it is 22.8% when $x=1$. This result can also be obtained from the original frequencies. We know that $y=1$ in 48 out of $(48+163)=211$ cells with $x=1$. Hence, $P(Y=1|x=1)=48/211$ or 23%. Likewise, $P(Y=1|x=0)=7\%$. Consequently, the results of the linear expression can be interpreted as probabilities. The regression approach can be extended to situations where Y and X are continuous random variables.

Prediction of gold values, South African gold mines

The theory of bivariate regression outlined in this section has other important applications. Fig. 37 is based on an illustration by Krige (1962, fig. 1). Along the X-axis, it shows the average value of gold concentration in inch-dwt. values for panel faces in a Witwatersrand gold mine. Along the Y-axis, similar values are plotted for panels which are located 30 ft. (9m) ahead of the panels whose values are plotted along the X-axis. Each value for a panel is the average of ten single values for narrow sections across the gold reef. The ten section values cover a distance of 150 ft. across the panel along which mining proceeds. Fig. 37 can be considered as a scattergram. Individual values are not shown; their frequency was counted per small block in the diagram. Since panel values are approximately lognormally distributed, a ratio scale is used for the coordinate axes rather than an ordinary scale. The correlation coefficient amounts to 0.59. The elliptic contour shown in Fig. 37 contains most data and is comparable to the contours shown in Fig. 36B.

The practical problem consists of predicting the value of a panel that will be 30 ft. ahead of the panel whose value is known at a given time. In order to solve this problem, we must estimate, as Krige (1962) has done:

$$E(Y|x)=\beta_0+\beta_1 x \qquad \text{[8.15, repeated]}$$

This linear relationship can be estimated from the means, standard deviations, and the correlation coefficient as before.

The result has been plotted as a straight line (A) in Fig. 37. Krige (1962) has pointed out that, in this manner, we avoid making the erroneous assumption that a given element-concentration value in a mine is representative for a larger region surrounding it. For example, it is shown that, if a panel value is $x=1102$, then the value predicted from it for a panel that is 30 ft. ahead is not 1102 but 688 which is considerably less. On the other hand, if $x=92$, $E(Y|x)$ is 158 which is more.

Consequently, relatively high values may overestimate the element concentration for other points in their immediate surroundings. On the other hand, relatively low values tend to underestimate. On the average, the overestimation of some values and the underestimation of others cancel out, and if the entire gold reef is mined, this statistical fact does not present practical problems. However, often element-concentration values, which are representative for relatively small blocks or areas in a mine, are used for making the decision of what should be mined out and what should not be mined.

Suppose that a mineral deposit is sampled in a number of places and that there exists a so-called cut-off level for the mine. This means that it will not be economical to mine blocks of ore where the concentration is below the cut-off level. Suppose now that we assume that each sample value is representative for a larger volume of ore around it and mine only those blocks for which the sample value is relatively large leaving the remainder unmined. In the past, mining engineers have found, to their surprise, that the average for a volume blocked out in this manner may systematically overestimate the average concentration for metal actually mined out. Systematic overestimation in the presence of a cut-off level can be avoided by using a method of "kriging". Fig. 37 provides a simple example of this. Another practical result is that the cut-off level should be applied to the predicted values $E(Y|x)$ rather than to the known values x. Otherwise, one would miss a certain amount of payable ore.

By using the method given in section 4.7, we calculated the major axis for the situation represented in Fig. 37. It plots as the line B which is approximately equal to a line with equation $y = x$. Line B indicates, that, on the average, there is no systematic difference between the value for a given panel and that for the panel which is 30 ft. ahead of it. Con-

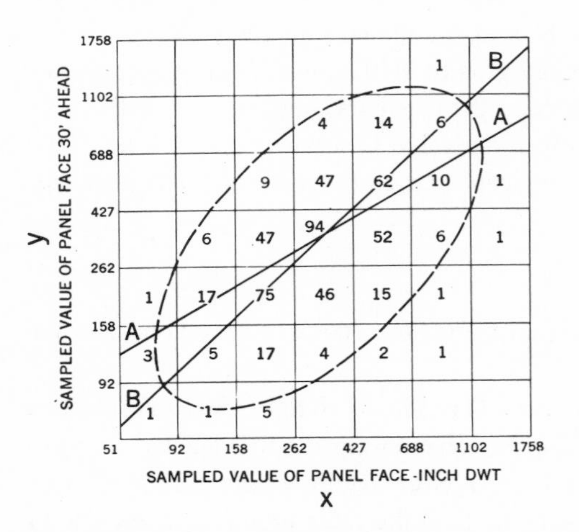

Fig.37. Contour ellipse contains most of values; frequencies shown for blocks. Krige's regression line (*A*) is used for prediction; line *B* represents the solution of Kummell's equation for linear functional relation between two random variables. It coincides approximately with the axis of the contour ellipse because, on average, y is about equal to x. (After Krige, 1960.)

sequently, Fig. 37 indicates that there are no systematic changes in average over a distance of 30 ft. or that there is no significant trend over this distance (see also next section).

In Chapter 10, we will make a distinction between "trend" and "signals". The signals are systematic variations in concentration of a limited spatial extent. This concept can be used to explain that the variables X and Y are positively correlated ($r_{xy} = 0.59$). We can also say that panel values, which are 30 ft. apart, are autocorrelated.

8.5. Bivariate regression analysis

If the parameters $\mu_x, \mu_y, \sigma_x, \sigma_y$ and ρ in [8.14] are replaced by the sample estimates $\bar{x}, \bar{y}, s_x, s_y$ and r_{xy}, we obtain a result previously represented by [4.52]. In the previous section, it was assumed that both X and Y are normally distributed. It has been shown that $(Y|x)$ then also is normally distributed. However, regression analysis can also be applied when Y is not normal. The Gauss-Markov theorem states that the least-square estimate of $(Y|x)$ always is a best unbiased estimate, regardless of the nature of the frequency distribution of Y. The only condition to be fulfilled is that the residuals for Y are random and uncorrelated.

Least-squares theory can be applied in situations when x is not a random variable but an ordinary variable, e.g., a function of distance or time. In fact, the variable x as it occurs in [8.15] is not a random variable. In Fig. 3, the method of least squares was applied in a situation where x denotes distance with its values equally spaced along the X-axis. From a theoretical statistical point of view, this situation is entirely different from regression analysis for a bivariate normal distribution. However, the equations used to obtain a solution are identical in the two situations.

Suppose that Y, given x, is normal. It does not matter whether x is the realization of a random variable X or not. When the residuals for Y are also normally distributed, methods of statistical inference can be developed as follows.

For convenience, we will make a change in notation and use matrix algebraic methods. When $\mathbf{Y}$ and $\mathbf{E}$ represent $(n \times 1)$ column vectors for observed values and residuals, the linear regression model can be written as:

$$\mathbf{Y} = \mathbf{X}\boldsymbol{\beta} + \mathbf{E} \qquad [8.18]$$

where:

$$\boldsymbol{\beta} = \begin{bmatrix} \beta_0 \\ \beta_1 \end{bmatrix}; \qquad \mathbf{X} = \begin{bmatrix} 1 & X_1 \\ 1 & X_2 \\ . & . \\ . & . \\ . & . \\ 1 & X_n \end{bmatrix}$$

The vectors Y, $\boldsymbol{\beta}$, and E are estimated by $\hat{\mathbf{Y}} = \mathbf{X}\hat{\boldsymbol{\beta}}$, $\hat{\boldsymbol{\beta}}$, and $\hat{\mathbf{E}}$, respectively. For [8.18], we obtain:

$$\mathrm{Y} = \mathbf{X}\hat{\boldsymbol{\beta}} + \hat{\mathrm{E}} \qquad [8.19]$$

Bivariate regression is a special application of multiple regression (see section 8.6). It consists of minimizing the sum of squared residuals $\Sigma \hat{E}_k^2 = \hat{\mathbf{E}}'\hat{\mathbf{E}}$ with:

$$\hat{\mathbf{E}}'\hat{\mathbf{E}} = \mathbf{Y}'\mathbf{Y} - \mathbf{Y}'\mathbf{X}\hat{\boldsymbol{\beta}} = \mathbf{Y}'\mathbf{Y} - \mathbf{Y}'\hat{\mathbf{Y}} \qquad [8.20]$$

If summation signs are used, [8.20] becomes:

$$\Sigma \hat{E}_k^2 = \Sigma Y_k^2 - \Sigma Y_k \hat{Y}_k \qquad [8.21]$$

If both Y_k and $\hat{Y}_k$ are corrected for the mean $\bar{Y}$ = Ave (Y_k) = Ave $(\hat{Y}_k)$, the sum of squared deviations from the mean can be decomposed as:

$$\underset{TSS}{\Sigma (Y_k - \bar{Y})^2} = \underset{SSR}{\Sigma (Y_k - \bar{Y})(\hat{Y}_k - \bar{Y})} + \underset{RSS}{\Sigma \hat{E}_k^2} \qquad [8.22]$$

It follows that the total sum of squares $TSS = \Sigma (Y_k - \bar{Y})^2$ is equal to the sum of squares due to regression $SSR = \Sigma (Y_k - \bar{Y})(\hat{Y}_k - \bar{Y}) = \Sigma (\hat{Y}_k - \bar{Y})^2$ plus the residual sum of squares $RSS = \Sigma \hat{E}_k^2$.

It should be kept in mind that the abbreviation "total sum of squares", which is written as TSS, usually refers to the sum of squares of the deviations of the Y_k from their mean $\bar{Y}$. Some of this variation is ascribed to the regression whereas the remainder (RSS) remains unexplained.

Suppose that the n observed values Y_k come from a normal random variable and that they are not significantly correlated with the X_k. In such a situation, TSS is distributed according to $\sigma^2\chi^2(n-1)$ with $(n-1)$ degrees of freedom and where σ^2 represents the variance of Y.

From [4.55], we know that:

$$SSR = \Sigma (\hat{Y}_k - \bar{Y})^2 = \hat{\beta}_1^2 \{\Sigma (X_k - \bar{X})^2\} \qquad [8.23]$$

From [4.57] it follows that $\hat{\beta}_1$ can be interpreted as a linear combination of the $(Y_k - \bar{Y})$. Since the $(Y_k - \bar{Y})$ are independent, $\hat{\beta}_1^2$ and SSR both are proportional to $\chi^2(1)$ with a single degree of freedom. By employing the addition theorem for χ^2, we find that RSS is proportional to χ^2 with $(n-1-1) = (n-2)$ degrees of freedom. In fact, we can write:

$$\underset{TSS}{\sigma^2\chi^2(n-1)} = \underset{SSR}{\sigma^2\chi^2(1)} + \underset{RSS}{\sigma^2\chi^2(n-2)} \qquad [8.24]$$

Consequently:

$$\frac{SSR}{RSS/(n-2)} = \frac{\chi^2(1)}{\chi^2(n-2)/(n-2)} = F(1, n-2) \qquad [8.25]$$

On this basis, the following analysis of variance table can be constructed:

Source	*Sum of squares*	*Degrees of freedom*	*Mean square*
Linear regression	*SSR*	1	*SSR*
Residuals	*RSS*	$n-2$	$RSS/(n-2)$
Total	*TSS*	$n-1$	

If, in an application, the ratio $\hat{F}(1, n-2)$ is significantly larger than 1, it may be assumed that there is a significant linear association between X and Y and that $\hat{\beta}_1$ differs from zero. On the other hand, if the ratio $\hat{F}(1, n-2)$ is sufficiently close to one, we assume that the values Y_k are independent random normal and that there is no need to use the values $\hat{Y}_k = \hat{\beta}_0 + \hat{\beta}_1 X_k$ since $\hat{\beta}_1$ can not be distinguished from $\beta_1 = 0$.

Suppose that we can prove that $\beta_1 \neq 0$, then the residuals $\hat{E}_k$ are normally distributed and $RSS = \Sigma\, \hat{E}_k^2$ has $(n-2)$ degrees of freedom. The so-called residual variance (variance of residuals) is estimated as:

$$s_{\mathrm{e}}^2 = RSS/(n-2) \qquad [8.26]$$

This is equivalent to writing:

$$s_{\mathrm{e}}^2 = \frac{n-1}{n-2} \cdot s_y^2 (1-r^2) \qquad [8.27]$$

which should be compared to [8.14]. In a bivariate situation:

$$r^2 = \frac{RSS}{TSS}$$

In practice, s_{e}^2 is often estimated by using the expression:

$$s_{\mathrm{e}}^2 = \frac{n-1}{n-2}(s_y^2 - b^2 s_x^2) \qquad [8.28]$$

Confidence belts

It can be shown (see section 8.7) that the variance of the calculated values $\hat{Y}_k$ satisfies:

$$s^2(\hat{Y}_k) = s_{\mathrm{e}}^2 \left[\frac{1}{n} + \frac{(X_k - \bar{X})^2}{\Sigma(X_k - \bar{X})^2}\right] \qquad [8.29]$$

Eq. [8.29] allows us to compute so-called confidence belts about the calculated values $\hat{Y}_k$ which lie on a straight line. In general, four types of confidence belt can be constructed, each of which has its own meaning. By setting:

$$R_k = \frac{1}{n} + \frac{(X_k - \bar{X})^2}{\Sigma(X_k - \bar{X})^2}$$

we may obtain for a best-fitting straight line:

$$(1) \quad \hat{Y}_k \pm t(n-2)\, s_e \sqrt{R_k} \qquad [8.30$$

This belt consists of two hyperbolas that enclose an area about the straight line $\hat{Y}_k = \hat{\beta}_0 + \hat{\beta}_1 X_k$. $t(n-2)$ is Student's t for $(n-2)$ degrees of freedom. A 95-% confidence belt has $t(n-2) = t_{0.975}(n-2)$. The purpose of the belt is to set confidence intervals on all single values of $\hat{Y}_k$ that can be calculated from given values of X_k.

$$(2) \quad \hat{Y}_k \pm \sqrt{2F(2, n-2)}\, s_e \sqrt{R_k} \qquad [8.31$$

The purpose of this belt which is wider than the previous one is to define a confidence region for the entire calculated regression line. It may be used in an accuracy test as discussed in section 6.1 where a line calculated from data was compared with a theoretical line $E(Y_k) = X_k$. The latter equation represents the hypothesis of complete accuracy of values Y_k which are being measured by a relatively imprecise method for known data X_k.

$$(3) \quad \hat{Y}_k \pm t(n-2)\, s_e \sqrt{1-R_k} \qquad [8.32$$

This represents a confidence belt for the residuals for data which have been used to calculate $\hat{Y}_k$.

In this way, it is possible to test the residuals for normality. In general, belt (3) and also belt (4) are considerably wider than belts (1) and (2), although all belts are hyperbolas.

$$(4) \quad \hat{Y}_k \pm t(n-2)\, s_e \sqrt{1+R_k} \qquad [8.33$$

Suppose that, after the calculation of the regression line, a new pair of values (X_k, Y_k) is observed. The new value should be compared to the fourth type of belt.

The subject of confidence belts for results obtained by regression is of considerable practical importance. Quenouille (1952) has given a discussion of belts of types (1), (3) and (4). The belt of type (2) first has been derived by Working and Hotelling (1929). In the following example, we will demonstrate the usage of confidence belts of types (3) and (4). For later applications, we will make use of type (2) belts generalized to a multivariate situation.

Determinative curve, olivine composition

We recall that in Chapter 5 the cell edge d_{174} of olivine was used to determine percentage forsterite. For this, a linear regression equation was calculated from 20 data with percentage Fo > 30 and actual percentages forsterite were determined by chemical analysis of olivine crystals.
The resulting equation is:

$$\hat{y} = 4145 - 3970\, x$$

where $\hat{y}$ represents estimated percentage forsterite and x is the measured cell edge d_{174}

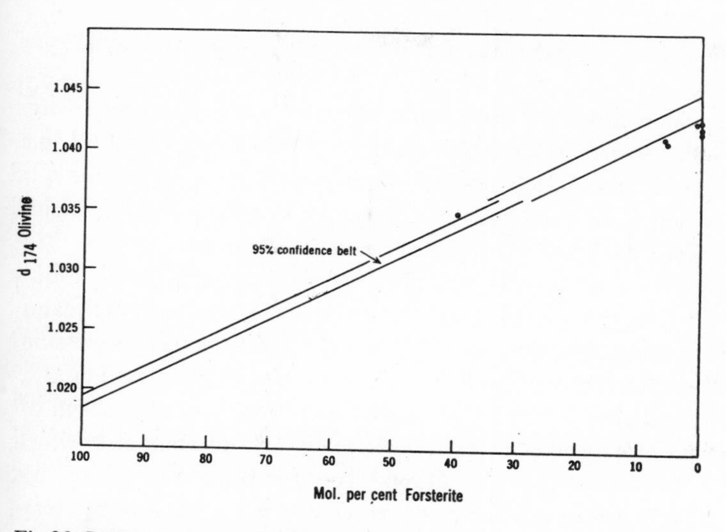

Fig.38. Practical use of confidence belts for a regression line. A line was fitted to data with more than 30% forsterite; only one of these falls outside the 95-% belt. All observation data with % Fo < 30 fall below the 95-% confidence belt for data not used for computing. (From Agterberg, 1964c.)

A belt of the third type is shown in Fig. 38 for this curve with original data in the range $30 < \%\,\text{Fo} \leqslant 100$. One of twenty values falls just outside the 95-% confidence belt. This indicates that the twenty original data may indeed be related to a single straight line. Six additional measurements were available for the range $0 \leqslant \%\,\text{Fo} < 30$. A belt of the fourth type was constructed for this range to see whether or not these six values are related to the regression line.

All six values fall below the 95-% confidence belt for new observations. This indicates a significant departure from linearity in the relationship between d_{174} and composition for the entire range ($0 \leqslant \%\,\text{Fo} \leqslant 100$). For practical determinations, we therefore used the straight line based on twenty data only. This statistical analysis was repeated for d_{130} cell edge measurements as used by Yoder and Sahama (1957) yielding a similar result. The d_{174} and d_{130} cell edges can be used for the same purpose of determining the composition of olivine crystals.

Regression and functional relationship

In problems of curve-fitting, one should distinguish between prediction equations and best estimates of the functional relationships between two variables.

In the preceding applications, the purpose was to derive an equation from data which is to be applied later to new data. In that situation, one should regress the variable which is to be predicted on the variable which will be measured. A different problem consists of estimating the functional relationship between two variables. If both variables are subject to error, the regression line does not provide an unbiased estimate of the functional relationship.

For the olivine composition curve, the resulting discrepancy is negligible (Agterberg, 1964c). On the other hand, it is large in the earlier example for gold values. In Fig. 37, line B approximates the functional relationship between the two variables, whereas line A is the line to be used for prediction. In some situations, the major axis or the reduced major axis can be used for estimating the linear relationship between two variables.

The problem of fitting best functions has been discussed in detail by Kendall (1965) and Deming (1943). If the ratio of the variances of the errors in the two variables remains constant and is known beforehand, one can construct what is called Kummell's regression line (Deming, 1943). When this ratio is equal to one, as in the situation of Fig. 37, Kummell's regression line and the major axis are identical. A more general problem of fitting a straight line to pairs of data whereby each single value has a different error which is known beforehand has been treated by York (1966). The best slope then is given by the solution of the so-called "least-squares cubic" equation, which may be done by iteration.

If the independent variable is free of error or nearly free of error, the regression line estimates the functional relationship. This condition will usually be satisfied if the independent variable is a function of time or distance.

8.6. Multiple regression

Results obtained for a bivariate situation are extended readily to the multivariate situation.

We consider a dependent variable Y whose n observations form a column vector $\mathbf{Y}$ and whose expectation is a function of p independent variables x_i each of which has n observed values X_{ik}. The matrix $\mathbf{X}$ is defined as the $n \times (p+1)$ matrix with rows equal to $[1 \quad X_{1k} \quad X_{2k} \quad \ldots X_{pk}]$ where k denotes a single observation. If $\boldsymbol{\beta}$ represents the $(p+1) \times 1$ vector of unknown coefficients, the general linear model, as before, can be written as:

$$\mathbf{Y} = \mathbf{X}\boldsymbol{\beta} + \mathbf{E} \qquad [8.34]$$

where $\mathbf{E}$ is the $(n \times 1)$ column vector for the residuals.

If the residual variance is σ^2, the frequency-distribution function of the conditional multivariate normal distribution for Y satisfies:

$$f_k(y | x_1, x_2, \ldots, x_p) = \frac{1}{\sigma\sqrt{2\pi}} \exp\left[-(1/2\sigma^2)(y_k - \mathbf{X}_k'\boldsymbol{\beta})^2\right] \qquad [8.35]$$

where f_k denotes frequency at point k where $y = y_k$; $\mathbf{X}_k'$ is the row vector consisting of values for the independent variables at point k. The maximum likelihood product for a sample of n observations (cf. section 7.11) satisfies:

$$\left(\frac{1}{\sigma\sqrt{2\pi}}\right)^n \exp\left[-(1/2\sigma^2)\,\mathbf{E}'\mathbf{E}\right] \tag{8.36}$$

By taking the logarithm it appears that this function will be a maximum if $\mathbf{E}'\mathbf{E}$ is a minimum. Consequently, the maximum-likelihood method for a multivariate distribution leads to the same results as the method of least squares. The maximum-likelihood estimator of the variance becomes:

$$\hat{\sigma}^2 = \hat{\mathbf{E}}'\hat{\mathbf{E}}/n \tag{8.37}$$

If the method of moments is applied, the result is the unbiased estimator:

$$\sigma^2(\hat{\mathbf{E}}_k) = \hat{\mathbf{E}}'\hat{\mathbf{E}}/(n-p-1) \tag{8.38}$$

which is used in statistical inference.

When [8.34] is written in full, we have:

$$\begin{bmatrix} Y_1 \\ Y_2 \\ \cdot \\ \cdot \\ \cdot \\ Y_n \end{bmatrix} = \begin{bmatrix} 1 & X_{11} & X_{21} & \cdot & \cdot & X_{p1} \\ 1 & X_{12} & X_{22} & \cdot & \cdot & X_{p2} \\ \cdot & \cdot & \cdot & \cdot & \cdot & \cdot \\ \cdot & \cdot & \cdot & \cdot & \cdot & \cdot \\ \cdot & \cdot & \cdot & \cdot & \cdot & \cdot \\ 1 & X_{1n} & X_{2n} & \cdot & \cdot & X_{pn} \end{bmatrix} \begin{bmatrix} \beta_0 \\ \beta_1 \\ \cdot \\ \cdot \\ \cdot \\ \beta_p \end{bmatrix} + \begin{bmatrix} E_1 \\ E_2 \\ \cdot \\ \cdot \\ \cdot \\ E_n \end{bmatrix}$$

For individual observations Y_k:

$$Y_k = \beta_0 + \beta_1 X_{1k} + \beta_2 X_{2k} + \ldots + \beta_p X_{pk} + E_k \tag{8.39}$$

From $\mathbf{Y} = \mathbf{X}\boldsymbol{\beta} + \mathbf{E}$, it follows that:

$$\mathbf{E}'\mathbf{E} = (\mathbf{Y} - \mathbf{X}\boldsymbol{\beta})'(\mathbf{Y} - \mathbf{X}\boldsymbol{\beta})$$

$$\mathbf{E}'\mathbf{E} = \mathbf{Y}'\mathbf{Y} + \boldsymbol{\beta}'\mathbf{X}'\mathbf{X}\boldsymbol{\beta} - \boldsymbol{\beta}'\mathbf{X}'\mathbf{Y} - \mathbf{Y}'\mathbf{X}\boldsymbol{\beta} \tag{8.40}$$

The terms $\boldsymbol{\beta}'\mathbf{X}'\mathbf{Y}$ and $\mathbf{Y}'\mathbf{X}\boldsymbol{\beta}$ are equal to the same scalar, as follows readily from the definitions of $\mathbf{Y}$, $\mathbf{X}$ and $\boldsymbol{\beta}$. Hence:

$$\mathbf{E}'\mathbf{E} = \mathbf{Y}'\mathbf{Y} - 2\boldsymbol{\beta}'\mathbf{X}'\mathbf{Y} + \boldsymbol{\beta}'\mathbf{X}'\mathbf{X}\boldsymbol{\beta} \tag{8.41}$$

This scalar can be minimized by partial differentiation with respect to the elements of $\boldsymbol{\beta}$. According to section 3.12, we have:

$$\frac{\partial(\mathbf{E}'\mathbf{E})}{\partial\boldsymbol{\beta}} = -2\mathbf{X}'\mathbf{Y} + 2\mathbf{X}'\mathbf{X}\boldsymbol{\beta} \tag{8.42}$$

If this $(p+1) \times 1$ column vector is set equal to zero, we obtain the $(p+1)$ normal equations:

$$\mathbf{X}'\mathbf{X}\hat{\boldsymbol{\beta}} = \mathbf{X}'\mathbf{Y} \tag{8.43}$$

with solution:

$$\hat{\boldsymbol{\beta}} = (\mathbf{X}'\mathbf{X})^{-1}\mathbf{X}'\mathbf{Y} \qquad [8.44]$$

The estimated values $\hat{\mathbf{Y}}$ satisfy:

$$\hat{\mathbf{Y}} = \mathbf{X}\hat{\boldsymbol{\beta}} = \mathbf{X}(\mathbf{X}'\mathbf{X})^{-1}\mathbf{X}'\mathbf{Y} \qquad [8.45]$$

Eq. [8.44] and [8.45] represent the generalization of a similar result obtained for $p = 2$ in Chapter 4.

An alternative method

$\mathbf{E}'\mathbf{E}$ also can be minimized by another method. This will provide an expression for $\hat{\mathbf{E}}'\hat{\mathbf{E}}$ and show that $\hat{\mathbf{E}}'\hat{\mathbf{E}}$ always is a minimum. By adding and subtracting the term $\mathbf{Y}'\mathbf{X}(\mathbf{X}'\mathbf{X})^{-1}\mathbf{X}'\mathbf{Y}$, [8.40] becomes:

$$\begin{aligned}\mathbf{E}'\mathbf{E} &= \boldsymbol{\beta}'\mathbf{X}'\mathbf{X}\boldsymbol{\beta} - \boldsymbol{\beta}'\mathbf{X}'\mathbf{Y} - \mathbf{Y}'\mathbf{X}\boldsymbol{\beta} + \mathbf{Y}'\mathbf{X}(\mathbf{X}'\mathbf{X})^{-1}\mathbf{X}'\mathbf{Y} + \mathbf{Y}'\mathbf{Y} - \mathbf{Y}'\mathbf{X}(\mathbf{X}'\mathbf{X})^{-1}\mathbf{X}'\mathbf{Y}\\ &= \boldsymbol{\beta}'\mathbf{X}'\mathbf{X}\boldsymbol{\beta} - \boldsymbol{\beta}'\mathbf{X}'\mathbf{X}(\mathbf{X}'\mathbf{X})^{-1}\mathbf{X}'\mathbf{Y} - \mathbf{Y}'\mathbf{X}(\mathbf{X}'\mathbf{X})^{-1}\mathbf{X}'\mathbf{X}\boldsymbol{\beta}\\ &\qquad + \mathbf{Y}'\mathbf{X}(\mathbf{X}'\mathbf{X})^{-1}\mathbf{X}'\mathbf{X}(\mathbf{X}'\mathbf{X})^{-1}\mathbf{X}'\mathbf{Y} + \mathbf{Y}'\mathbf{Y} - \mathbf{Y}'\mathbf{X}(\mathbf{X}'\mathbf{X})^{-1}\mathbf{X}'\mathbf{Y}\end{aligned}$$

Hence:

$$\mathbf{E}'\mathbf{E} = \{\boldsymbol{\beta} - (\mathbf{X}'\mathbf{X})^{-1}\mathbf{X}'\mathbf{Y}\}'\mathbf{X}'\mathbf{X}\{\boldsymbol{\beta} - (\mathbf{X}'\mathbf{X})^{-1}\mathbf{X}'\mathbf{Y}\} + \mathbf{Y}'\mathbf{Y} - \mathbf{Y}'\mathbf{X}(\mathbf{X}'\mathbf{X})^{-1}\mathbf{X}'\mathbf{Y} \qquad [8.46]$$

The vector $\boldsymbol{\beta}$ occurs in the first term on the right-hand side of this expression. This term is a quadratic form of the matrix $\mathbf{X}'\mathbf{X}$. According to section 3.12, $\mathbf{X}'\mathbf{X}$ is positive definite if its inverse exists, and the quadratic form can not be negative. $\mathbf{E}'\mathbf{E}$ will be a minimum if the quadratic form is equal to zero, which happens if:

$$\hat{\boldsymbol{\beta}} = (\mathbf{X}'\mathbf{X})^{-1}\mathbf{X}'\mathbf{Y} \qquad [8.44, \text{repeated}]$$

The minimum itself satisfies:

$$\begin{aligned}\hat{\mathbf{E}}'\hat{\mathbf{E}} &= \mathbf{Y}'\mathbf{Y} - \mathbf{Y}'\mathbf{X}(\mathbf{X}'\mathbf{X})^{-1}\mathbf{X}'\mathbf{Y}\\ &= \mathbf{Y}'\mathbf{Y} - \mathbf{Y}'\mathbf{X}\hat{\boldsymbol{\beta}} = \mathbf{Y}'\mathbf{Y} - \mathbf{Y}'\hat{\mathbf{Y}}\end{aligned} \qquad [8.47]$$

This shows that [8.20] for the bivariate situation is valid in the multivariate case.

Some basic properties of $\hat{\mathbf{Y}}$ and $\hat{\mathbf{E}}$

In section 4.6, the first column of $\mathbf{X}$ was written as $\mathbf{J}_n$ which consists of elements equal to unity only.

The first element of $\mathbf{X}'\mathbf{Y}$ is $\mathbf{J}_n'\mathbf{Y}$ and that of $\mathbf{X}'\mathbf{X}\hat{\boldsymbol{\beta}}$ is $\mathbf{J}_n'\mathbf{X}\hat{\boldsymbol{\beta}}$. Since $\mathbf{X}'\mathbf{X}\hat{\boldsymbol{\beta}} = \mathbf{X}'\hat{\mathbf{Y}}$, we have:

$$\mathbf{J}_n'\mathbf{X}\hat{\boldsymbol{\beta}} = \mathbf{J}_n'\mathbf{Y} \qquad [8.48]$$

This results in a linear relationship between the averages $\bar{Y}, \bar{X}_1, \bar{X}_2, \ldots, \bar{X}_p$, with:

$$\bar{Y} = \hat{\beta}_0 + \hat{\beta}_1 \bar{X}_1 + \ldots + \hat{\beta}_p \bar{X}_p$$

or: $$\hat{\beta}_0 = \bar{Y} - (\hat{\beta}_1 \bar{X}_1 + \ldots + \hat{\beta}_p \bar{X}_p) \qquad [8.49]$$

If this result is subtracted from:

$$\hat{Y}_k = \hat{\beta}_0 + \hat{\beta}_1 X_{1k} + \ldots + \hat{\beta}_p X_{pk}$$

we obtain:

$$(\hat{Y}_k - \bar{Y}) = \hat{\beta}_1 (X_{1k} - \bar{X}_1) + \ldots + \hat{\beta}_p (X_{pk} - \bar{X}_p) \qquad [8.50]$$

Consequently, the constant term $\hat{\beta}_0$ has been eliminated from the regression equation. Instead of solving $\boldsymbol{\beta}$ from $\mathbf{Y} = \mathbf{X}\boldsymbol{\beta} + \mathbf{E}$, we may solve $\boldsymbol{\alpha}$ from:

$$\mathbf{Y}_0 = \mathbf{X}_0 \boldsymbol{\alpha} + \mathbf{E} \qquad [8.51]$$

where:

$$\boldsymbol{\alpha} = \begin{bmatrix} \beta_1 \\ \beta_2 \\ \cdot \\ \cdot \\ \cdot \\ \beta_p \end{bmatrix}; \quad \mathbf{X}_0 = \begin{bmatrix} (X_{11} - \bar{X}_1) & \cdots & (X_{p1} - \bar{X}_p) \\ (X_{12} - \bar{X}_1) & \cdots & (X_{p2} - \bar{X}_p) \\ \cdot & \cdots & \cdot \\ \cdot & \cdots & \cdot \\ \cdot & \cdots & \cdot \\ (X_{1n} - \bar{X}_1) & \cdots & (X_{pn} - \bar{X}_p) \end{bmatrix}; \quad \mathbf{Y}_0 = \begin{bmatrix} Y_1 - \bar{Y} \\ Y_2 - \bar{Y} \\ \cdot \\ \cdot \\ \cdot \\ Y_n - \bar{Y} \end{bmatrix}$$

The least-squares solution for this reduced system of p variables is:

$$\hat{\boldsymbol{\alpha}} = (\mathbf{X}_0' \mathbf{X}_0)^{-1} \mathbf{X}_0' \mathbf{Y}_0 \qquad [8.52]$$

This gives all regression coefficients $\hat{\beta}_i$ except $\hat{\beta}_0$ which then is obtained by using [8.49]. From [8.48], it follows that:

$$\mathbf{J}_n' \hat{\mathbf{E}} = \mathbf{J}_n' \mathbf{Y} - \mathbf{J}_n' \hat{\mathbf{Y}} = 0 \qquad [8.53]$$

The residuals add to zero. Further:

$$\hat{\mathbf{E}}' \mathbf{X} = \mathbf{Y}' \mathbf{X} - \hat{\boldsymbol{\beta}}' \mathbf{X}' \mathbf{X} = [0 \ 0 \ldots 0]; \quad \hat{\mathbf{E}}' \hat{\mathbf{Y}} = 0$$

and: $$\hat{\mathbf{Y}}' \hat{\mathbf{Y}} = (\mathbf{X}\hat{\boldsymbol{\beta}} + \hat{\mathbf{E}})' \hat{\mathbf{Y}} = \mathbf{Y}' \hat{\mathbf{Y}} \qquad [8.54]$$

Consequently, [8.47] may be written as:

$$\mathbf{Y}' \mathbf{Y} = \hat{\mathbf{Y}}' \hat{\mathbf{Y}} + \hat{\mathbf{E}}' \hat{\mathbf{E}} \qquad [8.55]$$

We also have:

$$\underset{TSS}{\mathbf{Y}_0'\mathbf{Y}_0} = \underset{SSR}{\hat{\mathbf{Y}}_0'\hat{\mathbf{Y}}_0} + \underset{RSS}{\hat{\mathbf{E}}'\hat{\mathbf{E}}} \qquad [8.56]$$

If Y_0 is normally distributed and not correlated to one or more of the independent variables, we have $TSS = \sigma^2 \chi^2(n-1)$; $SSR = \sigma^2 \chi^2(p)$; $RSS = \sigma^2 \chi^2(n-p-1)$.
Hence:

$$\hat{F}(p, n-p-1) = \frac{SSR/p}{RSS/(n-p-1)} \qquad [8.57]$$

The analysis of variance table becomes:

Source of variation	*Sum of squares*	*Degrees of freedom*	*Mean square*
Regression	SSR	p	SSR/p
Residuals	RSS	$n-p-1$	$RSS/(n-p-1)$
Total	TSS	$n-1$	

The analysis-of-variance technique is more commonly applied in the following situation. Suppose that we have performed a multiple regression for p independent variables. Later, the regression analysis is carried out again by adding q new independent variables.

In total, we then have $(p+q)$ independent variables for the second regression. The problem consists of testing the q new terms for statistical significance. It is solved by forming the following table.

Source	*Sum of squares*	*Degrees of freedom*	*Mean square*
First regression (p var.)	SSR_1	p	
Difference between 1 and 2	$\Delta SSR = SSR_2 - SSR_1$	q	$\Delta SSR/q$
Second regression ($p+q$ var.)	SSR_2	$p+q$	
Residuals	RSS	$n-p-q-1$	$RSS/(n-p-q-1)$
Total	TSS	$n-1$	

The corresponding significance test consists of calculating:

$$\hat{F}(q, n-p-q-1) = \frac{\Delta SSR/q}{RSS/(n-p-q-1)} \qquad [8.58]$$

If this ratio is significantly larger than one, we may add the q new variables to the previous result. Otherwise, they will not be included in the regression equation.

Squared multiple correlation coefficient

An excellent measure for the overall fit of a regression is the so-called squared multiple correlation coefficient R^2 with:

$$R^2 = SSR/TSS \qquad [8.59]$$

If $p = 1$ (bivariate case), $R^2 = r_{xy}^2$. The so-called percentage explained sum of squares is obtained by multiplying R^2 by 100.

8.7. Geometric interpretation of least squares

In section 4.2 and 4.3, it was shown that fitting by the method of least squares can be interpreted as an orthogonal projection in n-space of the vector for the dependent variables. The situation is graphically represented in Fig. 39 for the case $n = 3$, $p = 2$. It is assumed that the three variables Y, x_1 and x_2 have been corrected for their means. We will use the geometrical approach to outline the concepts of multiple and partial correlation and for the construction of confidence regions on the entire regression equation.

If we drop the subscript 0, which has been used previously to indicate deviations from the mean, the data used for constructing Fig. 39 can be written as:

$$\mathbf{X} = \begin{bmatrix} X_{11} & X_{21} \\ X_{12} & X_{22} \\ X_{13} & X_{23} \end{bmatrix}; \quad \mathbf{Y} = \begin{bmatrix} Y_1 \\ Y_2 \\ Y_3 \end{bmatrix}$$

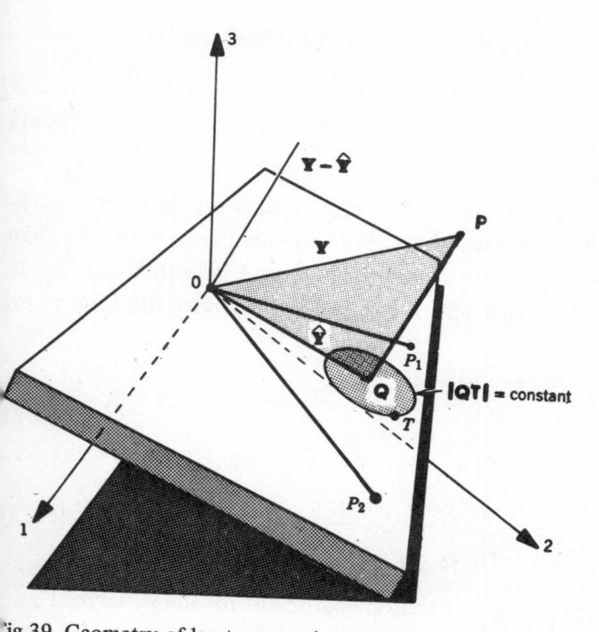

ig.39. Geometry of least squares in sample space when $n = 3$ and $p = 2$. For explanation, see text. After Draper and Smith, 1966.)

The two columns of **X** and the vector **Y** determine the three points P_1, P_2 and P. The vectors OP_1 and OP_2 define the plane onto which P is projected. Suppose that the perpendicular projection of P onto P_1OP_2 is Q. We then have $OP = \mathbf{Y}$; $OQ = \hat{\mathbf{Y}}$ and $QP = \hat{\mathbf{E}}$. We also see that $\mathbf{Y} = \hat{\mathbf{Y}} + \hat{\mathbf{E}}$ and, since $\hat{\mathbf{Y}}$ and $\hat{\mathbf{E}}$ are orthogonal, $\mathbf{Y}'\mathbf{Y} = \hat{\mathbf{Y}}'\hat{\mathbf{Y}} + \hat{\mathbf{E}}'\hat{\mathbf{E}}$ which is the Pythagorean theorem in matrix notation. We recall that $\mathbf{Y}'\mathbf{Y} = TSS$; $\hat{\mathbf{Y}}'\hat{\mathbf{Y}} = SSR$; $\hat{\mathbf{E}}'\hat{\mathbf{E}} = RSS$; and $TSS = SSR + RSS$.

A correlation coefficient can be interpreted as the cosine of the angle between two variables represented as vectors in n-space. If we write $r(y, x_1) = r_{y1}$, it follows that:

$$r_{y1} = \cos \angle POP_1$$

Likewise, $r_{y2} = \cos \angle POP_2$; $r_{12} = \cos \angle P_1OP_2$.
The multiple-correlation coefficient is defined as the positive square root of R^2. In Fig. 39, $R = \cos \angle POQ$. The so-called partial correlation coefficient $r_{y2.1}$ satisfies $r_{y2.1} = \cos \angle PQP_1$. Likewise, $r_{y1.2} = \cos \angle PQP_2$.

Ordinary and multiple-correlation coefficients also can be defined as follows: r^2_{y1} is the proportion of total sum of squares of **Y** explained by $\mathbf{X}_1$; and $R^2_{y.12}$ is the proportion of total sum of squares of **Y** explained by $\mathbf{X}_1$ and $\mathbf{X}_2$. The partial correlation coefficient $r_{y2.1}$ may be defined as: $r^2_{y2.1}$ is the proportion of remaining sum of squares of **Y** explained by $\mathbf{X}_2$ after elimination of the effect of $\mathbf{X}_1$.

In Fig. 39, $|PQ| = |PP_1| \sin \angle PQP_1$. If both sides are squared, it follows that:

$$1 - R^2 = (1 - r^2_{y1})(1 - r^2_{y2.1}) \qquad [8.60]$$

By repeated application of the Pythagorean theorem, it can be shown that:

$$r_{y2.1} = \frac{r_{y2} - r_{y1} \cdot r_{12}}{\sqrt{(1 - r^2_{y1})(1 - r^2_{12})}} \qquad [8.61]$$

The partial correlation coefficient may be treated as an ordinary correlation coefficient; $r_{y2.1}$ may provide a more useful measure of the association between **Y** and $\mathbf{X}_2$ than r_{y2}. For example, suppose that $\mathbf{X}_1$ represents a time trend present in both **Y** and $\mathbf{X}_2$. In that situation, $r_{y2.1}$ is the correlation coefficient for **Y** and $\mathbf{X}_2$ whereby the time trend has been removed.

Eq. [8.60] and [8.61] can be generalized to:

$$1 - R^2_p = (1 - r^2_{y1})(1 - r^2_{y2.1}) \ldots (1 - r^2_{yp.12\ldots(p-1)}) \qquad [8.62]$$

$$r_{yk.12\ldots p} = \frac{r_{yk.12\ldots(p-1)} - r_{yp.12\ldots(p-1)}\, r_{kp.12\ldots(p-1)}}{\sqrt{(1 - r^2_{yk.12\ldots(p-1)})(1 - r^2_{kp.12\ldots(p-1)})}} \qquad [8.63]$$

The expression $r_{yk.12\ldots p}$ denotes the partial correlation coefficient for the variables Y and x_k after elimination of the combined effect of a set of other variables $x_1, x_2, \ldots, x_p$

Proof of [8.61] – [8.63] is given by Cramér (1947, chapter 23) or Kendall and Stuart (1958, vol. 2, chapter 27).

These results are used in various methods of sequential regression analysis where the independent variables are successively included in the regression equation.

Confidence intervals

If [8.47] is subtracted from [8.46], we obtain:

$$\mathbf{E}'\mathbf{E} - \hat{\mathbf{E}}'\hat{\mathbf{E}} = [\boldsymbol{\beta} - (\mathbf{X}'\mathbf{X})^{-1}\mathbf{X}'\mathbf{Y}]'\mathbf{X}'\mathbf{X}[\boldsymbol{\beta} - (\mathbf{X}'\mathbf{X})^{-1}\mathbf{X}'\mathbf{Y}]$$
$$= (\boldsymbol{\beta} - \hat{\boldsymbol{\beta}})'\mathbf{X}'\mathbf{X}(\boldsymbol{\beta} - \hat{\boldsymbol{\beta}}). \qquad [8.64]$$

Writing $\mathbf{T} = \mathbf{X}(\boldsymbol{\beta} - \hat{\boldsymbol{\beta}}) = \mathbf{X}\boldsymbol{\beta} - \mathbf{X}\hat{\boldsymbol{\beta}}$:

$$\mathbf{E}'\mathbf{E} - \hat{\mathbf{E}}'\hat{\mathbf{E}} = \mathbf{T}'\mathbf{T} \qquad [8.65]$$

The vector $\mathbf{T}$ is equal to TQ in Fig. 39; we have $TQ = \mathbf{T}$, $QP = \hat{\mathbf{E}}$, and $TP = \mathbf{E}$.

It can be shown that the sum of squares $\mathbf{T}'\mathbf{T}$ is distributed according to χ^2 with $(p+1)$ degrees of freedom. On the other hand, $\hat{\mathbf{E}}'\hat{\mathbf{E}}$ is independent of $\mathbf{T}'\mathbf{T}$ and has $(n-p-1)$ degrees of freedom. If the residual variance σ^2 is estimated as $s^2 = \hat{\mathbf{E}}'\hat{\mathbf{E}}/(n-p-1)$:

$$\mathbf{T}'\mathbf{T} = (p+1)s^2 F(p+1, n-p-1) \qquad [8.66]$$

Suppose that $F(p+1, n-p-1)$ is set equal to the 95-% fractile of the cumulative F-distribution. The form:

$$\mathbf{T}'\mathbf{T} = \mathbf{E}'\mathbf{E} - \hat{\mathbf{E}}'\hat{\mathbf{E}} = (p+1)s^2 F_{0.95}(p+1, n-p-1) \qquad [8.67]$$

then defines a 95-% confidence region for the point T. If Q is given, the probability that T falls within the confidence region is 95 %. Since $\mathbf{T}'\mathbf{T}$ is equal to a constant,

$$|QT| = \text{constant}$$

which defines a circle in the plane P_1OP_2.
A circle of this type is shown in Fig. 39. The probability that the true solution $\mathbf{X}\boldsymbol{\beta}$ will plot as a point onto P_1OP_2, that is enclosed by the circle, is equal to 95%. We, therefore, have now developed a method for testing all $(p+1)$ parameters simultaneously.

In some applications, it is useful to determine a confidence interval for $\mathbf{X}_k'(\boldsymbol{\beta} - \hat{\boldsymbol{\beta}})$ where $\mathbf{X}_k'$ represents a single observation point with given values for the independent variables. In that case:

$$T_k^2 = (\boldsymbol{\beta} - \hat{\boldsymbol{\beta}})'\mathbf{X}_k\mathbf{X}_k'(\boldsymbol{\beta} - \hat{\boldsymbol{\beta}}) \qquad [8.68]$$

By using $\mathbf{T}'\mathbf{T} = (\boldsymbol{\beta} - \hat{\boldsymbol{\beta}})'\mathbf{X}'\mathbf{X}(\boldsymbol{\beta} - \hat{\boldsymbol{\beta}})$, it can be shown that:

$$T_k^2 = \mathbf{X}_k'(\mathbf{X}'\mathbf{X})^{-1}\mathbf{X}_k\,\mathbf{T}'\mathbf{T} \qquad [8.69]$$

By using [8.67], a 95-% confidence belt:

$$\hat{Y}_k \pm [\mathbf{X}'_k(\mathbf{X}'\mathbf{X})^{-1}\mathbf{X}_k(p+1)s^2 F_{0.95}(p+1, n-p-1)]^{1/2} \qquad [8.70]$$

may be constructed. If $p = 1$, this expression reduces to [8.31] for our second type of confidence belt on a straight line with $\hat{Y}_k = \hat{\beta}_0 + \hat{\beta}_1 X_k$.

It can be shown that the variance—covariance matrix of the $(p + 1)$ regression coefficients of $\hat{\boldsymbol{\beta}}$ satisfies:

$$\mathbf{V}(\hat{\boldsymbol{\beta}}) = (\mathbf{X}'\mathbf{X})^{-1}\sigma^2 \qquad [8.71]$$

The variance of a calculated value $\hat{Y}_k$ satisfies:

$$\sigma^2(\hat{Y}_k) = \mathbf{X}'_k\mathbf{V}(\hat{\boldsymbol{\beta}})\mathbf{X}_k = \mathbf{X}'_k(\mathbf{X}'\mathbf{X})^{-1}\mathbf{X}_k\sigma^2 \qquad [8.72]$$

If σ^2 is estimated as $s^2 = \hat{\mathbf{E}}'\hat{\mathbf{E}}/(n-p-1)$, one obtains, as a 95-% confidence belt for individual values $\hat{Y}_k$:

$$\hat{Y}_k \pm t_{0.975}\, s\sqrt{\mathbf{X}'_k(\mathbf{X}'\mathbf{X})^{-1}\mathbf{X}_k} \qquad [8.73]$$

In the bivariate situation, this reduces to [8.30].

Eq. [8.70] and [8.73] both determine confidence regions about $\hat{Y}_k$. The former is for the situation when we consider the regression equation as a whole. The latter applies if we wish to set 95-% confidence limits on a single calculated value $\hat{Y}_k$. The first belt [8.70] is always wider than the second belt [8.73].

Example: *Multiple regression of density of gold occurrences on mappable lithological units*

This example is for an area of 7200 sq. miles (18,500 km^2) in western Quebec (surroundings of Noranda and Val d'Or, cf. Fig. 79). A regular grid with spacing equal to 8 miles was superimposed on geologic maps with scale 1 inch = 4 miles (approx. 1 : 250,000) for this area. The vector **Y** from the dependent variable Y was obtained by counting the number of gold occurrences per unit cell of 8 × 8 sq. miles. In total, there are 113 cells or $n = 113$, and the total number of gold deposits in the area is equal to 444. Six lithological variables that could be distinguished everywhere in the area are listed in Table XXII.

TABLE XXII
Rock types in Abitibi area of western Quebec

1. Granitic rocks, granite gneisses, acidic intrusive rocks
2. Mafic intrusive rocks (includes gabbros and diorites)
3. Ultramafic rocks
4. Early Precambrian sedimentary rocks
5. Acidic volcanics (rhyolites and pyroclastic rocks)
6. Mafic volcanics (basalts and andesites)

TABLE XXIII

Percentage explained sum of squares (*ESS*) for 62 possible regressions; density of gold occurences regressed on six lithological variables

x_i's	*ESS* (%)	x_i's	*ESS* (%)	x_i's	*ESS* (%)	x_i's	*ESS* (%)
1	8.64	3, 5	16.29	2, 3, 5	16.43	1, 3, 4, 6	25.61
2	1.45	3, 6	0.33	2, 3, 6	1.79	1, 3, 5, 6	28.37
3	0.19	4, 5	28.15*	2, 4, 5	28.18	1, 4, 5, 6	28.61
4	10.48	4, 6	10.56	2, 4, 6	12.46	2, 3, 4, 5	29.40*
5	15.56*	5, 6	16.07	2, 5, 6	16.14	2, 3, 4, 6	12.78
6	0.17	1, 2, 3	8.95	3, 4, 5	29.32*	2, 3, 5, 6	16.84
1, 2	8.85	1, 2, 4	14.33	3, 4, 6	10.96	2, 4, 5, 6	28.18
1, 3	8.76	1, 2, 5	19.21	3, 5, 6	16.72	3, 4, 5, 6	29.33
1, 4	13.54	1, 2, 6	22.60	4, 5, 6	28.15	1, 2, 3, 4, 5	29.41
1, 5	18.86	1, 3, 4	13.79	1, 2, 3, 4	14.56	1, 2, 3, 4, 6	29.41
1, 6	22.52	1, 3, 5	14.40	1, 2, 3, 5	19.83	1, 2, 3, 5, 6	29.41
2, 3	1.58	1, 3, 6	22.58	1, 2, 3, 6	22.65	1, 2, 4, 5, 6	29.41
2, 4	12.42	1, 4, 5	28.19	1, 2, 4, 5	28.24	1, 2, 4, 5, 6	29.41
2, 5	15.65	1, 4, 6	24.87	1, 2, 4, 6	27.45	2, 3, 4, 5, 6	29.41
2, 6	1.68	1, 5, 6	28.34	1, 2, 5, 6	29.37		
3, 4	10.85	2, 3, 4	12.72	1, 3, 4, 5	29.32		

* For explanation see section 8.9.

The percentage of area occupied by each of these six rock types was determined for each of the 113 cells giving six independent variables $x_1 - x_6$ with vectors $\mathbf{X}_1 - \mathbf{X}_6$. The matrix $\mathbf{X}$ therefore has size (113 × 7). Because these data are percentage values, $\Sigma_i X_{ik} = 1$ for every cell (k). Because of this linear relationship between the variables, the matrix $\mathbf{X}'\mathbf{X}$ is singular and can not be inverted. However, multiple regression can be carried out for all possible combinations of one, two, three, four and five independent variables. In total, this gives 62 solutions which are shown in Table XXIII. We only present the percentage explained sum of squares (*ESS*) which is equal to the squared multiple-correlation coefficient multiplied by 100. Note that the largest value is reached for combinations of five variables all of which have *ESS* = 29.41 %. A proof of this property for closed number systems will be given in section 13.9. From [8.58] and [8.59], it is derived that:

$$\hat{F}(q, n-p-q-1) = \frac{(ESS - ESS_p)/q}{(100 - ESS)/(n-p-q-1)} \qquad [8.74]$$

where ESS_p is $100 \times R^2$ for p variables in regression, and *ESS* applies to $(p+q)$ variables. If five variables are considered, we may set $p = 0$, $q = 5$, whence:

$$\hat{F}(5{,}107) = \frac{29.41/5}{(100 - 29.41)/107} = 9.04$$

The calculated ratio can be compared to theoretical $F_{0.95}(5,107) = 2.31$. From $\hat{F} > F_{0.95}$, it is concluded that there exists a statistically significant relationship between Y and five lithological variables. Two of the six possible regression equations are as follows. Variables x_6 and x_5 have been deleted in the first and second equation, respectively.

$$\hat{\mathbf{Y}} = 0.108 - 0.0031\mathbf{X}_1 - 0.0475\,\mathbf{X}_2 + 0.1443\,\mathbf{X}_3 + 0.1148\mathbf{X}_4 + 0.3531\mathbf{X}_5$$

$$\hat{\mathbf{Y}} = 35.421 - 0.3563\,\mathbf{X}_1 - 0.4007\,\mathbf{X}_2 - 0.2089\,\mathbf{X}_3 - 0.2383\,\mathbf{X}_4 - 0.3531\,\mathbf{X}_6$$

This result illustrates that it may be dangerous to assign a physical meaning to individual coefficients in a regression equation. The two equations yield indentical values of $\hat{\mathbf{Y}}$ when the independent variables are given. For example, 2 out of the 113 cells consist of granitic material (x_1) only. For these cells, $X_1 = 100$ and $X_2 = X_3 = \ldots = X_6 = 0$. Allowing for round-off errors, the reader may verify that $\hat{Y} = -0.21$ for these cells, regardless of which equation is used.

In many problems of multiple regression, it is useful to identify those independent variables which are most important. It can happen that one or more of the variables are not at all related to the dependent variable. Such variables are called "irrelevant". Or, some variables may be almost linearly related to other independent variables, and would not increase the total strength of the regression as measured by *ESS*. These variables are termed "redundant".

A solution to this problem is given in Table XXIII. The application of the method of all possible regressions to geologic situations has been advocated by Krumbein et al. (1964). The single variable with largest *ESS* is x_5 (rhyolitic and pyroclastic rocks; ESS = 15.56 %). Setting $p = 0$, $q = 1$, in [8.74], we obtain:

$$\hat{F}(1,111) = \frac{15.56}{(100 - 15.56)/111} = 20.46$$

This value is large as compared to $F_{0.95}(1,111) = 3.94$. We can proceed and select the best combination of two variables. This gives the pair x_4 and x_5 with $ESS = 28.15$ %. By setting $p = 1$, $q = 1$, it is tested whether x_4 improves upon x_5 which already occurs in the regression equation:

$$\hat{F}(1,110) = \frac{28.15 - 15.56}{(100 - 28.15)/110} = 19.27$$

This computed $\hat{F}$-value is larger than $F_{0.95}(1,110) = 3.94$ indicating that x_4 should be added. The next variable that would be added is x_3 with $ESS = 29.32$ % (for x_3, x_4 and x_5). With $p = 2$, $q = 1$, $\hat{F}(1,109) = 1.80$ which is less that $F_{0.95}(1,109) = 3.94$ or $F_{0.90}(1,109) = 2.76$.

Consequently, variable x_3 should not be included in the regression equation. In fact, the pair x_4 and x_5 accounts for 96 % of the maximum value $ESS = 29.41$ % for five variables in regression.

Discussion

The regression equations for first and second step are:

$$\hat{\mathbf{Y}} = 2.550 + 0.3160\,\mathbf{X}_5\,; \qquad \hat{\mathbf{Y}} = 0.189 + 0.1151\,\mathbf{X}_4 + 0.3377\,\mathbf{X}_5$$

The first equation predicts that $\hat{Y}_k$ is relatively large if rhyolitic and pyroclastic rocks occur in a cell. The other equation states that in addition to x_5, the sedimentary rocks (x_4) constitute a favorable environment for the occurrence of gold deposits. However, if these results are evaluated in more detail, it appears that the sedimentary rocks in the map area are restricted to a number of relatively narrow belts with differences in geological age and facies, which could not be mapped in a systematic manner. One of these belts approximately coincides with a major gold belt in the area with a high density of gold deposits. It explains that x_4 enters the regression. However, $\hat{Y}_k$ now is predicted to be large in all cells that are situated on the sedimentary belts. In Chapter 15, a more flexible prediction model will be discussed where the possibility is left open that a geological environment coded as a single variable (e.g., sedimentary rocks) may be rich in mineral deposits in one place but barren in others.

8.8. Regression analysis and component analysis

In the previous example, the dependent variable could be regressed on any combination of five independent variables, giving the same vector $\hat{\mathbf{Y}}$ in all cases. However, the coefficients of the vectors $\hat{\boldsymbol{\beta}}$ showed little resemblance to each other indicating that they are subject to large variances. Many computer programs for multiple regression analysis also produce an estimate of the matrix $\mathbf{V}(\hat{\boldsymbol{\beta}})$ with $\mathbf{V}(\hat{\boldsymbol{\beta}}) = \sigma^2(\mathbf{X}'\mathbf{X})^{-1}$ whose diagonal elements $V_{ii}(\hat{\boldsymbol{\beta}})$ are the variances of the coefficients $\hat{\beta}_i$. The precision of $\hat{\boldsymbol{\beta}}$ can be described in general terms as follows.

Suppose that the independent variables are corrected for their mean. The vectors $\hat{\boldsymbol{\beta}}$ and $\boldsymbol{\beta}$ then fall in p-dimensional (coefficient-) space. The distance D between their end points in this space follows from the Pythagorean theorem or:

$$D^2 = (\hat{\boldsymbol{\beta}} - \boldsymbol{\beta})'\,(\hat{\boldsymbol{\beta}} - \boldsymbol{\beta}) \qquad [8.75]$$

The expected value of D^2 satisfies:

$$E(D^2) = \Sigma\, V_{ii}(\hat{\boldsymbol{\beta}}) = \text{Trace}\ [\mathbf{V}(\hat{\boldsymbol{\beta}})] \qquad [8.76]$$

Since $\mathbf{V}(\hat{\boldsymbol{\beta}}) = \sigma^2(\mathbf{X}'\mathbf{X})^{-1}$, we can write:

$$E(D^2) = \sigma^2\ \text{Trace}\ [(\mathbf{X}'\mathbf{X})^{-1}]$$

If $\lambda_1 \geqslant \lambda_2 \geqslant \ldots \geqslant \lambda_p$ are the eigenvalues of $\mathbf{X}'\mathbf{X}$, the eigenvalues of $(\mathbf{X}'\mathbf{X})^{-1}$ are:

$$1/\lambda_1 \leqslant 1/\lambda_2 \leqslant \ldots \leqslant 1/\lambda_p$$

Hence:

$$E(D^2) = \sigma^2 \sum_{i=1}^{p} 1/\lambda_i \qquad [8.77]$$

If one or more of the eigenvalues of $\mathbf{X}'\mathbf{X}$ are small, $E(D^2)$ will be large. If $\mathbf{X}'\mathbf{X}$ is singular, $\lambda_p = 0$ and $E(D^2)$ would be infinitely large.

We have calculated the eigenvalues of the (6×6) matrix $\mathbf{X}'\mathbf{X}$ for the six independent variables of the previous example after replacing the vectors $\mathbf{X}_i$ by $(\mathbf{X}_i - \overline{\mathbf{X}}_i)$ where $\overline{\mathbf{X}}_i = \overline{X}_i \mathbf{J}_n$.

This gave:

$$\lambda_1 = 1427; \quad \lambda_2 = 534; \quad \lambda_3 = 93; \quad \lambda_4 = 30; \quad \lambda_5 = 15; \quad \lambda_6 = 0$$

Since one of the eigenvalues is zero, $\mathbf{X}'\mathbf{X}$ can not be inverted and a regression equation containing all six variables can not be obtained.

It is convenient to multiply the eigenvalues by a constant so that they add to 100. This gives:

$$\lambda_1' = 68.0; \quad \lambda_2' = 25.4; \quad \lambda_3' = 4.4; \quad \lambda_4' = 1.4; \quad \lambda_5' = 0.7; \quad \lambda_6' = 0.0$$

These values may be considered as percentages of total sum of squares $TSS_x = \Sigma_{i,k}(X_{ik} - \overline{X}_i)^2$ explained by a principal component ξ_j for λ_j (cf. Chapter 5). The elements of the principal components satisfy $\xi_{jk} = \Sigma_i v_{n,ji}(X_{ik} - \overline{X}_i)$ where $v_{n,ji}$ are elements of the eigenvectors $\mathbf{V}_{n,j}$ which, for the example, are:

	$\mathbf{V}_{n,1}$	$\mathbf{V}_{n,2}$	$\mathbf{V}_{n,3}$	$\mathbf{V}_{n,4}$	$\mathbf{V}_{n,5}$	$\mathbf{V}_{n,6}$
1	0.79	−0.28	−0.30	−0.20	0.04	0.41
2	−0.02	0.00	0.26	0.19	−0.86	0.41
3	0.00	0.01	0.07	0.82	0.39	0.41
4	−0.13	0.80	−0.37	−0.20	0.05	0.41
5	−0.05	0.01	0.74	−0.42	0.33	0.41
6	−0.60	−0.54	−0.39	−0.19	0.05	0.41

Suppose that we regress on the principal components instead of on the original variables. This method is called canonical regression analysis and has been discussed by Kendall (1965). If ξ_6 is deleted as a variable, $\hat{\mathbf{Y}}$ will be the same as may be obtained by omitting any one of the vectors $(\mathbf{X}_i - \overline{\mathbf{X}}_i)$. The residual variance for these regressions is $s^2 = 519.005$. With [8.77], we obtain:

$$\hat{E}(D^2) = 519 \times \left(\frac{1}{1427} + \frac{1}{534} + \frac{1}{93} + \frac{1}{30} + \frac{1}{15}\right) = 58.8$$

On the other hand, the first two eigenvalues of $\mathbf{X}'\mathbf{X}$ account for $68.0 + 25.4 = 93.4$ % of the total sum of squares. This suggests that only two principal components should be used for the multiple regression. The average squared distance between $\hat{\boldsymbol{\beta}}$ and $\boldsymbol{\beta}$ then has:

$$\hat{E}(D^2) = 519 \times \left(\frac{1}{1427} + \frac{1}{534}\right) = 1.34$$

vhich is 44 times smaller than the previous value. Consequently, the estimate $\hat{\boldsymbol{\beta}}$ is rela-ively precise if only two principal components are used.

Discussion

The method of canonical analysis suggests that two instead of five independent vari-bles may be used in the multiple regression. In this respect, there is agreement with the esults reported in Table XXIII. This illustrates the usefulness of inspecting the eigenval-es of $\mathbf{X}'\mathbf{X}$. However, in general, regressing on a limited number of principal components as two drawbacks: (1) the final regression equation contains all independent variables be-ause none are eliminated completely; and (2) the association between dependent and in-ependent variables is not considered at the stage that the principal components are ormed. For example, from the coefficients of the eigenvectors derived here, it can be een that the first two principal components are mainly determined by the variables x_1, $_4$ and x_6. Although this combination of three in Table XXIII accounts for $ESS = 24.9\%$, which is a relatively large proportion, we failed to detect the association between Y and $_5$.

It should be kept in mind that possible difficulties encountered in obtaining precise stimates of $\hat{\beta}_i$ are not necessarily related to the absence of significant multiple associa-ions with regard to the dependent variable. Rather, they may be caused by problems of ampling the independent variables if these are strongly interrelated. Various transforma-ions such as those based on the eigenvalues can be useful in obtaining a numerical esti-nate of $\hat{\boldsymbol{\beta}}$ that is satisfactory. If $\mathbf{X}'\mathbf{X}$ has one or more small eigenvalues, it is ill-condi-ioned. It may not be possible to invert an ill-conditioned matrix by one of the conven-ional methods such as pivotal condensation (see section 3.6). This is because numerical alculations are done with limited precision. For example, so-called single precision on digital computer may be equivalent to eight or more digits, and double precision to sixteen or more digits depending on the type of computer in use. Errors arise because umbers are truncated during the calculations. It may happen that round-off errors begin o propagate at some point and distort the final result. Fortunately, such mishaps can be etected readily by using various checks. For example, the inverse $(\mathbf{X}'\mathbf{X})^{-1}$ can be post-r premultiplied with the original $\mathbf{X}'\mathbf{X}$, which should give an identity matrix. We will not e able to consider these problems in detail. An excellent introduction is given by Conté 1965). For a more advanced review, see Wilkinson (1965).

Because these problems are important in practical estimation, methods to solve them re being reviewed continuously in papers dealing with applied statistics (e.g., Beale, 1970; Vampler, 1970). Another method of obtaining good although slightly biased estimates of is ridge regression (Hoerl and Kennard, 1970).

In the next section, we will discuss three general and relatively simple methods by which redundant and irrelevant independent variables can be eliminated. In addition t leading to more meaningful coefficients, these methods also have the advantage that prob lems of propagation of round-off errors can be avoided. In this discussion, the terminolog will be as that used by Draper and Smith (1966).

8.9. Methods of sequential regression analysis

The best method of singling out variables and combinations of variables in regression i to study all possible regressions (see section 8.7). Unfortunately, this method can not b applied when the number of independent variables p is large. The number of possible re gressions N satisfies $N = 2^p$. This includes the calculation of the arithmetic mean which incidentally, provides the least-squares solution if $\mathbf{X} = \mathbf{J}_n$. When $p = 10, N = 1024$ whic is processable, but, e.g., $p = 100$ gives $N \approx 10^{30}$ which is beyond the capability of com puters. In that situation, a stepwise procedure may give useful results. A well-known method for this is forward selection.

Forward selection

Of p independent variables, the one which explains most of the dependent variable ha the largest correlation coefficient. This variable can be identified readily (e.g., x_5 i Table XXIII). Let it be referred to as x_1. By using [8.61], partial correlation coefficient $r_{yk.1}$ then can be calculated for the $(p-1)$ relationships between Y and the independent variables x_k after elimination of the effect of x_1. The variable with the largest partial-cor relation coefficient is called x_2. It is the next most important variable. We then calculat the $(p-2)$ coefficients $r_{yk.12}$ by using [8.63], select the largest one, and repeat the pro cedure. An F-test can be applied at each step. If the method is applied to the example o Table XXIII, the path followed is the one marked by asterisks.

We can define a cut-off level. For example, the procedure may be discontinued when $\hat{F} < F_{0.90}$. However, in doing this, the risk is taken that one or more variables, which ar relevant, are excluded. The method always is useful because it is relatively effective fo solving $\hat{\boldsymbol{\beta}}$ if no cut-off level is applied and all variables are included.

Strictly speaking, the F-test can not be applied for the statistical inference of thi method. The theoretical problems of developing a significance test for the method ar formidable as shown by Pope (1969) who has developed a test for the situation that $p = 3$

Stepwise regression

This method does the same as forward selection except for a significant refinement After completing a single step (selection of variable with largest partial-correlation coeffi cient), one goes back one step whereby the variables already included in the equation ar again checked for statistical significance. This can be done by using [8.62].

We can reject a variable if $\hat{F} < F_{0.90}$. However, in practice, it may be useful to employ other cut-off levels also, e,g., $\hat{F} < F_{0.50}$. For example, we have found in one application with $p = 22$ that for $\hat{F} > F_{0.90}$, only one variable was included, whereas several other significant independent variables were missed. The limitations of various methods of stepwise regression have been discussed by Hamaker (1962) and Beale et al. (1967).

For Table XXIII, this method gives the same result as forward selection. For an example where stepwise regression and forward selection do not give the same result as the method of all possible regressions, see section 15.8.

A computer algorithm for stepwise regression has been developed by Efroymson (1960). The method has been programmed for geologic applications by Miesch and Connor (1968).

Backward elimination

A third method consists of first including all variables in the regression equation and then eliminating them one by one, by using [8.62], until the F-ratio exceeds a predetermined level, e.g., $\hat{F} > F_{0.90}$.

If this method is applied to the example of Table XXIII, we run into a problem of how to begin. Because no solution can be obtained for six variables, one must be excluded. However, no matter which one is excluded initially, the *ESS*-value is the same for all combinations of five variables, and backward elimination can not make the choice for us. If x_1 is deleted initially, the path is the one marked by asterisks in Table XXIII. On the other hand, if x_4 is omitted, we have successively the combinations $(x_1, x_2, x_3, x_5, x_6)$, (x_1, x_2, x_5, x_6), (x_1, x_5, x_6), (x_1, x_6). At this point $\hat{F} > F_{0.90}$, and the final regression equation contains x_1 and x_6 with $ESS = 22.52$ instead of x_4 and x_5 with $ESS = 28.15\%$.

If the number of variables is large, it may be necessary to exclude a number of them in the beginning, because, otherwise, it may not be possible to invert $\mathbf{X}'\mathbf{X}$. Problems of this type constitute a drawback of the method. Of course, the problem does not occur if one of the previous two procedures is applied. On the other hand, backward elimination has the advantage that a large number of variables is considered initially, and it may be more difficult to exclude significant variables.

8.10. Dummy variables

Suppose that we wish to discriminate between the relationship of a dependent variable Y and a set of independent variables x_i in two or more different geologic environments or separate areas. This problem can be solved by the use of dummy variables whose purpose is to compare with one another sets of coefficients for different multiple regressions. Dummy variables are employed in several fields of applied statistics including econometrics. A good review was given by Gujarati (1970). Let us assume that a variable Y is linearly related to two variables x_1 and x_2 and that:

$$Y_k = \beta_0 + \beta_1 X_{1k} + \beta_2 X_{2k} + E_k \qquad [8.78]$$

where $k = 1, 2, \ldots, n$.

Suppose that measurements were made in two geographically distinct areas (1 and 2). We can then reformulate the model as follows:

$$\left.\begin{array}{ll} \text{Area 1:} \; Y_{1k} = \beta_{10} + \beta_{11} X_{11k} + \beta_{12} X_{12k} + E_{1k} & k = 1, 2, \ldots, n_1 \\ \text{Area 2:} \; Y_{2k} = \beta_{20} + \beta_{21} X_{21k} + \beta_{22} X_{22k} + E_{2k} & k = 1, 2, \ldots, n_2 \end{array}\right| \qquad [8.79]$$

The subscripts 1 and 2 denote the two areas. Eq. [8.79] will reduce to [8.78] if $\beta_{10} = \beta_{20} = \beta_0$; $\beta_{11} = \beta_{21} = \beta_1$; $\beta_{12} = \beta_{22} = \beta_2$.

The data matrix **X** and vector **Y** for the extended model are formed as follows:

$$\begin{array}{c} \quad\quad D_1 \quad X_1 \quad D_1X_1 \quad X_2 \quad D_1Y_2 \end{array}$$

$$\mathbf{X} = \begin{bmatrix} 1 & 0 & X_{111} & 0 & X_{121} & 0 \\ 1 & 0 & X_{112} & 0 & X_{122} & 0 \\ \cdot & \cdot & \cdot & \cdot & \cdot & \cdot \\ \cdot & \cdot & \cdot & \cdot & \cdot & \cdot \\ \cdot & \cdot & & \cdot & & \\ 1 & 0 & X_{11n_1} & 0 & X_{12n_1} & 0 \\ 1 & 1 & X_{211} & X_{211} & X_{221} & X_{221} \\ 1 & 1 & X_{212} & X_{212} & X_{222} & X_{222} \\ \cdot & \cdot & \cdot & \cdot & \cdot & \cdot \\ \cdot & \cdot & \cdot & \cdot & \cdot & \cdot \\ \cdot & & & \cdot & & \\ 1 & 1 & X_{21n_2} & X_{21n_2} & X_{22n_2} & X_{22n_2} \end{bmatrix} ; \quad \mathbf{Y} = \begin{bmatrix} Y_{11} \\ Y_{12} \\ \cdot \\ \cdot \\ \\ Y_{1n_1} \\ Y_{21} \\ Y_{22} \\ \cdot \\ \cdot \\ \\ Y_{2n_2} \end{bmatrix}$$

The resulting solution can be written as $\hat{\mathbf{Y}} = \mathbf{X}\hat{\boldsymbol{\alpha}}$ with $\hat{\boldsymbol{\alpha}} = (\mathbf{X}'\mathbf{X})^{-1}\mathbf{X}'\mathbf{Y}$ or:

$$\hat{Y} = \hat{\alpha}_0 + \hat{\alpha}_1 D_1 + \hat{\alpha}_2 X_1 + \hat{\alpha}_3 D_1 X_1 + \hat{\alpha}_4 X_2 + \hat{\alpha}_5 D_1 X_2 \qquad [8.80]$$

where the subscripts k are omitted. The additional independent variables are the dummy variable D_1 and its cross-products with the two independent variables. The coefficients $\hat{\alpha}$ can be interpreted as follows:

$\hat{\alpha}_0 = \hat{\beta}_{10}$ = intercept for area 1
$\hat{\alpha}_1 = \hat{\beta}_{20} - \hat{\beta}_{10}$ = differential intercept for area 2
$\hat{\alpha}_2 = \hat{\beta}_{11}$ = coefficient of X_1 for area 1
$\hat{\alpha}_3 = \hat{\beta}_{21} - \hat{\beta}_{11}$ = differential coefficient of X_1 for area 2
$\hat{\alpha}_4 = \hat{\beta}_{12}$ = coefficient of X_2 for area 1
$\hat{\alpha}_5 = \hat{\beta}_{22} - \hat{\beta}_{12}$ = differential coefficient of X_2 for area 2

The solution of [8.79] is obtained by a single regression using dummy variables because:

$$\begin{aligned}
&\text{Area 1:} \quad \hat{Y}_1 = \hat{\alpha}_0 + \hat{\alpha}_2 X_{11} + \hat{\alpha}_4 X_{12} \\
&\text{Area 2:} \quad \hat{Y}_2 = (\hat{\alpha}_0 + \hat{\alpha}_1) + (\hat{\alpha}_2 + \hat{\alpha}_3) X_{21} + (\hat{\alpha}_4 + \hat{\alpha}_5) X_{22}
\end{aligned} \qquad [8.81]$$

Dummy variables can be included in models of sequential regression analysis which were discussed in the previous section.

A simplified model that is used by econometricians to study relationships which may have been abnormal for certain, distinct periods of time, is as follows. Suppose that the cross-product terms $D_1 X_1$ and $D_1 X_2$ are deleted from the model. We retain:

$$\begin{aligned}
&\text{Area 1:} \quad Y_{1k} = \beta_{10} + \beta_1 X_{1k} + \beta_2 X_{2k} + E_{1k} \\
&\text{Area 2:} \quad Y_{2k} = \beta_{20} + \beta_1 X_{1k} + \beta_2 X_{2k} + E_{2k}
\end{aligned} \qquad [8.82]$$

The intercept for area 2 is allowed to differ from that in area 1, but the other coefficients are forced to be the same.

In general, both the number of independent variables and the number of areas or environments can be increased by considering additional dummy variables.

Chapter 9

TREND ANALYSIS

9.1. Introduction

The problem of describing large-scale, systematic variations of a variable or set of variables in space or with time can be solved by using the method of multiple regression discussed in the previous chapter. The method is then called trend analysis. It is named trend-surface analysis when the observation points for the dependent variable are areally distributed in a two-dimensional plane (map area).

It may be useful to divide the spatial variability of a geologic variable into two or more components. One of these components represents the "trend" if there exist systematic changes in the average or mathematical expectation of the variable for the area of study. Deterministic functions such as polynomials can be employed to describe a trend. The residuals, which are the deviations from the trend, are not necessarily free of systematic spatial variability. The term "trend" is used for convenience and does not cover all systematic variations in the data. For example, if a residual is sampled in detail, its systematic spatial variability may become the trend for the domain of sampling, which would be smaller in extent than that for the larger area where this particular residual is one of many events.

The concept of a trend is best illustrated by practical examples. We will complete the statistical analysis of the mineralogical variations in the Mount Albert intrusion which was initiated in Chapter 5. Another example to be discussed in this chapter concerns the spatial variability of the element copper in a mineral deposit (Whalesback copper deposit).

Methods of trend analysis are well-known in mathematical geology (cf. Krumbein and Graybill, 1965; Harbaugh and Merriam, 1968; Whitten, 1970). For applications in economic geology, see Forgotson (1963) and Link et al.(1964).

If the residuals are autocorrelated, the method of least squares, in general, may give unbiased estimators for the trend but these estimators do not have the property of minimum variance. Moreover, methods of statistical inference as discussed in the previous chapter may not be applicable if the residuals are autocorrelated. The subject of trends with autocorrelated residuals will be discussed in more detail in the next chapter.

A statistical model such as the one used for trend analysis is called linear if the dependent variable can be expresssed as a linear combination of unknown coefficients. In that situation, the coefficients are solved by matrix inversion as discussed in the previous chapter. However, one may choose to fit equations which do not have this property. This leads to nonlinear models which may be divided into intrinsically nonlinear models and intrinsically linear models. The former class (intrinsically nonlinear models) can not be changed into linear models and methods other than that of linear least squares must be used to obtain a solution. On the other hand, intrinsically linear models can be made into linear models by an appropriate transformation of the variables. In this chapter, we discuss a multiplicative model as an example of an intrinsically linear model. In section 11.6, we will use an intrinsically nonlinear model.

Finally, a geological variable may be subject to trend and, simultaneously, it may be correlated to other geological variables. This leads to a mixture of multiple regression and trend analysis as will be discussed in Chapter 15.

9.2. Empirical trend functions

Suppose that the variation along a line (X-axis) of a given geologic attribute can be described by a continuous function $f(x)$. Then the Taylor expansion of $f(x)$ about a point which $x = a$ satisfies:

$$f(x) = c_0 + c_1(x-a) + c_2(x-a)^2 + \dots$$

where the coefficients c_i satisfy [2.16]. By translating the origin along the X-axis to a point near the center of the range of values for x where measurements were made, and expanding about the new origin, we can use the Maclaurin form:

$$f(x) = f(0) + \left[\frac{\mathrm{d}f}{\mathrm{d}x}\right]_{x=0} x + \frac{1}{2}\left[\frac{\mathrm{d}^2 f}{\mathrm{d}x^2}\right]_{x=0} x^2 + \dots \tag{9.1}$$

If $f(x)$ is approximated by a limited sequence of this infinite series, the fit will be perfect in the origin. The degree of fit decreases outwards from the origin depending on the nature of the function and the number of terms used for approximation.

If $f(u, v)$ represents a two-dimensional function in u and v, then:

$$f(u,v) = f(0,0) + \left[\frac{\partial f}{\partial u}\right]_{u=0} u + \left[\frac{\partial f}{\partial v}\right]_{v=0} v + \frac{1}{2}\left[\frac{\partial^2 f}{\partial u^2}\right]_{u=0} u^2 + \left[\frac{\partial^2 f}{\partial u \partial v}\right]_{\substack{u=0\\v=0}} uv + \frac{1}{2}\left[\frac{\partial^2 f}{\partial v^2}\right]_{v=0} v^2 + \dots \tag{9.2}$$

Likewise, in a three-dimensional situation:

$$f(u,v,w) = f(0,0,0) + \left[\frac{\partial f}{\partial u}\right]_{u=0} u + \left[\frac{\partial f}{\partial v}\right]_{v=0} v + \left[\frac{\partial f}{\partial w}\right]_{w=0} w + \frac{1}{2}\left[\frac{\partial^2 f}{\partial u^2}\right]_{u=0} u^2 + \dots \tag{9.3}$$

We are using the notation u, v and w to denote distances along U-, V-, and W-axes of a hree-dimensional Cartesian coordinate system.

In trend analysis, it is assumed that the function $f(u, v, w)$ exists and can be approxi-nated by a limited sequence of terms such as those shown at the right-hand side of [9.1]–[9.3]. Because f is unknown, the coefficients can not be estimated by partial lifferentiation. They are estimated by using the method of multiple regression analysis liscussed in the previous chapter. If y represents the dependent variable, the following hree types of trend analysis can be distinguished, on the basis of number of coordinate xes used to define location in space:

Curve-fitting: $y = \beta_0 + \beta_1 u + \beta_2 u^2 + \ldots + \beta_p u^p + e$ [9.4]

Trend-surface analysis:

$$y = \beta_0 + \underbrace{(\beta_1 u + \beta_2 v)}_{\text{Linear}} + \underbrace{(\beta_3 u^2 + \beta_4 uv + \beta_5 v^2)}_{\text{Quadratic}} + \underbrace{(\beta_6 u^3 + \beta_7 u^2 v + \beta_8 uv^2 + \beta_9 v^3)}_{\text{Cubic}}$$
$$+ \underbrace{(\beta_{10} u^4 + \beta_{11} u^3 v + \beta_{12} u^2 v^2 + \beta_{13} uv^3 + \beta_{14} v^4)}_{\text{Quartic}} + \ldots + e \quad [9.5]$$

Three-dimensional trend analysis:

$$y = \beta_0 + (\beta_1 u + \beta_2 v + \beta_3 w)$$
$$+ (\beta_4 u^2 + \beta_5 uv + \beta_6 uw + \beta_7 v^2 + \beta_8 vw + \beta_9 w^2) + \ldots + e \quad [9.6]$$

s before, the variable e represents residuals whose sum of squares is being minimized. An xample of curve-fitting was discussed in section 4.6. An example of trend-surface analy-s is shown in Fig.40 according to Whitten (1966b). In this figure, the observed values are otted in three-dimensional space with the Y-axis for the dependent variable pointing in e vertical direction. The surface $y = f(u, v)$ is continuous and the observed values are attered around it. The vertical distance from these points to the surface are the residu-s.

Although [9.4]–[9.6] in form are analogous to [9.1]–[9.3], it should be kept in ind that the methods used for obtaining the coefficients are different. In fact, the aclaurin series does not have the least-squares property because the degree of fit de-eases moving away from a central point. Nevertheless, the coefficients of the best-fitting end equation can be interpreted as partial derivatives of the trend function at its center will be done later.

Another point of analogy between trend equations and truncated Maclaurin series is at the function which is being approximated must be analytic within the domain of

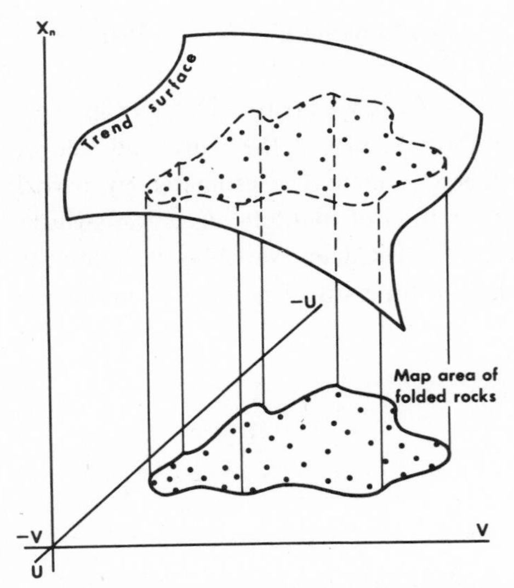

Fig.40. Map area lies in the horizontal UV-plane. A geological attribute of folded rocks, X_n, change systematically across the map area according to trend surface representing a function of u and v. The function also assumes values outside the map area which are usually not considered in practice. (From Whitten, 1966b.)

study. Discontinuities such as sharp boundaries or faults may introduce complications that can not be treated by using continuous functions. It is usually wise not to combine with one another data from both sides of a real or inferred break. One might distinguish between the areas separated by a discontinuity by using dummy variables as in section 8.10.

TABLE XXIV
Degree, number of independent variables and maximum number of extrema for trend equations in one-, two-, or three-dimensional space

Trend equations	*Degree*	*Number of independent variables*			*Maximum number of extrema*
		curve	*surface*	*hypersurface*	
Linear	1	1	2	3	0
Quadratic	2	2	5	9	1
Cubic	3	3	9	19	2
Quartic	4	4	14	34	3
Quintic	5	5	20	55	4
Sextic	6	6	27	83	5
.	.	.	.	.	.
.	.	.	.	.	.
.	.	.	.	.	.
k	k	k	$\frac{1}{2}k(k+3)$	$\frac{1}{6}k(k^2+6k+11)$	$k-1$

Some properties of the trend equation of the k-th degree in one, two, or three dimensions are shown in Table XXIV. It is noted that the degree of a trend equation and its geometrical representation are determined by the highest degree of individual terms occurring in it. For example, in [9.5] the quadratic terms are $\beta_3 u^2$, $\beta_4\, uv$, $\beta_5 v^2$. If the sequence is terminated after the term $\beta_5 v^2$, it is quadratic or of the second degree. The geometrical representation then is a quadratic trend surface. Sometimes, it is called the complete quadratic trend surface and the partial sum $(\beta_3 u^2+\beta_4 uv+\beta_5 v^2)$ is referred to as the quadratic component.

9.3. Applied trend analysis

Trend analysis is an empirical approach; it is based on the assumption that any function can be developed as a polynominal function. The fitting can be done in steps. For example, in trend-surface analysis (two-dimensional case), the linear surface may be fitted first, then the quadratic, cubic, quartic, and so on, by successively adding groups of terms for a higher degree. If a continuous trend function indeed does exist, it will be approximated by this stepwise procedure. The fitted trend function should converge to the "real" trend function. One must stop at a certain point, because including more and more terms in the equation brings one gradually close to a surface with 100% fit. For example, if there are ten observation points, there always exists a cubic surface that passes exactly through all observed values. This is because the cubic has ten coefficients and, therefore, ten normal equations. Consequently, $\hat{\mathbf{E}}'\hat{\mathbf{E}} = 0$ and the ten residuals $\hat{E}_k$ are all zero. The shape of such a surface may be completely unrealistic because strictly local fluctations are described perfectly and their influence is extended to the space between observation points where, in reality, other noisy values could be measured. The following three relatively simple methods are helpful for deciding on the cut-off point for the degree of a trend equation:

(1) We can construct contour maps for surfaces of different degrees and do a visual comparison of results. The trend maps are compared to one another and to maps with original data and residuals.

(2) One may calculate the squared multiple correlation coefficient R^2 for each surface or the percentage explained sum of squares $ESS = 100\, R^2$.

(3) Additional groups of terms of a given degree can be tested for statistical significance by using the F-test of [8.74].

The third procedure would be conclusive in the special situation that the residuals $\hat{E}_k$ are normally distributed and uncorrelated. In practice, these conditions usually are not realized for the residuals. The condition of normality can be checked readily by doing a χ^2-test on a histogram of the residuals. The condition of uncorrelated residuals is more difficult to verify especially if the observation points are irregularly distributed in the plane. However, if the calculated $\hat{F}$-value is less than the theoretical F for a given step, it may be concluded that additional groups of terms should not be added. On the other hand, a relatively large $\hat{F}$-value for a step is not necessarily related to the existence of a regional trend in the data for the area. It could mean that the residuals at observation points that are closely together, in one or more small subareas, are autocorrelated.

Other methods to decide if a calculated trend surface is realistic include dividing the data from which it was calculated into two or more subgroups and then comparing the trends for these subgroups to one another. This procedure is equivalent to adding a random variable to the independent variables whose values are either one or zero. The resulting dummy variable and its cross-products with the independent variables for terms of the polynomial provide a set of additional independent variables. Or, we may divide the study area into two or more geographically distinct subareas, and do a separate trend-surface analysis for every subarea for which the method of dummy variables also can be used.

If possible, one should avoid using trend equations of a high degree. Higher-degree surfaces may be too flexible, in that they tend to have unrealistic extrema in data voids where control points are missing, and near the boundaries of the area containing the control points. These edge effects sometimes can be avoided by constructing low-degree surfaces for a mosaic of smaller subareas. Because of their properties, the linear and quadratic surfaces are preferable in such situations.

From a numerical point of view, trend analysis may be difficult because the independent variables generated by a polynomial series usually are strongly correlated with one another and this can result in ill-conditioned matrices $\mathbf{X}'\mathbf{X}$. If, on the digital computer, a matrix $\mathbf{X}'\mathbf{X}$ can not be inverted in single precision, one may try double precision. Also, it can be helpful to replace the origin of the coordinate system to the center of the area of study and to change the scales along the coordinate axes. If this does not help, one may have to eliminate the linear or nearly linearly relationships between the independent variables by employing sequential or canonical regression (cf. sections 8.8 and 9).

Paleodelta on northwestern Melville Island, Arctic Archipelago

The data for this example were discussed in section 2.11; the observation points fall in the narrow zone with outcrops for the Bjorne Formation (Fig.13A). The 43 mean paleocurrent direction data were subjected to trend-surface analysis with the following results:

linear surface, $ESS = 78\%$; (complete) quadratic, $ESS = 80\%$; (complete) cubic, $ESS = 84\%$.

By using [8.74], we obtain for the three terms added by going from the linear to the quadratic surface:

$$\hat{F}(3,37) = \frac{(80-78)/3}{(100-80)/37} = 1.04$$

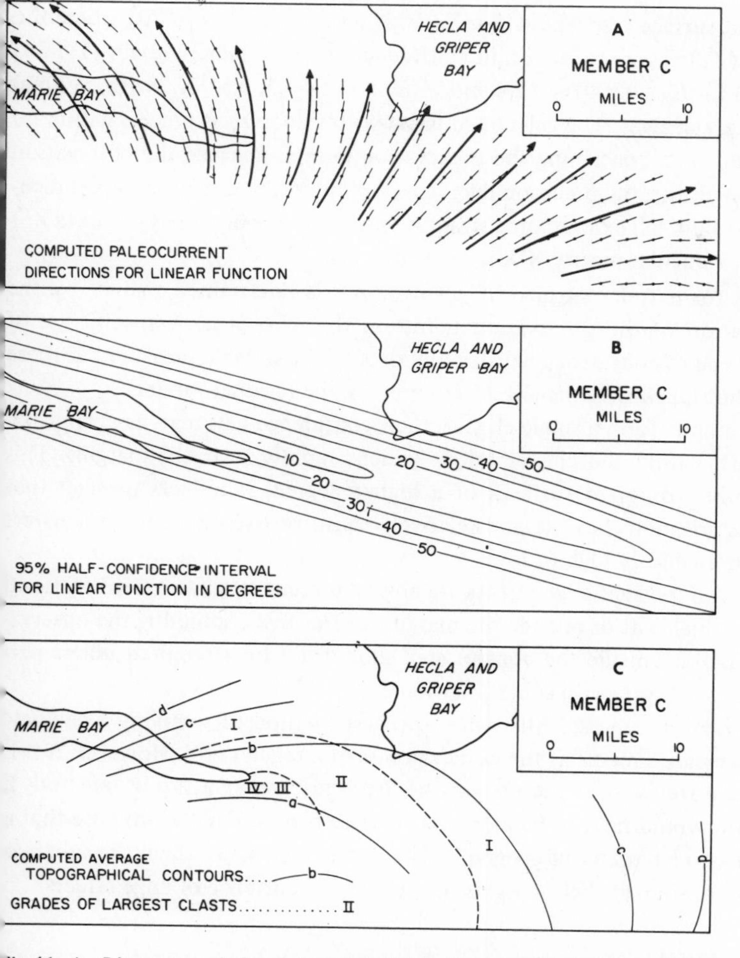

Fig.41. A. Linear trend for data of Fig.13A. Computed azimuths are shown by line segments for points on regular grid, both inside and outside exposed parts of Member C; flow lines are based on curves *1* and *2* of Fig.13B. B. Contour map of 95% half-confidence interval for A, confidence is greater for area supported by observation points. C. Computed topographic contours for delta obtained by shifting curve *1* of Fig.13B, 90° in the southeastern direction (perpendicular to isoazimuth lines), and then moving it into four arbitrary positions by changing the constant C (cf. section 2.11). Contoured grades of largest clasts provide independent information on shape of delta.

which is close to $F_{0.60}(3,37)$. Consequently, the linear trend surface as shown in Fig.41A is acceptable in this situation. This was suggested before by the graphical plot shown in Fig.12. The lines in the pattern of Fig.41C are perpendicular to those in Fig.41A.

Half-confidence intervals

The reliability of Fig. 41A can be determined by constructing a 95-% confidence interval on the linear trend surface (see Fig. 41B). For this, use is made of [8.70] where th term $[(p+1) \cdot F \cdot s^2(\hat{Y}_k)]^{\frac{1}{2}}$ was called a half-confidence interval by Krumbein (1963) It is shown in Fig. 41B for $F_{0.95}(3,40) = 2.84$ and $s^2(\hat{Y}_k) = s^2 \mathbf{X}'_k(\mathbf{X}'\mathbf{X})^{-1}\mathbf{X}_k$ with residua variance $s^2 = 380$ square degrees. The calculated trend with values $\hat{Y}_k$ is precise within 2(degrees for most of the area containing the observation points. Because the observatio points lie close to a single line, the precision decreases rapidly in the direction perpendicu lar to this line. If the points fall exactly on a straight line, trend-surface analysis could no be applied because of a singular matrix $\mathbf{X}'\mathbf{X}$ in that situation.

A confidence belt for a trend surface of given degree is determined mainly by th geometrical configuration of the observation points in the map area. A specific set o points give the same type of confidence belt for any dependent variable measured at thes points, except for a multiplicative constant determined by the residual variance.

If the observation points form a single cluster, their confidence belt may have approxi mately constant width within the cluster but it widens rapidly near the margins. Thi widening is particularly strong for surfaces of a higher degree. It reflects the fact tha higher-degree surfaces are more flexible and relatively sensitive to local values or cluster of values which are anomalously high or low.

It should be kept in mind that edge effects are always present to some extent. A tren surface usually is not reliable at or outside the margins of the area containing the observa tion points. Extrapolation outside the control area should not be attempted unless pe haps for a low-degree trend surface in some situations.

Unless prior information suggests otherwise, the best method of sampling a geologi environment is to do measurements at the corner points of a regular grid. However, if it i important to determine trends near the margins of the study area (e.g.,a plutonic body) the spacing of the grid would have to be narrower in these places. Finally, suppose that a larger area is divided into a number of geographically distinct subareas, then, the subarea may be given a certain overlap at their margins to avoid the occurrence of edge effects.

9.4. Mineralogical variations in the Mount Albert intrusion

The four variables for which measurements are available were discussed in detail ir Chapter 5. Trend surfaces for percentage forsterite in olivine, percentage enstatite ir orthopyroxene, and the chrome spinel cell edge are shown in Fig.42. The trends for thes three variables are similar in that they show an elongated minimum at approximately th same location. This phenomenon was called cryptic zoning by MacGregor and Smith (1963). Possible physico-chemical interpretations of the positive correlation between th three variables were discussed in Chapter 5.

Several aspects of trend-surface analysis and three-dimensional trend analysis will b discussed on the basis of the three mineralogical variables and rock density.

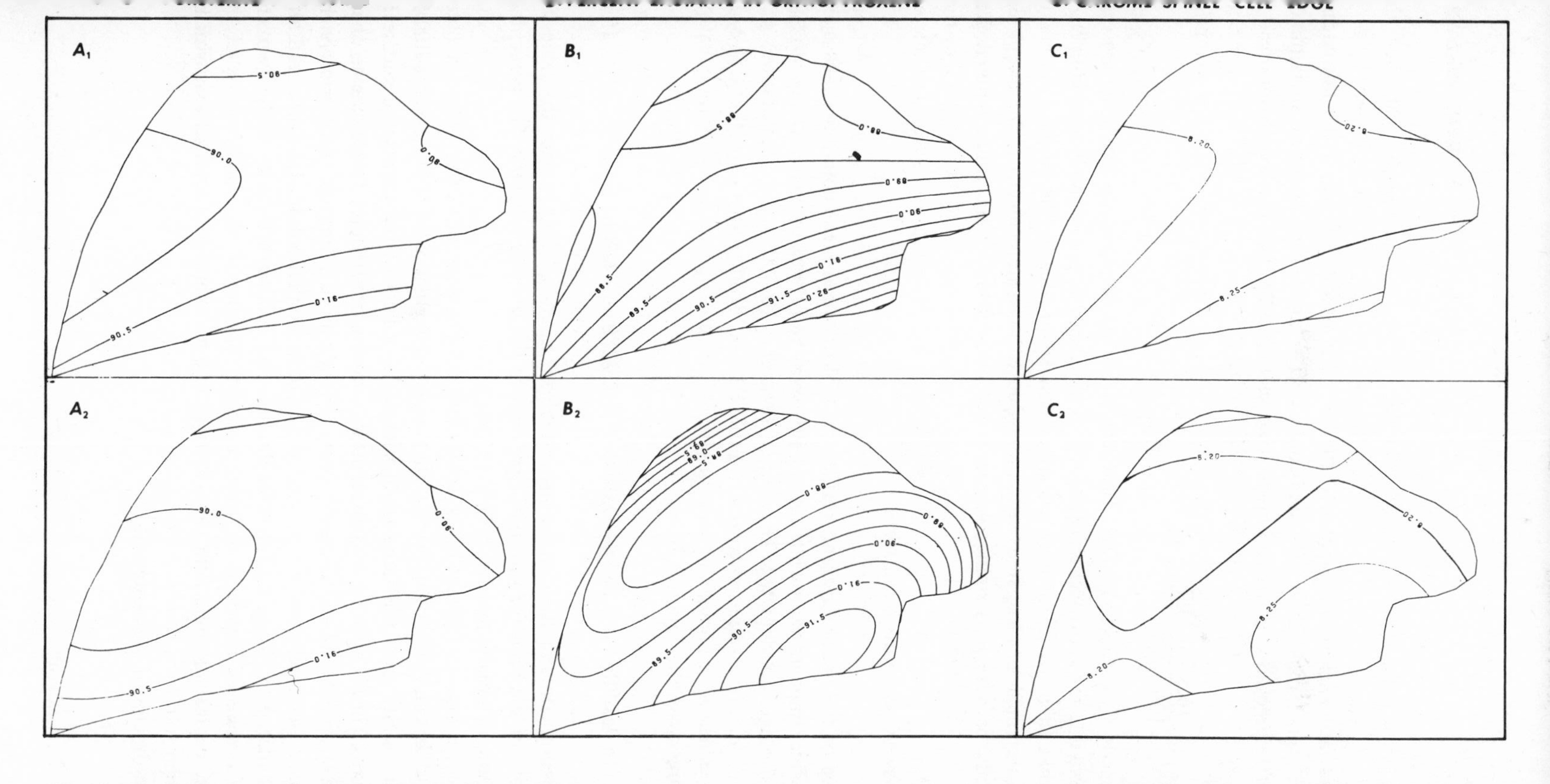

Fig. 42. Machine-constructed quadratic (1) and cubic (2) trend surfaces for (A) percent forsterite in olivines, (B) percent enstatite in orthopyroxenes, (C) cell edge, chrome spinel. Patterns for Mg content in olivine and orthopyroxene are similar in shape but variation is more pronounced in orthopyroxene (steeper gradients). For scale and grid, see Fig. 43.

Olivine

Trend-surface analysis as applied to the 167 measurements gave in the linear cas $ESS = 0.4\%$; quadratic, $ESS = 6.3\%$; and cubic, $ESS = 7.3\%$. These values are small mainl because of imprecision of measurements (see section 5.2). For the step from linear t quadratic:

$$\hat{F}(3,161) = \frac{(6.3 - 0.4)/3}{(100 - 6.3)/161} = 3.4$$

which exceeds $F_{0.95}(3,161) = 2.7$. However, if olivine composition would be the onl variable available for statistical analysis, this $\hat{F}$-value can not be considered as conclusiv evidence that there exists an olivine-composition trend because the inferred systemati variations are small as compared with those due to measurement error. We accept thi trend mainly because it is similar in shape to trends fitted to the other mineralogic variables.

Orthopyroxene, quadratic trend surface

The map of the quadratic trend surface for 174 orthopyroxene determinations is als shown in Fig.43, where the outline of the intrusion and the coordinate system used fc recording the data are depicted. The unit of distance is 10,000 ft. All computations wer done on coordinate readings as shown in Table III after dividing them by 10,000 in orde to obtain values of a comparable magnitude for the coefficients in the trend equatior The resulting quadratic equation is:

$$\hat{y} = 79.116 + 5.806\,u - 1.245\,v - 0.0496\,u^2 - 2.2289\,uv + 1.9661\,v^2 \qquad [9.7$$

This expression contains the continuous variables $\hat{y}$, u, and v which replace the values fc discrete data points used for multiple regression in the previous chapter. The percentag explained sum of squares satisfies $ESS = 39.2\%$.

It is noted that the choice of a coordinate system does not affect the calculate function $\hat{y}$. It can be shown readily that any rectangular coordinate system for th observation points gives the same trend surface regardless of its orientation and scale along the axes. The ESS-value also is invariant to rotation, translation or scale-change of th coordinate system. However, the coefficients depend on the definition of the grid system

The quadratic trend surface of Fig.43 is a hyperbolic paraboloid. Its center and axe can be calculated by using the method discussed in section 4.5. One of the two axe represents a zone of minimum enstatite content. Because of the saddle shape of thi surface, the center is a maximum for this axis but a minimum for the other axis which i perpendicular to it.

The coordinates of the center satisfy:

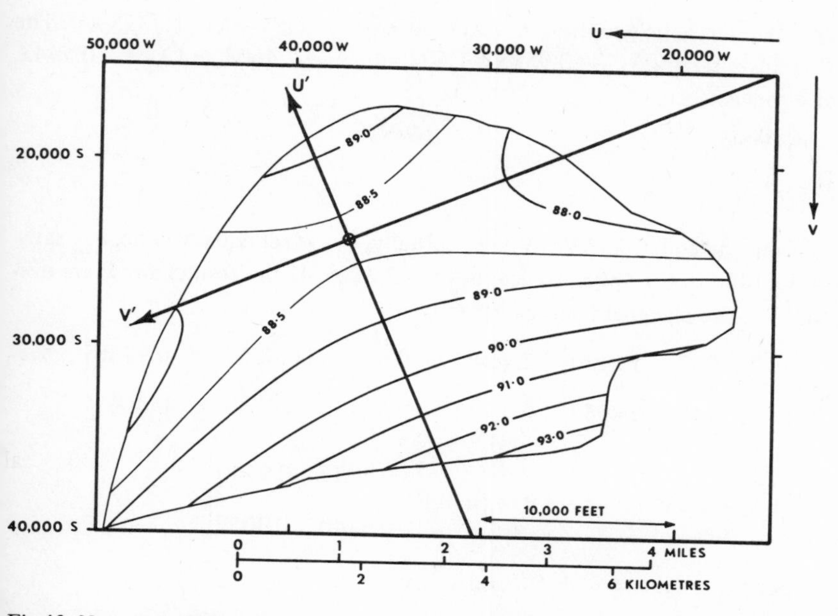

Fig.43. New coordinate system with U'- and V'- axes constructed for quadratic orthopyroxene surface; surface is hyperbolic paraboloid (cf. Fig.17E); individual contours are hyperbolas (cf. Fig.8).

$$\frac{\partial \hat{y}}{\partial u} = 0; \qquad \frac{\partial \hat{y}}{\partial v} = 0$$

This gives:

$$5.806 - 0.0992\,u_0 - 2.2289\,v_0 = 0$$

$$-1.245 - 2.2289\,u_0 + 3.9322\,v_0 = 0$$

with solution $u_0 = 3.743$; $v_0 = 2.438$. The center (u_0, v_0) is shown in Fig. 43. In order to determine the orientation of one of the axes, it is sufficient to calculate the angle φ between this axis and the U-direction. We consider the quadratic component:

$$\hat{y}^* = -0.0496\,u^2 - 2.2289\,uv + 1.9661\,v^2$$

This is a quadratic form; in matrix notation it can be written as:

$$\hat{y}^* = \mathbf{X}'\mathbf{A}\mathbf{X} \text{ where } \mathbf{X} = \begin{bmatrix} u \\ v \end{bmatrix}; \qquad \mathbf{A} = \begin{bmatrix} -0.0496 & -1.1145 \\ -1.1145 & 1.9661 \end{bmatrix}$$

The matrix A can be diagonalized. The matrix:

$$\mathbf{\Lambda} = \begin{bmatrix} \lambda_1 & 0 \\ 0 & \lambda_2 \end{bmatrix}$$

follows from the characteristic equation $|\mathbf{A} - \lambda\mathbf{I}| = 0$, or $\lambda^2 - 1.9165\lambda - 1.3395 = 0$. Th eigenvalues, therefore, satisfy $\lambda_{1,2} = 0.9583 \pm 1.5026$, or $\lambda_1 = 2.4608$ and $\lambda_2 = -0.544$ The matrix of eigenvectors

$$\mathbf{V} = \begin{bmatrix} v_{11} & v_{21} \\ v_{12} & v_{22} \end{bmatrix}$$

can be derived from the equation $\mathbf{AV} = \mathbf{V\Lambda}$. If, initially, v_{11} is set equal to one, v_{12} satis fies $-0.0496 - 1.1145\, v_{12} = 2.4608$. Hence, $v_{12} = -2.2526$. If the results for $\mathbf{V}$ are no malized by using $\mathbf{V}_n$, the diagonal form of $\mathbf{A}$ becomes

$$\mathbf{A} = \mathbf{V}_n \mathbf{\Lambda} \mathbf{V}_n' = \begin{bmatrix} 0.4057 & 0.0140 \\ -0.9140 & 0.4057 \end{bmatrix} \begin{bmatrix} 2.4608 & 0 \\ 0 & -0.5443 \end{bmatrix} \begin{bmatrix} 0.4057 & -0.9140 \\ 0.9140 & 0.4057 \end{bmatrix}$$

We can write $\hat{y}^* = \mathbf{X}'\mathbf{AX} = \mathbf{P}'\mathbf{\Lambda}\mathbf{P}$ where $\mathbf{P}' = [u \quad v] \begin{bmatrix} 0.4057 & 0.9140 \\ -0.9140 & 0.4057 \end{bmatrix}$

denotes a new coordinate system with U'- and V'-axis and:

$$u' = 0.4057\,u - 0.9140\,v$$

$$v' = 0.9140\,u - 0.4057\,v$$

Alternatively, $\mathbf{P} = \mathbf{V}_n'\mathbf{X} = \begin{bmatrix} u' \\ v' \end{bmatrix}$ with $\mathbf{X} = \begin{bmatrix} u \\ v \end{bmatrix} = \mathbf{V}_n\mathbf{P}$

or: $u = 0.0457\,u' + 0.9140\,v'$

$v = -0.9140\,u' + 0.4057\,v'$

Geometrically, this transformation represents the rotation of the U- and V-axes over th angle φ. The two elements of the first normalized eigenvector $\mathbf{V}_{1n}$ represent the cosine of the angles of the U'-axis with the U- and V-axes, respectively. Hence, $\varphi = \arccos 0.4057$ 66° and U' makes an angle of 66° with positive U. The second element of $\mathbf{V}_{1n}$ is negativ indicating that U' makes an angle of arccos 0.0140 = 24° with the negative V-axis. Th V'-axis is perpendicular to the U'-axis. Both new axes pass through the center and ar shown in Fig. 43.

Suppose that the origin is translated to the center and that the coordinate system i rotated by 66°. Eq. [9.7] than reduces to:

$$\hat{y} = 88.46 + 2.4608\,u'^2 - 0.5443\,v'^2 \qquad [9.8$$

The constant term is the value of $\hat{y}$ at the center with $u = u_0$ and $v = v_0$. The other tw coefficients control the shape of parabolas for the variation of $\hat{y}$ along the U'- and V' axis respectively.

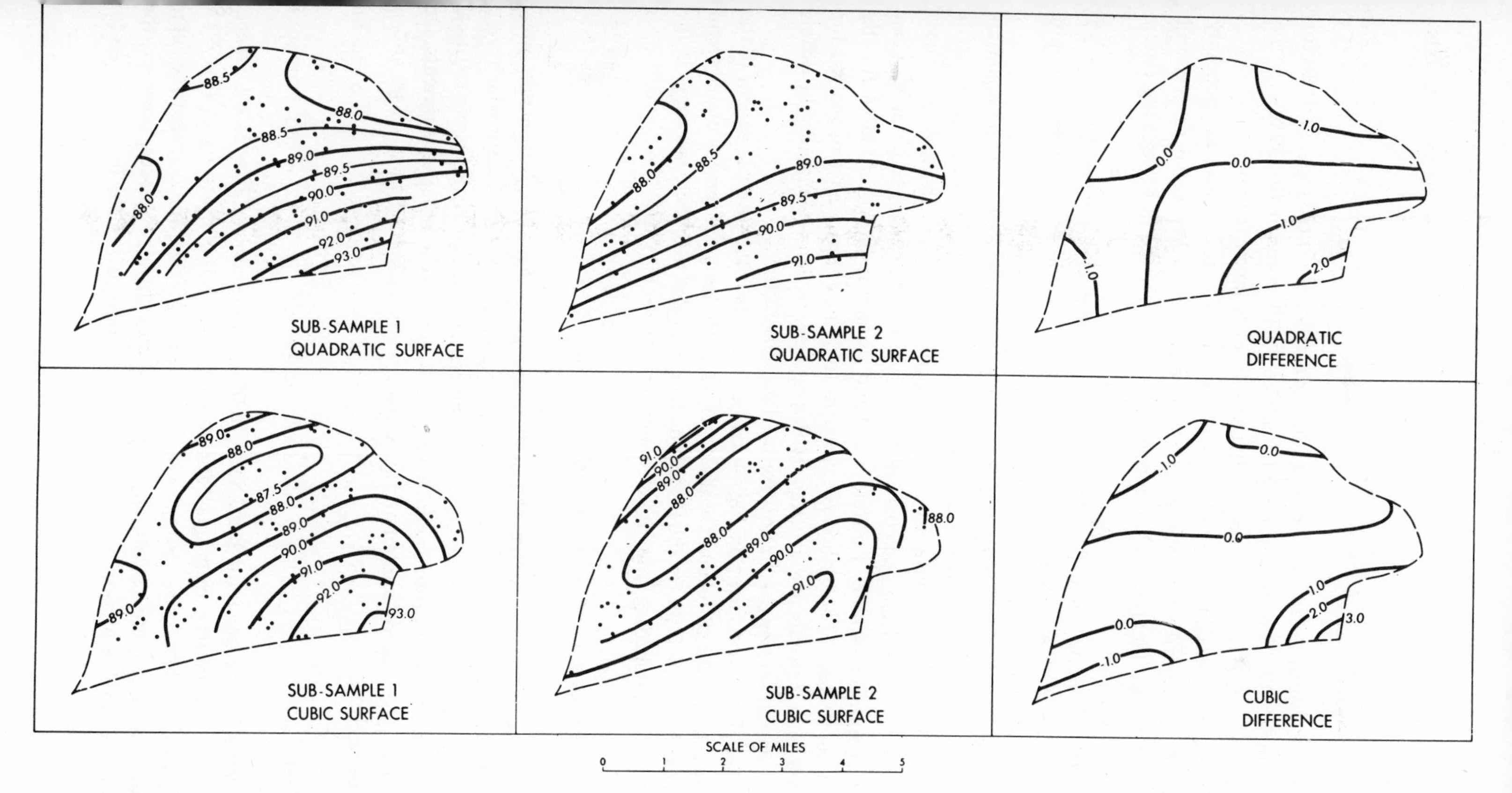

Fig.44. Evaluation of statistical significance of trend surfaces for orthopyroxene by computing differential surfaces based on two interpenetrating subsamples. (From Agterberg, 1964b.)

Differential trend surfaces

There is relatively little difference between the quadratic and cubic trend surfaces for orthopyroxene in Fig.42. The cubic surface has ESS = 49.1% compared to 39.2% for the quadratic. For this step, $\hat{F}(4,164) = 6.5$ which is significantly larger than one.

Another method of evaluating trend surfaces for significance consists of randomly dividing the set of original observations into two equal or nearly equal subsets and applying a separate trend analysis to the subsets. This can be done in a single regression by using a dummy variable.

The 174 determinations were divided into two subsamples each consisting of 87 values. The results are shown in Fig.44. The trend surfaces for the subsamples should duplicate each other apart from statistical fluctuations.

Suppose that we use the method of dummy variables of section 8.10. A trend equation then contains twice as many coefficients as before. As explained for [8.80], these coefficients are of two types. One set gives us the trend equation for one of the two subgroups. The other set consists of differential coefficients. This part of the trend equation also can be plotted in map-form yielding a differential trend surface as shown in Fig.44 for the quadratic and cubic case. This method is equivalent to doing a separate trend analysis for each subsample and then plotting a surface with coefficients that are equal to the differences between corresponding coefficients for the two subsamples.

The differential surfaces shown in Fig.44 are both less than 1% enstatite except in the vicinity of the edges of the intrusion. On the other hand, the range of the calculated trend values $\hat{Y}_k$ is about 4% enstatite for the intrusion. This indicates that there is a trend in the data which could be duplicated from independent subsamples of data.

Location of center and axes of quadratic surface

After obtaining the cubic surface and doing a duplicate trend analysis on two interpenetrating subsamples, we can evaluate the properties of the center and axes which were calculated for the quadratic surface. Obviously, the center is not an attribute of statistical significance, because this location could not be duplicated in the other experiments. On the other hand, the axis of minimum Mg content was emphasized in all cases and is thought to be realistic.

Chrome spinel

For this variable, the quadratic has ESS = 16.0%, and the cubic, ESS = 21.6%. Hence, $\hat{F}(4,179) = 3.20$ for the step from quadratic to cubic, which is between $F_{0.95}(4,179) = 2.42$ and $F_{0.99}(4,179) = 3.43$. Setting a cut-off value at 95%, we would accept the cubic trend surface which is more realistic in parts of the area. In other places, however, the quadratic surface is to be preferred (e.g., in the southeastern part of the intrusion). As for

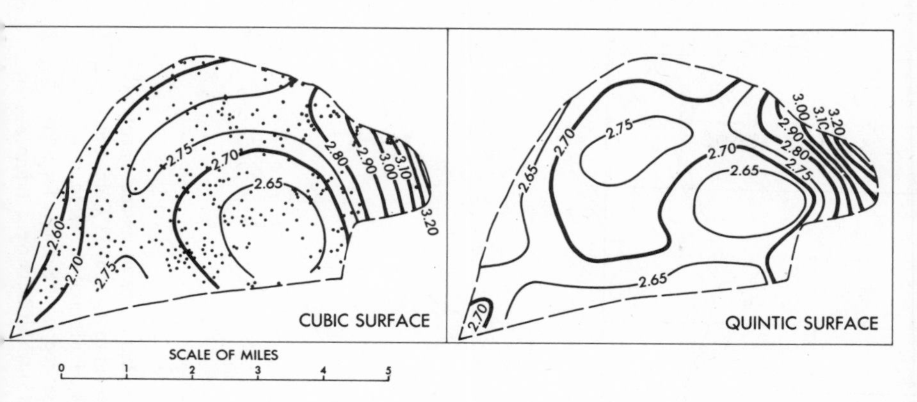

ig.45. Cubic and quintic trend surfaces for specific-gravity data, Mount Albert Peridotite intrusion dots denote positions of specimens). (From Agterberg, 1964b.)

rthopyroxene, there is relatively little difference in shape between quadratic and cubic urface and either one of them can be accepted as an approximation for the chrome pinel composition trend (cf. Fig.42).

.5. Specific gravity in Mount Albert intrusion

There are 359 observation points for specific gravity. The quadratic surface is not dequate for representing the trend of this variable. Cubic (*ESS* = 39.9%) and quintic *ESS* = 56.0%) surfaces are shown in Fig.45. The difference in pattern between these two urfaces is caused mainly by the occurrence of a pocket of practically unaltered peridotite n the eastern part of the body. The transition from high-density to lower-density mateial is relatively rapid and is poorly approximated by the cubic.

A schematic contour map for elevation is given in Fig.46A. A cross-section *CD* was onstructed and in Fig.46B all observations within 2500 ft. from the section line were rojected onto it. We then calculated average values for blocks measuring 5,000 ft. on a ide as indicated in Fig.46E. These block averages coincide with the intersection of the uintic surface with the cross-section *CD*. This shows that: (1) the quintic trend surface rovides a good fit to the specific gravity trend; and (2) a trend surface also can be btained by the relatively simple method of moving averages. The method of moving or unning averages consists of calculating arithmetic averages for a large numer of overlaping blocks and contouring the results. If few observations are available, trend-surface nalysis usually is the best method. If there are many data, moving average methods ecome increasingly applicable and may be preferable over trend-surface analysis.

Because as many as 359 data points are available for specific gravity, we have applied hree-dimensional trend analysis by considering the elevation of each observation point.

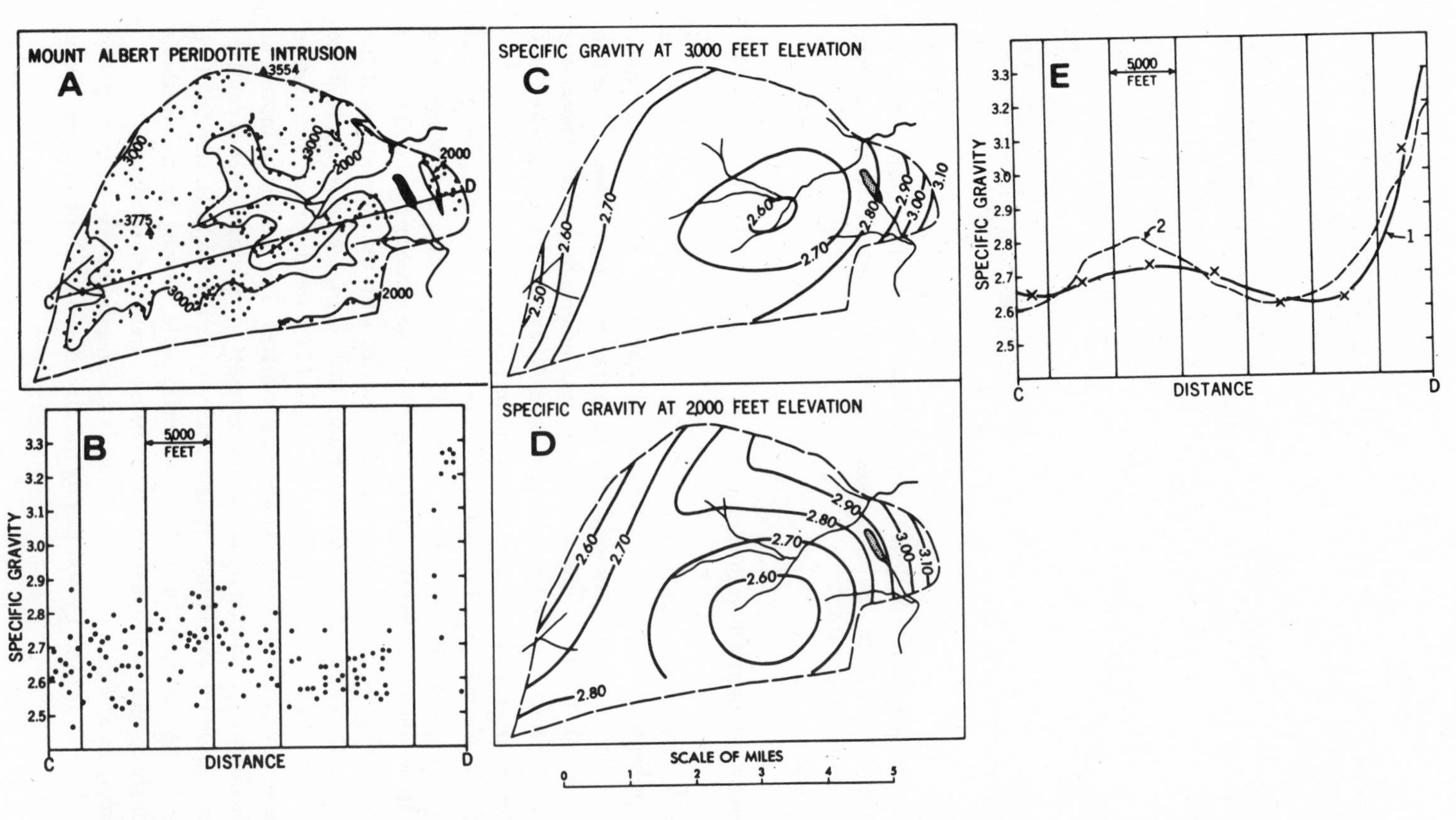

Fig. 46. A. Schematic contour map of the Mount Albert Peridotite; elevations are in feet above sealevel. B. Variations of specific gravity along section *CD* (see A for location). All values within 2500 ft. from section line were perpendicularly projected to it. C and D. Cubic trend surfaces for 3000-ft. and 2000-ft. levels obtained by contouring three-dimensional cubic hypersurface in two horizontal planes. In most of area, today's topography resembles the 3000-ft. pattern more closely than the 2000-ft. pattern. E. Crosses represent averages for values within blocks measuring 5000 ft. on a side (see B). Curve *1* is intersection of quintic trend surface (Fig.45) with *CD*; curve *2* represents

As shown in Table XXIV, the three-dimensional cubic trend equation contains 19 independent variables (u, v, w, u^2, uv, uw, v^2, vw, w^2, u^3, u^2v, u^2w, uv^2, uvw, uw^2, v^3, v^2w, vw^2, and w^3) and 20 coefficients. Its *ESS*-value amounts to 55.1% which is close to that for the quintic trend surface (21 coefficients). Calculated values for the three-dimensional cubic along the topographic surface are shown for the line *CD* in Fig.46E. Because all observation points lie in the topographic surface the elevation w can be approximated by a polynomial in u and v for horizontal distances. Consequently, a three-dimensional trend $y = f(u, v, w)$ becomes a two-dimensional trend $g(u, v)$ if w, as contained in $f(u, v, w)$, is replaced by its polynomial in u and v.

The reliability of the three-dimensional trend $f(u, v, w)$ as estimated from the data, decreases rapidly outside the topographic surface. We have attempted extrapolation in the vertical direction by setting w equal to 3,000 ft. and 2,000 ft., respectively. This gave two ordinary cubic trend surfaces for horizontal planes at these levels which are shown in Fig.46C and D, respectively.

The results may be compared with the contour map for the topographic surface. The following interpretation is suggested. In Mount Albert, specific gravity is, by approximation, linearly related to volume percentage serpentine. Thus, a low value indicates a soft rock relatively sensitive to erosion. Two of the three rivers originating on Mount Albert follow zones of weakness at the 3,000-ft. level rather than at the 2,000-ft. level (Fig.46). A further comparison with the contour map for the topographic surface suggests present-day topography was controlled by the distribution of less resistent rocks at higher levels.

9.6. Analysis of covariance

In analysis of variance, the total sum of squares *TSS* is divided into the components *SSR* (for trend) and *RSS*(for residuals). Similarly, an analysis of covariance can be developed for deviations of original data from their means. If y_1 and y_2 are two dependent variables, then:

$$\underset{\text{data}}{\mathbf{Y}_1'\mathbf{Y}_2} = \underset{\text{trend}}{\hat{\mathbf{Y}}_1'\hat{\mathbf{Y}}_2} + \underset{\text{residuals}}{(\mathbf{Y}_1 - \hat{\mathbf{Y}}_1)'(\mathbf{Y}_2 - \hat{\mathbf{Y}}_2)} \qquad [9.9]$$

Proof. Eq. [9.9] holds true if $\mathbf{Y}_1'\hat{\mathbf{Y}}_2 = \hat{\mathbf{Y}}_1'\mathbf{Y}_2 = \hat{\mathbf{Y}}_1'\hat{\mathbf{Y}}_2$. We have:

$$\mathbf{Y}_1'\hat{\mathbf{Y}}_2 = \mathbf{Y}_1'\mathbf{X}\hat{\boldsymbol{\beta}}_2 = \mathbf{Y}_1'\mathbf{X}(\mathbf{X}'\mathbf{X})^{-1}\mathbf{X}'\mathbf{Y}_2 = \hat{\mathbf{Y}}_1'\mathbf{Y}_2$$

$$\hat{\mathbf{Y}}_1'\hat{\mathbf{Y}}_2 = \hat{\boldsymbol{\beta}}_1'\mathbf{X}'\mathbf{X}\hat{\boldsymbol{\beta}}_2 = \mathbf{Y}_1'\mathbf{X}(\mathbf{X}'\mathbf{X})^{-1}(\mathbf{X}'\mathbf{X})(\mathbf{X}'\mathbf{X})^{-1}\mathbf{X}'\mathbf{Y}_2 = \mathbf{Y}_1'\mathbf{X}(\mathbf{X}'\mathbf{X})^{-1}\mathbf{X}'\mathbf{Y}_2 = \hat{\mathbf{Y}}_1'\mathbf{Y}_2$$

Consequently:

$$\mathbf{Y}_1'\hat{\mathbf{Y}}_2 = \hat{\mathbf{Y}}_1'\mathbf{Y}_2 = \hat{\mathbf{Y}}_1'\hat{\mathbf{Y}}_2 \qquad [9.10]$$

and [9.9] holds true.

Both calculated trend values $\hat{Y}_k$ and residuals $\hat{E}_k$ can be correlated with each other as if they were original data. The technique of analysis of covariance was discussed by Quenouille (1952). For an application to trend-surface analysis, reference is made to Krumbein and Jones (1970).

The correlation coefficient for observed values can be replaced by two correlation coefficients: one for the trend values and the other one for the residuals. Thus:

$$r_{12} = \mathbf{Y}_1' \mathbf{Y}_2 \{(\mathbf{Y}_1' \mathbf{Y}_1)(\mathbf{Y}_2' \mathbf{Y}_2)\}^{-1/2} \quad [9.11]$$

is replaced by:

$$r_{T12} = \hat{\mathbf{Y}}_1' \hat{\mathbf{Y}}_2 \{(\hat{\mathbf{Y}}_1' \hat{\mathbf{Y}}_1)(\hat{\mathbf{Y}}_2' \hat{\mathbf{Y}}_2)\}^{-1/2} \quad [9.12]$$

and a similar expression for the residuals.

Correlation matrices for original data and trend values were given in Chapter 5. The r_T values were based on the quadratic trend surface. Rather than using [9.12] for pairs of values, we substituted values from the quadratic surface as calculated from all data for each variable. This was done by placing a regular grid on the area and calculating trend values for 70 corner points. By an inspection of the trend maps for the four variables, it follows that the three mineralogical variables are positively correlated because they have a similar trend pattern. The specific-gravity trend is only in part related to those for the three mineralogical variables. A negative correlation is present in parts of the intrusion. However, an exception occurs in the eastern part of the intrusion where the rock is nearly unaltered peridotite with a specific gravity between 3.2 and 3.3. The correlation between trend patterns in most of the intrusion suggests that a single physico-chemical process was operative involving both serpentinization and changes in chemical composition of the minerals olivine, orthopyroxene and chrome spinel (cf. Chapter 5). This change was strongest in the southeastern corner of the intrusion and along the northwestern contact.

9.7. Spatial variation of copper in the Whalesback deposit, Newfoundland

In the remaining sections of this chapter, we discuss the Whalesback copper deposit near Springdale, Newfoundland, Canada. This example will also be used in later chapters. The orebody consists of chloritic altered volcanics containing pyrite and chalcopyrite mainly in disseminated form. However, these minerals also tend to cluster and form stringers, and occur in well-defined narrow veins with a maximum width of 4 ft.

The deposit coincides with a shear zone. It is about 1,200 ft. (400 m) long at the surface. Average width of the central copper zone which averages more than 1% copper is approximately 60 ft. and vertical depth is 800 ft. The copper zone is enveloped by a rather strongly altered "chlorite zone" with 0.35% copper on the average. The plate-shaped mineralized zone dips about 70° south-southwest. Before mining (by the British Newfoundland Exploration Company) commenced in 1965, ore reserves were estimated at 4 million short tons averaging 1.48% copper.

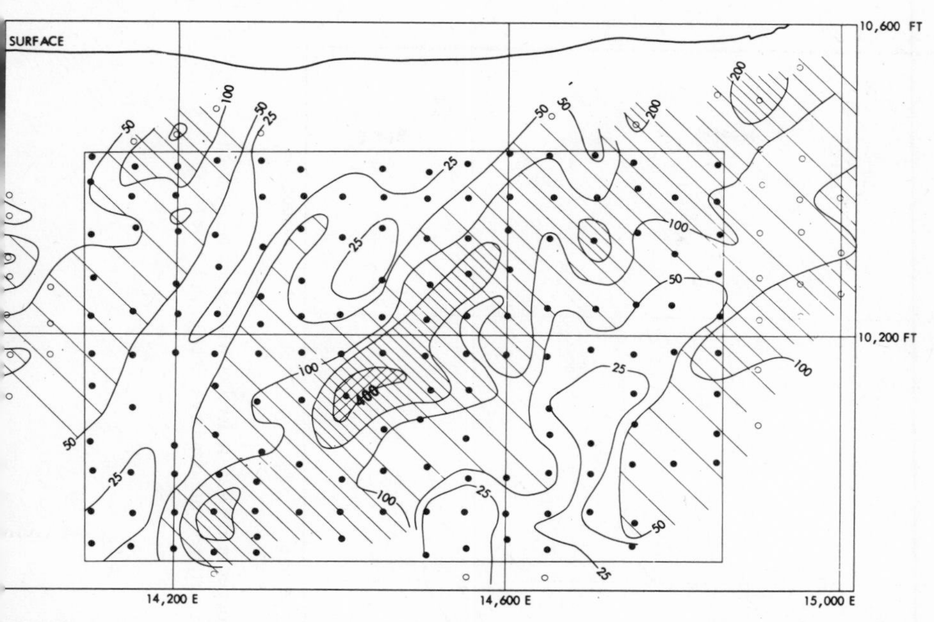

Fig.47. Manually contoured map of percent-foot values for copper in longitudinal section across Whalesback deposit, Newfoundland. Each dot represents the approximate intersection point of an underground borehole with the center of the mineralized zone. (After Agterberg, 1969.)

A longitudinal cross-section is shown in Fig.47. It contains the intersection points of 188 underground drillholes with the center of the ore zone. The average copper value was calculated for each hole over the width of the orebody. This value was multiplied by horizontal width yielding so-called percent-foot values. These percent-foot values were contoured by the mining staff with the result shown in Fig.47. In the center of this diagram, a relatively rich copper zone is shown dipping about 45° downward to the west. The mining grid which was used for location is shown in Fig.47 and 48.

9.8. Trend-surface analysis, 425-ft. level

This level occurs about 425 ft. below the surface and has a vertical coordinate equal to 10,175 ft. The core for underground drillholes, which are 50 ft. apart, was divided into pieces of 5 ft. length and assay values (percentages copper) were determined from these.

About 15 volume percent of the rocks in the deposit consists of dikes which do not contain copper. This yields zero assay values which were omitted from the data to be used for statistical analysis and trends will be projected across the dikes. Therefore, if average grade values are to be estimated for larger blocks of ore, these values, after calculation, can be reduced by 15% for dilution caused by barren dikes. Of course, more

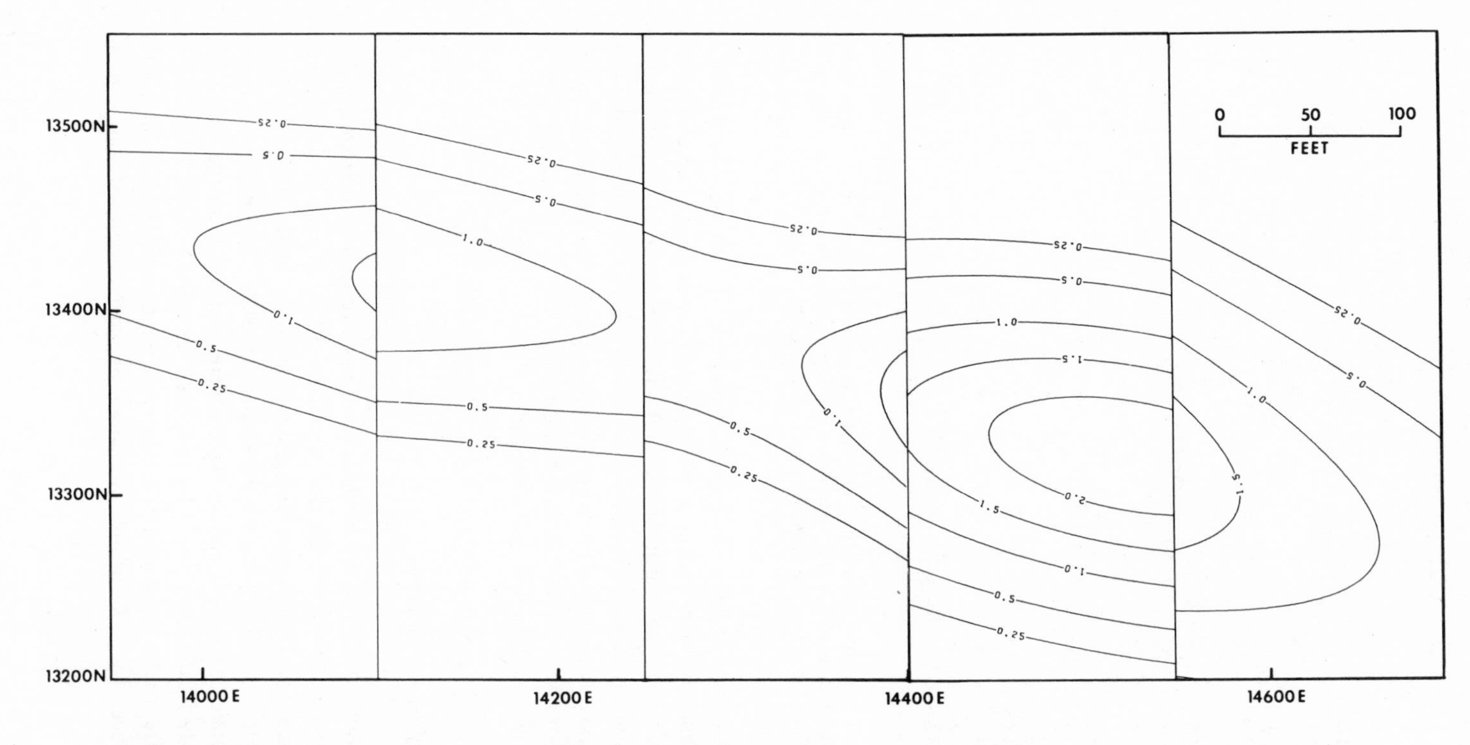

Fig. 48. Machine-constructed mosaic of exponential quadratic trend surfaces for copper in the 425-ft. level of Whalesback Mine; based on underground borehole data; holes were north–south oriented, approximately 50 ft. apart; western boundary coincides with hole No. 2.

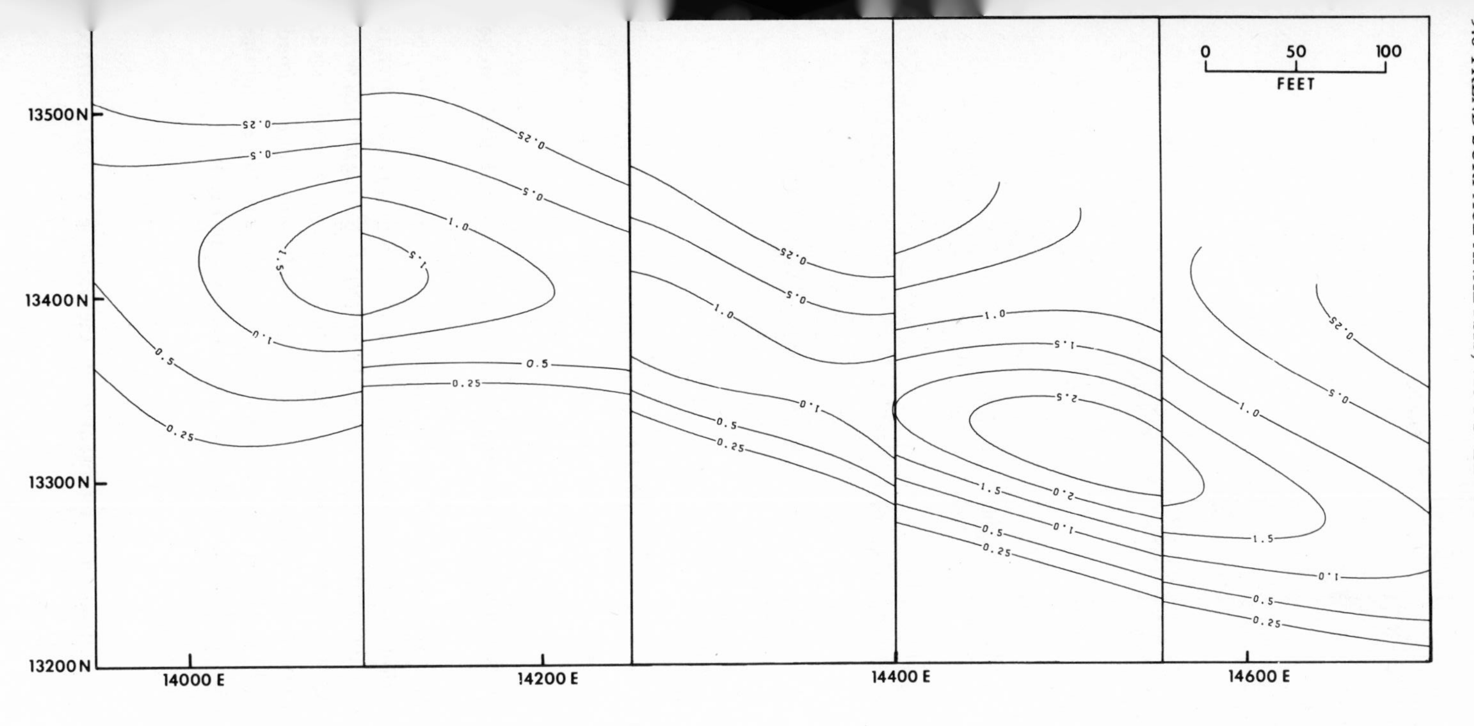

Fig.49. Same as Fig.48 for cubic trend surfaces.

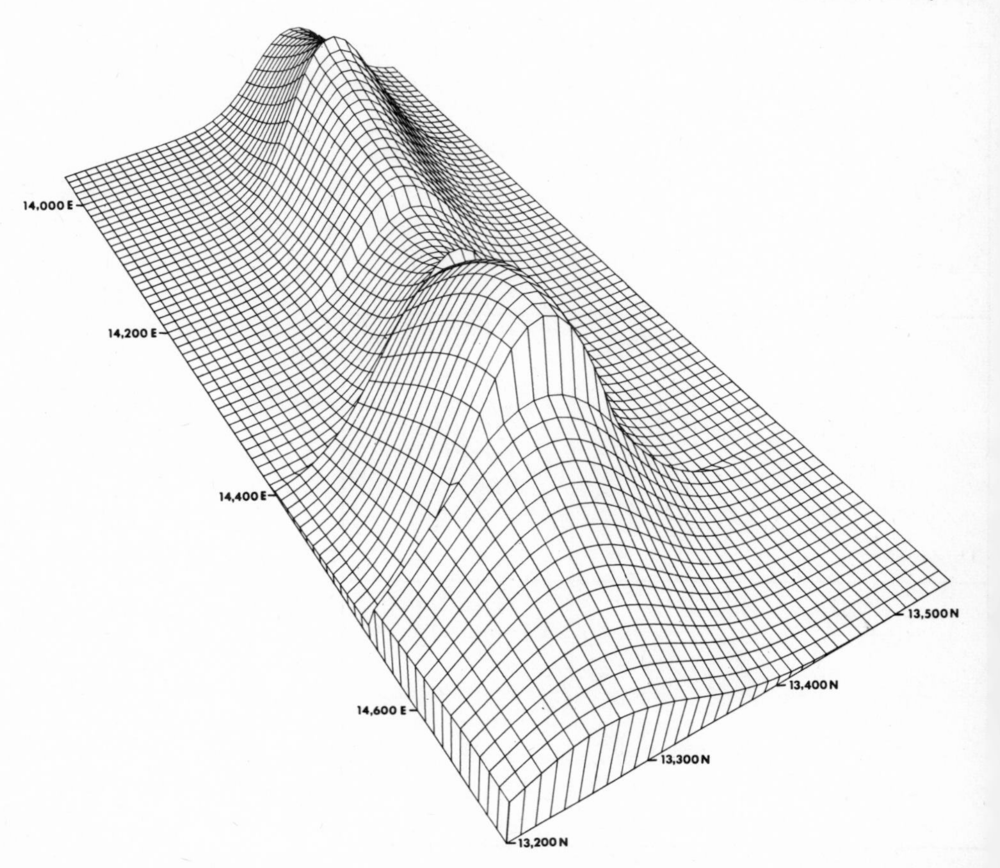

Fig.50. Three-dimensional block diagram for variation pattern of Fig.48; each cross-section is Gaussian curve.

precise corrections can be made in places where the approximate location of dikes can be predicted.

For trend-surface analysis, we took all data for overlapping sets of six drillholes, or cross-cuts which had been sampled in the same manner, and calculated quadratic trends for the natural logarithms of the original copper values. These results for logarithms were converted to ordinary percentage copper values by using a transformation to be discussed. The central part for each converted quadratic surface for a 150 ft. wide zone situated between second and fifth hole in each situation, is shown in the mosaic of Fig.48 and the block diagram of Fig.50. The corresponding cubic solutions are given in Fig.49.

Non-linear model

Suppose that copper concentration is represented by the variable x. Trend-surface analysis was accomplished by taking $y = {}^e\log x$ as the dependent variable. The statistical model can then be written as:

$$\mathbf{Y} = \mathbf{P}(u, v) + \mathbf{E} \tag{9.13}$$

where **Y**, **P**(u, v) and **E** are vectors for log-copper values Y_k, trend values, and residuals respectively. If the residuals are described by a frequency-distribution function whose parameters are independent of location in the orebody, then, according to the Gauss-Markov theorem, $\mathbf{Y} = \mathbf{P}(u,v)$ gives unbiased estimates of $E(Y)$ where Y is a random variable whose expectation $E(Y)$ is a function of u and v, but whose variance σ^2 is constant. However, we are more interested in unbiased estimates of $E(X)$ where X is a random variable corresponding to original copper values. Suppose that $E(X) = T(u, v)$ where $T(u,v)$ is the trend function in u and v which is of interest. Then, in general:

$$T(u, v) > \exp \{P(u, v)\}$$

This inequality at first may not be clear to the reader, especially because for all original data we have:

$$Y_k = {}^e\log X_k; \qquad X_k = \exp (Y_k)$$

Let us assume that the residuals E_k for the Y_k are normally distributed with constant variance σ^2. This implies that the X_k are lognormally distributed with both the expectation $E(X)$ and the variance $\sigma^2(X)$ depending on location in the orebody.

Because X is lognormal, its frequency distribution satisfies [7.7], or:

$$f(x) = \frac{1}{x\sigma\sqrt{2\pi}} \exp \{-(1/2\sigma^2)({}^e\log x - P)^2\} \quad (x > 0) \tag{9.14}$$

In this expression $P = P(u, v)$. According to [6.15]:

$$E(X) = (1/\sigma\sqrt{2\pi}) \int_0^\infty \exp \{-(1/2\sigma^2)({}^e\log x - P)^2\} \mathrm{d}x \tag{9.15}$$

After some manipulation, this gives:

$$E(X) = \exp \{P + \sigma^2/2\} \tag{9.16}$$

In fact, [9.16] is equivalent to the estimate for α given in [7.49] if μ is replaced by P. If the number of observations, n, is large, we have for the residual variance divided by two $\sigma^2/2 \approx \frac{1}{2}s^2(E_k)$. For small samples, the correction factor $\Psi_n(t)$ (see section 7.12) can be used.

Collecting the results, we obtain:

$$T(u, v) = \mathrm{e}^{P(u, v)}\, \mathrm{e}^{\sigma^2/2} \tag{9.17}$$

If σ^2 is small, the factor $e^{\sigma^2/2}$ can be neglected.

If the method is applied to data from holes 1–6 drilled from north to south and 50 ft apart (between 13,900 E and 14,150 E), we obtain for the quadratic trend surface of y a residual variance equal to $s^2 = 1.024$. In this example, $n = 177$; hence, $\Psi_n(t) \approx e^{s^2/2} = 1.67$.

Consequently, if we take the antilog of the trend surface for the logs and then multiply the result by 1.67, we obtain a trend surface for x. It is of the form:

$$\hat{x} = \exp(\hat{\sigma}^2/2 + \hat{\beta}_0 + \hat{\beta}_1 u + \hat{\beta}_2 v + \hat{\beta}_3 u^2 + \hat{\beta}_4 uv + \hat{\beta}_5 v^2) \qquad [9.18]$$

On the other hand, the trend surface for logarithmically transformed data satisfies:

$$\hat{y} = \hat{\beta}_0 + \hat{\beta}_1 u + \hat{\beta}_2 v + \hat{\beta}_3 u^2 + \hat{\beta}_4 uv + \hat{\beta}_5 v^2 \qquad [9.19]$$

A number of surfaces $\hat{x}$ were shown in Fig.48. One surface $\hat{y}$ (for holes 1–6) that corresponds to the westernmost surface $\hat{x}$ in Fig.48 (for holes 2–5) is shown in Fig.65 to be discussed later. Of course, x can also be expressed directly as a quadratic polynomial such as [9.19]. For the example (holes 1–6), quadratic trend-surface analysis on the logs of the copper values (Y_k) gave ESS = 37%. If a similar analysis is applied to the original copper values (X_k), ESS = 20%. Because the two models that underlie these results are different, the better fit for y does not necessarily mean that working with logarithmically transformed values is preferable. Both methods result in a trend surface with a maximum along the center of the ore zone.

The exponential form of [9.18] becomes a Gaussian curve for any line in the UV plane. For a profile perpendicular to the orebody, it shows inflection points at both sides of the maximum and decreases toward zero in the rocks surrounding the orebody. On the other hand, a quadratic trend surface for x decreases continuously and assumes negative values in the outer part of the chlorite zone and country rock. Vistelius and Janovskaya (1967), for a somewhat different situation, argued that if partial differential equations are used for modelling geologic processes, the solutions are likely to consist of exponential-type functions. For the Whalesback deposit, we may speculate that a mineralized solution penetrated from below along the shear zone and diffused sideways into the country rock. Diffusion from a point source into a homogeneous medium is described by a Gaussian curve if a Brownian motion model may be applied. In time, this curve would change according to the sequence shown in Fig.31A. If this process was discontinued at a specific time, the Gaussian curve (cf. Fig.50) may be a better basis for a statistical model than the parabola arising from ordinary surface-fitting. The statistical model that results in [9.18] is nonlinear because the trend function for x is not a linear combination of the coefficients. By using a logarithmic transformation on the dependent variable, however, the coefficients could be solved by matrix inversion. The model therefore is intrinsically linear.

Nonlinear models can be used for solving geologic problems if a specific theoretical equation (e.g., diffusion equation) is considered to provide a fair representation of the geometrical form expected in nature (cf. James, 1967); and this model is then confirmed by the subsequent observations.

leteroscedasticity

At a number of different levels in the mine, two types of drifts were sampled at an nterval of about 8 ft. Values for 1,110 samples from drifts in the chlorite zone gave an verage grade of 0.35%. On the other hand, 1,090 values from drifts in the copper zone ave 1.57% copper on the average. The variance for the second set of values is about eventeen times larger than that for the first set. The method of least squares, in theory, ives best linear unbiased estimates for the trend only if the residuals are uncorrelated and ome from the same population with constant variance. Little is known about how a trong heteroscedasticity, as observed in the Whalesback copper deposit, affects the trend. lowever, if the copper values are logarithmically transformed, the variances become 1.21 or the copper zone (1,090 values) and 1.32 for the chlorite zone (1,110 values), respecvely. The variance ratio is reduced to 1.09 which is not significantly larger than one.

Consequently, a trend analysis as applied to the logarithms of data (Y_k) would be referable from this point of view, because the variance of the residuals is made independent of location within the orebody. The problem of heteroscedasticity is avoided by the garithmic transformation. The least-squares model for [9.13] and [9.19] is additive, hereas that yielding [9.18] is multiplicative, because [9.13] can also be written as:

$$X_k = e^{Y_k} = e^{P_k(u,v)+E_k} = e^{P_k(u,v)} \cdot N_k \qquad [9.20]$$

here $N_k = e^{E_k}$ is a multiplicative noise component whose logarithmic variance is equal the parameter σ^2.

Little published work is available for least-squares models where the dependent variole is transformed. It should be kept in mind that, in general, it is not easy to calculate ends for original data from trends computed for transformed data. The problem was iscussed by Heien (1968) for linear regression with a logarithmically transformed ependent variable.

utocorrelation of residuals

Although this subject will be discussed in more detail in the next chapter, the trend ialysis for Whalesback is in part determined by the autocorrelation known to exist nong the residuals from the trend. A condition which should be fulfilled for the method f least squares to give unbiased, minimum variance estimators of the trend is that the siduals are not correlated to each other.

Along the drillholes in the Whalesback deposit, the midpoints of core samples are only ft. apart. Moreover, the core samples are adjacent to one another. Therefore, adjacent llues are likely to resemble each other more closely than values which are farther apart. his phenomenon is called autocorrelation which remains in existence after the eliminaon of a trend component from the data.

This phenomenon has several effects on the results of a trend analysis. Although trends calculated by the usual method of least squares are unbiased if the residuals are autocorrelated, they do not have the minimum variance property. It is possible to estimate a best minimum variance trend by a method to be discussed in section 10.13. However, for a sufficiently large set of Whalesback copper values, the pattern for the best estimator does not differ significantly from the pattern obtained by ordinary least squares. We concluded that the method of trend analysis used so far need not be modified because of autocorrelation of the residuals.

However, methods of statistical inference such as the F-test are strongly influenced by autocorrelation in the residuals. An analysis of variance for the step from a quadratic fit to a cubic fit is given in Table XXV for four nonoverlapping sets of six drillholes at the 425-ft. level. In general, the calculated $\hat{F}$-values are too large if autocorrelation is present. They were corrected by using the following approximate method.

The correlation coefficient (r_1) between residuals for data points which are one sampling interval apart can be computed readily. We can say that n autocorrelated data are equivalent to n' statistically independent values. Matalas (1963) has pointed out that, if a simple (first-order Markov) model can be used for the autocorrelation of discrete data, then:

$$n/n' = 1 + 2r_1 [n/(1-r_1) - 1/(1-r_1)^2]/n$$

TABLE XXV

Comparison of quadratic and cubic trend surfaces for four sets of six holes at 425-ft level; logarithmically transformed data. (After Agterberg, 1966b)

	Source of variation	*Sum of squares*	*Degrees of freedom*	*$\hat{F}$-ratio*	*Corrected df*	*Corrected $\hat{F}$*	*Confidence limi[ts]*	
							95%	*99%*
Holes 1–6	Complete quadratic	100.68	6	16.4	6	5.5	2.3	3.2
	Residuals	175.01	170		57			
	Total	275.69	176					
	Cubic minus quadratic	21.39	4	5.8	4	1.9	2.6	3.7
Holes 7–12	Complete quadratic	100.46	6	14.5	6	4.8	2.3	3.2
	Residuals	163.76	148		49			
	Total	264.21	154					
	Cubic minus quadratic	41.82	4	11.8	4	3.9	2.6	3.8
Holes 13–18	Complete quadratic	119.89	6	20.0	6	6.7	2.2	3.1
	Residuals	206.71	198		66			
	Total	326.60	204					
	Cubic minus quadratic	16.88	4	4.3	4	1.4	2.5	3.6
Holes 19–24	Complete quadratic	74.88	6	10.2	6	3.4	2.3	3.2
	Residuals	196.52	159		53			
	Total	271.40	165					
	Cubic minus quadratic	14.29	4	3.1	4	1.0	2.6	3.7

he F-ratio of [8.57] is the ratio of two variances for independent random variables. The ım of squares of n' independent data would be distributed according to $\sigma^2\chi^2(n')$ with n' egrees of freedom. In the situation of redundancy due to autocorrelation, we would btain $(n/n')\ \sigma^2\chi^2(n')$. Assume now that the sum of squares due to regression is not riously affected by the redundancy. This suggests that the $\hat{F}$-ratios, initially computed y [8.57], should be multiplied by (n'/n) in order to account for the redundancy. The umber of degrees of freedom for the denominator also should be corrected by this ctor. The modified $\hat{F}$-ratios of Table XXV indicate that the step from a quadratic to a ıbic fit should not be made except perhaps for holes 7–12 where the corrected $\hat{F}$-ratio significant at the 99-% level. The area of Fig.49 (quadratic) and 50 (cubic) is situated etween hole 2 to the west and hole 17 to the east. The difference between these two osaics is greatest in the central part (between 14,250E and 14,400E) computed from oles 7–12 (between 14,200E and 14,450E).

9. Three-dimensional trend analysis

A large part of the information on the distribution of the element copper in Whales-ıck is for points outside subhorizontal levels such as the 425-ft. level. We took all data 16 values) for a relatively small block of the Whalesback deposit extending from the rface to the 425-ft. level and situated between the 14,000E and 14,500E sections, corporating information from core samples and also from channel samples taken along ifts at the levels. Three-dimensional trend analysis on the logs of the copper values gave e following *ESS*-values: linear 18.0; quadratic 32.8; and cubic 43.3%. At the same time,

BLE XXVI
equency distribution of 516 Whalesback copper values

ıss limits % Cu)	No. values	Class limits (in % Cu)	No. values
0–0.25	188	3.00– 3.50	8
25–0.50	75	3.50– 4.00	11
50–0.75	47	4.00– 4.50	6
'5–1.00	31	4.50– 5.00	1
0–1.25	34	5.00– 6.00	5
25–1.50	18	6.00– 7.00	5
50–1.75	17	7.00– 8.00	1
'5–2.00	25	8.00– 9.00	2
0–2.25	13	9.00–10.00	0
25–2.50	6	10.00–12.00	1
50–2.75	6	12.00–14.00	6
'5–3.00	4	14.00–16.00	3
		16.00–18.00	1

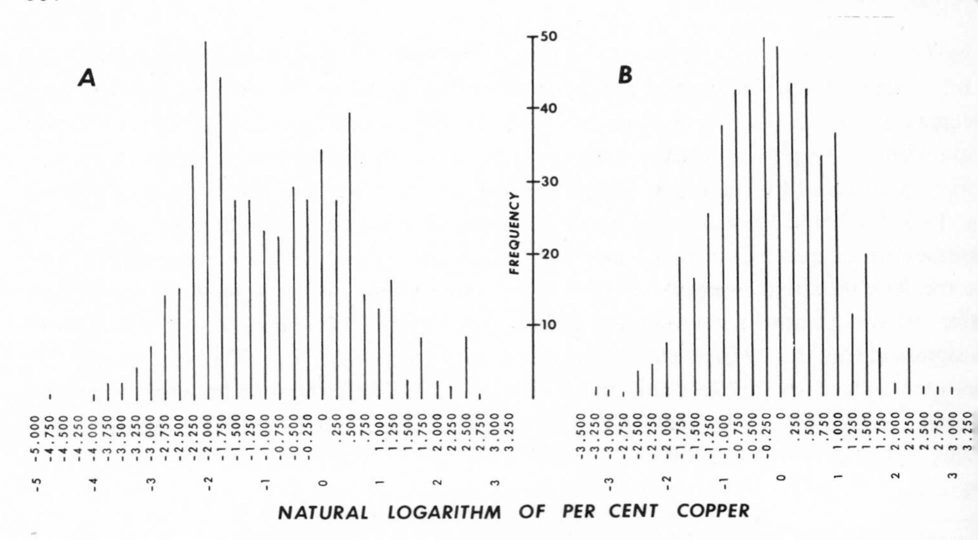

Fig.51. A. Histogram of natural logs of 516 copper values listed in Table XXVI. B. Ditto, for residual from cubic hypersurface; this curve is approximately Gaussian. Note that A is distinctly bimoda suggesting two populations for preferred types of concentration of chalcopyrite in small rock-sample (disseminated and more massive forms).

the residual variance was reduced as follows: original data 1.865; linear residuals 1.531 quadratic residuals 1.270; and cubic residuals 1.093.

A more detailed evaluation (Agterberg, 1968) showed that the cubic fit is better thar the quadratic in this example. However, the amount of information is not evenly dis tributed in space. In fact, because the underground holes were drilled from centers at th 225- and 425-ft. levels, little information is available for parts of the deposit betwee these centers. Nevertheless, it is interesting to study the effect of this trend analysis o the 516 original data of the example. Frequencies for original data (X_k) are given in Tabl XXVI; the positive skewness is obvious. Frequency histograms for logarithmically trans formed data (Y_k) and the corresponding residuals from the cubic trend are shown i Fig.51. The Y_k-values are not normally distributed. A χ^2-test gave $\hat{\chi}^2$ = 80.5 for 2 degrees of freedom versus $\chi^2_{0.95}$ = 32.7 and $\chi^2_{0.99}$ = 38.9. On the other hand, th residuals gave $\hat{\chi}^2$ = 17.3 for 17 degrees of freedom versus $\chi^2_{0.95}$ = 27.6 and $\chi^2_{0.99}$ 33.4. The results of this test are conservative because autocorrelation tends to increas $\hat{\chi}^2$-values. Consequently, the residuals are normally distributed by approximation whic justifies our usage of the nonlinear, multiplicative model for trend-fitting discussed in th previous section. We may argue that the trend component actually represents the dete ministic components of variations in the original data. The main trend shows that th relatively high copper values tend to be concentrated near the center of the mineralize zone. On the other hand, the residuals, although they are autocorrelated, conform to th frequency distribution of a random variable and, outside the spatial extent of the autocorrelation, can be treated as unpredictable, random events.

For logarithmically transformed data, 43% of the variation is accounted for by the deterministic component (trend) and 57% by the random component (residuals). Incidentally, the histogram of Fig.51A suggests bimodality for the original data. To a first approximation, the underlying population may be considered as a mixture of two normal populations with 42% coming from a normal distribution with its mean at – 1.7 and standard deviation 0.8, and 58% from another normal distribution with its mean at 0.3 and standard deviation 1.0. Because these results are for logarithmically transformed data, it would suggest that the original data X_k are a mixture of two lognormal distributions. The assumptions that underlie the latter model are not the same as those for the trend-analysis model. The copper occurs in two forms: (1) in a disseminated form, and (2) in more massive forms of sulphide mineralization. These two forms are mixed in the orebody with the disseminated form prevailing in the chlorite zone and more massive sulphide lenses in the copper zone. The bimodality is masked by the use of trend analysis. However, it may also be considered in its own right.

9.10. Three-dimensional trend analysis of surface-hole data

The 335 copper values used for this example come from 20 holes drilled from the topographic surface during an early stage of development of the orebody. They are for an 800 ft. wide zone between 14,100 E and 14,900 E which is 625 ft. deep. The holes were sampled at 10-ft. intervals, but mainly for material from the copper zone; relatively few values were obtained for the chlorite zone. Trend maps for the three-dimensional quadratic solution are shown for two levels in Fig.52. The trends were obtained as in the previous two sections by converting ordinary polynomial trends for logarithms of data to exponential-type trends. A schematic outline of the copper zone is also represented. The contours should not be extrapolated beyond these boundaries. In fact, the meaning of the calculated trends for this example is limited. It will be shown in section 11.6 that more meaningful results can be extracted from these data by a technique of harmonic analysis. The purpose of this example is to illustrate some general properties of three-dimensional quadratic trend functions.

The coordinate system has U-, V-, and W-axes, that point north, east and upward, respectively. If 100 ft. is taken as the unit of distance, and if, for convience, we delete the first two digits from every coordinate reading, the trend equation for logarithms of data is:

$$\hat{y} = -5.9397 + 2.7247u + 1.0172v - 1.1270w - 0.34353u^2 - 0.22897uv + 0.19687uw - 0.02277v^2 + 0.02159vw + 0.08329w^2 \qquad [9.21]$$

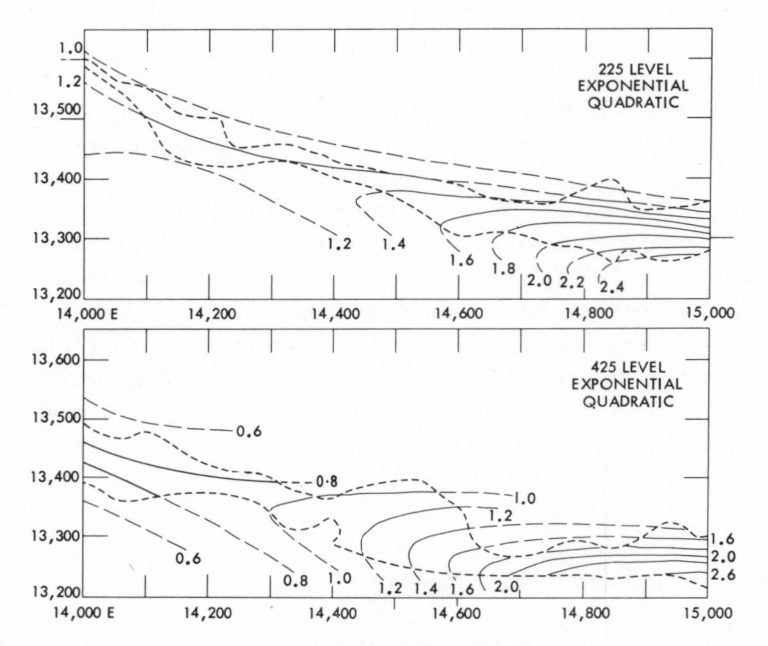

Fig.52. Contours from exponential quadratic hypersurface based on copper data from 20 explorator holes drilled from surface. Outline of orebody is approximately 1% copper contour based on late information. (After Agterberg, 1968.)

This is the equation for a quadratic hypersurface consisting of a family of hyperboloids o one sheet and two sheets (see section 4.4). There is a separate hyperboloid for ever constant value of $\hat{y}$. If w is set equal to a constant, [9.21] reduces to that of a singl hyperbolic paraboloid whose contours in the plane are hyperbolas. Although Fig.52 wa derived for the exponential-type trend for x, the shapes of the contours for $\hat{x}$ and $\hat{y}$ ar identical. In fact, the trend hypersurfaces for x and y have the same center and axes. W can calculate these from [9.21] as follows:

Centers for 225- and 425-ft. levels. The centers for the hyperbolic paraboloids show in Fig.52 can be calculated as in section 9.4. This procedure gave the centers (4.43, 1.82 3.75) for the 255-ft. level and (4.81, –1.03, 1.75) for the 425-ft. level. The distanc between these two points is 3.509 (X 100 ft.). The direction cosines of the line connect ing the points are (– 0.1082, 0.8145, 0.5700). It means that the line dips 55° wit azimuth 277° (N83°W). All horizontal planes have hyperbolic paraboloids with center falling on this line. However, the line does not represent an "axis" of minimum miner alization in the ore zone. In fact, any set of parallel planes gives an axis of this type bu because of the properties of the spatial configuration of a hypersurface, these axes diffe between sets of parallel planes. They have in common that all pass through the cente which is the same for all hyperboloids arising from the quadratic hypersurface of [9.21]

Center and axes of the hyperboloids

The coordinates for the center are calculated from [9.21] by solving the equations $\partial\hat{y}/\partial u = 0$; $\partial\hat{y}/\partial v = 0$; $\partial\hat{y}/\partial w = 0$, or:

$$2.725 - 0.687\,u_0 - 0.229\,v_0 + 0.197\,w_0 = 0$$

$$1.017 - 0.229\,u_0 - 0.0455\,v_0 + 0.0216\,w_0 = 0$$

$$-1.127 + 0.197\,u_0 + 0.0216\,v_0 + 0.1666\,w_0 = 0$$

By solving the three unknowns, we obtain $u_0 = 4.92$; $v_0 = -1.84$; $w_0 = 1.18$ for the center. A method to locate the center has also been given by Koch and Link (1971).

Every hyperboloid arising from [9.21] for a fixed value of $\hat{y}$ has the same set of three axes which are calculated by extending the method given in section 9.4 for quadratic trend surfaces.

The quadratic component of [9.21] satisfies $\hat{y}^* = \mathbf{X}'\mathbf{A}\mathbf{X}$ where:

$$\mathbf{X} = \begin{bmatrix} u \\ v \\ w \end{bmatrix}; \qquad \mathbf{A} = \begin{bmatrix} -0.3435 & -0.1145 & 0.0984 \\ -0.1145 & -0.0228 & 0.0108 \\ 0.0984 & 0.0108 & 0.0833 \end{bmatrix}$$

The matrix $\mathbf{A}$ satisfies $\mathbf{A} = \mathbf{V}_n \mathbf{\Lambda} \mathbf{V}_n'$.

A practical method to extract the eigenvalues and eigenvectors from $\mathbf{A}$ will be discussed in section 12.4. The result to be computed in section 12.4 is:

$$\mathbf{\Lambda} = \begin{bmatrix} -0.3997 & 0 & 0 \\ 0 & 0.1068 & 0 \\ 0 & 0 & 0.0099 \end{bmatrix}; \qquad \mathbf{V}_n = \begin{bmatrix} -0.9365 & 0.2444 & -0.2517 \\ -0.2900 & -0.1359 & 0.9473 \\ 0.1973 & 0.9600 & 0.1981 \end{bmatrix}$$

We are free to interchange columns in the matrix $\mathbf{V}_n$ or change the sign for all elements in one or more columns because these manipulations do not invalidate the equation $\mathbf{A} = \mathbf{V}_n\mathbf{\Lambda}\mathbf{V}_n'$ for the canonical form. Every method for extracting eigenvectors from a matrix has certain specifications assigned to it in order to avoid the ambiguity of the canonical form. The method used here has the property that the eigenvectors appear in order of decreasing absolute value of the corresponding eigenvalues. Further, the third element of each eigenvector was given a positive sign.

It can be shown (Albert, 1949) that the orthogonal matrix $\mathbf{V}_n$ corresponds to a true rotation only if its determinant is equal to one. For the example, the determinant is +1. A change of sign for elements in column one and, in addition to this, an interchange of columns two and three gives us the following modified result, which is more convenient from a geometrical point of view:

$$\mathbf{\Lambda} = \begin{bmatrix} -0.3997 & 0 & 0 \\ 0 & 0.0099 & 0 \\ 0 & 0 & 0.1063 \end{bmatrix} ; \quad \mathbf{V}_n = \begin{bmatrix} 0.9365 & -0.2517 & 0.2444 \\ 0.2900 & 0.9473 & -0.1359 \\ -0.1973 & 0.1981 & 0.9601 \end{bmatrix}$$

The new matrix $\mathbf{V}_n$ also has a determinant equal to one. Either one of the operations of changing the signs in a column or interchanging two columns has the effect of making the determinant negative. A combination of the two operations leaves it positive.

The three elements along the main diagonal of $\mathbf{V}_n$ are close to unity. This indicates that the three eigenvectors make small angles with the three coordinate axes U-, V-, and W.

The equation of the hypersurface can be described with respect to a new coordinate system with U'-, V'-, and W'-axes. The rotation satisfies:

$$u' = 0.9365\,u + 0.2900\,v - 0.1973\,w$$

$$v' = -0.2517\,u + 0.9473\,v + 0.1981\,w$$

$$w' = 0.2444\,u - 0.1359\,v + 0.9601\,w$$

If, by means of translation, the origin of the coordinate system is moved to the center (u_0, v_0, and w_0), and if the rotation is carried out, [9.21] becomes:

$$\hat{y} = -0.8435 - 0.3997\,u'^2 + 0.0099\,v'^2 + 0.1068\,w'^2 \qquad [9.22$$

For values of $\hat{y}$ which are larger than -0.8435, this equation corresponds to hyperboloids of one sheet (cf. Fig.17). For $\hat{y} < -0.8435$ (or $\hat{x} < 0.75\%$ Cu), we have hyperboloids of two sheets. For planes perpendicular to the U'-axis, the trend surface is an elliptic paraboloid. For the other two axial planes, we will find hyperbolic paraboloids. The horizontal plane makes only a slight angle with the axial plane perpendicular to the W' axis. The trend surfaces for the horizontal planes are hyperboloic paraboloids as show in Fig.52.

The three eigenvectors of $\mathbf{V}_n$ define three axes trough the center. The plane perpendicular to the first eigenvector has equation:

$$0.9365\,u + 0.2901\,v - 0.1973\,w = \text{constant}$$

Because the center (u_0, v_0, w_0) lies in the plane, the constant can be solved. It amounts t 3.838. This plane dips 78½° in the direction with azimuth 197° (S17°W). It provides a approximation for the central plane of the sheet-shaped orebody. The eigenvector for th U' direction, which is perpendicular to this plane, is considerably larger than the oth eigenvalues representing that the variation in grade is relatively rapid in the directio perpendicular to the orebody.

The other two axes lie within the plane of the orebody. The V'-axis dips 11½° WNW with the W'-axis perpendicular to it. The eigenvalue for the V'-direction is small (= 0.01) indicating that there is relatively little variation in this direction.

Critical evaluation. The position of the central plane of maximum mineralization ($V'W'$-plane) is in close agreement with the outline of the orebody shown in Fig.52 which was based on underground data. However, the existence of a subhorizontal axis of minimum mineralization within this plane was not confirmed by later underground drilling. Inspection of other results (e.g., Fig.47) shows that there exist several axes of minimum and maximum mineralization within the plane for the orebody. These axes dip approximately 45° towards the west. The quadratic hypersurface of [9.21] could represent only a single axis of this type and, in this regard, is inadequate as a model for trends within the plane of maximum mineralization.

'hapter 10

·TATIONARY RANDOM VARIABLES AND KRIGING

0.1. Introduction

In the previous chapter, spatial variability in rocks was assessed by the fitting of eterministic functions (polynomials). In this chapter, a different method of approach is ıken. We assume that variables which change in value from point to point obey sta- onary random functions. The values of a variable at different places in dimensional ›ace are related to one another. This gives relationships whereby a variable at a given lace is a function of the same variable but measured at different places.

The results can be used for interpolation and extrapolation. For practical usage, much epends on the validity of the assumption of stationarity. For example, if a random ıriable is stationary, we can determine its parameters for a small area and apply the ›lationship everywhere in a larger region.

Krige (1966a) argued as follows: In the Witwatersrand goldfields, mining is preceded y development. During both stages, chips of rock are sampled and assay values deter- ıined. However, more assays are obtained during the mining stage, and together, these ›says give a close approximation of the amounts of gold truly present in the gold reefs. rige divided a mined-out area into blocks. For each block, the true amount of gold resent was predicted from development values around the periphery of the same block ıd values from a set of adjacent blocks forming a specific geometrical configuration. his problem can be solved by multiple regression with the dependent variable, being the 'erage gold content per block, based on the mining assays. In other, nearby areas which ere developed but not mined out, the data are development assays and there are no ıining assays. The multiple regression equation can be applied to the development assays ı the other areas, if conditions there are the same as in the mined-out area.

Methods of this type were called "kriging" by Matheron (1962). Methods of autocor- ›lation and kriging have entered the earth sciences via problems in the field of mining. ther early work on the topic was done by De Wijs (1951) who argued as follows: Let x ›present the average concentration of a metallic element in the total block of study. sually, this block is an orebody, but Brinck (1971) went so far as to take the earth's pper crust to a depth of 2.5 km as the total block of study in an application of the ıodel of De Wijs.

The basic assumption is that, if a block of rock of any volume is divided into two parts f equal volume, the average concentration values of the two parts are $(1 + d)x$ and $- d)x$ where d is a constant regardless of the volume of the initial block with concentration x. The size-frequency distribution for a number of blocks, all of equal size, now i "logbinomial" as shown by De Wijs (1951). Suppose that the initial block with volume V is divided into 2^p blocks, then $\binom{p}{K}$ out of 2^p blocks have average concentration equal to

$$X(p,K) = x(1+d)^{p-K}(1-d)^K \qquad [10.1$$

K satisfies the binomial distribution.

The logbinomial frequency distribution $X(p, K)$ is positively skew and converge rapidly to a lognormal distribution for increasing p. In practice, only those blocks ar minable where $X(p, K)$ exceeds a specific cut-off value set by economic conditions.

The model does not account for either abrupt (discontinuous) changes or more grada tional changes (trends) within orebodies and rock units in general. Extrapolations on th basis of the De Wijsean model recently made by Brinck (1971) do not recognize the fac that rocks within a larger part of the earth's crust are of different types and do no constitute a continuum.

Nevertheless, the model is comprehensive in that both lognormality and autocorrela tion of assay values from orebodies are explained. In Brinck's application, the mode would account at the same time for all possible variations in size and grade of minera deposits of a metallic commodity and their clustering in three-dimensional space.

Matheron (1962) adopted principles of the work by De Wijs and Krige to develop general theory of regionalized random variables. Various practical methods of krigin resulted from Matheron's theoretical work. The De Wijsean model is not "stationary" i the sense that most random variables of stochastic process theory are stationary. One o the differences is that, in stationary models, the variance of measurements converges to constant value when the size of the sampled block or time interval is increased. In the D Wijsean model, the variance continues to increase with increasing block size.

An example is shown in Fig.53 (from Krige, 1966b). The logarithmic variance of gol

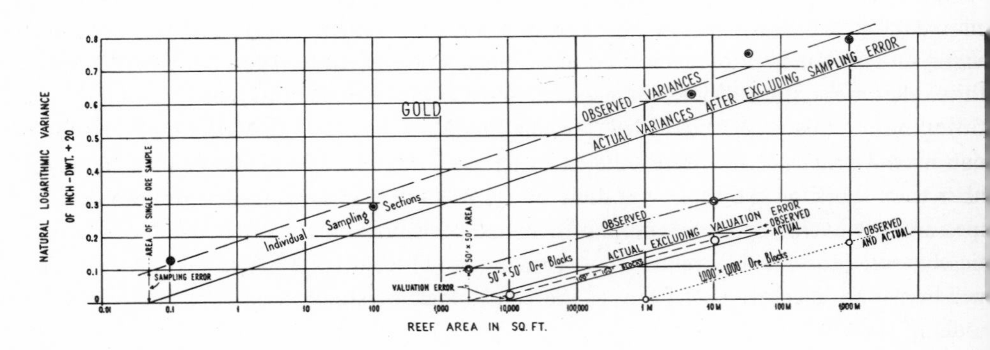

Fig.53. Logarithmic variance of gold values as a function of reef area sampled. The variance increase linearly with area if the area for gold values is kept constant. The relationship satisfies the model of D Wijs. (From Krige, 1966b.)

assays increases linearly with the logarithm of area of gold reef sampled from the smallest to the largest area. In the latter case, data from different mines in the Witwatersrand basin were lumped together.

The phenomenon of a variance depending on size of area sampled is not restricted to variations of ore mineral concentrations in deposits. In section 6.11, it was noted that the spread of age determinations in structural provinces in the Canadian shield increases with the size of area sampled. Likewise, the variance of paleocurrent flow directions measured from foresets may increase with the size of area sampled.

The models of De Wijs and Matheron are stationary for differences between values at equally spaced points rather than for individual values. A consequence is that the theory has a somewhat wider range of application than the more conventional theory of stationary random functions also to be discussed in this chapter, which is used in such fields as stochastic process theory, time-series analysis and statistical theory of communication. For example, the intrinsic models of Matheron can be applied when the variable is subject to a linear trend whereas stationary random variables could not be used in that case.

An important question to ask in specific situations is whether all variability can be expressed by a single random variable. When nonlinear trends occur forming distinct patterns at a more regional scale, deterministic functions must be used. However, deterministic functions could be fitted to random variables with stationary increments. The resulting patterns then would be highly artificial (Matheron, 1967).

Often the question of whether a variable is subject to trends cannot be answered in practice by methods of statistical inference for lack of sufficient data. Conceptual models of theoretical geology should also be considered. When trends can be interpreted geologically, and possibly reflect a phenomenon better observed in other, analogous situations, this must be considered as indirect evidence. For example, the quadratic trend for Mg content of olivine crystals in the Mount Albert intrusion (Fig.45) is weak but its pattern resembles that of stronger trends for other mineralogical data in this intrusion.

It is important to keep in mind that trend surfaces in geology with residuals that are mutually uncorrelated occur only rarely. More commonly, a variable subject to spatial variability has both random (or stochastic) and deterministic components. Until recently, there were two principal methods of approach to spatial variability. One consisted of fitting deterministic functions (as developed by Krumbein and Whitten), and the other one made use of stationary random functions (Matheron and Krige). Combined methods have been developed only recently and the subject is difficult theoretically. Nevertheless, methods such as "universal kriging" where use is made of both deterministic and stochastic components are likely to provide better answers than the earlier methods. This subject will be discussed at the end of the chapter.

Emphasis in this chapter is on autocorrelation functions and application of models from stochastic process-theory. The Fourier transform of the autocorrelation function is the spectral density function to be discussed in more detail in the next chapter. The methods of this chapter and the next one should not be applied separately from one another.

10.2. Space series

Suppose that measurements are made along a line that intersects a rock unit. In this way, we obtain an ordered set of values written as:

$$x_1, x_2, \ldots, x_n \text{ or } \{x_k\}, \qquad k = 1, 2, \ldots, n.$$

The sampling line may be an axis of distance or time. In fact, most work on series of this type was done for situations where the subscript k denotes discrete points in time. A description of various methods of time-series analysis is given in most textbooks of mathematical statistics; for a review of applications see Rosenblatt (1963). In geological situations, the observation points may be ordered not in time, but in space, even in situations where the distance might be interpreted as time (e.g., across a stratigraphic unit).

The analysis of space series can be extended to situations of observation points forming two- or three-dimensional arrays in the plane or in three-dimensional space.

It may be useful to assume that a given space series $\{x_k\}$ is part of an infinite series which, in turn, is a realization of an ordered set of random variables $\{X_k\}$, $k = -\infty, \ldots, -1, 0, 1, \ldots, \infty$. The theoretical set can be seen as a population that is doubly infinite, because any point out of an infinite number of points along the line can assume any one value in an infinite set of possible values.

Suppose that the random variables X_k all have the same probability distribution. This implies a constant population mean and variance. However, the random variables X_k may be correlated with one another. We recall (section 8.3) that two sets of South African gold values from panels 30 ft. apart had approximately the same mean and variance but were positively correlated by $r = 0.59$. If a variable is correlated with another variable relating to the same attribute but measured at locations with a given distance between them, the phenomenon is called autocorrelation.

In general, the autocorrelation coefficient $\rho(l)$ or ρ_l changes with distance or "lag" l between locations. Autocorrelation coefficients estimated from data or serial correlation coefficients are written as $r(l)$ or r_l. The coefficient $r(30 \text{ ft.}) = 0.59$ for the South African gold mine applies only if there is a distance of 30 ft. between the panels. If the distance l is increased, one finds smaller values for $r(l)$. On the other hand, if l is decreased, the autocorrelation coefficient becomes larger.

Suppose that we are measuring a variable which changes continuously in space. This results in a space series if the sampling is done along a line. If, for continuous variables, the interval between measurement points is decreased so that it approaches zero, the autocorrelation coefficient approaches its maximum value which is equal to one. A series is "weakly" stationary if the autocorrelation function for the random variables X_k is constant and does not depend on location along the series. A property of stationary series is that $r(l)$ must decrease to zero if l increases. If the variation along the series contains a yclical component, $r(l)$ can be negative even if the series is stationary. Many series are onstationary. For example, if the mean of the random variable X_k changes along the eries, it is nonstationary.

In general, the autocorrelation function for nonstationary series also decreases to zero vith increasing distance and later becomes negative. However, the shape of the autocor-elation function now depends on the length of the series being sampled. If this is so, the eries can not be characterized by an autocorrelation function although an apparent utocorrelation function can always be calculated. The autocorrelation function is mean-ngful only if the series is stationary or approximately stationary.

The so-called autocovariance function of an ordered sequence or series of random vari-bles X_k with $k = -\infty, ..., 0, 1, ..., \infty$, is defined as:

$$\Gamma_l = E[(X_k - \mu)(X_{k+l} - \mu)]$$

he sample autocovariance can be estimated for a finite series of n values x_k by:

$$C_l = \frac{1}{n-l} \sum_{k=1}^{n-l} (x_k - \bar{x})(x_{k+l} - \bar{x}) \qquad [10.2]$$

here

$$\bar{x} = \frac{1}{n} \sum_{k=1}^{n} x_k ; \qquad l = 0, 1, 2, ..., m .$$

Eq. [10.2] is one of three formulas more widely used to estimate autocovariances (see ection 10.5). If $l = 0$, C_0 is a biased estimate of the variance. The autocorrelation function atisfies:

$$\rho_l = \Gamma_l / \Gamma_0 = \Gamma_l / \sigma^2(X) \qquad [10.3]$$

he sample autocorrelation coefficient r_l satisfies:

$$r_l = C_l / C_0 \qquad [10.4]$$

tationarity

Weak stationarity means that all random variables X_k have the same mean, variance nd autocorrelation function. If, in addition to this, all higher-order moments remain qual, the series is "strictly" stationary. In most practical applications, it is sufficient to ssume weak stationarity or "second order" stationarity.

Terms such as "stationary" and "lag" originate from time-series analysis. Stationary ndom functions are also called "homogeneous" random functions. Stationarity implies at in all domains of the sampled assemblage, there existed an identical chance mecha-ism that governed the magnitudes of the values. In literature, considerable attention is ven to "ergodicity". Simply stated, this implies that the mean and autocovariances as timated from a small domain provide unbiased estimates for the entire assemblage. A

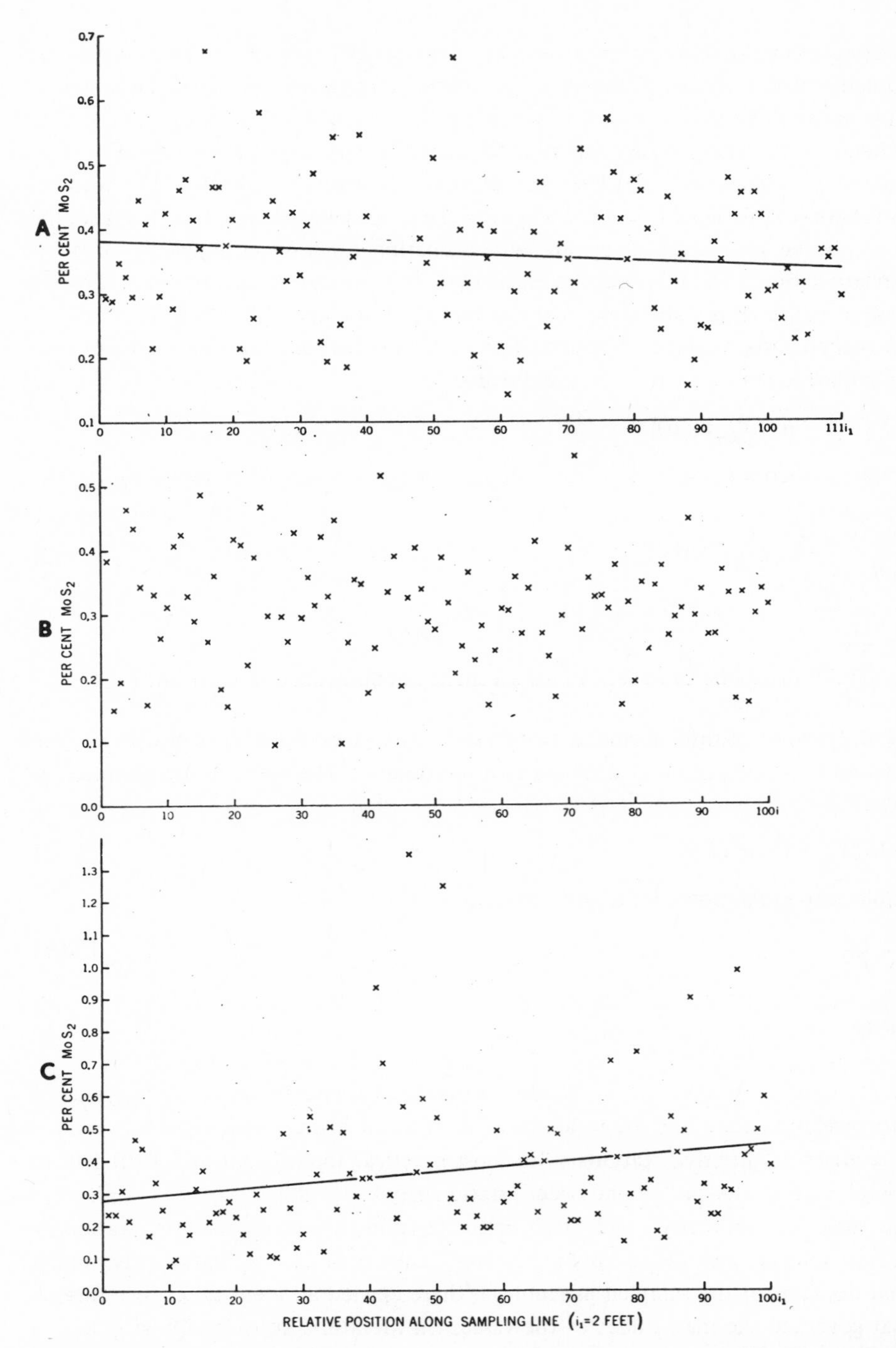

Fig.54. Space series of MoS_2 values from molybdenite orebody (porphyry type, Climax Mine, Denver, Colo., data for A and C from Hazen and Berkenkotter, 1962). A. 101 values from hole NX 636 are uncorrelated and satisfy normal distribution with mean 0.360 and S = 0.111%. B. 100 random normal numbers with mean 0.3 and S = 0.1%. C. 100 data from hole AX 649. Slope of trend line in A is not statistically significant contrary to that for C.

stationary random variable is always ergodic with respect to its mean and, in general, ergodic with respect to its autocovariance function. However, there are some special stationary random variables, not discussed here, which are nonergodic with respect to their autocovariance function (Pugachev, 1965, p.298).

Series of MoS_2 percent values from a molybdenite orebody near Denver, Colorado

Three series are shown in Fig.54. The original data for the first and third series are from Hazen and Berkenkotter (1962). The first series consists of 101 values. A straight line was fitted to it by least squares. The slope of this line is nearly zero and the series is stationary with regard to its mean and variance. The values are not significantly autocorrelated. In fact, the values r_l are close to zero for any lag l, except when $l = 0$, since $r_0 = C_0/C_0 = 1$.

For comparison, 100 random normal numbers are represented in the same manner in a second series. This artificial series has approximately the same mean and variance and its pattern is similar to that of the first series. Since autocorrelation is lacking, these series are said to consist of "noise" or white noise. Histograms for the data satisfy a Gaussian distribution function; hence, the first two series of Fig.54 are examples of Gaussian white noise.

On the other hand, the third series shows a slight but statistically significant increase in MoS_2 content with distance. This series is nonstationary because it contains a linear trend. If the trend is subtracted from the data, we retain residuals which form a stationary series with mean equal to zero. A histogram for this series suggests positive skewness (thin tail of relatively large values). The residuals have the property of white noise but not Gaussian white noise.

Zinc values from sphalerite deposit, Pulacayo, Bolivia

The original data which are shown in Fig.55 are from De Wijs (1951). A sphalerite-quartz vein was developed on the 446-m level (below the surface) by a mining drift with a length of 240 m. Along this drift, 118 channel samples were cut across the vein at 2-m intervals. The massive veinfilling averaged only 0.50 m in width but the wall rocks on both sides contained disseminated sphalerite, partly occurring in subparallel stringers. The channel samples were cut over a standard width of 1.30 m, corresponding to the expected stoping width. Consequently, the assays of Fig.55 represent average weight percentage zinc for rod-shaped samples of 1.30 m cut perpendicular to the vein.

When a straight line is fitted to the data, it is nearly horizontal and, in this sense, the series is stationary because a large-scale systematic variation is absent. However, in comparing Fig.55 to Fig.54, we note a difference. The variation along the series is less abrupt (more gradational) in Fig.55 because of autocorrelation. The first fifteen autocorrelation coefficients are shown in Table XXVII for both original data and logarithmically transformed data.

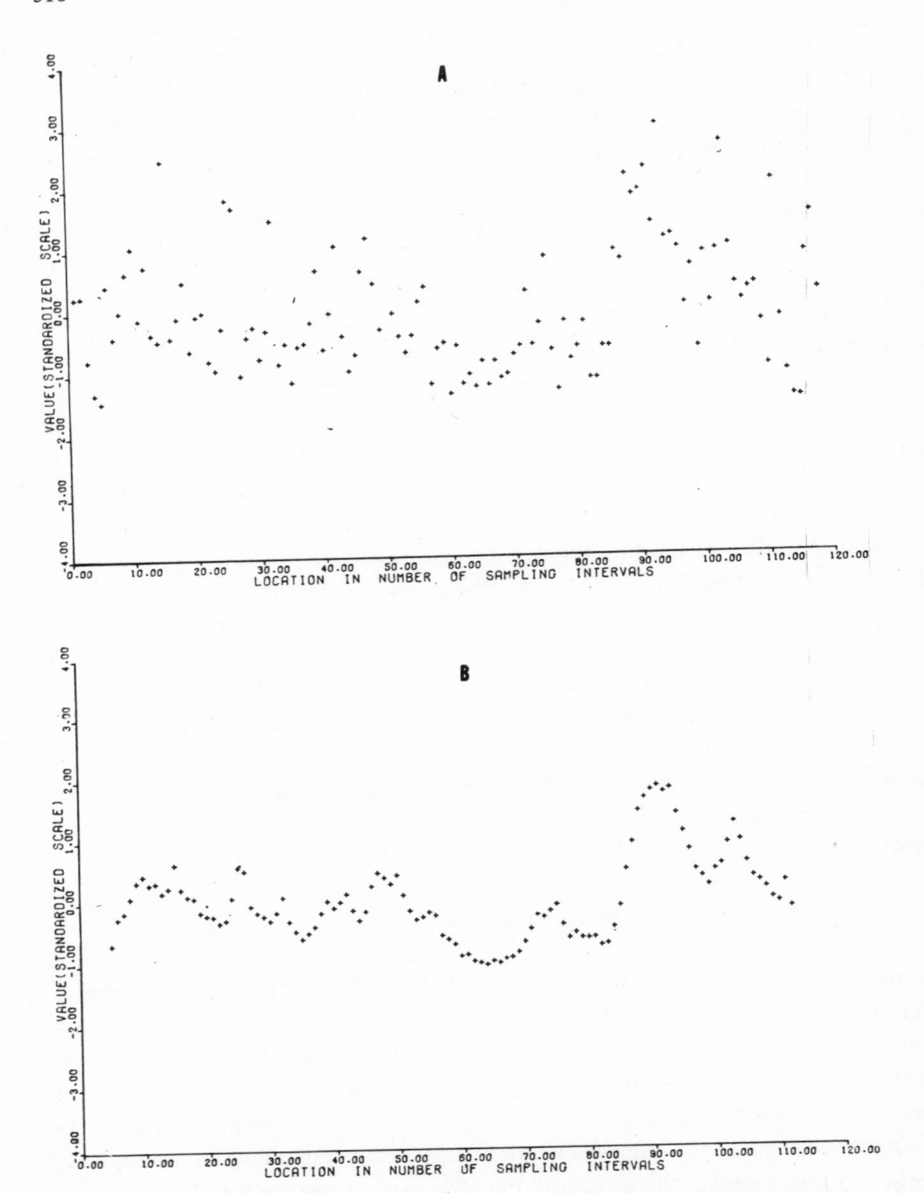

Fig.55. A. Space series of 118 zinc percent values from sphalerite orebody (vein type, Pulacay Bolivia, data from De Wijs, 1951); data are autocorrelated according to pattern of Table XXVII. "Signal" retained after extraction of a "white noise" component. Unit along vertical axis is s = 8.01 and 0.00 corresponds to $\bar{x}$ = 15.61% zinc in both diagrams; sampling interval is 2 m.

De Wijs (1951) showed that the 118 zinc values are lognormally distributed. Th logarithmic variance (base e) amounts to $s^2(\log x) = 0.29$. By taking logarithms, w therefore obtain a series whose values are normally distributed by approximation. Mo theory of autocorrelation was developed for series of this type. In section 10.7, it will b

TABLE XXVII

Autocorrelation coefficients ([10.4] and [10.2]) for first fifteen lags, sphalerite orebody, Pulacayo, Bolivia (for original data and logarithmically transformed zinc values)

Lag (l)	*Raw data of Zn, r (l)*	*Log Zn, r (l)*
0	1.000	1.000
1	0.453	0.465
2	0.290	0.258
3	0.302	0.236
4	0.180	0.175
5	0.246	0.227
6	0.172	0.177
7	0.110	0.088
8	0.129	0.129
9	0.075	0.091
10	0.018	−0.040
11	0.010	−0.014
12	0.075	−0.102
13	0.088	0.048
14	0.171	0.122
15	0.085	0.037

shown that if σ^2 $(\log x) < 1$, series of lognormally distributed data have approximately the same autocorrelation function as the same series after logarithmic transformation. However, when the positive skewness is large, or σ^2 $(\log x) > 1$, the two functions will be different.

Lognormality of data forming a space series in geology is a wide-spread phenomenon, particularly for chemical determinations. It also happens frequently that a few values in a series are much larger than the others. Such anomalous values may strongly influence the shape of the autocorrelation function and the power spectrum. A logarithmic transformation may overcome this difficulty. The drawback of a transformation is that it may be difficult to undo the transformation and return to the original scale of measurements after statistical treatment of transformed data (see section 9.8).

Autocorrelation constitutes a departure from randomness which can be observed as a clustering of points. Thus, if a value in Fig.55 is greater (or less) than the mean, its neighbors also are likely to be greater (or less) than the mean. Nevertheless, there is systematic variation in Fig.55 and we could fit a trend to describe it. Obviously, the trend line should have several maxima and minima. The method of trend-fitting discussed in the previous chapter is not satisfactory in situations of this type because one would have to use a long trend equation with many coefficients. On the other hand, we can assume that the series is weakly stationary and consists of two components: a systematic component (signal) and a white-noise component.

Then, by using a method of filtering, somewhat similar to methods in use by communication engineers, the noise component can be removed from the data which leaves the signal. The result is shown in Fig.55B. This method will be discussed in more detail in section 11.8.

It is sometimes difficult to make a clear-cut distinction between a stationary series and a nonstationary series that contains trends. In some situations, it is advantageous to perform a trend analysis first by using deterministic functions. Afterwards, the residuals can be treated as a stationary series.

Filtering is one reason for considering the autocorrelation function of a series. In section 9.8, we noted that several methods of statistical inference of trend functions are sensitive to autocorrelation. A third reason for considering the autocorrelation function is that it enables us to develop equations to estimate the precision of the mean of autocorrelated data. For example, the average zinc value as estimated from the 118 data of Fig.55 amounts to $\bar{x}$ = 15.61%. The standard deviation is $s(x)$ = 8.01%. An important practical problem consists of estimating the precision of the mean. By the method of section 6.8 (see also [2.29]), we could show that the standard deviation of the mean satisfies:

$$s(\bar{x}_{118}) = s(x)/\sqrt{118} = 0.737$$

Consequently, the probability that the population mean μ lies in the interval $\bar{x} \pm 0.737\, t_{0.975}$ amounts to 95 %. From $t_{0.975}(117) = 1.98$, it follows that:

$$P(14.15 < \mu < 17.07) = 95\ \%$$

This result is based on [2.29] which was derived for independent random variables. Eq. [2.29] also can be written as:

$$F_n = n \cdot s^2(\bar{x}_n)/s^2(x) = 1 \qquad [10.5]$$

The factor F_n is equal to the relative variance of the mean multiplied by the number of observations (n). For uncorrelated data it is equal to one. Except for random fluctuations, this equation is valid for the data shown in the first two series of Fig.54 but it is not valid for the present example (Fig.55A).

From the 118 zinc values, we can form 117 pairs consisting of two values. For each pair we can calculate a mean $\bar{x}_2$. The 117 values $\bar{x}_2$ have variance $s^2(\bar{x}_2)$ equal to 46.85. If [10.5] is applied, we obtain:

$$F_2 = 2s^2(\bar{x}_2)/s^2(x) = 2 \times \frac{46.85}{64.14} = 1.461$$

This factor is considerably larger than one and indicates that [10.5] is not valid for autocorrelated data.

Instead of this, we have:

$$s^2(\bar{x}_2) = F_2 \cdot s^2(x)/2$$

In order to estimate the variance of the mean for 118 data, we need:

$$s^2(\bar{x}_{118}) = F_{118} \cdot s^2(x)/118$$

or: $s(\bar{x}_{118}) = \sqrt{F_{118}} \cdot s(x)/\sqrt{118}$

It means that the above estimate of $s(\bar{x}_{118}) = 0.737$ must be corrected by the factor $\sqrt{F_{118}}$. Of course, F_{118} cannot be estimated from the data by the direct method used to estimate F_2.

However, by assuming that the series is stationary and consists of two components as in Fig. 55B, we have, by approximation:

$$F_n \simeq 1 - c + (2c/a)[1+(e^{-an} - 1)/an] \quad [10.6]$$

where a and c are two constants that characterize the stationary series. This result will be derived in section 10.10. For the example, $a = 0.1892$, $c = 0.5157$, $F_{118} \simeq 5.69$, and $s(\bar{x}_{118}) \simeq \sqrt{5.69} \times 0.737 = 1.769$. The above confidence interval for the mean must be replaced by:

$$P(12.1 < \mu < 19.1) = 95\,\%$$

The new interval is $\sqrt{5.45} = 2.4$ times as wide as the previous interval.

De Wijs (1951, p.21) has stated that at the 476-m level, which is 30 m deeper, the average zinc content was 14.30%, which is within the confidence interval for the average of 15.61% of the 446-m level.

By eliminating the noise from the data, we were able to obtain Fig.55B, indicating that the average changes from place to place along the drift. In fact, a 95% confidence interval for the "signal" can be calculated readily by using a result to be derived in section 11.8. The variance of the "moving" average of Fig.55B is $q \cdot s^2(x)$ with:

$$q = ac\,[a^2 + 2ac/(1-c)]^{-\frac{1}{2}} \quad [10.7]$$

where a and c, as before, are the two constants of the signal-plus-noise model.

Because $q = 0.3040$, a 95-% confidence belt has a constant width of approximately $2s(x)\sqrt{q} = 8.83\%$ zinc.

It is, therefore, likely that some of the maxima and minima of the signal in Fig.55B are real and that the average is not the same along the entire drift.

The problems discussed here are of practical importance in mining. Similar problems may occur in geochemistry and various fields of geology. For example, we may be interested in the average uranium content of a granite stock; in the average age of granitic stocks in a structural province; or in the preferred direction of paleocurrents transporting sediments in an ancient basin.

The example of the sphalerite vein of Pulacayo, Bolivia, was treated differently by Matheron (1962) who operated on logarithmically transformed data. The logarithmic variance is 0.29. From this we derive by the previous signal-plus-noise model that the variance of the mean of 118 logarithmically transformed data y_k is:

$$s^2(\bar{y}_{118}) \approx F_{118} \times 0.29/118 = 0.049$$

where F_{118} is now equal to 5.45, which is close to $F_{118} = 5.69$ used previously fo untransformed data. We assumed that each single assay provides an estimate for th average of the entire deposit, or at least the part where this average can be assumed t exist. Matheron (1962) used the model of De Wijs for analysis. According to this model it is not meaningful to estimate an average value for an assemblage; and an autocorrela tion function does not exist. Instead of this, any known concentration value can b interpreted as representing the average for a block of rock with specific shape an volume. A measured value for an initial block then has a so-called extension variance if i is applied to a larger block.

In section 10.11, it will be seen that Matheron estimated for the Pulacayo zinc bod that $3\alpha \simeq 0.050$ where α is the so-called absolute dispersion. Assuming that our average o 118 values is a good estimate for an elongated block of ore at the 446-m level, we ca determine the larger part of the orebody for which this average is representative with a extension variance equal to $s^2(\bar{y}_{118}) \approx 0.049$. Matheron's method gives that thi would be for about 340 m both upward and downward from the 240 m of drift sample at the 446-m level. Of course, both models are applicable only as long as the vein doe exist with a stationary variability for its zinc content. When abrupt or gradational change (trends) occur, these models can not be applied. However, the model of Matheron i detailed and formulas are available for many different geometrical configurations.

In this regard, Matheron's model is more flexible than a direct application of results o more conventional methods of stationary random variables such as the signal-plus-nois model. Unfortunately, much of Matheron's work was never fully explained in the Englis language, although detailed introductions to it were published by Matheron (1963), Blai and Carlier (1968), David (1969) and Watson (1971). It is beyond the scope of this boo to fully explain Matheron's methodology. We will give several examples, however, wher Matheron's method is applied to specific situations. At present, the most complete refe ence in English is a manual prepared by Matheron (1971a).

10.3. The model of De Wijs

We recall [10.1]. If the logarithm is taken:

$$\log X(p, K) = \log x + p \log (1 + d) + K \log \frac{1-d}{1+d} \qquad [10.8$$

where x, p and d are constants, whereas $\log X(p, K)$ and K are random variables. K ha the binomial distribution; for increasing p, its variance rapidly becomes $\sigma^2(K) \approx p/4$ Hence, the logarithmic variance of $X(p, K)$ approaches to:

$$\sigma^2 = (p/4) \left[\log \frac{1-d}{1+d}\right]^2 \qquad [10.9$$

For convenience, we will use natural logarithms (base e). When v represents the volume of a sample which is part of V(volume of total block), $V/v = 2^p$. Matheron (1962) has defined the "principle of similitude" by the equation:

$$\sigma^2 = \alpha \log \frac{V}{v} \qquad [10.10]$$

where α is the so-called "absolute dispersion". Eq. [10.9] and [10.10] are equivalent, and:

$$\alpha = \frac{1}{4 \log 2}\left[\log \frac{1-d}{1+d}\right]^2$$

Suppose that $\sigma^2(v_1, v_2)$ describes the variance of concentrations of samples with volume v_1 within a larger block with volume v_2 that is sampled; let $\sigma^2(v_2, V)$ be the variance of the average concentration of blocks with volume v_2 coming from a much larger block with volume V. Matheron (1962) has shown that the variance $\sigma^2(v_1, V)$ for samples within V, then satisfies:

$$\sigma^2(v_1, V) = \sigma^2(v_1, v_2) + \sigma^2(v_2, V) \qquad [10.11]$$

This additive property of variances for average concentration in blocks of different sizes applies when the spatial variability is stationary. If the model of De Wijs is satisfied, we have:

$$\sigma^2(v_1, V) = \alpha \log \frac{v_2}{v_1} + \alpha \log \frac{V}{v_2}$$

This rule for addition of variances was first discovered experimentally by Krige (1951) for assay values from South African gold mines. It holds true in general, provided that the variations are stationary. Krige (1966b) later has shown that the model of De Wijs can be applied to his data as shown in Fig.53.

A single sample groove across a gold reef for which a gold assay is determined has average cross-sectional area of 0.05 sq. ft. These assays are subject to a sampling error, which is a superimposed random distortion and can be eliminated by subtraction. If this noise is eliminated, samples of volume v_1 satisfy the equation:

$$\sigma^2(v_1, v) = 0.0295 \log \frac{v}{v_1}$$

being the equation of the straight line in Fig.53. In this expression, v is a variable determined at several points for increasingly larger parts of the reef. The linear relationship of Fig.53 holds true up to the large volume corresponding to a reef of the Witwatersrand gold field sampled in different mines which are far apart. The other straight lines in Fig.53 are for average gold grade in successively larger blocks.

Matheron (1962) has treated the 118 zinc values of Fig.55A by this method. The logarithmic variance is 0.29. In total, 240 m of drift was sampled, hence $R = 240$. Samples were 1.30 m long, but massive sulphides of the vein were only 0.50 m. Hence,

Matheron set $r = 0.5$, instead of 1.3, assuming that the sampling of wall rocks with disseminated sulphides had the effect of underestimating zinc content of massive sulphides by a constant factor. This would not affect the logarithmic variance for zinc variation of the vein. If it is assumed that variability is the same in all directions within the vein, then:

$$\sigma^2(r, R) = 3\log\frac{R}{r} \approx 3\alpha\log\frac{240}{0.5}$$

Hence, $3\alpha \approx 0.045$. The reason for using 3α is as follows.

Eq. [10.10] is generally applicable for blocks of exactly the same shape but of different volumes. For the example, the channel samples are rod-shaped with length r. The dimensions for the other two directions are negligibly small. Likewise, the drift is $R = 240$ m long and the dimensions for directions perpendicular to the drift are negligibly small. To a first approximation, we are comparing linear blocks with one another for which $\alpha\log v \approx \alpha\log(abc\cdot r^3)$ and $\alpha\log V = \alpha\log(abc\cdot R^3)$ where a, b, and c are the same. For V, all three linear dimensions are R/r greater than for v. Hence:

$$\sigma^2(v, V) = \alpha\log\frac{R^3}{r^3} = 3\alpha\log\frac{R}{r} \qquad [10.12]$$

When v is set equal to V, $\sigma^2(V, V) = 0$. It would imply that our estimate of average zinc content for the 446-level is exact. Its extension variance, however, will increase when we assume that the value is representative for a portion of the orebody extending upward and downward from the 446-level

10.4. Correlograms and variograms

A sequence of sample autocorrelation coefficients r_l with $l = 1, 2, \ldots m$, is called a correlogram. Several correlograms are shown in Fig.56. The first one is for the 118 zinc values, Pulacayo; Fig.56B is a composite correlogram obtained by averaging single correlograms for 24 space series along drifts of various levels of the Whalesback copper deposit. In Fig.56C, after Krige et al. (1969), logarithm of covariance is plotted against lag for two space series of gold values (see also Watson, 1968).

The data were subjected to a logarithmic transformation before computing the correlogram, although this is not necessary for the zinc values (see Table XXVII).

The three correlograms show trends approximated by curves of the type:

$$r(l) = c\,e^{-a|l|} \qquad [10.13]$$

When a logarithmic scale is used for the covariance as in Fig.56C, this exponential curve plots as a straight line.

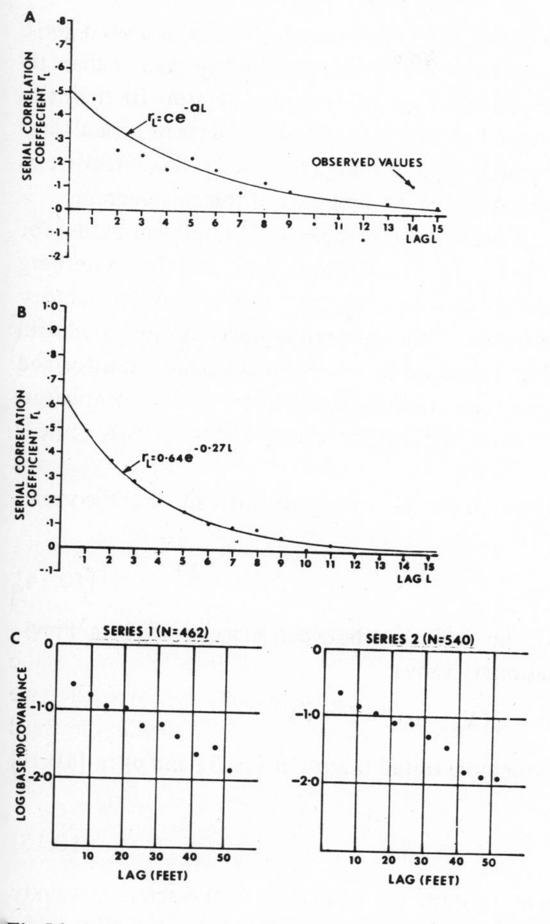

Fig.56. A. Correlogram for data of Table XXVII (2nd column) and best-fitting exponential curve. B. Composite correlogram; average of 24 correlograms for series of logarithmically transformed copper values (8 ft. apart) along drifts at various levels of Whalesback Mine, Newfoundland. C. Autocovariance of residuals from cubic trends fitted to two space series of log-gold values. (From Krige et al., 1969.) An exponential model would be according to a straight line in C because the logarithmic scale is used for autocovariance.

It will be shown later that [10.13] is the solution for a Markovian-type space series with superimposed noise (signal-plus-noise model). If the noise is absent, $c = 1$, and the trend lines of Fig.56A would pass through the point $r(0) = 1.0$ for zero lag. If the space series is without systematic variations (noise only, see Fig.54A), $c = 0$.

The ratio $c/(1-c)$ is called the signal-to-noise ratio in statistical theory of communication.

The De Wijsean model and the exponential model are basically different. However, for the data of Fig.56, both models provide an approximation that is satisfactory. Krige et al.

(1969) have shown that in the situation of Fig.56C, the exponential model gives a good fit until lag 50, whereas the De Wijsean model is not satisfactory for lags greater than 10 sample intervals. In that situation, the exponential model also gave a better fit than the so-called power-law covariance model proposed by Whittle (1954a) for data in agriculture.

In practice, it may not be important which type of model is used for the autocorrelation function provided that the fit is satisfactory for the first part of the correlogram.

The curves satisfying [10.13] in Fig.56 were fitted by means of a nonlinear model for exponentials developed by Deming (1943). The model was discussed by Agterberg (1967b). Later applications of this method have shown that it provides satisfactory solutions most of the time. In some situations, the approximations are not good and better results may be obtained by fitting a straight line to logarithmically transformed coefficients of the type shown in Fig.56C and undoing the transformation afterwards. This second method also has drawbacks, particularly if the scatter about the fitted line is large.

Matheron (1962, 1965) preferred to compute the variogram instead of the correlogram. The parameter estimated is:

$$\sigma_l^2 = E(X_{k+l} - X_k)^2 \qquad [10.14]$$

It is the expected value of the square of the difference between successive values. Previously, it was shown that for a weakly stationary series:

$$E(X_{k+l}) = E(X_k) = \mu; \quad E(X_{k+l} - \mu)^2 = E(X_k - \mu)^2 = \sigma^2(X)$$

If these conditions hold true, it is readily demonstrated that ρ_l in [10.3] and σ_l^2 in [10.14] are related by:

$$\sigma_l^2 = 2(1 - \rho_l)\,\sigma^2(X) \qquad [10.15]$$

Consequently, correlogram and variogram analysis are equivalent if the series is weakly stationary.

However, suppose that the data contain a linear trend. Then, $E(X_{k+l}) \neq E(X_k)$. Instead of this, $E(X_{k+l}) - E(X_k) = cl$ where c is a constant. In that situation, a constant correlogram does not exist since the shape of the correlogram depends on the length of the finite series from which it is estimated. On the other hand, the variogram can be calculated and will not be subject to this drawback.

Similarly, suppose that the mean is stationary and equal to zero but that the variance increases or decreases linearly along the series. Again, the shape of the variogram is independent of the length of the series, but the correlogram depends on it. In some practical situations, it may occur that, by approximation, both the mean and the variance increase linearly in positive or negative direction of the space series. The variogram then can be used but the correlogram should not be used. A number of results based on the variogram were derived by Matheron (1962, 1965), and outside the field of geostatistics by Jowett (1955) and Cramér and Leadbetter (1968).

If the mean and the variance show a linear increase, we would have to do two things before a correlogram may be calculated. First, all data are transformed, e.g., by taking their square roots. This is to stabilize the variance. Secondly, we would fit a linear trend and calculate the correlogram from the residuals.

In many practical situations, the mean is not stationary but there exist trends which are not necessarily linear. If the trend is not linear, correlogram or variogram analysis cannot be applied. However, we may eliminate the trend and calculate a correlogram from the residuals. The variance may also change with location in space. For example, in the Whalesback copper deposit, the variance is about seventeen times larger near the center of the mineralized zone than at the margins. This variance was stabilized by applying a logarithmic transformation (section 9.8).

10.5. Main properties of the autocorrelation function

Important properties are:

$$|\rho(l)| \leqslant 1 \quad \text{for all } l \qquad [10.16]$$

$$\rho(0) = 1 \qquad [10.17]$$

$$\rho(l) = \rho(-l) \qquad [10.18]$$

Eq. [10.16] is obvious if $\rho(l)$ is interpreted as an ordinary correlation coefficient for which the property holds true. Eq. [10.17] follows from setting $l = 0$ in [10.3]. Eq. [10.18] is a consequence of the assumption of stationarity according to which:

$$\Gamma(l) = \text{cov}\,[X(k), X(k+l)] = \text{cov}\,[X(k-l), X_{(k)}]$$

Further: $\text{cov}\,[X(k-l), X(k)] = \text{cov}\,[X(k), X(k-l)]$

Hence: $\Gamma(l) = \Gamma(-l)$ and [10.18] follows.

Eq. [10.16] – [10.18] also apply to the sample autocorrelation coefficients $r(l)$ or r_l.

A fourth property of $\rho(l)$ is that the so-called autocorrelation matrix:

$$\begin{bmatrix} 1 & \rho(1) & \rho(2) & \cdots \\ \rho(1) & 1 & \rho(1) & \cdots \\ \cdots & \cdots & \cdots & \cdots \end{bmatrix}$$

s nonnegative definite (see section 3.12). In this regard, it is equivalent to the ordinary orrelation matrix $\mathbf{R}$ for a set of p variables. We have $\mathbf{R} = (\mathbf{Z}'\mathbf{Z})/(n-1)$ where $\mathbf{Z}$ is an $n \times p)$ matrix consisting of standardized data. A matrix of the type $\mathbf{A}'\mathbf{A}$ is always non-egative definite. The fourth property of the coefficients $\rho(l)$ is not necessarily true for he $r(l)$ if [10.2] and [10.4] are used.

In the general case, the autocorrelation function $r(l)$ is estimated from residuals $\hat{E}_k$ of data X_k which form a series. Then:

$$\hat{E}_k = X_k - \hat{X}_k \tag{10.19}$$

where the $\hat{X}_k$ satisfy some deterministic function. The X_k are not necessarily original measurements since these may have been transformed in order to stabilize the variance. In many situations, the $\hat{E}_k$ are taken to be deviations from the arithmetic mean $\bar{X} = (1/n)\ \Sigma X_k$ or $\hat{E}_k = X_k - \bar{X}$. The estimation method then reduces to the one discussed in section 10.2. The covariance for residuals $\hat{E}_k$ satisfies:

$$C_l = \frac{1}{n-l} \sum_{k=1}^{n-l} \hat{E}_k \hat{E}_{k+l} \tag{10.20}$$

Consequently: $$r_l = C_l/C_0 = \frac{n}{n-l} \frac{\Sigma \hat{E}_k \hat{E}_{k+l}}{\Sigma \hat{E}_k^2} \tag{10.21}$$

If the residuals are deviations from the mean, [10.21] provides maximum likelihood estimates of ρ_l. The estimation can be slightly refined by using [13.28] in Chapter 13 which is useful for relatively short series. Estimates of this type are used by a number of authors including Kendall and Stuart (1958, vol. 3), Blackman and Tukey (1958) and Bartlett (1966).

A somewhat different approach was taken by Parzen (1962) and Jenkins and Watts (1968) who divided the sum in [10.20] by n instead of by $(n-l)$. This results in biased estimates of C_l and r_l (underestimation). However, the autocorrelation matrix for the sample:

$$\begin{bmatrix} 1 & r_1 & r_2 & \cdots & r_m \\ r_1 & 1 & r_1 & \cdots & r_{m-1} \\ \cdot & \cdot & \cdot & \cdots & \cdot \\ \cdot & \cdot & \cdot & \cdots & \cdot \\ \cdot & \cdot & \cdot & \cdots & \cdot \\ r_m & r_{m-1} & r_{m-2} & \cdots & 1 \end{bmatrix}$$

now always is nonnegative definite and the fourth property for ρ_l also holds true for r_l. If [10.20] is used for estimating r_l, this condition is usually not satisfied. We can interpret the Parzen-Jenkins approach as follows. Suppose that a set of m zeros is added to the series $\{\hat{E}_k\}$, $k = 1, \ldots, n$. We then calculate the covariance as:

$$C_l = \frac{1}{n} \sum_{k=1}^{n} \hat{E}_k \hat{E}_{k+l}, \quad l = 0, \ldots, m$$

from the series $\{\hat{E}_k\}$, $k = 1, \ldots, n, \ldots, n+m$, where $\hat{E}_k = 0$, if $k > n$. The autocorrelation matrix now is of the type **A′A** where **A** is a matrix consisting of residuals and zeros. Nonnegative definiteness is implied. The Fourier transform of a continuous autocorrelation function $\rho(l)$ satisfies:

$$\phi(\omega) = \int_{-\infty}^{\infty} e^{i\omega l} \rho(l)\, dl \qquad [10.22]$$

This is the spectral density function or power spectrum that will be discussed in more detail in Chapter 11. If $\phi(\omega)$ is estimated from discrete data, computational problems may arise when the autocorrelation function is not nonnegative definite. These problems (e.g., negative spectral density values which always should be positive) are minor when the maximum lag is kept considerably smaller than the length of the series and maximum-likihood estimators can be used.

In some applications, however, we will estimate, directly from the data, the complete power spectrum consisting of $n/2$ values. In that case, the Parzen-Jenkins approach will be used. The autocorrelation function can then be obtained from the spectrum by inverting [10.22]. Multiplication of its values by $n/(n-l)$ will give estimates identical to those of [10.21].

Traditionally, geologists have been more concerned with correlogram and variogram than with the power spectrum. When some data processing facilities are available, it is best to study both.

Correlogram and variogram suffer from a significant drawback in that they have correlated errors. On the other hand, adjacent values in the power spectrum tend to be statistically independent. It is difficult to separate real phenomena from random fluctuations in correlograms. Some authors go so far as not to believe patterns suggested by a correlogram. Admittedly, the power spectrum and harmonic analysis provide better results in a number of situations. However, the spectrum suffers from a disadvantage which is not present in the correlogram. One usually can obtain a precise estimate of a smoothed version of the power spectrum $\phi(\omega)$ but not necessarily of the true spectrum.

A detailed theory of the properties of the sample autocorrelation coefficient is not presented here. For this, the reader is referred to textbooks such as Bartlett (1966) or Jenkins and Watts (1968). If the series is long (large value for n), r_l is, by approximation, normally distributed with mean $\rho(l)$ and variance $1/n$. This large-sample property of the sampling fluctuations can be used to make rapid, though approximate, tests of significance. An advantage is that the method can also be used to test values of a two-dimensional autocorrelation function (cf. section10.8) for statistical significance. If the autocorrelation coefficients are to be compared to zeros as expected for Gaussian white noise, the distribution function of the ordinary correlation coefficient can be used for comparison. The results can then be refined as follows for series (Quenouille, 1952). Before the significance limit of the ordinary correlation coefficient is determined (e.g., by

[8.17]), 2 is added to the number of degrees of freedom. Afterwards, a constant $1/n$ is subtracted from the significance level. For a more rigorous test of randomness or non-randomness in space series, Durbin and Watson's (1950, 1951; with tables) method may be used.

10.6. Theoretical autocorrelation functions

In order to visualize theoretical autocorrelation, we return to the example of section 7.7 according to Krumbein and Dacey (1969). The data consist of a random sequence of lithologies. Randomness in this context means that if we are in a specific state (e.g., shale), there is a given probability that the next point measured along the section will be for the same state or that another state will occur. Suppose that instead of assigning letters to the states, we code +1 if a given state is present and −1 if another state occurs. The result is a series of discrete data which can be connected by horizontal lines giving a rectangular wave similar to the one shown in Fig.57. Several other processes can be represented in this fashion. For example, α-particles from a radioactive source are sometimes used to trigger an alternate-state device with values +1 and −1 (see Jenkins and Watts, 1968). The main statistical property of a wave of this type is that transitions from +1 to −1 or from −1 to +1 can occur any time with a probability that is independent of place of occurrence along the series.

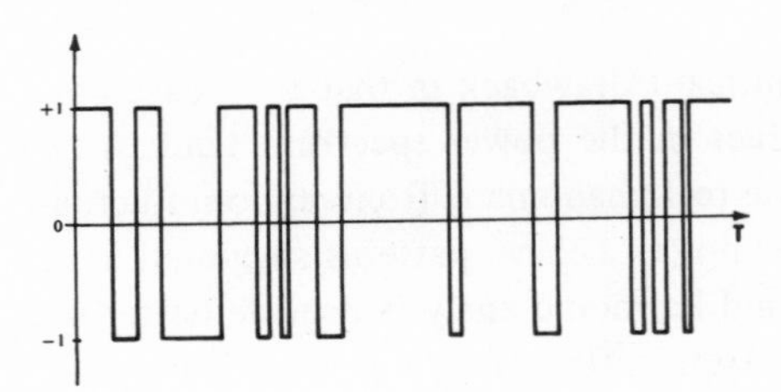

Fig.57. Random telegraph signal (after Jenkins and Watts, 1968). Zero crossings between two states +1 and −1 occur at random along the time axis.

In section 7.7, we derived that sampling at a regular interval results in a geometric distribution for the lengths of the sequences now coded as +1 or −1. The discrete frequencies of the geometric distribution lie on an exponential curve. This also holds true in the limit, if the sampling interval approaches zero, giving the exponential distribution as the continuous counterpart of the geometric distribution.

In order to derive the autocorrelation function for the process of Fig.57, we may take the following approach.

Assume that Fig.57 represents a process started at $T = -\infty$. For any discrete point k, the probabilities of being in state +1 or −1, are equal to constants adding to one. For simplicity, let us assume:

$$P(X_k = 1) = P(X_k = -1) = \tfrac{1}{2}$$

Consequently, $E(X_k) = 0$. The series is stationary and has expectation equal to zero. The autocovariance function satisfies:

$$\Gamma_l = E(X_k \cdot X_{k+l})$$

It turns out to be the sum of two terms: (1) probability that the number of changes in state is even during the interval $(k, k + 1)$ multiplied by +1; and (2) probability of an odd number of changes multiplied by −1. We may write this in the form:

$$\Gamma_l = \sum_{k=0}^{\infty} (P_{2k} - P_{2k+1}) \qquad [10.23]$$

where:

$$P_k = \frac{e^{-\lambda|l|}(\lambda|l|)^k}{k!}$$

denotes the Poisson-type probability of having exactly k changes in an interval of length l. According to [7.59], λ can be interpreted as the number of changes (both from +1 to −1 and −1 to +1) expected per unit of distance along the axis when $|l| = 1$.

From [10.23], it follows after some manipulation that:

$$\Gamma_l = e^{-2\lambda|l|}$$

Because $\Gamma_0 = 1$, we also have:

$$\rho_l = e^{-2\lambda|l|} \qquad [10.24]$$

The absolute value $|l|$ indicates that the lag l can be taken in positive direction or in negative direction of the series with the same outcome.

We will see presently that several other models for space series also result in exponential autocorrelation functions.

Time-dependent stochastic processes

Most of the theory of autocorrelation was developed for time series. Several important concepts of the theory of stochastic processes will be reviewed. Because of space limitations, this review is not comprehensive. For a more complete picture, the reader is referred to the books listed in section 5.6 and 5.7 of "Selected Reading".

The continuous m-th order autoregressive process is represented by the stochastic differential equation:

$$a_m \frac{d^m x}{dt^m} + a_{m-1} \frac{d^{m-1} x}{dt^{m-1}} + \ldots + a_0 x(t) = \eta(t) \qquad [10.25]$$

The stationary random variable $x(t)$ may be regarded as the response by a system to a white-noise function $\eta(t)$. Its autocorrelation function is of the type:

$$\rho(x) = A_1 e^{-\lambda_1 |x|} + A_2 e^{-\lambda_2 |x|} + \ldots + A_m e^{-\lambda_m |x|} \qquad [10.26]$$

where $A_1, A_2, \ldots, A_m$ are constants, and $\lambda_1, \ldots, \lambda_m$ represent the roots of a polynomial (cf. Jenkins and Watts, 1968). The roots are either real or they occur as pairs of conjugate complex roots. In the latter situation, a pair can be combined into a single term of the type $e^{-k_1 |x|} \cos(k_2 x + \phi)$ where k_1, k_2, and ϕ are constants.

First-order processes

The continuous first order process has equation:

$$dx/dt + \alpha x(t) = \eta(t) \qquad [10.27]$$

with:

$$\rho(x) = e^{-\alpha |x|} \qquad [10.28]$$

Strictly speaking, a white-noise function $\eta(t)$ may only be defined for a discrete series consisting of uncorrelated values. Some workers prefer not to use continuous process-models because $\eta(t)$ could not be continuous and uncorrelated at the same time. This subject was discussed in detail by Yaglom (1962). A way of resolving this dilemma is to define $\eta(t)$ as a continuous process with, e.g., an exponential autocorrelation function that dies out rapidly so that any sampling at regular time intervals gives uncorrelated values.

If the right-hand side of [10.27] is replaced by $\eta^*(t)$ for a stationary random process with uncorrelated increments, we obtain Langevin's equation, a famous physical application of which is to Brownian motion. If we take the first derivative of $dx/dt + \alpha x(t) = \eta^*(t)$, and set $v = dx/dt$ for velocity, the result is $dv/dt + \alpha v = \eta(t)$ which is comparable to [10.27]. The function $\eta(t)$ now is the derivative of a process with uncorrelated increments. It can be interpreted as a random force working on a particle. This force provokes both an acceleration dv/dt and a friction αv proportional to the velocity v.

First-order Markov process

Suppose that the continuous first-order process is sampled at a regular interval. This gives the discrete autoregressive process represented by the stochastic difference equation:

$$X_t = \alpha_1 X_{t-1} + \eta_t \qquad [10.29]$$

with:

$$\rho(x) = \alpha_1^{|x|}$$

which is also exponential and we may write:

$$\rho(x) = \mathrm{e}^{-a_1|x|}$$

If the sampling interval is sufficiently narrow $a_1 \approx \alpha$ with α as in [10.28]. Eq. [10.29] is also known as the first-order Markov process. The Markov property can be formulated as:

$$P\{X_t | X_\tau; \tau < t\} = P\{X_t | X_{t-1}\} \qquad [10.30]$$

indicating that X_t, apart from its random component given by η_t, depends on X_{t-1} only. This model is more general than that of Fig.57 which resulted in [10.24]. We may also write:

$$X_k = \rho_l X_{k-l} + \eta_{k,l}$$

where:

$$\rho_l = \rho_1^{|l|} = \mathrm{e}^{-a|l|} \qquad [10.31]$$

This conforms to the notation used previously for space series.

In [10.31], ρ_1 is the first autocorrelation coefficient. For the variance of $\eta = \eta_{k,l}$ we have:

$$\sigma^2(\eta)/\sigma^2(X) = 1 - \rho_1^{2|l|} \qquad [10.32]$$

which reduces to:

$$\sigma^2(\eta) = (1 - \rho_1^2)\,\sigma^2(X) \text{ if } |l| = 1$$

Second-order processes

The continuous process of the second order may be written in the form:

$$\frac{\mathrm{d}^2 x}{\mathrm{d}t^2} + 2\alpha \frac{\mathrm{d}x}{\mathrm{d}t} + (\omega_0^2 + \alpha^2)\, x(t) = \eta(t) \qquad [10.33]$$

This is equivalent to [10.25] if $m = 2$. The autocorrelation function satisfies [10.26] with $m = 2$. We will be interested in the situation that λ_1 and λ_2 form a pair of conjugate complex roots. This situation occurs if ${\omega_0}^2 > 0$. Then:

$$\rho(x) = \mathrm{e}^{-\alpha|x|}\,[\cos \omega_0 x + (\alpha/\omega_0) \sin \omega_0 |x|] \qquad [10.34]$$

The autocorrelation function of the discrete second-order process is of a different type (see Yule, 1927; Kendall and Stuart, 1958, vol.3). In the sequel, we will apply [10.34] t serial correlation coefficients estimated from discrete data. This is permitted only if th sampling interval is sufficiently small; then the theoretical function for the discret second-order process may be approximated by [10.34].

Cross-correlation

Autocorrelation can be seen as taking two identical copies of a series of discrete dat and sliding the two copies with respect to one another. The amount of sliding is the la and a correlation coefficient is computed for every lag considered. Because the two copie are identical, it does not matter which series is slid with respect to the other one. Th result will be the same as expressed by the property $\rho(-l) = \rho(l)$.

Suppose that we have two series of data (x and y) for different attributes measured a the same equidistant points. Sliding of these two series with respect to one another an correlation gives cross-correlation coefficients for different lags. If $\rho_{xy}(l)$ represents th cross-correlation function for x slid in the positive direction of lag with respect to y, an $\rho_{xy}(-l)$ for x slid in the negative direction, then $\rho_{xy}(-l)$ is generally not equal to $\rho_{xy}(l)$ On the other hand, we always have $\rho_{xy}(-l) = \rho_{yx}(l)$ and $\rho_{yx}(-l) = \rho_{xy}(l)$.

An example of cross-correlation will be given presently. Anderson and Kirkland (1966 used the method for correlating series of varve-thickness data from different places to on another. Other, related, statistical methods for correlating stratigraphic sections includ cross-association (Merriam, 1971) and teleconnection (Schove, 1971).

The largest possible value of a cross-correlation coefficient is +1 indicating a perfec match. If covariances are used instead of correlation coefficients, we have $\Gamma_{xy}(-l) = \Gamma_{yx}(l)$ and $\Gamma_{yx}(-l) = \Gamma_{xy}(l)$.

Signal-plus-noise models

The basic assumption for models of this type is that a series of observed values x_k i the sum of two series, signal (s_k) and white noise (η_k) with:

$$x_k = s_k + \eta_k \quad [10.35]$$

It is assumed that the three series are weakly stationary, e.g., the signal may satisfy an autoregressive process-model. The noise consists of uncorrelated values. For the three autocovariance functions we may write $\Gamma_{xx}(l)$, $\Gamma_{ss}(l)$ and $\Gamma_{nn}(l)$, respectively. We have

$$\Gamma_{nn}(l) = 0, \text{ if } l \neq 0 \text{ and } \Gamma_{xx}(l) = \Gamma_{ss}(l) + \Gamma_{nn}(l) \quad [10.36]$$

Because the noise is independent of the signal:

$$\Gamma_{sn}(l) = \Gamma_{ns}(l) = 0$$

Also:

$$\Gamma_{xs}(l) = \Gamma_{sx}(l) = \Gamma_{ss}(l) \qquad [10.37]$$

Suppose now that the data x have unity variance or:

$$\Gamma_{xx}(0) = 1, \quad \text{then} \quad \rho_{xx}(l) = \Gamma_{xx}(l)$$

but, in general:

$$\rho_{ss}(l) > \Gamma_{ss}(l) \quad \text{and} \quad \rho_{nn}(0) = 1 > \Gamma_{nn}(0)$$

We may define $\Gamma_{ss}(0) = c$ where c is a constant with $0 \leqslant c \leqslant 1$; then [10.36] gives:

$$\Gamma_{nn}(0) = 1 - c, \quad \text{because} \quad \Gamma_{xx}(0) = 1$$

Suppose that:

$$\Gamma_{ss}(l) = c\mathrm{e}^{-a|l|}$$

viz:, the signal satisfies a Markov process of the first order, then:

$$\left.\begin{aligned} \rho_{xx}(l) &= c\,\mathrm{e}^{-a|l|} \quad \text{if } l \neq 0 \\ \rho_{xx}(0) &= 1 \end{aligned}\right| \qquad [10.38]$$

This seems to be the model underlying the empirical patterns of Fig.56.

Approximate methods of calculating c when the signal is Markovian were mentioned in section 10.5. Walker (1960) has shown that the parameters of a series consisting of signal and noise can be estimated by treating the series as if it were generated by a so-called general mixed autoregressive-moving average process (cf. Bartlett, 1966). Other, more advanced methods for estimating the signal-to-noise ratio were given by Parzen (1967).

The presence of a noise component in space series of geochemical data from orebodies was discussed by Krige (1966) and Matheron (1962). It may be caused by sampling errors arising from the crushing and chemical analysis before assay values are obtained. Alternatively, it may be caused by random variations in concentration of ore minerals in small samples, e.g., at the sampling level, the ore grains may be randomly distributed through the rock. Matheron has referred to the phenomenon as a "nugget" effect. The sampling error was also noted in Fig.53 where its effect was eliminated before fitting trend lines for the De Wijsean model.

Varve-time series from glacial Lake Barlow–Ojibway, Ontario

A varve-time series consisting of 537 thickness measurements for silt and clay components was statistically analyzed by Agterberg and Banerjee (1969). These data (series 4) are used to exemplify several methods reviewed in the previous part of this section. They will also be used for example in the next Chapter (section 11.4).

Each varve consists of a thin silt layer overlain by a thin clay layer. It may be assumed that these varves were deposited with a frequency of one per year. It seems that the silty

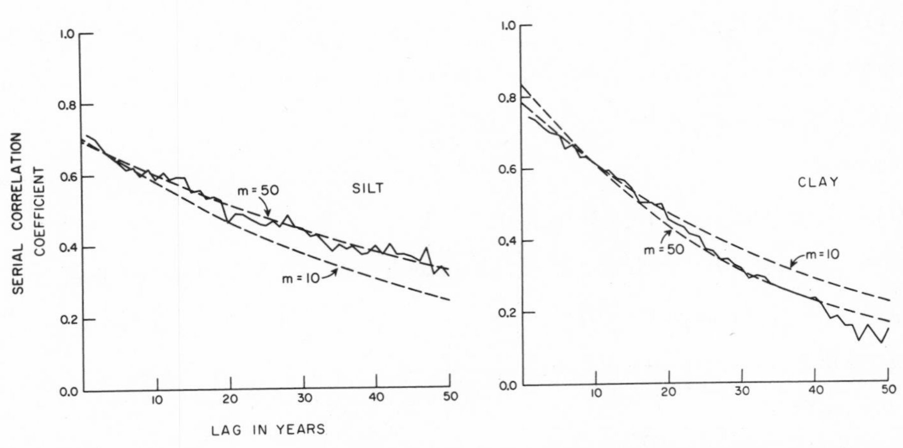

Fig.58. Correlograms of log-thickness data for silt and clay components in 537 varves (series 4), glacial Lake Barlow – Ojibway, Ontario. (From Agterberg and Banerjee, 1969.) Exponential curves fitted for $m = 10$ and 50 lags; fit for $m = 10$ is better close to the origin.

material was transported by turbidity currents during the summer. Series 4, consisting of 537 varves, is exposed near Twin Falls, Ontario. The average thickness of both silt and clay layers decreases from bottom to top with the silt decreasing more rapidly than the clay. Antevs (1925) explained the decrease by showing that varves were deposited continuously in the lake in front of a retreating ice-cap (Wisconsian ice sheet). He estimated the rate of retreat at about 500 m per year due to melting during the summer.

Before statistical analysis, the thickness data were logarithmically transformed. This was desirable because of heteroscedasticity of the variance for original data, in particular those of the silt component. For example, the total sum of squares of 537 original silt thickness data is 1037.2 cm^2. The first 100 varves near the bottom of series 4 account for 92.0% of this total but the last 100 varves near the top for only 0.02%. The ratio of the variances for these two sets of 100 values is 33.6 indicating heteroscedasticity. When logarithmically transformed data are used, this variance ratio is reduced to 1.5.

Correlograms for logarithmically transformed data are shown in Fig.58. We will take the clay correlogram for example. Two exponential curves were fitted to it conforming to the exponential model of the first-order Markov process. One curve was fitted on the first 10 serial correlation coefficients and the second one on the first 50 coefficients. Although the curve for $m = 50$ shows the best overall fit for the data shown in Fig.58, the other curve provides a better fit for the first part. The phenomenon of interest in this correlogram is that extrapolation toward zero-lag gives a value $c = 0.8$ which is obviously smaller than $r_{xx}(0) = 1$ indicating that about 20% of the variations in the series is white noise whereas 80% is a systematic variation. The logarithmic variance for the clay-thickness data (in cm) is 0.64. If low-degree polynomials are fitted by least squares, the percentage explained sum of squares *ESS* amounts to 38.0 (linear), 38.6 (quadratic), and 39.3% (cubic), respectively.

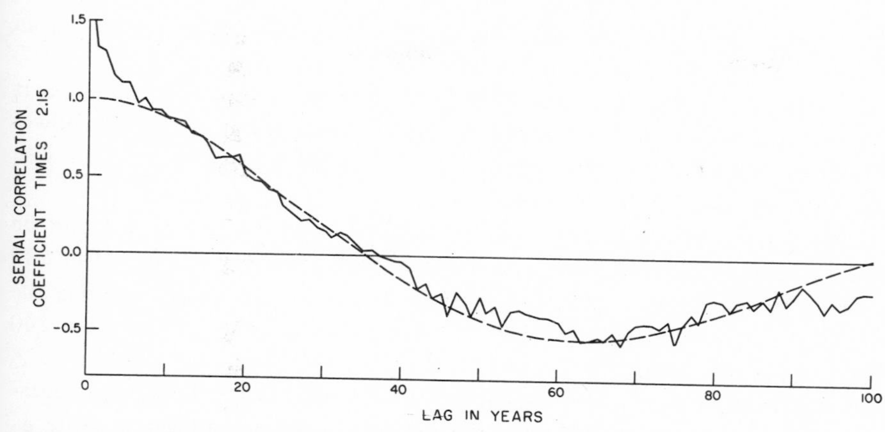

Fig.59. Correlogram for residuals from linear trend, clay log-thickness data, series 4; curve satisfying [10.34] provides good fit for lags between 10 and 90 years.

This indicates that 38% of the variation in the data is accounted for by a linear trend. A correlogram also was calculated for the residuals from the linear trend as shown in Fig.59. There exists a strong noise component in this correlogram which assumes negative values for lags longer than 37 (years). A curve with [10.34] was fitted to this correlogram. It is shown for $\alpha = 0.01$ and $\omega_0 = 0.05$. All autocorrelation coefficients were multiplied by 2.15 to let the theoretical curve begin at a point with $r_{ss}(0) = 1$. The model provides a good fit for lags between 10 and 90. The departure for lags < 10 can be explained by assuming that the noise is not completely "white" but extends over several lags.

If the theoretical curve is accepted for the residuals, we obtain the following interpretation:

(1) The log clay thickness decreases linearly in the course of time. A linear trend for logarithmically transformed data corresponds to an exponential trend for original data. This trend is probably related to a gradual increase in distance between the moving icefront and place of deposition.

(2) The noise component reflects superimposed random variations in clay thickness which are uncorrelated.

(3) The systematic variation that remains, as residuals, after elimination of the "trend" and the "noise" can be explained by the second-order stochastic differential equation of [10.34]. This solution also can be represented in the form:

$$x(t) = (1/\omega_0) \int_{-\infty}^{t} e^{-\alpha(t-s)} \sin \omega_0 (t-s) \cdot \eta(s) \, ds \qquad [10.39]$$

(see Yaglom, 1962). This facilitates the physical interpretation. The coefficient α (= 0.01) reflects the memory capacity of the process. When α is small the process has a good memory for the past. The constant ω_0 controls the period of a sinusoidal variation in thickness. Since $\omega_0 = 0.05$, the period is $T = 2\pi/\omega_0 = 126$ years.

Physically, this may mean the retreat of the icefront was not exactly linear but that there were periods of relatively rapid retreat alternating with periods of a slowing down. During one of the latter periods, the icefront actually may have readvanced as indicated by a study of the glacial till in the area (see section 11.4).

It is noted that, in several places along the series, one or more years may not be represented by varves. The total time span covered by the section is longer than 537 years, possibly by about 10%. These minor discontinuities were neglected during the statistical analysis. The estimated period $T = 126$ years could, in reality, be about 140 years because of missing varves.

The preceding analysis was for clay-thickness data only, which give a better representation of the mechanism of deposition than the silt data which are subject to more irregularities. However, residuals from a linear trend also were obtained for silt-thickness data and cross-correlated with the clay data. The result is shown in Fig.60. The cross-correlation function is not symmetrical about the origin. The first zero-crossing points for positive and negative lag-directions are $56\frac{1}{2}$ and $-25\frac{1}{2}$ years, respectively. Physically, this pattern probably represents the same long-term cyclical variation as that shown in Fig.59. However, the asymmetry suggests that the clay material responded on the average $\frac{1}{2} \times (56\frac{1}{2} - 25\frac{1}{2}) = 15$ years more rapidly to the periodic phenomenon than the silt. Our interpretation of this phenomenon is that the coarser material (silt), during periods of relatively rapid retreat (more meltwater) was initially deposited more closely to the icefront leaving the silt-thickness profile, at some distance from the source, unaffected. Later, the steeper profile of the material dumped near the source acted as a floor to turbidity currents for later years which helped the coarser material to move to more distal areas (cf. section 11.4).

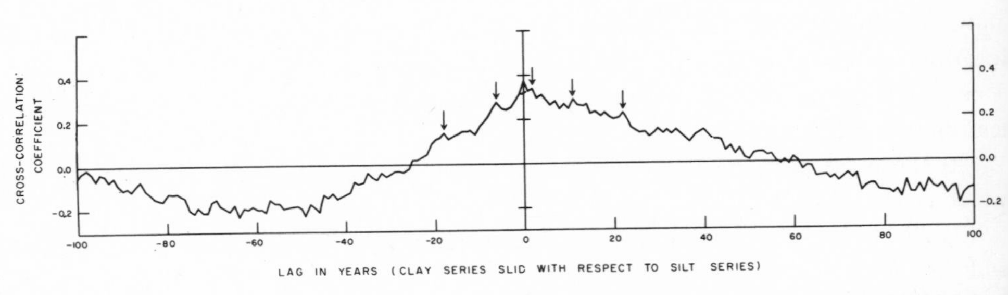

Fig.60. Cross-correlation function for residuals from linear trend, series 4; main pattern according to wave about 15 years out of phase with origin. Weak 10-year periodicity is indicated by arrows; this short-term cycle is better developed in cross-spectrum (cf. Fig.69).

10.7. Transformations of stationary series

Transformations have been discussed by Grenander and Rosenblatt (1957) and Granger and Hatanaka (1964). This summary is restricted to the logarithmic transformation. Suppose that the data in the space series are lognormally distributed. If the distribution is skew, large values occurring at random may strongly affect the autocorrelation function and the power spectrum. This effect can be avoided by operating on the logarithms of the data. However, it should be kept in mind here that if the distribution is skew, values near zero also are likely to occur and these values may assume large negative values when the logarithm is taken. This secondary drawback may be avoided by first adding a constant to all data before the logarithm is taken.

If X_k is lognormally distributed, $Y_k = {}^{e}\log X_k$ is normally distributed. Suppose that $E(X_k) = \alpha$ and $\Gamma_x(l) = \beta^2 \rho_x(l)$ where α and β^2 are mean and variance of the lognormal distribution. Further:

$$E(Y_k) = \mu; \qquad \Gamma_y(l) = \sigma^2 \rho_y(l)$$

Since: $\alpha = \exp\left(\tfrac{1}{2}\sigma^2 + \mu\right)$

and:
$$E(X_k X_{k+l}) = E[\exp(Y_k + Y_{k+l})] = \exp[\sigma^2 + 2\mu + \sigma^2 \rho_y(l)]$$

it follows that:

$$\beta^2 \rho_x(l) = \exp(\sigma^2 + 2\mu)\,[\exp\{\sigma^2 \rho_y(l)\} - 1]$$

and: $\sigma^2 \rho_y(l) = \log[1 + \gamma^2 \rho_x(l)]$ [10.40]

where $\gamma = \beta/\alpha$ is the coefficient of variation. If $\gamma \leqslant 1$, [10.40] can be expanded as the power series:

$$\sigma^2 \rho_y(l) = \sum_{j=1}^{\infty} (-1)^{j+1} \frac{\gamma^{2j}}{j} \{\rho_x(l)\}^j$$

$$= \sum_{j=1}^{\infty} (-1)^{j+1} \frac{\{\beta^2 \rho_x(l)\}^j}{j\alpha^{2j}}$$

Suppose that: $\alpha \gg \beta$, then:

$$\sigma^2 \rho_y(l) \approx \gamma^2 \rho_x(l) \qquad [10.41]$$

Since then, at the same time, $\sigma^2 \approx \gamma^2$ (see section 7.4), we have:

$$\rho_y(l) \approx \rho_x(l)$$

Consequently, the expected autocorrelation functions for the series $\{X_k\}$ and $\{Y_k\}$ are the same.

De Wijs (1951) has shown that the 118 zinc values of Fig.55A are lognormally distributed. The coefficient γ amounts to about 0.51 for these data. From Table XXVII, we may conclude that the sample autocorrelation functions for original data and normalized values are nearly the same. This result is in accordance with [10.41].

10.8. Two-dimensional autocorrelation functions

In several situations, it is possible to obtain measurements in the plane at the corner points of a regular grid. A two-dimensional (2D) autocorrelation function can be computed and represented in map form. For geological applications, see Hempkins (1970) and Horton et al. (1964). A 2D-autocorrelation function has a central point or origin. A profile across it that passes through the origin gives the autocorrelation function of a space series for a sampling line in the plane that is parallel to the line of the profile.

For example, the data for underground drillholes shown in Fig.47 were written as the elements of a (16 × 11) matrix shown in Table XXVIII.

TABLE XXVIII

Matrix of underground drillhole data for the area depicted in Fig. 47. Values represent products of horizontal width of orebody (in feet) and average concentration of copper in percent

29	146	116	76	22	6	(23)	11	18	48	60	94	9	32	137
51	47	91	(60)	12	21	43	32	40	52	158	139	94	77	118
43	60	94	34	19	67	31	22	65	112	126	78	258	80	(86)
91	(70)	60	34	66	29	(41)	32	164	286	160	(126)	(97)	(97)	74
91	73	68	21	75	44	35	73	297	89	140	52	66	49	52
105	43	36	42	40	77	83	272	209	107	177	40	29	42	52
66	32	(47)	58	128	114	354	(194)	284	169	(91)	25	(34)	9	(61)
47	(38)	31	86	76	(159)	(150)	113	43	78	(69)	40	21	64	(51)
41	8	41	125	196	128	118	74	76	33	70	65	29	64	100
19	19	62	260	135	174	135	163	23	18	33	24	36	95	(67)
20	35	59	168	199	(146)	91	(85)	16	13	58	30	(40)	23	(59)

The rectangle in Fig.47 encloses 151 solid circles whose values occur in Table XXVIII. In total, there should be 16 × 11 = 176 elements in the matrix. The 176 – 151 = 25 elements which are missing were inserted by averaging the eight or less surrounding values that are known. An exception was made for the value 68 in the lower right-hand corner which is the average of the surrounding three interpolation values. Suppose that the elements of the $(m \times n)$ data matrix $\mathbf{X}$ are written as $X_{i,j}$ $(i = 0,1,..., m-1; j = 0,1,..., n-1)$. The two-dimensional autocovariance function $C(r, s)$ is defined as:

$$C(r, s) = \frac{1}{(m-r)(n-s)} \sum_{i=0}^{m-r-1} \sum_{j=0}^{n-s-1} (X_{i,j} - \bar{X})(X_{i+r,j+s} - \bar{X})$$

and:

$$C(-r, s) = \frac{1}{(m-r)(n-s)} \sum_{i=r}^{m-1} \sum_{j=0}^{n-s-1} (X_{i,j} - \bar{X})(X_{i-r,j+s} - \bar{X}) \qquad [10.42]$$

where $r = 0, 1, \ldots, M_1$ and $s = 0, 1, \ldots, N_1$.

Eq. [10.42] is a direct extension of [10.2]. For example, since $s = 0$ for the horizontal direction, $C(r, 0)$ is the average of 11 autocovariance functions for the 11 rows of matrix.

If all the elements of $C(r, s)$ are divided by $C(0, 0)$, the result is $R(r, s)$ which is shown as a map in Fig.61. Although individual values $R(r, s)$ are subject to sampling fluctuations, it appears to be possible to contour them. A 0-contour is shown. If a two-dimensional autocorrelation function is isotropic, it would plot as circular contours and the autocorrelation of space series as obtained by line-sampling would be independent of direction. The pattern in Fig.61 is far from being circular.

10.9. Theoretical spatial autocorrelation functions

Space series are obtained by sampling along a line through a three-dimensional medium. There is a fundamental difference between space series and time series. As pointed out by Whittle (1954a), time is a dimension that provides a direction to the causal relation of a process. One may attempt to predict the future from the present and the past of a

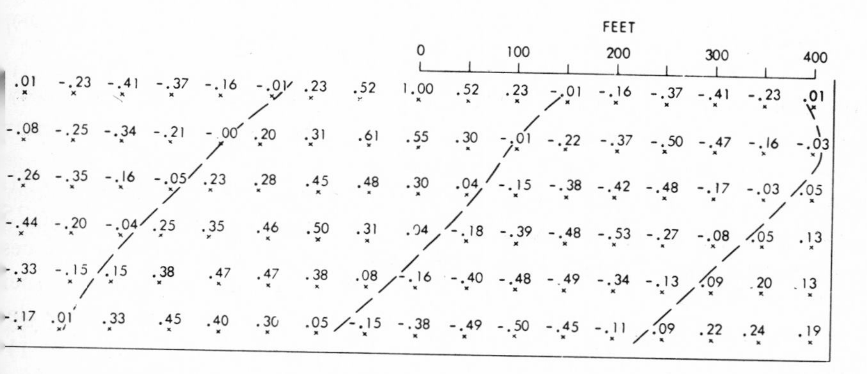

Fig.61. Two-dimensional autocorrelation function for data from Table XXVIII, Whalesback copper deposit, Newfoundland.

time series. In the analysis of space series, we are more interested in interpolation between points. In case of a homogeneous medium, there is no preference for a positive or negative direction along the sampling line.

Suppose that for a space series, we want to determine the value of points situated midway between points with known values. Suppose that the model can be written as:

$$X_k = \beta_1 X_{k-1} + \beta_2 X_{k+1} + \epsilon_k \qquad [10.43]$$

where ϵ_k is a random variable (white noise); X_{k-1} and X_{k+1} are known values; and X_k is the unknown value.

The coefficients β_1 and β_2 can be solved by least squares. For a finite series of n values:

$$\begin{bmatrix} \hat{\beta}_1 \\ \hat{\beta}_2 \end{bmatrix} = \begin{bmatrix} \Sigma X_{k-1}^2 & \Sigma X_{k-1} X_{k+1} \\ \Sigma X_{k-1} X_{k+1} & \Sigma X_{k+1}^2 \end{bmatrix}^{-1} \begin{bmatrix} \Sigma X_k X_{k-1} \\ \Sigma X_k X_{k+1} \end{bmatrix}$$

For the underlying population (infinite series):

$$\begin{bmatrix} \beta_1 \\ \beta_2 \end{bmatrix} = \begin{bmatrix} 1 & \rho_2 \\ \rho_2 & 1 \end{bmatrix}^{-1} \begin{bmatrix} \rho_1 \\ \rho_1 \end{bmatrix} \qquad [10.44]$$

or: $\beta_1 = \beta_2 = \dfrac{\rho_1}{1 + \rho_2}$

Intuitively, this result is satisfactory. If $\beta_1 = \beta_2 = \beta$, there is no preference for one of the two directions along the line. Eq. [10.43] expresses the so-called nearest-neighbor property of space series. A more general expression is:

$$P\{X_k \mid X_j; j \neq k\} = P\{X_k \mid X_{k+l}, X_{k-l}\} \qquad [10.45]$$

We state that the probability of a random variable X assuming value X_k when it is given that in other points along the series it assumes values X_j ($j \neq k$), is equal to a probability which is solely determined by the two neighbors at distances $+l$ and $-l$. Although theoretically l can have any value, one considers the two nearest neighbors if practical estimation is attempted. In fact, a series of distance data which has the nearest-neighbor property, also has the Markov property (cf. [10.30]):

$$P\{X_k \mid X_j; j < k\} = P\{X_k \mid X_{j-l}\} \qquad [10.46]$$

and consequently, an exponential autocorrelation function.

However, [10.45] can be expanded to a multi-dimensional situation whereas this is not possible for [10.46].

Writing $X_k = \beta(X_{k-1} + X_{k+1}) + \eta_k$, it follows that:

$$(1 - 2\beta)\, X_k = \beta(X_{k-1} - 2X_k + X_{k+1}) + \eta_k$$

Suppose that the space series X_k was obtained by sampling a continuous series $X(s)$ at regular interval Δs. Then

$$X(s + \Delta s) - 2X(s) + X(s - \Delta s) - \frac{1 - 2\beta}{\beta} X(s) = \eta_k$$

If both sides are divided by Δs^2 and if Δs is replaced by infinitesimal ds, then:

$$\frac{\mathrm{d}^2 X(s)}{\mathrm{d}s^2} - \alpha^2 X(s) = \eta(s) \qquad \text{where} \qquad \alpha^2 = \frac{1 - 2\beta}{\beta \mathrm{d}s^2} \qquad [10.47]$$

This approach can be generalized to the p-dimensional case where $X(s)$ are values in p-space.

Whittle (1963b) has generalized [10.47] to:

$$(\nabla^2 - \alpha^2)\, X(u) = \eta(u) \qquad [10.48]$$

where u is a vector representing location in p-space with coordinates $u_1, u_2, \ldots, u_p$ and:

$$\nabla^2 = \sum_{i=1}^{p} \frac{\partial^2}{\partial^2 u_i}$$

For $p = 1$, [10.48] reduces to [10.47]. We are interested in the cases $p = 2$ and $p = 3$ where the data have positions in the plane and in 3D-space, respectively.

The autocovariance function $\Gamma(s)$ for the general process satisfies:

$$\Gamma_p(s) = \frac{1}{(2\pi)^p} \int_{-\infty}^{\infty} e^{i\omega s} \cdot \frac{1}{(\omega^2 + \alpha^2)^p}\, \mathrm{d}\omega \qquad [10.49]$$

which is the Fourier transform of the p-dimensional spectral density function:

$$f_p(\omega) = \frac{1}{(\omega^2 + \alpha^2)^p} \qquad [10.50]$$

Eq. [10.49] is a Hankel-Nicholson type integral with general solution:

$$\Gamma_p(s) = \tfrac{1}{2}(s/\alpha)^{2-p/2}\, K_{2-p/2}(\alpha s) \qquad [10.51]$$

The correponding autocorrelation function satisfies:

$$\rho_p(s) = \Gamma_p(s)/\Gamma_p(0)$$

If $p = 2$, we have:

$$\rho_2(s) = \alpha s K_1(\alpha s) \qquad [10.52]$$

where $K_1(\alpha s)$ is a modified Bessel function of the second kind. This solution was discussed and applied by Whittle (1963b) and Bartlett (1966). If $p = 3$, we obtain simply:

$$\rho_3(s) = e^{-\alpha|s|} \qquad [10.53]$$

The relationship between [10.48] and [10.49] will not be discussed in more detail. Reference is made to Bartlett (1966, chapter 6), Yaglom (1962) and to the work of Whittle (1954a, 1962, 1963b). Eq. [10.51] and [10.52] were discussed by Whittle and Bartlett. Eq. [10.53] was discussed by Agterberg (1970a). Solutions of [10.49] are given by Abramowitz and Stegun (1965).

The implications of [10.53] are that, if a variable assumes values in three-dimensional space and if it is subject to the property that a value is determined by its neighbors on all sides, the autocorrelation function is exponental for all directions.

If the spatial dependence is the same for all directions, any space series would have the autocorrelation function:

$$\rho_s = e^{-\alpha|s|}$$

Suppose now that the spatial autocorrelation is dependent upon direction. The following model can then be defined:

$$(\nabla^2 - \alpha^2)\, X(u, v, w) = \eta(u, v, w) \qquad [10.54]$$

where u, v, and w denote location in 3D-space and:

$$\nabla^2 = a\frac{\partial^2}{\partial u^2} + b\frac{\partial^2}{\partial v^2} + c\frac{\partial^2}{\partial w^2} + 2d\frac{\partial^2}{\partial u\,\partial v} + 2e\frac{\partial^2}{\partial u\,\partial w} + 2f\frac{\partial^2}{\partial v\,\partial w} \qquad [10.55]$$

∇^2 is a basic operator in the theory of partial differential equations (Courant and Hilbert, 1962, vol. 2). By a rotation of the coordinate system (u, v, w) similar to that discussed in Chapter 4, the last three terms of [10.55] can be eliminated. Secondly, by changing the scale along the three axes, the new coefficients can be set equal to one and [10.54] reduces to [10.48]. This can be done for any general elliptic partial differential equation. The solution of [10.54], in this more general case, is a hypersurface whose contours in 3D-space are ellipsoids. Line-sampling always gives an exponential correlogram but the coefficient α will depend on the direction of the sampling line. The two-dimensional autocorrelation, in general, will have elliptic contours.

By adopting a modified Markovian property for spatial variability ([10.45] instead of [10.46]) and changing from a directed process to an undirected process, we obtain a result that makes space series comparable with time series with a first-order Markov property.

We have not found a stochastic process leading to spatial variability of this type. Instead of this, we have shown that it is physically possible to have a solid in which the space series for any line of intersection has an exponential autocorrelation function. It was known that the first-order Markov processes could be used for space series in practice. However, objections have been raised against the model in that it would not be physically possible for a medium.

In the exponential model, the mutual association between any two points in the medium, decreases gradually to zero with increasing distance. On the other hand, if the medium satisfies a De Wijsean model, the difference between two values continues to increase with increasing distance. The use of these two models for spatial variability (exponential model and De Wijsean model) will be further explored in the next two sections.

10.10. Variance of average concentrations for volumes of rock

Suppose that a rock volume is divided into n pieces of equal size and that a specific attribute can be measured for each piece. If the n values are statistically independent, we have:

$$s^2(\bar{x}_n) = s^2(x)/n \quad \text{or} \quad F_n = \frac{n \cdot s^2(\bar{x}_n)}{s^2(x)} = 1 \qquad [10.56]$$

When the volume of individual pieces is set equal to 1, we have for the block with average $\bar{x}_n$, $V = n$. The model results in the equation: variance × volume = constant, as proposed by Hazen and Berkenkotter (1962). It can be used for blocks of rock for which assays are uncorrelated (cf. Fig.54). Also, it can be used for the sampling of broken ore where possible gradational spatial changes of rock in situ have been destroyed. The model does not apply when the spatial variability follows the De Wijsean model. However, as we will see, it does apply if spatial change satisfies the exponential model provided that the volumes are large.

In section 10.2, it was shown that [10.56] is not satisfied for averages of two adjacent values of a space series. Suppose that $\bar{x}_2$ represents an average of two values; then:

$$s^2(\bar{x}_2) = s^2\left(\frac{(x_k + x_{k+1})}{2}\right) = \tfrac{1}{4}[s^2(x_k) + s(x_k, x_{k+1}) + s(x_{k+1}, x_k) + s^2(x_{k+1})]$$

where $s(x_k, x_{k+1})$ and $s(x_{k+1}, x_k)$ are covariances. If the series is stationary $s^2(\bar{x}_2) = \tfrac{1}{2}s^2(x)[1 + r_1]$ or $F_2 = 1 + r_1$ where r_1 is the first autocorrelation coefficient.

This result may be extended to the mean of n adjacent data, so that:

$$s^2(\bar{x}_n) = s^2(x)\,[1/n + (2/n^2) \sum_{j=1}^{n-1} (n-j)\, r_j]$$

or:

$$F_n = 1 + (2/n) \sum_{j=1}^{n-1} (n-j)\, r_j \qquad [10.57]$$

When we set $r_j = r_1^j$, it can be shown that $F_n \approx n/n'$ where n' is the equivalent number of statistically independent values as used in section 9.8 (Table XXV). Assume that, in general, the situation can be described by using a signal-plus-noise model as discussed in section 10.5. Then:

$$F_n = 1 + (2/n) \sum_{j=1}^{n-1} (n-j)\, r_s(j) \qquad [10.58]$$

where the $r_s(j)$ are autocorrelation coefficients for the signal.

We may say that [10.57] is based on the more general expression:

$$F_n = 1 + (1/n) \sum_{j=1}^{n-1} (n-j)\, r_s(j) + (1/n) \sum_{j=1}^{n-1} (n-j)\, r_s(-j)$$

which reduces to [10.58] because $r_s(j) = r_s(-j)$. If $j = 0$, $r_s(0) = c$ (cf. [10.45]).

If the summation is to include the term for $j = 0$, it becomes larger and the term c must be subtracted from the new sum, or:

$$F_n = 1 - c + (1/n) \sum_{j=-n+1}^{n-1} (n - |j|)\, r_s(j) \qquad [10.59]$$

$$\text{and: } F_l \approx 1 - c + (1/l) \int_{-l}^{l} (l - |\lambda|)\, r_s(\lambda)\, \mathrm{d}\lambda \qquad [10.60]$$

where l denotes length of the block sampled by means of the space series and λ is the distance between two points on the line segment.

Setting $r_s(\lambda) = c\mathrm{e}^{-a|\lambda|}$, we obtain, after some manipulation:

$$F_l = 1 - c + (2c/a)\, \{1 + (\mathrm{e}^{-al} - 1)/al\} \qquad [10.61]$$

with $\lim_{l \to \infty} F_l = 1 - c + 2c/a$. This indicates that for large values of l, we have constant F_l or variance × volume = constant.

The function F_l was determined experimentally by averaging an increasingly greater number of adjacent zinc values for Pulacayo in Fig.62. From the correlogram of Fig.56A calculated from logarithmically transformed data, we estimated $a = 0.1962$ and $c = 0.5078$. These values were used for [10.61] and the theoretical curve is also shown. The correspondence between experimental and theoretical values is good in this situation. For larger values of n or l, the curve tends to become horizontal; in the limit, $F_n = 5.68$

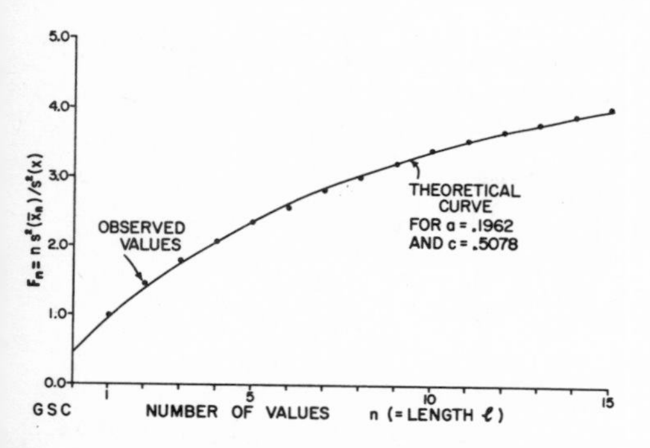

Fig.62. Variance–volume relationship for logZn in blocks elongated in direction of drift. (From Agterberg, 1967b.) Observed values based on variance of average of n adjacent values; theoretical curve satisfies [10.61].

10.11. Matheron's geostatistical theory

Suppose that $f(x)$ represents concentration of an element at a specific point x, which may be a vector (x_1, x_2, x_3) in 3D-space.

The average m for a volume v satisfies:

$$m = (1/v) \int_v f(x)\,\mathrm{d}x \qquad [10.62]$$

Suppose that M is the assemblage average for the entire block of study with volume V; then:

$$M = (1/V) \int_V f(x)\,\mathrm{d}x = (1/V)\,\Sigma\, m_i v_i \qquad [10.63]$$

where the m_i are for smaller blocks with volumes v_i.

On the basis of $f(x)$ and V, we can define two intrinsic functions:

$$\text{Covariance: } g(y) = (1/V) \int_V [f(x) - M]\,[f(x+y) - M]\,\mathrm{d}x \qquad [10.64]$$

$$\text{Semivariogram: } \gamma(y) = (1/2V) \int_V [f(x+y) - f(x)]^2\,\mathrm{d}x \qquad [10.65]$$

The variance is $g(0)$; we derive immediately:

$$\gamma(y) = g(0) - g(y) \qquad [10.66]$$

An important problem is to determine the covariance $\sigma(m_1,m_2)$ of averages m_1 and m_2 for two volumes v_1 and v_2.

$$\sigma(m_1, m_2) = (1/V)\int_V (m_1 - M)(m_2 - M)\,\mathrm{d}x$$

$$= \frac{1}{Vv_1v_2} \int_V\int_{v_1} [f(x + x_1) - M]\,\mathrm{d}x_1 \int_{v_2} [f(x + x_2) - M]\,\mathrm{d}x_2\,\mathrm{d}x$$

$$= \frac{1}{Vv_1v_2} \int_{v_1} \mathrm{d}x_1 \int_{v_2} \mathrm{d}x_2 \int_V [f(x + x_1) - M]\,[f(x + x_2) - M]\,\mathrm{d}x \qquad [10.67]$$

From [10.64], we obtain:

$$\sigma(m_1, m_2) = \frac{1}{v_1v_2} \int_{v_1} \mathrm{d}x_1 \int_{v_2} g(x_2 - x_1)\,\mathrm{d}x_2$$

or, when y represents the distance between all possible pairs of points in the two volumes:

$$\sigma(m_1, m_2) = \frac{1}{v_1v_2} \int_{v_1} \mathrm{d}x_1 \int_{v_2} g(y)\,\mathrm{d}x_2 \qquad [10.68]$$

Thus $\sigma(m_1, m_2)$ is an average of the covariance function $g(y)$.

When v_1 and v_2 coincide, it reduces to:

$$\sigma^2(m) = \frac{1}{v^2} \int_v \mathrm{d}x_1 \int_v g(y)\,\mathrm{d}x_2 \qquad [10.69]$$

representing the variance of average concentration m for a block with volume v.

Suppose that we calculate the variance of M for the volume V. Because M was taken as the average in [10.64] and [10.67], it follows that $\sigma^2(M) = 0$. Integration over V of both sides of [10.66] now gives:

$$g(0) = \frac{1}{V^2} \int_V \mathrm{d}x_1 \int_V \gamma(y)\,\mathrm{d}x_2 \qquad [10.70]$$

At this point we define an auxiliary function $F(m)$ with:

$$F(m) = \frac{1}{v^2} \int_v \mathrm{d}x_1 \int_v \gamma(y)\,\mathrm{d}x_2 \qquad [10.71]$$

From [10.66] and [10.70] it then follows that:

$$\sigma^2(m) = F(M) - F(m) \qquad [10.72]$$

Likewise, $G(m_1, m_2)$ is defined as:

$$G(m_1,m_2) = \frac{1}{v_1 v_2} \int_{v_1} \mathrm{d}x_1 \int_{v_2} \gamma(y)\,\mathrm{d}x_2 \qquad [10.73]$$

Then, from [10.68]:

$$\sigma(m_1,m_2) = F(M) - G(m_1,m_2) \qquad [10.74]$$

Suppose that we want to use m_1 of v_1 to estimate m_2 of v_2, which may surround v_1. We are interested in the so-called "extension variance" σ_E^2 with:

$$\sigma_E^2 = \sigma^2(m_1 - m_2)$$

or:

$$\sigma_E^2 = \sigma^2(m_1) - 2\sigma(m_1,m_2) + \sigma^2(m_2)$$

Substitution of [10.72] and [10.74] gives:

$$\sigma_E^2 = 2G(m_1,m_2) - F(m_1) - F(m_2) \qquad [10.75]$$

It should be kept in mind that $g(y)$ of [10.64] is not a covariance function in the usual sense. If $f(x)$ is stationary with expectation $E[f(x)] = \mu$, M can be replaced by μ. However, the present theory can be applied when $\gamma(y)$ exists, even when μ can not be defined such as for the model of De Wijs. It is also noted that m (of v) is considered in relation to M (of V). Previously (see [10.11]), we have defined the variance $\sigma^2(v, V)$. Thus, [10.72] can be written as:

$$\sigma^2(v, V) = F(V) - F(v)$$

The model of De Wijs

Matheron (1962), for most of this earlier models, assumed:

$$\gamma(y) = 3\alpha \log y \qquad [10.76]$$

where α is the absolute dispersion. Suppose that the samples are "linear", viz. rod-shaped with length equal to l. The field of study also is linear with length L.

Because our two volumes have been reduced to a point and a line segment:

$$F(l) = \frac{2}{l^2} \int_0^l (l - x)\,\gamma(x)\,\mathrm{d}x$$

By substituting [10.76], it follows that:

$$F(l) = 3\alpha\,[\log l - 3/2] \qquad [10.77]$$

In section 10.3, using a somewhat different notation, we derived [10.12] which now follows from [10.77] because:

$$\sigma^2(l, L) = F(L) - F(l) = 3\alpha \log \frac{L}{l} \qquad [10.78]$$

Variogram of Pulacayo zinc deposit

Although [10.76] is simple and all variability for the model of De Wijs is determined by a single constant α, the approach rapidly leads to more complicated integrations. The following example is geometrically simpler than most practical situations. Matheron (1962) and his colleagues have prepared tables and nomograms that cover the more common geometrical configurations of mine sampling.

In the Pulacayo deposit, the channel samples are $l = 0.50$ m long and a distance $h = 2$ m apart (see Fig.63). To every sample, a zone of influence may be assigned that extends over a distance of $h/2$ on both sides.

We will limit the problem to a calculation of the variogram (cf. [10.14]) for these samples which satisfies:

$$\sigma_h^2 = 2[\sigma^2(m) - \sigma(m_1, m_2)]$$

where $\sigma^2(m)$ denotes the variance of the assay values and $\sigma(m_1, m_2)$ their covariance for sampling interval h. This expression is equivalent to [10.74] with $\sigma^2(m_1) = \sigma^2(m_2) = \sigma^2(m)$. Employing [10.75] and setting $G(m_1, m_2) = F(h)$; $F(m_1) = F(m_2) = F(0)$, we obtain:

$$\sigma_h^2 = 2[F(h) - F(0)]$$

$F(0) = 3\alpha\,[\log l - \frac{3}{2}]$ would arise when $d = 0$ and is given by [10.77]. Let us define a rectangle $AA'BB'$ with sides $AA' = BB' = l$ and $AB = A'B' = h$. Further, let $m = \sqrt{l^2 + h^2}$ and $\theta = \arctan(l/h) = \arcsin(l/m)$. The problem now consists of calculating the average logarithm of the distance between an arbitrary point on AA' and an arbitrary point on BB'.

First, we can estimate the average log of r connecting B to a point $x \epsilon [0, l]$ on AA'. We assume that x is a random variable with a uniform frequency distribution $f(x) = 1/l$. Since $x^2 = r^2 - h^2$, $\mathrm{d}x/\mathrm{d}r = r(r^2 - h^2)^{-\frac{1}{2}}$. According to [7.3] in section 7.2:

$$f(r) = \frac{r}{l} \frac{1}{\sqrt{r^2 - h^2}}$$

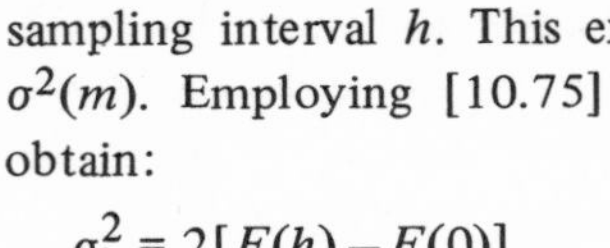

Fig.63. Sampling scheme, Pulacayo Mine; length of the massive sulphide part of channel sample $l = 0.50$ m; sampling interval $h = 2$ m. Complete channel samples were 1.30 m long, also cutting across wall-rock rich in disseminated sphalerite. (From Matheron, 1962.)

We seek the expected value:

$$E(\log r) = \frac{1}{l} \int_{l}^{m} \frac{r \ln r}{\sqrt{r^2 - h^2}} \, \mathrm{d}r$$

After some manipulation, we derive:

$$E(\log r) = \log l - \log(l/m) - 1 + \frac{h \arctan(l/h)}{l}$$

or: $E(\log r) = \log l - \log \sin \theta - 1 + \dfrac{\theta}{\tan \theta}$

We now seek:

$$F(h) = 3\alpha E(\log y)$$

where y is the distance between two points, one on AA' and the other one on BB'. $E(\log y)$ is obtained by integrating $E(\log r)$ over BB', giving:

$$F(h) = 3\alpha \left[\log l - \log \sin \theta - 3/2 + \frac{2\theta}{\tan \theta} + \frac{\log \cos \theta}{\tan^2 \theta}\right]$$

Hence:

$$\sigma_h^2 = 6\alpha \left[-\log \sin \theta + \frac{2\theta}{\tan \theta} + \frac{\log \cos \theta}{\tan^2 \theta}\right] \qquad [10.79]$$

In the Pulacayo zinc deposit, the sampling interval is 2 m. The experimental variogram σ_h^2 can be calculated by using [10.14]. We can evaluate $\sigma_h^2/6\alpha$ by [10.79] using θ = arctan (l/h) with l = 0.5 m and h = 2 m. Multiples of 2 meter can also be employed for this comparison. The results are shown in Table XXIX.

TABLE XXIX

Variogram of Pulacayo zinc deposit. (From Matheron, 1962, p. 180).

h *(2-m units)*	σ_h^2 [10.14]	$\sigma_h^2/6\alpha$ [10.79]	6α
1	0.303	2.89	0.106
2	0.420	3.58	0.117
3	0.436	3.98	0.110
4	0.465	4.27	0.109
5	0.408	4.50	0.091
6	0.412	4.68	0.088
7	0.464	4.83	0.096
8	0.452	4.96	0.091
9	0.472	5.08	0.093
10	0.545	5.18	0.105

The ten estimates of 6α have an average equal to 0.1006. Hence, $3\alpha \approx 0.050$, which is close to the estimate obtained previously in section 10.3 by a simpler method.

It is interesting to compute the extension variance of the assay values for the 2-m intervals. This can be accomplished by [10.75] but we must know the expected value of logarithm of distance between any two points within the rectangle $AA'BB'$. The solution can be written as an infinite series:

$$\sigma_E^2/3\alpha = -0.3863 + \log\frac{h}{l} + 1.0472\frac{l}{h} + \ldots$$

This amounts to 1.26 when $l/h = 0.25$. Since $3\alpha = 0.050$, $\sigma_E^2 = 0.063$.

Matheron has shown that the extension variance for a sequence of n equal domains of this type is:

$$\sigma_E^2(n) = \sigma_E^2/n$$

Hence, the logarithmic variance for the mean of the 118 zinc values would amount to 0.0005 which is practically zero. In turn, we can compute the extension variance of the average for 118 data in the 240 m of drift at the 446-level. For a 240 m wide block extending 340 m upward and downward from this level, the extension variance would amount to 0.049 which is equal to the result obtained by our signal-plus-noise model as applied to logarithmically transformed data at the end of section 10.2.

10.12. Kriging

Linear prediction in time-series analysis can be done as follows. Suppose that the series $\{X_t\}$ with $t = \ldots, -2, -1, 0, 1, \ldots, s, \ldots$ is weakly stationary and has mean equal to zero. A value $X_s (s > 0)$ in the future can be predicted linearly by least squares from the past $(s \leq 0)$. Let $\hat{X}_{s,n}$ be the predictor for X_s based on the history $X_{-n}, \ldots, X_{-1}, X_0$. Then:

$$\hat{X}_{s,n} = \sum_{i=-n}^{0} \hat{A}_{i,n} X_i \qquad [10.80]$$

where $\hat{A}_{i,n}$ are coefficients estimated by minimizing:

$$E(X_s - \hat{X}_{s,n})^2$$

Because $\hat{X}_{s,n}$ is equivalent to a calculated value $\hat{Y}_k$, the normal equations (cf. [8.43] may be written as $\mathbf{C}_{xx}\hat{\mathbf{A}} = \mathbf{C}_{xy}$ or $\mathbf{R}_{xx}\hat{\mathbf{A}} = \mathbf{R}_{xy}$, with solution $\hat{\mathbf{A}} = \mathbf{C}_{xx}^{-1}\mathbf{C}_{xy}$ or $\hat{\mathbf{A}} = \mathbf{R}_{xx}^{-1}\mathbf{R}_{xy}$.

The meaning of these matrix equations is illustrated by writing $\hat{\mathbf{A}} = \mathbf{R}_{xx}^{-1}\mathbf{R}_{xy}$ in full; or

$$\begin{bmatrix} \hat{A}_{0,n} \\ \hat{A}_{1,n} \\ \cdot \\ \cdot \\ \cdot \\ \hat{A}_{n,n} \end{bmatrix} = \begin{bmatrix} 1 & \rho_1 & \cdots & \rho_n \\ \rho_1 & 1 & \cdots & \rho_{n-1} \\ \cdot & \cdot & \cdots & \cdot \\ \cdot & \cdot & \cdots & \cdot \\ \cdot & \cdot & \cdots & \cdot \\ \rho_n & \rho_{n-1} & \cdots & 1 \end{bmatrix}^{-1} \begin{bmatrix} \rho_s \\ \rho_{1+s} \\ \cdot \\ \cdot \\ \cdot \\ \rho_{n+s} \end{bmatrix} \quad [10.81]$$

where ρ_i, $i = 1, 2, \ldots, n + s$, are the autocorrelation coefficients with:

$\rho_i = \Gamma_i/\Gamma_0 = E(X_k X_{k+i})/\sigma^2(X)$

A typical geologic prediction problem is illustrated in Fig.64. The control points P_i, $i = 1, 2, \ldots, n$, are irregularly distributed in the map area. Values X_i for a given attribute are known for these points. P_0 is a point with arbitrary coordinates. The problem is to predict a value $\hat{X}_0$ for point P_0 from the known values in the vicinity.

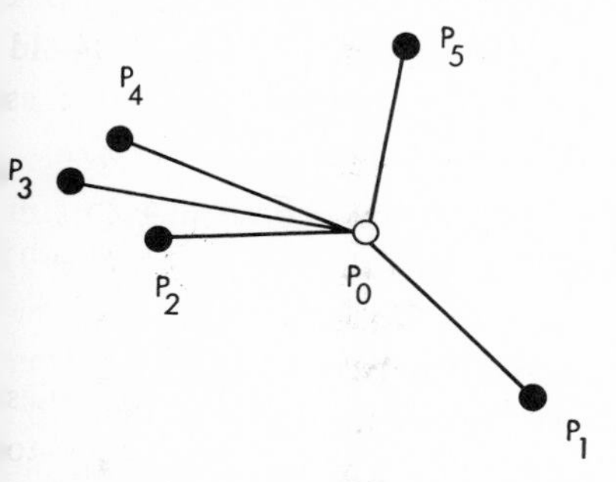

Fig.64. Typical kriging problem; values are known at five points. Problem is to estimate value at point P_0 from the known values at P_1–P_5. (From Agterberg, 1971.)

The method of linear prediction in time series can be adapted to the situation of Fig.64 by defining a two-dimensional autocorrelation function ρ_{ij} for the linear relationship between all possible pairs of points P_i and P_j. The value X_0 at point P_0 is estimated by:

$$\hat{X}_0 = \sum_{i=1}^{n} \hat{A}_{0,i} X_i \quad [10.82]$$

with solution:

$$\begin{bmatrix} \hat{A}_{0,1} \\ \hat{A}_{0,2} \\ \cdot \\ \cdot \\ \cdot \\ \hat{A}_{0,n} \end{bmatrix} = \begin{bmatrix} 1 & \rho_{12} & \rho_{13} & \cdots & \rho_{1n} \\ \rho_{21} & 1 & \rho_{23} & \cdots & \rho_{2n} \\ \cdot & \cdot & \cdot & \cdots & \cdot \\ \cdot & \cdot & \cdot & \cdots & \cdot \\ \cdot & \cdot & \cdot & \cdots & \cdot \\ \rho_{n1} & \rho_{n2} & \rho_{n3} & \cdots & 1 \end{bmatrix}^{-1} \begin{bmatrix} \rho_{01} \\ \rho_{02} \\ \cdot \\ \cdot \\ \cdot \\ \rho_{0n} \end{bmatrix} \quad [10.83]$$

The method only can be applied when ρ_{ij} is known. A practical procedure is as follows:

(1) Correct all values for the regional mean $\bar{X}$ or remove a deterministic trend $X(u, v)$ where u and v are geographical coordinates.

(2) For pairs of values X_i and X_j with $i = 0, 1, \ldots, n, j = 0, 1, \ldots, n; i \neq j$, determine $(u_i - u_j)$ and $(v_i - v_j)$ and the distance s_{ij} between points with:

$$s_{ij} = \sqrt{(u_i - u_j)^2 + (v_i - v_j)^2} \quad [10.84]$$

If ρ_{ij} is isotropic (circular contours), it depends on the value of s_{ij} only.

(3) From the known autocorrelation function, determine all relevant values of ρ_{ij}

(4) Solve the coefficients $\hat{A}_{0,i}$ by using [10.83] and determine $\hat{X}_0$ in terms of the X ([10.82])

(5) Add the regional mean $\bar{X}$ or $X(u, v)$ for the trend.

This method closely resembles that used by Krige (1966) discussed in section 10.1. It is equivalent to Matheron's method of punctual kriging except that Matheron (1969) imposes a constraint by letting the coefficients $\hat{A}_{0,i}$ add to one. Kriging can then be applied in situations that the regional mean can not be estimated by the arithmetic mean $\bar{X}$. Unless there are few data, this seems to make little difference in practice. Matheron (1971a) has emphasized that $\bar{X}$ is generally not the minimum variance estimate of the mean of data which are autocorrelated. It is noted that the variogram with [10.14], contrary to the various expressions in use for the autocorrelation coefficient, is estimated without use of $\bar{X}$. For this reason, Jowett (1955) proposed variogram analysis as an alternative method of approach.

Top of the Arbuckle group (Cambrian – Ordovician) in Kansas, U.S.A.

Good (1964) performed trend-surface analysis on data supplied by D.F. Merriam on the tops of this group from 200 wells irregularly distributed throughout the state of Kansas. Agterberg (1970a) used the same data to compare predictions based on the methods of kriging and trend-surface analysis with one another. For this experiment, the 200 data were divided randomly into three groups. Two sets of 75 data were used for control, and the remaining 50 data were set aside for testing predictions based on the two control samples.

TABLE XXX

Values of check sample predicted by equations based on control samples 1 and 2 (unit for sum of squared differences is 10^8)

		Mean	*Linear*	*Quadratic*	*Cubic*	*Quartic*
Sum of squares	1	0.46	0.32	0.12	0.18	0.11
	2	0.40	0.27	0.13	0.10	0.11

For example, polynomial trend surfaces were computed on the basis of the control samples and values in the trend surface were calculated at the 50 check points. The sum of squares for differences between predicted values and known values at the check points was calculated (see Table XXX). For sample 1, the best result was obtained for the quartic but it is almost the same as that for the quadratic whereas the cubic shows greater deviations. For sample 2, the cubic solution is best but close to that of the quartic and the quadratic.

In order to apply kriging, we must know the two-dimensional autocorrelation function. A crude estimate was obtained as follows: We assumed isotropy, so that ρ_{ij} depends on s_{ij} only, with:

$$\rho(s) = a + bs + cs^2 \qquad [10.85]$$

Thus the autocorrelation function is approximated by a quadratic polynomial. It would have circular contours which is somewhat unrealistic since anisotropy could be demonstrated by other methods (Agterberg, 1970a).

Each value of a control sample was surrounded by a circle with radius equal to 40 miles. On average, this unit circle contained 4.0 other control points of sample 1 and 4.8 other points of sample 2.

For each of the 75 values per control sample X_0, the distance s_i to other points X_i within the unit circle was calculated and $Y = X_0$ was regressed on three auxiliary variables: $Z_1 = X_i$, $Z_2 = s_i X_i$ and $Z_3 = s_i^2 X_i$.

The solution was of the type:

$$\hat{Y} = aZ_1 + bZ_2 + cZ_3$$

providing estimates for a, b and c in [10.85].

The underlying reasoning is that for any pair of values X_0 and X_i, we have:

$$X_0 = \rho(s_i)\, X_i + \epsilon_i$$

By accepting these estimates for [10.85], kriging can be applied. The complete method was carried out for deviations from the mean but also for residuals from the four trend surfaces of Table XXX. The results are shown in Table XXXI.

The first estimates in Table XXXI are based on deviations from the mean. That for sample 1 is as good as the best estimates of Table XXX. However, we note that the result

TABLE XXXI
Values of check sample predicted by kriging on deviations from mean and residuals

		Mean	*Linear*	*Quadratic*	*Cubic*	*Quartic*
Sum of	1	0.11	0.21	0.09	0.14	0.09
squares	2	0.19	0.16	0.08	0.07	0.09

is poorer for sample 1 if residuals from the linear trend-surface are used.

On the whole, best results are obtained by first developing a quadratic trend surfac and then kriging on the residuals. We note that the quartic in Table XXXI is almost a good as the quadratic but the cubic is less satisfactory for sample 1.

This experiment indicates that, if trends do exist, prediction may be done by poly nomial interpolation or by kriging. However, it may be dangerous to use polynomials of higher degree because of their increased flexibility. Likewise, kriging can be dangerous i situations where the condition of stationarity is not satisfied. Best results are obtained b using both deterministic and stochastic components for prediction. This subject will b explored further in the next section. It is emphasized, however, that kriging can be don by various methods. It is unlikely that we have chosen the best possible method for th above example.

10.13. Least-squares model with deterministic and stochastic components

We have seen that the problem of prediction can be solved by: (1) trend analysis or th fitting of empirical deterministic functions to data, and (2) kriging or linear predictio assuming that the spatial variability is weakly stationary.

The basic assumptions underlying these two methods are different. Nevertheless, th two methods may give similar results in the same situation. In the examples of Chapter 9 trend analysis readily gave a good result. The result of Fig.55B can be regarded as a forn of kriging and this method is to be preferred here. The two methods can be combine into a single, more general model, with both deterministic and stochastic components. I is discussed in several textbooks notably by Whittle (1963a) and publications by Krige e al. (1969), Matheron (1971a), and Watson (1971). The use of this model in the eartl sciences has been limited. In applications of universal kriging (Huijbregts and Matheron 1971), space series are analyzed and divided into a "drift" and a stochastic component Polynomial functions are used for the "drift" which is not a polynomial trend for a entire space series or a large part of it. Instead of this, Matheron (1971a) uses low-degre polynomials for relatively short segments to cover the nonstationary variations. The resul is that the variation pattern is broken into two components, one for large-scale variation and one for more local random fluctuations. This is useful for the processing of spac series of many types, for example in geophysics, where anomalies must be separated fron regional trends.

Suppose that, for a single variable, we wish to predict m values written as an $(m \times 1)$ column vector $\mathbf{Y}$ from n known values with $(n \times 1)$ vector $\mathbf{X}$ by $\hat{\mathbf{Y}} = \mathbf{AX}$ where $\mathbf{A}$ is an $(m \times n)$ matrix of coefficients. Suppose further that $\mathbf{X}$ and $\mathbf{Y}$ are both subject to deterministic trends or that:

$$\begin{aligned} E(\mathbf{X}) &= \mathbf{G}\boldsymbol{\beta} \\ E(\mathbf{Y}) &= \mathbf{H}\boldsymbol{\beta} \end{aligned} \qquad [10.86]$$

where $\boldsymbol{\beta}$ is a $(p \times 1)$ vector of regression coefficients. The columns of the $(n \times p)$ matrix $\mathbf{G}$ and $(m \times p)$ matrix $\mathbf{H}$ consist of known values for the same deterministic function in space.

The model can be illustrated by the example of areally distributed data, e.g., those for the Whalesback 425-ft. level (section 9.8), with:

$$X(u, v) = \Sigma\, \beta_j G_j(u, v) + S(u, v)$$

where $E(\mathbf{X}) = \Sigma\beta_j G_j\,(u, v)$ represents a polynomial for a trend surface. $S\,(u, v)$ is a stationary component (with zero expectation) in the (U, V)-plane. For Whalesback, we know that the residuals are autocorrelated along the boreholes. Their variation would be according to $S\,(u, v)$. In Chapter 9, we have assumed that $S\,(u, v)$ is white noise. We now use the notation $Y\,(u, v)$ for values that do not coincide with the $X\,(u, v)$ although they share the same polynomial trend $G\,(u, v)$. In terms of the problem of Fig.64, P_0 is the location of one unknown value of $Y\,(u, v)$ and $P_1 - P_5$ those of the known values $X\,(u, v)$. In that case, $\mathbf{Y}$ has only one element and $\mathbf{H}$ is equal to the value of $G\,(u, v)$ at point P_0.

We may attempt to find the unknown coefficients by minimizing the $(m \times m)$ variance–covariance matrix of the residuals:

$$E(\hat{\mathbf{Y}} - \mathbf{Y})(\hat{\mathbf{Y}} - \mathbf{Y})' = E(\mathbf{AX} - \mathbf{Y})(\mathbf{AX} - \mathbf{Y})' \qquad [10.87]$$

This approach is more general than that used in section 10.12 because now we are also optimizing the covariances. However, the following results remain valid when the elements of $\mathbf{Y}$ are predicted separately. Hence, if the variances of the residuals are minimized, their covariances are minimized simultaneously.

Eq. [10.87] is the expectation of a quadratic form with $(m \times m)$ elements which also can be written as $Q(\mathbf{AX} - \mathbf{Y})$. In general, the following relation holds true for the expectation of a quadratic form $Q(\mathbf{X})$:

$$E\,[Q(\mathbf{X})] \doteq EQ[\mathbf{X} - E(\mathbf{X})] + Q[E(\mathbf{X})] \qquad [10.88]$$

This is a generalization of [6.20] and [8.3] obtained previously for variances and covariances, respectively.

Use of this result gives:

$$E(\mathbf{AX}-\mathbf{Y})(\mathbf{AX}-\mathbf{Y})' = \mathbf{AC}_{xx}\mathbf{A}' - \mathbf{AC}_{xy} - \mathbf{C}_{yx}\mathbf{A}' + \mathbf{C}_{yy} + E(\mathbf{AX}-\mathbf{Y})E(\mathbf{AX}-\mathbf{Y})' \qquad [10.89]$$

Substitution of [10.86] gives:

$$E(\mathbf{AX}-\mathbf{Y})\,E(\mathbf{AX}-\mathbf{Y})' = (\mathbf{AG}-\mathbf{H})\boldsymbol{\beta}\boldsymbol{\beta}'(\mathbf{AG}-\mathbf{H})' \qquad [10.90]$$

The minimum with respect to **A** cannot be found by the method of section 10.12 unless we neglect the trends.

For solution Whittle (1963a) has used the so-called minimax criterion which is a powerful tool in advanced methods of mathematical analysis (Courant and Hilbert, 1962). According to Whittle we seek the optimum predictor $\hat{\mathbf{Y}}$ determined by:

$$E(\hat{\mathbf{Y}}-\mathbf{Y})(\hat{\mathbf{Y}}-\mathbf{Y})' = \underset{\mathbf{A}}{\text{Min}}\ \underset{\beta}{\text{Max}}\ E(\mathbf{AX}-\mathbf{Y})(\mathbf{AX}-\mathbf{Y})' \qquad [10.91]$$

It means that the coefficients of **A** are optimized as before but, at the same time, we want the trend $\mathbf{G}\boldsymbol{\beta}$ (and $\mathbf{H}\boldsymbol{\beta}$) to be of maximum strength.

The last term of [10.89] contains elements which we can make as large as we want, no matter what **A** is. Consequently, unless $\mathbf{AG} = \mathbf{H}$, the minimax criterion does not give a finite solution. This imposes a condition upon the minimization of $E(\mathbf{AX}-\mathbf{Y})(\mathbf{AX}-\mathbf{Y})'$ with respect to **A**; otherwise, the procedure is the same as that employed before. An interpretation of this condition is that when the stochastic component is disregarded and $\mathbf{X} = \mathbf{G}\boldsymbol{\beta}$, $\mathbf{Y} = \mathbf{H}\boldsymbol{\beta}$ exactly. **A** should be such that **Y** and **X** satisfy the same deterministic functions, no matter what $\boldsymbol{\beta}$ is.

Consequently, we minimize:

$$E(\hat{\mathbf{Y}}-\mathbf{Y})(\hat{\mathbf{Y}}-\mathbf{Y})' = \mathbf{AC}_{xx}\mathbf{A}' - \mathbf{AC}_{xy} - \mathbf{C}_{yx}\mathbf{A}' + \mathbf{C}_{yy} + (\mathbf{H}-\mathbf{AG})\mathbf{P} + \mathbf{P}'(\mathbf{H}-\mathbf{AG})' \qquad [10.92]$$

where **P** is a matrix consisting of Lagrangian multipliers. If a single value is predicted, **Y** is a scalar. Then **P** is a $(p \times 1)$ column vector and $(\mathbf{H}-\mathbf{AG})\,\mathbf{P} = \mathbf{P}'\,(\mathbf{H}-\mathbf{AG})'$ is a scalar. If **Y** is an $(m \times 1)$ vector, **P** is a $(p \times m)$ matrix ensuring that the condition holds at all points.

By treating the first four terms as was done for [8.46], we find that the minimum is now reached when:

$$\mathbf{A} = (\mathbf{C}_{yx} + \mathbf{P}'\mathbf{G}')\,\mathbf{C}_{xx}^{-1} \qquad [10.93]$$

Substituting [10.93] in $\mathbf{AG} = \mathbf{H}$ gives:

$$(\mathbf{C}_{yx} + \mathbf{P}'\mathbf{G}')\,\mathbf{C}_{xx}^{-1}\mathbf{G} = \mathbf{H}$$

or: $$\mathbf{P}' = (\mathbf{H} - \mathbf{C}_{yx}\mathbf{C}_{xx}^{-1}\mathbf{G})\,(\mathbf{G}'\mathbf{C}_{xx}^{-1}\mathbf{G})^{-1}$$

and: $$\mathbf{A} = [\mathbf{C}_{yx} + (\mathbf{H} - \mathbf{C}_{yx}\mathbf{C}_{xx}^{-1}\mathbf{G})\,(\mathbf{G}'\mathbf{C}_{xx}^{-1}\mathbf{G})^{-1}\mathbf{G}']\,\mathbf{C}_{xx}^{-1}$$

Hence: $$\hat{\mathbf{Y}} = \mathbf{C}_{yx}\mathbf{C}_{xx}^{-1}\mathbf{X} + (\mathbf{H} - \mathbf{C}_{yx}\mathbf{C}_{xx}^{-1}\mathbf{G})\hat{\boldsymbol{\beta}}_t$$

or: $$\hat{\mathbf{Y}} = \mathbf{H}\boldsymbol{\beta}_t + \mathbf{C}_{yx}\mathbf{C}_{xx}^{-1}(\mathbf{X} - \mathbf{G}\hat{\boldsymbol{\beta}}_t) \quad [10.94]$$

where: $$\hat{\boldsymbol{\beta}}_t = (\mathbf{G}'\mathbf{C}_{xx}^{-1}\mathbf{G})^{-1}\mathbf{G}'\mathbf{C}_{xx}^{-1}\mathbf{X} \quad [10.95]$$

If $\mathbf{Y}$ and $\mathbf{X}$ are both corrected for their deterministic components $\mathbf{H}\hat{\boldsymbol{\beta}}_t$ and $\mathbf{G}\hat{\boldsymbol{\beta}}_t$ [10.94] is equivalent to [8.95]. When $\mathbf{C}_{xx} = \sigma^2\mathbf{I}$, where σ^2 represents the variance of the data $\mathbf{X}$ and $\mathbf{I}$ is the identity matrix, [10.95] reduces to:

$$\hat{\boldsymbol{\beta}} = (\mathbf{G}'\mathbf{G})^{-1}\,\mathbf{G}'\mathbf{X} \quad [10.96]$$

which is the ordinary least-squares solution for the coefficients in the case of the random residuals.

Intuitively, it is clear that the solution by [10.96] is unbiased because the stochastic component has an expected value equal to zero at all places. The positive and negative anomalies will cancel out. A proof of this is given in several textbooks, e.g., Grenander and Rosenblatt (1957). Watson (1967, 1971) discussed the problem in detail. If $\boldsymbol{\beta}_t$ is estimated by ordinary least squares, $\mathbf{C}_{xx}$ as calculated from the residuals will be biased. Of course, the bias will be small when $\hat{\boldsymbol{\beta}}_t$ can be replaced by $\hat{\boldsymbol{\beta}}$.

Practical example for the Whalesback deposit

In Fig.65A, a quadratic trend surface is shown which was fitted to the logarithm of percent copper from six N–S directed boreholes (holes 1–6) at the 425-ft. level of the Whalesback deposit. Part of the antilog for this surface has been shown in Fig.48 (see section 9.8).

Along the six holes, there were 177 observation points, 5 ft. apart, for assays from contiguous core samples. For convenience, one longer space series was formed by placing the 177 data points for the six separate holes in sequence. The first serial correlation coefficient for the residuals from the quadratic trend of Fig.65A at these points amounts to 0.50039. Clearly, this is an example of where we have used ordinary least squares for estimating the trend in a situation of autocorrelated residuals.

In order to evaluate $\hat{\boldsymbol{\beta}}_t$, we assumed that the stochastic component for variations in

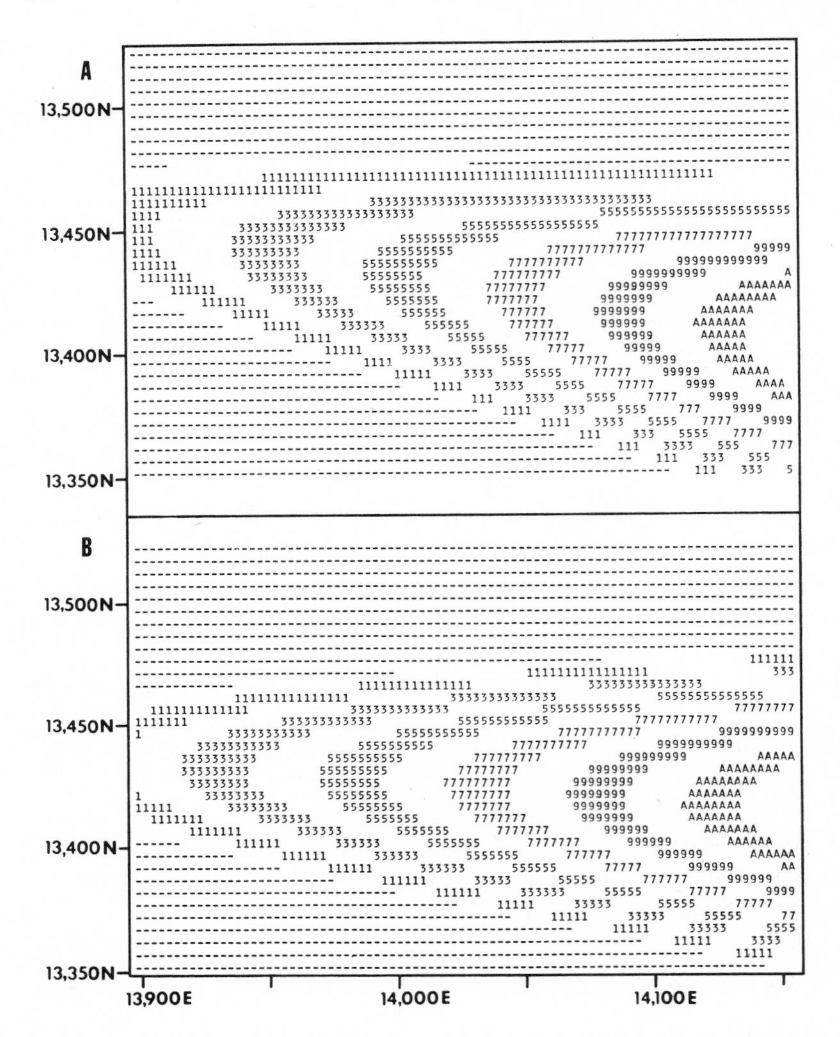

Fig.65. Quadratic trend surface fitted to 177 log-copper data from 6 boreholes (holes 1 – 6) at 425-ft. level, Whalesback deposit. Fitted log-copper value is printed number divided by 5 (A = 11, B = 13). A. Ordinary least-squares fit (cf. westernmost segment of mosaic in Fig.48). B. Best linear unbiased estimate (B.L.U.E.) based on generalized least-squares method.

residuals has an exponential autocorrelation function. The objective of this example is to estimate $\hat{\boldsymbol{\beta}}_t$ by approximation and compare its trend surface with the one for $\hat{\boldsymbol{\beta}}$ obtained previously. Adopting the notation of previous sections, [10.95] can be written as:

$$\hat{\boldsymbol{\beta}}_t = (\mathbf{X}'\mathbf{R}^{-1}\mathbf{X})^{-1}\mathbf{X}'\mathbf{R}^{-1}\mathbf{Y} \qquad [10.97]$$

with:

$$\mathbf{R} = \begin{bmatrix} 1 & \rho & \rho^2 & \ldots & \rho^{n-1} \\ \rho & 1 & \rho & \ldots & \rho^{n-2} \\ \rho^2 & \rho & 1 & \ldots & \rho^{n-3} \\ \cdot & \cdot & \cdot & \ldots & \cdot \\ \rho^{n-1} & \cdot & \cdot & \ldots & \cdot \end{bmatrix}$$

where ρ is the first autocorrelation coefficient to be approximated by $r_1 = 0.50$ for the example. It is readily confirmed that:

$$\mathbf{R}^{-1} = [1/(1-\rho^2)] \begin{bmatrix} 1 & -\rho & 0 & \ldots & 0 \\ -\rho & (1+\rho^2) & -\rho & \ldots & 0 \\ 0 & -\rho & (1+\rho^2) & \ldots & 0 \\ \cdot & \cdot & \cdot & \ldots & \cdot \\ 0 & 0 & 0 & \ldots & 1 \end{bmatrix}$$

It follows that $\mathbf{R}^{-1} \approx \mathbf{Q}'\mathbf{Q}/(1-\rho^2)$ with:

$$\mathbf{Q} = \begin{bmatrix} -\rho & 1 & 0 & \ldots & 0 \\ 0 & -\rho & 1 & \ldots & 0 \\ \cdot & \cdot & \cdot & \ldots & \cdot \\ 0 & 0 & 0 & \ldots & 1 \end{bmatrix}$$

Hence: $\hat{\boldsymbol{\beta}}_t = (\mathbf{X}'\mathbf{Q}'\mathbf{Q}\mathbf{X})^{-1}\mathbf{X}'\mathbf{Q}'\mathbf{Q}\mathbf{Y}$

or: $\hat{\boldsymbol{\beta}}_t = [(\mathbf{QX})'(\mathbf{QX})]^{-1}\,[(\mathbf{QX})'(\mathbf{QY})]$ [10.98]

In form, this is equivalent to [8.44] where $\hat{\boldsymbol{\beta}} = (\mathbf{X}'\mathbf{X})^{-1}(\mathbf{X}'\mathbf{Y})$.

Thus $\hat{\boldsymbol{\beta}}_t$ is obtained by computing $\hat{\boldsymbol{\beta}}$ for transformed data $X^*_{ik} = X_{ik} - r_1 X_{i,k-1}$ and $Y^*_k = Y_k - r_1 Y_{k-1}$.

The trend surface based on the resulting vector $\hat{\boldsymbol{\beta}}_t$ is shown in Fig.65B. The difference between values in Fig.65A and B is very small, indicating that $\hat{\boldsymbol{\beta}}_t$ can be approximated by $\hat{\boldsymbol{\beta}}$ in this type of application. As a rule, it is only if sets of relatively few data are considered that $\hat{\boldsymbol{\beta}}_t$ differs significantly from $\hat{\boldsymbol{\beta}}$ as in Matheron's (1971a) method of universal kriging.

It is noted that in the previous derivation, $\mathbf{Q}$ can be interchanged with $\mathbf{Q}'$, since we also have:

$$\mathbf{R}^{-1} \approx \mathbf{Q}\mathbf{Q}'/(1-\rho^2).$$

Chapter 11

HARMONIC ANALYSIS AND POWER SPECTRA

11.1. Introduction

In previous chapters, considerable use was made of polynomial series for the fitting of trends. Fourier series whose terms are cosine and sine functions provide a tool which is approximately as powerful as polynomial series. Not only can Fourier series be used for fitting trends, they also provide a natural approach for describing periodical phenomena. These two applications are called harmonic analysis.

In some situations, polynomial series may be better for describing trends, since the periodicity of Fourier approximations can be a disadvantage where periodicity is absent in nature. Also, the linear and quadratic polynomials have simple geometrical representations. Linear or gently curved trends which occur frequently in nature should not be approximated by a Fourier series.

Harmonic analysis leads to the power spectrum. It has already been shown that the autocorrelation function and the power spectrum are Fourier transforms of one another. Techniques of Fourier analysis are important in the study of stationary stochastic processes. The spectrum has more desirable properties than the autocorrelation function. Also, it is easier to describe specific techniques for filtering (see Fig.55), if this is done in the frequency domain by using Fourier transforms (see section 11.8).

In Chapter 7, we made use of this technique by forming characteristic functions for deriving equations for the moments of frequency distributions.

It may seem that matters are more complicated if the time or distance domain is replaced by the frequency domain; however, after the mathematical derivations are made in the frequency domain, one usually returns to the time or distance domain. In practice, the solution to a problem may be obtained much faster this way.

We begin this chapter with a simple exposition of Fourier series. Any continuous function or series of discrete data can be expressed by a Fourier series. Individual terms of the harmonic expressions have both an amplitude and a phase. For applications, in stochastic process theory, the phase of the individual sine-waves is often not considered.

The power spectrum consists of values equivalent to the squares of the amplitudes of the sine-waves. The phase is considered only if the location of the waves in space is of significance. Methods of harmonic analysis are readily extended into two dimensions giving double Fourier series. Our main use of Fourier harmonic analysis will be to study periodic phenomena.

In section 11.8, which may be skipped at first reading, we derive the simple method of filtering used previously to construct Fig.55B. By applying the results of section 11.8 to the signal-plus-noise model where the signal is characterized by a stationary exponential autocorrelation function the equations for best-fit can be obtained and confidence intervals computed by applying more general equations, such as those formulated by Yaglom (1962).

11.2. One-dimensional Fourier series

Suppose that a variable y assumes the values $y_k (k = 0, 1, \ldots, n-1)$ for n equally spaced points. We can define an auxiliary variable x that ranges in value from 0 to 360°. If we assume that the space series is cyclical with period equal to n sampling intervals, the value of y for $x = 360°$ is equal to that for $x = 0°$.

To every value y_k, there corresponds a value $x_k = k \cdot 360°/n$. The finite Fourier series for y now satisfies:

$$y = a_0 + (a_1 \cos x + b_1 \sin x) + (a_2 \cos 2x + b_2 \sin 2x) + \ldots + (a_m \cos mx + b_m \sin mx) \quad [11.1]$$

where $m = \frac{1}{2}(n-1)$ if n = odd and $m = \frac{1}{2}n$ if n = even. It can be shown that:

$$\left.\begin{aligned} a_0 &= (1/n) \sum_{k=0}^{n-1} y_k = \text{mean} \\ a_i &= (2/n) \sum_{k=0}^{n-1} y_k \cos ix_k \\ b_i &= (2/n) \sum_{k=0}^{n-1} y_k \sin ix_k \end{aligned}\right| \quad [11.2]$$

There are as many coefficients as there are data. If n is even, b_m vanishes and:

$$a_m = (1/n) \sum_{k=0}^{n-1} y_k \cos mx_k$$

A simple example may illustrate the calculation of the Fourier coefficients a_i and b_i. Suppose that a space series consists of four values: {3, 2, 3, 5}. We calculate:

x_k	y_k	$\cos x_k$	$y_k \cos x_k$	$\sin x_k$	$y_k \sin x_k$	$\cos 2x_k$	$y_k \cos 2x_k$
0°	3	1	3	0	0	1	3
90°	2	0	0	1	2	–1	–2
180°	3	–1	–3	0	0	1	3
270°	5	0	0	–1	–5	–1	–5
Sum	13	0	0	0	–3	0	–1

According to [11.2], $a_0 = 13/4 = 3.25, a_1 = 0, b_1 = -3/2 = -1.5$ and $a_2 = -1/4$. Hence, the fitted function is:

$$y = 3.25 - 1.5 \sin x - 0.25 \cos 2x.$$

It is readily verified that this expression gives exactly the series $\{3, 2, 3, 5\}$ for $x = 0°, 90°, 180°, 270°$, respectively. On the other hand, one can represent a series by the first set of terms of the Fourier series only:

$$\hat{y}_p = a_0 + (a_1 \cos x + b_1 \sin x) + \ldots + (a_p \cos px + b_p \sin px) \qquad [11.3]$$

where $p < m$. $\hat{y}_p$ sometimes is called the harmonic trend. It consists of low-frequency waves only and the higher-frequency waves have been deleted from the representation. The function $\hat{y}_p$ has the least-squares property. This is illustrated numerically for the previous example. When the series is $\{3, 2, 3, 5\}$, the function $\hat{y} = 3.25 - 1.5 \sin x$ is the least-squares solution for the expression $y = a_0 + a_1x_1 + b_1x_2$ which contains three coefficients and two independent variables $x_1 = \cos x$ and $x_2 = \sin x$.

When the variables are corrected for the mean, we can define the two matrices:

$$\mathbf{X} = \begin{bmatrix} 1 & 0 \\ 0 & 1 \\ -1 & 0 \\ 0 & -1 \end{bmatrix}; \quad \mathbf{Y} = \begin{bmatrix} -0.25 \\ -1.25 \\ -0.25 \\ 1.75 \end{bmatrix}$$

According to [8.44]:

$$\begin{bmatrix} a_1 \\ b_1 \end{bmatrix} = (\mathbf{X}'\mathbf{X})^{-1}\mathbf{X}'\mathbf{Y} = \begin{bmatrix} 0 \\ -1.5 \end{bmatrix}$$

Since $\bar{y} = 3.25$, we have $\hat{y} = 3.25 - 1.5 \sin x$ constituting the first part of $y = 3.25 - 1.5 \times \sin x - 0.25 \cos 2x$. The calculated values for the function $\hat{y}$ form the series $\{3.25, 1.75, 3.25, 4.75\}$.

Continuous functions

If $y = f(x)$ is continuous in a range of width 2π, it can be represented by [11.1] with m going to infinity. In this situation:

$$a_0 = \frac{1}{2\pi} \int_{-\pi}^{\pi} f(x)\, dx$$

$$a_i = \frac{1}{\pi} \int_{-\pi}^{\pi} f(x) \cos ix\, dx \qquad b_i = \frac{1}{\pi} \int_{-\pi}^{\pi} f(x) \sin ix\, dx \qquad [11.4]$$

We are mainly interested in series of discrete data. If n is even, and the origin is chosen in the center of the series, [11.1] can be written as:

$$y_k = a_0 + 2 \sum_{i=1}^{\frac{1}{2}n-1} [a_i \cos(2\pi ik/n) + b_i \sin(2\pi ik/n)] + a_{\frac{1}{2}n} \cos \pi k \qquad [11.5]$$

with $k = -n/2, \ldots, 0, 1, \ldots, (n/2-1)$, where:

$$a_i = (1/n) \sum_{k=-\frac{1}{2}n}^{\frac{1}{2}n-1} y_k \cos(2\pi ik/n)$$

$$b_i = (1/n) \sum_{k=-\frac{1}{2}n}^{\frac{1}{2}n-1} y_k \sin(2\pi ik/n) \qquad [11.6]$$

with $i = 0, 1, 2, \ldots, \frac{1}{2}n$.

According to [2.2], we can write:

$$y_k = R_0 + 2 \sum_{i=1}^{\frac{1}{2}n-1} R_i \cos(2\pi ik/n + \phi_i) + R_{\frac{1}{2}n} \cos \pi k \qquad [11.7]$$

where $R_i = \sqrt{a_i^2 + b_i^2}$ and $\phi_i = \arctan(-b_i/a_i)$.

This is the so-called amplitude and phase representation of a Fourier series. Using results of section 2.12, we can write:

$$S_i = R_i \, e^{I\phi_i} = a_i - Ib_i \text{ where } I^2 = -1$$

This leads to the Fourier transform with:

$$y_k = \sum_{i=-\frac{1}{2}n}^{\frac{1}{2}n-1} S_i \, e^{I(2\pi ik)/n} \qquad [11.8]$$

and the inverse Fourier transform:

$$S_i = (1/n) \sum_{k=-\frac{1}{2}n}^{\frac{1}{2}n-1} y_k \, e^{-I(2\pi ik)/n} \qquad [11.9]$$

where $-\frac{1}{2}n \leqslant i \leqslant \frac{1}{2}n - 1$. These summations also can be carried out by letting i and k go from 0 to $(n-1)$ or from 1 to n.

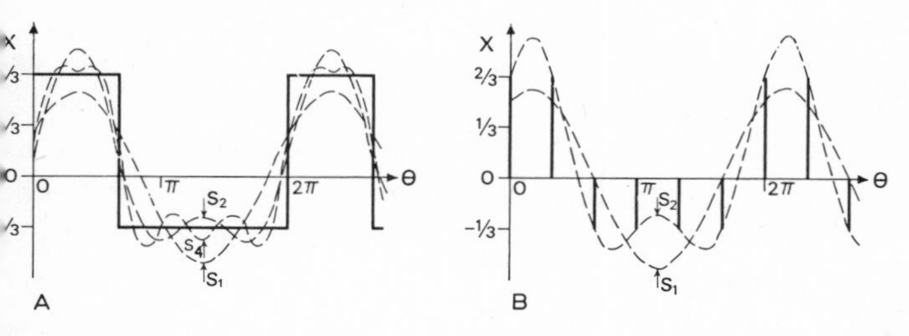

ig.66. A. Truncated Fourier series expansion of a continuous space series (rectangular wave). B. itto, for a discrete series derived by sampling A at regular intervals equal to $\pi/6$.

1.3. Spectral analysis

An example of one-dimensional Fourier analysis is shown in Fig.66. Suppose that a tratigraphic section consists of alternate layers of two types which are assigned the values ne and zero, respectively. The periodicity is assumed to be perfect and one type of layer half as thick as the other one.

From [11.4], it follows immediately that $a_o = 1/3$ and this value was subtracted from ne data represented in Fig.66. We computed the curves S_j with:

$$S_j = \sum_{i=1}^{j} (a_i \cos i\theta + b_i \sin i\theta) \qquad [11.10]$$

nown for $j = 1, 2, 3$, and 4 in Fig.66. For this example, $S_3 = S_2$. Every component (a_i os $i\theta + b_i \sin i\theta$) of S_j is a sine curve with period $2\pi/i$, amplitude $A_i = \sqrt{a_i^2 + b_i^2}$ and phase ngle $\phi_i = \arcsin (a_i/A_i)$. The original pattern is a continuous rectangular wave which is pproximated by the S_j with the fit improving with increasing j. The fundamental wave ength for the periodicity was set equal to 2π. Other components ($i > 1$) are "harmonics" f this fundamental wavelength. On the other hand, suppose that the measurements were nade at a regular sampling interval equal to $2\pi/6$, beginning at the point with $\theta = 0$ and xtending over $n = 6$ measurements. In this situation (discrete data), Fourier analysis is one by using [11.2]. The new curves S_1 and S_2 (see Fig.66B) differ from those in ig.66A and S_2 gives a 100% fit.

Suppose that in either one of these two examples, A_i^2 is plotted against i. This ives the so-called Schuster periodogram. In the situation of Fig.66B, there are only values in the periodogram regardless of the size of n. These are a large value corre- ponding to fundamental wave length and a small value for the first harmonic. None of ne values in a periodogram can be negative. They provide information about the relative trength of the sine-waves fitted to the data.

If the technique is applied to series generated by stationary random functions, we find nat $P_i = A_i^2$ is distributed as a multiple of χ^2 with 2 degrees of freedom (Jenkins, 1961).

Adjacent values in the periodogram can be averaged; the average of q values bein distributed as χ^2 with approximately $2q$ degrees of freedom. We then obtain a smoothe version of the so-called power spectrum $E\ (P_i)$. If the series is long, it may be to laborious to compute the periodogram by the direct method. Then, the Fast Fourie Transform method by Cooley and Tukey (1965) may be utilized if the number of data i the series (n) is equal to an integer power of 2, or $n = 2^p$ where p is an integer numbe The series may be enlarged by zeros until this condition is fulfilled. However, this ma cause distortions which may be troublesome, in particular when the method is used fo fitting trends to data. It is assumed that the series has zero-mean.

Conventional spectral analysis consists of first calculating the autocovariance functio for m lags:

$$C_k = \frac{1}{n-k} \sum_{j=1}^{n-k} x_j x_{j+k} \quad (k = 0, 1, \ldots, m) \tag{11.11}$$

Next the discrete Fourier transform is taken giving:

$$P_r = C_0 + 2 \sum_{k=1}^{m-1} C_k \cos\left(\frac{kr\pi}{m}\right) + C_m \cos(r\pi) \tag{11.12}$$

P_r is the "raw" estimate of the spectral density for frequency $f = r/2m$. Blackman an Tukey (1958) showed that the estimates P_r are subject to distortions which can b eliminated largely by using "refined" spectral density estimates U_r. The latter can b obtained either by applying a cosine transformation to the autocovariance functio before its Fourier transform is taken or by afterwards applying a simple moving averag scheme to the P_r-values of [11.12].

In Fig.67 we show the 15 values U_r computed from the first 15 values of the autoco variance function for Pulayaco. We used Tukey's method of "hanning".

$$U(f) = C_0 + 2 \sum_{k=1}^{m-1} W(k)\, C_k \cos 2\pi k f \tag{11.13}$$

with $W(k) = \frac{1}{2}(1 + \cos \pi k/m)$. Other types of weighting function $W(k)$ can be used (see e.g., Jenkins and Watts, 1968). The spectral power spectrum of a continuous autocovariance function is:

$$P(f) = s^2(x) \int_{-\infty}^{\infty} r_l \cos 2\pi f l \, \mathrm{d}l \tag{11.14}$$

If r_l satisfies [10.13], then:

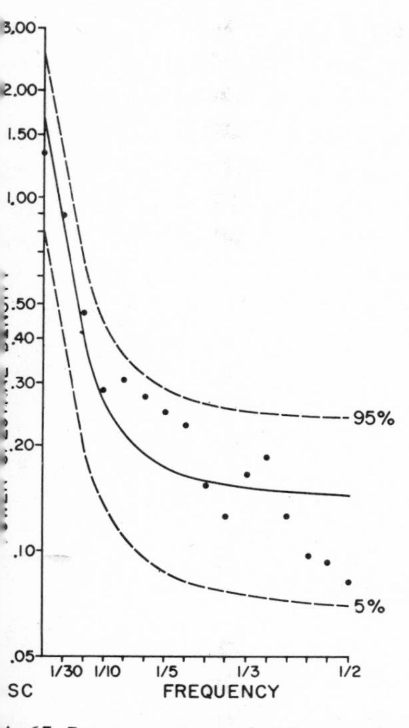

ig.67. Power spectrum of 118 zinc values, Pulacayo, obtained by a Fourier transformation of both ata points and curve in Fig.56A; smoothed spectral density values (dots) derived by hanning; dots are ithin the 90% confidence belt on the theoretical curve. Nevertheless, the dots suggest a somewhat ifferent pattern.

$$P(f) = (1-c)\, s^2(x) + \frac{cs^2(x)/\pi f_c}{1 + (f/f_c)^2} \qquad [11.15]$$

here $f_c = a/2\pi$. This curve is also shown in Fig.67 with a and c as estimated from ig.56A.

Confidence bands can be readily computed by using a method outlined by Blackman nd Tukey (1958). It is noted that all dots in Fig.67 fall within the 90% confidence band ut a departure between dots and curve is indicated.

1.4. Stochastic model for varve deposition in glacial Lake Barlow – Ojibway

In section 10.6, we have analyzed two sets of silt and clay component thickness mea-ırements for series 4 consisting of 537 varves. The power spectrum of this series is ıown in Fig.68, together with spectra for other series from the same area, each with

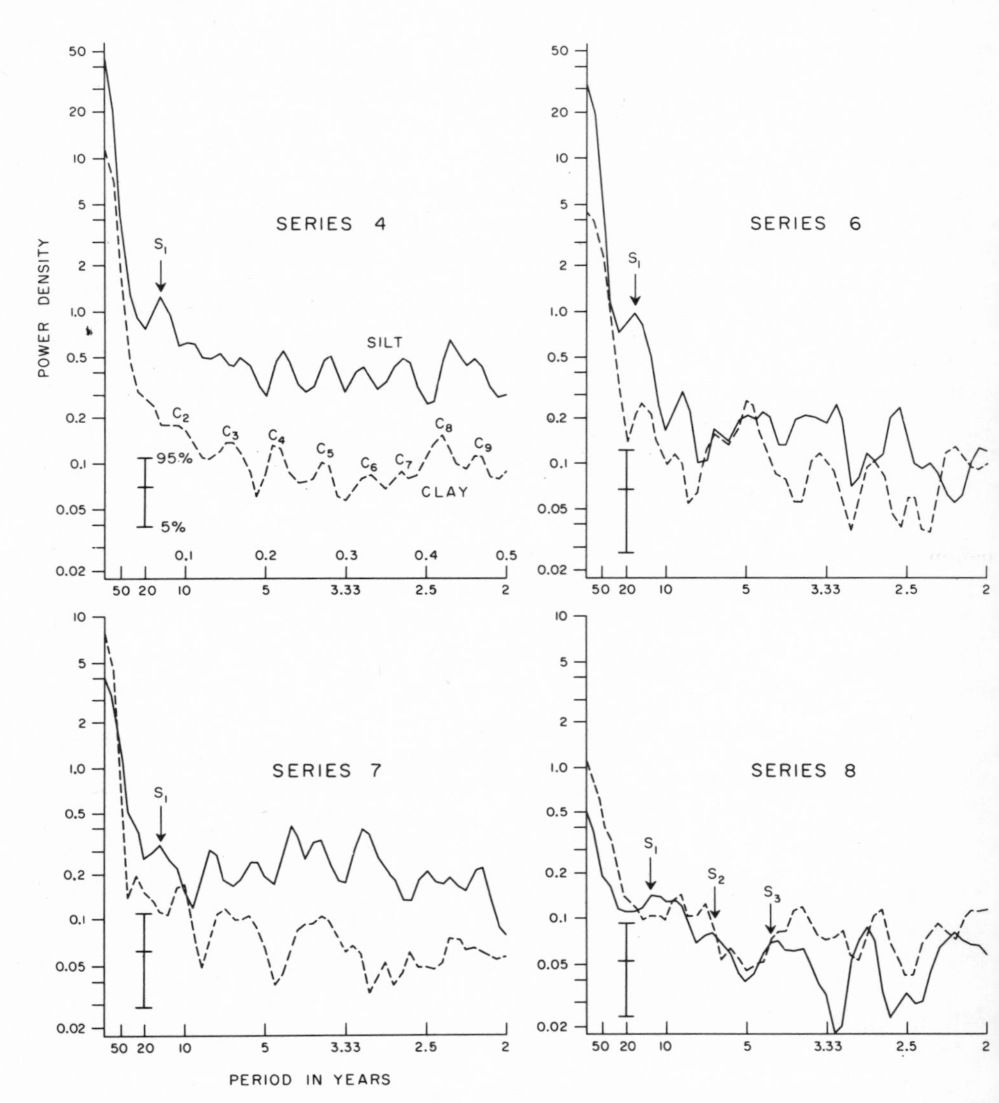

Fig.68. Power spectra for log-thickness data of silt and clay components from series with more than 250 varves each; measured at different locations in glacial Lake Barlow – Ojibway. (From Agterberg and Banerjee, 1969.) Confidence belts also shown. Note S_1-peak developed in the four series.

more than 250 data. Eq.[11.13] was used to obtain these spectra. For other applications of this method to varve time series, see Anderson and Koopmans (1965).

It is interesting that the silt component shows a peak for periods of about 14 years in the four spectra of Fig.68. The four series are for localities at some distance apart, with the distance between localities 4 and 8 exceeding 100 km. This suggests that the 14-year

eriodicity reflects a process that was operative in large parts of Lake Barlow – Ojibway.
Schove (1972) suggested climatological variations due to soli-lunar cycles as an ex-
lanation. Brier (1968) showed that soli-lunar cycles produce significant tidal effects in
ne atmosphere at $13\frac{1}{2}$ and 27 years. According to Schove (1972), silt components of
arves are suitable for study of soli-lunar cycles as affect the summer months of June and
uly. The $13\frac{1}{2}$-year (163 calendar months) cyclicity is the so-called beat period between
ne calendar month (30.44 days) and the synodic month (29.53 days). These two cycles
re in phase with one another every 163 months, and both are then approximately in
hase with the cycle arising from the anomalistic month (27.55 days). The synodic cycle
the period from one new moon to the next and the anomalistic cycle that from one
erigee to the next. These two cycles are most important in determining the magnitude of
ne soli-lunar gravitational tide, which has significant effects on monthly weather condi-
ons as shown by Brier (1968).

Another nonrandom feature in the arrangement of maxima and minima in the spectra
f Fig.68 is that they tend to lie at equal intervals along the frequency scale. This
henomenon is conspicuous in the clay-spectrum for series 4. Other geological time series
ith similar arrangements in the positions of peaks have been described by Carrs and
eidell (1966) and Schwarzacher (1967). Phenomena of this type can be: (1) harmonics
f a fundamental wavelength; (2) "side lobe" effects from a large peak for long periods;
r (3) "ripples" associated with a relatively small number of extreme values in the original

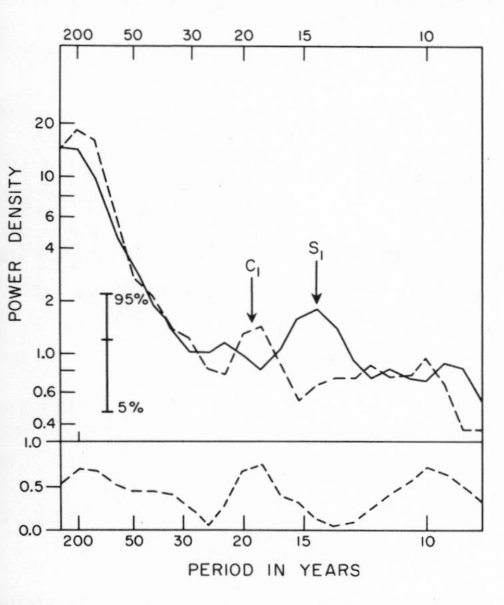

g.69. Spectra for residuals from linear trend, series 4 (537 varves); coherence (COH.) shown at the
ttom. Note that peak C_1 was not developed in Fig.68, and agreement in location between peaks in
y spectrum and coherence.

data. The latter two factors can be evaluated to some extent by rerunning the spectral analysis for different values of m and transforming the original data in various ways so that clusters of extreme values are eliminated from the original data. We used logarithmically transformed thickness data to counter the effect of larger values.

Power spectra with $m = 100$ for residuals from a linear trend are shown in Fig.69 (for series 4). The S_1-peak for silt is well developed in this diagram; there is a C_1-peak for clay at about 20 years. It is likely to be the fundamental period of the harmonics shown in Fig.68 (series 4).

In the remainder of this section, we will apply several more advanced methods of spectral analysis restricting the mathematical treatment to a listing of equations used, with brief explanations and references for further reading. The emphasis will be on geological application and interpretation.

Cross-spectral analysis

A particularly powerful tool for data analysis of bivariate series is cross-spectral analysis (see Koopmans, 1967; Jenkins and Watts, 1968). The so-called cospectrum $C_{sc}(f)$ and quadrature spectrum $Q_{sc}(f)$ for silt (s) and clay (c) are estimated by:

$$C_{sc}(f) = r_{sc}(0) + \sum_{k=1}^{m} W(k) \cos 2\pi kf \, [r_{sc}(k) + r_{cs}(k)] \quad [11.16]$$

and:

$$Q_{sc}(f) = \sum_{k=1}^{m} W(k) \sin 2\pi kf \, [r_{sc}(k) - r_{cs}(k)] \quad [11.17]$$

In these expressions $W(k)$ is as in [11.13]. The data were standardized and $r_{sc}(k)$ and $r_{cs}(k)$ together constitute the cross-correlation function shown previously in Fig.60. From [11.17], the coherence $R(f)$ and phase $\phi(f)$ are computed by:

$$R(f) = [\{C^2(f) + Q^2(f)\}/P_s(f)\,P_c(f)]^{\frac{1}{2}} \quad [11.18]$$

and:

$$\phi(f) = \arctan\,[Q(f)/C(f)] \quad [11.19]$$

$R(f)$ is a measure of the amount of linear relationship between two series for frequency bands around f. It is equivalent to the correlation coefficient between two variables as a function of frequency (Koopmans, 1967).

The power spectrum can be regarded as a decomposition of all variability in a series

TABLE XXXII
Phase angles for largest coherences of Fig.69

Period	*Coherence*	*Observed phase*	
(years)		*(degrees)*	*(years)*
200	0.71	−27	−15
100	0.70	−20	− 6
20	0.67	− 4	0
18.2	0.75	− 1	0
10	0.71	−72	− 2
9.5	0.62	−91	− 2

terms of components of the variance for small frequency bands. Likewise, the coherence is a decomposition of the total correlation coefficient between two variables. For example, Anderson (1967) has shown that the data of two time series can be uncorrelated when time is disregarded whereas, in reality, the long-term fluctuations are negatively correlated and the short-term fluctuations positively correlated (or vice versa). By disregarding the nature of the variability in time, the overall correlation coefficient can be a meaningless quantity. Partial correlation with trend elimination was used previously to give a solution to this problem (see section 8.7). Cross-spectral analysis may provide a more refined answer.

The coherence is positive in all cases and should not be interpreted separately from the phase $\phi(f)$ which can be either positive or negative and lies between $-180°$ and $180°$. In other words, if the phase is zero, a high coherence means positive correlation. When the phase is close to $-180°$ or $180°$, the variations in time of the two variables are $180°$ out of phase implying a negative correlation.

Coherences for the silt–clay series are shown in Fig.69. They can be tested for statistical significance by consulting the tables prepared by Amos and Koopmans (1963). In the present situation, coherences greater than 0.63 are significantly different from zero at the 95-% confidence level. These values are also shown in Table XXXII with the corresponding phase angle which is negative in all cases, indicating that the variations in the clay lead those of the silt by a variable amount.

The phase angles can be tested for statistical significance by using a method given by Jenkins and Watts (1968). Table XXXII was modified after Agterberg and Banerjee (1969) who used Goodman's (1957) earlier method for testing the phase angles for significance. For the Jenkins-Watts approximate test, we first estimate an approximate number of degrees of freedom (d). For the Tukey window (hanning method):

$$d = \frac{8}{3}\frac{n}{m} = \frac{8}{3}\frac{537}{100} \approx 14 \qquad [11.20]$$

We then use the equation (cf. Jenkins and Watts, 1968, p.435):

$$\phi(f) \pm \arcsin \left[\frac{2}{d-2} F_{1-\alpha}(2, d-2) \frac{1-R^2(f)}{R^2(f)}\right]^{\frac{1}{2}} \qquad [11.21$$

For the example, $R^2(f) \approx 0.5$, and $F_{0.95}$ (2, 12) = 3.89. It follows that the 95-9 confidence interval is approximately 54° wide. Consequently, only two of the six phas lags listed in Table XXXII are statistically significant according to the approximate test

The pattern of the coherence in Fig.69 resembles the clay spectrum rather than the sil spectrum. A 10-year and 20-year periodicity is well developed and may correspond to th 11- and 22-year sunspot cycle. We accepted cyclical climatological variation in tempera ture related to sunspot activity as the most likely explanation for this phenomenon. Th clay would reflect slight increases and decreases in the annual supply of meltwater bette and earlier than the silt because of the following difference in transportation mechanisr for the two materials.

The silt was deposited during the summer by one or more sudden events resulting i turbidity currents. On the other hand, clay remained dispersed in the lake water and wa deposited all year around. A greater rate of retreat of the land ice due to an increase i temperature would produce more meltwater and consequently more sediments to b deposited. The clay was transported across the lake almost immediately, but the silt wa delayed. Initially, the silty material was dumped close to the ice front leaving the thick ness profile of varves in more distal areas relatively unaffected. However, later the steepe profile of underlying varves, acting as a floor to turbidity currents for later years, helpe to spread the silt farther with the thicker parts moving to more distal areas.

The 10- and 20-year cycle can be observed as a slight fluctuation in the cross-correla tion of Fig.60. The phase lead of the clay with respect to the silt, about two years, can b seen in this diagram. The cross-correlation function suggests that the 10-year cycle is phenomenon in its own right and not purely a harmonic of the 20-year cycle.

The main variation in the correlogram of Fig.59 was interpreted in terms of a second order Markovian process (section 10.6). The cross-correlation function suggests that fo this process the clay leads the silt about 15 years.

Fourier transformation of [10.34] for $c = 1$ by [11.14], gives:

$$f(\omega) = \frac{2\alpha(\omega_0^2 + \alpha^2)/\pi}{(\omega^2 - \alpha^2 - \omega_0^2)^2 + 4\alpha^2\omega^2} \qquad [11.22$$

where $\omega = 2\pi f$ is angular frequency. Since $\alpha \approx 0.03$ for silt and $\alpha \approx 0.01$ for clay, thei theoretical functions satisfying [11.22] are different in appearance (see Agterberg and Banerjee, 1969, fig.13). By solving [10.41] completely and considering $\eta(t)$ as the input for a process with $x(t)$ as output, we not only derive [11.22] for the square of amplitude but for the phase lag:

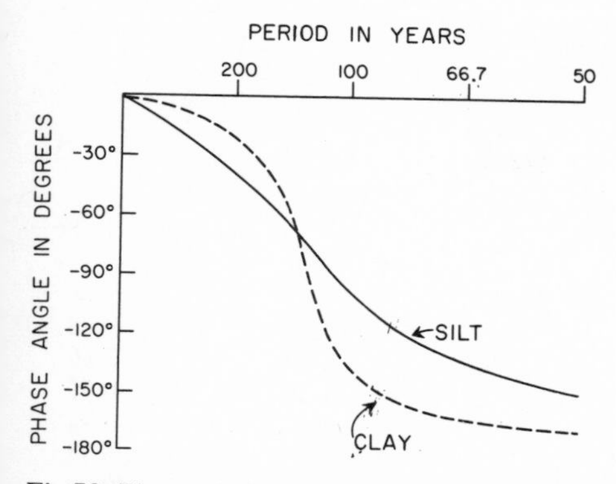

Fig.70. Theoretical phase difference between silt and clay as a function of period if silt and clay thickness-variation patterns are responses to same second-order Markov process with differences in damping coefficient α. Clay would lead silt for long-period waves where most of the power is concentrated (Fig.69).

$$\phi^*(\omega) = \frac{-2\alpha\omega}{\omega_0^2 + \alpha^2 - \omega^2} \qquad [11.23]$$

Eq. [11.23] can be established by using derivations given by Parzen (1962, p.112) and Sommerfeld (1948, p.101). Consequently, the response $x(t)$ lags behind the initial random process $\eta(t)$. Setting $\omega_0 = 0.05$ as before (see section 10.6), we obtain the curves of $\phi^*(\omega)$ for $\alpha = 0.03$ (silt) and $\alpha = 0.01$ (clay) which are shown in Fig.70. According to this diagram, the response function of clay leads that of silt for periods greater than 130 years. This corresponds to frequency bands where most of the power is concentrated in (Fig.69) for both silt and clay. Consequently, the phase lag of the silt, which averages to about 15 years, is explained, at least in a qualitative manner, by the stochastic model.

The silt and clay thickness data for series 4 were filtered by the method used previously to obtain Fig.55B from 55A. In the vicinity of year 140, counted from the bottom of the section, there is a relatively sudden increase in thickness. Original data and filtered values of silt and clay components for years 130–159 from the beginning of series 4 are shown in Table XXXIII.

Several small peaks are removed from the original data by the filtering. A phase lag of silt thickness with regard to clay of about 6 years is evident. The sudden increase in thickness can be cross-correlated between different localities in the area and was referred to as the "datum" by Antevs (1925) and Hughes (1965).

In total, we now have discussed three separate indications that the silt (with coarser particles) lagged behind the clay during the process of sedimentation. The phenomenon may be explained by adopting the Kuenen-De Geer model for varve sedimentation (De Geer, 1940; Kuenen, 1951). According to this model, the silt was transported by

TABLE XXXIII

Method of filtering applied to silt and clay data for years 130–159 of series 4. The "datum" occurs between years 141 and 142. Significant maxima and minima indicated by asterisk. (From Agterberg and Banerjee, 1969.)

Year	*Thickness (in cm)*		*Log thickness* (x_k)		*Signal* (y_k)	
	silt	*clay*	*silt*	*clay*	*silt*	*clay*
159	0.20	1.75	−1.61	0.56	−0.18	0.71
158	1.10	1.90	0.10	0.64	−0.10	0.69
157	1.90	1.90	0.64	0.64	−0.05*	0.70
156	0.90	1.50	−0.11	0.41	−0.08	0.71
155	1.80	3.00	0.59	1.10	−0.11	0.74
154	0.45	2.20	−0.80	0.79	−0.21	0.83
153	0.70	2.40	−0.36	0.88	−0.25	0.87
152	0.60	2.60	−0.51	0.96	−0.27	0.92
151	0.65	5.80	−0.43	1.76	−0.26	0.99
150	1.70	2.20	0.53	0.79	−0.25	1.05*
149	0.60	2.50	−0.51	0.92	−0.33	0.98
148	2.20	2.20	0.79	0.79	−0.38	0.93
147	0.30	2.00	−1.20	0.69	−0.58	0.89
146	0.35	3.00	−1.05	1.10	−0.70	0.86
145	0.30	3.20	−1.20	1.16	−0.78	0.85
144	0.70	2.50	−0.36	0.92	−0.82	0.79
143	0.55	1.80	−0.60	0.59	−0.90	0.65
142	0.90	1.30	−0.11	0.26	−1.02	0.46
141	0.15	0.70	−1.90	−0.36	−1.25	0.25
140	0.20	0.60	−1.61	−0.51	−1.39	0.04
139	0.20	1.00	−1.61	0.00	−1.52	−0.11
138	0.05	0.60	−3.00	−0.51	−1.62*	−0.17
137	0.20	0.60	−1.61	−0.51	−1.57	−0.27
136	0.30	1.00	−1.20	0.00	−1.51	−0.32
135	0.25	0.70	−1.40	−0.36	−1.48	−0.38
134	0.30	0.70	−1.20	−0.36	−1.46	−0.44
133	0.25	0.75	−1.40	−0.29	−1.47	−0.51
132	0.10	0.20	−2.30	−1.61	−1.49	−0.62*
131	0.35	0.60	−1.05	−0.51	−1.42	−0.53
130	0.30	0.75	−1.20	−0.29	−1.38	−0.46

turbidity currents whereas most of the clay settled from the lake water after being transported in suspension in the meltwater.

The deposition of varves in Lake Barlow–Ojibway took place about 11,000 years ago during the retreat of the Late Wisconsian ice-sheet in North America. This relatively recent process can be studied in more detail than most other geological processes since, in general, uncertainties related to cause and effect increase with age of the geological formation of study. We should be careful not to attach too much significance to the limited data available for this study and the results of statistical analysis based on these data. Nevertheless, the example indicates clearly that methods such as filtering and cross-spectral analysis have great potential as tools for the study of processes of sedimentation.

Results (also see section 10.6) for major changes in the varve-thickness data suggest the following interpretations. The exponential thickness decrease in most varve time series for this glacial lake suggests a rapid linear retreat of the land-ice. Superimposed on this trend, there occurred two relatively rapid increases in thickness. One of these falls at the "datum" (cf. Table XXXIII); the other one near the end of series 4. In time, the second increase coincides with the so-called Cochrane readvance of the land-ice (cf. Antevs, 1925; Hughes, 1965). The stochastic model indicated a 126-year cyclicity superimposed on the linear retreat. Thus, the rate of retreat was accelerated and decelerated periodically. It is possible that at the end of a deceleration, the ice-sheet not only came to a stand-still but occasionaly readvanced relatively rapidly during a short period of time. This phenomenon is referred to as "surging".

In considering causes for the periodicity and possible "surging", Agterberg and Banerjee (1969) pointed out that long-term climatological cycles of this order are known to exist. They are perhaps related to the motion of the sun about the center of mass of the solar system (Jose, 1965). However, in 1969, the proceedings of two seminars on the causes and mechanics of glacier surges were published (Ambrose, 1969). In view of the results described in these 27 papers dealing mainly with glaciers, the possibility must be considered that periodic accelerations and decelerations of the retreating Wisconsian ice sheet, with occasionally associated surging, were intrinsically related to the mechanics of land-ice movement.

Theoretical models as developed by Weertman (1969) and Lliboutry (1969) for surging glaciers admit the possibility of cyclicity for this process. Several authors assumed that ice-sheets may also have been subject to periodical surging.

Wilson (1969) assumed that Carboniferous cyclothems which occur in different places of the world are related to sea-level fluctuations. These would be caused by periodical surging into the sea of continental ice-sheets. In turn, this would lead to melting and a raising sea-level. Afterwards, the ice-sheets would build up again until a new surge occurred according to Wilson. On the other hand, Hollin (1969) pointed out that, until now, there is no conclusive evidence that ice-sheets have surged.

Periodic arrangements of layers in stratigraphic sequences involving more than two components also can be studied by means of Markov-chain techniques as will be done for some Carboniferous cyclothems in Chapter 12. A third method of approach to the mathematical analysis of layered sequences is to formulate random-genetic models for simulation (Matheron, 1970b; Jacod and Joathon, 1971).

11.5. Spectral analysis of minerals in thin section

Serra (1966) has analyzed a thin section of Lorraine oolitic iron ore, France, by coding occurrences of three minerals at 6,000 points forming a regular grid with spacing equal to

0.05 mm. The minerals were: (1) matrix consisting of chlorite; (2) limonitic oolithes; an (3) quartz.

The coded array was shifted with respect to itself in four directions (N, E, SE and NE By machine, the following frequencies (or probabilities) were computed:

$$\begin{bmatrix} P_{11} & P_{12} & P_{13} \\ P_{21} & P_{22} & P_{23} \\ P_{31} & P_{32} & P_{33} \end{bmatrix}$$

P_{ij} denotes the relative number of times mineral i is followed by j after a single shift In order to study minerals separately, Serra calculated:

$$\gamma_i = \tfrac{1}{2} \Sigma (P_{ik} + P_{ki}), \quad k \neq i \qquad [11.24$$

For example, $\gamma_1 = \frac{1}{2}(P_{12} + P_{13} + P_{21} + P_{31})$. This is the first value of the semivariogran for mineral 1.

We can write for each mineral:

$$\begin{bmatrix} P_{11}(l) & \gamma_1(l) \\ \gamma_1(l) & P_{RR}(l) \end{bmatrix}$$

where l denotes the number of shifts and P_{RR} is for minerals other than mineral 1

Suppose that, when l increases to infinity, randomness prevails with $\gamma_1(\infty) = P_1 \cdot P_R$ The totals $P_1 = P_{11}(l) + \gamma_1(l)$; $P_R = P_{RR}(l) + \gamma_1(l)$ are independent of the value of l They represent volume percentage values of mineral 1 and of the sum of other mineral within the domain of analysis. We have $P_R = 1 - P_1$.

The experimental variograms of Serra (1966) assume horizontal position for large value of l. Different directions have different variograms. The schematic diagram o Fig.71 shows the characteristic features. The oolithes occur frequently in groups of tw or three (separated by chlorite). This would explain the waves in Fig.71A. Serra's inte pretation of Fig.71B is that larger chlorite crystals may envelop grains of other mineral Hence, the "hole effect" (effect du trou) in this diagram. On the other hand, the quart grains are separated from one another and occur randomly in the rock (Fig. 71C).

The author has carried out similar experiments on four thin sections from rocks of th Muskox layered intrusion, District of Mackenzie, Canada. Some results were reported b Agterberg (1967a). T.N.Irvine of the Geological Survey of Canada has made photomicro graphs available for this study. A grid with spacing equal to 0.1 inch was overlain on 10 enlarged prints and 4,900 observations were made per thin section for areas of 7 X inches (or 0.7 X 0.7 inch in true distance). The dark grains were coded as ones and th light grains as zeros. The two-dimensional autocovariance function $C(r, s)$ was compute by [10.42].

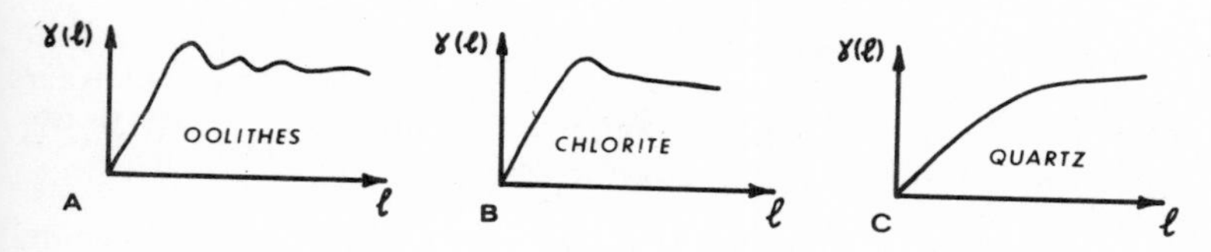

Fig.71. Schematic variograms for presence–absence data of three minerals in thin section of oolitic iron ore, Lorraine, France. (From Serra, 1966.)

The maximum lag $m = M_1 = N_1$ was changed (cf. [10.42]). Results to be shown are for $m = 10$ and $m = 15$. Any profile through the centre of a 2D-autocovariance function gives the 1D-autocovariance function for that direction. In order to compare our results with those of Serra, a profile for the horizontal direction parallel to foliation in a gabbro (layer 33) is shown in Fig.72. The autocovariance is related to the semivariogram by:

$$C(0,\, l) = (1 - P_1)\, P_1 - \gamma_1(l) \qquad [11.25]$$

When the rock is homogeneous for larger domains, P_1 is a constant, and $C(0, l)$ approaches zero for increasing l.

Agterberg (1967a) assumed that, to a first approximation, the theoretical autocovariance function satisfies:

$$C(0,\, l) = C(0,\, 0)\, \mathrm{e}^{-2n_0|l|} \qquad [11.26]$$

where $C(0, 0) = (1 - P_1)\, P_1$ is the variance and n_0 the average number of boundaries between different minerals per unit of distance. This result was derived earlier (cf. [10.24]) for the Poisson model. n_0 can be estimated independently as follows. The

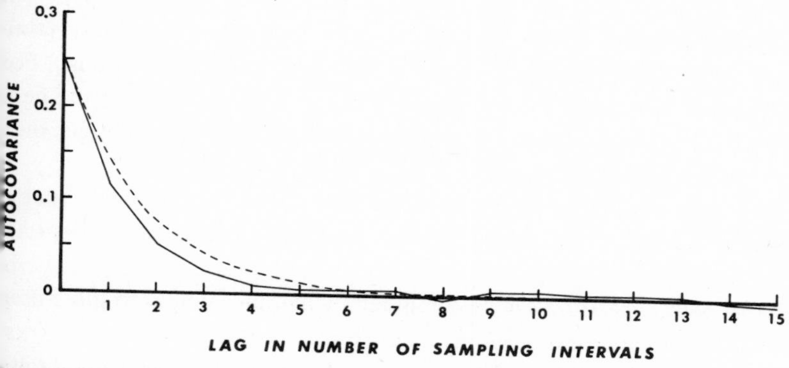

Fig.72. Observed autocovariance function compared to exponential curve calculated from frequency of transitions between dark and light minerals in direction parallel to foliation of gabbro, layer 33, Muskox intrusion; sampling interval is 0.01 inch.

TABLE XXXIV
Modal analyses (in percent) of three rock types, Muskox layered intrusion. Generalized results fc gabbro and peridotite after Smith et al. (1963). Result for troctolitic peridotite after T.N.Irvine (perso nal communication, 1966). A and B indicate groupings made for coding.

	Gabbro (general)		*Troctolitic peridotite*		*Peridotit (general)*
Olivine (+ alteration)			43	} A	69–90 }
Orthopyroxene	5–20	} A			5–10
Clinopyroxene	20–35		12½	} B	10–20
Plagioclase	45–55	B	44½		5–15

average number of points in strings of ones and zeros coded for black and white minera along lines in the direction used for Fig.72 amounts to 3.46 and 3.52, respectively. I true distance, this is equivalent to 0.0346 and 0.0352 inch. It follows that:

$$n_0 = 2/(0.0352 + 0.0346) = 28.7 \text{ boundaries per inch.}$$

Since $C(0, 0) = 0.25$, [11.26] becomes $C(0, l) = 0.25\, e^{-57.4l}$. This curve is also show in Fig.72. The fit is remarkably good. It is readily verified that $C(r, s)$ is independent c which mineral (light or dark) is coded as one or zero.

In total, two gabbros (from layers 33 and 35), a troctolitic peridotite and a peridotit were analyzed by this method. Modal analyses from these rock types of the Musko intrusion are shown in Table XXXIV.

The gabbros are foliated and the sections were cut perpendicular to foliation which subhorizontal in the photomicrographs. Each thin section was coded as ones and zeros b dividing the minerals into two groups (A and B) in Table XXXIV. The foliation of th gabbros is a planar lamination perhaps due to gravity settling. For this rock type, a minerals are both cumulus (settled from the magma by gravity) and postcumulus (crysta lized from an interstitial liquid trapped between the larger crystals of the cumulus). Fc the troctolitic peridotite, the cumulus consists of olivine and plagioclase; for the per dotite the cumulus is olivine only. Other minerals may be present in small amounts; the are not listed in the Table XXXIV.

The purpose of this statistical analysis is to illustrate methods of studying spati relationships between mineral groups rather than to contribute to petrological interpret tion.

A two-dimensional power spectrum can be computed by methods similar to those use for 1D-space series. We computed raw spectral estimates by Fourier transforming $C(r,$ and the results were smoothed by using a 2D-conversion of Tukey's hanning method (als see Preston and Esler, 1967). The result for $m = 15$ (gabbro, layer 33) is shown i Fig.73A. By doubling the grid-spacing to 0.02 inch, a slightly larger area could be used fc

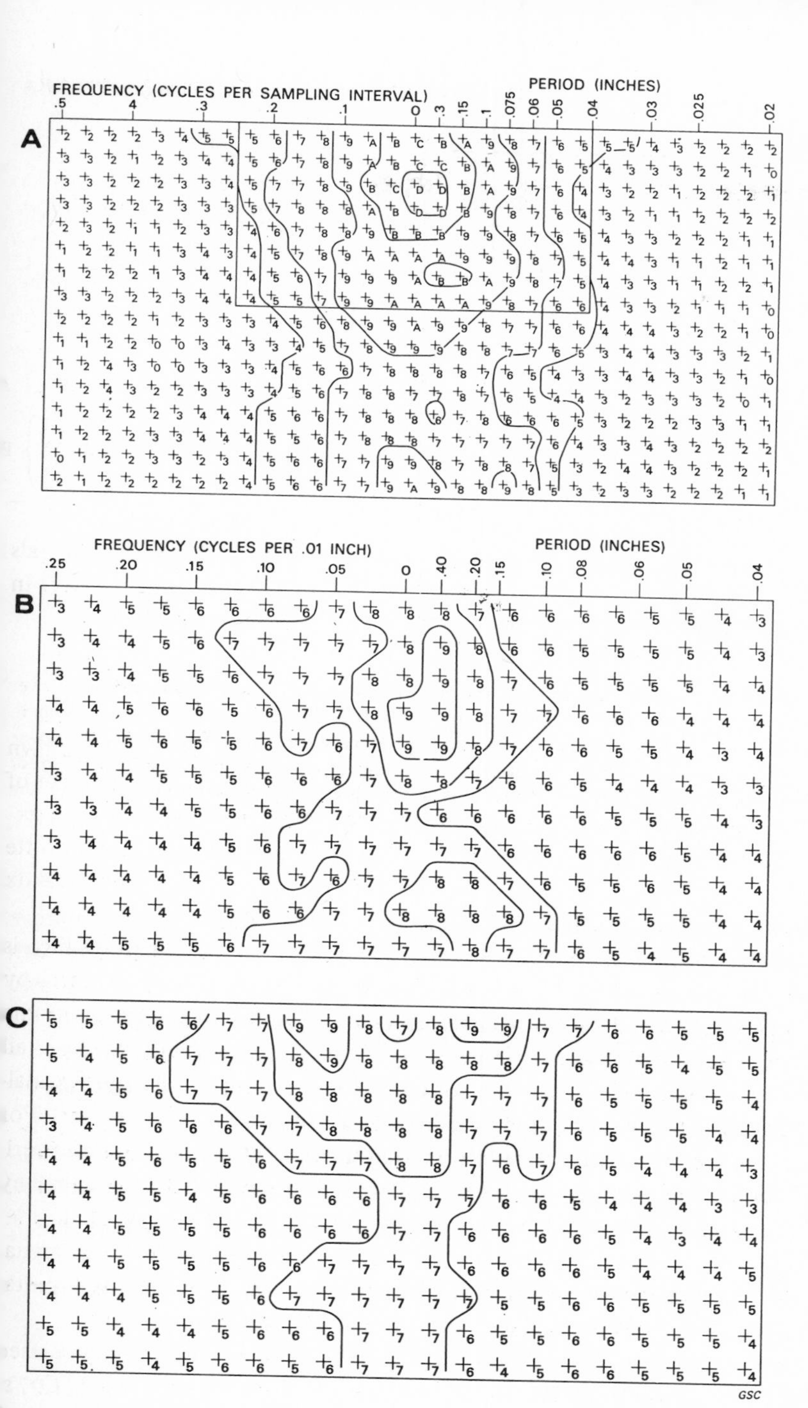

Fig.73. Two-dimensional power spectra calculated by hanning method for thin sections of gabbros of layer 33 (A and B) and layer 35 (C). Horizontal direction is parallel to foliation; lower half of complete spectrum represented only. Frame in A for grid spacing 0.01 inch corresponds to B for grid spacing 0.02 inch. Highest peak for layer 33 with value $D = 13$ in A and 9 in B corresponds to layering; that for layer 35 in C is ring-shaped and represents clustering of pyroxene crystals. Other peaks probably are related to size and shape of crystals. (From Agterberg, 1967a.)

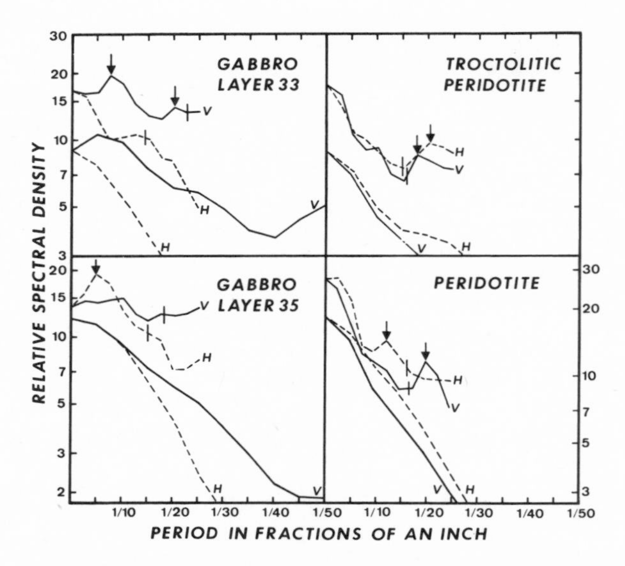

Fig.74. Profiles in horizontal (parallel to foliation) and vertical direction across two-dimensional power spectra of thin sections for four rock types. Curves for gabbros 33 and 35, up to period 1/25 inch correspond to Fig.73B and C, respectively. Peaks marked by arrows. In these four rocks, the space between crystals of one type (e.g., pyroxenes in gabbro) is filled in by many smaller crystals of other type (e.g., plagioclase in gabbro). This may have caused ordering in the spatial arrangement of the grains and would be represented by relatively weak, oval-shaped maxima in the power spectra; predicted locations of these maxima are indicated by vertical bars.

computing the autocovariance function and spectrum with $m = 10$. In order to improve the estimates, refined spectral densities were computed for four overlapping areas with the 0.02-grid and the average of four values was taken at each point. The result is shown for both gabbros in Fig.73B and for horizontal and vertical profiles to the center of the power spectrum for the four thin sections in Fig.74. For comparison, similar profiles are represented for runs with $m = 10$ based on the original 0.01-inch grid. In all diagrams, the refined spectral-density values were plotted by using a logarithmic scale.

On the whole, the spectra provide more information than the 2D-autocovariance functions. Several peaks are superimposed on a general decrease in power from the center of the spectrum. The latter is illustrated by the profiles for the $m = 10$, 0.01-grid runs in Fig.74.

It is noted that the theoretical equation for this trend is not given by the Fourier transform of [11.26]. For the Poisson model, the 2-D autocovariance function is:

$$C(u, v) = C(0, 0) \exp [\alpha\sqrt{u^2 + (v/\beta)^2}] \qquad [11.27]$$

which reduces to an exponential curve for every profile. In the isotropic case, $\beta = 1$.

The coefficient β gives a measure of anisotropy. When the crystals are not round, contours of $C(u, v)$ may be ellipses instead of circles. The two-dimensional spectrum then is proportional to $[\omega_1^2 + (\omega_2/\beta)^2 - \alpha^2]^{-3/2}$ as can be derived from results by Papoulis (1968) or Matérn (1960). The trend line for any profile such as those of Fig.74 then is proportional to $(\omega^2 - \alpha^2)^{-3/2}$ instead of $(\omega^2 - \alpha^2)$, with α depending on grid-spacing and orientation of the profile.

Possibly significant peaks are marked by arrows in Fig.74. For comparison, we also show cyclicities corresponding to the sum of average lengths of strings of ones and zeros. For the gabbro, layer 33 (horizontal direction), a mark was plotted at period 0.0698 which is the sum of 0.0346 and 0.0352 inch for average lengths of dark and light minerals respectively. Theoretically, a periodicity of this type could arise when grains or clusters of grains of one coded mineral group tend to be surrounded by those of the other group and vice versa.

The two peridotites have periodicities of this type mainly because the olivine crystals lie in a matrix of other minerals. The latter are smaller although their clusters (which were coded) are of the same size as the olivine crystals. This type of periodicity would form a ring or oval-shaped maximum at considerable distance from the origin of the spectrum. Since a maximum of this type is not concentrated at a point, it usually is not clearly developed in the spectrum.

Certainly, for the gabbros, it is difficult to see maxima of this type in Fig.74, but their existence may be inferred from the contours in Fig.73 for the vertical direction.

In Fig.73A, this periodicity has a harmonic at twice the distance from the origin. Both gabbros show another peak closer to the origin.

For the gabbro of layer 33, its period is about 0.15 inch. This may be a single peak concentrated at a point in the vertical direction, and layering parallel to the foliation is indicated. On the other hand, for the gabbro of layer 35, the peak seems to be oval-shaped and is more strongly developed in the horizontal direction. It indicates the existence of clusters of pyroxene crystals separated by clusters of plagioclase crystals.

In general, information is gained by adopting a two-dimensional instead of a one-dimensional approach. The spectra can contribute information with regard to hidden layering and clustering. The semivariogram and autocovariance function may provide additional data for average shape and texture of individual crystals or clusters of crystals.

Spectra for thin sections with dark and light grains were constructed optically by Davis (1971).

A drawback of dividing the minerals into two groups by coding is that boundaries between grains of the same mineral are not considered. In order to study these transitions, another approach is needed. Vistelius (1966a) and Kretz (1969) have developed methods based on counting boundaries between individual crystals along traverses across a thin section. These procedures are similar to the Markov-chain approach to be discussed in Chapter 12.

In addition to having the advantage that boundaries between grains of the same mineral can be considered, Markov-chain methods provide a multivariate approach to the problem. A drawback is that, until now, these methods are one-dimensional and restricte to one or a few transitions.

In recent years, different mathematical methods have been developed for the analys of textures of mineral grains in rocks. Reference is made to work by Engel (1957), Gig (1968), Giger and Erkan (1968), Lafeber (1966), Matheron (1971b), Serra (1971), an Switzer (1965). Several workers in this field used advanced methods of geometric probability. It is beyond the scope of this book to review this interesting topic in mo detail.

11.6. Two-dimensional harmonic analysis

The inverse Fourier transform $A(p, q)$ of a two-dimensional array of gridded $(m \times r$ data, $\mathbf{X}(i, j)$ satisfies:

$$A(p, q) = (1/mn) \sum_{i=0}^{m-1} \sum_{j=0}^{n-1} \mathbf{X}(i, j) \exp\,[-2\pi I(ip/m + jq/n)] \qquad [11.28$$

where I denotes $\sqrt{-1}$. Without loss of generality, it can be assumed that the average $\bar{X}$ zero. Every value $A(p,q)$ consists of a real part $\mathrm{Re}\,(A_{p,q})$ and an imaginary part I $(A_{p,q})$. It can be represented by a single sine-wave which is determined by four paran eters: (1) direction of axis, (2) period, (3) amplitude, and (4) phase angle with respe to an origin. Examples will be given below. Each wave can be described by a continuo function:

$$X^*(p, q; u, v) = \mathrm{Re}\,(A_{p,q}) \cos\,[2\pi(pu/m + qv/n)] - \mathrm{Im}\,(A_{p,q}) \sin\,[2\pi(pu/m + qv/n)] \qquad [11.29$$

where u and v are geographical coordinates. When i and j denote columns and rows o $\mathbf{X}(i, j)$ respectively, the U-axis points eastward and the V-axis southward.

The square of amplitude is given by:

$$P^*(p, q) = \mathrm{Re}^2(A_{p,q}) + \mathrm{Im}^2(A_{p,q}) \qquad [11.30$$

The values $P^*(p, q)$ form a two-dimensional array equivalent to the periodogram. The can be standardized by computing $P(p,q) = m \cdot n \cdot P^*(p,q)/s^2$, where s^2 is the mean squar of all $(m \times n)$ data $\mathbf{X}(i, j)$.

If $\mathbf{X}(i, j)$ is a realization of a two-dimensional stationary random variable, $P(p, q)$ i distributed as $\chi^2/2$ with two degrees of freedom. $\mathbf{X}(i, j)$ is the Fourier transform c $A(p, q)$ with:

$$\mathbf{X}(i, j) = \sum_{p=0}^{m-1} \sum_{q=0}^{n-1} A(p, q) \exp\,[2\pi I\,(ip/m + jq/n)] \qquad [11.31$$

The following three numerical problems can be solved by using [11.29] and [11.31]: 1) representation of $\mathbf{X}(i, j)$ by a double Fourier series, (2) computation of a 2-D power pectrum, (3) computation of the 2-D autocovariance and autocorrelation functions.

The first problem results in the harmonic trend surfaces discussed by Bhattacharyya 1965), Harbaugh and Preston (1965), Krumbein (1966), James (1966, 1967) and Har- augh and Merriam (1968). Using present notation, we would develop X^* of [11.29] for block of values (p, q) and sum the individual sine-waves. For example, a harmonic trend urface of the second degree satisfies:

$$\hat{Y}(u, v) = \sum_{p=-2}^{2} \sum_{q=-2}^{2} X^*(p, q; u, v) \qquad [11.32]$$

For the second problem, the computation of $P(p, q)$ gives an answer but values for locks of P-values can be averaged in order to obtain a smoothed version of the spectrum. we average four values together, $\bar{P}(p, q)$ is distributed in proportion to χ^2 with X 2 = 8 degrees of freedom. It may be useful to work with the moving average (over- pping blocks of four or more values) in that situation.

A 2-D autocovariance function is obtained by taking the Fourier transform of $^*(p, q)$:

$$C^*(r, s) = \sum_{p=0}^{m-1} \sum_{q=0}^{n-1} P^*(p, q) \exp\,[2\pi I(pr/m + qs/n)] \qquad [11.33]$$

owever, this result is for a cyclical scheme where the array $\mathbf{X}(i, j)$ repeats itself in both e U- and V-directions. Other consequences of this artificial cyclicity with fundamental avelengths determined by m and n were discussed by Krumbein (1966). For example, a ırmonic trend surface based on a double Fourier series must have identical values at its estern and eastern boundaries and also at its northern and southern boundaries.

In section 10.8 we obtained an autocovariance function satisfying [10.42]. If $\bar{X} = 0$ which can be done in practice by subtracting $\bar{X}$ from all data X), the array $\mathbf{X}(i, j)$ n be artificially enlarged by zeros. When the maximum lags amount to M_1 and N_1 spectively, we will need at least M_1 and N_1 extra columns and rows for $\mathbf{X}(i, j)$ con- sting of zeros.

When $P^*(p, q)$ is developed for the enlarged array, and the Fourier transform taken, e result is $C^*(r, s)$ from which $C(r, s)$ satisfying [10.42] is computed by:

$$C(r, s) = m'n'\,C^*(r, s)/[(m - |r|)(n - |s|)] \qquad [11.34]$$

here m' and n' are the dimensions of the enlarged array $\mathbf{X}(i, j)$ with $m' \geqslant m + M_1$ and $\geqslant n + N_1$.

An advantage of the present approach is that the Fast Fourier Transform method ca be used for computing (Cooley, 1966; Cooley and Tukey, 1965; Tukey, 1967; Gentle man and Sande, 1966).

In order to use the F.F.T.-method, $\mathbf{X}(i, j)$ is expanded by zeros so that the size of th enlarged array $(m' \times n')$ becomes an even power of 2. Hence, $m' = 2^{k_1}$ and $n' = 2^{k}$ where k_1 and k_2 are integer numbers. In practice, it may be convenient to choose $m' = n$ and work with square matrices.

It is noted that two-dimensional data arrays also can be treated by methods of spatia filtering well-known in geophysical data processing (cf. Spector and Grant, 1970). Thes methods were introduced into geology by Robinson (1970) and Robinson and Merriar (1972).

The two-dimensional power spectrum of yes–no data is equivalent to the diffractio patterns in optics and X-ray analysis. There is a close relationship between spectra analysis and optical data processing as applied by Davis (1971) and Pincus and Dobri (1966).

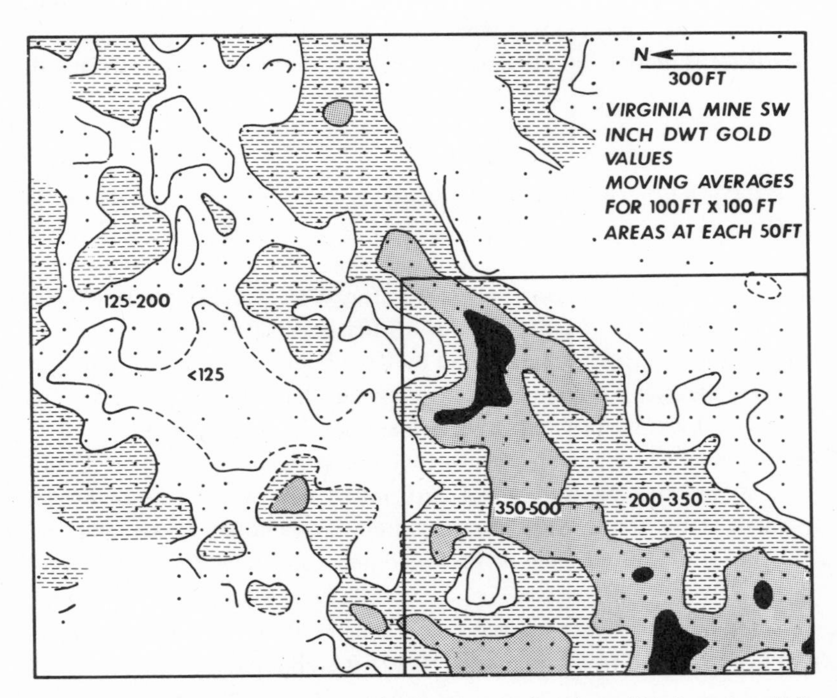

Fig.75. Manually contoured map of moving-average data for gold in inch-dwt. (After unpublished ma by Dr. D.G.Krige; also fig.1 in Whitten, 1966.) Each value of 50-ft. grid is average for enclosin 100 × 100 ft. cell; outline of area selected for harmonic analysis.

Gold values from the Virginia Mine, South Africa

A map of average gold values for the southwestern portion of the Virginia Mine is shown in Fig.75. Each contoured value is the average of all individual values in the surrounding 100 X 100 ft. area.

FEET

								PERIOD	1600	800	533	400	320	267	229	200
1	0	2	1	6	1	12	5	0	5	12	1	6	1	2	0	1
0	2	1	0	1	2	34	38	10	12	9	1	2	0	1	1	1
0	5	4	0	3	2	12	13	9	13	11	0	2	1	4	1	0
0	1	2	0	3	0	3	2	1	1	0	1	1	0	1	0	1
0	1	4	1	5	2	2	0	2	1	4	0	4	3	1	1	0
0	1	4	4	1	0	3	4	0	3	2	1	4	1	1	2	0
2	0	2	2	1	2	1	2	5	2	2	2	0	1	2	1	0
1	0	0	0	1	1	4	1	1	1	0	1	1	0	2	0	0
0	0	1	0	0	0	0	1	2	1	0	1	0	1	1	0	0

A

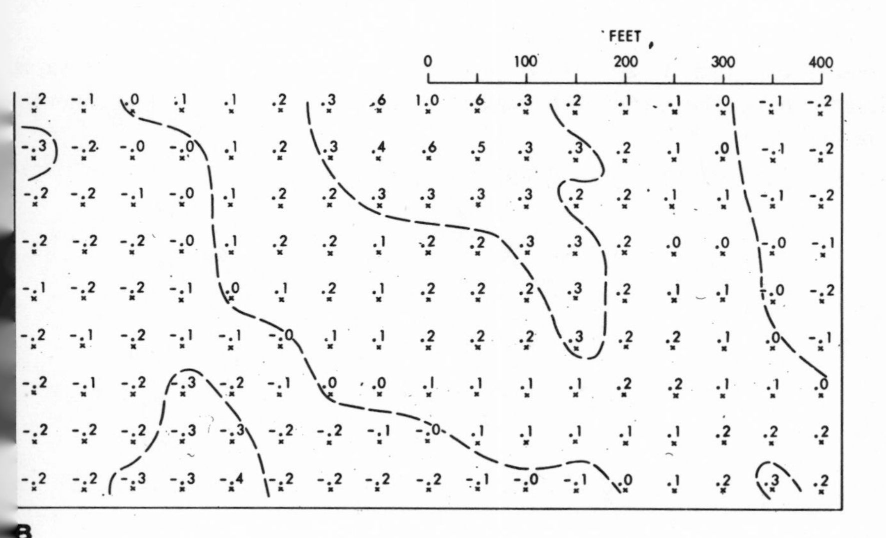

B

Fig.76. Part of two-dimensional power spectrum (A) and autocorrelation (B) functions for array depicted in Fig.75. Note that largest values in A are concentrated in block around the origin, suggesting a trend which may be described by harmonic trend surface of second degree. Autocorrelation surface is elongated in the direction of the trend.

Moving average values for a 50-ft. grid were subjected to various types of statistical analysis (Krige, 1966a; Krige and Ueckermann, 1963; Whitten, 1966b).

We have taken (18 X 18) values for a 900 X 900 ft. area in the lower right-hand side corner of Fig.75. Some gaps in the array were filled in by using interpolation values. The spectrum $P(p, q)$ and autocorrelation function are shown in Fig.76. In order to use the F.F.T.-method, the array was enlarged to size (32 X 32). Only part of the full (32 X 32) spectrum and autocorrelation function are shown. Both maps are symmetrical with respect to the central point (origin) and only part of the lower half is shown.

It is noted that the relatively large values of $P(p, q)$ are located within a block around the origin. This indicates that the second-degree harmonic trend surface provides a reasonable approximation. The autocorrelation function was contoured by hand. It shows a pattern comparable with those derived for the Hartebeestfontein Mine by Krige et al. (1969).

The values of the autocorrelation function for points at a distance of less than 100 ft. from the origin in Fig.76B are biased because the original values for the 50-ft. grid are averages for overlapping cells measuring 100 ft. on a side.

The Whalesback copper deposit

The array of Table XXVIII, which was based on Fig.47 and consists of 176 data, was enlarged to size (32 X 32). Part of the power spectrum $P(p, q)$ is shown in Fig.77. The corresponding autocorrelation function was discussed previously (see Fig.61).

The spectral-density values can be contoured. They form a peak for a direction dipping about 45° toward the lower right-hand side corner of the spectrum. If the period-scale is rotated around the origin an average period of 460 ft. is indicated for the sine-waves constituting the peak. The axis of these waves is perpendicular to the line that connects the peak to the origin.

The contoured pattern for the underground drillhole data of Fig.47 indicates a ridge

FEET

								PERIOD	1600	800	533	400	320	267	229	200
0	0	1	1	0	5	5	1	0	1	5	5	0	1	1	0	0
1	0	1	2	0	1	1	0	0	6	18	16	5	0	0	0	0
0	0	0	1	1	1	1	0	1	11	29	28	12	2	0	1	0
1	0	0	0	0	0	0	0	2	10	23	27	15	3	0	1	2
1	1	0	1	1	1	0	0	2	4	8	13	11	3	0	0	1
1	1	0	1	2	1	1	0	1	1	1	3	5	3	1	1	1

Fig.77. Part of two-dimensional power spectrum for data in Table XXVIII; sharply defined peak reflects zoning of copper percent-foot values shown in Fig.47.

FEET

PERIOD 1600 800 533 400 320

1	1	3	5	9	10	9	9	8	7	0	7	8	9	9	10	9	5	3	1	1
1	1	1	2	4	5	4	4	3	1	4	5	9	14	18	20	18	13	8	6	5
4	3	3	6	7	7	3	1	0	1	4	4	10	20	28	30	30	23	13	12	11
9	7	7	14	16	12	5	1	1	2	4	5	12	25	34	38	40	31	16	16	15
12	10	13	23	24	16	8	4	2	3	5	6	14	28	37	43	45	32	17	17	15
13	13	17	26	27	17	9	5	4	5	7	8	15	25	35	44	42	27	17	13	11
10	15	18	22	24	14	7	5	5	8	10	11	15	19	28	36	29	18	13	8	5
10	17	17	18	18	9	4	4	7	11	15	15	15	14	17	27	13	8	7	3	2
17	23	17	16	12	5	2	4	7	13	19	17	14	10	8	7	3	3	2	1	1
32	33	20	14	7	2	0	2	6	14	19	18	14	7	2	1	3	5	2	2	3
42	39	22	11	4	2	0	1	5	14	17	17	13	4	1	3	8	9	4	5	8

A

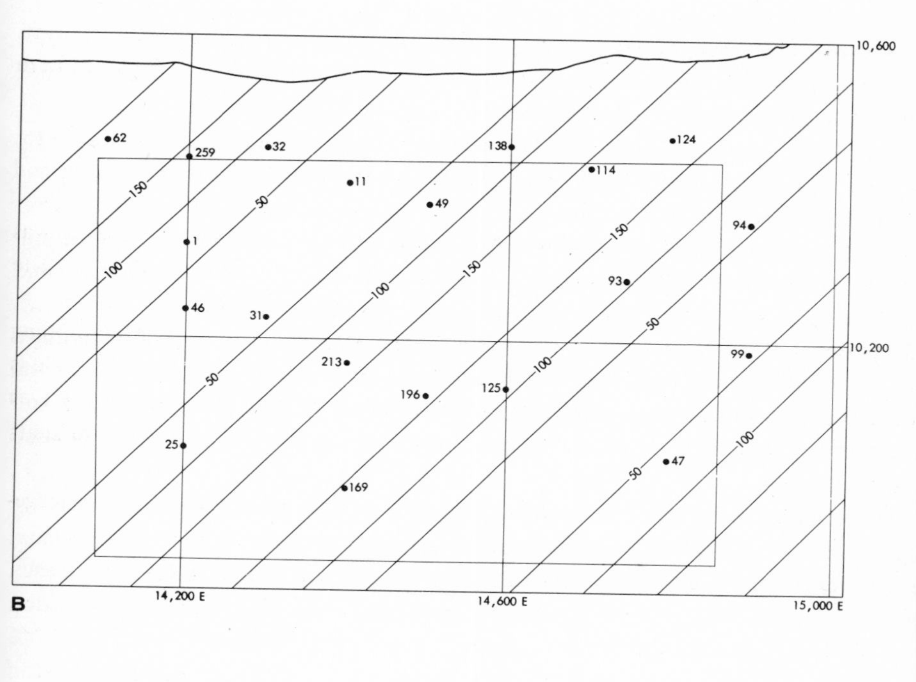

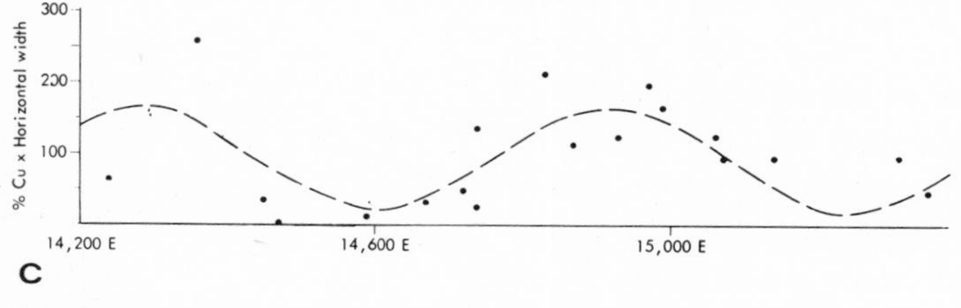

Fig.78. Harmonic analysis of irregularly spaced data. (From Agterberg, 1969.) Intersection points of twenty exploratory holes drilled from surface with orezone shown in B. Sine-waves satisfying [11.29] were fitted to twenty values by least-squares method; percentage explained sum of squares for each fitted wave shown in A. Wave for value 44 on peak in A is shown by straight-line contours in B, and in cross-section for horizontal line (V = 10,600) near topographic surface in C.

shaped maximum with axis dipping in the western direction. This central zone is flanked by two elongated minima and at about 460 ft. from it there occur two other relatively copper-rich zones, although their maxima are not as high as that for the central zone. The same configuration is clearly expressed in Fig.77.

The following experiment indicates that harmonic analysis can provide good estimates of spatial variation patterns in practice.

Development of the Whalesback copper mine was done in two separate stages. First holes were drilled from the surface. Copper percent-foot values for twenty surface-holes are shown in Fig.78B. The underground-holes of Fig.47 were drilled later. They give a fair representation of the true amount of copper in the mineralized zone.

The question can be asked of how well the pattern of Fig.47 can be predicted by a statistical analysis of the data of Fig.78B. This problem is analogous to that of constructing a topographical contour map of a mountain chain when only a few elevations are given. Trend-surface analysis produces patterns which show no resemblance to Fig.47. In fact, analysis of variance indicated that the fit of linear and quadratic surfaces was not statistically significant. ($R_l^2 = 0.16$, and $R_q^2 = 0.32$, respectively). The reader may attempt other contouring methods. We fitted [11.29] and estimated the coefficients b_1 = Re [$A_{p,q}$] and b_2 = $-$Im [$A_{p,q}$] by least squares for many possible directions of axes and periods for the waves. Ideally, we should have a least-squares model by which the four parameters (direction of axis, period, amplitude and phase) would be estimated simultaneously. However, [11.29] is nonlinear. A linear model can only be used when p/m and q/n are assumed to be known. This is the reason for trying many solutions.

This method has been discussed by Draper and Smith (1966) as one way of obtaining a best solution by least squares when the model is nonlinear. Other methods such as the steepest-descent method (Draper and Smith, 1966; James, 1967) can also be used and give more precise estimates provided that crude estimates of the p/m and q/n to be used are initially available.

For each sine-wave, we plotted the R^2-value (measure of degree of fit) as a percentage in Fig.78A. The result can be regarded as a power spectrum for irregularly spaced data. Since three coefficients were fitted (including a constant term related to the mean), each R^2-value can be converted to an F-value at the upper tail of the theoretical F-distribution with 3 and 17 degrees of freedom. The $F_{0.95}$- and $F_{0.99}$-values correspond to 27- and 37-% values for R^2 and these were contoured. We note that two contoured peaks are present in Fig.78A. One of these corresponds to that of Fig.77.

In addition, there are several other maxima and minima of lesser magnitude. Of course we know that by gradually increasing the number of drillhole data for the ore zone, and repeating the present statistical analysis many times, patterns similar to that of Fig.78A would gradually approach that of Fig.77 which was based on data from 150 underground holes. The "true" distribution would be approached as the number of data increased.

The contour map for a sine-wave whose amplitude falls on the peak of Fig.78A is shown in Fig.78B. Its profile for a line along the surface (V = 10,600) is also shown

(Fig.78C) and the twenty observations were plotted by projecting them along lines parallel to the contours. Obviously, the contours in Fig.78B are an oversimplification of those in Fig.47 but the zones with alternately high and low copper content are correctly predicted.

11.7. Harmonic analysis of copper and gold occurrences in east-central Ontario

The areal distribution of copper deposits in part of the Superior province of the Canadian shield (see Fig.79) is shown in Fig.80. It comes from a compilation made by Agterberg et al. (1972) and criteria for selecting deposits are discussed in that publication and by Agterberg and Fabbri (1973). The large copper deposits with more than 1,000 tons of copper at minable grade (0.8–6.0% Cu) are indicated by circles.

Similar patterns of deposits were analyzed by Leymarie (1970) who used two-dimensional autocorrelation. In that case, two identical copies of the two-dimensional original map are shifted with respect to one another in all possible directions. Every time that a point on the first map coincides with a point on the second map, this event is recorded as a point on a third map. When the number of point falling within a small area on the new map is divided by total number of points on this map, the result is the probability that, on the original map, a point will fall in a small area of the same size and shape at a given distance in a given direction from another point on the original map.

If in nature, there exists a tendency for the deposits to cluster along equally spaced lines, this may show up as a similar but better-defined set of lines in this type of autocorrelation pattern.

Several methods of approach can be taken in "dot-map" analysis. One can assume that the dots obey a two-dimensional stochastic process. The reader interested in that particular problem may consult Cox and Miller (1965) and Bartlett (1964). Our analysis of Fig.80 consisted of putting a grid on top of the area and counting numbers of dots falling in 4 × 4 miles cells. This gave a basic array of numbers for harmonic analysis. The methodology is comparable to that used by Bartlett (1964).

Kutina (1968) has emphasized that endogenic mineral deposits do form regular patterns in some areas. He related the phenomenon to the existence of deep-seated fractures in the crust of the earth which can be both approximately parallel and spaced at approximately equal intervals. If this type of regularity can be established from data in a region, it can be used for further prospecting in that region.

The data from Fig.80 were coded as an array with 28 columns and 20 rows. As done in previous examples, all data were corrected for the mean $\bar{X}$ and the array was augmented to size (32 × 32) by zeros. The inverse Fourier transform was calculated by [11.28]. The resulting spectrum is shown in Fig.81A. A moving average for four P-values was contoured on the basis of the 95- and 99-% fractiles of the theoretical χ^2-distribution. It is noted that the number of degrees of freedom (df) must be corrected by the factor

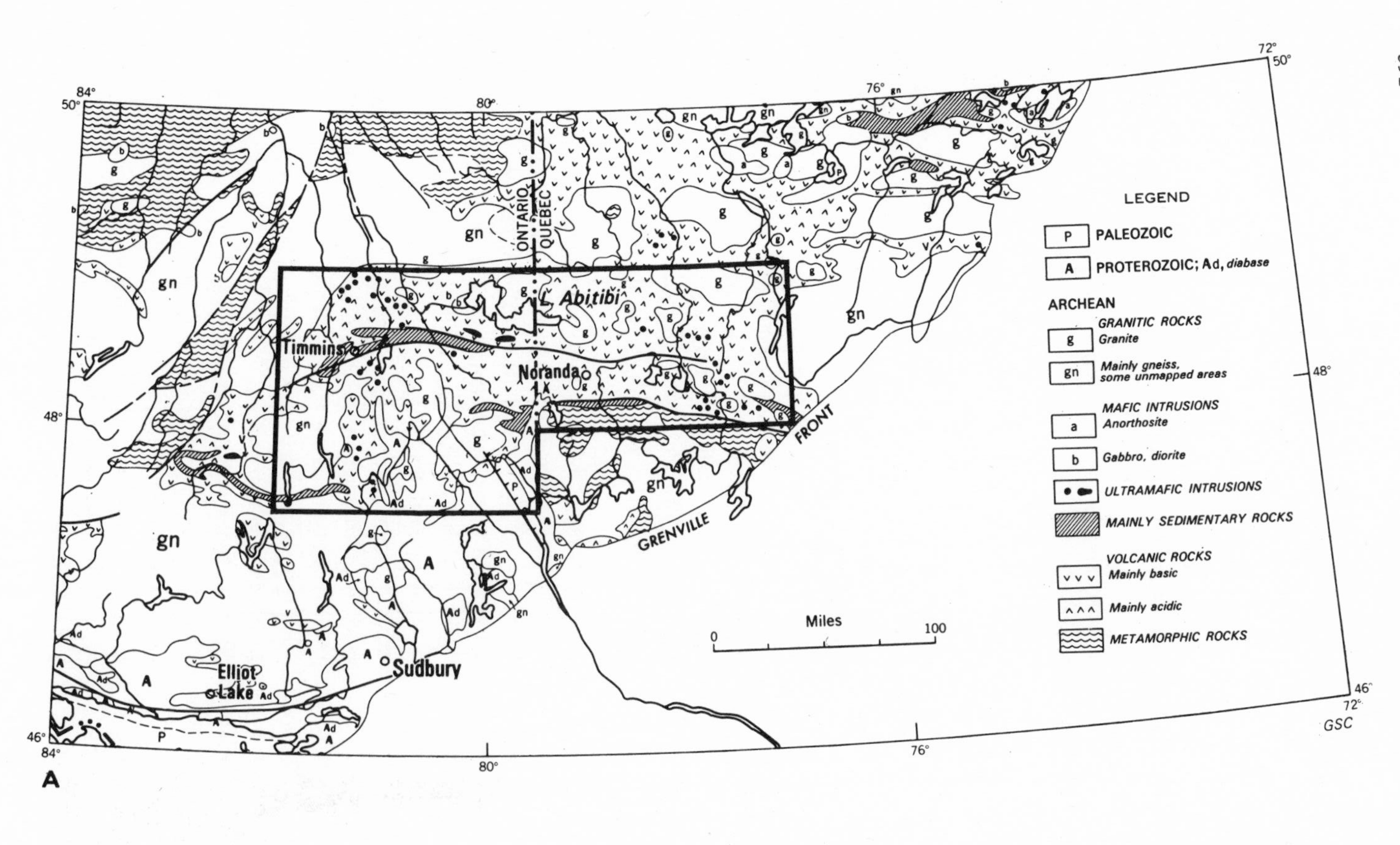

Fig. 79.A. Geological map of part of Superior province, Canadian shield. (After G.S.C.-Map no. 1250A by R.J.W. Douglas.) Area of B is outlined.

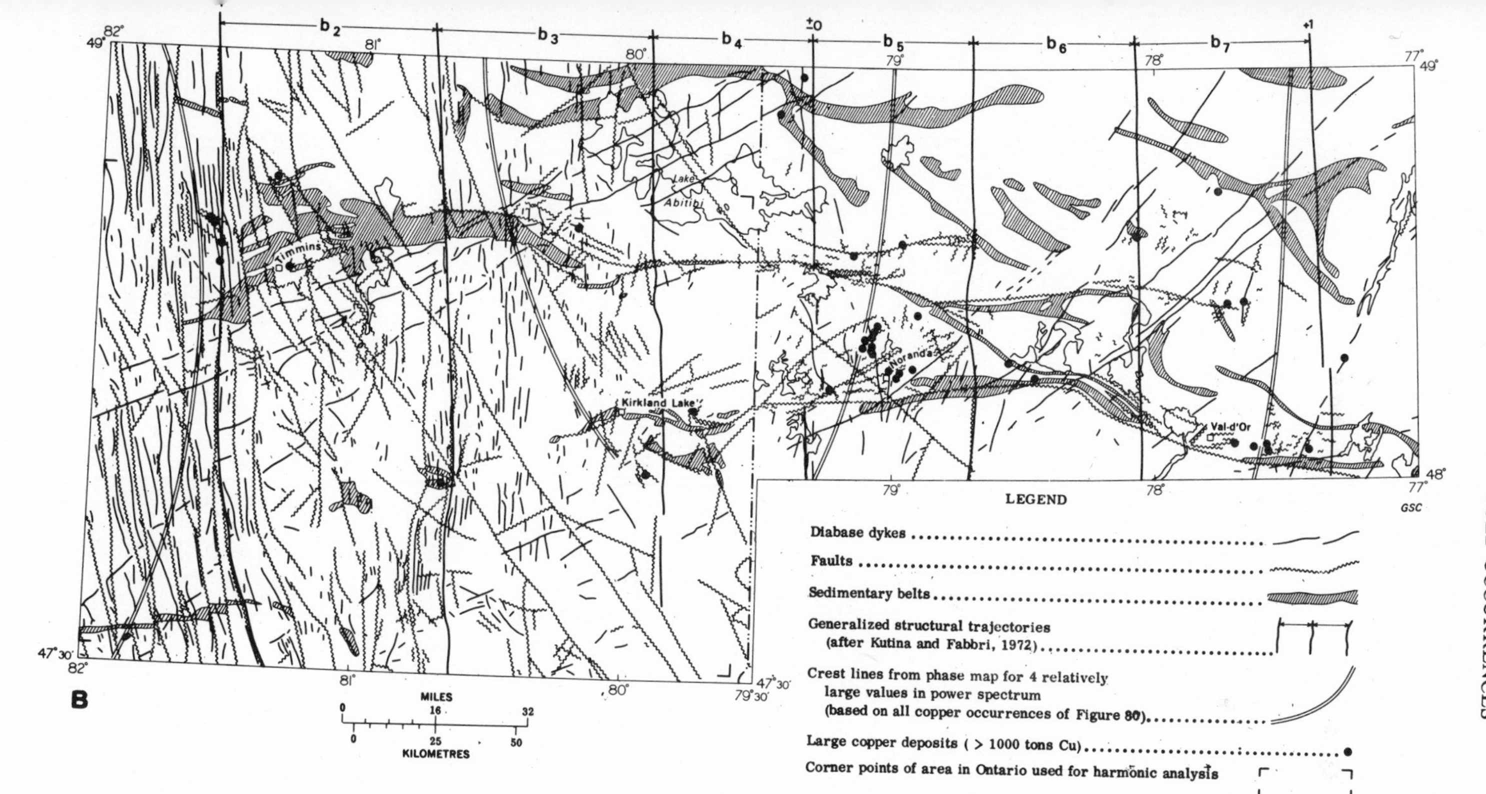

Fig.79.B. Generalized structural trends in the Abitibi area.

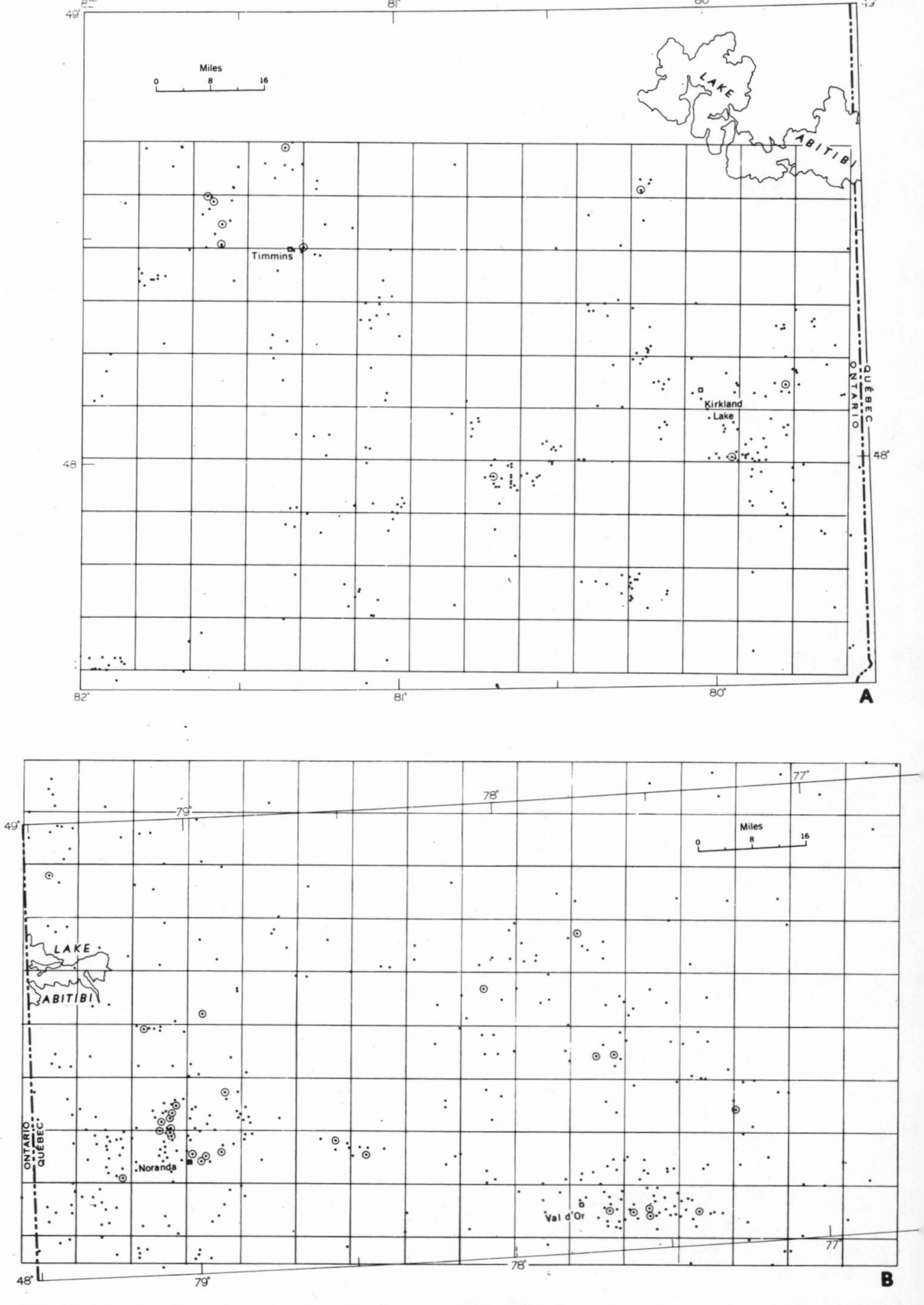

Fig.80. Distribution of copper deposits in area of Fig.79B; deposits with more than 1000 tons copper are circled. Alternate lines (8 miles apart) are shown for grid (4 X 4 miles cells) used for harmonic analysis.

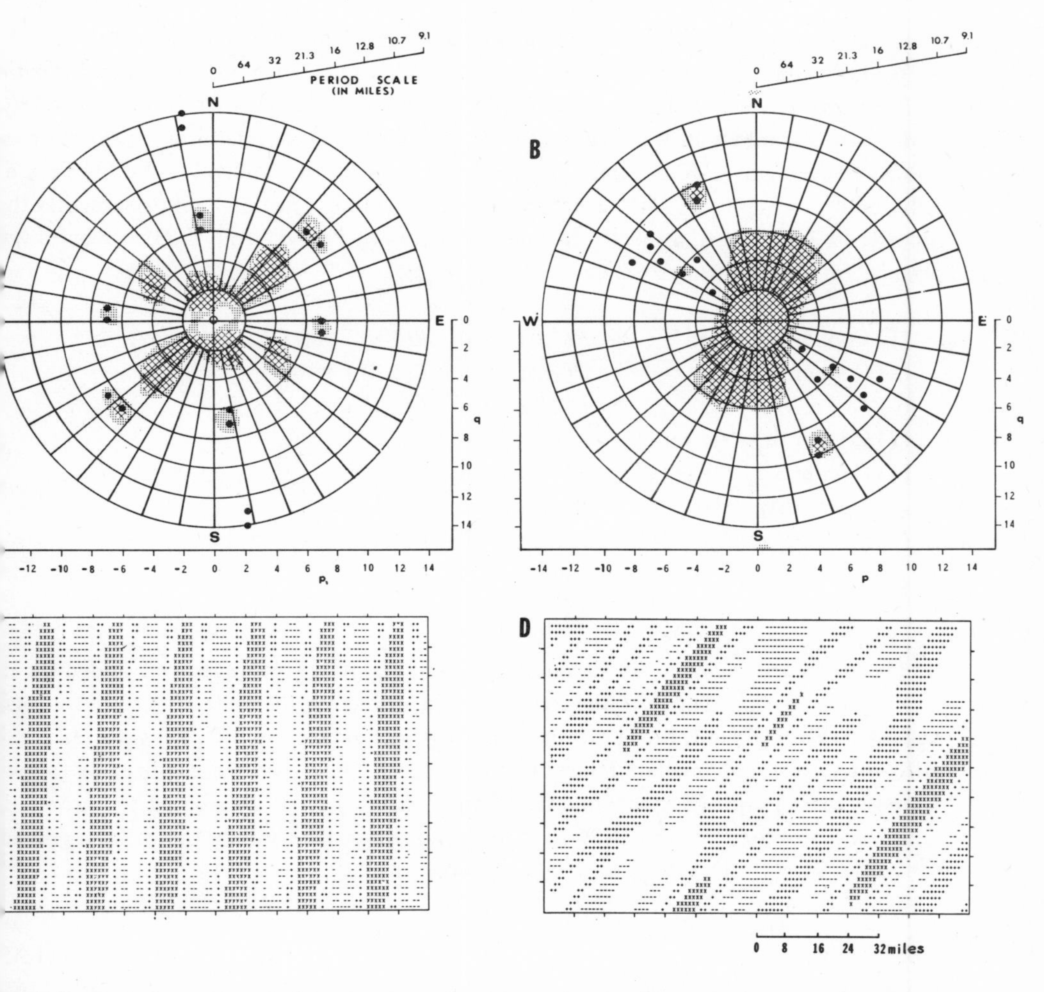

g.81. Power spectra (A and B) and phase maps (C and D) for distribution of copper and gold :currences in east-central Ontario (gridded area in Fig.80A). Dots in A and B denote P-values selected r constructing phase maps. (From Agterberg and Fabbri, 1973.)

$\times m/n' \times m'$ when zeros are added to the basic array of coded data. For individual P-val-es $df \approx (20 \times 28)/(32 \times 32) \times 2 \approx 1$. Hence, averages of four values will be distributed :cording to $\chi^2(4)/4$ by approximation.

The P-values were standardized so that their average amounts to one. The χ^2-test now related to a two-dimensional random variable whose clustering effects, if present, would ot extend beyond the boundaries of the cells used for coding. This condition is not tisfied in Fig.80. It implies that the present test is conservative, viz. values are more gnificant than indicated by the contours. Gold deposits in the same area were subjected to the same treatment as the copper deposits with the result shown in Fig.81B.

We have selected relatively large P-values at some distance from the origin (with peri $\leqslant$ 20 miles). These values are marked by dots in Fig.81A. They may be accompanied by contoured maximum in the diagram but this is not always so, because the contours a based on averages of four values. As mentioned previously, individual P-values are d tributed according to χ^2 with a single degree of freedom. The 95- and 99-% fractiles th amount to 3.8 and 6.6, respectively. Selected values of $P(p, q)$ from the two spectra we arranged in the following six sets:

(1) Copper N–S set: $P(7,0) = 4$; $P(7,1) = 5$
(2) Copper E–W set: $P(1,6) = 4$; $P(1,7) = 5$
$P(2,13) = 3$; $P(2,14) = 5$
(3) Copper SE–NW set (1): $P(-6,6) = 7$
(4) Copper SE–NW set (2): $P(-6,6) = 7$; $P(-7,5) = 4$
(5) Gold WSW–ENE set: $P(4,8) = 7$; $P(4,9) = 4$
(6) Gold SW–NE set: $P(3,2) = 4$; $P(4,4) = 4$;
$P(5,3) = 3$; $P(6,4) = 3$; $P(7,5) = 2$;
$P(7,6) = 5$; $P(8,4) = 2$

Eq.[11.29] was developed for each value but the resulting sine-waves were added for t P-values within a set. The resulting continuous functions are shown in map-form for sets and 6 in Fig.81C and D. These maps are for interfering sets of sine-waves and ha different largest and smallest values. For this reason, they were standardized as follow Suppose that $X^*(u, v)$ is the function representing a set of interfering sine-waves. I value was calculated for the $28 \times 20 = 560$ centers of cells forming the original grid a the values X^*_{max} and X^*_{min} were taken to compute:

$$X(u, v) = \frac{X^*(u, v) + |X^*_{min}|}{X^*_{max} + |X^*_{min}|} \qquad [11.3$$

The standardized version $X(u, v)$ then varies between zero (new minimum at c centers) and one (new maximum). The interval [0,1] can be divided into equal segmen For Fig.81, we took seven segments to which the symbols: b, –, b, +, b, X, X we assigned. Here b represents a blank. Consequently, the largest 2/7 part of $X(u, v)$ marked by a band of X-es surrounding the crest-lines of the patterns.

The crest-lines of sets 1, 2, 3 and 6 are shown in Fig.82. A combined phase map w computed for sets 1, 2, and 4, and is shown as a three-dimensional block-diagram Fig.83.

The selection of the above sets from the power spectra was rather evident for sets 1, and 5. The P-values in these sets are markedly higher than those for surrounding poin

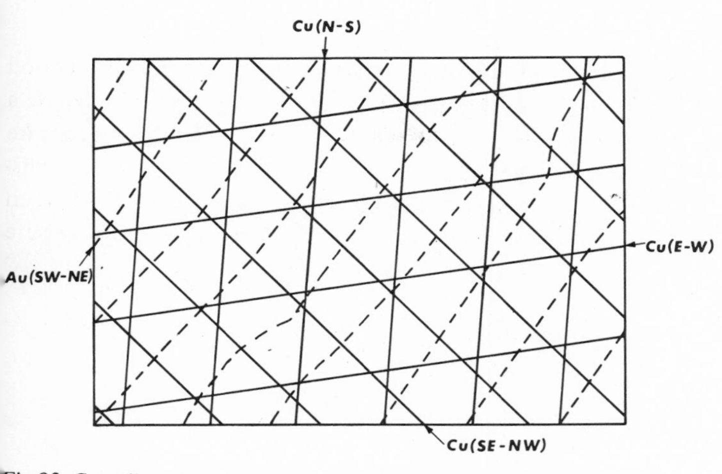

Fig.82. Crest-lines for some phase maps of copper and gold occurrences.

Peaks in the same directions but closer to the origin are not present indicating that the selected peaks represent fundamental periodicities. The waves corresponding to different values of the same set are in phase and the addition of smaller values surrounding the peaks does not change the patterns of the crest-lines. Set 2 consists of waves for two separate peaks. The higher frequency peak represents a harmonic of the lower frequency peak. It has the effect of flattening the synclines between the peaks for the first set as can be seen in Fig.83.

Set 3 is based on a single value of *P* which is equal to 7. However, if we add the largest adjacent value, which results in set 4, the two waves are not exactly in phase. A possible reason for this phenomenon is that between the peak of sets 3 and 4 and the origin, there occurs another peak (see Fig.81A); sets 3 and 4 may in part be a harmonic of this more prominent peak at lower frequencies.

Set 6 is not based on contiguous *P*-values that form a single peak, but on seven relatively large values which form an ill-defined ridge.

From Fig.82 and 83, it can be seen that the three sets for copper pass through approximately the same intersection points where they reenforce one another. Set 6 for gold also seems to be related to this pattern. Set 5 is independent of the other sets.

A comparison of the phase maps with the original dot maps indicates that the crest-lines and their intersection points tend to coincide with clusters of deposits but there are several exceptions. By consulting geological maps of the area, it was found that, if the crest-lines cross granitic stocks or belts of glacial cover, they are not accompanied by deposits. The patterns of Fig.82 and 83 may be used for prediction of occurrence of unknown deposits in places where the bedrock is not exposed because of abundant glacial debris. An exploration strategy would be to assume that the undecovered deposits are described by the pattern derived from the known deposits in the region.

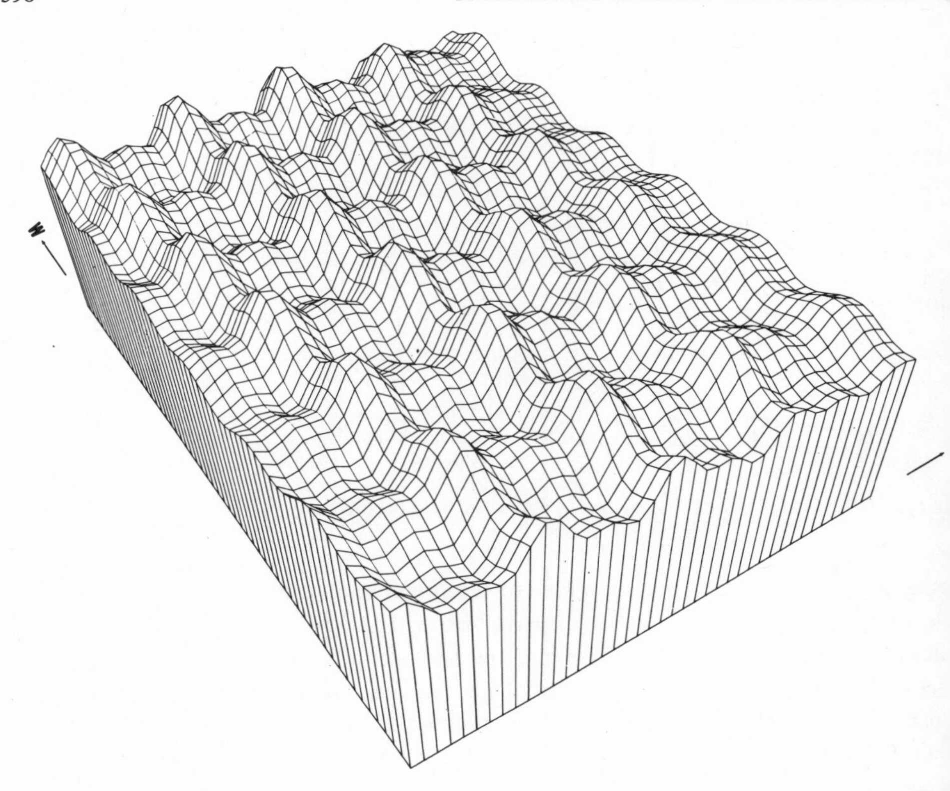

Fig.83. Three-dimensional block diagram depicting composite phase map based on eight P-values in cated by dots in Fig.81A; area is same as gridded area in Fig.80A. Note presence of the harmonic f copper E–W set.

A comparison with structural lineaments in the area indicated several coincidences illustrated in Fig.79B where the features were copied from map No. 2198 of the Ontari Department of Mines and Northern Affairs (1971). All sets correspond to a structur trend in the area.

The copper N–S set coincides with N–S trending faults in several places.Kutina an Fabbri (1972) constructed so-called structural "trajectories" by Kutina's (1968) metho These would indicate deep-seated fractures in the crust of the earth. It is interesting th the copper N–S set with 20-mile spacing tends to coincide with these trajectories whic are spaced at about 40 miles.

The copper SE–NW set also coincides with several faults, notably the Montreal-Riv fault extending to the southeast from Timmins and the major fault crossing Lake Abitit

The copper E–W set tends to coincide with occurrences of sedimentary belts in th area. If the number of copper deposits per cell is related to rock types for these cells b stepwise regression (cf. section 8.9), Archean sediments is the only variable selected significantly correlated to the copper mineralization. The sedimentary belts occur in the centers of large synclines or in graben-structures bounded by faults or shear zones.

The previous three structural trends were developed mainly during the Kenoran orogeny (about 2,480 m.y. ago, see section 6.13), although a number of faults remained active afterwards. Most copper deposits are relatively small bodies occurring within shear zones. They originated during the Kenoran orogeny or somewhat later.

It is noted that the very large copper deposits (circled in Fig.80) occur in specific lithological environments marked by the presence of Archean rhyolites and felsic tuffs indicating ancient volcanic centers (cf. Goodwin and Ridler, 1970). This subject will be discussed further in Chapter 15. They are believed to be older than the Kenoran orogeny. The large deposits are of economic significance but they are few in comparison with the total number of mineralized zones rich in chalcopyrite (principal copper mineral).

It could also be established that most of the gold deposits (gold-bearing quartz veins) occur in specific lithological environments. These are, in part, the same as where the large copper deposits occur (ancient volcanic centers with rhyolites) but gold-veins occur also elsewhere, e.g., near sedimentary rocks of a specific facies, in particular those containing layered iron-formations. Nevertheless, it would appear from the harmonic analysis, that part of the quartz-gold veins was formed by structural processes similar to those related to most smaller copper deposits. The gold SW–NE set is geometrically related to the three copper sets. However, there are few important faults with this direction in the area.

The copper N–S set is parallel to the oldest diabase swarm in the area (N–S trending Matachewan swarm dated at approximately 2,485 m.y. by Fahrig et al., 1965). The gold SW–NE set is parallel to the Preissac swarm which is approximately 2000 m.y. old (W.F. Fahrig, personal communication, 1971). It suggests that part of the gold deposits were formed during Proterozoic time which followed the Kenoran orogeny (2,300–2,600 m.y. ago).

The gold WSW–ENE set is not related to the other four sets. It is parallel to the so-called Grenville front that originated some 955 m.y. ago. It is also parallel to the Abitibi diabase dike swarm which trends WSW–ENE and is 1230 m.y. old. This suggests that the gold WSW–ENE set was formed relatively late in the sequence of geological events for the Abitibi area.

In conclusion, we suggest that methods of harmonic analysis can be used to recognize sets of deposits which formed at different times and were controlled by fracturing of the earth's crust. In nature, these processes were similar to those which led to the formation of diabase dike swarms. Most gold deposits in the area, however, were formed by other processes reflected by the block-shaped central maximum in Fig.81B.

By using different geological arguments, several authors including Dugas (1966), Latulippe (1966) and Hutchinson et al. (1971) also concluded that part of the gold-bearing quartz veins in the Abitibi volcanic belt are relatively young in age.

11.8. Filtering

In this section, we present the derivation of the optimum double exponential filter used to derive the patterns of Fig.55B and Table XXXIII. Filtering provides a useful alternative to trend analysis (fitting deterministic functions) and kriging, when observed data can be separated into signal and noise components. Use will be made of the cross-correlation theorem of the statistical theory of communication. Readers interested in this subject are referred to Lee (1960) or Yaglom (1962).

Suppose that $\rho_{xx}(l)$ represents the autocorrelation function of a continuous random variable, then the power spectrum satisfies:

$$\phi_{xx}(\omega) = \int_{-\infty}^{\infty} \rho_{xx}(l)\, e^{i\omega l}\, dl \qquad [11.36]$$

Since $\rho_{xx}(l)$ is even, it follows that:

$$\phi_{xx}(\omega) = \int_{-\infty}^{\infty} \rho_{xx}(l) \cos \omega l \, dl \qquad [11.37]$$

Eq. [11.36] is also known as Wiener's theorem of autocorrelation.

Suppose further that in the situation of [10.80] values X_s are located at the same points as the X_i. These unknown values X_s may be called Y_i here. All rows of the autocorrelation matrix consist of values of the autocorrelation function. The cross-correlation function of X and Y is given by the column vector at the right-hand side of [10.81]. The normal equations may be written as:

$$\rho_{xy}(l) = \Sigma\, h(l + \lambda)\, \rho_{xx}(\lambda) \qquad [11.38]$$

or, in the continuous case:

$$\rho_{xs}(l) = \int_{-\infty}^{\infty} h(l + \lambda)\, \rho_{xx}(\lambda)\, d\lambda \qquad [11.39]$$

where the subscript s denotes a continuous signal to be estimated. This is the so-called input–output cross-correlation theorem of a linear system in statistical theory of communication.

$\rho_{xs}(l)$ denotes the cross-correlation of a record (= input) and signal (= output) and $\rho_{xx}(l)$ is the autocorrelation function of the record. The function $h(l)$ represents the filter to which the record must be subjected in order to obtain the signal:

$$s(k) = \int_{-\infty}^{\infty} h(l)\, x(k + l)\, dl \qquad [11.40]$$

Suppose that the signal-plus-noise model of section 10.6 is satisfied. We recall $\rho_{xs}(l) = \Gamma_{ss}(l)$ and $\rho_{xx}(l) = \Gamma_{ss}(l) + \Gamma_{nn}(l)$. It follows that:

$$\Gamma_{ss}(l) = \int_{-\infty}^{\infty} h(l+\lambda)\,\Gamma_{ss}(\lambda)\,\mathrm{d}\lambda + \int_{-\infty}^{\infty} h(l+\lambda)\,\Gamma_{nn}(l)\,\mathrm{d}\lambda$$

The function $\Gamma_{nn}(l)$ is zero everywhere except at $l = 0$ where $\Gamma_{nn}(0) = 1 - c$. Hence:

$$\Gamma_{ss}(l) = \int_{-\infty}^{\infty} h(l+\lambda)\,\Gamma_{ss}(\lambda)\,\mathrm{d}\lambda + (1-c)\,h(l) \qquad [11.41]$$

Because of stationarity, there is no preferred direction on the l-axis and $h(l)$ is an even function with Fourier transform:

$$H(\omega) = \int_{-\infty}^{\infty} h(l)\cos\omega l\,\mathrm{d}l$$

If $G(\omega) = \int_{-\infty}^{\infty} \Gamma_{ss}(l)\cos\omega l\,\mathrm{d}l$, Fourier transformation of both sides of [11.41] gives:

$$G(\omega) = H(\omega)\cdot G(\omega) + (1-c)\,H(\omega)$$

It follows that:

$$H(\omega) = \frac{G(\omega)}{G(\omega) + (1-c)} \qquad [11.42]$$

The filter $h(l)$ can be found by taking the inverse Fourier transform:

$$h(l) = (1/2\pi)\int_{-\infty}^{\infty} H(\omega)\cos\omega l\,\mathrm{d}\omega \qquad [11.43]$$

When $\Gamma_{ss} = c\,\mathrm{e}^{-a|l|}$, then:

$$G(\omega) = \frac{2ac}{a^2+\omega^2} \qquad [11.44]$$

From [11.42], [11.43], and [11.44], we derive:

$$h(l) = \frac{ac}{\pi p^2(1-c)}\int_{-\infty}^{\infty} \frac{1}{\omega^2/p^2+1}\cos\omega l\,\mathrm{d}\omega \qquad [11.45]$$

where $p = \sqrt{a^2 + 2ac/(1-c)}$ [11.46]

After some further manipulation:

$$h(l) = q\, e^{-p|l|}$$ [11.47]

where $q = \dfrac{ac}{p(1-c)}$ [11.48]

For the Pulacayo zinc percent values, $a = 0.1892$ and $c = 0.5157$.
It follows from [11.46] and [11.48] that $p = 0.6624$ and $q = 0.3042$. The coefficients of this bilateral exponential filter decrease rapidly for increasing $|l|$.

In order to apply the result to the observations, which are discrete, we can compare values of $h(l)$ for $l = 0, \pm 1, \pm 2, \ldots$. It is desirable to define a cut-off value for $|l|$. For the example, we selected four values at each side of points with $l = 0$, leading to:

$$\sum_{l=-4}^{l=4} h(l) = 0.9060$$

to be compared with:

$$\sum_{l=-\infty}^{\infty} h(l) = q + \frac{2qe^{-p}}{1-e^{-p}} = 0.9517$$

There is only a slight difference between these two sums which is less than 0.05. In practice, the estimated values

$$s_k = \sum_{l=-m}^{m} h(l)\, x_{k+l}$$

can be corrected for this cut-off value. For the example, $m = 4$ and we multiplied the weights $h(l)$ by $0.957/0.9060 = 1.0504$ which gave corrected weights $h'(l)$. The result is

$$s_k = 0.3195 x_k + 0.1648\,(x_{k+1} + x_{k-1}) + 0.0849\,(x_{k+2} + x_{k-2})$$

$$+ 0.438\,(x_{k+3} + x_{k-3}) + 0.0226\,(x_{k+4} + x_{k-4})$$ [11.49]

We recall that this derivation was done for values with mean equal to zero. In reality both signal S_k and data X_k have an average approximately equal to $\bar{X}$. We may set $s_k = S_k - \bar{X}$ and $x_{k+l} = X_{k+l} - \bar{X}$. Hence:

$$S_k = \left[1 - \sum_{l=-m}^{m} h'(l)\right] \bar{X} + \sum_{l=-m}^{m} h'(l) X_{k+l}$$ [11.50]

or, for the example:

$$S_k = 0.0483\bar{X} + \sum_{l=-m}^{m} h'(l) X_{k+l} \qquad [11.51]$$

where the $h'(l)$ are as in [11.49] and $\bar{X} = 15.61\%$ Zn.

The weights of [11.51] add to one whereas, in [11.49], they were slightly less than one. If this filter is applied to the data of Fig.55A, the result is as in Fig.55B. It is noted that Fig.55B was derived by automatic processing of the original data of Fig.55A without using a logarithmic transformation. In fact, the difference between filters derived from original and logarithmically transformed data, respectively, is negligible in this case. This was explained by the results derived in section 10.7 (also, see Table XXVII). In general, the method has a wider range of applicability when logarithmically transformed data are used. For example, the silt data of the varve series, part of which is shown in Table XXXIII, have logarithmic variance equal to 1.47 which is greater than one. In that situation, the original data should not be used. However, if the objective is to estimate average values of a variable (e.g., mining grade), one should apply the method to original data when this is possible. This is because results for transformed data often can not be readily recomputed to apply to the original variable.

Since [11.38] was derived from [10.80], filtering is closely related to kriging. In practice, one would use these more elaborate methods only when the object is to do a detailed analysis of data subject to considerable noise. Frequently, good results are obtained by methods with predetermined weighting factors (cf. Vistelius, 1961).

Mean square filtering error

The estimated values of the signal S_k are subject to error. Since these errors are strongly autocorrelated, it is useful to compute their precision. The derivation of the mean square filtering error σ_f^2 is laborious although the result is simply:

$$\sigma_f^2 = (1-c)\, h(0)\, \sigma^2(x) \qquad [11.52]$$

where $h(0)$ satisfies [11.43]. When $\rho_{ss}(l) = c \exp(-a|l|)$, it follows from [11.47] that $h(0) = q$ and [10.7] is satisfied. For the derivation of [11.52], the reader is referred to Lee (1960, chapter 14) or Yaglom (1962, chapter 5).

Chapter 12

POWERS OF MATRICES AND MARKOV CHAINS

12.1. Introduction

This chapter is a continuation of the matrix algebra presented in Chapter 3. In particular, Sylvester's theorem will be introduced. As will be seen later, this theorem has direct application to cyclical stratigraphic sequences. Sylvester's result applies to power series of matrices which are formed as follows.

If $\mathbf{A}^l$ and $\mathbf{A}^m$ represent the l-th and m-th power of a square matrix $\mathbf{A}$ respectively, then:

$$\mathbf{A}^l \mathbf{A}^m = \mathbf{A}^m \mathbf{A}^l = \mathbf{A}^{m+l}$$

This relationship follows directly from the definition of matrix multiplication. If $\mathbf{A}^0$ is interpreted as the unit matrix $\mathbf{I}$, a polynomial of degree m of $\mathbf{A}$ is:

$$P(\mathbf{A}) = p_0\mathbf{I} + p_1\mathbf{A} + p_2\mathbf{A}^2 + \ldots + p_m\mathbf{A}^m$$

This matrix polynomial can be compared to the polynomial of a scalar variable x:

$$P(x) = p_0 + p_1x + p_2x^2 + \ldots$$

The algebras of scalar polynomials of a single variable and of polynomials that involve a single square matrix are completely analogous.

For example, the analogue of $P(x) = \prod_{r=1}^{m} (x - c_r)$ is $P(\mathbf{A}) = \prod_{r=1}^{m} (\mathbf{A} - c_r\mathbf{I})$.

If $P(\mathbf{A}) = \mathbf{A}^2 - \mathbf{A} - 2\mathbf{I}$ with:

$$\mathbf{A} = \begin{bmatrix} 1 & 2 \\ 3 & 2 \end{bmatrix} \quad \text{and} \quad \mathbf{A}^2 = \begin{bmatrix} 7 & 6 \\ 9 & 10 \end{bmatrix}$$

then:

$$P(\mathbf{A}) = \begin{bmatrix} 7 & 6 \\ 9 & 10 \end{bmatrix} - \begin{bmatrix} 1 & 2 \\ 3 & 2 \end{bmatrix} - 2\begin{bmatrix} 1 & 0 \\ 0 & 1 \end{bmatrix} = \begin{bmatrix} 4 & 4 \\ 6 & 6 \end{bmatrix}$$

We also can write $P(\mathbf{A}) = (\mathbf{A} - 2\mathbf{I})(\mathbf{A} + \mathbf{I})$ which gives the same result:

$$P(\mathbf{A}) = \begin{bmatrix} -1 & 2 \\ 3 & 0 \end{bmatrix} \begin{bmatrix} 2 & 2 \\ 3 & 3 \end{bmatrix} = \begin{bmatrix} 4 & 4 \\ 6 & 6 \end{bmatrix}$$

The exponential function of a square matrix is defined by the same power series as the exponential function of a scalar, or:

$$\begin{aligned} &\mathrm{e}^{\mathbf{A}} = \mathbf{I} + \mathbf{A} + \mathbf{A}^2/2! + \mathbf{A}^3/3! + \ldots \\ \text{and: } &\mathrm{e}^{-\mathbf{A}} = \mathbf{I} - \mathbf{A} + \mathbf{A}^2/2! - \mathbf{A}^3/3! + \ldots \end{aligned} \qquad [12.1]$$

These equations are examples of infinite series of matrices. As in the case of a scalar, an infinite series of a matrix may or may not converge to a constant form when the power increases.

Another infinite series is:

$$(\mathbf{I} - \mathbf{A})^{-1} = \mathbf{I} + \mathbf{A} + \mathbf{A}^2 + \mathbf{A}^3 + \ldots \qquad [12.2]$$

This series will converge to a constant form if the inverse of $(\mathbf{I} - \mathbf{A})$ exists.

12.2. Sylvester's theorem

We will use this important theorem for several purposes and will prove it by first deriving an analogous theorem for the scalar variable x.

Sylvester's theorem states that if the n roots λ_r of the square $(n \times n)$ matrix $\mathbf{A}$ are all distinct, and $P(\mathbf{A})$ is any polynomial of $\mathbf{A}$, then:

$$P(\mathbf{A}) = \sum_{r=1}^{n} P(\lambda_r)\,\mathbf{Z}(\lambda_r) \qquad [12.3]$$

where $\mathbf{Z}(\lambda_r) = \dfrac{\prod\limits_{j \neq r} (\lambda_j \mathbf{I} - \mathbf{A})}{\prod\limits_{j \neq r} (\lambda_j - \lambda_r)}$

The matrices $\mathbf{Z}(\lambda_r)$ are the spectral components of $\mathbf{A}$. Together they form the spectral set of $\mathbf{A}$.

The proof of Sylvester's theorem is based on Lagrange's well-known interpolation formula. If $P(x)$ is a polynomial of x with a degree that does not exceed $(n - 1)$, then:

$$P(x) = \sum_{r=1}^{n} P(a_r) \frac{\prod\limits_{s \neq r} (a_s - x)}{\prod\limits_{s \neq r} (a_s - a_r)} \qquad [12.4]$$

where $a_1, a_2, \ldots, a_n$ are distinct, arbitrary constants.

Proof. $P(x) = p_1 + p_2 x + \ldots + p_n x^{n-1}$
Suppose that $P(x)$ is expressed as the sum of n polynomials $Q_r(x)$ with:

$$P(x) = P(a_1)Q_1(x) + P(a_2)Q_2(x) + \ldots$$

It follows that, when x is replaced by the constants a_r $(r = 1, 2, \ldots, n)$, the result is n equations of the type:

$$P(a_1)Q_1(a_r) + P(a_2)Q_2(a_r) + \ldots = P(a_r).$$

These equations are satisfied if:

$$Q_r(a_r) = 1 \quad \text{and} \quad Q_r(a_s) = 0 \qquad (s \neq r)$$

Each polynomial $Q_r(x)$ has $(n-1)$ distinct roots and can be written in the form:

$$Q_r(x) = C_r(a_1 - x)(a_2 - x)(a_{r-1} - x)(a_{r+1} - x)\ldots(a_n - x)$$

$$= C_r \prod_{s \neq r} (a_s - x)$$

The constants C_r can be found by evaluating $Q_r(x)$ at the points with $x = a_r$. That is:

$$Q_r(a_r) = C_r \prod_{s \neq r} (a_s - a_r) \quad \text{or:} \quad C_r = \frac{1}{\prod_{s \neq r} (a_s - a_r)}$$

Hence:

$$Q_r(x) = \frac{\prod_{s \neq r} (a_s - x)}{\prod_{s \neq r} (a_s - a_r)} \qquad \text{and} \qquad P(x) = \sum_{r=1}^{n} P(a_r) \frac{\prod_{s \neq r} (a_s - x)}{\prod_{s \neq r} (a_s - a_r)} \qquad \text{[12.4, repeated]}$$

This is Lagrange's interpolation formula.
When the scalar variable x is replaced by the matrix $\mathbf{A}$, it follows that:

$$P(\mathbf{A}) = \sum_{r=1}^{n} P(a_r) \frac{\prod_{s \neq r} (a_s \mathbf{I} - \mathbf{A})}{\prod_{s \neq r} (a_s - a_r)}$$

When the n values a_r are replaced by the eigenvalues λ_r, this identity immediately results in Sylvester's theorem [12.3].

Application of Lagrangian interpolation

In the skewness test of section 6.13, we were faced with an elementary interpolation problem reformulated as follows. $P(x)$ was known for $x_1 = 70$ and $x_2 = 80$, with $P(70) =$ 0.459 and $P(80) = 0.432$, respectively. The problem was to estimate $P(x)$ for $x = 76$.

If two values of $P(x)$ are used, a linear polynomial can be fitted. Lagrangrian interpolation [12.4] gives the straight-line equation:

$$P(x) = \frac{70 - x}{70 - 80} \times 0.432 + \frac{80 - x}{80 - 70} \times 0.459$$

For $x = 76$, it follows that $P(76) = 0.443$.

Suppose now that a third value is also considered, e.g., $P(90) = 0.409$ (from Pearson-Hartley (1958) tables). Eq. [12.4] now gives the quadratic polynomial:

$$P(x) = \frac{(80 - x)(90 - x)}{(70 - 80)(70 - 90)} \times 0.459 + \frac{(70 - x)(90 - x)}{(80 - 70)(80 - 90)} \times 0.432 + \frac{(70 - x)(80 - x)}{(90 - 70)(90 - 80)} \times 0.409$$

which becomes $P(76) = 0.442$ for $x = 76$.

In fact, we have constructed a parabola passing through three known points. Lagrangian interpolation is used in several computer contour programs. For a general description of contouring by machine, see Tomlinson (1972) and Walters (1969). For other applications of Lagrangian interpolation, the reader is referred to textbooks on numerical analysis (e.g., Ralston, 1965).

12.3. Applications of Sylvester's theorem

If [12.3] is applied to the matrix $\mathbf{A}$, with $P(\mathbf{A}) = \mathbf{A}$, the result is:

$$\mathbf{A} = \sum_{i=1}^{n} \lambda_i \mathbf{Z}_0(\lambda_i) \qquad [12.5]$$

When the canonical form of $\mathbf{A}$ is written in the form:

$$\mathbf{A} = \sum_{i=1}^{n} \lambda_i \mathbf{V}_i \mathbf{T}_i'$$

where $\mathbf{V}_i$ and $\mathbf{T}_i'$ are the right and left eigenvectors of $\mathbf{A}$, respectively, as in section 3.11 (see [3.40]), then:

$$\mathbf{Z}_0(\lambda_i) = \mathbf{V}_i \mathbf{T}_i' \qquad [12.6]$$

Likewise, if $P(\mathbf{A}) = \mathbf{A}^s$, the result is:

$$\mathbf{A}^s = \sum_{i=1}^{n} \lambda_i^s \mathbf{Z}_0(\lambda_i) \qquad [12.7]$$

which is comparable to the expression:

$$\mathbf{A}^s = \sum_{i=1}^{n} \lambda_i^s \mathbf{V}_i \mathbf{T}_i' \qquad [12.8]$$

These two expressions can only be valid for different values of s if [12.6] holds true.

An arbitrary square matrix $\mathbf{A}$ can have one or more pairs of eigenvalues forming conjugate pairs of complex numbers. In practice, this situation may arise in the case of stratigraphic sequences with cyclical components. Suppose that the matrix $\mathbf{A}$ has dominant roots which form a conjugate complex pair with:

$$\lambda_1 = \mu + i\omega$$

$$\lambda_2 = \mu - i\omega$$

$P(\mathbf{A})$ in [12.3] can be chosen as:

$$P(\mathbf{A}) = \mathbf{A}^s P_0(\mathbf{A})$$

where $P_0(\mathbf{A})$ is independent of s.
It follows that:

$$\mathbf{A}^s P_0(\mathbf{A}) \approx \lambda_1^s P_0(\lambda_1)\, \mathbf{Z}_0(\lambda_1) + \lambda_2^s P_0(\lambda_2)\, \mathbf{Z}_0(\lambda_2)$$

for large values of s.

If $P_0(\mathbf{A}) = (\lambda_1 \mathbf{I} - \mathbf{A})\,(\lambda_2 \mathbf{I} - \mathbf{A})$, the polynomials $P_0(\lambda_1)$ and $P_0(\lambda_2)$ are both equal to $\mathbf{N}$, which is a so-called null matrix consisting of zeros only. It follows that:

$$\mathbf{A}^s (\lambda_1 \mathbf{I} - \mathbf{A})\,(\lambda_2 \mathbf{I} - \mathbf{A}) = \mathbf{N}$$

If a_s denotes any individual element of $\mathbf{A}^s$, it follows that:

$$a_s(\lambda_1 - a_s)\,(\lambda_2 - a_s) = 0$$

$$a_{s+2} - (\lambda_1 + \lambda_2)\, a_{s+1} + \lambda_1 \lambda_2 a_s = 0$$

$$a_{s+2} - 2\mu a_{s+1} + (\mu^2 + \omega^2)\, a_s = 0 \qquad [12.9]$$

The elements a_{s+2} and a_{s+1} take the same position in $\mathbf{A}^{s+2}$ and $\mathbf{A}^{s+1}$ as a_s in $\mathbf{A}^s$. If b represents another element of $\mathbf{A}^s$, then:

$$b_{s+2} - 2\mu b_{s+1} + (\mu^2 + \omega^2)\, b_s = 0 \qquad [12.10]$$

It follows that:

$$\mu^2 + \omega^2 = \frac{a_{s+1}b_{s+2} - a_{s+2}b_{s+1}}{a_s b_{s+1} - a_{s+1}b_s} \qquad [12.11]$$

The square root of this expression $r = \sqrt{\mu^2+\omega^2}$ is the modulus of the complex pair λ and λ_2.

When $\mathbf{A}$ is powered, the modulus as calculated from three successive powers of $\mathbf{A}$ wi differ when s is small. For large values of s, however, it assumes a constant value. Onc convergence has been reached, μ and ω can be solved from [12.9] and [12.10].

The components $\mathbf{Z}_0(\lambda_1)$ and $\mathbf{Z}_0(\lambda_2)$ can be found as follows. If $P_0(\mathbf{A})$ is put equal t $(\lambda_2\mathbf{I} - \mathbf{A})$ in [12.3], then:

$$\mathbf{A}^s(\lambda_2\mathbf{I} - \mathbf{A}) \approx \lambda_1^s(\lambda_2 - \lambda_1)\, \mathbf{Z}_0(\lambda_1)$$

or, approximately:

$$\mathbf{Z}_0(\lambda_1) = \frac{\mathbf{A}^s(\lambda_2\mathbf{I} - \mathbf{A})}{\lambda_1^s(\lambda_2 - \lambda_1)} \qquad [12.12]$$

Likewise:

$$\mathbf{Z}_0(\lambda_2) = \frac{\mathbf{A}^s(\lambda_1\mathbf{I} - \mathbf{A})}{\lambda_2^s(\lambda_1 - \lambda_2)} \qquad [12.13]$$

The elements of $\mathbf{Z}_0(\lambda_2)$ are the conjugates of those of $\mathbf{Z}_0(\lambda_1)$. $\mathbf{Z}_0(\lambda_2)$ is said to be th conjugate matrix of $\mathbf{Z}_0(\lambda_1)$.

The joint component for a conjugate pair of eigenvalues

The results obtained in the previous section can be used to explain the behavior o matrices subjected to powering. We also have derived practical methods for extractin dominant roots and spectral components of a matrix. Once a dominant root (or pair o complex roots) and the corresponding component (or pair of components) has bee found, the product of the root and the component (or the sum of two of these product for a complex pair) can be subtracted from the original matrix and the method o powering can be repeated until all eigenvalues and components are known. Complex root always occur as a conjugate pair and they are being solved simultaneously.

Suppose that $\mathbf{Z}_0(\lambda_1)$ and $\mathbf{Z}_0(\lambda_2)$ are the components for a complex pair. Their joint omponent $\mathbf{Z}^{(1,2)}_{(0)}$ with:

$$\mathbf{Z}^{(1,2)}_{(0)} = \mathbf{Z}_0(\lambda_1) + \mathbf{Z}_0(\lambda_2) \qquad [12.14]$$

onsists of real values only, because $\mathbf{Z}_0(\lambda_2)$ is the conjugate matrix of $\mathbf{Z}_0(\lambda_1)$. If the atrix $\mathbf{A}$ has a dominant pair of roots, $\mathbf{A}^s$ does not converge for large values of s. owever, it approaches the form:

$$\mathbf{Z}^{(1,2)}_{(s)} = \lambda_1^s \mathbf{Z}_0(\lambda_1) + \lambda_2^s \mathbf{Z}_0(\lambda_2) \qquad [12.15]$$

hich also consists of real values only. This expression may be modified as follows. We n write:

$$\mathbf{Z}_0(\lambda_1) = \mathbf{V}_1\mathbf{T}_1' \quad \text{and} \quad \mathbf{Z}_0(\lambda_2) = \mathbf{V}_2\mathbf{T}_2'$$

here $\mathbf{V}_1$ and $\mathbf{V}_2$ are the (complex) eigenvectors for λ_1 and λ_2. $\mathbf{T}_1'$ and $\mathbf{T}_2'$ are the rresponding left eigenvectors. The vector $\mathbf{V}_1$ is uniquely determined except for a com- ex constant. It is convenient to set the first element of $\mathbf{V}_1$ equal to one. The first row $\mathbf{Z}_0(\lambda_1)$ then is equal to the vector $\mathbf{T}_1'$. The other elements of $\mathbf{V}_1$ can be obtained by viding them by the first elements of $\mathbf{T}_1'$. By using the polar form, we can write:

$$\left.\begin{aligned} \lambda_1 &= r(\cos\phi + i\sin\phi) \\ v_{1j} &= p_j(\cos\alpha_j + i\sin\alpha_j) \qquad (j = 1, 2, \ldots, p) \\ t_{1k} &= q_k(\cos\beta_k + i\sin\beta_k) \qquad (k = 1, 2, \ldots, p) \end{aligned}\right| \qquad [12.16]$$

ie vectors $\mathbf{V}_2$ and $\mathbf{T}_2'$ are defined by $\mathbf{V}_2 = \overline{\mathbf{V}}_1$ and $\mathbf{T}_2' = \overline{\mathbf{T}}_1'$, with the bar denoting the mplex conjugate. The individual elements of the matrix $\mathbf{Z}^{(1,2)}_{(s)}$ with:

$$\mathbf{Z}^{(1,2)}_{(s)} = \begin{bmatrix} z_{11}(s) & z_{12}(s) & \ldots \\ z_{21}(s) & z_{22}(s) & \ldots \\ \ldots & \ldots & \ldots \\ \ldots & \ldots & \ldots \end{bmatrix}$$

e:

$$z_{jk}(s) = 2r^s p_j q_k \cos(\phi s + \alpha_j + \beta_k) \qquad [12.17]$$

th j and $k = 1, 2, ..., p$. In each case, the values of $z_{jk}(s)$ therefore describe a damped sine with period $2\pi/\phi$, phase angle $(\alpha_j + \beta_k)$, and amplitude equal to $2r^s p_j q_k$.

Power series and functions of matrices

From Sylvester's theorem, it follows that a power series of a matrix always can b represented as the sum of a number of power series for the eigenvalues. Another impo tant consequence is that if $F(\mathbf{A})$ represents a function of the matrix $\mathbf{A}$, that can b expressed as an infinite series, then P can be replaced by F in [12.3].

Example. The matrix exponential $e^{\mathbf{A}}$ can be expressed as an infinite series (cf. [12.1] Consequently:

$$e^{\mathbf{A}} = \sum_{r=1}^{n} e^{\lambda_r} \mathbf{Z}_0(\lambda_r) \qquad [12.18$$

where the $\mathbf{Z}_0(\lambda_r)$ are the spectral components of $\mathbf{A}$.

12.4. An iterative method for extracting eigenvectors

The following method can be used on the desk calculator. For example we take th (3 X 3) symmetrical matrix of section 9.10. The method also can be used for asymmetri cal matrices with real eigenvalues but then both a matrix $\mathbf{A}$ and its transpose must b subjected to the method.

The proof of this method is of interest since it will be used in Chapter 13 in th context of transition matrices for multivariate Markovian series.

For this example:

$$\mathbf{A} = \begin{bmatrix} -0.3435 & -0.1145 & 0.0984 \\ -0.1145 & -0.0228 & 0.0108 \\ 0.0984 & 0.0108 & 0.0833 \end{bmatrix}$$

This matrix is symmetrical, and we know beforehand that the eigenvalues will be real

The matrix $\mathbf{A}$ is postmultiplied by the vector $\begin{bmatrix} 1 \\ 0 \\ 0 \end{bmatrix}$, giving:

$$\mathbf{A} \begin{bmatrix} 1 \\ 0 \\ 0 \end{bmatrix} = \begin{bmatrix} -0.3435 \\ -0.1145 \\ 0.0984 \end{bmatrix}$$

The first element of the resulting column vector is made equal to unity by dividing by th scalar -0.3435, or:

$$1/(-0.3435) \begin{bmatrix} -0.3435 \\ -0.1145 \\ 0.0984 \end{bmatrix} = \begin{bmatrix} 1 \\ 0.333333 \\ -0.286463 \end{bmatrix}$$

ext compute:

$$\mathbf{A}\begin{bmatrix}1\\0.333333\\-0.286463\end{bmatrix}=\begin{bmatrix}-0.409855\\-0.125194\\0.078138\end{bmatrix}$$

he first element of the resulting vector is again made equal to one. This procedure is epeated and leads to the iteration results shown in the following table.

ep 1		step 2		step 3	
−0.3435	1	−0.409855	1	−0.397235	1
−0.1145	0.333333	−0.125194	0.305459	−0.123523	0.310957
0.0984	−0.286463	0.078138	−0.190648	0.085818	−0.216038

ep 4		step 5		step 6	
0.400363	1	−0.399528	1	−0.399751	1
0.123923	0.309527	−0.123817	0.309908	−0.123845	0.309805
0.083762	−0.209215	0.084315	−0.211307	0.084168	−0.210551

ep 7	
0.399691	1
0.123838	0.309834
0.084207	−0.210680

ne result after step 7 is close to that after step 6 and the column vector $\begin{bmatrix}1\\0.3098\\-0.2107\end{bmatrix}$ proximates the first eigenvector $\mathbf{V}_1$, while the largest eigenvalue of $\mathbf{A}$ is $\lambda_1 = -0.3997$.

Proof. A general proof of this method can be based on [12.8] giving that $\mathbf{A}^s \approx \lambda_1^s \mathbf{V}_1 \mathbf{T}_1'$ r large values of s.

onsequently, $\mathbf{A}^s\mathbf{X} = \lambda_1^s(t_{11}x_1 + t_{12}x_2 + t_{13}x_3)\begin{bmatrix}v_{11}\\v_{12}\\v_{13}\end{bmatrix}$ where $\mathbf{X} = \begin{bmatrix}x_1\\x_2\\x_3\end{bmatrix}$ is an arbitrary column ctor.

$c = \lambda_1^s(t_{11}x_1 + t_{12}x_2 + t_{13}x_3)$ denotes a scalar factor, which has been extracted at step then:

$$(1/c)\,\mathbf{A}^s\mathbf{X} \approx \mathbf{V}_1$$

d at step $(s+1)$:

$$(1/c)\,\mathbf{A}^{s+1}\mathbf{X} \approx (1/\lambda_1)\mathbf{V}_1$$

In a similar fashion, the row vector $\mathbf{T}_1'$ can be extracted from $\mathbf{A}$ by repeated premulti- cation with an arbitrary row vector and extracting a scalar at each step. In the present

example, **A** is symmetric and $\mathbf{T}'_1 = \mathbf{V}_1$ and it is not necessary to determine $\mathbf{T}'_1$ by ite-
tion

We now proceed as follows to determine the second largest eigenvalue. First t
product

$$\begin{bmatrix} -0.3435 \\ -0.1145 \\ 0.0984 \end{bmatrix} [1 \quad 0.3098 \quad -0.2107]$$

is calculated, which is the product of the first column of **A** and the row vector $\mathbf{T}'_1$. In
case of general, nonsymmetrical matrices, the elements of $\mathbf{T}'_1$ would be divided by t_{11},
that the first element of the row vector is one. The resulting matrix is subtracted from
yielding:

$$\mathbf{C} = \begin{bmatrix} 0 & -0.008084 & 0.026025 \\ 0 & 0.012672 & -0.013325 \\ 0 & -0.019684 & 0.104033 \end{bmatrix}$$

This matrix **C** is given the same treatment as **A**, except that now the multiplication is do
by $\begin{bmatrix} 0 \\ 0 \\ 1 \end{bmatrix}$ instead of $\begin{bmatrix} 1 \\ 0 \\ 0 \end{bmatrix}$. We obtain:

	step 1		step 2		step 3	
0	0.026025					
0	−0.013325	−0.128084	−0.014948	−0.140286	−0.015103	−0.1414
1	0.104033	1	0.106554	1	0.106794	1

step 4	
0.027168	1
−0.015117	−0.5564
0.106817	3.9317

The first element need only be calculated after that convergence has been reached (af
step 4). The elements then are divided by it, yielding:

$$\mathbf{V}_2 = \begin{bmatrix} 1 \\ -0.5564 \\ 3.9317 \end{bmatrix}$$

For the corresponding eigenvalue, $\lambda_2 = 0.1068$.

Proof. Because of orthogonality, $\mathbf{T}'_1 \mathbf{V}_i = 0 \quad \text{if } i \neq 1$.

The first element of $\mathbf{T}_1'$ is t_{11}. Let **B** be a matrix that consists of zeros except for its first row which is equal to $\mathbf{T}_1'/t_{11}$. Then:

$$\mathbf{V}_i = (\mathbf{I} - \mathbf{B})\,\mathbf{V}_i \quad \text{if } i \neq 1$$

In general, $(\lambda_i \mathbf{I} - \mathbf{A})\mathbf{V}_i = 0$. Hence:

$$(\lambda_i \mathbf{I} - \mathbf{C})\,\mathbf{V}_i = 0 \qquad [12.19]$$

where $\mathbf{C} = \mathbf{A}(\mathbf{I} - \mathbf{B})$.

It follows that the λ_i are eigenvalues of **C** and the $\mathbf{V}_i$ its eigenvectors. The remaining eigenvalue of **C** is zero, since the first column of this auxiliary matrix consists of zeros. This completes the proof for the second step. It could be repeated for later steps if the matrix is larger than (3 X 3). For the example, **A** is (3 X 3) and the first two eigenvalues and eigenvectors have now been found. The third eigenvector is orthogonal to the other two and can be solved from the equations $\mathbf{V}_1'\mathbf{V}_3 = 0$ and $\mathbf{V}_2'\mathbf{V}_3 = 0$.

Putting $v_{31} = 1$, the solution of the resulting two linear equations is:

$$v_{32} = -3.7642; \quad v_{33} = -0.7873$$

and the third eigenvector is known.

The corresponding eigenvalue λ_3 satisfies:

$$a_{11}v_{31} + a_{12}v_{32} + a_{13}v_{33} = \lambda_3 v_{31}$$

or $\lambda_3 = -0.3435 + (-0.1145)(-3.7642) + 0.0984(-0.7873) = 0.0099$. The precision of these calculations can be checked by calculating the trace of **A** with $\text{tr}(\mathbf{A}) = -0.3435 - 0.0228 + 0.0833 = -0.2830$. On the other hand, $\text{tr}(\mathbf{A}) = \lambda_1 + \lambda_2 + \lambda_3 = -0.3997 + 0.1068 + 0.0099$ which happens to be exactly equal to -0.2830.

12.5. Iterative procedures, review of powering methods

Earlier in this chapter, a discussion was given of Lagrange's interpolation formula. The purpose of introducing it here was to use it for an easy derivation of Sylvester's theorem. Likewise, Sylvester's method is not used directly to obtain the spectral components of a matrix, but as a means to understand the behavior of matrices subjected to powering. In principle, we now have at our disposal a method by which the eigenvalues and eigenvectors can be determined for any square matrix that consists of real numbers. An exception is the special case of a matrix with nonzero eigenvalues that are not distinct. For example, a matrix with two eigenvalues that are exactly equal to one another will not converge to one of the forms discussed in this chapter when it is subjected to powering. If two eigenvalues are almost equal, the previous version of Syl-

vester's theorem applies, but the rate of convergence can be slow. This is a reason th
there are differences in efficiency of iterative methods for digital computers. In fac
when a matrix is large, it may not be possible to obtain the required solution by con
puter methods based on the methods of powering that have been discussed so far.

Confluent form of Sylvester's theorem

If a matrix has one or more sets of two or more nonzero eigenvalues which are equ
to one another, Sylvester's theorem requires modification. This leads to the so-calle
confluent form of Sylvester's theorem which is applicable without exceptions. For
proof of this theorem, see Frazer et al. (1960). The discussion here is restricted to
presentation of the general equation, since in our practical situations we have not found
situation where two eigenvalues were exactly equal to one another.

Some remarks on notation are given first. The cofactor matrix $\mathbf{A}^{c}$ of a square matrix
was discussed in section 3.4. Its transpose $(\mathbf{A}^{c})'$ also is called the adjoint matrix of **A** an
can be written as adj **A**. In section 3.12, the first derivative $\mathrm{d}\mathbf{A}/\mathrm{d}t$ of a matrix **A** wit
respect to a variable t has been discussed. Higher derivatives $\mathrm{d}^{k}\mathbf{A}/\mathrm{d}t^{k}$ with $k = 2, 3, \ldots$ ca
be obtained by repeating the process of differentiation with respect to t.

Suppose that the eigenvalues of an $(n \times n)$ matrix **A** are written as $\lambda_1, \lambda_2, \ldots, \lambda_i, \ldots$
λ_n. Further, that there are one or more sets of s_j eigenvalues which are equal to λ_j. In th
case, s_j is called the multiplicity of λ_j. The confluent form of Sylvester's theorem the
satisfies:

$$P(\mathbf{A}) = \Sigma_j \frac{1}{(s_j - 1)!} \left[\frac{\mathrm{d}^{s_j - 1}}{\mathrm{d}\lambda^{s_j - 1}} \left\{ \frac{P(\lambda)\ \mathrm{adj}\ (\lambda\mathbf{I} - \mathbf{A})}{\prod_{i \neq j} (\lambda - \lambda_i)^{s_i}} \right\} \right]_{\lambda = \lambda_j} \qquad [12.2$$

There are as many terms in the sum as there are distinct eigenvalues λ_j. If $s_j = 1$ for a
roots, the confluent form reduces to [12.3]. When **P** represents a transition matrix that
raised to a high power n, [12.20] applies with **A** replaced by **P** and the polynomials $P($
and $P(\lambda)$ by $\mathbf{P}^{n}$ and λ^{n}, respectively.

The methods for extracting eigenvalues by powering have been programmed for digit
computer. However, it has become clear that these methods do not always provide a
answer that is satisfactory. The main reason for this is that numerous precise multiplic
tions and additions must be performed to raise a larger matrix to a high power. Th
precision of digital computers is restricted. Although in practical applications, the roo
of a matrix usually are distinct, some of them may be almost equal to one another ar
convergence may then be slow. During the calculations, all numbers are truncated at
certain point, e.g., after eight or sixteen significant digits. An iterative procedure is calle
unstable if it is sensitive to rounding off to the extent that numerical errors are intr
duced and propagated.

An effective method to treat symmetric matrices, which always seems to give good results, has been known for some time. This is the Jacobi method for eigenvalues and eigenvectors of symmetric matrices (cf. Wilkinson, 1965).

Extracting the eigenvalues from a general, unsymmetric matrix has remained a numerical problem until recently, when the so-called *QR*-method of Francis (1961) has been found to yield satisfactory results.

The Jacobi method and the *QR*-method have been used by the author to obtain the eigenvalues of the larger matrices in examples to be discussed later. These two procedures are briefly introduced in the next section. A full discussion with proofs is beyond the scope of this book. In particular, the *QR*-method is based on concepts of advanced matrix algebra.

12.6. Computer algorithms for extracting eigenvalues

Jacobi method for symmetric matrices

This discussion is restricted to an outline of the procedure. A proof that the matrices obtained by iteration actually converge toward the correct solution can be found in Fröberg (1969). Computer algorithms for the Jacobi method are available at most computer centers.

For a real symmetric matrix $\mathbf{A}$, there always exists an orthogonal matrix $\mathbf{V}$ so that $\mathbf{V}^{-1}\mathbf{AV} = \Lambda$. In the Jacobi method, $\mathbf{V}$ is formed as the product of a set of special orthogonal matrices $\mathbf{O}_k$, $k = 1, 2, ..., n$. $\mathbf{O}_1$ is formed as follows.

The off-diagonal elements of $\mathbf{A}$ are scanned to find the element with largest absolute value, or:

$$|a_{ij}| = \max.$$

A (2×2) submatrix $\mathbf{A}_s$ then is formed with the elements a_{ii}, $a_{ij} = a_{ji}$, and a_{jj}:

$$\mathbf{A}_s = \begin{bmatrix} a_{ii} & a_{ij} \\ a_{ji} & a_{jj} \end{bmatrix}$$

$\mathbf{A}_s$ can be diagonalized as follows. Its characteristic equation is:

$$\lambda^2 - (a_{ii} + a_{jj})\lambda + a_{ii}a_{jj} - a_{ij}^2 = 0$$

If $R = \sqrt{(a_{ii} - a_{jj})^2 + 4a_{ij}^2}$, then:

$$\lambda_{1,2} = \tfrac{1}{2}[a_{ii} + a_{jj} \pm R]$$

There are two ways of writing the diagonal matrix of $\mathbf{A}_s$:

$$\begin{bmatrix} \lambda_1 & 0 \\ 0 & \lambda_2 \end{bmatrix} \quad \text{or} \quad \begin{bmatrix} \lambda_2 & 0 \\ 0 & \lambda_1 \end{bmatrix}$$

One of these two possibilities is selected according to the following principle. Let us in troduce a variable k with $k = 1$ if $a_{ii} > a_{jj}$, and $k = -1$ if $a_{ii} < a_{jj}$, and let

$$\lambda_1^* = \tfrac{1}{2}[a_{ii} + a_{jj} + kR] \quad \text{and} \quad \lambda_2^* = \tfrac{1}{2}[a_{ii} + a_{jj} - kR]$$

Consequently, $\lambda_1^* = \lambda_1; \lambda_2^* = \lambda_2$, if $k = 1$, and $\lambda_1^* = \lambda_2; \lambda_2^* = \lambda_1$, if $k = -1$.

The orthogonal matrix $\mathbf{O}^*$ of $\mathbf{A}_s$ satisfies:

$$\mathbf{O}^{*-1}\mathbf{A}_s\mathbf{O}^* = \mathbf{\Lambda}_s^*$$

or:
$$\begin{bmatrix} \cos\varphi & \sin\varphi \\ -\sin\varphi & \cos\varphi \end{bmatrix} \begin{bmatrix} a_{ii} & a_{ij} \\ a_{ji} & a_{jj} \end{bmatrix} \begin{bmatrix} \cos\varphi & -\sin\varphi \\ \sin\varphi & \cos\varphi \end{bmatrix} = \begin{bmatrix} \lambda_1^* & 0 \\ 0 & \lambda_2^* \end{bmatrix}$$

After a few manipulations, it follows that:

$$\sin 2\varphi = 2ka_{ij}/R \quad \text{and} \quad \cos 2\varphi = k(a_{ii} - a_{jj})/R$$

From the definition of k, it follows that $\cos 2\varphi$ cannot be negative which means:

$$-90° \leqslant 2\varphi \leqslant 90°$$

In fact, we also can write:

$$\varphi = \tfrac{1}{2}\arctan\{2a_{ij}/(a_{ii} - a_{jj})\} \text{ with } -45° < \varphi < 45°, \text{ if } a_{ii} \neq a_{jj}, \text{ and:}$$

$$\left.\begin{array}{l} \varphi = 45° \text{ when } a_{ij} > 0 \\ \varphi = 45° \text{ when } a_{ij} < 0 \end{array}\right\} \text{ if } a_{ii} = a_{jj}$$

This result suggests why the auxiliary variable k was introduced in the first place. Th angle of rotation φ has been made as small as possible by enforcing the conditio $-45° \leqslant \varphi \leqslant 45°$.

The matrix $\mathbf{O}_1$ is defined as a matrix equal to the identity matrix with the same size a $\mathbf{A}$ in all its elements except $O_{1,ii} = O_{1,jj} = \cos\varphi$ and $-O_{1,ij} = O_{1,ji} = \sin\varphi$. The matri $\mathbf{A}$ is transformed by the two-dimensional rotation $\mathbf{O}_1^{-1}\mathbf{A}\mathbf{O}_1$. This result is subjecte to the same procedure as $\mathbf{A}$ which yields an orthogonal matrix $\mathbf{O}_2$ and the produc $(\mathbf{O}_1\mathbf{O}_2)^{-1}\mathbf{A}\mathbf{O}_1\mathbf{O}_2$. If the procedure is repeated, $\mathbf{P}_n = \mathbf{O}_1\mathbf{O}_2 \ldots \mathbf{O}_n$ approaches the form when n increases. Simultaneously, $\mathbf{P}_n^{-1}\mathbf{A}\mathbf{P}_n$ approaches $\mathbf{\Lambda}$.

The Jacobi method has been proven to be efficient on digital computers.

QR-method of Francis

It is relatively easy to check for precision a set of eigenvalues and eigenvectors. Fc example, this check can be based on the identity $\mathbf{AV} = \mathbf{\Lambda V}$. If $\mathbf{V}$ has been found, $\mathbf{V}^{-1}$ can be obtained by inversion of $\mathbf{V}$. The complete canonical form of $\mathbf{A}$ satisfies $\mathbf{A} = \mathbf{V \Lambda V}^{-1}$. If the eigenvectors are extracted from the transpose of $\mathbf{A}$, this yields $(\mathbf{V}^{-1})'$, because:

$$\mathbf{A}' = (\mathbf{V \Lambda V}^{-1})' = (\mathbf{V}^{-1})' \mathbf{\Lambda V}' \qquad [12.21]$$

These fundamental identities can be helpful if a subroutine is available on the computer to calculate $\mathbf{\Lambda}$ and $\mathbf{V}$ from a general, nonsymmetric matrix.

A real nonsymmetric matrix $\mathbf{A}$ can be considered as a special case of an arbitrary complex matrix with the imaginary parts of its elements equal to zero. The *QR*-method is based on a theorem that every square complex matrix $\mathbf{A}$ can be written as $\mathbf{A} = \mathbf{QR}$, where $\mathbf{Q}$ is a unitary matrix and $\mathbf{R}$ an upper triangular matrix.

A matrix $\mathbf{Q}$ is called unitary when it has the property that its complex conjugate $\mathbf{Q}^*$ is equal to its inverse $\mathbf{Q}^{-1}$. An upper triangular matrix $\mathbf{R}$ has elements below its diagonal equal to zero, or $r_{ij} = 0$ for $i - j > 0$.

Francis (1961) has found a method to write a matrix in the form $\mathbf{QR}$. In general, a computer algorithm based on the *QR*-method is too laborious for arbitrary matrices but it seems to be excellent for some special matrices, including the so-called Hessenberg matrix.

A matrix $\mathbf{A}$ is of upper Hessenberg form if $a_{ij} = 0$ for $i - j > 1$. This is almost upper triangular. However, until recently, it has been relatively easy to convert an arbitrary matrix to Hessenberg form by a sequence of orthogonal rotations. On the contrary, it was difficult to make the step from a Hessenberg matrix to a triangular matrix by computer. This step now can be made by using the *QR*-method.

The resulting triangular matrix $\mathbf{R}$ has the same eigenvalues as $\mathbf{A}$. It is relatively easy to extract the eigenvalues from a triangular matrix.

For a detailed discussion of the *QR*-method with proof see Francis (1961), Wilkinson (1965), or Fröberg (1969). The latter two references can also be consulted for algorithms in addition to the *QR*-method necessary for diagonalizing a nonsymmetric matrix. A FORTRAN IV version of the complete method was published by Grad and Brebner (1968).

12.7. Markov chains; explicit form of transition matrix

Example of a transition matrix

Vistelius (1949) has derived the following matrix for a sequence of lithologies from the Klev-grdzeli section of Cretaceous flysch.

$$\mathbf{P} = \begin{array}{c} \\ \text{sand} \\ \text{silt} \\ \text{clay} \end{array} \begin{array}{c} \begin{array}{ccc} \text{sand} & \text{silt} & \text{clay} \end{array} \\ \left[\begin{array}{ccc} 0.002 & 0.998 & 0.000 \\ 0.388 & 0.014 & 0.598 \\ 0.712 & 0.288 & 0.000 \end{array}\right] \end{array}$$

P is called a transition matrix. Its elements p_{ij} represent the probabilities by which lithology i is followed by lithology j. **P** has the following two specific properties: (1) its elements p_{ij} cannot be negative or $p_{ij} \geqslant 0$; (2) the elements for each row of P sum up to one or:

$$\sum_{j=1}^{n} p_{ij} = 1$$

Both properties are related to the physical meaning of this matrix. For example, in all cases, sand must be followed by either sand, silt or clay, with probabilities that are positive (or zero) and add to one.

In practice, the rows of the matrix **P** are calculated by counting how many times a given state is followed by itself or the other states. The resulting values are then divided by the total number of one-step transitions for that state.

The Chapman-Kolmogorov equation

A Markov process is defined as a stochastic process in which knowledge about the state of the process at a given time t_2 can be deduced from knowledge of its state at any earlier time t_1 and is independent of the history of the system before t_1. A Markov chain is a process for which development in time can be treated as a series of transitions between distinct states (e.g., lithologies).

The Markov property can be formulated as:

$$p_{ik}(t_2) = \sum_j p_{ij}(t_1)\, p_{jk}(t_2 - t_1) \qquad [12.22]$$

This is the Chapman-Kolmogorov equation for homogeneous processes. It can be explained as follows. Let the system be in state i at time 0 and in state k at time t_2. The course followed by the process must have gone via some other state j at time t_1 $(0 < t_1 < t_2)$. Since the course of the process between t_1 and t_2 is independent of the past $t < t_1$, the probability that the process is in state j at time t_1 and in state k at time t_2 is the product of the probabilities for two independent random events, or:

$$p_{ij}(t_1)\, p_{jk}(t_2 - t_1)$$

The probability of state k at time t_2, given i at time 0, is obtained by summing all possible values of j. In the case of Markov chains, we can write:

$$p_{ik}^{(m+n)} = \sum_j p_{ij}^{(m)} p_{jk}^{(n)} \qquad [12.23]$$

Here $p_{ij}^{(m)}$ denotes the probability of a transition from state i to j in exactly m steps. If the first m steps lead from i to j, the probability of a subsequent passage from j to k in n steps does not depend on the manner in which the intermediate state j was reached.

In the previous section, the probabilities p_{ij} were arranged in a matrix denoted by **P**. In the same way, the probabilities $p_{ij}^{(n)}$ can be arranged in the matrix $\mathbf{P}^n$. The Chapman-Kolmogorov equation then reduces to the identity:

$$\mathbf{P}^{m+n} = \mathbf{P}^m \mathbf{P}^n \qquad [12.24]$$

By raising the matrix **P** for the Klev-grdzeli section to the second power, Griffiths (1966a) has found:

$$\mathbf{P}^2 = \begin{array}{c} \begin{array}{ccc} \text{sand} & \text{silt} & \text{clay} \end{array} \\ \begin{bmatrix} 0.432 & 0.559 & 0.009 \\ 0.224 & 0.442 & 0.335 \\ 0.400 & 0.173 & 0.427 \end{bmatrix} \end{array}$$

Also:

$$\mathbf{P}^{16} = \begin{array}{c} \begin{array}{ccc} \text{sand} & \text{silt} & \text{clay} \end{array} \\ \begin{bmatrix} 0.338 & 0.415 & 0.247 \\ 0.338 & 0.414 & 0.248 \\ 0.338 & 0.413 & 0.248 \end{bmatrix} \end{array}$$

If the lithological sequence has the property of a Markov chain, then the matrix $\mathbf{P}^2$ denotes the probabilities that state i is followed by j after two steps. If **P** is raised to a high power, its rows tend to become equal to the same row vector, which is equal to [0.338, 0.414, 0.248] for the example. This row vector is called the fixed vector. For lithological sequences, its elements simply are estimates of the proportions for the lithologies in the section. Thus, approximate percentage values for frequency of sand, silt, and clay layers in this flysch section are 34, 41, and 25%, respectively.

If the Markov chain was used as a prediction device for unknown parts of the section, the probability for a particular state after n steps (n large) is the proportion by which the state occurs in the known part of the section. This is equivalent to saying that the prediction is a random guess based on the proportion values. If, in a section, the states follow each other at random, the fixed vector is reached after a single step, except for random fluctuations.

A transition matrix **P** can be tested for this type of randomness in the sequence by applying a χ^2-test (cf. [12.28], section 12.9). The following analysis is only meaningful if the succession of states is nonrandom.

Explicit form of Markov chains

If **P** represents a $(p \times p)$ transition matrix, we can write:

$$\mathbf{P}^n = \sum_{i=1}^{p} \lambda_i^n \mathbf{Z}_0(\lambda_i) \qquad [12.25]$$

where the λ_i's are the eigenvalues of **P**, and $\mathbf{Z}_0(\lambda_i)$ the spectral components:

$$\mathbf{Z}_0(\lambda_i) = \frac{\prod_{j \neq i} (\lambda_j \mathbf{I} - \mathbf{P})}{\prod_{j \neq i} (\lambda_j - \lambda_i)} \qquad \text{(cf. [12.3])}$$

It will be assumed that the p eigenvalues are distinct.

It can be shown that the largest eigenvalue is equal to one. If λ_1 represents the largest eigenvalue, we have in the limiting case (large n):

$$\mathbf{P}^n = \lambda_1^n \mathbf{Z}_0(\lambda_1) = \lambda_1^n \mathbf{V}_1 \mathbf{T}_1' \qquad [12.26]$$

All rows of $\mathbf{P}^n$ have elements proportional to the elements of $\mathbf{T}_1'$. Because $\mathbf{P}^n$ is a transition matrix, the sum of the elements for each of its rows is equal to one. This is only possible if: (1) $\lambda_1 = 1$; and (2) all elements of $\mathbf{V}_1$ are equal to one. The row vector $\mathbf{T}_1'$ is the fixed vector discussed previously.

The remaining $(p - 1)$ roots λ_i are either real or they occur as conjugate complex pairs. The eigenvalues and corresponding spectral components can be solved by various methods. (For a method based on Sylvester's theorem, see section 12.3.) The explicit form of the matrix **P** for the flysch section will be derived in the next section. Inasmuch as this example involves a (3×3) transition matrix, a relatively simple procedure can be followed to obtain the solution.

Explicit form of the matrix **P** *for the Klev-grdzeli flysch section*

The eigenvalues of:

$$\mathbf{P} = \begin{bmatrix} 0.002 & 0.998 & 0.000 \\ 0.388 & 0.014 & 0.598 \\ 0.712 & 0.288 & 0.000 \end{bmatrix}$$

can be found by solving the characteristic equation:

$$|\mathbf{P} - \lambda \mathbf{I}| = 0$$

which becomes:

$$(\lambda - 1)(\lambda^2 + 0.984\lambda + 0.42458) = 0$$

Consequently:

$$\lambda_1 = 1 \quad \text{and} \quad \lambda_{2,3} = -0.492 + 0.427i$$

Let the matrix $\mathbf{V}$ be of the form:

$$\mathbf{V} = \begin{bmatrix} 1 & 1 & 1 \\ 1 & a_1 + b_1 i & a_1 - b_1 i \\ 1 & a_2 + b_2 i & a_2 - b_2 i \end{bmatrix}$$

From $\mathbf{PV}_2 = \lambda_2 \mathbf{V}_2$, it follows that:

$$p_{11} + p_{12}(a_1 + b_1 i) + p_{13}(a_2 + b_2 i) = \lambda_2$$

or: $0.002 + 0.998\,(a_1 + b_1 i) = -0.429 + 0.427i$
Hence, $a_1 = -0.495$ and $b_1 = 0.428$.
Further:

$$p_{21} + p_{22}(a_1 + b_1 i) + p_{23}(a_2 + b_2 i) = \lambda_2(a_1 + b_1 i)$$

This results in $a_2 = -0.536$ and $b_2 = -0.716$.

Because $\mathbf{V}$ now is known, $\mathbf{T}'$ can be solved from the relationship $\mathbf{T} = \mathbf{V}^{-1}$. However, let us first solve $\mathbf{T}'_1$ by using $\mathbf{T}'_1\mathbf{P} = \lambda_1 \mathbf{T}'_1$. This standard method for calculating the fixed vector results in the equations:

$$\begin{aligned} 0.002t_{11} + 0.388t_{12} + 0.712t_{13} &= t_{11} \\ 0.998t_{11} + 0.014t_{12} + 0.288t_{13} &= t_{12} \\ 0.598t_{12} \qquad\qquad &= t_{13} \end{aligned}$$

Also: $t_{11} + t_{12} + t_{13} = 1$
Note that one of the first three equations is superfluous. The resulting fixed vector is:

$$\mathbf{T}'_1 = [0.388 \quad 0.414 \quad 0.248]$$

The vectors $\mathbf{T}'_2$ and $\mathbf{T}'_3$ will consist of complex numbers. Let $\mathbf{T}'_2$ be of the form:

$$\mathbf{T}'_2 = [p_1 + q_1 i \quad p_2 + q_2 i \quad p_3 + q_3 i]$$

$\mathbf{T}'_3$ is the conjugate of $\mathbf{T}'_2$. From $\mathbf{T} = \mathbf{V}^{-1}$, it follows that:

$$t_{11} + 2p_1 = 1$$

$$t_{12} + 2p_2 = 0$$

$$t_{13} + 2p_3 = 0$$

Hence, $p_1 = 0.331$, $p_2 = -0.207$, $p_3 = -0.124$.

Also:

$$t_{11} + 2a_1p_1 - 2b_1q_1 = 0$$

$$t_{12} + 2a_1p_2 - 2b_1q_2 = 1$$

$$t_{13} + 2a_1p_3 - 2b_1q_3 = 0$$

with solution $q_1 = 0.012$, $q_2 = -0.445$, $q_3 = 0.433$.

Collecting the results, we can write:

$$\mathbf{P}^n = \begin{bmatrix} 1 & 1 & 1 \\ 1 & -0.495 + 0.428i & -0.495 - 0.428i \\ 1 & -0.536 - 0.716i & -0.536 + 0.716i \end{bmatrix}$$

$$\times \begin{bmatrix} 1 & 0 & 0 \\ 0 & (-0.492 + 0.427i)^n & 0 \\ 0 & 0 & (-0.492 - 0.427i)^n \end{bmatrix}$$

$$\times \begin{bmatrix} 0.338 & 0.414 & 0.248 \\ 0.331 + 0.012i & -0.207 - 0.445i & -0.124 + 0.433i \\ 0.331 - 0.012i & -0.207 + 0.445i & -0.124 - 0.433i \end{bmatrix}$$

The interpretation of the explicit form is greatly facilitated by using real numbers. We can write:

$$\mathbf{P}^n = \mathbf{Z}_0(1) + \mathbf{Z}_{(n)}^{(2,3)}$$

where $\mathbf{Z}_{(n)}^{(2,3)}$ represents the joint spectral component for the pair of complex roots. By using [12.16] of section 12.3, the final result is:

$$\mathbf{P}^n = \begin{bmatrix} 0.338 & 0.414 & 0.248 \\ 0.338 & 0.414 & 0.248 \\ 0.338 & 0.414 & 0.248 \end{bmatrix} +$$

$$+ 0.651^n \begin{bmatrix} 0.662 \cos(139n + 2) & 0.982 \cos(139n + 245) & 0.900 \cos(139n + 106) \\ 0.216 \cos(139n + 141) & 0.321 \cos(139n + 24) & 0.294 \cos(139n + 245) \\ 0.218 \cos(139n + 235) & 0.417 \cos(139n + 98) & 0.382 \cos(139n + 339) \end{bmatrix} \quad [12.27]$$

The angles in the second matrix are given in degrees. Clearly, if **P** is raised to a high power, the second matrix will disappear. We are close to this stage when $n = 16$ (see before). The second matrix describes the cyclicity in the section. Sand tends to be followed by silt, silt by clay, and clay by sand, but this regularity rapidly dies out when n increases. The period of the system is $360/139 = 2.6$ which is less than the periodicity of 3 for an idealized cyclicity. The individual elements of the second matrix all describe damped cosine curves when n increases.

It can be verified readily that when $n = 1$, [12.27] gives the original matrix **P**. The interpretation of the explicit form is facilitated by the graphical representation of Fig.84.

In the first diagram, the largest transition probabilities are indicated by arrows. The cycle 1–2–3 is indicated but chances of making a backward step are significant for states 2 and 3.

Suppose that the cyclical component of [12.27] is represented by a circle. The angles α_j of [12.16] can be used to divide this circle into three segments whereas the p_j are used to represent relative strength of amplitude (see Fig.84B). For example, the angle between arrows for states 1 and 2 is $141° - 2° = 139°$. Likewise, the angles β_k and coefficients q_k can be represented on a circle (see Fig.84C). The second circle would become the first one when the series is reversed before the forward transition matrix is determined. All three diagrams indicate that the preferred cycle for the forward direction is –sand–silt–clay–sand–.

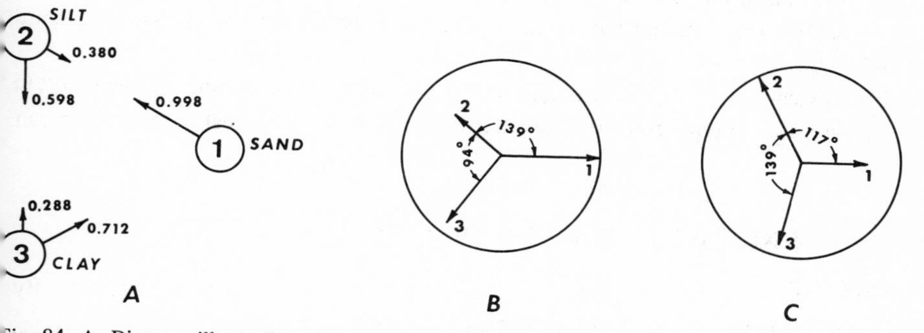

Fig. 84. A. Diagram illustrating observed upward transition probabilities between lithologies of Klevrdzeli flysch section (original data from Vistelius, 1949). B. Circle represents cyclical component of forward transition matrix; anticlockwise direction shows preferred cycle of lithologies from forward transition matrix. C. Ditto, for backward transition matrix.

12.8. Embedded Markov chains

Let us momentarily return to Krumbein and Dacey's matrix of section 7.9. This Markov chain has the property that after each increment there is a constant probability s_i of remaining in the same state i. Otherwise, the system is random.

The probabilities p_{ii} can be set equal to zero. At the same time, off-diagonal element can be made to add to one again by computing:

$$p_{ij}^* = \frac{p_{ij}}{(1-p_{ii})}$$

The resulting matrix is called an embedded Markov chain. For the example:

	A	B	C	D
A	0	0.333	0.352	0.315
B	0.226	0	0.288	0.486
C	0.184	0.554	0	0.262
D	0.269	0.553	0.188	0

The embedded Markov chain was used for studying lithologic transitions in strat graphic sequences by several authors including Carr et al. (1966), Potter and Blakel (1968), Gingerich (1969) and Read (1969).

An additional constraint has been forced on the chain in that a given state is no followed by itself. The fixed vector will represent the number of beds for a litholog rather than its relative thickness in the stratigraphic section of study.

Read (1969) has studied an embedded Markov chain for six lithologies in a boreho section through paralic Namurian sediments (Limestone Coal Group) east of Stirlin central Scotland.

The lithological states are: (1) mudstone; (2) siltstone; (3) sandstone; (4) seatclay; (5 silty and sandy rooty beds; (6) coal. The transition matrix is:

	1	2	3	4	5	6
1	0	0.606	0.212	0.030	0.091	0.061
2	0.196	0	0.647	0.039	0.098	0.020
3	0.083	0.375	0	0.104	0.417	0.021
4	0.156	0.063	0.031	0	0.313	0.438
5	0.125	0.150	0.125	0.375	0	0.225
6	0.385	0.154	0.038	0.346	0.077	0

A matrix of this type also can be expressed in explicit form. Some results are shown Fig.85.

Fig.85A is from Read who recognized a preferred cyclicity in the upward transition The sequence 1–2–3–5–4–6–1 etc. is broadly similar to the idealized cyclothems pr posed by Duff et al. (1967) for the Limestone Coal Group. Read (1969) pointed out th

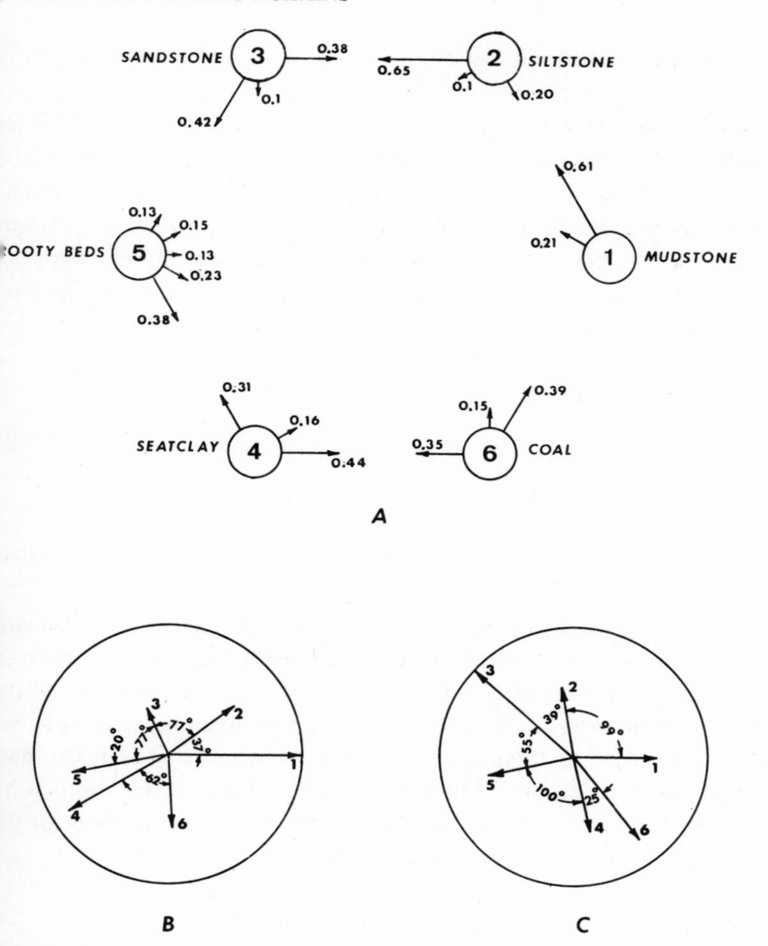

g.85. A. Upward transitions with probabilities of 0.10 or more between lithologies of Limestone al Group section, Scotland (from Read, 1969). B. Circle represents a cyclical component of forward ansition matrix. C. Ditto, for backward transition matrix.

vo subsets of lithological states, once entered, may be difficult to leave. These are the ternation of siltstone and sandstone and a subset consisting of three states (4, 5 and 6) aracterized by the growth of land vegetation in situ.

The preferred cyclicity also is reflected in the circles of Fig.85B and C. These were nstructed in the same manner as Fig.84. They are for a pair of complex roots ($0.177 \pm$ $238i$) with modulus 0.300. The periodicity of this component is 6.75. The other roots r Read's embedded chain were $1, -0.540$, and ($-0.408 \pm 0.075i$). The negative root d the other pair of complex rocks have cyclicities with periods equal to 2 and 2.1 states spectively.

One of the problems of interpretation of these results is that statistical tests of signifi- nce for the explicit form do not exist.

12.9. Upper Pennsylvanian cyclothems in Kansas

Schwarzacher (1967, 1969) studied sedimentary cycles by Markov-chain models fo part of the Upper Pennsylvanian sequence of northeastern Kansas. The Kansas cyclo thems provide an example of a remarkable cyclicity persisting over long distances laterall and of a considerable stratigraphic thickness. They were studied in great detail by Moor (1936) who proposed ten members of an ideal cyclothem comprising different types o shale and limestone. For field descriptions, this classification was summarized by Merrian (1963) to a cyclothem consisting of:

Outside shale	
Limestone 5	(Clay Creek)
Outside shale	
Limestone 4	(Kereford)
Limestone 3	(Ervine Creek)
Limestone 2	(Leavenworth)
Limestone 1	(Toronto)
Outside shale	

Limestones 1 through 4 are separated by inside shales, of which the black shale followin limestone 2 is the most conspicuous. Part of this ideal cyclothem is present in the thirtee cyclothems of the schematic diagram of Fig.86. This was based on measurements of th dominant lithology at 5-ft. intervals. A distinction was made between limestone (1), in side shale (2), and outside shale (3). Cumulative thickness of limestone and outside shal for each cyclothem is shown in Fig.86C. Elimination of the trends in this diagram b Schwarzacher resulted in the transformed section of Fig.86B. Transition matrices for th 5-ft. sampling interval before and after transformation are:

$$\text{Before transformation} \qquad \begin{matrix}1\\2\\3\end{matrix}\begin{bmatrix}0.6190 & 0.0952 & 0.2857\\0.6923 & 0.2308 & 0.0769\\0.2173 & 0.0264 & 0.8364\end{bmatrix} \qquad \text{After transformation} \qquad \begin{bmatrix}0.6400 & 0.1200 & 0.2400\\0.6000 & 0.3333 & 0.0667\\0.1579 & 0.0526 & 0.7895\end{bmatrix}$$

Both matrices have real roots only. Schwarzacher concluded that this non-oscillatin behavior of the chain is due to underrepresentation in that either the number of states i too few or the vertical sampling interval is too small. Schwarzacher (1969, p.33) nex reduced the system to two states (limestone, L; and shale, S), simultaneously expanding i by forming a fourth-order chain. This can be done by making all possible orderings c four of the two states (L and S) and defining these as the new states. There are 1 different orderings of this type but 4 of 16 were not found in the section of study. transition matrix was formed for the 12 remaining states:

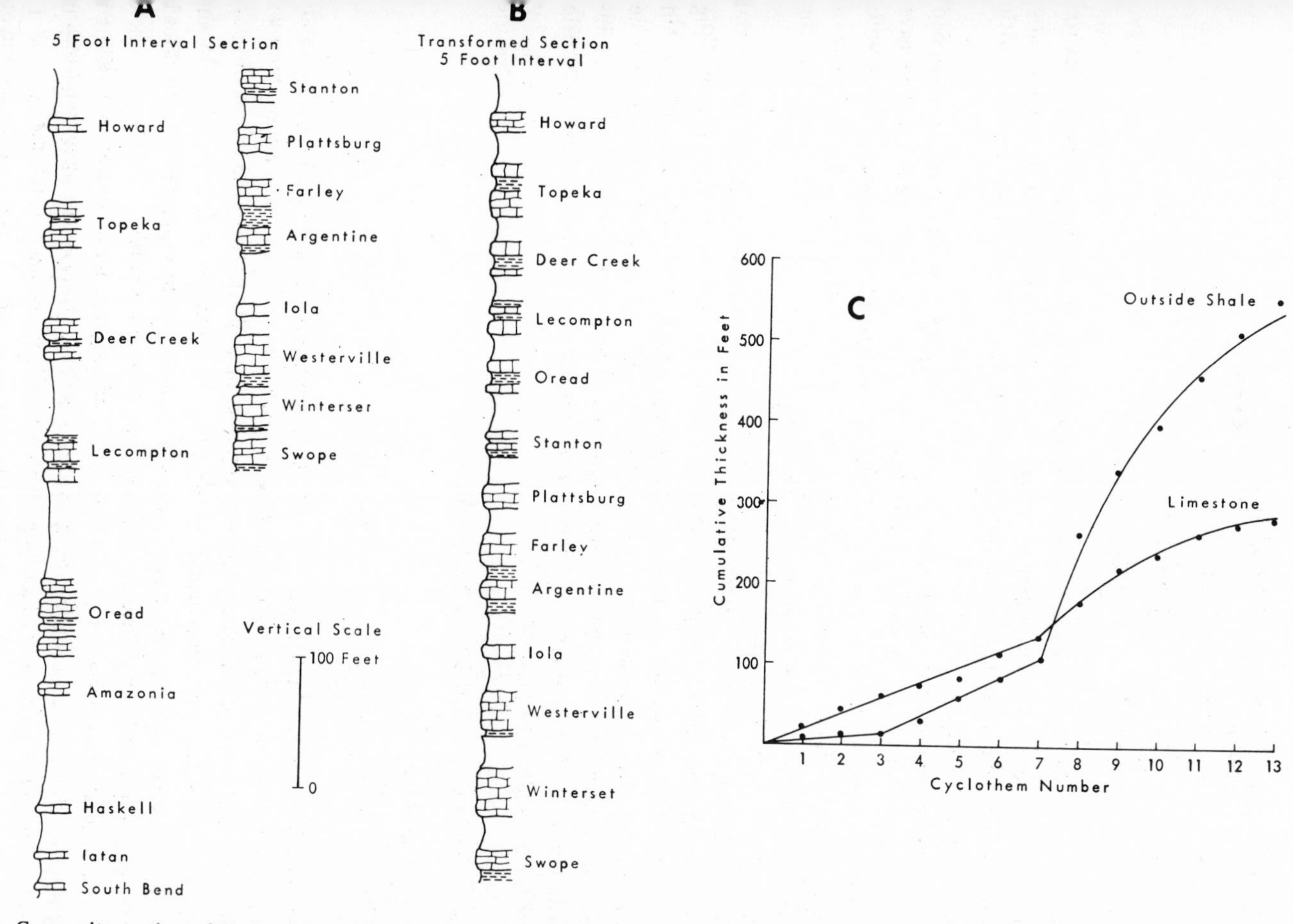

Fig.86. A. Composite section of Upper Pennsylvanian rocks in north-eastern Kansas coded according to dominant lithology at 5-ft. intervals. B. Same section after removal of sedimentation rate trend. C. Cumulative thickness of limestone and outside shale for cyclothems. Fitted curves represent sedimentation rate trend removed in B. (From Schwarzacher, 1969.)

1. LLLL
2. LLLS
3. LLSL
4. LLSS
5. LSLS
6. LSSS
7. SLLL
8. SLLS
9. SLSS
10. SSLL
11. SSSL
12. SSSS

As in the case of first-order chains, the largest eigenvalue of the resulting transition matri is equal to one, but next comes a pair of complex roots with modulus equal to 0.868. Th periodicity for the corresponding component amounts to 45 ft. which is comparable wit the cyclicity of Fig.86B.

In an earlier publication, Schwarzacher (1967) analyzed a 2000-ft. section compile from a generalized graphic column of the Pennsylvanian sequence in Kansas. This gav 400 percentage values for 5-ft. intervals of four lithologies: (1) sandstone; (2) shale wit coal; (3) shale without coal; and (4) limestone. It is noted that the "outside" shale c Fig.86 generally is silty or sandy and includes increasing amounts of sandstone in th more southern parts of Kansas.

The percentage values for three states 1, (2 + 3) and 4 were subjected to spectr analysis by Schwarzacher using the "hanning" method described in Chapter 11. Bot shale (2 + 3) and sandstone (1) showed pronounced peaks at periods equal to 166 ft. Th limestone peak was shifted to a 125-ft. period having a minimum coinciding with th shale-sandstone peak. Schwarzacher fitted various autoregressive schemes to the data an concluded that at least three coefficients were needed to provide a realistic stochast model. For sandstone the scheme is:

$$\hat{X}_t = 0.826X_{t-1} - 0.301X_{t-2} + 0.072X_{t-30}$$

and for limestone:

$$\hat{X}_t = 0.534X_{t-1} + 0.002X_{t-2} + 0.066X_{t-20}$$

The third term is for higher lags 30 and 20, respectively, in accordance with a mod proposed by Whittle (1954b) for analyzing sunspot numbers. For shale, a third term w not statistically significant. There are several points of similarity with the results obtaine for silt–clay thickness components of varves in the example of section 11.4. In th situation, we also needed three coefficients (c, α and ω_0) to represent the process ar peaks in the spectra of two different lithologies at different periods.

The problem of whether the sandstone and limestone cyclicities in Kansas reflect tw separate mechanisms or a single would require a detailed study of the phase relationshi and this could not be done for lack of suitable data (Schwarzacher, 1967, p.11).

Schwarzacher determined the dominant lithologies at 5-ft. intervals for the section a formed the transition matrix:

$$
\begin{array}{c}
1 \\ 2 \\ 3 \\ 4
\end{array}
\begin{bmatrix}
0.6456 & 0.1266 & 0.1519 & 0.0759 \\
0.0625 & 0.3125 & 0.3125 & 0.3125 \\
0.1272 & 0.0060 & 0.7212 & 0.1456 \\
0.0476 & 0.0119 & 0.7212 & 0.5953
\end{bmatrix}
$$

The author obtained the explicit form of this matrix. The fixed vector can be recomputed to percentage values for the entire section of study giving 22% sandstone (1); 5% shale with coal (2); 48% shale without coal (3); and 25% limestone (4). The four eigenvalues are 1, $(0.493 \pm 0.059i)$ and 0.289.

The periodicity corresponding to the complex pair of rocks amounts to 52.9 sampling intervals or about 265 ft. The components are represented by circles in Fig.87. The transition matrix suffers from underrepresentation but a long-term periodicity is indicated. Limestone (4) would follow sandstone (1) which in turn follows shale without coal (3). There were few data for shale with coal (2) and this would explain the erratic behavior of state 2 in Fig.87. In terms of distances, Fig.87A shows that sandstone lags $(360/129) \times 265 = 95$ ft. behind shale; and shale 111 ft. behind limestone. Schwarzacher (1967, p.8) has determined phase angles by cross-spectral analysis and obtained estimates of 74 and 77 ft. for these phase lags at period 166 ft.

The previous results indicate that the explicit form of Markov chains can provide information on cyclicity in stratigraphic sequences. Some of the problems include lack of methods of statistical inference. To some extent, the precision of estimates could be assessed by Monte Carlo simulation. Other problems are possible, such as underrepresentation but, then, the analysis of high-order Markov chains may give better answers. Another problem is possible lack of stationarity. This can sometimes be resolved by trend elimination as was done by Schwarzacher in the case of Fig.86B.

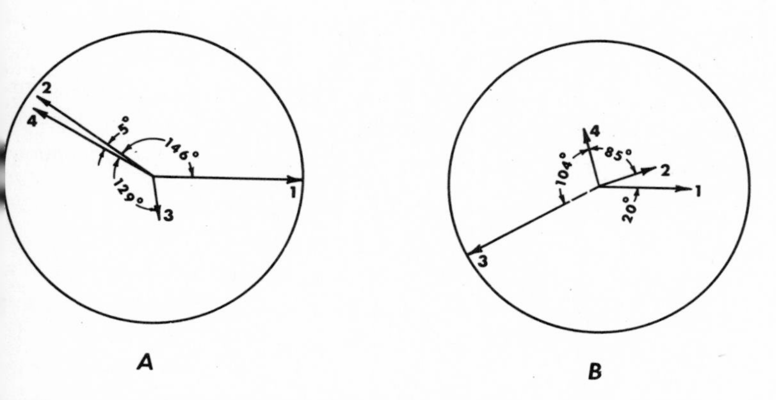

Fig.87. Unit circles representing cyclical component in forward and backward transition matrices compiled by Schwarzacher (1967) from a generalized column of Pennsylvanian rocks in Kansas.

The emphasis in this section was on the potential use of Markov chains for studyi cyclicities and the approach was exploratory. Other applications of Markov chains ha been reviewed by Krumbein (1967) and Krumbein and Dacey (1969). A semi-Mark process was used by Schwarzacher (1972).

In some applications, the existence of any regularity in the transitions is in doub Then, the transition matrix can be tested for nonrandomness. The test statistic:

$$\hat{\chi}^2 \approx 2 \Sigma n_{ij} \, {}^{e}\log \hat{p}_{ij}/\hat{p}_j \qquad [12.2$$

is distributed according to χ^2 with $(m-1)^2$ degrees of freedom if the states are arrange randomly (Anderson and Goodman, 1957; Kullback, 1959; Krumbein, 1967). In th expression, m denotes the number of states, the n_{ij} are absolute numbers for transitio $\hat{p}_{ij}$, and each $\hat{p}_j = (1/n) \sum_i n_{ij}\hat{p}_{ij}$ denotes the total probability for the j-th colum $n = \sum_{ij} n_{ij}$ is total number of transitions. Of course, if the sampling interval is at regul distances and less than the thickness of the layers, we must have nonrandomness, becau then the diagonal elements $\hat{p}_{ii}$ will be relatively large. Eq. [12.28] involves a comparis of the transition matrix with a matrix whose rows are identical and which would arise case of purely random changes in state. When nonrandomness is indicated by the test, t second matrix may be subtracted from the transition matrix. This procedure leads almost the same result as the elimination of the first component of the explicit for whose rows are given by the fixed vector.

For embedded Markov chains, [12.28] can be applied but m additional degrees freedom are lost because the diagonal elements are zeros in both matrices compared wi one another (Potter and Blakely, 1968). The transition matrices used in this chapter a all far from being random.

It is interesting to speculate on the reasons of preferred cyclicities in stratigraph sequences such as the Carboniferous rocks of Scotland and Kansas. Possible explanatio were reviewed by Duff et al. (1967) and, for the Kansas cyclothems, by Weller (1964 The problem was reviewed in several papers of a symposium on cyclic sedimentati (Merriam, 1964). Most explanations are climatic, tectonic, or a combination. An examp of a climatic explanation by Wilson (surging land ice) was mentioned in section 11. Twenhofel (1950) has warned that it is a mistake to relate all cyclicities in sediments to single cause.

Chapter 13

MULTIVARIATE STOCHASTIC PROCESS-MODELS WITH APPLICATIONS TO THE PETROLOGY OF BASALTS

13.1. Introduction

In the course of geologic time, basaltic magma has poured out on the surface of the arth in many places. In addition, it has formed dike swarms (cf. Fig.79B) and con- ordant intrusions such as sills inserted between the layers of sedimentary sequences.

According to Barth (1962), approximately 98% of effusive rocks of all ages on land is basaltic–andesitic in composition (cf. Daly, 1968). Baragar and Goodwin (1969) deter- mined that in four volcanic belts of Precambrian age, the abundance of volcanic classes is ather uniform with, on the average, basaltic (60%), andesitic (28%) and rhyolitic (12%). It seems that the proportion of rhyolite flows, which include ignimbrites, has decreased since Archean time.

The theory of basaltic magma will be discussed in some detail. Petrologists have constructed a variety of approximate models for observable variations in the fractional rystallization of basaltic rocks from magma and for the interrelationships between the omponents of rocks and magma. As for granitic rocks, the models are mainly based on xperimental phase diagrams. It may be possible to calculate a characteristic index on the asis of a model. A set of values for this index as determined from an assemblage of basalt amples can provide information on the composition of the original magma and the esulting series of fractionation products. Several petrological indices will be discussed. At ne same time, multivariate statistical techniques will be applied and the results compared o those for the indices.

The main practical example is for the Yellowknife volcanic belt in northern Canada. The geochemistry of this belt, which is Archean in age, has been studied by Baragar (1966). Total thickness exceeds 40,000 ft. (12 km) which amounts to about a third of ne present thickness of the earth's crust. Gradual changes in magma composition during me also will be studied for this belt by using techniques based on the theory of stochas- c processes. The oxides for successive observations are seen as a multivariate time series. The Kolmogorov forward differential equations for this series may be solved by assuming hat a simple process of time-continuous change in magma composition has been opera- ve.

13.2. Petrological models

Basalts mainly consist of two minerals: plagioclase and pyroxene. Barth (1962) discussed the following simplified model. Basalt magma can be defined as a silicate melt that on cooling yields plagioclase and the pyroxenes. These may be considered in terms of two main reaction series: (1) the series of plagioclase feldspars going from calcic to sodic composition (Ca → Na); and (2) a series of clinopyroxenes developing from diopside to hypersthene (Mg → Fe). Although this is an oversimplification of pyroxene crystallization relationships, it will be helpful for our discussion to use this combination of the two processes for representing the principal crystallization process of a basaltic magma. At specific compositions of magma, the second series can begin with initial precipitation of olivine that, on further cooling, is converted into pyroxene.

The relationship between the two processes is shown graphically in Fig.88. There is a boundary line *OP* in the center of this illustration. If the composition of the original basaltic liquid lies to the right of *OP,* crystallization will start with precipitation of only pyroxene. If it is located to the left of the boundary, only plagioclase will be formed. Because of the precipitation of crystals, the magma composition will change describing a path that is directed to the center. Once the line *OP* is reached, simultaneous precipitation of both mineral phases takes place. A melt of composition *O* crystallizes by simultaneous formation of pyroxene of composition *A* and plagioclase of composition *Q*. As crystallization proceeds, the melt changes from *O* to *P* and the two solid phases change from *A* to *B* and from *Q* to *R*. Fig.88 is only approximate and the model which has four constituents can not be completely represented by it. For a more complex representation, a tetrahedron with four corner points can be used for projection (cf. Barth, 1962).

One of the objectives of a model of this type is to classify a given basaltic rock. After the mode or the norm has been calculated for a basalt sample, the result can be projected

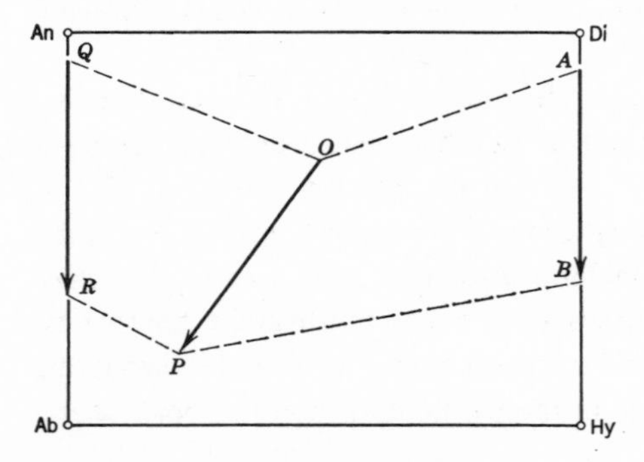

Fig.88. Schematic presentation of the crystallization process of basaltic lavas (from Barth, 1962). A melt of composition *O* crystallizes by simultaneous precipitation of pyroxene of composition *A* and plagioclase of composition *Q*. During crystallization, melt changes from *O* to *P*, and the two solid phases from *A* to *B* and *Q* to *R*, respectively.

on a phase diagram and from the position of the point with respect to one or more boundaries such as *OP* in Fig.88, it can be concluded which solid phase has crystallized first: plagioclase or pyroxene.

Another practical application is as follows. Suppose that basaltic magma occurs in an underground magma chamber which is subject to cooling and where continuous precipitation of crystals takes place in the manner described above. Suppose further that periodically some magma escapes from the chamber to the surface of the earth where it forms basalt lavas. An assemblage of samples from these lavas may show some regularities in the relationship between components. Ca-rich plagioclase will tend to coexist with Mg-rich pyroxene. If the major oxides for a suite of basalts are correlated to one another, regardless of time of deposition, the following signs may occur for the coefficients in the correlation matrix:

	FeO	MgO	CaO	Na_2O
FeO	1			
MgO	–	1		
CaO	–	+	1	
Na_2O	+	–	–	1

Likewise, if the major oxides are plotted against time, the cooling of an underground magma chamber may be reflected at the surface where CaO and MgO decrease whereas FeO and Na_2O increase with time. In this example, the time of deposition can be entered as a variable by sampling successive flows in a volcanic pile from bottom to top.

Two methods of presentation of data

Petrological data commonly are graphically presented by using one of the following methods:

(1) The relationship between the constituents of the system is regarded without consideration of the position in time or space of the individual rock samples. Examples include the triangular diagram (see below) and the Harker diagram where the major oxides are plotted against percent silica (SiO_2).

(2) The position in time or space of the rock samples is considered.

In many practical situations, the first method is the only possible one. Usually, when the dimensions of time or space also can be used as one or more coordinate axes in the graphical representation, the parameters that are plotted have been determined by the first method. The mathematical analogue of the first method is to correlate the constituents with one another. This method will be discussed in more detail. The constituents commonly are expressed as percent values (by weight or volume) and therefore add to a constant when summed for an individual rock sample. This introduces strong correlations into the system. For example, silica frequently is negatively correlated with most other

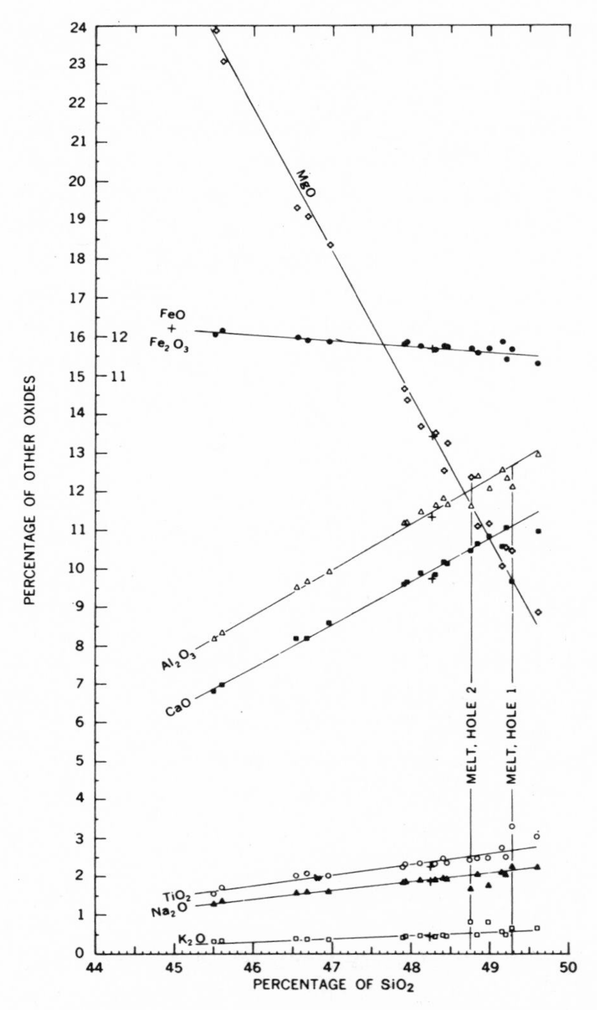

Fig.89. Harker diagram of crust rocks and melt samples from Kilauea Iki lava lake, Hawaii (1959–1960 eruption). Trend lines are olivine (85% forsterite) control lines. (After Murata and Richter, 1966.)

components. SiO_2 is the most abundant constituent. If it increases, a number of other components must decrease to maintain the constant sum. Problems associated with closed number systems have been studied in detail by Chayes (1970 and 1971), Chayes and Kruskal (1966), Miesch (1969), Smith (1972), and Vistelius (1968). Suppose that in a statistical simulation experiment a random variability is assigned to SiO_2, and that there are no systematic variations of petrological significance in the system. Yet trends may occur in the Harker diagram because of the closure of the system. Great care should be

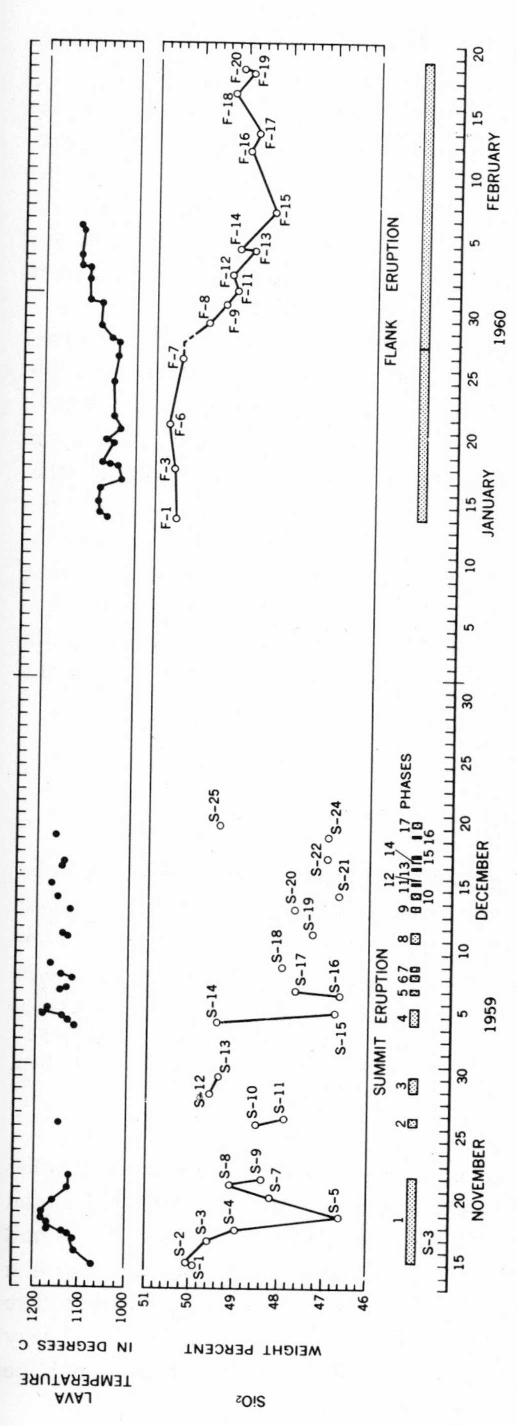

Fig.90. Temperature and bulk silica content of lavas for summit and flank eruptions as a function of time (from Murata and Richter, 1966); hotter lavas carried more (silica-poor) olivine crystals in suspension.

taken to eliminate these artificial trends when the attempt is made to detect petrological trends from data plotted in triangular diagrams, tetrahedrons or Harker diagrams.

An example of a Harker diagram is shown in Fig.89 which is based on work by Murata and Richter (1966) in relation to the 1959–60 eruption of the Kilauea Volcano, Hawaii. Both rock and melt samples are represented. The pattern that emerges is much simpler than results usually obtained for basaltic rocks. Trends are caused by variations in the amount of olivine crystals suspended in the magma. Other petrological trends are largely absent. The system is characterized by a lack of random noise. The chemical variation can be approximated by a set of straight lines through the points in Fig.89.

An example of the second method of presentation of data is shown in Fig.90. In the lower part, the variation in bulk silica content is plotted against time (November, 1959; January–February, 1960). The variation of lava temperature is also shown. There is negative correlation between temperature and silica content of the lava. Murata and Richter (1966) have shown that if the lava is hotter, it carries more olivine crystals. Olivine is poor in SiO_2 as compared to the other constituents of the magma such as hot fluid, plagioclase and pyroxene crystals. Although the data shown in Fig.89 are not the same as those shown in Fig.90, the trend in Fig.89 can be interpreted in the same manner. The variation in Fig.89 is characterized by the line for MgO with a steep dip to the right corresponding to a decrease in the number of Mg-rich olivine crystals carried by the magma. Simultaneously, most other elements have lines with a positive dip. In this example, noise is virtually absent.

13.3. Projection of a p-dimensional situation onto the Harker diagram

The Harker diagram of Fig.89 can be regarded as a projection of an eight-dimensional relationship onto the plane. The mechanism of this projection can be explained as follows by using a simpler three-component system as represented in Fig.91. The three components A, B, and C add to unity (100%), and any value must fall in the triangle ABC with corner points (1, 0, 0), (0, 1, 0), and (0, 0, 1). The so-called triangular diagram is

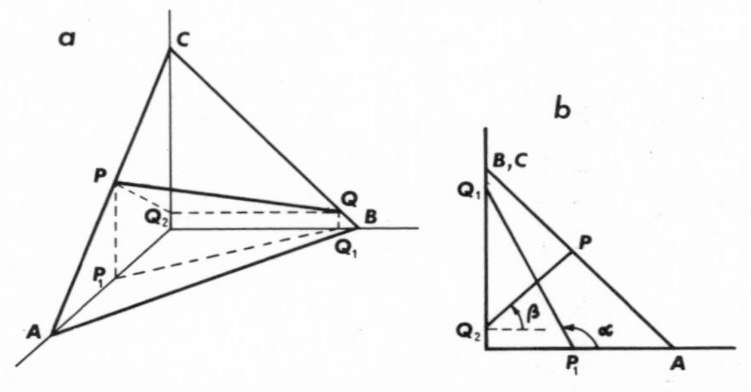

Fig.91. Harker-type diagram for three-component system (for explanation, see text).

obtained by representing the triangle ABC without the corresponding Cartesian coordinate system. Suppose that the observations in the triangle lie on a single straight line PQ. The line PQ can be projected onto both the horizontal AB-plane and the vertical AC-plane. This results in the lines P_1Q_1 and PQ_2 in Fig.91b. Representation of the new lines in a single diagram gives a simplified version of a Harker diagram.

By methods of analytical geometry, the slopes of the lines in Fig.91b can be converted into the direction cosines of the line PQ in Fig.91a.

If α represents the angle of slope of P_1Q_1 in the AB-plane, the slope of a line normal to P_1Q_1 can be defined as $(\alpha - 90°)$. This line is normal to the plane PQQ_1P_1 and has direction cosines:

$$\{\cos(\alpha - 90°), \sin(\alpha - 90°), 0\}$$

or $\{\sin\alpha, -\cos\alpha, 0\}$

The normal of the triangle ABC has direction cosines:

$$\{1/\sqrt{3}, 1/\sqrt{3}, 1/\sqrt{3}\}$$

PQ is perpendicular to these two normals. If its direction cosines are labelled $\{x, y, z\}$ the unknowns x, y and z can be solved from the system of equations:

$$x \sin\alpha - y\cos\alpha = 0$$

$$x/\sqrt{3} + y/\sqrt{3} + z/\sqrt{3} = 0$$

and: $x^2 + y^2 + z^2 = 1$

By putting $x' = 1$, it follows that:

$$y' = \tan\alpha \quad \text{and} \quad z' = -(\tan\alpha + 1)$$

Division of x', y', and z' by $\sqrt{x'^2 + y'^2 + z'^2}$ gives $\{x, y, z\}$.

By using the relationship $A + B + C = 1$, or alternately, $x + y + z = 0$, it seems that the problem can be solved without using β which is the slope of PQ_2 in the AC-plane. If the system is closed, β is fully determined by α and vice versa.

Suppose now that the line PQ does not necessarily fall in the triangle ABC. In that situation, PQ will be perpendicular to the two lines with $\{\sin\alpha, -\cos\alpha, 0\}$ and $\{\sin\beta, 0, -\cos\beta\}$. Consequently:

$$x\sin\alpha - y\cos\alpha = 0$$
$$x\sin\beta - z\cos\beta = 0$$
$$x^2 + y^2 + z^2 = 1$$

with solution:

$$x' = 1$$
$$y' = \tan\alpha$$
$$z' = \tan\beta$$

and:

$$x = (1 + \tan^2\alpha + \tan^2\beta)^{-1/2}$$
$$y = \tan\alpha(1 + \tan^2\alpha + \tan^2\beta)^{-1/2}$$
$$z = \tan\beta(1 + \tan^2\alpha + \tan^2\beta)^{-1/2}$$

This result is readily extended to a situation in p-space. For example, in Fig.89 there are eight components. The slopes of the seven lines are approximately:

K_2O	: 0.053	Al_2O_3	: 0.807
Na_2O	: 0.159	Iron	: −0.106
TiO_2	: 0.159	MgO	: −2.527
CaO	: 0.754		

Except for a normalizing factor, these numbers represent the direction cosines of a single line in 8-space. As in the three-dimensional example, the scaling is done by calculating the sum of squares (7.667), addition of 1 for silica (→ 8.667), taking the square root (→ 2.944) and multiplication of all slopes by the inverse of the square root (0.340). The direction cosines of the trend line in 8-space then are:

K_2O	: 0.02	Al_2O_3	: 0.27
Na_2O	: 0.05	Iron	: −0.04
TiO_2	: 0.05	MgO	: −0.86
CaO	: 0.26	SiO_2	: 0.34

This line can be given its proper position in 8-space by letting it pass through a point fo which the eight oxides are known. In the sequel, this point will be the one whose coordinates are equal to the arithmetic averages for the oxides. The direction cosine indicate that the trend line makes the smallest angle (arccos 0.86 = 31°) with the MgO axis. The signs show that an increase in MgO, and to a minor extent iron, corresponds to decrease in other oxides and vice versa.

Given a set of chemical analyses for an assemblage of rock samples, the direction cosines of the best-fitting line in p-space can be estimated by component analysis (see Chapter 5). This method can be useful if the scatter is large, viz. when the trends in the Harker diagram are obscured by noise. Also, more than a single principal component can be calculated. This method will be applied to the following practical example.

13.4. The Yellowknife volcanic belt

The Yellowknife volcanic belt is a homoclinal succession of flows and minor pyro

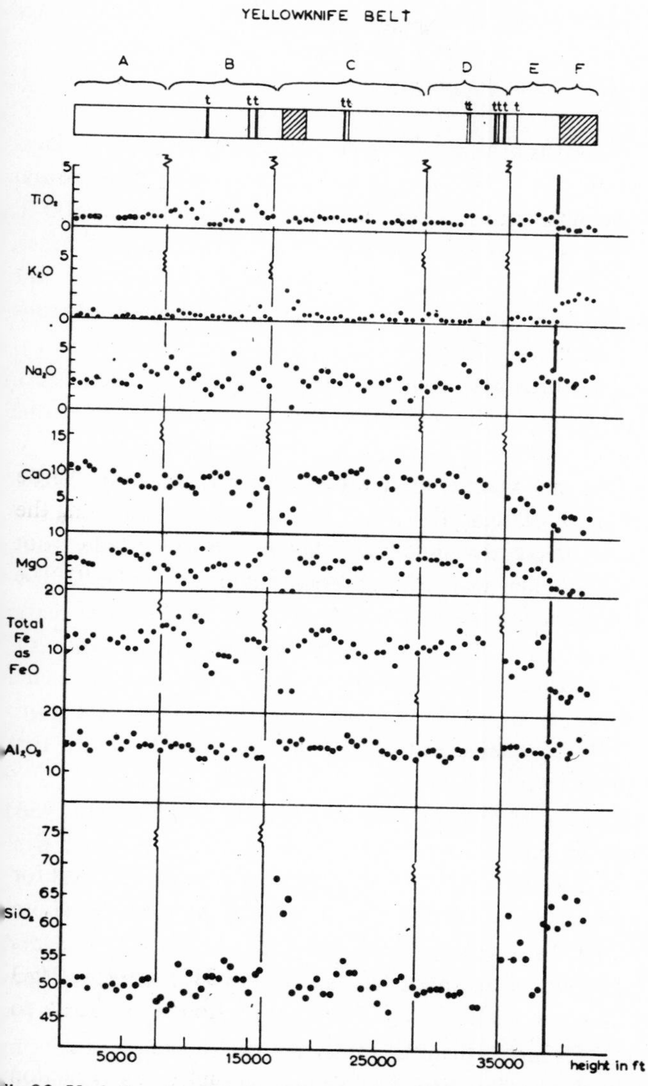

'ig.92. Variation in major oxides with stratigraphic height in the Yellowknife volcanic belt (from ʾaragar, 1966); acidic rocks, hatched; mafic rocks, plain; tuffs, *t*.

lastic rocks that generally dip steeply in southeastern direction. It is part of the Yellow-nife Group which is widely distributed in the Slave province of the northern Canadian hield. It occurs in irregularly shaped areas that may measure as much as 100 miles 160 km) across and are separated and perforated by bodies of granite and granitic gneiss. ʾommonly, the volcanic member occurs as discontinuous strips at the margins of predominantly sedimentary units which it separates from granitic gneisses. The averag potassium/argon age of the granitic rocks is about 2400 m.y., indicating an Archean ag as found in the Superior province of Ontario and Quebec whose average age is 2490 m.y (cf. Chapter 15).

The Yellowknife volcanic pile consists mainly of dark-green massive and pillowed lava with acidic layers at about the midpoint and top of the section (see Fig.92). Baraga (1966) collected specimens for chemical analysis at stratigraphic intervals of about 500 ft (150 m). Some variation from the 500-ft. interval was permitted to avoid altered zones tuffs, and pyroclastic beds. Gabbro sills were sampled but because their position in th time sequence is uncertain they were omitted from the stratigraphic variation diagram and have not been included here in our statistical analyses.

As indicated in Fig.92, the data originate from six separate sampling lines (A–F) These lines cross different parts of the belt but can be readily correlated with one another. Combined, they provide a complete stratigraphic section through the belt.

The variation of the major elements against stratigraphic height is shown in Fig.92 Not all these data will be used for the statistical analysis. The first three observations o sampling interval C were made on the central acidic layer; they have been excluded fron the numerical calculations because of their markedly different composition. The 31 ob servations from sections A and B provide a fairly complete sampling of the lower mafic member and will be referred to as series 1. The remaining 29 observations of sections C and D represent the lower three-quarters of the second mafic member; they are callec series 2. Sections E and F have not been used for statistical analysis. There is a strati graphic gap of approximately 2000 ft. between the last specimen from line D and the firs from line E. Section F is for the upper acidic member.

The chemical analyses shown in Fig.92 and used for computations by Baragar (1966 and Agterberg (1966a) were done by rapid methods in the Geological Survey's analytica laboratories. Water was not determined as a rock constituent. For our statistical analysis the raw data for the major elements were recomputed to add up to 100%. FeO anc Fe_2O_3 have not been combined. MnO also was determined separately but has been addec to FeO. Subsequent analysis by classical methods has led to improved chemical data (Baragar, 1966 and personal communication, 1970). For example, the new values fo Al_2O_3 are on the average about 0.5% higher than the earlier results. This improvement in accuracy and precision is too small to alter previous conclusions based on petrological indices and component analysis. However, the multivariate time-series analysis discussed at the end of this chapter was completed in 1970 on the revised data supplied by Baraga (personal communication, 1970). Our preprocessing of the new data is the same as that performed previously (Agterberg, 1966a).

13.5. Calculation of trend lines in the triangular diagram

Kuno et al. (1957) stated that the fractionation trend of basaltic magma is bes

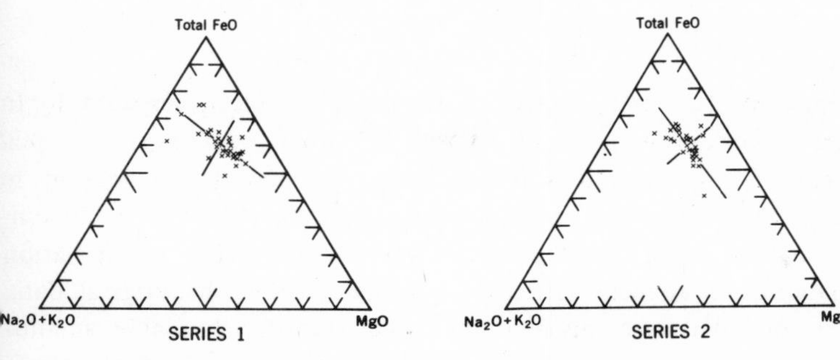

Fig.93. Geometrical representation of the first two principal components derived from the correlation matrix in a closed system of three components (after Agterberg, 1966a). Crosses represent ratio MgO/total iron as FeO/total alkalies in series 1 and 2. PC_1 (pointing upwards) approximately coincides with first principal component derived from variance–covariance matrix.

illustrated by plotting the ratio (MgO/total iron as FeO/total alkalies) in a triangular diagram. These three oxides or combinations of oxides are extracted from the total system by recalculating them to 100%. The triangle *ABC* as shown in Fig.91 can then be used for graphical representation. In Fig.93, the ratios for the chemical analyses of series 1 and 2 are plotted in this fashion. Means, standard deviations and correlation coefficients are shown in Table XXXV. The variances and covariances are not shown but can be readily calculated from these data.

TABLE XXXV

Correlation matrix $\mathbf{R}_0$, mean and standard deviation for MgO/FeO (total)/alkali ratios. (From Agterberg, 1966a, table 2.)

	Series 1			*Series 2*		
	MgO	FeO	alk.	MgO	FeO	alk.
MgO	1			1		
FeO	−0.823	1		−0.851	1	
alk.	−0.638	0.087	1	−0.322	−0.222	1
Mean	24.29	60.59	15.12	26.53	58.70	14.77
S.D	7.71	5.96	4.40	5.77	5.60	3.11

As a working hypothesis, it can be assumed that the irregular clusters of dots in Fig.93 contain a linear trend that has been obscured by noise. The statistical problem then is to extract this trend from the data. Component analysis can be used to calculate the direction cosines of a best-fitting line for which the sum of squares of perpendicular distances of points from the line is a minimum (cf. Le Maitre, 1968).

Multivariate statistical analysis is most frequently performed on the basis of correlation matrices. We recall from section 5.1 that the correlation matrix is the variance–covariance matrix of standardized data $z_{i,k}$ with:

$$z_{i,k} = (x_{i,k} - \bar{x}_i)/s(x_i)$$

where $\bar{x}_i$ and $s(x_i)$ represent mean and standard deviation for variable x_i, respectively. In general, an advantage of working with standardized data is that, with respect to total variation in the system, an equal weight is assigned to all variables. Thus, variations in minor constituents are deemed as important as those in major constituents which commonly have a larger variance. In many situations, it is better to operate on correlation matrices than on variance–covariance matrices as calculated from the original data.

The best-fitting lines obtained by component analysis are sensitive to scaling and not independent of linear transformations applied to the original data (e.g., standardization). Therefore, a trend line extracted from the variance–covariance matrix will differ from one extracted from the correlation matrix even when the effects of the transformation are removed. This aspect of component analysis will now be investigated for the triangular diagrams of Fig.93.

First, component analysis is applied to the variance–covariance matrix. The resulting eigenvalues and eigenvectors are shown in Table XXXVI. The smallest eigenvalue is zero and has eigenvector $\{1/\sqrt{3}, 1/\sqrt{3}, 1/\sqrt{3}\}$. This line can be interpreted by inspection of Fig.91. It is the normal of the triangle *ABC.* The eigenvector for the largest root in Table XXXVI (series 1) has elements $\{0.799, -0.544, -0.255\}$. It defines a line of best fit or trend line and can be plotted in a triangular diagram as follows.

TABLE XXXVI

Eigenvalues and eigenvectors for MgO/FeO (total)/alkali ratios

		Variance–covariance matrix			*Correlation matrix*		
		1	2	3	1	2	3
Series 1							
Eigenvalue		92.12	22.23	0.00	2.084	0.916	0.00
	MgO	0.799	0.167	0.577	0.693	0.021	0.577
Eigenvector	FeO	– 0.544	0.608	0.577	–0.562	–0.611	0.577
	alk.	– 0.255	– 0.776	0.577	–0.452	0.791	0.577
Series 2							
Eigenvalue		59.88	14.43	0.00	1.858	1.142	0.000
	MgO	0.722	0.382	0.577	0.719	0.189	0.670
Eigenvector	FeO	– 0.692	0.433	0.577	–0.690	0.320	0.649
	alk.	– 0.030	– 0.826	0.577	–0.091	–0.928	0.359

The line *AB* in Fig.91 has direction cosines $\{1/\sqrt{2}, -1/\sqrt{2}, 0\}$. This is also the direction of the side MgO–total iron in the triangle of Fig.93. Consequently, the angle ϕ between the trend line and this side satisfies:

$$\cos\phi = 0.799/\sqrt{2} + 0.544/\sqrt{2} = 0.950$$

ith $\phi = 18°$. By plotting the average (24.29, 60.59, 15.12) in the triangular diagram and onstructing the line which makes angle ϕ with the MgO–iron line, the trend line can be onstructed. The second eigenvector also defines a line that passes through the average nd is perpendicular to the trend line. For series 2, the angle ϕ amounts to 2°.

The eigenvalues and eigenvectors of the correlation matrix are also shown in Table XXVI. These eigenvectors are orthogonal but the coordinate system which is now used ffers from that shown in Fig.91. The three axes have been scaled by standardizing the iginal data. Suppose that the unit of distance before the ratios are divided by their andard deviations is 1% along each of the three axes. The unit is changed into 1/7.71, 5.96, and 1/4.40 for MgO, iron, and alkali (series 1), respectively. Consequently, the angular diagram becomes distorted. The first two eigenvectors define two lines for incipal components (PC_1 and PC_2) that are mutually perpendicular and can be plotted this distorted diagram. If the distortion is removed from the diagram by plotting the me results in the common triangular diagram, it appears that PC_1 and PC_2 for the rrelation matrix do not exactly coincide with the previous results for the variance–variance matrix. The difference is small for PC_1 but for PC_2 it is considerable. This is e reason that PC_1 and PC_2 in Fig.93 are not orthogonal, these principal components ing based on the correlation matrix.

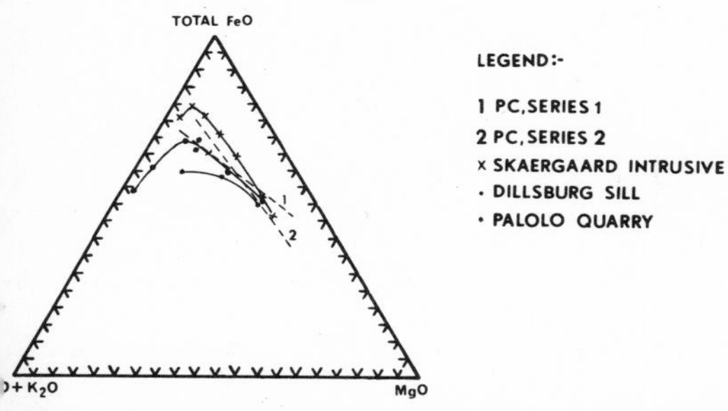

g.94. Well-defined fractionation trends of several tholeiitic magmas compared with trends derived in g.93. (After Baragar, 1966.)

In Fig.94, the calculated trends for the Yellowknife mafic members are compared to me other fractionation trends (cf. Baragar, 1966) constructed from data which are pposed to be free of noise. These trends represent (1) Skaergaard intrusion (Wager, 60) with extreme iron-enrichment; (2) Dillsburg sill (Hotz, 1953) with moderate iron-richment; and (3) Palolo quarry (Kuno et al., 1957) with low iron-enrichment. It is oted that these trends show a sharp bend for Mg-poor rocks not represented in our series and 2.

It is concluded that considerations on the type of trend line to be fitted are important component analysis as applied to petrological phase diagrams. The methods discussed re are readily extended to a four-component system. For example, the first three

TABLE XXXVII
Correlation matrix $\mathbf{R}_0$

	Series 1								*Series 2*							
	SiO_2	Al_2O_3	Fe_2O_3	FeO	MgO	CaO	Na_2O	K_2O	SiO_2	Al_2O_2	Fe_2O_3	FeO	MgO	CaO	Na_2O	K_2O
Al_2O_3	−0.20								0.08							
Fe_2O_3	−0.20	−0.03							−0.27	−0.28						
FeO	−0.63	−0.25	−0.12						−0.32	−0.19	0.33					
MgO	−0.35	−0.06	0.17	−0.18					−0.48	−0.35	−0.18	−0.21				
CaO	−0.08	0.17	−0.58	−0.15	0.14				−0.03	−0.21	−0.27	−0.58	0.16			
Na_2O	0.04	−0.05	0.26	0.13	−0.47	−0.59			−0.46	0.16	0.03	0.15	−0.12	−0.15		
K_2O	−0.15	−0.01	0.12	0.14	−0.19	−0.07	0.07		−0.44	0.11	−0.15	−0.16	0.57	0.07	−0.12	
TiO_2	−0.07	−0.16	0.36	0.42	−0.57	−0.64	0.57	0.23	−0.19	0.11	0.47	0.58	−0.49	−0.57	0.47	−0.26

igenvectors of the variance–covariance matrix then represent three perpendicular lines in tetrahedron. The fourth eigenvector corresponds to zero eigenvalue. It is equal to {1/2, /2, 1/2, 1/2} and represents all values that must lie in the tetrahedron.

If the observations lie within a single plane through the tetrahedron, there will be two igenvalues equal to zero. The third eigenvector now provides the direction cosines of the ormal of the plane that contains the data. This suggests that in applications of com-onent analysis to petrological problems, the eigenvectors for small eigenvalues can con-ain important information on "best-fitting" planes for the system.

3.6. Best-fitting trends in a nine-component system

Correlation matrices for the mafic members of Yellowknife (series 1 and 2) are shown n Table XXXVII. Note that the signs of the correlations between FeO, MgO, CaO and Na_2O satisfy those expected for Barth's simplified basalt model (section 13.2). Another eature of Table XXXVII is that SiO_2 is negatively correlated with nearly all other oxides. his reflects the closure of the system as discussed by Chayes (1960, 1971).

The 95-% confidence interval for testing if a correlation coefficient is significantly ifferent from zero is ±0.36 for series 1 (31 data) and ±0.37 for series 2 (29 data). This uggests that few coefficients in Table XXXVII are statistically significant. Moreover, ecause of the closure of the system, this statistical test which has been developed for andom variables is not applicable and provides a crude guideline only. For a detailed iscussion of these problems, the reader is referred to Chayes (1971).

The eigenvalues and eigenvectors of the correlation matrices are shown in Tables XXVIII and XXXIX. The coefficients of the eigenvectors have been normalized and epresent direction cosines of lines in 9-space. The standardized variables will be abbre-iated as follows:

SiO_2, z_1	FeO, z_4	Na_2O, z_7
Al_2O_3, z_2	MgO, z_5	K_2O, z_8
Fe_2O_3, z_3	CaO, z_6	TiO_2, z_9

he oxide TiO_2 was included in the analysis although it is a minor constituent. Chayes 964) has shown that TiO_2 is important for the classification of basaltic rocks. He has mpared a group of 360 circum-oceanic basalts to a group of 579 "oceanic" basalts. It is oted that Chayes' analyses for oceanic basalts represent basalts from the oceanic islands, d are not to be confused with ocean-floor basalts which have a low TiO_2 value; 93% of hayes' basalts with more than 1.75% TiO_2 are oceanic basalts, proving that TiO_2 is an

TABLE XXXVIII
Eigenvalues of $\mathbf{R}_0$

i	λ_i		i	λ_i	
	Series 1	*Series 2*		*Series 1*	*Series 2*
1	2.883	2.887	6	0.448	0.458
2	1.703	2.029	7	0.322	0.306
3	1.447	1.413	8	0.144	0.160
4	1.124	1.035	9	0.000	0.000
5	0.928	0.712			

TABLE XXXIX
Matrix of eigenvectors from $\mathbf{R}_0$

V_1	V_2	V_3	V_4	V_5	V_6	V_7	V_8	V_9
Series 1								
0.05*	0.73	0.03	0.18	0.16	0.11	0.22	0.10	0.58
0.12	−0.02	−0.03	−0.86	−0.31	0.10	0.28	0.02	0.24
−0.28	−0.10	−0.65	−0.10	0.13	0.29	−0.45	0.35	0.22
−0.25	−0.53	0.39	0.24	−0.17	0.08	0.18	0.37	0.50
0.31	−0.33	−0.52	0.21	0.03	−0.26	0.27	−0.43	0.38
0.47	−0.09	0.34	−0.10	0.07	0.02	−0.69	−0.20	0.34
−0.45	0.17	0.00	−0.11	−0.25	−0.77	−0.26	−0.07	0.18
−0.17	−0.17	0.15	−0.30	0.87	−0.21	0.14	−0.01	0.05
−0.54	0.00	0.10	−0.02	−0.03	0.43	−0.04	−0.71	0.11
Series 2								
0.09	0.64	0.14	0.21	0.10	0.26	0.35	0.17	0.54
−0.05	0.21	−0.71	0.21	−0.31	−0.11	−0.43	−0.13	0.29
−0.33	−0.17	0.41	−0.06	−0.71	0.22	−0.20	0.22	0.25
−0.44	−0.19	0.18	0.25	0.40	−0.55	−0.16	0.17	0.40
0.33	−0.50	0.08	0.13	0.22	0.42	−0.19	−0.38	0.46
0.40	0.03	0.09	−0.59	−0.20	−0.52	0.10	−0.17	0.37
−0.27	−0.20	−0.43	−0.57	0.24	0.30	0.13	0.41	0.22
0.24	−0.44	−0.25	0.39	−0.28	−0.17	0.56	0.31	0.07
−0.54	−0.04	−0.09	−0.04	−0.11	0.01	0.50	−0.66	0.08

* Entries for v_{i1} were chosen positive.

excellent diagnostic index for discriminating between circum-oceanic and oceanic basal The Yellowknife rocks have a frequency curve for TiO_2 that corresponds closely Chayes' circum-oceanic curve for TiO_2. However, Baragar (1966) made a more extensi comparison. His conclusion was that the Yellowknife Group shows mixed affinities, e. in Al_2O_3 content it was more similar to oceanic than to circum-oceanic basalts. Chay

(1968) has developed methods for multivariate discrimination between groups of rocks in petrology.

Comparison of eigenvectors for the two mafic members

The eigenvalues and eigenvectors in Tables XXXVIII and XXXIX are comparable to each other in the following respects:

(1) The dominant eigenvalue is 2.88 (series 1) and 2.89 (series 2). It means that 32% of the total variation is accounted for by the first principal component for both mafic members. The first eigenvector has comparable coefficients except for K_2O.

Note that Na, Fe^{3+} and Fe^{2+} have coefficients with magnitudes approximately equal but signs opposite to those of Mg and Ca. This is in agreement with Barth's simplified model for basaltic magma.

(2) The second eigenvector accounts for 19% of total variation in series 1 and 23% in series 2. It is characterized by a high coefficient for Si in both mafic members. An independent variation trend which involves silica therefore is suggested.

(3) The remainder of the coefficients for the second eigenvector and also the coefficients for the third, fourth, fifth, sixth, and seventh eigenvector do not duplicate each other in a systematic manner for the two mafic members.

(4) The ninth eigenvector represents the closure of the system. Its coefficients are proportional to the standard deviations of the nine oxides.

(5) The eighth eigenvalue is small. It accounts for 1.6% of total variation in series 1 and 1.8% in series 2. Correspondence is fair except for the alkalies Na and K. The eighth eigenvector is perpendicular to a plane in which the observations lie by close approximation; the first eigenvector also lies in this plane. Note that Ti now is at the Mg, Ca side of the origin whereas in the first eigenvector, it occurs at the Fe^{3+}, Fe^{2+}, Na end.

Principal components

The results of component analysis indicate a complex system where more than a single trend is present. The interpretation will be tentative. Nevertheless, the two series have a number of features in common and the application of component analysis to complex systems of this type is warranted. By means of this method, a set of linear relationships between all variables is established which are extracted in order of importance.

Principal components (PC's) are derived from the eigenvectors by summation of products of coefficients and oxides. For example, the first principal component (PC_1) for series 1 satisfies:

$$PC_1 = 0.05z_1 + 0.12z_2 - 0.28z_3 - 0.25z_4 + 0.31z_5 + 0.47z_6 - 0.45z_7 - 0.17z_8 - 0.54z_9$$

Values of PC_1 can be calculated for the individual observations. This yields 31 com ponent scores for series 1. Geometrically, the calculation of one of these componen scores is equivalent to projection of the observations in 9-space onto the best-fitting lin given by the first eigenvector. The variance of the 31 component scores is equal to th largest eigenvalue (= 2.88).

If the eigenvalue is zero, the scores for the corresponding component (PC_9 in th example) all are equal to zero.

When the scores of a component are graphically represented in a diagram, it is custom ary to standardize them so that their variance becomes one. This is a possible source o confusion. For example, all scores for PC_1 (series 1) have been divided by the square roo of their variance ($=\sqrt{2.88}$) before plotting them in Fig.96A which is to be discusse later.

13.7. Other petrological models of basalt magma; comparison of principal components t indices

The results obtained by component analysis can be summarized as follows: Fe^{3+} an Fe^{2+} behave similarly; Mg follows Fe but has opposite sign; Ca tends to follow Mg; S describes a trend that seems to be independent; Na tends to follow Fe in the main tren (PC_1) but it also is affiliated to K which shows an independent behavior.

The properties of Ti in relation to the other elements are not clear. Although it i relatively strongly correlated to other oxides, this may in part be caused by the fact tha it is a minor constituent with a positively skew frequency distribution.

Until now, the following petrological theories for basalts that are relevant for th Yellowknife volcanic rocks have been discussed: (1) Barth's simplified model for basalt in general; (2) Kuno's method of using a triangular diagram for the (MgO/iron/alkali ratio; (3) Chayes' use of TiO_2 as an index to discriminate between circum-oceanic an oceanic basalts.

In this section, some further aspects of petrologic theory for basalts will be discusse and background will be provided for the following three petrological indices: (1) solidifi cation index (Kuno et al., 1957); (2) differentiation index (Thornton and Tuttle, 1960) (3) Poldervaart's alkalinity index (Poldervaart, 1964).

Two fundamental magma types are generally recognized: tholeiitic and alkali basal magma. Each gives a series of fractionation products. This leads to silica-oversaturatio for tholeiitic magma and silica-undersaturation for alkali basalts. In addition to this, th calc-alkali series is commonly recognized which, amongst other properties, may have hig initial silica content relative to tholeiitic and alkali magmas (Nockolds and Allen, 1953 and a differentiation course noted by a generally high CaO/alkali ratio. Baragar (1966 has pointed out that the Yellowknife volcanic rocks have mixed tholeiitic and calc-alkal characteristics.

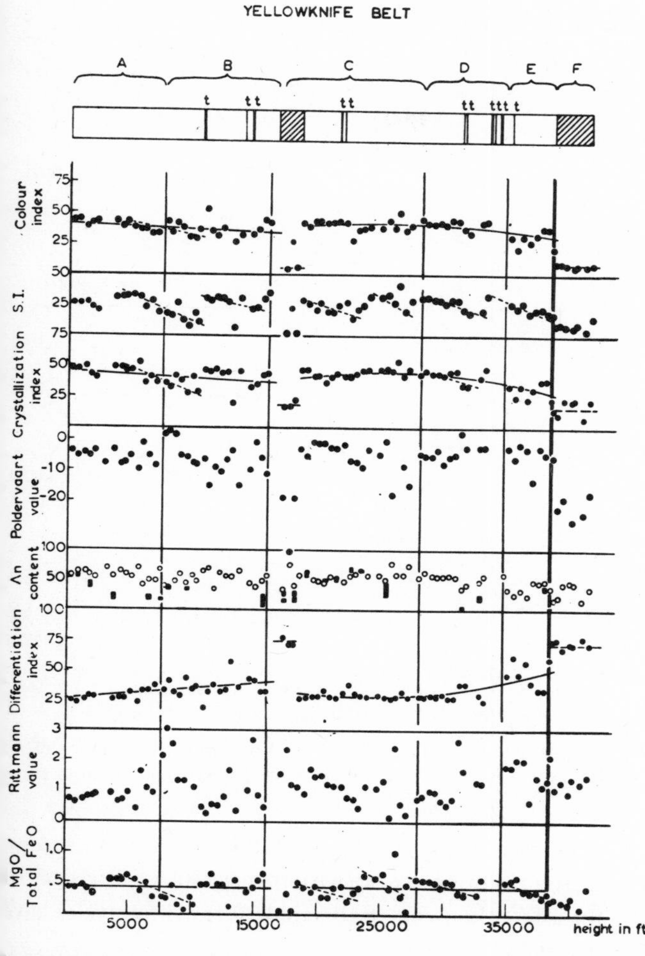

Fig.95. Variation in petrological indices with stratigraphic height in Yellowknife volcanic belt. (From Baragar, 1966.) Color index, crystallization index, and differentiation index exhibit same type of trend; short-range trends in solidification index and MgO/total FeO ratio fitted by Baragar (1966).

The tholeiitic trend is mainly characterized by iron-enrichment. Its presence is shown in the triangular diagrams of Fig.93. The iron-enrichment is mainly at the expense of MgO. Kuno et al. (1957) have proposed to use the so-called solidification index (*S.I.*) which is directly based on the ratio plotted in their triangular diagram. The equation is:

$$S.I. = \frac{MgO \times 100}{MgO + FeO + Fe_2O_3 + Na_2O + K_2O}$$

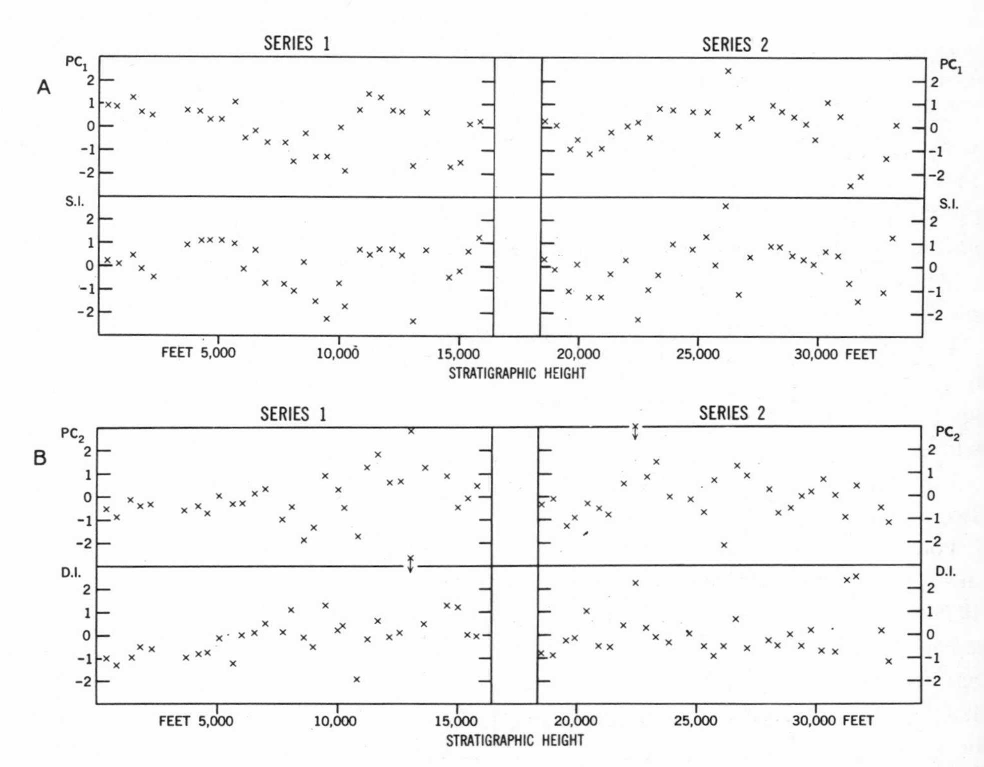

Fig.96. Comparison of the first two principal components (PC_1 and PC_2) to solidification index (*S.I.* and differentiation index (*D.I.*); unit along vertical axes is one standard deviation; means equal to zer

The stratigraphic variation of *S.I.* is shown in Fig.95. Baragar (1966) has fitted a succe sion of straight lines to the data. The resulting zig-zag pattern suggests a succession c iron-enrichments each persisting for a few thousand stratigraphic feet. Three cycles hav been recognized for both our series 1 and 2.

In Fig.96A, standardized values for the solidification index are compared to corre sponding values for PC_1. The agreement is fair. The linear correlation coefficient is 0.7 (series 1) and 0.73 (series 2).

The second trend recognized by Baragar is a slight increase in acidity towards the to in both members (see Fig.95). This trend can be brought out by the differentiation inde (*D.I.*) developed by Thornton and Tuttle (1960) with equation:

$$D.I. = q + ab + or(+ne + kp + lc)$$

where the symbols denote normative quartz, albite, orthoclase, nepheline, kaliophilit and leucite, respectively. The latter three minerals are not present in the norms calculate for the Yellowknife volcanic rocks with an exception for nepheline which occurs in fe norms.

It is noted that the *D.I.* has been developed to study a vast assemblage of igneous and extrusive rocks including granitic rocks. Its use is not restricted to basalt samples.

Standardized values for the *D.I.* are shown in Fig.96B. They can be compared to the scores for PC_2. The linear correlation coefficient amounts to 0.67 (series 1) and 0.38 (series 2). In Fig.95 and 96B, it is indicated that the *D.I.* gradually increases in series 1 but remains constant in series 2. The PC_2 has a relatively high coefficient for silica in both series but is relatively poorly correlated to the *D.I.* in series 2.

Baragar (1966) has interpreted the second trend (raise in acidity with height in the lava pile) as caused by an increased contamination by sialic material of the crust.

Inspection of the original data (Fig.92) suggests that TiO_2 also conforms to this trend in that, on the average, its content increases with height. This feature has not been borne out by the statistical analysis indicating that the correspondence in pattern is of a general nature.

In view of results to be obtained in subsequent sections (e.g., see 13.11), a third index also will be discussed.

Poldervaart (1964) has designed an index to measure the alkalinity of basalts which can be determined from norms. The main purpose of the Poldervaart index is to separate alkali-basalts from tholeiitic basalts. A simplified basaltic magma is used from which nepheline, clinopyroxene, olivine and quartz may be precipitated. According to Poldervaart, the tetrahedron formed by these four components has a plane in it that separates alkaline and tholeiitic basalts. The equation of this plane is $0.53\ ne' - 0.47\ qu' = 0$ where the symbols ne′ and qu′ indicate norms after a recalculation to 100% for the system ne, di, ol, and qu. If the index $A = 0.53\ ne' - 0.47\ qu'$ for a rock sample is negative, a tholeiite is indicated. Positive values correspond to alkali-basalts. The variation of the Poldervaart alkalinity index is shown in Fig.95. It assumes a negative value for nearly all samples. An alkalinity trend, if present, has been obscured by a large amount of noise. The following interpretation is therefore tentative at this point. It is suggested from Fig.95 that, on the average, the Poldervaart index decreases towards the right for sections *A*, *B*, *C*, and $(D + E)$. This alkalinity trend tends toward a sinusoidal pattern rather than to the zig-zag pattern exhibited by the iron-enrichment trends.

Summary of results by time-independent methods

Time has not been considered as a variable until now except that the results obtained from the interrelationships between major oxides have been plotted against stratigraphic height which is a measure of time. The calculations made until now indicate that the Yellowknife volcanic pile has been formed by a complex process of magma composition change. The existence of at least three more or less independent processes is indicated:

(1) The solidification index and first principal component indicate a succession of iron enrichments. The pattern is zig-zag rather than sinusoidal. Three cycles occur in both series 1 and 2.

(2) The differentiation index and second principal component indicate a slow raise i acidity towards the tops of the mafic members each of which is capped by an acidic laye

(3) The Poldervaart index possibly indicates an oscillation in alkalinity of the magm with two cycles for each of the mafic members.

The first two trends have been explained by Baragar (1966) and their existence ha also been indicated by component analysis. The third trend is based on scanty data an its existence is in doubt.

In the remainder of this chapter, the data for series 1 and 2 will be subjected to method of multivariate time-series analysis. The methods discussed in Chapter 12 wher use is made of Markov chains will be extended to the present situation of a multivariat series of data. First, background information will be presented on the Kolmogorov diffe ential equations and the statistical analysis of multivariate time series.

13.8. The Kolmogorov differential equations

As in Chapter 12, the starting point is the Chapman-Kolmogorov equation for discret data:

$$p_{ik}(s+t) = \sum_j p_{ij}(s)\, p_{jk}(t) \qquad [13.1$$

An analogous, more general expression of this equation is:

$$p_{ik}(\tau, t) = \sum_j p_{ij}(\tau, s)\, p_{jk}(s, t) \qquad [13.2$$

In this notation, $\tau < s < t$ and $p_{ik}(\tau, t)$ signifies the probability that the system has move from state i at time τ to state k at time t. The first equation is restricted to processe where direct transitions from a state at time t are possible only to the neighboring state at time $(t+1)$ and $(t-1)$. Also, the process has to be time-homogeneous in that th transition probabilities $p_{ik}(t)$ are the same for all time intervals of length t. These cond tions are met in the case of Markov chains but were dropped in [13.2]. However, a before, a direct transition from any state i to any state k can occur. Also, as before, ther is the side condition:

$$\sum_k p_{ik}(\tau, t) = 1 \qquad [13.3$$

Eq. [13.2] can be used as a starting point for deriving the Kolmogorov differenti equations in their general form. They are the continuous-time analogue of the Chapmar Kolmogorov equation for the discrete-time process.

A process that takes place in continuous time with gradual change can also be subje to the Markov property.

Suppose that to every state i there corresponds a continuous function $c_{ii}(t) \leqslant 0$, such that, as $\Delta t \to 0$, then:

$$\frac{1 - p_{ii}(t, t+\Delta t)}{\Delta t} \to -c_{ii}(t) \qquad [13.4]$$

Also, to every pair of states (i, k) with $i \neq k$, there corresponds a continuous function $c_{ik}(t)$ such that, as $\Delta t \to 0$, then:

$$\frac{p_{ik}(t, t+\Delta t)}{\Delta t} \to c_{ik}(t) \qquad [13.5]$$

Eq. [13.4] and [13.5] are called the regularity assumptions. Their meaning becomes clear if they are written as:

$$\left.\begin{aligned} p_{ii}(t, t+\Delta t) &\to 1 + c_{ii}(t)\Delta t \\ p_{ik}(t, t+\Delta t) &\to c_{ik}(t)\Delta t \end{aligned}\right| \qquad [13.6]$$

The transition probabilities p_{ik} are continuous in time and for every fixed (t, i):

$$\sum_k p_{ik}(t) = 1$$

or: $$\sum_k c_{ik}(t) = 0 \qquad [13.7]$$

If the Chapman-Kolmogorov equation [13.2] is satisfied, then:

$$p_{ik}(\tau, t+\Delta t) = \sum_j p_{ij}(\tau, t)\, p_{jk}(t, t+\Delta t)$$

$$= p_{ik}(\tau, t)\, p_{kk}(t, \Delta t) + \sum_{j \neq k} p_{ij}(\tau, t)\, p_{jk}(t+\Delta t) \qquad [13.8]$$

By using the regularity conditions of [13.4] and [13.5], it follows that, as $\Delta t \to 0$:

$$\frac{p_{ik}(\tau, t+\Delta t) - p_{ik}(\tau, t)}{\Delta t} \to p_{ik}(\tau, t)\, c_{ii} + \sum_{j \neq k} p(\tau, t) \frac{p_{jk}(t, t+\Delta t)}{\Delta t}$$

or: $$\frac{\partial}{\partial t} p_{ik}(\tau, t) = \sum_j p_{ij}(\tau, t)\, c_{jk}(t) \qquad [13.9]$$

This is the system of Kolmogorov forward differential equations. The parameters i and τ appear only in the initial condition:

$$p_{ik}(\tau, \tau) \begin{cases} = 1 & \text{for } k = i \\ = 0 & \text{otherwise} \end{cases}$$

In an analogous manner, the following relationship may be derived:

$$\frac{\partial}{\partial t} p_{ik}(\tau, t) = -\sum_j c_{ij}(\tau)\, p_{jk}(\tau, t) \qquad [13.10]$$

with: $p_{ik}(t, t) \begin{cases} = 1 & \text{for } k = i \\ = 0 & \text{otherwise} \end{cases}$

These are the Kolmogorov backward differential equations. For a detailed discussion of the difference between the forward and backward equations, see Feller (1968).

In matrix form, the Chapman-Kolmogorov equation becomes:

$$\mathbf{P}(\tau, t) = \mathbf{P}(\tau, s)\,\mathbf{P}(s, t) \qquad [13.11]$$

The forward differential equations [13.9] can be written as:

$$\frac{\partial}{\partial t}\mathbf{P}(\tau, t) = \mathbf{P}(\tau, t)\,\mathbf{C}(t) \qquad [13.12]$$

with the initial condition $\mathbf{P}(\tau, \tau) = \mathbf{I}$ where $\mathbf{I}$ represents the identity matrix. The convention has been followed that elements of rows of $\mathbf{P}(\tau, t)$ and $\mathbf{C}(t)$ add to one and zero, respectively.

The backward differential equations [13.10] become:

$$\frac{\partial}{\partial \tau}\mathbf{P}(\tau, t) = -\mathbf{C}(\tau)\,\mathbf{P}(\tau, t) \qquad [13.13]$$

with $\mathbf{P}(t, t) = \mathbf{I}$.

If it now is assumed that the process is stationary, or that $\mathbf{C}(t)$ remains the same regardless of t, the system reduces to:

$$\frac{\mathrm{d}\mathbf{P}(t)}{\mathrm{d}t} = \mathbf{P}(t)\,\mathbf{C} \qquad [13.14]$$

and: $$\frac{\mathrm{d}\mathbf{P}(t)}{\mathrm{d}t} = \mathbf{C}\mathbf{P}(t) \qquad [13.15]$$

with $\mathbf{P}(0) = \mathbf{I}$.

The Chapman-Kolmogorov equation becomes:

$$\mathbf{P}(\tau + t) = \mathbf{P}(\tau)\,\mathbf{P}(t) \qquad [13.16]$$

In general, there exists a common solution to the Kolmogorov forward and backward differential equations for time-homogeneous processes that satisfies the initial condition and also the Chapman-Kolmogorov equation. This general solution is:

$$\mathbf{P}(t) = e^{\mathbf{C}t} \qquad [13.17]$$

In explicit form the solution of the Kolmogorov equation is:

$$\mathbf{P}(t) = \sum_{i=1}^{p} e^{\pi_i t}\, \mathbf{Z}_0(\pi_i) \qquad [13.18]$$

This equation is obtained by writing **C** in its canonical form:

$$\mathbf{C} = \sum_{i=1}^{p} \pi_i \mathbf{Z}_0(\pi_i) \qquad [13.19]$$

If t is made equal to one, [13.17] reduces to:

$$\mathbf{P}(1) = e^{\mathbf{C}} \qquad [13.20]$$

In practical applications, the unit time interval is made equal to the sampling interval. If t is expressed in unit intervals and **P**(1) is written as **P**, [13.17] becomes:

$$\mathbf{P}(s) = \mathbf{P}^s$$

or: $\mathbf{P}(s) = \Sigma \lambda_i^s \mathbf{Z}_0(\lambda_i)$ [13.21]

This is equivalent to [12.7] obtained in section 12.3. A Markov chain in discrete time whose solution is given by [13.18] always can be regarded as a sequence extracted from the complete chain in continuous time. The spectral components $\mathbf{Z}_0(\pi_i)$ for **C** and $\mathbf{Z}_0(\lambda_i)$ for **P** are identical. Their eigenvalues are related by:

$$e^{\pi_i} = \lambda_i \quad \text{or:} \quad \pi_i = \ln \lambda_i \qquad [13.22]$$

Practical considerations

The practical example is the previous multivariate series of observations taken at regular stratigraphic intervals through the Yellowknife volcanic pile. It may be assumed that the sampling interval of approximately 500 ft. specifies a unit time interval. We will be able to establish a matrix **P** for this sequence. It can also be assumed that the changes in lava composition which have been recorded took place in continuous time. Therefore, the matrix **C** can be determined; it has the same eigenvectors and spectral components as **P** and its eigenvalues are equal to those of **P** after a logarithmic transformation. This procedure will enable us to calculate the transition matrix $\mathbf{P}(t) = e^{\mathbf{C}t}$ for any given value of t. In particular, extrapolation from **P** to $\lim_{t \to 0} \mathbf{P}(t)$ will be interesting.

Suppose that a multivariate observation made at time t is represented by the rov vector $\mathbf{X}(t)$. The forward differential equations then reduce to:

$$\frac{\mathrm{d}\mathbf{X}(t)}{\mathrm{d}t} = \mathbf{X}(t)\,\mathbf{C} \qquad [13.23]$$

In this section, the Kolmogorov differential equations first were derived for nonstationar series. Later the condition of stationarity was imposed on the solution for a nonstation ary situation. In the nonhomogeneous case, the matrix $\mathbf{C}(t)$ changes with time t; in thi case, the forward and backward differential equations have different solutions. Use of th forward equations seems more natural in that the changes of the process take place in th positive time direction. The backward equations specify a process that takes place in th negative time direction. However, by rigorous mathematical arguments as applied b Kolmogorov (1931) and Feller (1968), it is shown that the specifications leading to th forward and backward equations are somewhat different. Moreover, the backward equa tions provide a second powerful tool for studying the solution of Markov processes i general. It will be shown in the next section that we will be able to obtain a solution fo the transition matrix in practice, if and only if the assumption that the process is time homogeneous is satisfied. In this situation, the forward and backward equations shoul theoretically have a common solution. It is unlikely that any geological process is strictl time-homogeneous. Changes in the transition matrix through time are to be expected. A in the case of Markov chains, the following approximate method will therefore be follow ed: (1) the transition matrix is calculated on the assumption that the chain is stationary and (2) the forward Kolmogorov equations are used for further statistical analysis.

13.9. The transition matrix of multivariate series

Series 1 and 2 for the mafic members of the Yellowknife belt were analyzed separatel and combined with one another (combined series) whereby the discontinuity betwee series 1 and 2, where an acidic layer occurs, has been neglected.

Every observation in these series may be represented by a row vector $\mathbf{X}_k$ with elements corresponding to the oxides. The subscript k increases in the positive tim direction (cf. Fig.92); it goes from 1 to 31 in series 1; from 1 to 29 in series 2; and from to (31 + 29 =) 60 in the combined series.

Notation can be a source of confusion in practical applications. The convention use here is analogous to that generally used by geologists in the application of Markov-chai models.

An element p_{ij} in the transition matrix for Markov chains indicates the probabilit that state i is followed by state j after one time shift in the positive direction. Although multivariate series, such as the one represented in Fig.92, is for discrete time intervals, th variables (oxides) are continuous. The procedure of constructing a tally matrix and deriv ing the transition matrix from it (see section 12.7) cannot be followed. If the time serie consists of n multivariate observations on p variables and has the Markov property, the predictable parts of observations $x_{i,k}$ at time k are determined by a transition matrix and the observations $x_{i,k-1}$ at time $(k-1)$. It will be shown later that, without loss in generality, we may use $(p-1)$ standardized variables z_i instead of p original variables x_i.

In matrix notation, our model then becomes:

$$\mathbf{Z}_k = \mathbf{Z}_{k-1}\mathbf{U} + \mathbf{E}_k \qquad [13.24]$$

where $\mathbf{Z}_k$, $\mathbf{Z}_{k-1}$ and $\mathbf{E}_k$ are row vectors with $q(=p-1)$ elements and $\mathbf{U}$ is a $(q \times q)$ matrix. The elements of $\mathbf{E}_k$ are random variables and not correlated with the elements of $\mathbf{E}_{k-1}$. The elements of $\mathbf{U}$ can be written in the form:

$$\mathbf{U} = \begin{bmatrix} u_{11} & u_{12} & \cdots & u_{1q} \\ u_{21} & u_{22} & \cdots & u_{2q} \\ \cdot & \cdot & \cdots & \cdot \\ \cdot & \cdot & \cdots & \cdot \\ \cdot & \cdot & \cdots & \cdot \\ u_{q1} & u_{q2} & \cdots & u_{qq} \end{bmatrix}$$

Let $z_{k,i}$ denote the value assumed by variable $z_i (i = 1, 2, \ldots, q)$ at time k. This individual value then satisfies the equation:

$$z_{k,i} = \mathbf{Z}_{k-1}\mathbf{U}_i + e_{k,i}$$

where $\mathbf{U}_i$ denotes the i-th column of $\mathbf{U}$, $\mathbf{Z}_{k-1}$ is a p-dimensional row vector as before, and $e_{k,i}$ is a scalar.

The method of least squares can be used to find the coefficients of $\mathbf{U}_i$ from the series of observations; the coefficients then represent a column vector of regression coefficients (cf. section 8.6).

Consequently:

$$\hat{\mathbf{U}}_i = \mathbf{A}^{-1}\mathbf{B}_i$$

where $\mathbf{A}$ is a $(q \times q)$ matrix consisting of the elements:

$$a_{ij} = \sum_{k=2}^{n} z_{k-1,i} z_{k-1,j}$$

and $\mathbf{B}_i$ a column vector with elements:

$$b_{ij} = \sum_{k=2}^{n} z_{k-1,i} z_{k,j}$$

In this manner, the q columns of $\mathbf{U}$ can be solved by using the matrix $\mathbf{A}$ and vectors $\mathbf{B}_i$ with i going from 1 to q. This is an application of multivariate least-squares theory. The resulting matrix $\hat{\mathbf{U}}$ consists of maximum-likelihood estimators as, e.g., shown by Anderson (1958).

The complete matrix $\hat{\mathbf{U}}$ is estimated by:

$$\hat{\mathbf{U}} = \mathbf{A}^{-1}\mathbf{B} \tag{13.25}$$

where $\mathbf{B}$ consists of q columns $\mathbf{B}_i (i = 1, 2, \ldots, q)$. The matrix $\mathbf{A}$ may be replaced by the correlation matrix $\mathbf{R}_0$ computed from one observation point $(k = n)$ more than $\mathbf{A}$. Writing $\mathbf{R}_1$ instead of $\mathbf{B}$, we obtain:

$$\hat{\mathbf{U}} = \mathbf{R}_0^{-1}\mathbf{R}_1 \tag{13.26}$$

where $\mathbf{R}_0$ is the correlation matrix for the observations, and $\mathbf{R}_1$ is the lagged correlation matrix. From here on, the hat on $\mathbf{U}$ will be omitted.

Alternative method

It is also instructive to solve the transition matrix by means of another method. Let both sides of [13.24] be premultiplied by $\mathbf{Z}_{k-1}$. When expectations are taken, this results in:

$$E(\mathbf{Z}_{k-1}\,\mathbf{Z}_k) = E(\mathbf{Z}_{k-1}\,\mathbf{Z}_{k-1})\,\mathbf{U}$$

Because the variables have unit variance, $E(\mathbf{Z}_{k-1}\mathbf{Z}_{k-1})$ represents the correlation matrix and $E(\mathbf{Z}_{k-1}\mathbf{Z}_k)$ the lagged correlation matrix for the population. This procedure also leads to [13.26].

Instead of using $\mathbf{Z}_{k-1}$ for premultiplication in this method, we can take $\mathbf{Z}_{k-s}$ where s is an arbitrary integer with $s > 0$. Hence, $\mathbf{U}$ also can be estimated by using the more general equation:

$$\mathbf{U} = \mathbf{R}_{s-1}^{-1}\mathbf{R}_s \qquad s = 1, 2, \ldots \tag{13.27}$$

When $s = 1$, this equation reduces to [13.26]. The elements of $\mathbf{R}_0$ are correlation coefficients and can be estimated in the usual way. In estimating $\mathbf{R}_s$ $(s = 1, 2, \ldots)$ end corrections have to be made. The problem of end corrections has been discussed by Quenouille (1957). In practical applications, we have used his formula:

$$r_{(ij)s} = \frac{\sum_{k=1}^{n-s} x_{i,k+s}x_{j,k} - \left(\sum_{k=1}^{n-s} x_{i,k+s}\right)\left(\sum_{k=1}^{n-s} x_{j,k}\right) \Big/ (n-s)}{\sqrt{\left[\left\{\sum_{k=1}^{n-s} x_{i,k+s}^2 - \left(\sum_{k=1}^{n-s} x_{i,k+s}\right)^2 \Big/ (n-s)\right\}\left\{\sum_{k=1}^{n-s} x_{j,k}^2 - \left(\sum_{k=1}^{n-s} x_{j,k}\right)^2 \Big/ (n-s)\right\}\right]}} \tag{13.28}$$

with $s = 1$ or 2.

In this equation, $r_{ij,s}$ denotes a single element of the matrix $\mathbf{R}_s$.

Estimation by least-squares method

The equation $\mathbf{U} = \mathbf{R}_0^{-1}\mathbf{R}_1$ derived in the previous section cannot be applied when the matrix $\mathbf{R}_0$ does not have an inverse. It has been discussed in section 13.6 that the nine oxides for series 1 and 2 form a closed-number system so that the matrix $\mathbf{R}_0$ has at least one eigenvalue equal to zero. A modification in $\mathbf{R}_0$ therefore was required before $\mathbf{U}$ can be solved. In the practical example, silica has been omitted from the system. This results in a $(p-1) \times (p-1)$ matrix $\mathbf{U}$ with the properties of the $(p \times p)$ transition matrix $\mathbf{P}$ discussed in section 13.8 from which the first component has been removed.

First, the general problem of multiple regression of a variable on a set of independent variables which form a closed-number system will be discussed.

Suppose that a set of p geological variables expressed as $(n \times 1)$ column vectors $\mathbf{X}_i$ form a closed-number system with $\mathbf{J}_n$ consisting of n ones. Then:

$$\sum_{i=1}^{p} \mathbf{X}_i = \mathbf{J}_n \qquad [13.29]$$

Let the column matrix $\mathbf{X}$ with p columns of observed values be postmultiplied by an arbitrary $(p \times p)$ matrix $\mathbf{T}$. This is equivalent to transforming the variables X_i into new variables X_i^* with $\mathbf{X}^* = \mathbf{XT}$. If a dependent variable Y with zero mean is regressed on the new variable, the solution is:

$$\hat{\mathbf{Y}}^* = \mathbf{X}^*\hat{\boldsymbol{\beta}}^* = \mathbf{XT}\hat{\boldsymbol{\beta}}^*$$

with $\hat{\boldsymbol{\beta}}^* = (\mathbf{T}'\mathbf{X}'\mathbf{XT})^{-1}\mathbf{T}'\mathbf{X}'\mathbf{Y} = \mathbf{T}^{-1}(\mathbf{X}'\mathbf{X})^{-1}\mathbf{X}'\mathbf{Y}$.

In general, $\hat{\boldsymbol{\beta}}^*$ will not be equal to $\hat{\boldsymbol{\beta}}$. However:

$$\mathbf{T}\hat{\boldsymbol{\beta}}^* = \hat{\boldsymbol{\beta}}$$

and, consequently, $\hat{\mathbf{Y}}^* = \hat{\mathbf{Y}}$.

The only condition to be satisfied by $\mathbf{T}$ is that its inverse $\mathbf{T}^{-1}$ exists. Probably the simplest transformation consists of making $\mathbf{T}$ equal to a p-dimensional identity matrix for which the zeros in the last column are replaced by ones.

When the original regression equation is:

$$\hat{\mathbf{Y}} = \sum_{i=1}^{p} \hat{\beta}_i \mathbf{X}_i \qquad [13.30]$$

then:

$$\hat{\mathbf{Y}} = \sum_{i=1}^{p-1} \hat{\beta}_i^* \mathbf{X}_i + \hat{\beta}_p \sum_{i=1}^{p} \mathbf{X}_i \qquad [13.31]$$

The last term in this expression is a constant, say $\hat{\beta}_0^*$. Hence:

$$\hat{\mathbf{Y}} = \hat{\beta}_0^* + \sum_{i=1}^{p-1} \hat{\beta}_i^* \mathbf{X}_i \qquad [13.32]$$

The relationship between $\hat{\boldsymbol{\beta}}$ and $\hat{\boldsymbol{\beta}}^*$ is:

$$\left.\begin{aligned} &\hat{\beta}_i = \hat{\beta}_i + \hat{\beta}_0^* \qquad (i = 1, 2, \ldots, p-1) \\ &\hat{\beta}_p = \hat{\beta}_0^* \end{aligned}\right| \qquad [13.33]$$

The regression of $\mathbf{Y}$ on the p variables $\mathbf{X}_i$ can be interpreted in two ways. The direct method consists of applying the method of least squares by evaluating [13.30] where the constant term is forced to equal zero. A second method consists of correcting all variables $\mathbf{X}_i$ for their respective means. Then, a constant term $\hat{\beta}_0$* will arise if ordinary multiple regression is applied after deleting one of the independent variables. If $\mathbf{Y}$ has a mean $\bar{Y}$ which is larger than zero, the constant term $\hat{\beta}_0^*$ will be augmented by $\bar{Y}$.

In estimating the matrix $\mathbf{P}$ for a Markov chain consisting of values that add to unity at any time, the constant term should be suppressed if the calculation is done on the original data. This is equivalent to estimating $\mathbf{U}$ by multiple regression, where a constant term also is calculated, for a reduced system from which one of the variables has been deleted. By means of [13.33], a $(p \times p)$ matrix $\mathbf{P}^*$ can be calculated from the $(p-1) \times (p-1)$ matrix $\mathbf{U}$ to approximate $\mathbf{P}$. The elements in the rows of $\mathbf{P}^*$ add to unity. Its largest eigenvalue, therefore is equal to one. In this respect, $\mathbf{P}^*$ is comparable to a common transition matrix $\mathbf{P}$. However, the elements of $\mathbf{P}^*$ can be negative because its elements are obtained by regression analysis. As pointed out in section 8.8, regression coefficients can be imprecise if there are approximate linear relationships in the system which are characterized by small eigenvalues of the correlation matrix. In section 13.7, it was shown that the matrices $\mathbf{R}_0$ for series 1 and 2 of the practical example have several small eigenvalues. Consequently, $\mathbf{R}_0$ is ill-conditioned if used for multiple-regression analysis.

Calculation of transition matrix for combined series, Yellowknife volcanic belt

The combined series of 60 observations will be subjected to analysis. Means and standard deviations are shown in Table XL. The matrices $\mathbf{R}_0$, $\mathbf{R}_1$, and $\mathbf{U} = \mathbf{R}_0^{-1}\mathbf{R}_1$ are represented in Tables XLI and XLII.

The matrix $\mathbf{P}^*$ can be calculated from $\mathbf{U}$ as follows. The oxide Al_2O_3 (x_2) is taken for example. Since $\mathbf{U}$ has been obtained from standardized variables, its first column results in:

TABLE XL
Means and standard deviations for 8 oxides (in percent)

	Mean	*S.D*
Al_2O_3	15.441	0.999
Fe_2O_3	2.517	0.939
FeO	9.644	1.729
MgO	6.405	1.759
CaO	9.784	1.534
Na_2O	2.405	0.774
K_2O	0.343	0.218
TiO_2	1.161	0.481

TABLE XLI
Correlation matrix ($\mathbf{R_0}$)

Al_2O_3	1.00							
Fe_2O_3	−0.10	1.00						
FeO	−0.13	0.04	1.00					
MgO	−0.16	−0.05	−0.14	1.00				
CaO	0.03	−0.56	−0.18	0.18	1.00			
Na_2O	0.05	0.21	0.15	−0.36	−0.45	1.00		
K_2O	0.06	−0.11	0.09	0.13	0.07	−0.05	1.00	
TiO_2	−0.04	0.47	0.43	−0.54	−0.61	0.54	0.06	1.00

TABLE XLII
Lagged correlation matrix ($\mathbf{R_1}$) and forward transition matrix ($\mathbf{U} = \mathbf{R_0^{-1} R_1}$)

	0.22	−0.23	−0.02	−0.01	0.14	0.14	−0.10	−0.03
	0.05	0.25	0.02	−0.11	−0.46	0.41	−0.08	0.29
	0.31	0.16	0.42	−0.09	−0.03	0.19	0.15	0.26
$\mathbf{R_1}$ =	−0.18	0.03	−0.01	0.26	−0.14	−0.19	−0.04	−0.06
	−0.14	−0.36	0.14	0.05	0.36	−0.33	0.11	−0.25
	0.15	0.24	0.07	−0.12	−0.19	0.15	−0.01	0.23
	0.09	−0.09	0.11	−0.03	−0.12	−0.01	0.15	0.19
	0.31	0.24	0.02	−0.10	−0.20	0.24	0.08	0.18
	0.23	−0.27	−0.04	0.22	0.22	0.21	−0.19	0.27
	0.04	0.02	−0.05	0.03	−0.37	0.29	−0.04	−0.06
	0.39	0.30	0.47	0.08	0,31	0.17	0.09	0.06
$\mathbf{U}$ =	−0.13	−0.15	−0.04	0.26	−0.31	−0.26	−0.07	0.12
	−0.15	−0.20	0.21	−0.14	0.12	−0.23	0.09	−0.14
	0.18	0.21	0.03	−0.01	0.01	0.03	−0.01	0.11
	0.11	−0.32	−0.05	0.17	−0.15	−0.21	0.06	0.49
	0.33	0.26	0.04	0.00	−0.01	0.20	0.10	−0.06

$$z_{2,k} = 0.23z_{2,k-1} - 0.27z_{3,k-1} - 0.04z_{4,k-1} + 0.22z_{5,k-1}$$
$$+ 0.22z_{6,k-1} + 0.21z_{7,k-1} - 0.19z_{8,k-1} + 0.27z_{9,k-1}$$

From:

$$z_{i,k} = (x_{i,k} - \bar{x}_i)/s(x_i)$$

it follows, by using the values shown in **Table** XL, that:

$$x_{2,k} = 9.65 + 0.23x_{2,k-1} - 0.28x_{3,k-1} - 0.02x_{4,k-1} + 0.13x_{5,k-1}$$
$$+ 0.14x_{6,k-1} + 0.27x_{7,k-1} - 0.89x_{8,k-1} + 0.56x_{9,k-1}$$

From $\sum_{i=1}^{9} x_i = 100$, it follows that the constant term 9.65 can be replaced by:

$$0.0965 \sum_{i=1}^{9} x_i$$

Hence:

$$x_{2,k} = 0.10x_{1,k-1} + 0.33x_{2,k-1} - 0.19x_{3,k-1} + 0.07x_{4,k-1}$$
$$+ 0.22x_{5,k-1} + 0.24x_{6,k-1} + 0.38x_{7,k-1} - 0.80x_{8,k-1} + 0.66x_{9,k-1}$$

The coefficients of this equation constitute the second column of the matrix **P***. Th coefficients for $x_{i,k}$ ($i = 3, ..., 9$) can be obtained by the same method; and those for silic ($i = 1$) by using the fact that the elements for each row of **P*** add to one.

A number of the elements of **P*** are negative. It is likely that the true values of **P** which is estimated by **P***, are positive. The main reason for the appearance of negativ

TABLE XLIII
Eigenvalues of transition matrix **U** for combined series and series 1 and 2

	Eigenvalues		
	Combined series	*Series 1*	*Series 2*
1	0.60722	0.68559	0.65900
2	0.42142	}−0.44509 ± 0.18190 *i*	−0.53726
3	} 0.28406 ± 0.19711 *i*		0.48733
4		} 0.32979 ± 0.30468 *i*	}−0.36196 ± 0.27201 *i*
5	}−0.31069 ∓ 0.09776 *i*		
6		0.42512	−0.33037
7	0.31479	} 0.14352 ∓ 0.08212 *i*	0.15150
8	−0.16043		0.02898

elements is that there are approximate linear relationships in the system.

Perhaps, the system can be improved by using geochemical considerations; for example, the original values x_{ik} are concentration values in weight percent for the oxides. Although these values add to unity for each sample, the total number of cations per unity weight of rock will not add to the same constant for different samples. The main drawback of the present method, however, is that multiple regression was used for estimating the elements of the transition matrix which therefore are imprecise. A more detailed discussion of this problem was given in section 8.8. For this reason, interpretations will be based on groups of coefficients rather than on individual values.

The eigenvalues for the matrix **U** are represented in Table XLIII.

Relationship between transition matrices U and ***P******

It will now be proved that the matrix **U** contains the same information as the $(p \times p)$ matrix **P***.

Any transition matrix **P**, with elements that add to unity for each row, can be reduced in rank by one by: (1) subtracting one of its rows from the other rows; and (2) omitting the row of zeros from the result together with the corresponding column. Let the resulting reduced matrix be denoted by **S**. The relationship between **P** and **S** can be derived from the method discussed in section 12.4 for the calculation of the nondominant eigenvalues of a matrix. If **P**′ is the transpose of **P**, these two matrices will have the same eigenvalues. Also, the row eigenvectors of **P**′ are the column eigenvectors of **P** and vice versa. The largest eigenvalue of **P** and **P**′ is equal to one and the first row eigenvector of **P**′ consists of ones. This row can be used for $\mathbf{T}_1'/t_{11}$ in section 12.4. It follows that the reduced matrix **S**′ that corresponds to **P**′ has the same eigenvalues as **P**′ except for the eigenvalue one which is missing in **S**′. The eigenvectors of **S**′ also are the same as those of **P**′ except that one value is missing that can be obtained by using [12.19] of section 12.4. An analogous relation exists between the matrices **P** and **S**. The relation between the matrix **U** and the matrix **S** can be written as:

$$\mathbf{U} = \mathbf{Q}^{-1}\mathbf{S}\mathbf{Q} \qquad [13.34]$$

where **Q** denotes a diagonal matrix with elements $q_{ii} = 1/s(x_i)$, $i = 2, 3, ..., p$.

The matrices **U** and **S** share the same eigenvalues but the eigenvectors are different. By writing **U** in its canonical form:

$$\mathbf{U} = \mathbf{V}\mathbf{\Lambda}\mathbf{V}^{-1}$$

it follows that:

$$\mathbf{S} = (\mathbf{QV})\mathbf{\Lambda}(\mathbf{QV})^{-1} \qquad [13.35]$$

The eigenvectors of **S** (and those of **P**), therefore, can be readily obtained from the matr **V**.

The matrix **S** satisfies the equation $\mathbf{S} = \mathbf{C}_0^{-1}\mathbf{C}_1$ where the covariance matrices $\mathbf{C}_0$ ar $\mathbf{C}_1$ are related to the correlation matrices by $\mathbf{R}_0 = \mathbf{QC}_0\mathbf{Q}$ and $\mathbf{R}_1 = \mathbf{QC}_1\mathbf{Q}$. This result interesting in that it illustrates that the eigenvectors of **U** and **S** are identical after correction for scaling. In this respect, it does not matter which matrix is used for extra tion of components: **U** or **S**. It was pointed out in section 13.5 that the components $\mathbf{R}_0$ and $\mathbf{C}_0$ are not linearly related in this manner.

13.10. Properties of the first spectral component

Successive observations of the series of standardized data $\mathbf{Z}_k$ are related by the equ tion:

$$\mathbf{Z}_k = \mathbf{Z}_{k-1}\mathbf{U} + \mathbf{E}_k \qquad [13.3$$

Let $\mathbf{Z}_k\mathbf{Y}$ denote a linear combination of the variables with **Y** being a column vector th consists of arbitrary coefficients. It will be interesting to optimize **Y** in such a way th $\mathbf{Z}_k\mathbf{Y}$ ($k = 1, 2, ..., n$) is a new variable that shows the most gradual variation or mo "trend".

The geological meaning of this model can be illustrated on the basis of Fig.90. Th variations in bulk silica content in this situation are directly related to the variations temperature of the lava. It can be said that temperature is the underlying factor th controls the variations in the major oxides. All oxides describe a pattern which is linearl related to that of the factor.

The following experiment can be performed. Let noise be added to all oxides in th system in such a manner that the trends in their time-variation diagrams are obscure The time-variation pattern of the factor (temperature in the example) will be present i time-variation diagrams for the oxides. Calculation of the transition matrix **U** by lea squares helps to eliminate part of the noise. However, an additional step can be taken t improve the filtering process. Since the variation of the underlying factor is continuous i time, it will be useful to calculate a pattern $\mathbf{Z}_k\mathbf{Y}$ from the system with a variation that more continuous than that of any other of the infinite number of patterns which can b constructed. The end product is then a single systematic variation in time which described by all oxides and possibly also by the underlying controlling agency. Th method, of course, is most useful if the latter cannot be directly measured.

This approach is comparable to that of fitting a line to the data in p-space whic results in the first principal component. However, in the new method the position samples in time is considered.

First, the desirable property of $\mathbf{Z}_k\mathbf{Y}$ described as "most trend" has to be quantified. We will look for the $\mathbf{Z}_k\mathbf{Y}$ with maximum autocorrelation and maximize the first autocorrelation coefficient α in:

$$\mathbf{Z}_k\mathbf{Y} = \alpha\mathbf{Z}_{k-1}\mathbf{Y} + e_k^* \qquad [13.37]$$

where $e_k^*(k = 2, 3, \ldots, n)$ represents a residual.

If both sides of [13.36] are postmultiplied by $\mathbf{Y}$, the result is

$$\mathbf{Z}_k\mathbf{Y} = \mathbf{Z}_{k-1}\mathbf{U}\mathbf{Y} + e_k \qquad [13.38]$$

For a given value of k, the terms in this expression are scalars. The residual e_k satisfies:

$$e_k = \mathbf{E}_k\mathbf{Y}$$

The problem may be solved by using the following minimax criterion. The vector $\mathbf{Y}$ should have coefficients such that the difference between $\alpha\mathbf{Z}_{k-1}\mathbf{Y}$ in [13.37] and $\mathbf{Z}_{k-1}\mathbf{U}\mathbf{Y}$ in [13.38] is a minimum whereas α is a maximum. This difference is zero if:

$$\mathbf{U}\mathbf{Y} = \alpha\mathbf{Y} \qquad [13.39]$$

In that case, the residuals $e_k{}^*$ and e_k are identical. Eq. [13.39] only holds true if α is an eigenvalue of $\mathbf{U}$ and $\mathbf{Y}$ the corresponding eigenvector. Since α should be as large as possible, the dominant eigenvalue $\lambda_1 (= \alpha)$ and corresponding eigenvector $\mathbf{V}_1 (= \mathbf{Y})$ provide the required solution to our problem.

Application to combined series

The values $\mathbf{Z}_k\mathbf{V}_1$ as calculated for a multivariate series have a first autocorrelation coefficient which is equal to λ_1. In the practical example (combined series), $\lambda_1 = 0.607$ (see Table XLIII). In Table XLII, it is shown that the first autocorrelation coefficients for the individual oxides (along the main diagonal of $\mathbf{R}_1$) have the following values:

Al_2O_3 : 0.22	MgO : 0.26	K_2O : 0.15
Fe_2O_3 : 0.25	CaO : 0.36	TiO_2 : 0.18
FeO : 0.42	Na_2O : 0.15	

The value for SiO_2 (not shown in Table XLII) is 0.30.

For comparison, some additional values for indices used in section 13.7 are:

S.I. : 0.25	PC_1 : 0.36	Poldervaart Index : 0.29
D.I. : 0.30	PC_2 : 0.25	

The maximum value $\lambda_1 = 0.61$ is larger than anyone of these values. The corresponding

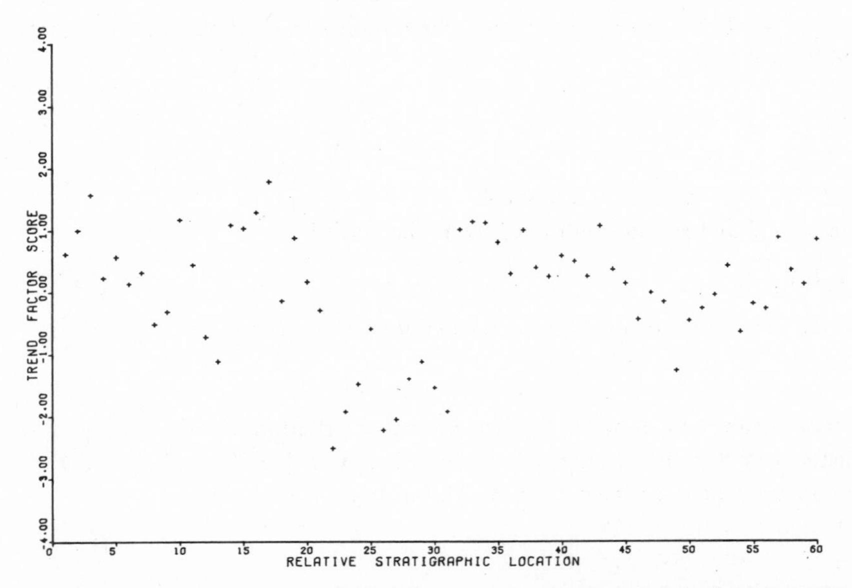

Fig.97. Trend factor or linear combination of all oxides with maximum first serial correlation coeffi cient; for combined series. Acidic volcanics excluded. Series 1 ends after point no. 31; unit alon vertical axis is one standard deviation.

linear combination $\mathbf{Z}_k\mathbf{V}_1$ is:

$$\mathbf{Z}_k\mathbf{V}_1 = 0.23z_2 - 0.35z_3 + 0.66z_4 + 0.42z_5 + 0.46z_6 + 0.00z_7 - 0.06z_8 + 0.01z_9$$

It is shown in Fig.97 for the 60 points of the combined series, with one standar deviation as the unit of distance along the vertical axis.

A comparison of Fig.97 with Fig.95 and 96 indicates that the new pattern resemble that of Poldervaart's alkalinity index. This is surprising because the alkalinity index ha not been emphasized by the component analysis of section 13.7. The trend in the diffe entiation index is also contained in the pattern of Fig.97. Since there exist approximat linear relationships in the system, it is not possible to interpret the individual coefficient of the first eigenvector $\mathbf{V}_1$.

Computing methods

The results shown in Table XLIII were obtained in 1970 by using the compute program by Grad and Brebner (1968). The earlier computer program of Agterberg an Cameron (1967) which includes a routine for obtaining eigenvalues and eigenvectors f an asymmetrical matrix such as **U** gave nearly the same results. The eigenvalues of Table XLIII were also obtained by this technique. However, the results for the last three corresponding eigenvectors appeared to be relatively inaccurate. Grad and Brebner's eigenvector routine which is based on the *QR* double-step iterative process (cf. section 12.6) is more efficient than our earlier routine which is based on large powers of matrices. Convergence for the first eigenvalue λ_1 in Table XLIII was reached after power 32; λ_2 after power 64; and $\lambda_{3,4}$ which is a pair of complex eigenvalues after power 512.

The additional components contain relevant information, and, in general, it may be dangerous to base conclusions solely on the dominant component; this is particularly true if the second dominant root is close to the first one. In the present example, $\lambda_2 = 0.42$ which is less than $\lambda_1 = 0.61$ but as large as the largest autocorrelation coefficient for individual oxides (FeO: 0.42). The next real eigenvalue is $\lambda_7 = 0.31$. These results will be further interpreted below. First, the method will be discussed in more detail.

Geometrical interpretation

The n multivariate observations denote n points in p-space. The vectors that point from each $\mathbf{X}_k$ to its following $\mathbf{X}_{k+1}$ are the observed changes in composition of $\mathbf{X}_k$. The multivariate series, therefore, can be seen as a chain in p-space which is controlled by a transition matrix **P**. We were able to estimate **P** by a matrix **P*** whose elements are subject to imprecision because of approximate linear relationships between the oxides. It was accomplished by estimating the matrix **U** from standardized data in a reduced space of $(p-1)$ dimensions. A geometrical interpretation for the modified chain in $(p-1)$-space is shown in Fig.98.

In Fig.98A, the points $\mathbf{Z}_k$, $\mathbf{Z}_{k+1}$, and $\mathbf{Z}_{k+2}$ form part of an observed chain. The points $\mathbf{Z}_k\mathbf{U}$ and $\mathbf{Z}_{k+1}\mathbf{U}$ are the predicted positions for $\mathbf{Z}_{k+1}$ and $\mathbf{Z}_{k+2}$, respectively. Differences between predicted and observed points are according to a random vector. Also represented in Fig.98A is the point $\mathbf{Z}_k\mathbf{U}^2$ which is the prediction of $\mathbf{Z}_{k+2}$ from $\mathbf{Z}_k$.

When the chain of values $\mathbf{Z}_k$ ($k = 1, 2, ..., n$) is short, it can be expected that not all randomness in the system has been eliminated by forming the transition matrix **U**. A dominant tendency for the process of change may be determined by extraction of the dominant component $\mathbf{U}_1 = \lambda_1\mathbf{V}_1\mathbf{T}_1'$ from **U**. This second step is represented in Fig.98B.

The transition matrix $\mathbf{U}_1$ has the property that, for a given point $\mathbf{Z}_k$, it specifies the direction for all following points $\mathbf{Z}_{k+s}$ with $s \geqslant 1$. For example, $\mathbf{Z}_k\mathbf{U}_1^2$ in Fig.98B lies on the line between $\mathbf{Z}_k$ and $\mathbf{Z}_k\mathbf{U}_1$, as do $\mathbf{Z}_k\mathbf{U}_1^3, ..., \mathbf{Z}_k\mathbf{U}_1^s, ...$ The expression $\mathbf{U}_1/\lambda_1^s$ is invariant to powering. On the average, the angle between the vectors connecting $\mathbf{Z}_k$ to $\mathbf{Z}_k\mathbf{U}_1^2$ and $\mathbf{Z}_k\mathbf{U}$ is smaller than the angle of the vectors which connect $\mathbf{Z}_k$ to $\mathbf{Z}_k\mathbf{U}$ and $\mathbf{Z}_k\mathbf{U}_1$, respectively. The angle between the vectors from $\mathbf{Z}_k$ to $\mathbf{Z}_k\mathbf{U}^s$ and $\mathbf{Z}_k\mathbf{U}_1$ approaches zero when s increases.

In the previous section, it has been shown algebraically that $\mathbf{U}_1$ specifies a direction

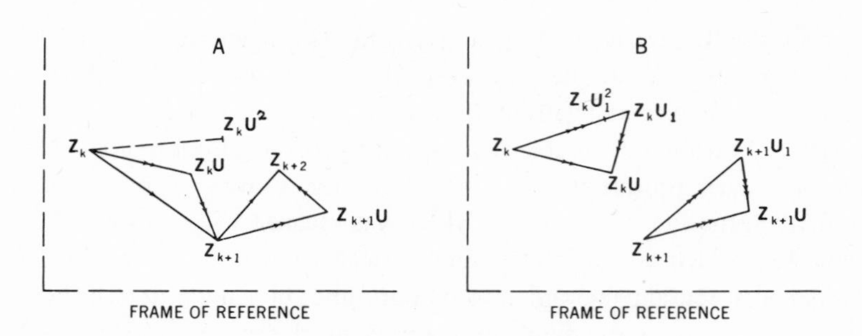

Fig.98. Schematic representation of the effect of the forward transition matrix **U** and its dominant component $\mathbf{U}_1$ on the observations $\mathbf{Z}_k$ and $\mathbf{Z}_{k+1}$. A. Single arrows denote two observed changes; double arrows represent inferred changes due to **U**. B. Triple arrows denote changes due to **U** extracted from **U**.

which accounts for most of the transition to which $\mathbf{Z}_k$ is subjected by **U**.

Evaluation of **U** in its canonical form can be considered as dividing the transformation by **U** into p individual transitions along p axes which are at right angles to one another and which are parallel to themselves regardless of the position of $\mathbf{Z}_k$. The direction cosines of these p axes are given by the vectors $\mathbf{T}_i'$ extracted from **U**. The row vector $\mathbf{T}_1$ that corresponds to the maximum root λ_1 defines the individual direction which accounts for most of the transition. Its coefficients provide a measure of the intensity by which the original variables follow the trend pattern $\mathbf{Z}_k\mathbf{V}_1$. Unfortunately, the estimation of the coefficients of $\mathbf{T}_1'$ also is affected by approximate linear relationships in the system.

Backward transition matrix

The matrix **U** is the forward transition matrix. A backward transition matrix **W** that corresponds to the backward Kolmogorov differential equations for the negative time direction can also be determined. The matrix **W** is equal to **U** for the series in reversed order. Hence:

$$\mathbf{W} = \mathbf{R}_0^{-1}\mathbf{R}_1' \qquad [13.40]$$

where $\mathbf{R}_1'$ is the transpose of **R**. **W** will be equal to **U** if $\mathbf{R}_1$ is a symmetrical matrix. The matrix **W** always has the same eigenvalues as **U** but its eigenvectors generally differ from those for **U**.

Other estimates of **U**

A result of section 13.9 is that **U** also can be estimated by the expression:

$$\mathbf{U} = \mathbf{R}_{s-1}^{-1}\mathbf{R}_s \qquad s > 1$$

Quenouille (1957) recommends the use of these matrix quotients and a comparison of successive results for the following reason. It is possible that the residuals $\mathbf{E}_k$ in [13.24] are not solely dependent upon $\mathbf{Z}_k$. For instance, the scheme:

$$\mathbf{Z}_k = \mathbf{Z}_{k-1}\mathbf{U} + \mathbf{E}_k + \mathbf{E}_{k-1}$$

where $\mathbf{E}_k$ and $\mathbf{E}_{k-1}$ are different random variables, may be a closer approximation to reality. In that case, the estimate $\mathbf{R}_0^{-1}\mathbf{R}_1$ will be biased but the estimates $\mathbf{R}_{s-1}^{-1}\mathbf{R}_s$ $(s > 1)$ are unbiased as autocorrelation of residuals is restricted to adjacent values. $\mathbf{R}_0^{-1}\mathbf{R}_1$ also would provide a biased estimate if the original variables individually follow a signal-plus-noise model (see section 10.6). The estimate $\mathbf{U} = \mathbf{R}_1^{-1}\mathbf{R}_2$ has been obtained for the practical example. It could not be used because of the approximate linear relationships between the variables. The precision of $\mathbf{R}_{s-1}^{-1}\mathbf{R}_s$ decreases rapidly for increasing s if these are present. This phenomenon can be illustrated by using a univariate series for example. The matrix $\mathbf{R}_{s-1}^{-1}\mathbf{R}_s$ then is reduced to a scalar $r_{s-1}^{-1}r_s$. For long univariate series, the expected variance of $r_{s-1}^{-1}r_s$ is:

$$\sigma^2(r_{s-1}^{-1}r_s) = \frac{\sigma^2(r_1)}{\rho_1^{2(s-1)}} \qquad [13.41]$$

where r_1 (for $s = 1$) represents the estimated first autocorrelation coefficient with expectation ρ_1. If, for example, $\rho_1 = 0.01$, [13.41] indicates that each successive estimate is hundred times as imprecise as the previous one for increasing s. Similar considerations hold true for multivariate series if one or more of the eigenvalues of $\mathbf{U}$ are close to zero. This is the case when the variables are subject to one or more linear constraints (Quenouille, 1957, p.38). The solutions of $\mathbf{U}$ by [13.27] with $s > 1$ may then be dominated by the approximate linear constraints in the system.

13.11. Extrapolation by Kolmogorov differential equations

The determination of $\mathbf{U}$ corresponds to a solution for the forward Kolmogorov differential equations. The matrix $\mathbf{C}$ in [13.19] has the same spectral components as $\mathbf{U}$. Its eigenvalues λ_i are equal to ln λ_i. There is no need to calculate $\mathbf{C}$ in explicit form. In fact, the underlying continuous process is fully described by the expression:

$$\hat{\mathbf{Z}}_{k+s} = \mathbf{Z}_k\mathbf{U}^s \qquad [13.42]$$

f s is changed from an integer variable to a nonnegative continuous variable. The results or $s = 1$ then are for the unit time interval. $s = 0$ means extrapolation toward a zero-time nterval.

The continuous process which is constructed by using the Kolmogorov forward equations corresponds exactly to the observed values in the limit $s = 0$. Since it is likely that the observed values (see Fig.92) are subject to a considerable amount of noise, this model is not realistic if the entire matrix **U** is used for the extrapolation. It therefore is practical to divide **U** into its spectral components and to disregard those components of **U** which have small eigenvalues or otherwise seem to indicate systematic variations which die out rapidly if the time interval s increases from zero.

In the limit $s = 0$, [13.42] reduces to $\hat{\mathbf{Z}}_k = \mathbf{Z}_k$ since $\mathbf{U}^s$ becomes an identity matrix. If **U** is evaluated in its canonical form, we can write:

$$\hat{\mathbf{Z}}_k = \sum_{i=1}^{p-1} \mathbf{Z}_k \mathbf{Z}_0(\lambda_i)$$

or: $$\hat{\mathbf{Z}}_k = \sum_{i=1}^{p-1} \mathbf{Z}_k \mathbf{V}_i \mathbf{T}_i'$$

Because of the properties of the spectral components $\mathbf{Z}_0(\lambda_i) = \mathbf{V}_i\mathbf{T}_i'$, this also reduces to $\hat{\mathbf{Z}}_k = \mathbf{Z}_k$. However, if a number of components is eliminated, $\hat{\mathbf{Z}}_k \neq \mathbf{Z}_k$.

As for Markov chains (see section 12.7), the eigenvalues of the transition matrix are positive, negative, or they form conjugate pairs of complex values.

These three types are represented in Table XLIII for the practical example. The positive values $\lambda_1 = 0.61$, $\lambda_2 = 0.42$, and $\lambda_7 = 0.32$ already have been considered in the previous section. They correspond to autocorrelated linear combinations. The autocorrelation would show exponential decay. For $\mathbf{Z}_k\mathbf{V}_1$, the first seven autocorrelation coefficients would be: 0.61, 0.37, 0.22, 0.14, 0.08, 0.05 and 0.03. These values were obtained by raising $\lambda_1 = 0.61$ to successive powers. It is interesting to compare the theoretical values with empirical autocorrelation coefficients computed for the data plotted in Fig.97, which are 0.61, 0.41, 0.30, 0.29, 0.15, – 0.12, and – 0.15. The agreement between the two sequences is fair in this part of the correlogram. The dependence between values is practically zero (< 0.05) for a time interval equal to seven units. The first few exponential autocorrelations for $\mathbf{Z}_k\mathbf{V}_2$ are: 0.42, 0.18, 0.07 and 0.01; and for $\mathbf{Z}_k\mathbf{V}_7$: 0.32, 0.10, and 0.03. They die out after four and three units, respectively.

A negative eigenvalue λ_i corresponds to a cyclical $\mathbf{Z}_k\mathbf{V}_i$ with period equal to two sampling intervals. In the example, $\lambda_8 = -0.16$ which dies out rapidly. The moduli of the complex pairs $\lambda_{3,4}$ and $\lambda_{5,6}$ are equal to 0.35 and 0.33, respectively. The corresponding periods are 10.4 and 3.3. The real part of the pair $\lambda_{5,6}$ has a negative value, indicating a short period. This rapid oscillation probably has no petrological meaning. The pair $\lambda_{3,4}$, on the other hand, is interesting in that its period is 10.4. Since there are 60 observations in the combined series, this eigenvalue may correspond to $60/10 = 6$ oscillations. This is equal to the six iron enrichments in the zig-zag pattern shown by the solidification index (Fig.95). Statistical tests to determine the significance of these results are not available.

that situation, a good procedure is to apply the same technique to different sets of data and to compare the results. Series 1 and 2 have been separately analyzed. The resulting eigenvalues of **U** were also shown in Table XLIII.

The largest eigenvalue for series 1 is equal to 0.69. The complex pair $\lambda_{2,3}$ has modulus 0.48 and period 2.3. This rapid fluctuation has not been detected in the other series.

The complex pair $\lambda_{4,5}$ has modulus 0.45 and period 8.4. This result is fairly close to that of $\lambda_{3,4}$ for the combined series. It also occurs in series 2 as $\lambda_{4,5}$ with modulus 0.45 and period 9.7. Clearly, the periodicity averaging about ten sampling intervals in the combined series, also is present in the two subseries.

It may be concluded that the solution of the Kolmogorov forward differential equation for the Yellowknife mafic volcanic members shows results that are comparable to those obtained by the use of petrological indexes and principal components. The existence of six iron-enrichment trends has been confirmed. A surprising result is the cyclical variation for the linear combination of all variables that shows most trend. This pattern resembles a variation in alkalinity suggested by the Poldervaart index. The linear correlation coefficient between these two patterns amounts to 0.58.

An increasing number of petrologists is adopting the approach of subjecting a set of data to a variety of statistical techniques. Many indices or principal components are computed in the hope that patterns will emerge that can be interpreted. In the last part of this chapter, we have introduced a potentially powerful technique for the statistical treatment of chemical analyses of samples forming an ordered sequence.

Chapter 14

CALCULATION OF PREFERRED ORIENTATIONS FROM VECTORIAL DATA

14.1. Introduction

A number of geological attributes may be approximated by lines or planes. Measuring them results in "angular" data consisting of azimuths for lines in the horizontal plane or azimuths and dips for lines in 3D-space. Although a plane is usually represented by its strike and dip (cf. section 4.3), it is fully determined by the line perpendicular to it, and statistical analyses of data sets for lines and planes are analogous.

Examples of angular data are strike and dip of bedding, banding, and planes of schistosity, cleavage, fractures or faults. Azimuth readings with or without dip are widely used for sedimentary features such as axes of elongated pebbles, ripple marks, foresets of cross-bedding and indicators of turbidity current flow directions (sole markings). Then there are the B-lineations in tectonites; problems at the microscopic level include that of finding the preferred orientation of crystals in a matrix (e.g., quartz axes in petrofabrics). Another example is the direction of magnetization in rocks.

It may be useful to treat lines of these types as unit vectors which point in directions assigned to them. In some situations (e.g., direction of magnetization), it is obvious that the lines are directed and a positive direction makes good sense. In other situations (e.g., B-lineation in gneiss), the lines may be undirected in that a distinction between a positive and a negative direction is arbitrary.

We begin this chapter with a brief review of some of the methods in use to treat vectorial data in various fields of geology. Next, several methods to compute the preferred direction of data in the horizontal plane are discussed and examplified. This raises the problem of how to compare means from different areas; lines in space will be treated similarly. Finally, the more advanced subject of trend analysis of vectorial data is introduced in the last part of the chapter. It consists of fitting unit vector fields to data from many localities with preferred orientations that are not parallel to one another.

14.2. Methods of treating angular data

It is useful to plot the angular data under study in a diagram before a statistical calculation is attempted. Azimuth readings may be plotted on various types of rose

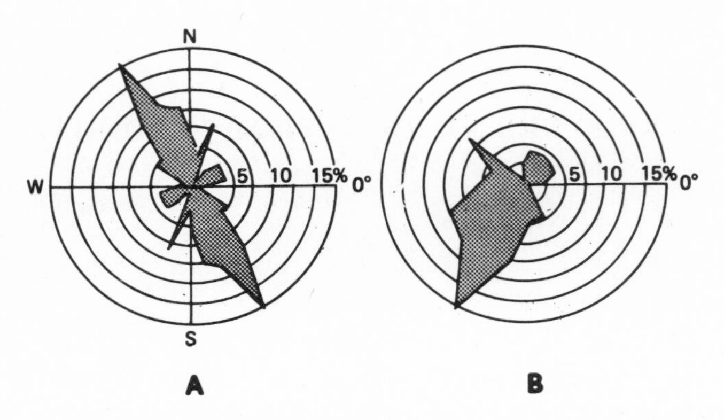

Fig.99. Axial symmetric rose diagram of the direction of pebbles in glacial drift in Sweden according to Köster (1964). Step from A to B is accomplished by doubling the angles. (From Koch and Link, 1971; after Batschelet, 1965.)

diagram (Potter and Pettijohn, 1963). If the lines are directed, the azimuths can be plotted from the center of a circle and data within the same class intervals may be aggregated. Suppose that the lines are undirected, then a rose diagram with axial symmetry may be used such as the one shown in Fig.99A.

Lines in space can be projected onto the surface of a sphere. The equal-area Schmidt net and the Wulff stereographic projection are most commonly used for this purpose. The data plotted on a net may be contoured. Methods to accomplish this are discussed in most textbooks of structural geology (e.g. Billings, 1972). Comprehensive summaries of graphical methods of treating lines in space are given by Vistelius (1966b) and Phillips (1972). These methods were also reviewed by Stauffer (1966). Some textbooks of structural geology contain methods for calculating the best-fitting plane passing through a given set of lines (Whitten, 1966a), or treat more advanced problems such as computing the best fitting cone of a set of lines (Ramsay, 1967).

It is important to realize that most methods for plotting, contouring and computing are applicable only if the directional features occur within a single "domain". Turner and Weiss (1963, p.20) specified a domain as any three-dimensional portion of a rock body that is statistically homogeneous on the scale of the domain.

Graphical contouring of the data from domains may be done automatically by computer (Adler and Pilger, 1968; Bonyun and Stevens, 1971).

A set of angular data should be carefully inspected before statistical treatment is undertaken. It is useful to calculate the average or preferred orientation if: (1) a single population of unit vectors can be defined; (2) the scatter about the mean is relatively large; and (3) a preferred orientation can not be obtained by graphical contouring for lack of sufficient data.

If the scatter of the data is large, it may be questionable whether preferred orientations really exist, although contouring usually provides maxima in the diagrams. Tests to decide between randomness or nonrandomness include the Tukey χ^2-test for azimuth readings (Middleton, 1965) and several methods for lines in space including those listed by Hopwood (1968).

The body of publications dealing with statistics of orientation data is large. Introductory papers on unit vectors in the plane include Pincus (1956), Steinmetz (1962), Jones (1968) and Rao and Sengupta (1970).

Methods for unit vectors in space are used extensively in paleomagnetism (Irving, 1964). Several important methods such as calculating the cone of confidence for an average pole (Fisher, 1953) and comparing poles from different sites by analysis of variance (Watson, 1966) were originally developed using paleomagnetic data for example.

It happens frequently, particularly in problems of structural geology and sedimentology that the data from a single domain correspond to two or more populations of features. Allocation of single observations to one of two known populations may be accomplished by using a discriminant function for vectorial data as discussed by Watson (1966). In some situations, the allocation of individual data can be avoided by using more flexible statistical methods (Jones and James, 1969).

An example of a special problem solved by using statistical methods for orientation data is provided by Cruden and Charlesworth (1966). These workers were able to establish a statistically significant angular unconformity of 4° between bedding on the two sides of the contact between Mississippian and Jurassic rocks near Nordegg, Rocky Mountain Foothills of central Alberta.

In this chapter, we are mainly concerned with features that scatter considerably about a mean so that an individual measurement is not representative for a domain extending beyond the place where this measurement was taken. Some linear features are representative of large domains and, therefore, may be plotted on a small-scale map without further statistical treatment. For example, sole markings, which are indicators of turbidity currents, may point in the same direction both at different localities in a large part of a basin and in successive layers (Potter and Pettijohn, 1963, p.130). On the other hand, foresets and ripple marks, formed by slower currents, under other conditions, may differ strongly in direction between layers at the same outcrop and between outcrops. Likewise, B-lineations for gneiss may be mappable whereas those for more strongly deformed micaschist or phyllite may change rapidly in orientation from outcrop to outcrop.

14.3. Methods of calculating the mean for vectorial data in the plane

This section consists of the following parts. First, there is a discussion of elementary methods to obtain means from directed and undirected lines in the horizontal plane. A brief review of computing the more widely used types of mean is next, followed by a practical example with supplementary discussions.

Graphical methods

An example of undirected lines was given in Fig.99A where each line is plotted in the two opposite directions. If the angles for azimuth are doubled, the result is as in Fig.99B with a single preferred direction. We may divide the angle for this direction by two to obtain the preferred orientation on the circle used for presentation in Fig.99A. The advantage of this method is that, during statistical manipulation, the two directions of a line are treated equally, without preference to one direction which possibly can not be defined consistently for all measurements of a set of undirected lines, especially if the deviations from the mean are large.

The mean of unit vectors may be constructed graphically as shown in Fig.100. Vectors of equal length, one for each measurement, are added as was done previously in Fig.15A; the direction of the resultant vector is the preferred orientation. The resulting vector mean is a sample estimate of a population mean shown schematically in Fig.100. Its length R is a measure of the scatter of the original data about the mean. R will be longer if the scatter is less, e.g., if the unit vectors are subparallel.

Suppose that we use this method to compute the vector mean after doubling the angles as in Fig.99B, and then divide the angle for the resultant by two. This method was originally developed by Krumbein (1939) for the preferred orientation of pebbles in sedimentary deposits. It gives the same result as when a major axis (cf. section 4.7) is fitted to the points of intersection between the undirected lines and a circle. We recall that this method consisted of minimizing the sum of squares of perpendicular distances from all points to the axis. The procedure is illustrated for an artificial example in Fig.101. Axial symmetry is preserved when the major axis is constructed. Note that the result is independent of which type of azimuth (e.g., 10° or 190°) is originally coded for any undirected line.

This method may be extended to the situation of undirected lines in 3D-space. Then the first principal component is computed using the coordinates of the intersection points of the lines, which pass through the center of a sphere, and the surface of that sphere.

If taking the arithmetic mean of a set of azimuths also is considered as a method to obtain a preferred direction, we now have three procedures for computing: (1) arithmetic

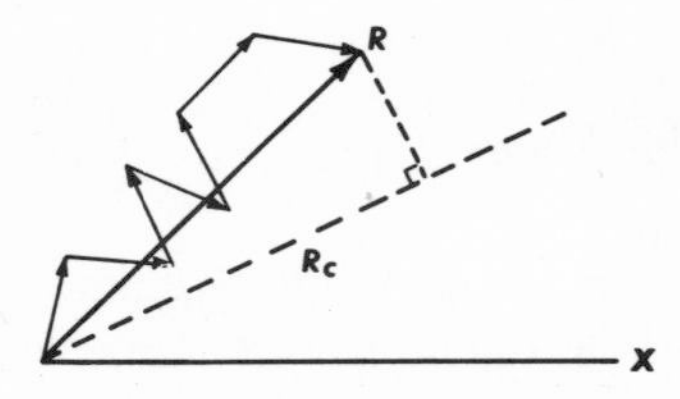

Fig.100. Vector sum or resultant **R** of seven unit vectors and its projection $\mathbf{R}_c$ on true mean direction relative length of **R**, provides a measure of scatter; vector mean is direction parallel to **R**. (After Watson, 1966.)

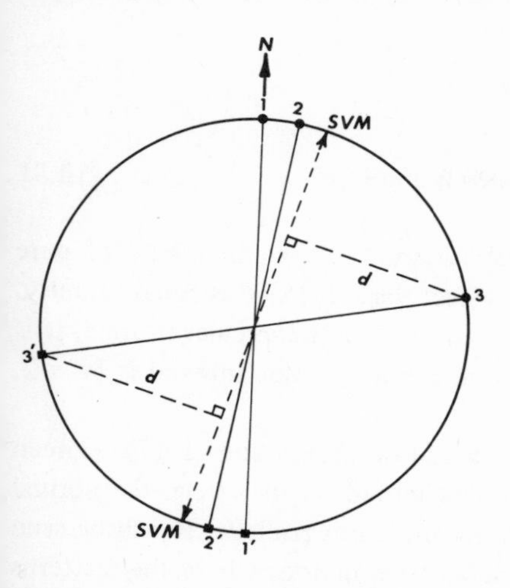

Fig.101. Calculation of semicircular vector mean. This is equivalent to minimizing sum of squares of perpendicular distance (d) as illustrated for azimuth reading no. 3.

mean; (2) vector mean (Fig.100); and (3) "semicircular" vector mean (Fig.101). If the azimuths are written as x_i, these means satisfy the equations:

AM (arithmetic mean) : $\bar{x} = \frac{1}{N} \Sigma x_i$

VM (vector mean) : $\hat{\gamma} = \arctan \frac{\Sigma \sin x_i}{\Sigma \cos x_i}$

SVM (semicircular vector mean) : $\hat{\gamma}_2 = \frac{1}{2} \arctan \frac{\Sigma \sin 2x_i}{\Sigma \cos 2x_i}$

From a statistical point of view, each mean provides the answer to a different question. If deviations from the mean are written as θ_i, specific conditions are satisfied by each estimator:

AM : $\Sigma \theta_i = 0$; $\Sigma \theta_i^2$ = minimum

VM : $\Sigma \sin \theta_i = 0$; $\Sigma \cos \theta_i$ = maximum

SVM: $\Sigma \sin 2\theta_i = 0$; $\Sigma \cos^2 \theta_i$ = maximum

Each mean is the maximum-likelihood estimator for a specific type of frequency distribution with density $f(\theta)$:

AM : normal (Gaussian) with $f_n(\theta) = \frac{1}{\sigma\sqrt{2\pi}} \exp(-\theta^2/2\sigma^2)$ [14.1]

VM : circular normal (Von Mises) with $f(\theta) = \dfrac{1}{2\pi I_0(\kappa)} \exp(\kappa \cos\theta)$ [14.2]

SVM: semicircular normal with $f_2(\theta) = \dfrac{1}{\pi I_0(\kappa_2)} \exp(\kappa_2 \cos 2\theta)$ [14.3]

In the latter two expressions, $I_0(\kappa)$ is the Bessel function of the first kind of pur imaginary argument. It is a scaling factor with the effect that $\int_R f(\theta)d\theta$ is equal to unity This integration is done for the relevant range of θ-values. For the circular normal, it i the interval $[-\pi, +\pi]$ or $[0, 2\pi]$. For the semicircular normal, this interval is $[-\pi/2$ $+\pi/2]$ or $[0, \pi]$.

The scaling factor for the normal distribution is based on integration of $f(\theta)$ betwee the limits $-\infty$ and $+\infty$. Since orientation data are distributed on the circle, the norma distribution model, which underlies the arithmetic mean, is not realistic. It will be see that the usefulness of the arithmetic mean is restricted to situations where the scatter i relatively small ($\sigma < 30°$), because the other two types of distribution converge fairl rapidly to the normal one when κ and κ_2 become larger.

Paleocurrent indicators, Melville Island

The purpose of this example is to compare the preceding three methods to on another by applying them to the same data and to introduce further the underlyin frequency distributions.

The areal variation of average paleocurrent directions in sandstones from the Bjorn Formation (Lower Triassic) Melville Island, Arctic Archipelago, was discussed in sectio 2.11.

Each value shown in Fig.13A is a locality average based on N_k readings for two type of directional features: cross-bedding foresets and scoop axes. The "scoops" are spoon shaped troughs elongated in the current direction. Detailed information on the origina data is given by Trettin and Hills (1966) and Agterberg, Hills and Trettin (1967). Th preferred orientation of the scoop axes does not differ significantly from that of th foresets, although the scatter of the scoop axes may be slightly less (difference no statistically significant at the 95-% level).

The three types of mean were computed for the 43 localities with, in total, 49 readings (see Table XLIV). The difference between arithmetic mean and vector mean i small for all localities. The largest difference amounts to 5° (Station 23). A simila comparison between these two types of mean was done by Krumbein and Graybill (1965 for slopes of valley walls to demonstrate the problem when the range of angles is narro (20°).

The semicircular means in Table XLIV deviate from the two other means by somewha larger amounts, about 20° at a maximum (Stations 2 and 23). Inspection of rose diagram for these two localities shows that their range of angles amounts to 230° and 160°

TABLE XLIV

Three types of mean and vector strength ($\bar{a}$) for 43 localities, Bjorne Formation, Member C. (Locality averages in degrees)

No.	N_k	*AM*	*VM*	*SVM*	$\bar{a}$	*No.*	N_k	*AM*	*VM*	*SVM*	$\bar{a}$
1	4	343	345	357	0.82	29	15	351	353	359	0.82
2	15	359	356	334	0.56	30	5	29	29	29	0.90
3	4	304	305	311	0.88	31	15	353	353	351	0.80
6	10	336	335	331	0.89	32	7	8	8	8	0.92
7	6	288	288	288	0.92	33	7	348	349	351	0.91
9	17	318	319	325	0.84	34	12	232	322	315	0.68
11	13	359	359	358	0.88	35	12	8	8	7	0.94
12	7	337	341	351	0.84	36	10	348	347	344	0.88
14	15	330	330	332	0.86	37	12	34	34	33	0.95
15	14	335	335	337	0.78	39	14	46	48	54	0.84
16	5	347	347	347	0.98	40	10	42	40	40	0.93
17	12	349	349	349	0.93	41	12	48	48	51	0.84
18	12	331	330	329	0.94	44	11	53	52	49	0.86
19	14	332	330	325	0.85	46	12	39	42	51	0.53
20	11	15	15	15	0.87	47	13	61	61	61	0.90
21	13	16	16	19	0.92	48	10	64	64	65	0.94
23	17	7	12	30	0.72	50	12	51	53	59	0.85
24	16	339	339	341	0.82	52	7	72	72	74	0.95
25	14	15	15	15	0.96	53	5	79	79	80	0.97
26	14	5	5	6	0.93	54	9	120	120	120	0.95
27	23	356	356	357	0.90	55	13	88	89	93	0.90
28	17	358	357	353	0.87						

respectively, which is greater than that for the other localities. It is possible that the populations for these samples are not homogeneous, in which case the semicircular mean may be a better indicator of preferred orientation since it is closer to the suggested maximum of density in the rose diagram.

Pincus (1956) and Wood and Wood (1966) have done experiments on sets of artificial data in order to illustrate the effect of outlying observations on arithmetic and vector mean, respectively. Wood and Wood (1966) formed a set of nine readings equal to zero (pointing northward) and a variable tenth (x_{10}) increasing stepwise from 0° to 180°. The arithmetic mean of the set increases linearly to 18° (when x_{10} = 180°). The vectorial mean, however, first increases to a maximum value of 6°20′ (x_{10} = 90°), then decreases to 0° (x_{10} = 180°). We may add to this experiment the calculation of the semicircular mean. It increases to 3°30′ (x_{10} = 45°), then decreases to 0° (x_{10} = 90°) and −3°30′ (x_{10} = 135°) and increases again to 0° for x_{10} = 180°. The effect of outlying observations on the mean is largest for the arithmetic mean. It is less for the vectorial mean and, in turn, less for the semicircular mean.

The three types of density distribution [14.1–3] were fitted to 180 scoop axes from 27 sets (with at least 5 data) and 388 foresets from 39 of the stations listed in Table XLIV.

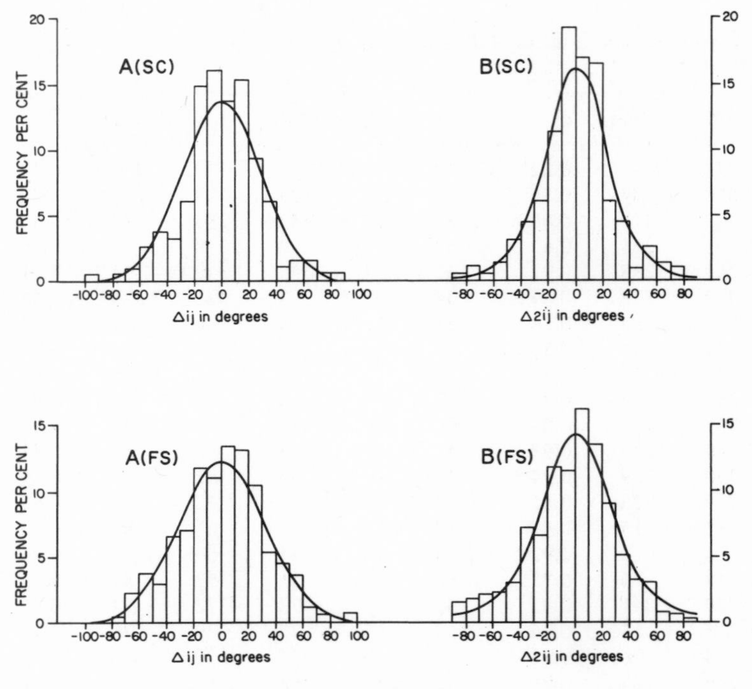

Fig.102.A. Histograms and best-fitting Gaussian curves for scoop axes (SC) and foreset-dip azimuth (FS); Δ_{ij}- values represent deviations from arithmetic means . . localities; difference between Gaussian curves (A) and circular normal curves (not shown) fitted to deviations from vector means is negligibly small. B. Ditto, for semicircular normal curves fitted to deviations (Δ_{2ij}) from semicircular vector means.

We used deviations from the locality means for two types of feature. The results are shown in Fig.102. In this situation, the difference in shape between the normal and the circular normal distribution is negligibly small.

A χ^2-test for goodness of fit gave significance levels of 88% for scoop axes ($\hat{\chi}^2$ = 14.2; 9 degrees of freedom) and 50% for foresets ($\hat{\chi}^2$ = 12.4; 13 d.f.). The histograms for deviations from the semicircular means for localities differ in shape from those for the other two types of mean. The best-fitting semicircular normal distributions also differ in shape from the normal or circular normal ones. The significance levels now are 72% for scoop axes ($\hat{\chi}^2$ = 11.0; 9 d.f.) and 57% for foresets ($\hat{\chi}^2$ = 16.3; 13 d.f.). The experiment suggests that each of the three methods of approach gives satisfactory results in this situation.

The method of fitting a circular normal distribution to data was discussed by Gumbel et al. (1953) who gave tables. These tables can also be used for fitting a semicircular normal distribution if the angles are doubled as before. It is pointed out again that, in the semicircular model, no distinction is made between the positive and negative direction of a line. It is less suitable for situations where the readings are directed, particularly if the total range of data exceeds 90° at either side of the mean. This model is recommended for situations where: (1) the lines are undirected; and (2) where the scatter is large. A further drawback of the semicircular mean is that, until now, detailed methods of statistical significance have not been developed for this model.

The term "semicircular normal" distribution was used by Arnold (1941) who derived its properties by solving specific heat-flow equations on a circle. Arnold demonstrated that the distribution can arise from a stochastic process on the circle which is analogous to one giving a normal distribution on the straight line. Later, Stephens (1962) showed that the circular normal distribution can be interpreted as the result of another type of stochastic process on the circle, also leading to a normal distribution on the line.

The previous two arguments extend to the spherical normal (or Fisher) and the hemispherical normal distribution on the sphere which are equivalent to the two distributions for the circle discussed in this section. Thus, the vector (or Fisher) mean of lines in 3D-space is equivalent to the vector mean of lines in the plane, and the hemispherical vector mean is the extension of the semicircular vector mean. The vector mean may be used for directed lines and also for undirected lines if the deviations from the mean do not exceed 90°. The semicircular mean (and hemispherical mean) should be used for undirected lines if many of these deviate more than 90° from the mean. It may be instructive to calculate the three types of mean for data on the circle (and two means on the sphere) in order to compare the results with one another. Other recommendations for directed lines were made by Wood and Wood (1966) who preferred the arithmetic mean above the vector mean and Scheidegger (1965) who used hemispherical instead of Fisher mean.

In Table XLIV, vector strengths ($\bar{a}_k$) for the 43 localities are also shown. This parameter is obtained by dividing the length R_k of the sum of N_k unit vectors for a set by N_k. In general, the magnitude R of the sum of N unit vectors satisfies:

$$R^2 = (\Sigma \sin x_i)^2 + (\Sigma \cos x_i)^2 \qquad [14.4]$$

and $\bar{a} = R/N$. If all unit vectors are parallel (no scatter), $R = N$ and $\bar{a} = 1$. The scatter increases when $\bar{a}$ decreases. The maximum-likelihood estimator of the parameter κ (circular normal distribution) satisfies the equation:

$$\frac{I_1(\hat{\kappa})}{I_0(\hat{\kappa})} = \bar{a} \qquad [14.5]$$

Gumbel et al. (1953) have tabulated this function so that $\hat{\kappa}$ can be determined if $\bar{a}$ is given. Series expansion of $\cos\theta$ in the exponent ($\kappa \cos\theta$) of the circular normal distribution ([14.2]), indicates that, by approximation $\cos\theta \approx 1 - \theta^2/2$ when $|\theta| < 1$. Since θ is in radians for this expansion it means that, when a deviation is less than 57° on either side of the vector mean, the circular normal can be approximated by the normal curve.

The average vector strength for the 43 sets of Table XLIV amounts to $\bar{a} = 0.863$ with corresponding value of $\hat{\kappa} = 4.0$. In this situation, 93% of area under the circular norma density curve falls within the range $\pm$ 57° at either side of the mean. We are approxi mately at the limit of where the normal and circular normal distributions have the sam shape.

The arithmetic mean has the advantage that it is calculated readily, and differen means can be compared with one another by conventional analysis of variance.

As a rule, it may be used instead of the vector mean if the standard deviation is les than 30° or if the range of observations does not exceed 114° (Agterberg and Briggs 1963).

14.4 Comparison of means

If the scatter increases further, the circular normal becomes markedly different an this model should then be used. It was fitted to data with $\hat{\kappa}$ ranging from 2.5 to 3.3 b Trettin (1971) for sole markings (turbidity current indicators) in the Imina Formation Lower Paleozoic on Ellesmere Island, Arctic Archipelago.

Sengupta and Rao (1966) have found that in units of the Kamthi Formation (Uppe Permian), India, cross-bedding foresets have $\hat{\kappa}$ ranging from 1.2 to 2.3.

In these two situations (Imina and Kamthi Formations), all readings from a large are were lumped together. In general, the scatter will increase with the size of area (an length of time interval) that is sampled, as established during the original work on cross bedding directions in basal Pennsylvanian sandstones in central U.S.A. by Potter an Olson (1954). Means from different localities may originate from a single population fo the entire area of study or the parameters of the population may change from place t place. Analysis of variance can be used to test for the existence of significant difference between means for localities (Watson and Williams, 1956; Stephens, 1962; Watson, 1966)

Rao and Sengupta (1970) sampled $n = 14$ outcrops of the upper Kamthi unit an collected 10 observations per outcrop; hence, there were $N = 140$ measurements in total If R is vector resultant length, then the quantity $2\kappa\,(N - R)$ is approximately distribute as χ^2 with $(N - 1)$ degrees of freedom.

Suppose that R is the resultant for all unit vectors, and R_i an outcrop resultant. Th total variation $(N - R)$ can be divided into two parts: one for variation between outcrop $(\Sigma R_i - R)$ and one for variation within outcrops $(N - \Sigma R_i)$. ΣR_i is simply the sum o lengths for the n outcrop resultants. We have that:

$$F = \frac{(\Sigma R_i - R)}{(N - \Sigma R_i)} \frac{N - n}{n - 1} \qquad [14.6$$

is F-distributed with $(n - 1)$ and $(N - n)$ degrees of freedom. For the example, Tabl XLV can be constructed.

TABLE XLV
Analysis of variance for 14 outcrops means, foresets, upper Kamthi unit, India. (From Rao and Sengupta, 1970.)

Variation	*df*	*SS*	*MS*	*F*(13,126)
Between outcrops	$n - 1 = 13$	$\Sigma R_i - R = 6.26$	0.48	$\hat{F} = 1.26$
Within outcrops	$N - n = 126$	$N - \Sigma R_i = 48.19$	0.38	
Total	$N - 1 = 139$	$N - R$		

The $\hat{F}$-ratio computed in Table XLV is not significant at the 95-% level with $F_{0.95}(13,126) = 1.83$. It may be concluded that, in the upper Kamthi unit, the same preferred paleocurrent direction was valid over a large region. This situation is exceptional. Usually, outcrop means or locality means have preferred directions different from one another. The averages may show patterns of systematic regional variation as in Fig.13.

14.5. Systematic variation of locality means

In section 2.11, we saw how a pattern could be fitted to the locality averages plotted in Fig.13. The method consisted of fitting a linear trend surface to the arithmetic means (also noted in Table XLIV) and solving the resulting differential equation.

Fröberg (1969) has stated that explicit solutions can be obtained for a negligible minority of differential equations only. He has summarized more general methods of numerical analysis by which a solution is obtained without the use of calculus for solving the differential equation.

A method of this type has been used by Roberts and Mark (1970). Suppose that the lines of equal azimuth are equidistant (Fig.103). A flow line can then be constructed as a succession of a large number of short straight-line segments. The flowlines in this situation will satisfy [2.42]. The method can also be applied when the isoazimuth lines are not parallel and equidistant. Roberts and Mark have used the technique to reconstruct ice-flow lines from the average orientation of stones in till. The original data for this example were reported by Wright (1957) for the Wadena drumlin field of west-central Minnesota. The first step of the procedure consisted of calculating semicircular vector means for localities. A linear trend surface was fitted to these averages. Each of the resulting curves satisfies [2.42] and [2.44] but there is no need to evalute [2.44]. These ice-flow lines, which are based on till fabric samples, are shown in Fig.104. Their pattern is confirmed by that of the drumlins in the same area since drumlins are also ice-flow indicators.

When the method is applied, care should be taken that the line segments (see Fig.103) are not too long. Otherwise, the constructed flow lines may diverge from their correct

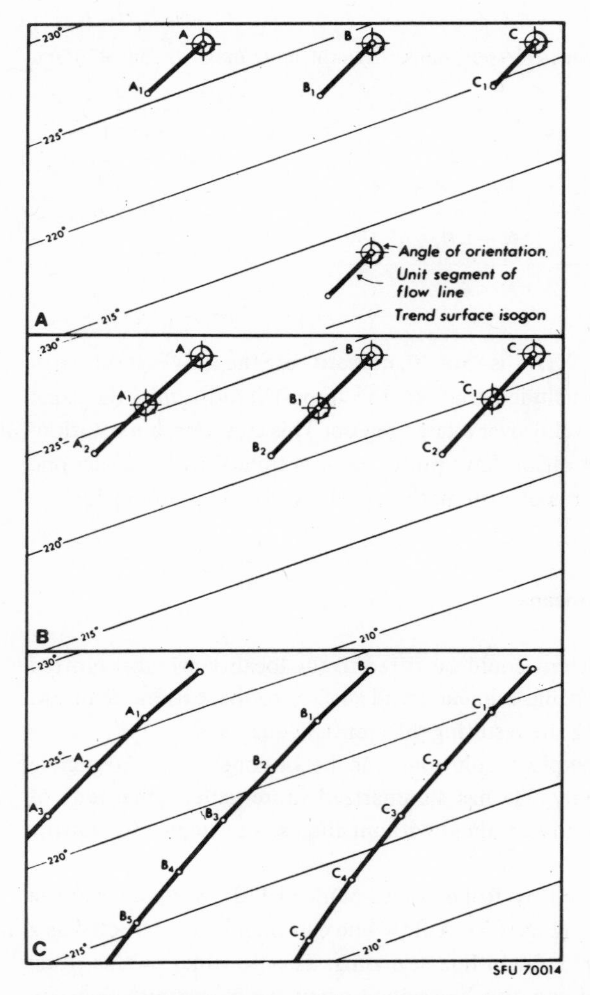

Fig.103. Steps in graphical construction of flow lines from isoazimuth lines for preferred flow direction (describing a linear surface in example); curves are approximated by succession of straight-line segments whose direction is given by isoazimuth line pattern. (From Roberts and Mark, 1970.)

course. An easy method to check for errors of this type is to do the complete numerical analysis choosing different lengths of segments and comparing the results with one another. If the segments are too long this should show as a difference in the patterns.

The method can be readily computerized. We obtained a graphical plot by fitting a linear trend surface to the vector means of Table XLIV. The result is shown in Fig.105.

The result is approximately equal to that shown previously in Fig.41A. This is because the vector means are nearly equal to the arithmetic means used previously.

It is noted that the flowlines in Fig.105 are equidistant along a horizontal line passing

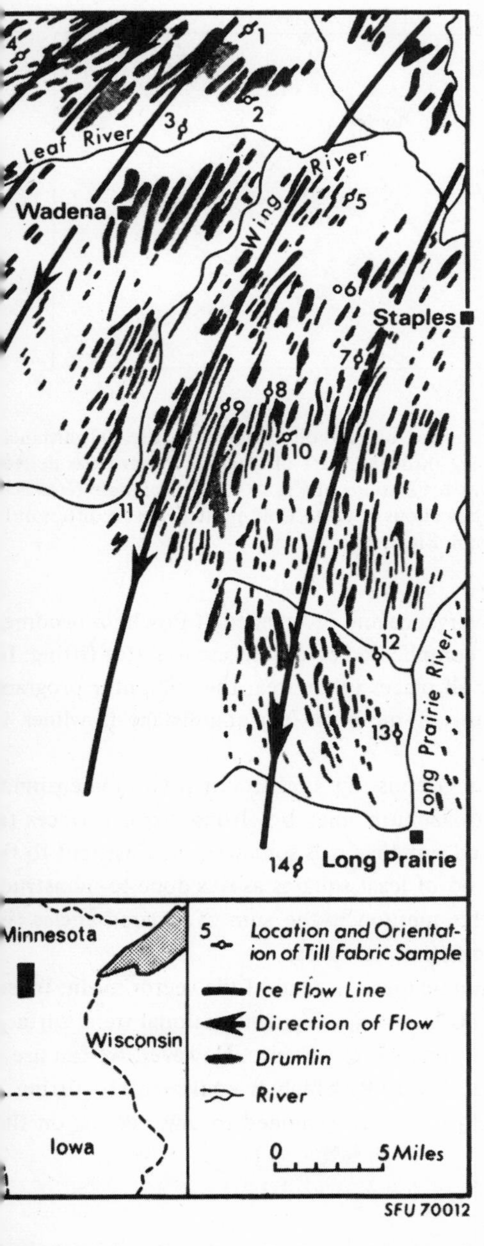

ig.104. Ice-flow lines from linear trend surface fitted to averages for till fabric samples from Wright 1957); sample averages were semicircular vector means. (From Roberts and Mark, 1970.)

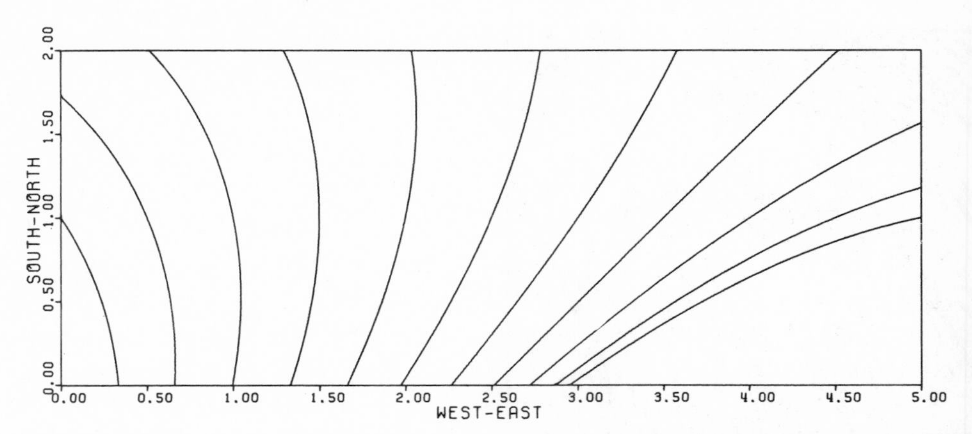

Fig.105. Machine-produced output for flow lines derived by method explained in Fig.103; distance between curves are equal for horizontal line passing through center of diagram. Pattern was derive from a linear trend surface fitted to vector means for stations of Fig.13A. Vertical scale (*U*-axis i Fig.13A) was reversed in order to conform to specifications of CALCOMP-plotter. Result differs only slightly from that based on arithmetic means shown in Fig.41A.

through the center of the diagram. The convergence and divergence of flowlines become increasingly difficult to predict when more complicated equations are used for fitting. In order to obtain a pattern representative for all places in an area, the computer program may be run several times so that a pattern of more or less equidistant flowlines is obtained.

Pelletier (1958) has fitted paleocurrent trends by constructing two-dimensiona moving averages. Fox (1967) constructed isoazimuth lines by fitting trend surfaces to vector means for Pelletier's data. It is pointed out that it is somewhat inconsistent to fit trend surfaces to vector means by the method of least squares as was done for construc ting Fig.105. The quantity minimized in this method is the sum of squared angles (in degrees) between observed azimuths and best-fitting flowlines.

This is equivalent to calculating the arithmetic mean instead of the vector mean. When the fit is good (standard direction of residuals less than 30°), conventional trend-surface analysis can be used and be supplemented by analysis of variance. However, we can use a more general method to be discussed in section 14.9, which is equivalent to fitting a vector mean to dimension-free data and which can be extended to unit vectors on the sphere.

14.6. Unit vectors in space

The two types of mean most frequently used for lines in space are the vector mean (VM) and the hemispherical mean (HVM).

These means are maximum-likelihood estimators for the density functions:

$$\text{VM} \; : f(\theta, \phi) = \frac{\kappa}{4\pi \sinh \kappa} \, e^{\kappa \cos\theta} \quad \text{(Fisher distribution)} \qquad [14.7]$$

$$\text{HVM}: f_2(\theta, \phi) = e^{\kappa_2} \left[2\pi \int_{-1}^{1} e^{2\kappa_2 x^2} dx \right]^{-1} e^{\kappa_2 \cos 2\theta} \qquad [14.8]$$

As before, κ and κ_2 are non-negative parameters which control the scatter; θ is the angle between the preferred orientation and an observation. When the mean forms the pole of a sphere, θ is the latitude as measured from the pole, and ϕ is the longitude of an observation.

Watson (1965, 1966) investigated the properties of $f_2(\theta, \phi)$. The exponent can be expressed in terms of $\cos^2 \theta$, since $\cos 2\theta = 2 \cos^2 \theta - 1$. If $\kappa_2 < 0$, a new type of distribution occurs with a density that is least at the poles and with a girdle-shaped maximum around the equator.

A vector mean was originally calculated by Reiche (1938) but the method became used generally after the work by Fisher (1953) and Watson (1966, 1970) by whom its properties were established in great detail.

The vector mean (or Fisher mean) has direction cosines (λ, μ, ν) which are estimated as:

$$\lambda = \Sigma l_i/R; \quad \mu = \Sigma m_i/R; \quad \nu = \Sigma n_i/R \qquad [14.9]$$

with $R = (\Sigma l_i^2 + \Sigma m_i^2 + \Sigma n_i^2)^{\frac{1}{2}}$

In [14.9], the direction cosines of the observations are written as (l_i, m_i, n_i). As before, R represents the magnitude of the resultant vector. It is the direction for which $\Sigma \cos \theta_i$ is a maximum.

The method of calculating VM will be extended in section 14.8 for the purpose of computing best-fitting unit vector fields. For this reason, we present two equivalent methods by which [14.9] is derived.

Method 1. For the i-th observation, we have $\cos \theta_i = \lambda l_i + \mu m_i + \nu n_i$. A necessary condition for (λ, μ, ν) is $\lambda^2 + \mu^2 + \nu^2 = 1$. Hence, $\Sigma \cos \theta_i$ reaches its maximum when:

$$S_1 = \lambda \Sigma l_i + \mu \Sigma m_i + \nu \Sigma n_i - (\lambda^2 + \mu^2 + \nu^2 - 1)\, p/2 \qquad [14.10]$$

is partially differentiated with regard to λ, μ and ν, where p is a Lagrangian multiplier. Equating the three results to zero gives immediately [14.9] with $p = R$.

Method 2. We may write the direction cosines for the vector mean as $(x/R, y/R, z/R)$ where R is the magnitude of the sum vector and x, y, and z its projections onto three mutually perpendicular axes. Then, $\cos \theta_i = (l_i x + m_i y + n_i z)/R$. The sum:

$$S_2 = \frac{\Sigma(l_i x + m_i y + n_i z)}{\sqrt{x^2 + y^2 + z^2}} \qquad [14.11]$$

is maximized by partial differentiation which regard to x, y and z, and by equating the results to zero. It follows that $x = \Sigma\, l_i$, $y = \Sigma\, m_i$, $z = \Sigma\, n_i$. This solution is also equivalent to [14.9].

The second method will be extended in section 14.8. In order to make calculations for axes in space, a Cartesian coordinate system must be defined. For example, Whitten (1966) proceeded as follows (see Fig.106). An *U*-, *V*-, and *W*-axis point southward, eastward and upward, respectively. An observation makes angles α_i, β_i and γ_i with these three axes. Two examples are shown in Fig.106. The cosines of the three angles are l_i, m_i and n_i. Addition for each direction and normalization gives the desired vector mean also shown in Fig.106.

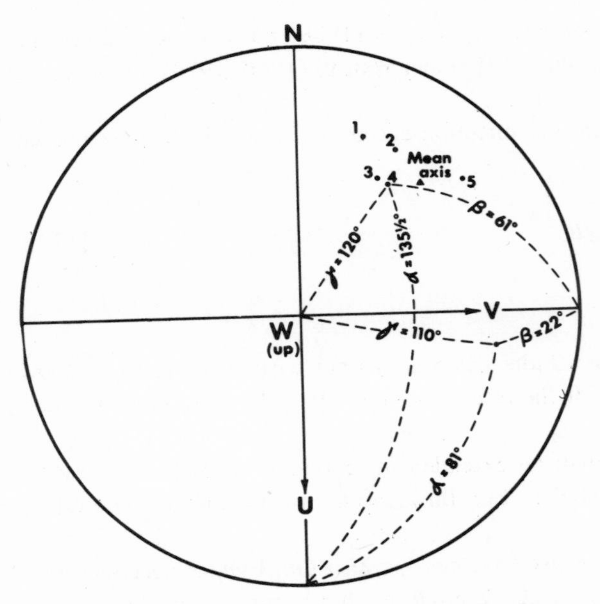

Fig.106. Stereogram (Wulff net) showing vector mean computed from direction cosines of six measured fold axes. (From Whitten, 1966.)

Suppose that instead of $\Sigma \cos \theta_i$, the expression $\Sigma \cos^2 \theta_i$ is maximized. This is equivalent to minimizing the sum of squares of the distances of points on the sphere from the axis. By using the method of component analysis (cf. Chapter 5) it follows that the hemispherical mean is given by the normalized eigenvector corresponding to the largest eigenvalue of a matrix **M** with:

$$\mathbf{M} = \begin{bmatrix} \Sigma l_i^2 & \Sigma l_i m_i & \Sigma l_i n_i \\ \Sigma m_i l_i & \Sigma m_i^2 & \Sigma m_i n_i \\ \Sigma n_i l_i & \Sigma n_i m_i & \Sigma n_i^2 \end{bmatrix} \quad [14.12]$$

Scheidegger (1964) used this method in connection with the analysis of fault-plane olutions of earthquakes and applied it to axes normal to cross-bedding measurements Scheidegger, 1965). Loudon (1964) used it for orientation data in structural geology. 'he eigenvector approach also provides a rapid solution to the problem of finding a est-fitting axis which is perpendicular to a set of normals for measured planes. This ıcludes Sander's (1948) π-axis problem. Whitten (1966) used the method for this purose. Ramsay (1967) has solved the same problem by minimizing a somewhat different um of squares which results in an algebraically simpler result. The eigenvector approach best applied when some data-processing facilities are available although one can directly se the method of section 12.4 on a desk calculator. Watson (1966) and Koch and Link 1971) discussed the method and gave additional references.

Additional properties of the vector mean will be introduced next in relation to a ractical example.

'hirty-five axes of minor folds in quartz phyllites near Monguelfo, Pusteria region

The azimuth and dip were measured for the axes of minor folds (B-lineations) in 35 ifferent outcrops within a small area of approximately 0.15 km^2 near Monguelfo, usteria region, northern Italy. This example was discussed previously by Agterberg 1959, 1961a). The lines were plotted on a stereographic (Wulff) net that has the propery that a circle on the sphere is projected as a circle in the net. The original data and their ector mean are shown in Fig.107A and C. Each circle around the mean in Fig.107C coresponds to a cone with a different (semi-) topangle. The length of the resultant vector mounts to $R = 33.85$.

Fisher (1953) has shown that the parameter κ of [14.7] can be estimated by the pproximate formula $\hat{\kappa} = (N-1)/(N-R)$ which is valid for $\kappa > 3$. For the example, $= 29$. In order to illustrate the meaning of $\hat{\kappa}$, Watson (1966) has constructed a set of urves (see Fig.107B) for topangle of cone as a function of κ such that 50, 75 and 95% of ie observations fall within ψ_0 degrees from the mean direction of the Fisher distribution. . visual comparison of Fig.107B and C suggests that the fit of the Fisher model is xcellent.

The 35 axes were originally contoured by using a unit circle (1% of total net area) to epict density. The contours are shown in Fig.107A, where the number of points per unit ircle is expressed as a percentage of total number of points ($N = 35$). The density ontours are irregular in shape but can be approximated by circles corresponding to cones

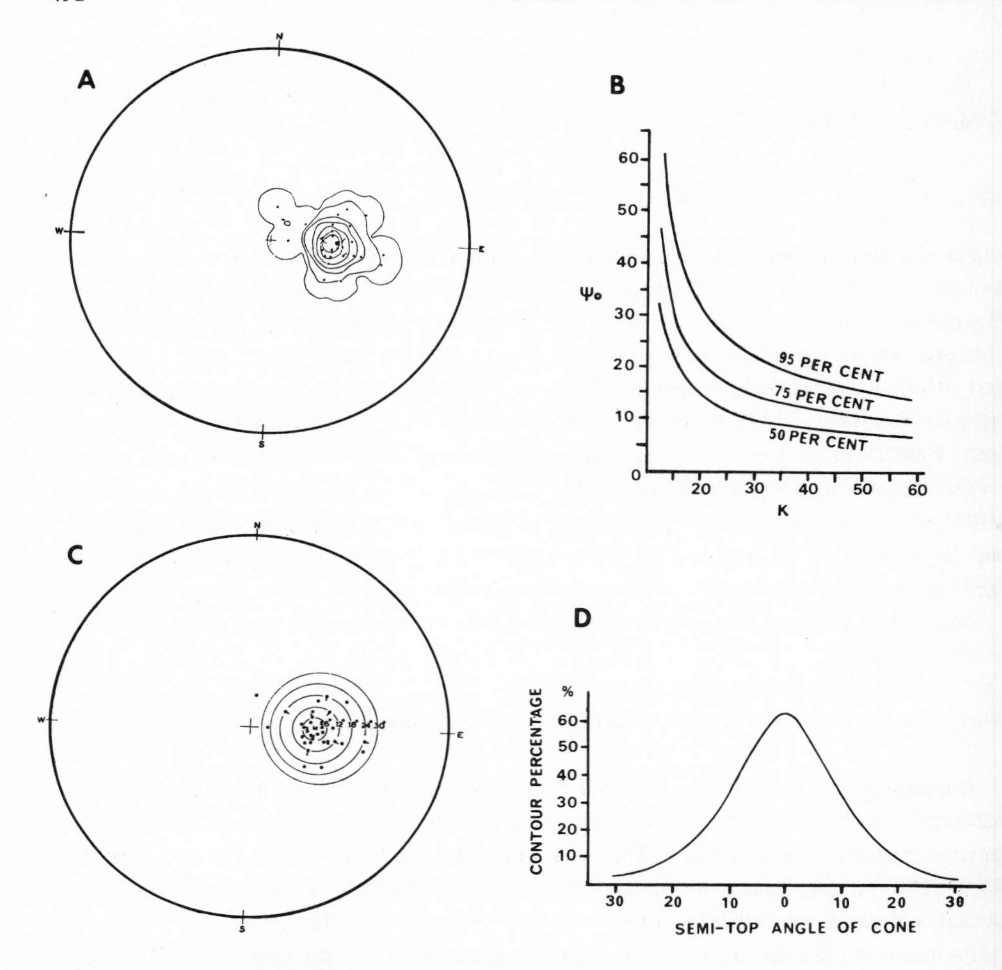

Fig.107. A. Stereogram (Wulff net, lower hemisphere) for 35 axes of minor folds from localities ne Monguelfo; 3-, 9-, 17-, 26-, 34-, 43- and 57-% contours drawn by using 1-% unit circle; 63-% conto indicated by black dot. B. Theoretical curves for angle ψ_o as a function of κ such that 50, 75 and 95 of structures will fall within ψ_o degrees from mean direction. (After Watson, 1966.) C. Same data as A with vector mean M in direction 89–54°; the semi-top angles of cones around M also shown. I Empirical frequency curve obtained by plotting areas bounded by contours in A against semi-t angles of cones with same areas in C.

in 3D-space. The contoured density value was plotted against topangle of these auxiliar cones in Fig.107D. The result is an empirical density profile through the mean axis of th distribution which is assumed to be isotropic. This curve is Gaussian by close approxima tion ($\sigma \approx 12°$).

It was pointed out previously that, for large values of κ, there is no difference betwee a circular normal distribution and a Gaussian distribution. Likewise, if κ is large, th

Fisher and two-dimensional Gaussian distributions are the same.

The so-called cone of confidence for the vector mean can be calculated by using Fisher's (1953) formula:

$$\cos\theta = 1 - \frac{N-R}{R}\,[(1/P)^{1/(N-1)} - 1] \qquad [14.13]$$

where θ = topangle of cone of confidence, N = number of data, R = length of resultant vector, P = probability of mean being located within cone of confidence with top angle θ. For the example $\theta = 4\frac{1}{2}°$ for $P = 0.95$.

The previous method is widely used in paleomagnetism. It is subject to the condition that the Fisher distribution applies. In some situations, the lines tend to be distributed around a plane in space (or along a great circle in the net) which indicates a nonisotropic density distribution. Eq. [14.13] should not then be applied.

The presentation of methods related to the Fisher distribution, as given in this section, was kept simple. The serious student should study the extensive work by Watson (1966) and Stephens (1962) on this subject. The accuracy of the approximate treatment has been examined in great detail by these authors and goodness-of-fit tests were developed.

4.7. Comparison of vector means

Application of the χ^2-test of Table XIV to the frequencies in Fig.107 gave $\hat{\chi}^2$ (1) = .27 with $P(\chi^2 > 0.27) = 0.61$, indicating normality. For this reason, the curve may be regarded as a profile of an isotropic two-dimensional normal distribution as shown in Fig.36A. For uncorrelated standard normal variables Z_1 and Z_2, $Z^2 = (Z_1^2 + Z_2^2)$ is distributed as χ^2 with 2 degrees of freedom.

A point with coordinates (Z_1, Z_2) falls at a distance Z from the origin. If θ is the angle between an observation and the true mean direction (λ, μ, ν) we have for large κ and small θ, $Z^2 = \theta^2/\sigma^2 \approx 2\,\kappa(1 - \cos\theta)$.

This approximation is valid when the Fisher distribution can be approximated by the two-dimensional normal one. Hence, $2\kappa\,(1 - \cos\theta) \approx \chi_2^2$. It also follows that:

$$2\kappa\,(N - X) \approx \chi_{2N}^2;\quad 2\kappa\,(R - X) \approx \chi_2^2;\quad 2\kappa\,(N - R) \approx \chi_{2(N-1)}^2 \qquad [14.14]$$

where X is the projection of an estimated resultant vector for N data onto the true mean direction. These are the approximations on which Watson (1960) has based his method of analysis of variance for vectorial data. It is pointed out that Watson, by more rigorous methods, proved that these approximations are good for a wider range than suggested by the present analogy which is based on comparing the Fisher with the two-dimensional normal distribution for large κ.

When there are n outcrops or sites, and N_i represents the number of measurements pe site (labelled i), the following analysis of variance (Table XLVI) may now be performe as shown by Watson and Irving (1957):

TABLE XLVI
Analysis of variance for unit vectors in space

Variation	*df*	*SS*	*F(2n – 2, 2N – 2n)*
Between sites	$2(n-1)$	$\Sigma R_i - R$	$F = \frac{\Sigma R_i - R}{N - \Sigma R_i} \cdot \frac{(N - n}{(n - 1}$
Within sites	$2(N-n)$	$N - \Sigma R_i$	
Total	$2(N-1)$	$N - R$	

The analogy between Tables XLV and XLVI is obvious. The difference is in the numbe of degrees of freedom (df). According to Watson's approximations ([14.14]), we have fo the vector mean in space that $2\kappa\ (N - R)$ is distributed as χ^2 with $(2N - 2)$ degrees o freedom, instead of $(N - 1)$ degrees of freedom for the vector mean in the plane.

14.8. Geological introduction to unit vector fields

The method to be discussed in the remainder of this chapter may contribute to sever problems of structural geology, particularly in the study of interference patterns pro duced by two or more successive periods of folding in micaschists or phyllites. The practic examples to be discussed are based on measurements collected by the author durin 1957–1960 for a study of the crystalline basement of the Dolomites in northern Ital All data to be used here were published (Agterberg, 1961a).

The data and previous methods of statistical analysis are introduced briefly in th section. A model of Ramsay (1967) is used for interpretation. In the next section, method for fitting unit vector fields is developed and applied to these data.

The basement of the Dolomites consists of quartz phyllites which were intensivel deformed during several phases of Hercynian and Alpine orogenies. Minor folds or micro folds can be observed in most outcrops of these tectonites. The attitudes of the B-axe change between outcrops as exemplified by Fig.107 which is for measurements from domain of 300 m × 500 m. The vector mean changes systematically from place to pla in the area. This can be established by defining domains and calculating a vector mean fo each domain. However, this method is at best approximate because the systematic are change in the vector mean may be rapid. This implies that the domains should be small i order to be homogeneous by approximation. Unfortunately, the number of outcrops i the area is too restricted to calculate vector means for small homogeneous domains.

One method of avoiding the difficulty of defining domains, is to fit curved vecto

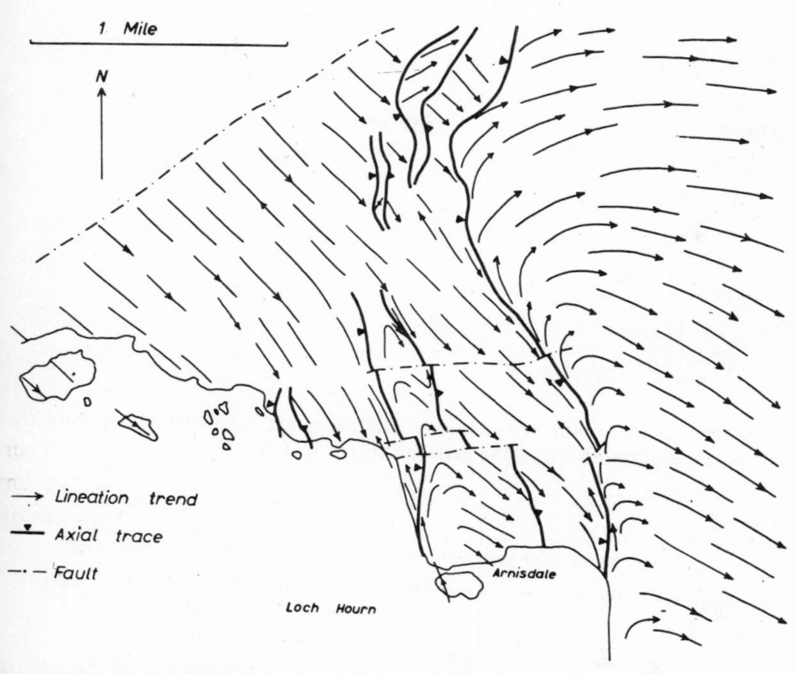

Fig.108. Traces of deformed lineations in gneisses at Arnisdale, western Highlands of Scotland. (From Ramsay, 1967.)

means that change continuously from place to place. This method then replaces that of subdividing the area into a mosaic of smaller areas each of which has a constant vector mean.

The style of folding in the basement of the Dolomites is analogous to that described by Ramsay (1960, 1967) for gneisses of the Moine Series at Arnisdale, western Highlands of Scotland. The latter example will be discussed first because in these gneisses the B-lineations are mappable and local variations do not obscure the regional trends.

Ramsay (1967) has pointed out that if similar folds develop in gneisses with preexisting subvertical foliation and subhorizontal B-lineation, the latter can be deformed leading to extremely complex variation over larger areas. An example is shown in Fig.108 where the pattern is for the azimuth of deformed B-lineations. Axial traces (intersections of axial planes with surface) for the late folds are indicated. A schematic explanation for this style of folding is given in Fig.109. During the late folding, the foliation, which is according to parallel original surfaces r, s, t and u, did not change in attitude. The axial plane of the late folding (ab) makes a small angle with the original surfaces. The a-direction is subvertical and is contained in the original surfaces.

Patterns such as the one shown in Fig.108 can become almost completely obscured when the surfaces, on which the B-lineation was developed were themselves variably

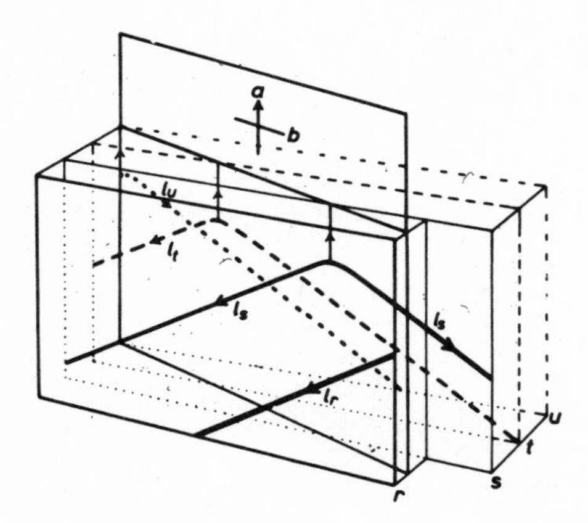

Fig.109. Variation in pattern of deformed lineations on surfaces *r*, *s*, *t*, and *u* resulting from the oblique intersection of the shear plane *ab* with these surfaces. (From Ramsay, 1967).

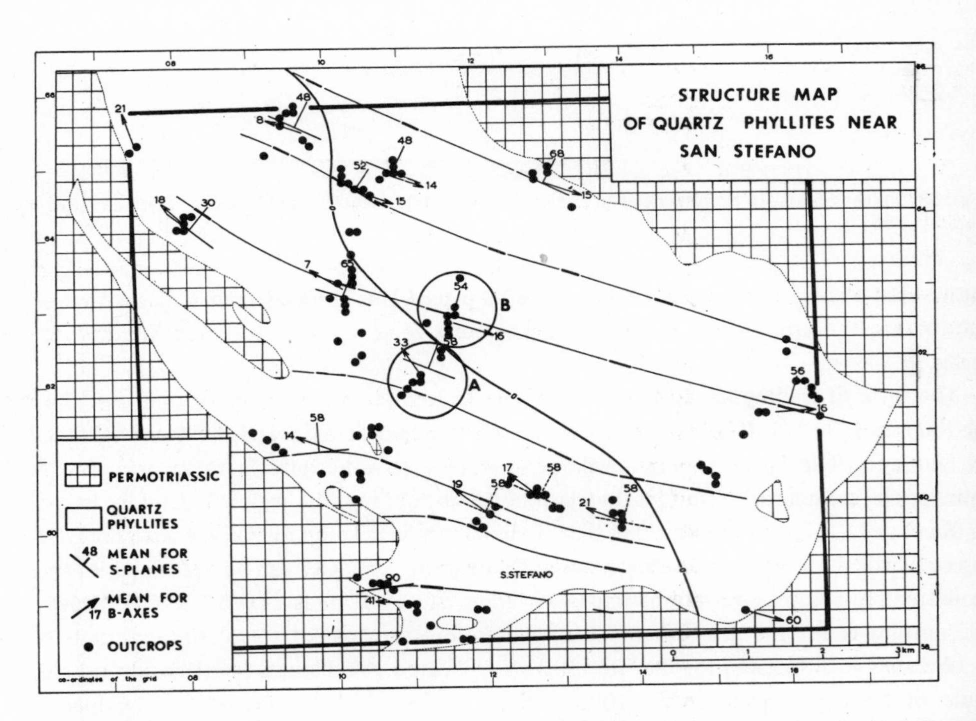

Fig.110. Preferred attitudes of schistosity-planes and B-lineations with extrapolated regional patte[rn] for crystalline basement of eastern Dolomites near San Stefano. (After Agterberg, 1961a). Cells [o] 100 m on a side, where one or more B-lineations were measured, are shown by dots. Domains f[or] Fig.111 indicated by circles (*A* and *B*); outline of boundary of Fig.112.

oriented or when the amount of compressive strain accompanying the original folding was variable in space. In such situations we may never be able to reconstruct a pattern of deformation in an accurate manner. However, by developing the vector mean for the lineation, trends can be established providing a broad outline of the geometry of the folds and the underlying process. It should be kept in mind that once a statistical model has been formulated for a complex situation, the calculations will give a result which is regular. Its validity depends on the validity of the model that is used.

In more complex situations, it becomes increasingly difficult to test the goodness of fit of a model. The methods to be discussed in the next examples are thought to be useful but they should be applied with discretion.

The San Stefano tectonites

The rocks of the Dolomites in northern Italy are of Permo-Triassic age. They are part of the Southern Limestone Alps. The Dolomites are surrounded by an intermittent rim of strongly deformed quartz phyllites. The schistosity planes and minor folds in this crystalline basement are likely to be of pre-Permian age because clear-cut structural disconformities at the base of the Permo-Triassic series can be observed locally and boulders of quartz phyllites similar to basement in situ may be contained in the basal conglomerate (Verrucano) of the Permo-Triassic stratigraphic column.

The San Stefano area (Fig.110) is located to the northeast of the Dolomites. Measurements on attitudes of S-planes and B-lineations were done in most outcrops of this area. Care was taken to maintain a sufficient distance between measurement points to avoid redundancy as much as possible and to exclude cross-folds from the population since these were developed locally.

The B-lineation ranges from microscopic ripples on the S-planes to outcrop-size folds. Several problems of the quantitative mapping of directional minor structures were discussed by Elliott (1965). It seems that in the basement of the Dolomites, the B-lineation can be mapped consistently but the schistosity, as measured in the field, may not represent the same type of structure everywhere (also see D'Amico, 1964). The statistical methods to be discussed here will be restricted to strike and plunge of the minor folds.

In total, 379 B-lineations were measured. Each measurement point was located to the nearest 100 m by using the rectangular grid on Italian topographic maps at scale 1 : 25,000 according to the Universal Tranverse Mercator projection system.

Grid-cells measuring 100 m on a side, where at least one measurement was done, are shown in Fig.110. Large parts of the area are not exposed. The dots for outcrops have a tendency to fall along lines following the course of streams. This limits the extent to which the data are representative for larger areas. On the other hand, exposure along streams is more or less continuous so that structural data for different types of phyllites were recorded.

Previously (see Fig.110), the measurements were arbitrarily grouped into 16 sets with the number of data on B-axes per set ranging from 9 to 52 (about 24 data on the average). For each domain, the data were plotted on the equal-area Schmidt net. An example for two domains is shown in Fig.111. These occur near the center of Fig.110 and fall along a stream (Rio Larice) providing a continuous sequence of outcrops perpendicular to the mean strike direction.

This method was applied in other areas also (e.g., Fig.113) and the following statistical analysis was done. Contouring of the data plotted on the Schmidt net was not always possible because the scatter can be large. An average attitude per domain was calculated by averaging strikes and dips for individual measurements. This rapid method is approximate but the resulting mean will not differ strongly from the vector mean and it tends to coincide with maximum density on the net in situations where contours can be drawn. These approximate averages were plotted on maps (e.g., Fig.110) and various types of lines were fitted, partly by hand, and partly by using an adaptation of Busk's (1929) method of profile construction for similar folds. Approximate patterns were constructed by this method for: (1) mean-strike lines for schistosity; (2) mean-azimuth lines for minor folds (not shown in Fig.110); (3) isolines for mean dips of minor folds (zero-dip line shown in Fig.110).

The Alpine trend in Fig.110 is southeast–northwest as reflected by the contact between tectonites and overlying Permo-Triassic rocks. The mean-strike lines for Hercynian S-planes trend ESE–WNW. The pattern is characterized by the existence of an isoline of zero mean dip trending SE–NW. On average, the B-axes at the two sides of this line plunge in opposite directions. The interpretation was that the San Stefano tectonites constitute the core of an Alpine anticlinal structure with northeasterly dipping axial plane. Before the late (Alpine) deformation, the Hercynian B-lineations in the area were subhorizontal. During a late phase of Alpine orogeny, deformation took place along the

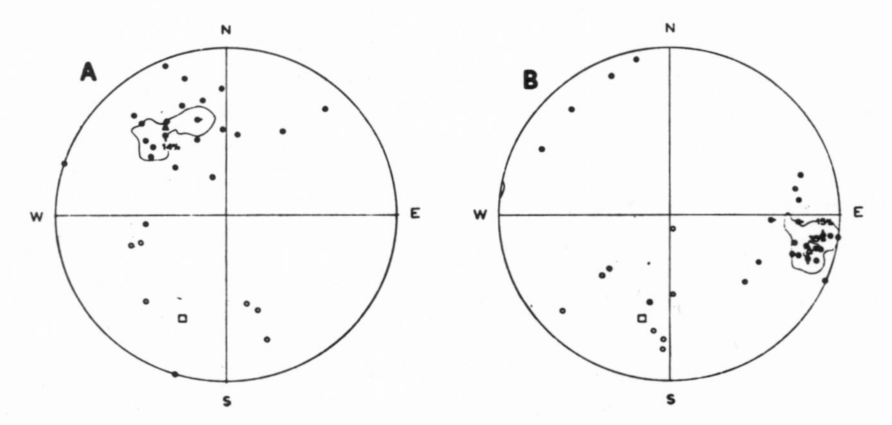

Fig.111. Lineations (dots) and S-poles (circles) plotted in equal-area Schmidt set, lower hemisphere for domains *A* and *B* in central part of Fig.110. Averages plotted in Fig.110 indicated by triangles and open squares, respectively.

existing Hercynian planes of schistosity. There was an angle between the axis of the Alpine anticline and the strike of the Hercynian schistosity. The situation is analogous to Ramsay's diagram (Fig.109). The previous model is supported by the attitudes of Permian strata at their contact with the crystalline core. These strata may be overturned toward the southwest along the southwestern boundary of the tectonites and, in general, dip gently northeast at the northeastern border.

Fig.111 is for domains separated by the axial trace of the Alpine anticline. There are few measurements for S-planes but the calculated averages for the two domains fall at nearly the same point. By approximation this point is also the pole of the plane in which the B-axes tend to fall. If a great circle is constructed through the calculated mean poles for B-axes in the two domains, if follows that the angle between these two lines amounts to 63°.

The preceding statistical analysis depends heavily on the rather arbitrary definition of domains combining outcrops which are thought to have approximately the same vector mean. The calculations for this mean were approximate; an alternative method of statistical analysis is as follows.

14.9 Method of fitting unit vector fields

We can assume that, at any given place, the observed minor fold-axis direction deviates by an angle θ from a curved line which is part of a set of continuous lines constituting a vector field. The angle θ is thought to be a random variable satisfying the Fisher distribution.

As before, observations can be described by their direction cosines (l_i, m_i, n_i) with $l_i^2 + m_i^2 + n_i^2 = 1$. The direction cosines for the unit vector field are $(\lambda_i, \mu_i, \nu_i)$. For convenience, these will be written as:

$$\left.\begin{aligned} x_i(u, v)/R_i(u, v) &= \lambda_i \\ y_i(u, v)/R_i(u, v) &= \mu_i \\ z_i(u, v)/R_i(u, v) &= \nu_i \end{aligned}\right| \qquad [14.15]$$

where $x_i^2(u, v) + y_i^2(u, v) + z_i^2(u, v) = R_i^2(u, v)$. The parameters $x(u, v)$, $y(u, v)$ and $z(u, v)$ change from place to place in the area according to continuous functions of the geographical grid coordinates u and v.

The deviation θ_i of an observation from the unit vector field satisfies:

$$\cos\theta_i = \frac{l_i x_i(u, v) + m_i y_i(u, v) + n_i z_i(u, v)}{R_i(u, v)} \qquad [14.16]$$

In analogy with the second procedure used in section 14.6 to calculate the vector mea[n] we maximize the sum of all (N) $\cos \theta_i$ in the area or:

$$S = \sum_{i=1}^{N} \frac{l_i x_i(u, v) + m_i y_i(u, v) + n_i z_i(u, v)}{R_i(u, v)} \qquad [14.1\text{'}$$

The nature of the continuous functions $x(u, v)$, $y(u, v)$ and $z(u, v)$ must be specified f[or] each situation. We may use polynomial functions. The simplest model then is:

$$\left. \begin{aligned} x(u, v) &= b_{10} + b_{11}u + b_{12}v \\ y(u, v) &= b_{20} + b_{21}u + b_{22}v \\ z(u, v) &= b_{30} + b_{31}u + b_{32}v \end{aligned} \right| \qquad [14.1\text{8}$$

This set of linear equations contains nine unknowns to be estimated by partial differenti[a]tion of S and equating the nine partial derivatives to zero. This gives:

$$\left. \begin{aligned} b_{10}N + b_{11}\Sigma u_i + b_{12}\Sigma v_i &= \Sigma l_i \\ b_{10}\Sigma u_i + b_{11}\Sigma u_i^2 + b_{12}\Sigma u_i v_i &= \Sigma u_i l_i \\ b_{10}\Sigma v_i + b_{11}\Sigma v_i u_i + b_{12}\Sigma v_i^2 &= \Sigma v_i l_i \end{aligned} \right| \qquad [14.19a$$

$$\left. \begin{aligned} b_{20}N + b_{21}\Sigma u_i + b_{22}\Sigma v_i &= \Sigma m_i \\ b_{20}\Sigma u_i + b_{21}\Sigma u_i^2 + b_{22}\Sigma u_i v_i &= \Sigma u_i m_i \\ b_{20}\Sigma v_i + b_{21}\Sigma v_i u_i + b_{22}\Sigma v_i^2 &= \Sigma v_i m_i \end{aligned} \right| \qquad [14.19b$$

$$\left. \begin{aligned} b_{30}N + b_{31}\Sigma u_i + b_{32}\Sigma v_i &= \Sigma n_i \\ b_{30}\Sigma u_i + b_{31}\Sigma u_i^2 + b_{32}\Sigma u_i v_i &= \Sigma u_i n_i \\ b_{30}\Sigma v_i + b_{31}\Sigma v_i u_i + b_{32}\Sigma v_i^2 &= \Sigma v_i n_i \end{aligned} \right| \qquad [14.19c$$

The nine equations of [14.19] are grouped into three sets of three each. Every set can b[e] solved separately to give three coefficients, and the best-fitting vector field is obtained b[y] applying a trend surface analysis to each of the three direction cosines l_i, m_i, n_i. F[or] [14.18], the solution of the direction cosines $\lambda(u, v)$, $\mu(u, v)$, and $\nu(u, v)$ is found b[y] dividing by the normalizing function:

$$R(u, v) = [x^2(u, v) + y^2(u, v) + z^2(u, v)]^{\frac{1}{2}} \qquad [14.20$$

For graphical representation, the aximuth $\delta(u, v)$ and dip $\phi(u, v)$ of the vector field ca[n] be computed from $\lambda(u, v)$, $\mu(u, v)$ and $\nu(u, v)$. For example, if U-, V- and W-axes are defined as in Fig.106, we have:

$$\left.\begin{aligned} \tan \delta(u, v) &= -\mu(u, v)/\lambda(u, v) \\ \sin \phi(u, v) &= -\nu(u, v) \end{aligned}\right| \quad [14.21]$$

Allowance must be made for the octants in which the unit vectors of the field fall. These are determined by the signs of their direction cosines; $\delta(u, v)$ and $\phi(u, v)$ may be plotted in degrees.

When the method is applied, care must be taken to compensate for small departures between grid north and true north. Also, the direction cosines for the observations should be defined in a manner which is consistent for the entire area.

In the previous derivation, linear functions of u and v were used for example. Eq. [14.18] can be replaced by any set of three functions in u and v. Results for quadratic and cubic polynomials as applied to the San Stefano tectonites are shown in Fig.112.

The azimuth of the vector field (δ) was initially calculated as a set of isoazimuth lines. From these, curved mean-azimuth lines were constructed by the same method as used for Fig.105. The isodip lines may be plotted as calculated. The isoline of zero dip in Fig.112 is situated in the central part of the area and, in location, shows good correspondence to the hand-constructed line in Fig.110. Agreement is better for the cubic solution (Fig. 112B).

The example also illustrates that the new patterns should be interpreted with moderation. The calculated functions are flexible. They provide valid approximations only at places where control points occur.

In Fig.112, the quadratic solution (112A) suggests the existence of a single anticline. On the other hand, the cubic (112B) shows an anticline flanked by "synclines" at both sides. These "synclines" in Fig.112B would continue under the Permo-Triassic northeast and southwest of the tectonites. This is unlikely and is caused by a combination of flexibility of the vector field and lack of observation points near the borders of the crystalline rocks. The quadratic solution is more realistic in these parts of the area. On the other hand, the cubic is probably better in the central parts of the area. For example, it indicates that the central anticline is more pronounced than suggested in Fig.112A and this is more in agreement with the relatively rapid variation in average orientation of the B-axes along the Rio Larice (Fig.111).

The San Stefano area contains several isolated blocks of Permo-Triassic rocks some of which have been strongly deformed. These younger rocks tend to lie near the axial planes of the synclines suggested by Fig.112B and a structural relationship is suggested.

Analysis of variance

It is interesting to do a statistical study on the angles of deviation (θ) between the observations and the best-fitting unit vector fields.

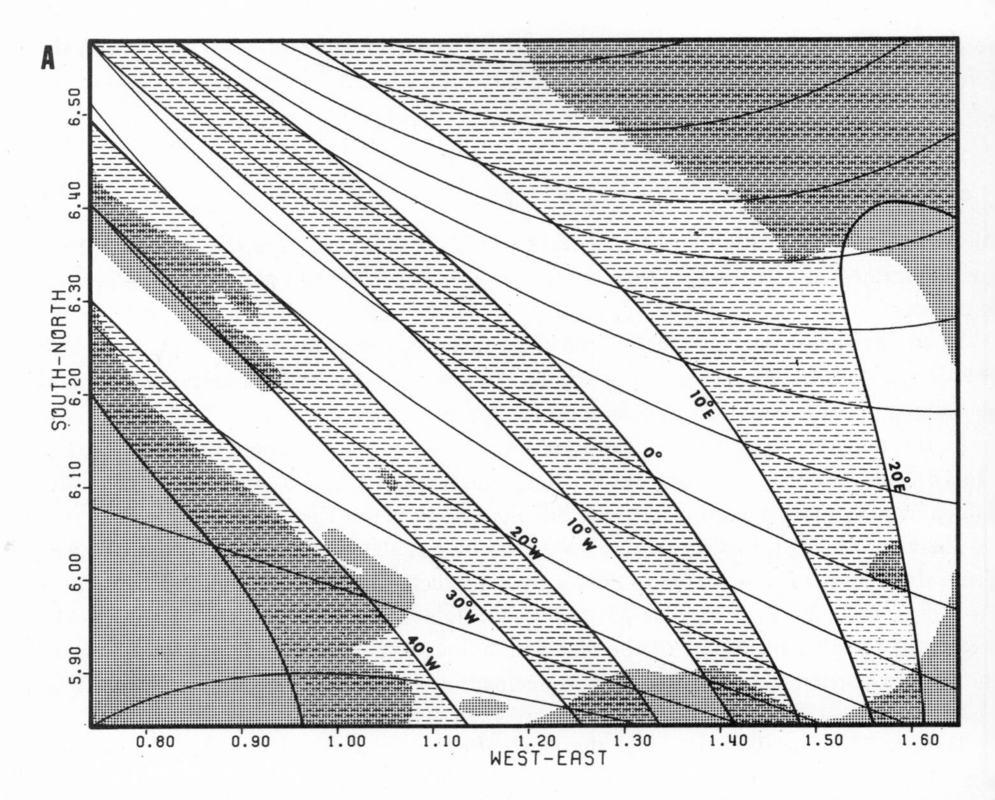

Fig.112. Azimuth (lines) and dip (solid lines marked with amount of dip) of best-fitting unit vector fields based on (A) quadratic and (B) cubic solution, for area outlined in Fig.111. Zero-dip line in B corresponds to similar line drawn by hand in Fig.110. Note that axes of "synclines" on limbs of superimposed anticline tend to coincide with isolated segments of Permo-Triassic (dotted pattern) enclosed by crystalline rocks in B.

The resultant vector of all 379 unit vectors in the area dips 8° in direction 296 (N64°W). The length of this vector amounts to 296.87. From Fisher's formula $\hat{\kappa} = (N - 1)/(N - R)$, it would follow that $\hat{\kappa} = 4.60$. The original data do not satisfy an isotropic Fisher distribution. They tend to fall within the average plane of schistosity as illustrated by Fig.111.

Similar statistics can be obtained for the deviations from the vector fields. They are calculated from the sums of squares due to regression (not corrected for the mean) which are generated when trend surfaces of the same degree are fitted to each of the three direction cosines for the observations. The sum of these three totals is equal to R_p^2/N where R_p is the vector length due to regression. The quantity R_p is equivalent to the sum ΣR_i for addition of lengths of vectors fitted separately to data for sites in Table XLVI

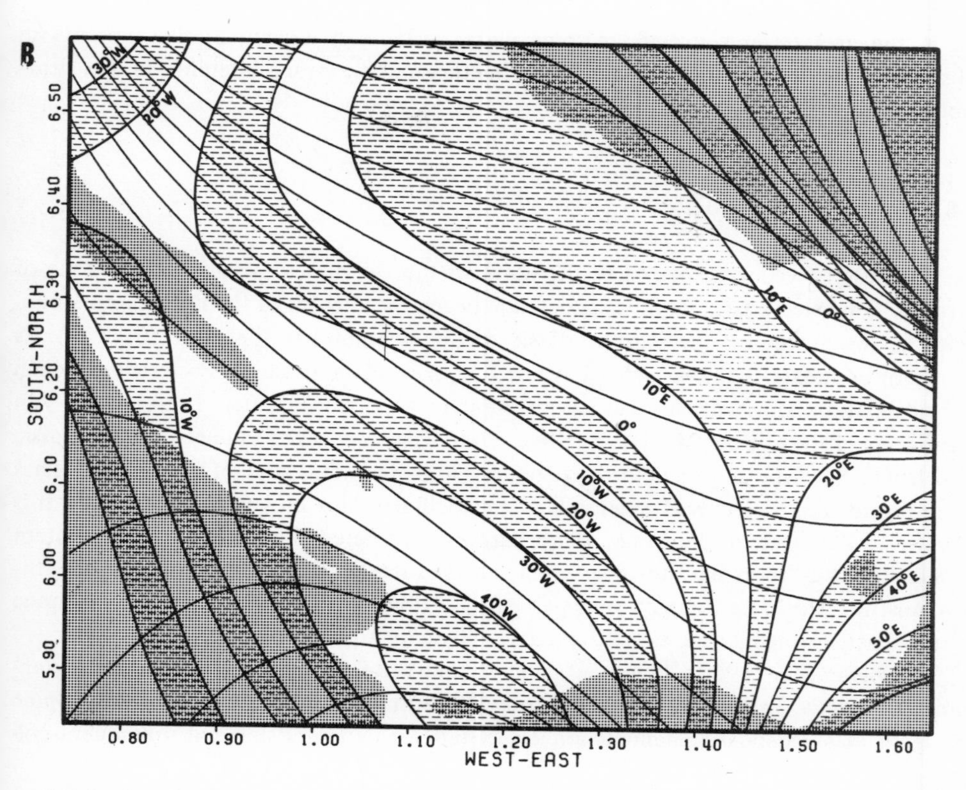

In the San Stefano area, $R_1 = 314.17$; $R_q = 318.99$; $R_c = 327.85$, with the subscripts denoting linear (l), quadratic (q) and cubic (c) fits, respectively. Consequently, the parameter $\hat{\kappa}$ increases gradually with increasing degree, from $\hat{\kappa} = 4.60$ (overall vector mean) to $\hat{\kappa}_1 = 5.83$; $\hat{\kappa}_q = 6.30$; $\hat{\kappa}_c = 7.39$. In principle, an analysis of variance can be done to compare overall fits of vector fields to one another (Table XLVII). The $\hat{F}$-ratio of Table XLVII would indicate that the step from quadratic to cubic fit is statistically significant. However, it is not permitted to assume that all deviations from the best-fitting cubic vector field are uncorrelated, especially since, in the field, the changes in attitude of minor folds are known to be gradational. The erratic way in which outcrops are distribut-

TABLE XLVII
Analysis of variance for comparison of quadratic with cubic solution

Variation	*df*	*SS*	*F (8,738)*
Cubic–quadratic	2 × 4	$R_c - R_q = 8.87$	$\hat{F} = 16.0$
Deviations from cubic	2 × 369	$N - R_c = 51.15$	
Deviations from quadratic	2 × 373	$N - R_q = 60.01$	

ed in the area (strong clusters in some places, wide gaps elsewhere) also presents an obstacle to the application of more formal methods of statistical inference in this situation.

14.10. The Pusteria tectonites

The Pusteria valley is located to the north of the Dolomites. It consists of a zone of strongly deformed quartz phyllites situated between the gneisses of the Defereggen Mountains to the north and the Permo-Triassic of the Dolomites to the south (Fig.113). Its northern boundary is the so-called Pusteria line, part of the Insubric–Pusteria line which is one of the major tectonic features of Alpine structure (De Sitter, 1964, p.163; Van Bemmelen, 1966). Along the Pusteria line, there occur wedges of younger rocks (Triassic and Upper Silurian) and small intrusive bodies (tonalites). This zone of younger rocks has been regarded as a westward extension of the Gailtaler Alps in Austria (Drauzone). It is known as the western Drauzone. Many authors considered the Drauzone (and western Drauzone) as a "rootzone" for overthrust sheets of the Northern Limestone Alps (northern Austria) which are thought to be part of the extensive upper Austrides or East-Alpine thrust sheets formed during a Mid-Cretaceous phase of Alpine orogeny.

Van Bemmelen (1960), however, assumed that the Drauzone, and its western extension in Pusteria, originated as graben-structures during a Mid-Tertiary phase of Alpine orogeny, as a secondary phenomenon associated with the uprising and subsequent collapse of the eastern Central Alps.

It is beyond the scope of our presentation to give an extensive discussion of Alpine orogeny. Although many facts on structures in the Alps are well-established, a general consensus among geologists as to the sequence of events leading to the present configura-

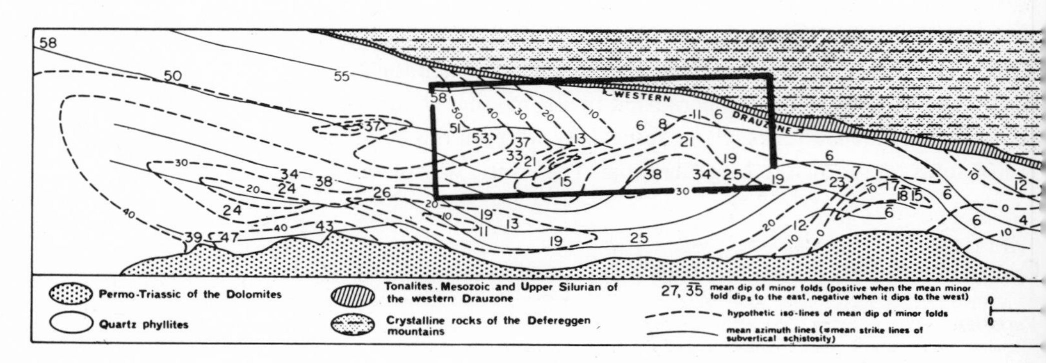

Fig.113. Complex structures of superimposed folding in Pusteria tectonites, northern Italy. (After Agterberg, 1961a.) Pattern of mean azimuth and dip of minor fold axes were constructed by averaging for domains and manual contouring. Frame of Fig.115 is outlined.

tions has never been reached. This is illustrated in Fig.114 compiled by K.A. de Jong and published with discussion by Van Bemmelen (1966). Emphasis in this section is on illustrating mathematical methodology but it is not possible to detach the geomathematical analysis from geological theory.

As a framework for discussion, we will use the interpretation of Fig.114E by Van Bemmelen (1960, 1966) and Fallot (1955). These workers assumed two main phases of Alpine orogeny: (1) a Mid-Cretaceous, unilateral phase with northward directed transportation of material (see scheme at bottom of Fig.114E); and (2) a Mid-Tertiary, bilateral phase. The Insubric–Pusteria line separates Eastern Alps from Southern Limestone Alps in Fig.114.

According to Van Bemmelen (1960), the Insubric–Pusteria fault was active during the Mid-Tertiary phase of Alpine orogeny which commenced with uplift of the Central Alps. It would be one of a series of step faults on the southern flank of the late central Alpine uplift (geanticline) leading to Po Plain and Adriatic Sea.

On the basis of the structural contours shown in Fig.113, Agterberg (1961a) concluded that the Pusteria tectonites were strongly compressed locally during the latest phase of Alpine orogeny with the compression increasing toward the east. It was assumed that the tectonites had been folded twice during Hercynian orogeny, leading at first to a regular pattern of an isoclinal series with subvertical schistosity. Superimposed on this structure was a secondary Hercynian B-lineation dipping steeply eastward in the western

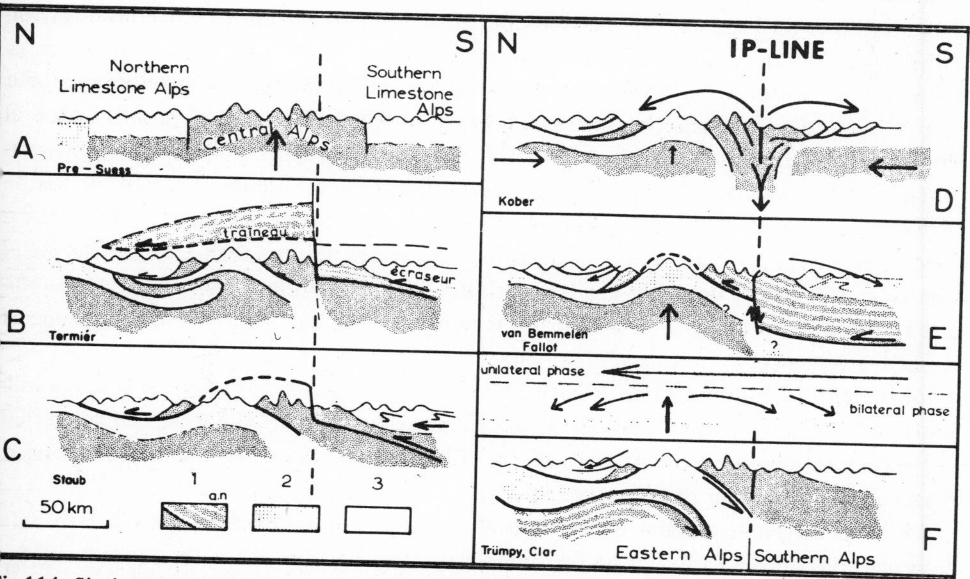

ig.114. Six interpretations of eastern Alpine orogeny as distinguished by Van Bemmelen (1966); for ubsequent discussion here, use is made of E. *1* = basement complex, striped (a.n.), if Alpine overrust sheet is assumed; *2* = parautochthonous units (flysch and Pennides); *3* = Northern and Southern imestone Alps. Insubric–Pusteria line (IP-line) indicated by broken line. (After Van Bemmelen, 966; based on original diagram compiled by K.A. de Jong.)

parts of Pusteria and of subhorizontal attitude in the eastern parts, from where the pattern continues into the Carnian Alps to the east of the area. The San Stefano tectonites (Fig.110) lie to the south of the Carnian Alps.

During Alpine orogeny, several anticlines were formed in Pusteria whose cores coincide with the areas of relatively low eastward mean dip of minor folds in Fig.113. In these zones, the alpine rotation of the *B*-axes within the planes of schistosity was stronger than elsewhere in Pusteria. Leonardi (1967) also assumed the existence of an anticlinal structure in the part of Pusteria which is enclosed by a rectangle in Fig.113. On his structural map of the Dolomites, it is part of the same Alpine anticlinal structure as the San Stefano anticline (Fig.111). According to Leonardi, this structure would envelop the northern and northeastern borders of the Dolomites.

Unit vector field, Pusteria

This statistical analysis was restricted to 461 measurements on minor folds in the part of Pusteria within the rectangle of Fig.113. A large part of this subarea is covered by Quaternary deposits related to the Rienza River flowing from east to west through the region. The 461 measurement points lie to the north of the Rienza River. Initially, average orientations were estimated for domains and the mean dips of minor folds shown in Fig.113 fall at the centers of these domains. This provides a generalized measure of the location of control points in this area. Quadratic and cubic vector fields fitted to the 461 data are shown in Fig.115. The patterns are more or less in agreement with those derived earlier by graphical methods (Fig.113). However, the eastward convergence of the mean azimuth lines seems to be strong in both Fig.115A and B. It suggests the existence of a narrow zone along which the tectonites were squeezed out with a rapid increase in compression toward the east.

The length of the resultant R_p for vector fields of different degree fitted to the ($N = 461$) data is $R = 369$; $R_1 = 405$; $R_q = 413$ and $R_c = 426$. In terms of κ, we have $\hat{\kappa} = 5.02$; $\hat{\kappa}_1 = 8.24$; $\hat{\kappa}_q = 9.61$ and $\hat{\kappa}_c = 13.03$. These values are less than that for the small area near Monguelfo used for Fig.107 with $\hat{\kappa} = 29.4$. This domain of 0.15 km^2 is indicated by the mean dip of 53° within the rectangle of Fig.113. It lies west of the zone of strong convergence.

Although the rapid variation in attitude of the mean *B*-axes in Pusteria was previously explained as a late Alpine deformation increasing eastward but, otherwise, similar to that in the San Stefano area, the patterns of Fig.113 and 115 suggest that the deformations in Pusteria are by an order of magnitude stronger than those near San Stefano.

The Defereggen Mountains of the Central Alps north of the Pusteria line were strongly deformed during the earlier phase of Alpine orogeny. They constitute the "rootzone" for northward directed thrustsheets. The Pusteria tectonites may be related to the "rootzone" north of the Pusteria line. This would imply that they were deformed during the Mid-Cretaceous Unilateral phase of Alpine orogeny. Van Bemmelen (1960) has assumed

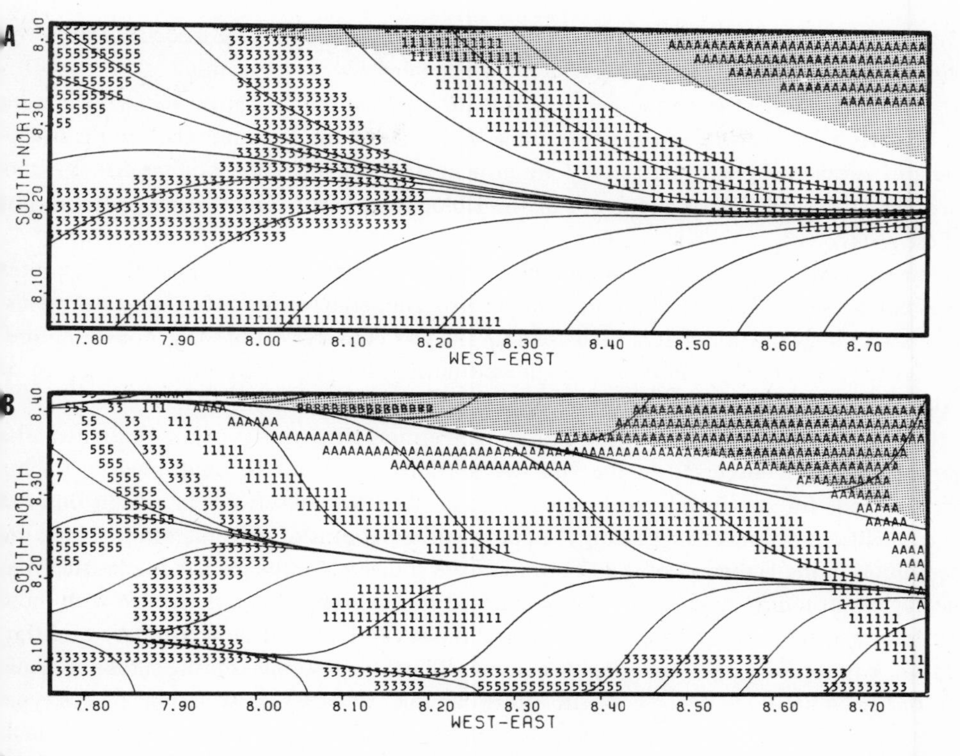

Fig.115. Azimuth (lines) and dip (in symbols) for (A) quadratic and (B) cubic unit vector field fitted to data within rectangle of Fig.113; printed symbols for dip are *1* (10–20°E), *3* (30–40°E), *5* (50–60°E), *A* (0–10°W). Strong eastward convergence of azimuth lines suggests existence of Alpine lineament or zone through which material moved with considerable mobility but with preservation of pre-Alpine *S*-planes and B-lineations.

that, at that time, parts of the Permo-Triassic cover slid northward toward the Northern Limestone Alps where they occur now. Large slices of the underlying crystalline basement participated in this movement as shown schematically in Fig.114E. At deeper levels, the Pennine-type nappes now exposed in the large Tauern window in Austria were injected into parautochthonous, mainly Mesozoic sediments which, in metamorphic form, occur in this window, and, to the west, in association with the Pennine nappes in Switzerland.

The zone along which the upper East Alpine nappes became dislocated from the southern Alps may have been located in the vicinity of the Insubric–Pusteria line. We might assume that the Pusteria tectonites participated in this large-scale formation of thrustsheets with extensive movements toward the north.

Later, during the Mid-Tertiary phase of Alpine orogeny, the "rootzone", including the Pusteria tectonites, was deformed into a large flexure resulting in its present subvertical attitude. The Permo-Triassic layers at the southern boundary of the Pusteria crystalline rocks participated in this flexuring but the deformation decreased rapidly in a southern direction. The rifting and faulting along the Pusteria line also were associated with the late Alpine differential vertical movements which persisted until recently.

An implication of this modified interpretation is that the Hercynian minor folds were subhorizontal everywhere in the crystalline basement of the Dolomites before the beginning of Alpine orogeny; just as they are now in the crystalline cores of the San Stefano area and those northwest and south of the Dolomites except for late Alpine rotations of these rocks.

In summary, the existence of complex structural patterns in the Pusteria tectonites was confirmed by the present statistical analysis consisting of the fitting of vector fields. The strike-lines in the subarea used for this analysis converge more than thought before. Structures of this type may have originated during the earlier Mid-Cretaceous phase of Alpine orogeny when large northward directed thrustsheets were formed.

Except for the structures at the base of the Permo-Triassic series of the Dolomites, the structural trends discussed in the last two sections could not be confirmed by tracing marker beds for stratigraphic control. Individual structures observable in a single outcrop provide little information regarding the complicated patterns of deformation to which the tectonites in the basement of the Dolomites were subjected in the past. It is admitted that geological structures here can not be established with a probability comparable with those in deformed sedimentary rocks where correlations can be based on relative age of different rock types forming a stratigraphic column. B-lineations formed during earlier orogenic deformation are a relatively poor substitute for this. However, in situations of this type, the collection of hundreds or thousands of measurements and subsequent statistical analysis can contribute significantly to the solution of structural problems of geology.

Chapter 15

SPATIAL VARIABILITY OF MULTIVARIATE SYSTEMS

15.1. Introduction

It can be said that in geology there are three important types of relationship amenable to mathematical treatment: (1) spatial variability; (2) variations in the course of time; and (3) interrelationships between many variables. Since observations usually are restricted to a record of past events expressed by features of a different age that occur at different places, variations in time and space often cannot be clearly separated from one another. The dimension of time usually is embedded in three-dimensional space.

In this final chapter, methods will be discussed to analyze data from multivariate systems which are subject to spatial variability across large regions. A specific problem on the estimation of the probability of occurrence of specific types of mineral deposits from geological and geophysical measurements obtained systematically for larger regions is taken to exemplify this approach. By analogy, the procedures can be applied to other types of multivariate systems in geology. Some specific problems arise in applying mathematics to the prediction of unknown mineral occurrences.

The practical importance of doing regional evaluations of mineral potential has been discussed by several authors including Derry (1968), who stated that the number of orebodies which could be readily detected is limited and the rate of discovery of new resources (future mines) is decreasing. Griffiths (1967) and Pretorius (1968) also made this point and argued that geomathematical methods can contribute significantly to problems of exploration strategy and decision-making in the search for hidden deposits.

This field of applied geology is tied to mineral-economic conditions. For example, although copper is rather abundant on earth (the earth's crust contains on average 70 p.p.m. Cu), it was not readily concentrated by geological processes into masses with a grade suitable for mining (Sullivan, 1970). The supply of high-grade copper orebodies is nearly exhausted and larger lower-grade deposits must be mined. In the United States, the average grade for all copper mines has fallen from 2.6% in 1900 to 0.65% in 1967 (cf. Sullivan, 1970).

The factors related to the occurrence of a specific type of mineral deposit can be evaluated in a systematic manner and probabilities for this can be assigned to single factors or combinations of many factors. Original contributions to this field include those by Allais (1957) and Harris (1965). These authors superimposed networks of equal-area cells across large regions and developed probabilistic models for occurrence of mineralization. Both attempted to predict economic measures such as total dollar value of metal mined per cell without distinguishing between different types of mineralization Allais did not consider geological factors. His method of appraisal was discussed in section 7.9. On the other hand, Harris (1965) systematically coded a number of geological factors for his cells, establishing by multivariate statistical methods that total amount of metal per cell, weighted according to price, depends upon geological factors in parts of the western United States. The measures for amount of metal used by these authors are rather general and weighting according to price of metal is an arbitrary procedure from a geological point of view. However, giant mineral deposits are rare events and it is difficult to classify them according to distinct types. By using value, these authors obtained a measure of mineral wealth that was of direct economic significance and the number of data was sufficiently large to be treated by the statistical methods they employed. Brinck (1971) analyzed data on size and grade for mines per commodity (e.g., uranium); his populations were fairly large because deposits from all over the world had been lumped together. Lack of data for the variable that is to be estimated (occurrence of deposits) presents a problem in all statistical work on mineral deposits. One can attempt to enlarge the population of deposits for a given commodity by including information on the numerous small deposits (occurrences) that have been located for most elements. However, most of these are poorly documented, mainly because they were usually not explored with great detail in three dimensions. Also, the small deposits of an element do not necessarily occur in the same places as the giants.

In problems of prediction of occurrence of mineralization, we will mainly use measurements made on the geological framework. In several other methods (Griffiths, 1966b; Griffiths and Drew, 1966; Hazen and Meyer, 1966; and Griffiths and Singer, 1973), more use is made of specific properties of the deposits such as their physical dimensions or changes in grade.

Griffiths and Singer (1973) by using a simulation model due to Engel (1957) did a two-stage search of a region for large prices. A sensor system was used which responded to natural resource targets with signals of two types: true responses to resource targets and false responses. Because the former tended to cluster around targets, it was possible to distinguish between potential targets and barren ground.

15.2. Problems related to complexity of the geological framework

It has been argued that it may not be practical to subject geological data from large regions, enclosing different types of environments, to mathematical procedures because of the complexity of the problems (heterogeneous populations) and difficulties inherent to establishing facts, especially quantitative measurements, in a consistent manner. In the methods to be discussed and applied here, the attempt is made to be sufficiently flexible to cope with part of this complexity. Nevertheless, the facts that will be used are limited in scope and not always well established because extensive use is made of scale-model such as geological maps which are partly based on subjective interpretation.

A few examples may illustrate several of the factors to be considered when geo-mathematical methods are applied to evaluate the association between occurrence of de-posits and geological framework.

Suppose that N deposits have been found in a given area, A, and that quantitative in-formation is available for p different geological attributes X_i (i=1,2,...,p), such as age of formations, abundance of specific rock types, contacts between different rock types, structural parameters, geochemical element-concentration values, and geophysical data. If the deposits are small, it can be assumed that to each feature, there is associated a probability π_i that it indicates the occurrence of one or more mineral deposits. Of course, the problem must be specified in greater detail for practical applications. Also, the features X_i are interrelated and usually the probabilities for combinations of features will differ from those for single features. Hopefully, numerical estimates of the π_i can be obtained and will prove useful for prediction in geologically similar environments within the region where the features have been mapped at the surface but more detailed ex-ploration for deposits was never done.

In some areas (control areas), it may be possible to compute proportions, $\hat{\pi}_i$, which can be used as approximations for the "true" probabilities π_i. They can be used for extra-polation into poorly explored areas (target areas). Instead of estimating the probability that a small area of fixed size (a "cell") contains one deposit, we can estimate other, related, quantities such as the expected number of deposits in a larger cell and the chances that a deposit, if found, has certain properties such as a size between specific limits.

Suppose that out of N observed deposits, a fraction π_1 (say 10%) is associated with attribute X_1; for example, these deposits occur close to east–west trending faults. Although this number can be obtained rapidly by counting, it may be difficult to inter-pret. Ten percent of all locations in the area A could occur in the vicinity of east–west trending faults. In that situation, the proportion, although it is a realistic number, has no value for prediction. A more meaningful interpretation can only be arrived at when ex-pressions such as "in the vicinity of" are specified numerically. We also must obtain in-formation on the relative abundance of the feature X_1 in the area; otherwise, the pro-portion cannot be evaluated statistically. In fact, we need at least two probabilities for evaluating a single feature, one for places where the feature is present and one for where it absent. Several types of feature are well documented only in places where explora-tion was intense. Because these are the places where deposits have been found, these features cannot be used for extrapolation outside say the mining camp.

Another potential difficulty is illustrated by the following example. Suppose that the area A consists of two subareas, A_1 and A_2, both of which contain the feature X_1. Sup-pose further that A_1 comprises 10% of A and that deposits in A_1 always are associated with X_1 and vice versa. In A_2 (90% of A), none are associated with X_1. A proportion π_1 = 10% for area A then represents the weighted averaged of two more meaningful probabilities, amounting to 100% for A_1 and 0% for A_2, respectively. Hence, if a new east–west fault is discovered in A_1, a new deposit also is indicated, but the chances are nil when the new fault is located in subarea A_2.

It is possible in the previous example that the subareas A_1 and A_2 can be distinguished on the basis of another geological feature X_j ($j \neq 1$), e.g., the age could be different. This would compel us to use two features instead of one to compute probabilities of occurrence, leading to use of multivariate methods.

Erroneous correlations can result in the following situation. Suppose that many features X_i were quantified systematically for a region that contains a mineralized belt with many deposits whose occurrence is not associated with any of the geological features. In general, the X_i are not equally abundant everywhere in a map area. Of many features some may happen to be relatively abundant in the mineralized belt. This can result in a strong, but purely fortuitous, positive correlation between occurrence of deposits and features. Using this relationship for prediction in other areas could lead to invalid results. This problem is analogous to a familiar problem of time-series analysis arising when two time series are correlated to one another in a situation where both variables are subject to time-trends. In that field, the problem may be solved by doing a partial correlation after elimination of the time-trends or by working with coherences of the cross-spectrum.

Regional trends such as mineralized belts may be related to other larger-scale features such as paleoclimates, metamorphic gradients, geosynclines, orogenic belts or continental drift. Even when these features cannot be measured in the study area, we can by partial correlation, to some extent, separate their effect from the relationship with the features that can be measured.

Some of the problems listed above can be overcome by using more flexible methods of least squares. One problem, however, is considered as too severe for this. We will recognize that the variable to be predicted (occurrence of deposits) is biased due to incompleteness of information, and consider methods to eliminate this bias. However, if the features X_i are not known everywhere in the total areas of study consisting of both control and target areas, a bias may arise which cannot be eliminated. Suppose that information on deposits and certain features is restricted to those places which are well explored (e.g., mining camps). When a correlation analysis is done for the entire area strong but meaningless positive associations must arise because the variables correlated with one another will both be underestimated in the places that were not fully explored. In general, this is the main reason that the features which can be used are limited in scope. It is, however, possible to have gaps in the area of study consisting of subareas for which information is not available. The previous discussion dealt with cells from all parts of a study area. A different method of approach consists of restricting the data collection to those places where mineral deposits have been found. The result of this can be a classification of orebodies into groups as illustrated by Collyer and Merriam (in preparation). These workers used cluster analysis to classify tungsten mines in North America in terms of a number of geological features coexisting with the mineralization. In studies of this ype, more so than in regional studies with cells, use can be made of the fact that existing lata on mineral deposits are relatively abundant although dispersed and of varying quality.

5.3. Methods of multivariate analysis

Several important multivariate methods were listed in Table I. For the study of geolog- cal environments, considerable use has been made of factor analysis (both Q-mode and R-mode), cluster analysis and discriminant analysis. Factor analysis was discussed in Chapter 6. General references to books on multivariate analysis are given in the list of elected reading at the end of the book. Geological applications of multivariate methods ave been reviewed by Parks (1972), Gower (1970) and McCammon (1969). For a useful ntroduction with listings of computer programs for several basic techniques, see Cooley nd Lohnes (1962). Methods of geological classification by discriminant analysis were xplained by Klovan and Billings (1967). A multivariate method not often applied in eology is canonical correlation. For an interesting model based on this technique, see ee (1969).

In both R- and Q-mode factor analyses, the data points are assumed to form a single luster (in p- or n-space, respectively) whose shape is investigated by fitting axes to it. he approach may become difficult when the data occur in several clusters instead of one. n that situation, cluster analysis may become a more appropriate tool for data analysis. his technique is extensively used by numerical taxonomists (Sokal and Sneath, 1963). he elements of a matrix $\mathbf{R}$ or $\mathbf{Q}$ can be ordered and represented in a two-dimensional iagram or dendrogram. New methods of cluster analysis were developed recently by witzer (1970) arguing that we cannot visualize p dimensions but can look at large umbers of differently oriented projections of the points in one or two dimensions. lthough some groupings may not survive the process of projection, the method may be seful for pattern recognition.

If the original data are arranged in the form of a $(n \times p)$ matrix $\mathbf{X}$, it can be said that ne previous methods operate from either the $(p \times p)$ matrix $\mathbf{X}'\mathbf{X}$ or the $(n \times n)$ matrix $\mathbf{X}\mathbf{X}'$, not considering the transformations of the coordinate systems which are applied. nother approach consists of operating from the $(n \times n)$ matrix $\mathbf{D} = \mathbf{X}(\mathbf{X}'\mathbf{X})^{-1}\mathbf{X}'$. The eader will recognize that this matrix was used previously to obtain calculated values $\hat{Y}$ n multiple-regression analysis. It is also used for computing a discriminant function to eparate between two populations in discriminant analysis. A column of n ones may be dded to the data matrix $\mathbf{X}$ for a dummy variable in order to correct all variables for heir means. If the dummy variable is not used, the coordinate system in sample space is quivalent to that in Imbrie's (1963) vector analysis where the n data points are repre- ented as vectors all beginning at the origin.

The significance of the matrix $\mathbf{D}$ will be illustrated by means of the following ex- mple.

Maps for one or more columns of the matrix **D** *for the Abitibi area*

The Abitibi area in the Superior province of the Canadian shield has been used fo several previous examples (sections 8.7 and 11.6). This region was selected in 1967 fo various studies in the Geomathematics Program of the Geological Survey of Canada o the systematic quantification of geological variables for larger regions and to correlat the occurrence of mineral deposits to digitized versions of the geological framework Most practical examples in this chapter are for the occurrence of copper and gold de posits in this region.

For the present example, a grid of equal-area 10 × 10 km cells was projected o various geological maps for this area at scales 1 inch to 4 miles (approx. 1:250,000) o larger. This grid corresponds to the Universal Transverse Mercator map projectio method used for Canadian topographic maps at scales 1:250,000 and larger. Ever 10 × 10 km cell was subdivided into four hundred $\frac{1}{2} \times \frac{1}{2}$ km cells and for each subcell th predominant rock type was measured from the geological maps. A subcell measurin 500 m on a side is reduced to about 2 × 2 mm on a 4-mile geological map and contact between map units are approximately straight line segments if they occur in a subcell. A the reconnaissance scale of mapping (1:250,000), the procedure therefore is equivalen to point-counting as used in petrographic modal analysis.

Geological maps are scale-models and contain a considerable amount of interpreta tion. The surface of the earth is shown as a mosaic with fields of different colors repre senting rock types or formations of different age. In many cases, the scale-model provide a consistent representation of reality. For example, an intrusive stock with subvertica boundaries and fairly homogeneous composition is easily made part of the mosaic. I other instances, contacts may have complicated shapes, especially if the topographi surface is subject to relief; or layers of a different composition may be interbedded in sedimentary volcanic pile. Structures (folds and faults) cause additional complications Representation in the mosaic at a smaller scale then must involve considerable generaliza tion. In fact, results for subcells as measured from the geological map individually may b meaningless. Their average for larger cells, however, becomes increasingly more meaning ful.

Larger regions may contain different geological environments and usually were mappe by many different geologists. A rock type called quartz-monzonite on one map shee may be called granite on an adjacent map (cf. section 1.1). For consistency, one may hav to be satisfied with a lowest common denominator for classification of the features in larger region. In the Abitibi area, all mapped Archean rocks could be classified accordin to the scheme shown previously in Table XXII. Rock type 4 in this list could be sub divided further on the basis of metamorphic grade giving sedimentary rocks and meta morphic rocks. The greater metamorphism in the latter is marked by the abundance o quartz-biotite schists which are missing in the lower grade metamorphic "sedimentar rocks".

The area used for example here consists of 834 cells of 10 km on a side. It contains the areas outlined in Fig.79. Proterozoic and Paleozoic rocks also occur in this area but the statistical analysis will be restricted to Archean rocks. In addition, some parts of the area are covered by a relatively thick layer of overburden; other parts were not mapped in the detail required for our analysis. An arbitrary cut-off value was used to compute percentage values of Archean rocks in the cells. If more than 100 subcells of 500 m on a side could be classified according to one of the seven rock types, a 10 × 10 km cell was used for analysis. Of the 834 cells, 644 met this arbitrary condition.

At this point, we recapitulate the seven lithological variables, adding three new variables to be explained later. The ten variables measured for the 644 cells were: (1) granitic rocks (acid intrusives and gneisses); (2) mafic intrusions; (3) ultramafics; (4) acidic volcanics; (5) mafic volcanics; (6) Archean sedimentary rocks; (7) metamorphic rocks of sedimentary origin; (8) approximate combined length of layered iron formations; (9) average Bouguer anomaly; and (10) aeromagnetic anomaly at cell center after removal of effect of earth's total magnetic field.

The iron formations are layered and of the Algoma type (Gross, 1965). They are useful for separating between Archean sedimentary rocks on a regional basis. These hard rocks are reasonably well exposed and have high local aeromagnetic anomalies providing indirect evidence of their occurrence in places with overburden.

The two geophysical variables (9 and 10) provide indirect information on rocks underlying those exposed at the surface. A relatively high Bouguer anomaly in this area is related to the occurrence of a relatively thick pile of mafic volcanic rocks because this type of rock has a specific gravity which is larger than that of the other rock types (Gibb et al., 1969; Moorhouse, 1965). It is noted that ultramafics would have a higher density but in the study area those rocks usually occur in altered form as low-density serpentinites.

Aeromagnetic anomalies are related to the magnetic properties of bedrock to some depth but the penetration is less than that of the gravimetric methods. A high regional anomaly over volcanic rocks, for example, may reflect that these rocks are rich in magnetite (Roscoe, 1965).

A geological environment is not only characterized by the individual attributes which are present but the different types of coexistences between attributes provide important additional information. For example, the coexistence of Archean sedimentary rocks and iron formations in a cell may define another facies of sediments. Coexistence of acidic volcanics and a relatively high Bouguer anomaly may indicate an ancient volcanic center where rhyolites and tuffs cap a rather thick pile of andesites and basalts. Contacts between rock types in a cell are another form of coexistence, e.g., a number of mineral deposits occur at contacts between mafic and acidic volcanics. In order to consider combinations of features, we have formed 45 additional variables for all possible pairs of the 10 measurement variables. These new variables are cross-products of the 10 variables and will be used in multivariate models with equations that are linear in the variables.

The Archean rocks are approximately three billion years old and have been subjecte to intensive deformation and intrusion of plutonic stocks. Faults and folds are abundan in the area (cf. Fig.79B) but these structures are known in specific places only and wer not coded for this analysis. Mapping of bedrock is hampered by glacial debris but tens c thousands of shallow holes were drilled in the region, through the overburden, amplifyin outcrop information on rock types. Diabase dikes also are abundant (Fig.79B) but thes are Proterozoic in age and were not considered during our analysis.

The data matrix **X** for the Abitibi area consists of 644 rows and 56 columns. The latte number was obtained by adding the 45 combination variables to the 10 measuremen variables listed previously and using a dummy variable with value equal to one in all cell Because the cells are small (10 km on a side), a number of variables for rock types ha zero value in any one cell if these rock types are absent.

We can now imagine forming the (644×644) matrix $\mathbf{D} = \mathbf{X}(\mathbf{X}'\mathbf{X})^{-1}\mathbf{X}'$. This matrix ha more than 400,000 elements and these are not all computed in practice. Instead of thi we select individual columns or linear combinations of columns from **D**. It is noted tha **D** is symmetric. At most, we could consider 52 out of 56 variables to construct **D** becaus otherwise, the part $\mathbf{X}'\mathbf{X}$ cannot be inverted. This subject will be discussed further i section 15.8. Every row or column of **D** corresponds to one of the 644 cells and can b represented as a map for that cell. An example is shown in Fig.116 for cell (20,4) ne Timmins. This cell is unique in the area in that it contains the Kidd Creek deposit, sulphide mine rich in copper, zinc, lead and silver. There exist other large sulphide bodi in the area. Those containing at least 1000 tons of copper or zinc are represented i Fig.116. The 644 elements of the selected column of **D** were grouped into three classe Large values (black) are for cells whose geological composition in terms of the variables closest to that of the reference cell. Framed cells are somewhat similar (more so than o average) and blank cells are more different. A more detailed interpretation of "similarity in this context will be given below. It is not surprising that the reference cell is surrounde by a cluster of similar cells. This holds true for most other reference cells since geologic environments usually have more or less continuous patterns, e.g., if mafic intrusions occu in one cell, they are likely to occur also in the surrounding cells, or, if a cell is underlai by granites and gneisses, it may be part of a larger belt consisting of these rock type

This method is useful in the search for other orebodies because one might first chec the other places in a region where the geological setting is similar to that where a lar deposit has been found.

Technically, this measure of similarity can be explained as a special application multiple regression analysis. Fig.116 is a map of the column vector **DY** where **Y** is column vector with 643 zero elements and a single one for the reference cell.

Another way of interpreting what was done is in terms of discriminant analysis. W may define two multivariate populations, one for the 643 other cells and one from whic the reference cell was drawn. These two populations are assumed to have an identic variance–covariance matrix but different means. The cluster of 644 points for cells

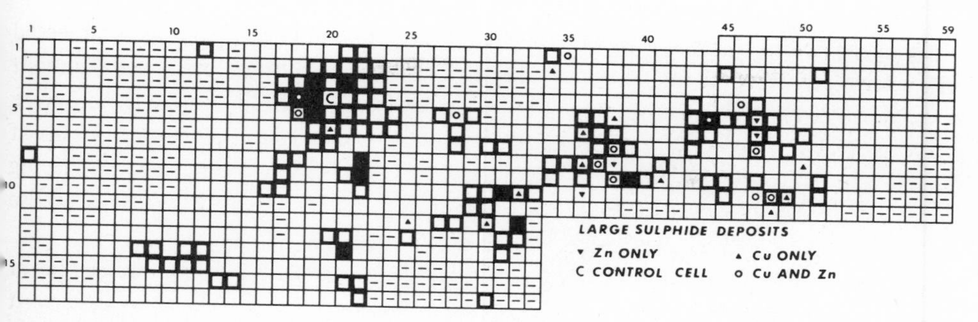

Fig.116. Abitibi area, Canadian shield (see Fig.79 and 119 for location); 643 cells measuring 10 km on a side were compared with single control cell (*C*) containing a large (Cu, Zn, Pb, Ag) sulphide deposit (Kidd Creek Mine); 12 black cells have geological setting similar to that of *C*; 123 other cells with black frame have similarity index greater than average. Systematic comparisons of this type are useful in problems of exploration strategy.

p-space was divided by a plane providing the best separation between the two populations. Next, the distances from the plane are computed for all 644 points including that for the reference cell. These distances are represented by the elements of the vector **DY**.

Discriminant analysis was developed for situations in which individual multivariate observations belong to two or more different populations. In fact, the method is commonly used to assign objects to one out of several distinct classes. In the present application, we are dealing with an illdefined continuum comprising geological environments of different types. These environments pass into one another either gradationally or abruptly. The formal model underlying discriminant analysis cannot be applied under these circumstances but the geometrical projection methods in *p*-space are analogous. Each point in *p*-space has a specific location with regard to the cluster for all points. Points reflecting similar cells will occur close to the point of reference.

Instead of comparing one cell with all others, we can select a group of cells which have certain characteristics in common. For example, there are 27 cells in the Abitibi area with at least one sulphide deposit known to contain more than 1000 tons of copper. These 27 cells are shown in Fig.116 and also in Fig.119. A vector **Y** can be defined with 27 elements equal to one. Instead of ones, we can assign weights to the cells where a significant amount of copper is present. For this, we have used the logarithm (base 10) of total tonnage copper which ranges from about 3 to 6 in the Abitibi region (see Fig.119). Selection of a group of cells does not imply that there will not be other cells with similar characteristics in the region. In fact, the occurrence of a copper deposit is, in the first place, a random event and empty cells adjacent to one with significant copper may be more similar than other cells with copper, occurring elsewhere in the region.

For simplified interpretation, the distances computed in this experiment for 10×10 km cells were averaged for square-shaped blocks of sixteen cells simultaneously. These results form the approximately continuous pattern shown in Fig.117A.

It may be assumed that not all large copper deposits have been found in the region and

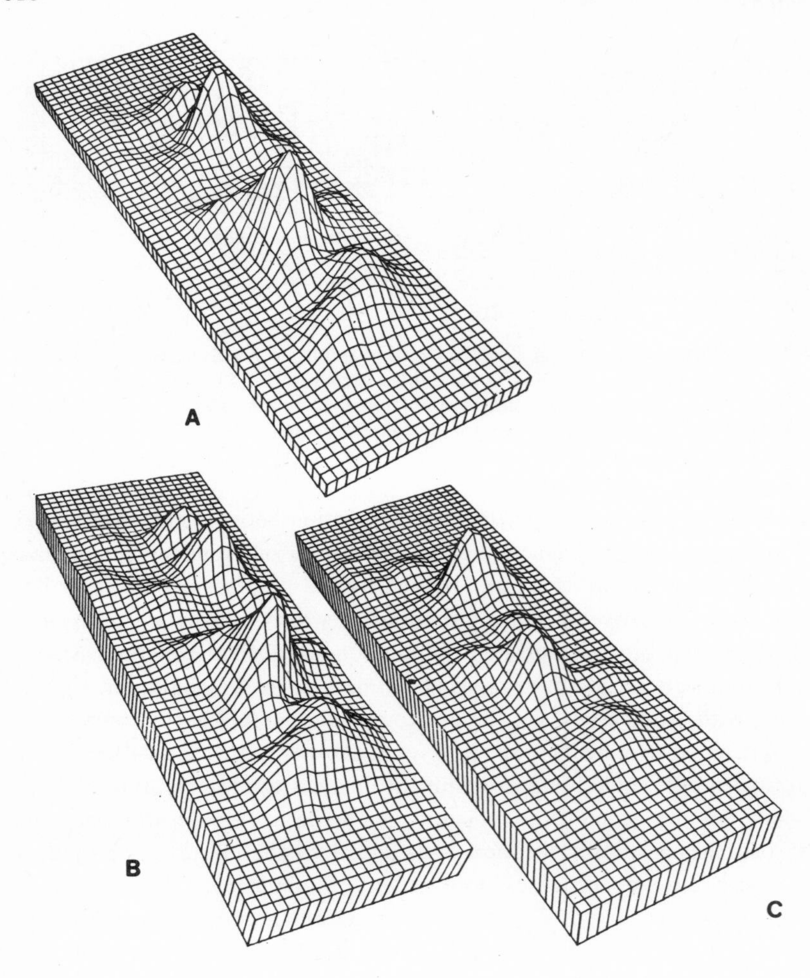

Fig.117. Abitibi area (boundaries as outer frame in Fig.119); occurrence of copper mines (for 27 control cells) per 40 X 40 km area (= 16 smaller 10 X 10 km cells) expressed in terms of geological an geophysical measurements for regional framework of Archean rocks (A); a group of 27 control cells was divided into two subgroups: 19 east of 600 km E line in Fig.119 and 8 west of it. Patterns for the entire area as calculated for these subgroups are shown in B and C, respectively. Note that B resembles A but C underestimates A in the eastern part of the region.

a pattern of this type can be rescaled with regard to a control area (see later). The pattern then would represent both known and unknown deposits in the region. The potency of methods of this type can be checked by comparing predicted patterns based on a larger set of reference or "control" cells with patterns based on subsets of cells The simplest test consists of constructing a pattern based on one control cell only (Fig.116). We also separated the 19 control cells east of the north–south directed line at

600 km E in Fig.119 from the total set of 27 cells used for Fig.117A. These 19 cells also provide a pattern for the entire region (see Fig.117B) but the western part of it is a prediction that can be checked against the locations of the (27–19=)8 control cells not used for analysis. The pattern of Fig.117B closely resembles that of Fig.117A. A similar prediction based on the eight westernmost control cells in the region is shown in Fig.117C. This pattern underestimates the occurrence of large copper deposits in the eastern part of the region but the pattern of Fig.117A is indicated in a general manner.

15.4. Discriminant analysis

Suppose that two p-dimensional populations have different means but the same variance–covariance matrix. In p-space, the situation can be represented as two clusters of points which may in part overlap. Schematically, this is represented in Fig.118. In discriminant analysis, all observations must be classified beforehand and assigned to one of the two populations. The so-called discriminant function (DF) is a plane which divides the space into two parts, one for each population.

This plane should be perpendicular to the line connecting the two population means. It cuts a slice from each population and a decision rule for classification is arrived at by stating that these two slices should have equal size. One admits the possibility of misclassification but the probability of this happening is kept the same for the two populations.

The mathematical solution of this problem results in a plane with equation:

$$DF = \sum_i \sum_j \alpha^{ij}(\bar{X}_{1j} - \bar{X}_{2j})x_i = \text{constant} \quad [15.1]$$

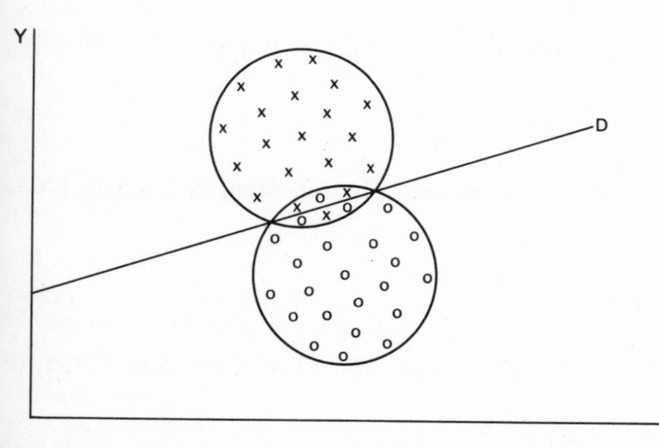

Fig.118. Schematic representation of discriminant analysis; if the points are projected on a line perpendicular to D (discriminant function), the amount of overlap is a minimum.

where α^{ij} denotes the elements of the inverse of the variance–covariance matrix which the two populations have in common; $\overline{X}_{1j}$ and $\overline{X}_{2j}$ are the means of the j-th variable in the two populations; and x_i denotes the i-th variable with coordinate axis in p-space.

For the derivation of [15.1], originally due to Fisher (1936), the reader is referred to such textbooks as Wilks (1962), Anderson (1958) and Rao (1952). It is noted that population means have been replaced by sample means in [15.1] for practical estimation.

It is convenient to set the overall mean for the two populations equal to zero for all variables so that $DF = 0$ represents a plane that passes through the origin. A value of DF can be computed for every observation to represent its distance from the plane $DF = 0$ in p-space (cf. [4.12]). Positive values are assigned to one population and negative values to the other. Because DF is used to separate between two populations and for measuring relative distances rather than absolute distances, it can be multiplied by any convenient constant (Kendall, 1965).

A point of some consideration is the estimation of the variance–covariance matrix **A** which, after inversion, gives the coefficients α^{ij}. Sums of squares and cross-products for deviations from the means can be determined for both populations, and the resulting matrices may be summed. In order to obtain an unbiased estimate of the pooled variance–covariance matrix, we would have to divide by $(n-2)$ instead of n (total number of observations). In practice, this distinction is immaterial since all coefficients of DF may be rescaled later by using an arbitrary constant.

Suppose that we use a multiple regression as discussed in the previous section but that, instead of using a dummy variable, all independent variables and the dependent variables are corrected for their overall means (weighted means for two populations). We recall that this procedure leads to the same end result as using a dummy variable.

The mean of the dependent variable then is $\overline{Y} = n_1/n$ where n_1 represents the number of ones (for population 1) in the vector **Y**. Consequently, the ones in **Y** are replaced by $(1-n_1/n=)n_2/n$ and the zeros by $-n_1/n$.

In terms of the means for two populations, the overall mean of a variable x_j satisfies:

$$\overline{X}_j = \frac{n_1\overline{X}_{1j} + n_2\overline{X}_{2j}}{n}$$

The problem now has become identical to a problem solved in detail by Kendall (1965) who obtained for the regression coefficients:

$$\hat{\beta}_i = (n_1 n_2/n^2)(1-c) \sum \alpha^{ij}(\overline{X}_{1j}-\overline{X}_{2j}) \qquad [15.2]$$

where c is a constant satisfying $c = \Sigma\,\hat{\beta}_i(\overline{X}_{1j}-\overline{X}_{2j})$. The calculated values $\hat{Y}$ satisfy:

$$(\hat{Y}-\overline{Y}) = \sum \hat{\beta}_i(X_i-\overline{X}_i) \qquad [15.3]$$

Except for a constant factor, [15.3] is equivalent to [15.1]. We have completed the proof that performing a regression analysis using a dependent variable with values equal to one and zero is, from a computational point of view, the same as doing a discriminant analysis to separate between two populations. Kendall used this result to evaluate a discriminant function for statistical significance. He pointed out that conventional analysis of variance for multiple regression can be used to test for homogeneity; that is, whether it is meaningful to separate two populations (1 and 2) instead of having one population for all observations. The only requirement for this test is that the independent variables are multivariate normal. For more refined testing, e.g., to assess individual coefficients of the discriminant function for statistical significance, the approach generally breaks down because of the dichotomous nature of the dependent variable.

Extension to k discriminant functions

When the objective is to separate between k populations, the space may be partitioned into k parts, one for each of the clusters around the different population means. The method resembles component analysis. At first, a single discriminant function DF is computed to separate between the k populations by means of a set of $(k-1)$ parallel planes perpendicular to a straight line in p-space on which the k population means would fall.

Multiple discriminant functions were used by Harris (1965) to distinguish between the geological compositions of cells containing variable amounts of base and precious metals concentrated in orebodies. As mentioned above, Harris using geological map systematically measured a number of features including stratigraphic age of formation, length of contacts, age and types of igneous intrusions, and structural parameters. His cells were 20 miles on a side (approximately 32 × 32 km); 243 were situated in Arizona and New Mexico (control area) and 144 in Utah (target area). As discussed in section 6.8, the measurements for each variable were transformed so that the transformed variables approximately satisfy a multivariate normal distribution model. Populations were defined on the basis of classes of total dollar value per cell. For example, the highest value class covered the range of 100 million to 5 billion dollars. Of 243 cells in the control area, 11 were known to belong to this class. By computing multiple discriminant functions and assigning cells to the populations on this basis, Harris found that 8 of these 11 were correctly identified by the model, whereas 3 were assigned to lower value classes. Two other cells, not known to belong there, also were assigned to the highest value class. It is noted that for this identification, Harris made use of a Bayesian decision rule (cf. Cooley and Lohnes, 1962).

The previous analysis was restricted to the control area. Next, Harris used data from the control area for extrapolation into the target area (144 cells in Utah). For this purpose, the probabilities for all membership in each of six groups were aggregated to two groups, separating two multivariate populations only; those cells with value greater than and those less than \$1 million, respectively. An arbitrary decision rule was applied to the results. Harris assumed that any cell in Utah, computed to have a probability greater than 20% for having a value greater than \$1 million, should be retained for more detailed exploration. This separated 19 from the set of 144 cells. In total, 17 cells in Utah were known to belong to this class. Ten of these 17 were correctly indicated by the model and 7 were missed. This left 9 cells indicated as belonging to the greater value class but where lesser value was actually found. Later, Harris (1969) used the (243+144=)387 cells of the earlier study as a control area for assessing the mineral potential of a target area of 493 cells in Alaska. The values expected by using the geostatistical model are shown in a map and in a table (Harris, 1969, fig.110 and table 11). 126 cells are predicted to be more valuable than \$1 million. On the other hand, 26 cells are known to be that valuable. The model found six of these. This result is not as satisfactory as the previous one. It illustrates that the potency of geostatistical methods may decrease significantly when the target area is far removed from the control area. Nevertheless, it can be fruitful to evaluate mineral potential for an area by using control areas in other parts of the world, especially when it can be speculated that the area was never thoroughly explored for types of deposits known to exist in the control areas.

The method of selecting a control area to extrapolate from there to a target area that occurs elsewhere was used in similar studies by Kelly and Sheriff (1969), Sinclair and Woodsworth (1970) and DeGeoffroy and Wignall (1971). These workers used multiple regression instead of discriminant analysis. It was pointed out previously in this section that these two methods are to some extent similar in this type of problem because of the dichotomous nature of the dependent variable used in multiple regression.

15.5. Calculation of a probability index

A measure that expresses quantitatively the likelihood that one or more deposits occur within a specific cell will be called a probability index. For example, by multivariate assessment of the geological features mapped for a cell, we may be able to say that, on the basis of these measured features, it is expected that the cell contains 4 deposits of a given type. The number 4 then is the probability index. Usually, it defines a sequence of probabilities. For example, if the deposits can be considered as points which are randomly distributed in a larger region, the index is the expected value for a Poisson distribution with individual terms $e^{-\lambda}\lambda^x/x!$ for the probabilities that the cell contains exactly x deposits. If $\lambda = 4$, the probability that $x = 4$ is about 20% and there is, for example, a 2-% probability that no deposits will occur. Likewise, if $\lambda = 0.2$, $P(X{=}0){=}0.82$, $P(X{=}1){=}0.16$, and $P(X{=}2){=}0.02$.

On a local scale, it may be difficult to define exactly what is meant by a mineral deposit. It seems logical to take mines or developed prospects (deposits which are not mined but have been delineated by drilling) as the objects of investigation. These objects usually have several well defined parameters such as a total size in tons of various metals mined and average grade values. However, a mine is partly defined by economic factors. It may consist of several separate orebodies exploited simultaneously, or several mines with different owners may fall in the same mineralized zone. In order to construct Fig.117A, we used 27 control cells, each of which contains at least one sulphide deposit with more than 1000 tons of copper. The object of study, therefore, was a 10 × 10 km cell with significant high-grade mineralization rather than a mine. In fact, the 27 cells contain a total of 41 copper mines or prospects: 19 have only 1, 6 have 2, 1 has 3, and 1 has 7 large copper deposits.

When the object of study is a cell, the probability index becomes the expected value of a binomial distribution for values greater than one; and values less than one can be directly interpreted as a Bernoulli variable. For example, if $P(X{=}1){=}0.2$, $P(X{=}0){=}0.8$ and $P(X{=}2){=}0.0$, which is close to the previous results based on the Poisson model with $\lambda = 0.2$.

In section 8.2, we discussed a method for relating the occurrence of mineralization (Y) to that of a mappable geological feature (X). The two required probabilities $P(Y|x{=}1)$ and $P(Y|x{=}0)$ could be obtained by linear regression analysis. Suppose now that not all deposits have been discovered in the region. We then can replace the matrix **A** of section 8.2 by that of Table XLVIII. The factor f represents the bias in P_y that occurs if not all deposits have been found in the area. We require that P_x is not affected by a bias of this type but that it is known in all parts of the area. This distinction between the properties of X and Y is possible because the feature X is measured at the surface of the earth during reconnaissance-type mapping, whereas information on Y can only be obtained by more detailed exploration with drilling. For considering the bias in Y, we can regard Table XLVIII as a "sample" drawn from the "population" with $f = 1$.

The conditional probability $P(Y_p|x)$ for the population (subscript p) is underestimated by the factor f if $P(Y_s|x)$ for the sample (subscript s) is used, because:

$$P(Y_p|x) = \hat{\beta}_{0p} + \hat{\beta}_{1p}x$$

and for the sample:

$$P(Y_s|x) = f\hat{\beta}_{0p} + f\hat{\beta}_{1p}x$$

TABLE XLVIII

2 × 2 contingency table with missing data for random variable Y

	$x = 0$	$x = 1$	
$y = 1$	fa	fb	fP_y
$y = 0$	$c + (1 - f)\,a$	$d + (1 - f)\,b$	$1 - fP_y$
	$1 - P_x$	P_x	1

Consequently:

$$\hat{\beta}_{0s} = f\hat{\beta}_{0p}\ ; \qquad \hat{\beta}_{1s} = f\hat{\beta}_{1p} \qquad [15.4]$$

Eq.[15.4] remains valid if X becomes a continuous random variable. The bias can be removed only if f can be estimated by comparison with a "control area" where P_y is supposedly unbiased. Because deposits are relatively small objects which may be hidden at greater depths, we usually cannot claim that all deposits were found in the control area. However, bias can be removed with regard to a control area which then functions as a norm to which other, lesser explored areas are compared.

Practical estimation of f is illustrated by the following example. The 27 control cells for copper in the Abitibi area are shown in Fig.119A. Three of the 55 geological variables used for this area are represented in simplified form in Fig.119**B**.

For illustration, rock types were reduced to binary form (presence–absence data). The same was done for the two geophysical variables, for which positive deviations from their overall mean (for 644 cells) were set equal to one and negative deviations equal to zero. The resulting presence–absence patterns are shown for acidic volcanics and Bouguer anomaly in Fig.119**B**. The meaning of the combination variable (acidic volcanics × Bouguer anomaly) can be visualized from this example. This variable is one if both measurement variables are present (since $1 \times 1 = 1$); it is zero in all cells where one or both measurement variables are absent (since $0 \times 1 = 1 \times 0 = 0 \times 0 = 0$). By simplifying the problem, it can now be treated by methods of Boolean algebra. In fact, a combination variable corresponds to the intersection of two sets defined for presence of the features in a cell.

If all 55 variables are reduced to this simple form, cells with acidic volcanics present and Bouguer anomaly above average are most strongly correlated with copper mineralization as represented in Fig.119A. The measure of strength of association in this experiment was the simple product-moment correlation coefficient.

Suppose that we disregard the weight assigned to control cells in Fig.119A and compare the copper mineralization pattern with that for the 120 black cells in Fig.119**B**. The two patterns have coinciding black cells in 18 places. If $f = 1$, it would mean that the probability that a black cell in Fig.119B has significant copper mineralization amounts to 18/120 or 15.0%. For other cells, the probability would be 9/524 = 1.7%. Table XLVIII becomes:

9/644	18/644	27/644
515/644	102/644	617/644
524/644	120/644	1

The two areas of 25 cells near Timmins, Ont. and Noranda, Que., are relatively well explored and can be used as a control area. Their combined area of 50 cells has 32 black

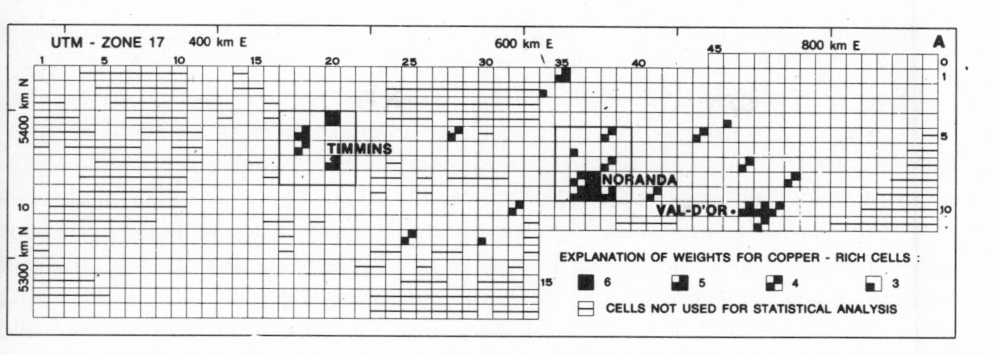

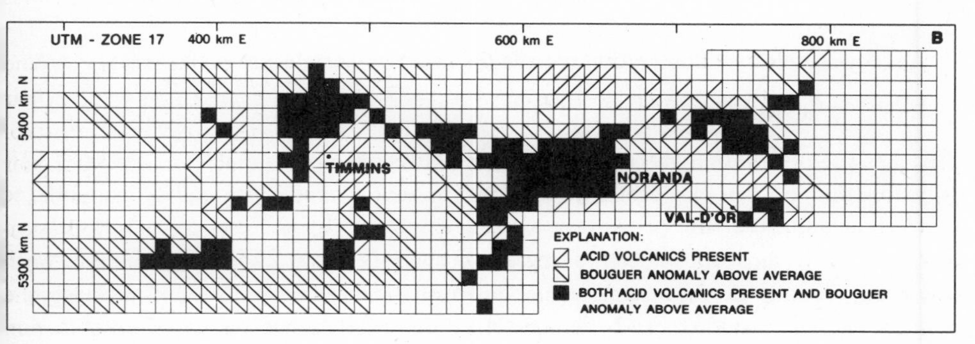

Fig.119. Abitibi area, Canadian shield; cells based on Universal Transverse Mercator grid, and measure 10 km on a side. A. Weight of 27 control cells is rounded value of logarithm (base 10) of total tonnage of copper per cell; control areas of 25 cells around Timmins and Noranda to be used for scaling. B. Of 55 binary variables, the one indicating both acid volcanics present and Bouguer anomaly above average is most strongly correlated to copper weights in A.

cells in Fig.119B and 18 other cells. If the biased probabilities are used for computing the expected number of copper-rich cells in the control area, we obtain ($32 \times 0.150 + 18 \times 0.017 =$)5.1. In reality, 12 copper-rich cells are present in the control area (4 near Timmins and 8 near Noranda). This results in $f = 5.1/12 = 1/2.4$. The probabilities for the population are derived by dividing by f which is equivalent to multiplication by the factor 2.4. The corrected probabilities become $2.4 \times 15.0 = 35.2\%$ for black cells and $2.4 \times 1.7 = 4.1\%$ for other cells.

The probability index for individual 10×10 km cells assumes one of these two values only. For larger areas, individual probabilities can be added, giving an expected value. For the control area, this value (=12) amounts to number of discovered copper-rich cells. In the entire region, there are 120 black cells and the expectation would amount to ($120 \times 0.352 + 524 \times 0.041 =$)63. Since 27 copper-rich cells are known to exist, this would leave ($63 - 27 =$)36 cells to be found. A prediction of this type should be further evaluated. For example, the probability of 4% for the 524 other cells is probably too high because many of these are underlain by granites and gneisses mainly. These rock types are not known to contain large copper deposits in the region, suggesting that more geological variables should be considered in the model.

Before extending the approach to the multivariate case, two further remarks are made on the basis of this example. We could have used a simpler model consisting of an immediate extrapolation from the control area. Because 11 and 12 control cells coincide with 11 of 32 black cells, this would give a probability of 11/32 = 34% for the black cells. This result is close to that obtained previously by the more elaborate method. The corresponding probability for other cells is 1/18 or 6%. The two procedures would give identical answers only when P_x in the control area is exactly equal to P_x in the entire region. We will see later that, in general, the more elaborate procedure is to be preferred in a multivariate situation if the control area does not provide a representative sample of the entire area.

A second additional observation regards the ordinary correlation coefficient:

$$r_{yx} = (b-P_xP_y)[(P_x-P_x^2)(P_y f^{-1}-P_y^2)]^{-\frac{1}{2}} \,. \qquad [15.5]$$

This measure of the strength of the association between the two variables of Table XLVIII is a function of the bias factor f. For the present example, its square amounts to 0.067 indicating that the percentage explained sum of squares due to regression is as low as 6.7%. In most applications of regression analysis, this would be regarded as a weak linear association. However, a visual comparison between Fig.119A and B suggests a connection. This paradox can be explained by the difference in nature of the two variables related to one another. Y expresses presence or absence of deposits which are discrete points on the small-scale map. Suppose that the spacing of the grid is decreased. The effect would be that P_y decreases rapidly because it is the ratio of number of copper-rich cells to total number of cells. In total, there will not be more than 41 copper-rich cells because this is the total number of large copper deposits in the area. On the other hand the total number of cells can be made large. Contrary to P_y, P_x will remain more or less constant, because it is related to features which are more widespread in the area. The total effect of decreasing cell size on [15.5] is that the numerator $(b-P_xP_y)$ approaches zero more rapidly than $(P_y f^{-1}-P_y^2)^{\frac{1}{2}}$ in the denominator. Clearly, the correlation coefficient depends strongly on the size of the cells which are used. Its absolute value, therefore, has relatively little meaning. Use of larger cells would give higher correlation coefficients but at the same time, the resolving power of the method would be reduced.

15.6. Generalization to multivariate situations

Suppose that to each independent variable X_i, there is associated a probability that a mineral deposit will occur. A simple example is that each color on the geological map representing a specific rock type, has a different probability π_i of containing a deposit per unit of area. Then, for a specific cell labelled k, we have:

$$\pi_k = \sum_{i=1}^{p} \pi_i X_{ik} \qquad [15.6]$$

When there are more cells than variables, the π_i can be estimated by multivariate least squares. A slight modification of this method consists of estimating:

$$n = \mu(n)\pi_k \quad \text{and} \quad n_i = \mu(n)\pi_i$$

where $\mu(n)$ is average number of deposits per cell with the effect of differences in X_i removed.

In practice, complications arise when the variables X_i are closely interrelated. The π_i of [15.6] are directly estimated by regression coefficients $\hat{\beta}_i$ if these have a small variance. This is rarely the case when p is greater than 1 or 2.

The effect of bias can be exemplified as follows. Suppose that the region is poorly exposed and that the last independent variable X_p denotes percentage of cell covered by unconsolidated overburden. If the probability π_p is zero, [15.6] becomes $\pi = \Sigma_{i=1}^{p-1} \pi_i X_i$. If π_t represents the probability that the bedrock contains a deposit, we would have, ideally, $\pi_t = (1/f)\pi = (1/f)\Sigma\pi_i X_i$ where $f < 1$ is a bias factor which is exactly equal to percentage of bedrock exposed in the cell, or $f = 1 - X_p$. However, this simple relationship is valid only if all rock types are equally well or poorly exposed. When the softer rocks are underrepresented in the exposed portion of bedrock, a more complex model must be used.

Another example is as follows. Suppose that, in a larger region, observational estimates of π_k are available for many cells. These estimates will depend on the so-called intensity of exploration I_k representing the total amount of information with regard to presence or absence of deposits. In general, I_k is a composite of such factors as observations made in available outcrops, results of geochemical and geophysical surveys, and quantity of drillhole data. In addition to this, the subjective interpretations made of these data by exploration geologists must be weighted strongly. It may be difficult or impossible to quantify the variable I_k in a precise manner. In fact, a cell would be fully explored only if composition of bedrock is known to a depth of say $2\frac{1}{2}$ km. The variable I_k also depends on geological factors. For example, if the bedrock consists of a thick pile of plateau basalts, this could be established with near certainty by limited observations at the surface (e.g., a geophysical survey) and perhaps one drillhole.

The intensity of exploration changes from place to place in a region. If u and v represent geographical coordinates and $I(u,v)$ ranges from 0 to 1, we can write:

$$\pi(u,v) = I(u,v)\Sigma\pi_i X_i \qquad [15.7]$$

Another possibility is that the probabilities π_i depend on where the variables X_i are measured in a region. A given rock type or contact is likely to contain more deposits if it occurs in a mineralized belt whose presence in the region could be related to large-scale features such as a zone of weakness in the crust along which ore-bearing solutions could ascend more easily. In that situation, we can write:

$$\pi(u,v) = \Sigma\pi_i(u,v)X_i \ . \qquad [15.8]$$

In general, [15.8] can not be solved by least squares unless some additional assumptions are made. A simplification that is suggested here is $\pi_i(u,v) = \pi_i F(u,v)$, where $F(u,v)$ is a continuous function depending on geographical location only. It would account for the situation that all features have higher probabilities in parts of the region (e.g., on the mineralized belts) provided that this areal change in the probabilities can be described by a continuous function (e.g., a two-dimensional polynomial). Eq.[15.8] then reduces to the form:

$$\pi(u,v) = F(u,v)\Sigma\pi_i X_i \ . \qquad [15.9]$$

We note that [15.9] is of the same type as [15.7] if $I(u,v)$ can be approximated by a continuous function. This illustrates that it may not be possible in practical applications to distinguish clearly between regional variations in intensity of mineralization and intensity of exploration.

Unfortunately, [15.9] is nonlinear in the coefficients unless $F(u,v)$ is fully known. Suppose that $F(u,v)$ is a polynomial consisting of q functions of the geographical coordinates.

An approximate model:

$$\pi(u,v) = F(u,v) + \Sigma\pi_i X_i \qquad [15.10]$$

consists of the first $(p+q+1)$ terms of the Taylor expansion of [15.9] and results in a linear system of $(p+q+1)$ normal equations if the method of least squares is used for solution. The properties of systems of equations related to [15.10] will be studied in more detail in this chapter.

First, the extension of [15.4] to multivariate situations is considered. In general, each independent variable X_i will have its own, specific bias factor f_i. After some manipulation we can obtain that multiple regression applied to the "sample" then gives the coefficients

$$\hat{\boldsymbol{\beta}}_s = (\mathbf{X'X})^{-1}\mathbf{X'Y}_s = (\mathbf{X'X})^{-1}\mathbf{FEX'Y}_p \qquad [15.11]$$

where $\mathbf{Y}_p$ consists of unbiased data; $\mathbf{F}$ is a column vector with elements f_i, and $\mathbf{E}$ is the p-dimensional identity matrix. $\hat{\boldsymbol{\beta}}_s$ has p elements, one for each variable X_i. This situation arises when the constant term of regression is found to be zero as it should be for some systems including the one of [15.6]. When a constant term is admitted, [15.11] applies if all variables are corrected for their means.

Except in the special situation that all bias factors f_i are equal to one another (=f), the coefficients $\hat{\beta}_{is}$ will be biased by different amounts and not proportional to the unbiased estimates for the population $\hat{\beta}_{ip}$. Also, the calculated values for the sample, $\hat{Y}_s$ will be biased by an amount that differs from cell to cell in the region. It seems that, in multivariate situations, there is no simple way of correcting for bias as we had in [15.4] with a single independent variable. Consequently, we must be careful if a single bias factor f is computed as before by comparing calculated values $\hat{Y}_s$ with known values Y in a control area. We can correct the estimated probabilities for variables and cells by multiplication by $(1/f)$ only if the control area is representative for the entire region with respect to all variables X_i. In some applications, it is useful to replace all variables X_i by new variables whose expected value is constant everywhere in the region. This ensures that any part of the region (e.g., a control area) is representative with regard to the average values for the variables X_i. If the changes in X_i are gradational across the region (e.g., for a metamorphic gradient or a regional change in density of intrusions of a given type), the original data for the X_i can be replaced by residuals from deterministic functions (e.g., polynomial trends) expressing these regional changes.

Gold occurrences in Abitibi

Gold-bearing quartz veins are abundant in the Abitibi area and the region has been thoroughly explored for gold in the past. The areal distribution of 1257 deposits is shown

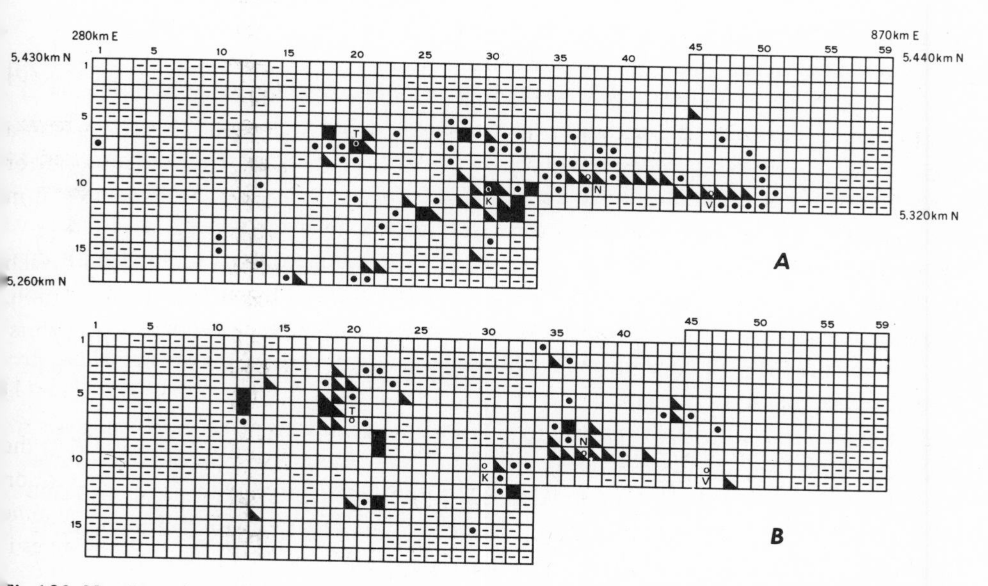

Fig.120. Number of gold occurrences per U.T.M.-cell; symbols in order of intensity indicate presence of 5–9, 10–19 and ⩾ 20 occurrences, respectively. B. Ditto, as predicted from 21 lithological variables; largest calculated values arbitrarily grouped into three classes. Note that favorable lithology in B tends to lie north of zone with most occurrences in A. T = Timmins; K = Kirkland Lake; N = Noranda; and V = Val d'Or.

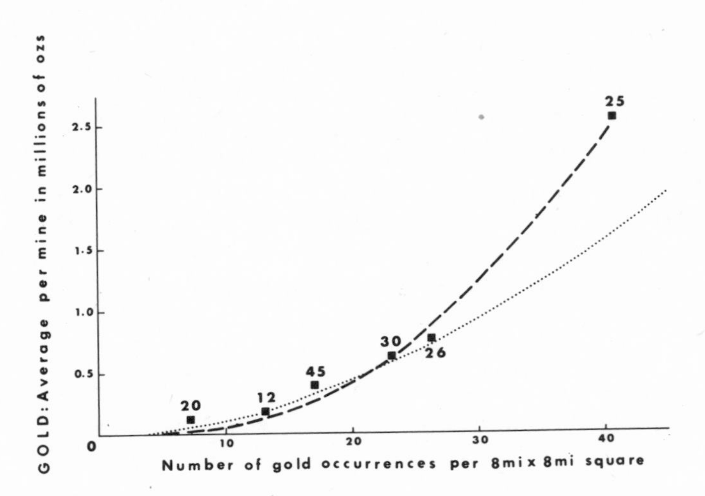

Fig.121. Average size of gold deposits containing more than 1000 ounces of gold plotted against total number of (both large and small) gold occurrences per 8 × 8 mile cell; numbers of mines, on which averages are based, are also shown. Broken line is least-squares curve fitted to data as shown, but some cells were relocated to correct for geological discontinuities. If this correction is not made, the least-squares curve is according to the dotted line which corresponds to the regular grid with 8 × 8 mile cells shown in Fig.80.

schematically in Fig.120A. The number of gold occurrences per 10 × 10 km cell will be taken as the dependent variable for statistical analysis.

The original data for this study were compiled from the literature (see Agterberg et al. 1972). Although it can be ascertained by assaying whether a given quartz vein is a gold occurrence, it is difficult to develop consistent rules for their representation on a map at a small scale because, locally, a gold deposit may consist of many separate veins penetrating a considerable volume of rock. In some gold mines, gold-vein systems were mined to depths of more than one mile below the surface. We adopted as a rule that individual gold "deposits" should be at least 500 m apart on the 1:250,000 maps used for the compilation. Regionally, these generalized gold occurrences occur as swarms. It is shown in Fig.121 that the size of larger deposits (mines) in the swarms increases with total number of occurrences per unit of area.

The dependent variable shown in Fig.120A was regressed on percentage values for six rock types, the same as those listed in Table XV, and on combination variables for pairs of these rock types. This gave 21 lithological variables in total. The result for 644 cells is shown schematically in Fig.120B. This pattern is not unrealistic but rather incomplete in that several gold-rich areas seem to be missing in Fig.120B.

Next, two subareas were defined which fall within the larger region of Fig.120 and density of gold occurrences was run on the same set of 21 lithological variables. The result is shown in Fig.122A where two separate solutions were combined into a single diagram. This pattern is unrealistic in several places: In Ontario, the Timmins area was missed and

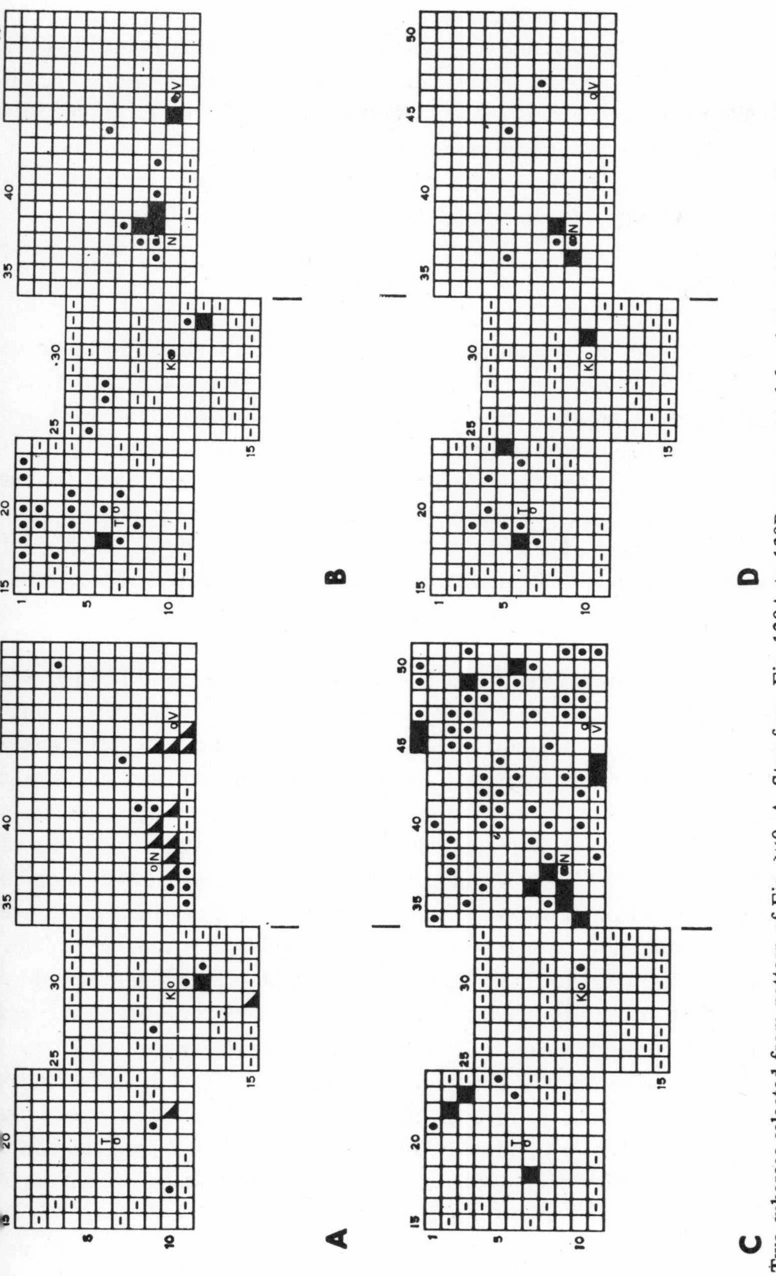

Fig. 122. Two subareas selected from pattern of Fig. 120. A. Step from Fig. 120A to 120B was repeated for data restricted to these subareas. Now favorable lithology would fall south of where gold was actually found, mainly in environments known to be barren. Occurrence of gold deposits is mainly independent of local lithology; however, within gold "belts", local gold mineralization is also controlled by favorableness of lithology. This is brought out in B where a cubic trend was first eliminated for each subarea. The new patterns (B) are more similar to those of Fig. 120A and B. C. Equations used to construct the two patterns of B were also used for extrapolation into the other subarea in each case; resulting patterns are rather unrealistic with local overestimation of values, particularly in the northeastern part of the area. D. Extrapolation patterns are more realistic if lithological data in target areas are considered when prediction equation is calculated. (From Agterberg and Kelly, 1971.)

in Quebec, the largest values occur to the south of Noranda, in an area underlain by metamorphic rocks known to contain few gold deposits. It is interesting that this bias is not present in Fig.120B where the same technique was used for a larger region. Also, it does not occur in Fig.122B representing the components for $\Sigma \pi_i X_i$ in a solution corresponding to [15.10] where $F(u,v)$ was approximated by a cubic polynomial for both subareas. This example illustrates that if there are strong trends in the area, failing to consider these may result in patterns that are strongly biased.

Extrapolation and interpolation

It was pointed out that for a regional prediction of the occurrence of deposits, we need a "control area" for comparison. An important question is whether the information on the variables X_i in the study area but outside the control area should be regarded when the prediction equations are being developed.

The patterns in Fig.122B obtained from data for the two subareas were extrapolated laterally toward the east and west, respectively (Fig.122C). The two predictions were again joined into a single diagram. Thus, the pattern for subarea 2 in Fig.122C is based on the linear equation also used for the pattern for subarea 1 in Fig.122B, and vice versa. This method of extrapolation resulted in several unrealistically high values, particularly in the northeastern corner of subarea 2.

Another type of lateral prediction is shown in Fig.122D. The model of [15.10] was applied to the data on X_i from both subareas (1 and 2), this time using a six degree polynomial for $F(u,v)$ to ensure greater flexibility. The pattern for subarea 2 in Fig.122D now represents part of a solution for the combined area using data from subarea 1 (but not from subarea 2), and vice versa.

Clearly the pattern of Fig.122D is more realistic than that of 122C. This second method of prediction may be referred to as "interpolation" because geological data on all cells of the study area are being considered from the beginning. Estimated values are biased when not all deposits in the region can be considered. For this reason, we have plotted relatively high and low values in Fig.122. If required, the bias can be removed by rescaling the dependent variable afterwards. The choice of a control area seems less critical if interpolation is used. A more important reason, however, why interpolation may be better than extrapolation is that the final estimated coefficients $\hat{\beta}_i$, individually, are subject to considerable imprecision for both methods. If the control area is not fully representative for the entire area, a number of cells outside the control area can easily assume unrealistically high or low values. The possibility of such "blown up" values is suppressed in interpolation.

Several details for the previous experiments were not fully explained. For example the method of stepwise regression was used with the level of significance set at 0.5 for individual steps. The graphical representation was obtained by defining several classes of high and low values in a rather arbitrary manner. In the next section, more formal proce dures will be discussed with reference to the occurrence of the large copper deposits in the same area.

15.7. Large copper deposits in the Abitibi area

The following geological discussion provides a further background for interpreting the statistical results. For more detailed general descriptions, reference is made to Baer (1970), Lang (1970), and McGlynn (1970). Most large sulphide bodies in the region are lenticular, massive to disseminated, stratabound deposits enclosed by volcanic and sedimentary rocks of Archean age. They contain mixtures, in various proportions, of pyrite, pyrrhotite, sphalerite, chalcopyrite and, in some instances, galena. They also may contain significant (minable) amounts of gold and silver.

Copper from chalcopyrite is economically most significant; zinc from sphalerite is recovered from a number of deposits, in some of which significant copper is absent.

Several authors believe that these deposits are related genetically to the Archean volcanism and would be only slightly younger than the Archean rocks in their immediate vicinity. Both deposits and enclosing rocks were folded and tilted to their present attitudes during the Kenoran orogeny which closed Archean time. Goodwin and Ridler (1970) took this view, pointing out that most large sulphide deposits occur in the vicinity of the sites of ancient volcanic centers whose main characteristic feature is the relative abundance of various types of acidic volcanic rocks (rhyolites and pyroclastics) in the upper parts of volcanic sequences, otherwise mainly consisting of mafic volcanics (basalts and andesites). The conclusion by these authors and also by Hutchinson et al. (1971) would be in agreement with our statistical results. For example, the rock type which singly is most strongly correlated to large copper deposits is acidic volcanics (cf. Fig.119B); in the simplified model discussed previously, the first variable selected during stepwise regression was the combination of acidic volcanics present and Bouguer anomaly above average. It suggests that the preferred environment consists of a thicker than average pile of andesites and basalts capped by rhyolites and tuffs. This could constitute an ancient volcanic center.

It is not clear whether certain stratigraphic horizons, especially some related to acidic volcanics, in the Abitibi volcanic belt are just favorable host rocks or whether there is a genetic connection between the deposits and the hostrock (Pettijohn, 1970).

The association also can be explained by Boyle's (1968) theory stating that metal in "epigenetic" mineral deposits probably comes from the country rocks that enclose, lie below, or lie lateral to the deposits. The elements were mobilized during regional metamorphism and taken in solution by deep-seated groundwater or brines according to this theory. They were later concentrated in favorable geological environments such as specific stratigraphic horizons and dilatant structures. A fact in favor of Boyle's concept is that the mafic volcanics are relatively rich in copper, zinc, gold and silver which also occur in the mineral deposits of the Abitibi volcanic belt. A slight reduction in overall copper content of the mafic volcanics would account for all copper concentrated in the large sulphide deposits. A further result favoring Boyle's concept is that the second variable selected by stepwise regression is the combination of acidic volcanics and granitic rocks. The latter are mainly smaller plutonic stocks intruded during the Kenoran orogeny. They may have provided the conditions required to mobilize the trace elements in the mafic volcanic rocks.

It is known that mafic and ultramafic intrusions can be a source of sulphide mineralization including chalcopyrite. In a single diagram for the Abitibi belt of western Quebec, Wilson (1967) plotted both copper occurrences and mafic and ultramafic intrusions. A spatial relationship is suggested by this diagram.

On the Canadian shield, nickel–copper deposits usually are markedly different from other types of sulphide deposits in that they occur in close association with ultramafic and mafic stocks which were their source. The area of Fig.119 contains one large nickel–copper deposit only (Zalupa Mine, Quebec) situated close to a diorite intrusion. Cameron et al. (1971) found that ultramafics with nickel–copper deposits in their vicinity usually have a larger nickel content than ultramafics without deposits.

Kutina and Fabbri (1972) have emphasized that both copper and gold mineralization in the Abitibi area are in part controlled by faults. They suggested that deep-seated fractures in the crust of the earth provided channels along which mineralized solutions could ascend.

Some of the previous hypotheses and suggested interrelationships may seem mutually exclusive. All offer explanations for the same data, although the emphasis on certain facts may differ.

Clearly, the geological and geophysical variables used in our statistical model are strongly interrelated. Unless the model is kept simple by basing it on few variables, this interaction may blur the possible relevance of single factors or combinations of factors.

Another difficulty for statistical analysis is whether the mineral deposits treated as a group are really comparable with one another, i.e., are a homogeneous population. Lang (1970) stated that the large Archean sulphide deposits in the Abitibi region are essentially of one type. Exceptions are: (1) a copper deposit at the McIntyre gold mine near Timmins, Ont., where disseminated chalcopyrite occurs in a quartz-feldspar porphyry stock, and (2) the nickel–copper deposit (Zalupa Mine) mentioned previously. We have lumped these two deposits with the other large copper deposits.

For statistical analysis, the 27 control cells of Fig.119A were weighted according to the logarithm (base 10) of their size in tons of copper and related to the 55 variables discussed previously. This gave 644 calculated values $\hat{Y}$ (subscript k deleted). These were added for the control area consisting of 50 cells outlined in Fig.119A, giving $\Sigma\hat{Y}_c = 38.4$. There are $N_c = 12$ control cells in the control area and the factor $\Sigma\hat{Y}_c/N_c = 3.20$ was used for scaling. If all values $\hat{Y}$ are divided by 3.20, the results can be interpreted as probabilities that a cell contains significant copper.

These estimated probabilities range from slightly negative to as high as 88% for one cell near Timmins, Ont. Because the method is approximate, negative values and values greater than 100% may occur but this did not present a serious problem in this study.

If the calculated values $\hat{Y}$ are rescaled by a single constant, two assumptions are implied: (1) for comparison, it is supposed that all minable deposits have been found in the control area; and (2) the calculated values $\hat{Y}$ are proportional to the probabilities of occurrence so that size can be neglected as a factor.

The second assumption may be tested by correlating the values Y to the $\hat{Y}$ for control cells. We would expect a positive correlation between these quantities if size increases together with probability of occurrence. For the example, the correlation coefficient amounts to 0.31 which is not statistically significant at the 5-% level of significance for the ordinary correlation coefficient (27 observations only). The 27 values plot as a cluster that does not suggest a functional relationship. For this reason, it was decided to use a constant correction factor rather than one which would depend on size.

The resulting pattern of 644 probabilities shows strong autocorrelation reflecting gradational changes in composition of the geological framework. These changes are not according to a continuous pattern but reflect the artificial boundaries between the 10 × 10 km cells used for coding. For further simplification, values were added for overlapping blocks of 16 cells simultaneously. The result is shown graphically in Fig.117A. During the smoothing operation, cells not used for statistical analysis were counted as zeros.

The final probability index can be contoured as shown in Fig.123A. It represents the number of copper-rich cells expected per surrounding 40 × 40 km cell.

Each of the calculated values $\hat{Y}$, and the values derived from these, is subject to uncertainty. Because of the dichotomous and distinctly nonnormal nature of the variable Y, results based on conventional methods of statistical analysis are difficult to interpret. In order to test the method, an indirect approach was taken by predicting areas of which the known values of Y are not used but saved for checking the predictions. A test of this type was presented earlier in Fig.117B and C where a vertical scale was not given. By using the present method, we may scale the results for Fig.117B with regard to the control area consisting of the 25 cells near Noranda, and those for Fig.117C by using the 25 cells near Timmins. The outcome is shown in Fig.123B and C. The scaling factors, by which the calculated values $\hat{Y}$ were divided, amounted to 2.45 and 2.25, respectively and are less than the factor 3.20 used for Fig.123A. Note that the westernmost peak in the area is missing in Fig.123C indicating that it is mainly caused by similarities of this site with the Noranda–Val d'Or area rather than with the nearby Timmins area.

Indirect tests of this type suggest that the method has predictive potency although precise methods to evaluate the results statistically have not been developed.

It seems that the greatest source of uncertainty arises from the estimated probabilities themselves. This uncertainty can be evaluated by approximate methods. As a first approximation, we can assume that the model of Allais (see section 7.10) is satisfied for

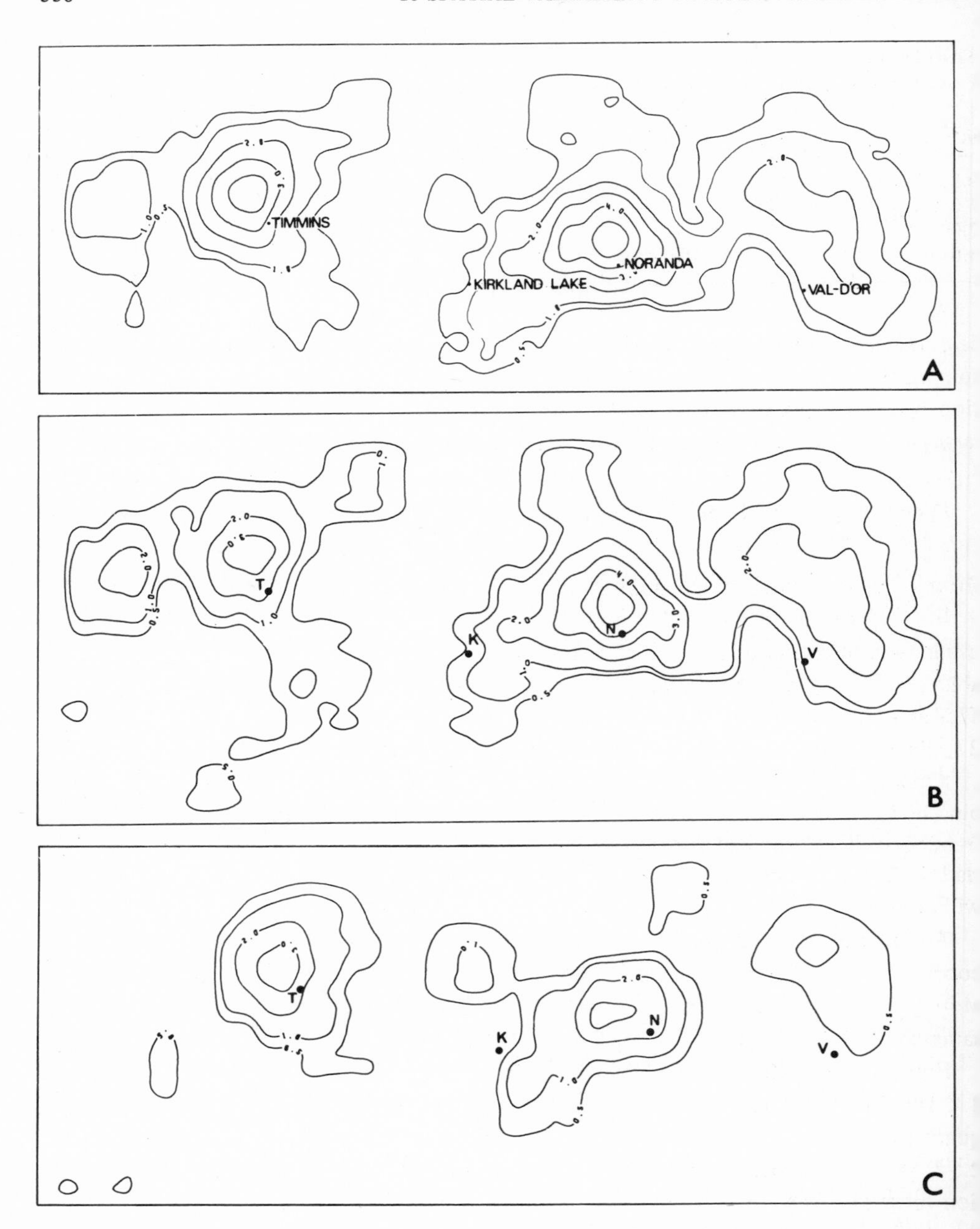

Fig.123. Copper probability maps corresponding to Fig.117. A. Probability index (based on 27 control cells) scaled with respect to control area of 50 cells shown in Fig.119A; result based on stepwise regression with $Q = 0.5$; 26 variables were included. B. Ditto, as computed from 19 control cells east of 600 km E line of Fig.119; control area consisted of 25 cells near Noranda. C. Ditto, from 8 control cells in western part of area; control area near Timmins.

TABLE XLIX
Probability (in percent) of x copper-rich 10 × 10 km cells per 40 × 40 km cell when m is given

m \ x	0	1	2	3	4	5	6	7	8	9
0.5	60	31	8	1						
1.0	36	38	19	6	1					
2.0	12	27	29	19	9	3	1			
4.0	1	5	13	21	23	18	11	5	2	1

each cell measuring 40 km on a side. As discussed in section 15.5, this is equivalent to assuming that a value for the probability index (m) represents the mathematical expectation of a binomial distribution. The exact number of copper-rich 10 × 10 km cells per larger 40 × 40 km cell then satisfies the terms of the binomial formula:

$$P(x) = \binom{n}{x} (m/n)^x (1-m/n)^{n-x} \qquad [15.12]$$

for $n = 16$. Sequences of probabilities for selected values of m are shown in Table XLIX.

Finally, we may estimate the total amount of copper in tons for any subarea, again by using a modification of the Allais model. Estimates of this type are subject to considerable uncertainty. On the average, a control cell contains about 140,000 tons of copper. This amount is the mean of a wide, positively skewed frequency distribution consisting of 27 values ranging from 1084 to 1,179,900 tons of copper.

Amount of copper per area satisfies a compound random variable consisting of a binomial distribution for number of copper-rich smaller cells in the area, and their size-frequency distribution. The latter may be approximated by a discrete distribution consisting of the 27 known values to each of which we assign a probability of 1/27 that it will occur.

An approximate solution for the confidence limits of expected values for amount of copper per area can be obtained by random sampling of this compound distribution. A Monte Carlo technique was used to generate the random numbers required for this sampling.

For example, a 40 × 40 km cell with $m = 4$ has an expected amount of copper equal to 4 × 140,000 = 560,000 tons but the median falls at 300,000 tons. The lower and upper quartiles (25 and 75% fractiles) are 50,000 and 850,000 tons, respectively. There is a 2.5-% probability that the actual amount is less than 3,500 tons or greater than 2.2 million tons. The latter two values define a 95-% confidence interval. The great width of this interval is comparable with those sometimes encountered when the amount of oil for an area is predicted in barrels (Kaufman, 1963; also see section 7.12). Clearly, the total amount of copper is more difficult to predict than the occurrence of large deposits since we have x = 2, 3, 4, 5 or 6 with a probability of 86% if m = 4 (Table XLIX).

Computing methods and other probabilistic maps

Standard methods of multiple regression were used for numerical analysis. Commonly the matrix $\mathbf{X}'\mathbf{X}$ has a rank which is less than its order. For example, the seven rock type add to one for every cell implying that $\mathbf{X}'\mathbf{X}$ must be singular if a dummy variable is usec There may be other linear or approximately linear relationships between the 55 variable used for this example, particularly because 45 of these are combinations (cross-products of the 10 measurement variables. For matrix inversion, we used the method of pivota condensation with interchange of rows which is part of Efroymson's (1960) algorithr for stepwise regression. This procedure is tied to the precision by which computing i being done. It was possible, at a maximum, to obtain the inverse of $\mathbf{X}'\mathbf{X}$ using 51 of th 55 variables.

For the following examples, a larger control area was used consisting of 145 cell surrounding the mining camps of Timmins, Kirkland Lake, Noranda and Val d'Or. Thi larger control area (see Agterberg et al., 1972) contained 24 out of 27 control cells fc copper. The result for 51 variables is shown in Fig.124G which resembles Fig.123A rathe closely. Not all 51 variables are needed. In order to see how many are really required, th method of forward selection was employed with results as shown in Fig.124 for termina tions after steps 1, 2, 4, 8, 16, 32 and 51. The squared multiple-correlation coefficient R increases from 0.073 (1 variable) to 0.248 (51 variables) but already has reached the valu 0.246 after 32 steps indicating that almost nothing was gained by going on to the en The final pattern of contours was reached even earlier (after 8 steps). This is because closer fit by using more variables also leads to an increase in $\Sigma \hat{Y}_c$ which is the sum of th calculated values in the control area; and for scaling, all values $\hat{Y}$ were divided by $\Sigma\hat{Y}_c/N$ where $N_c(=24)$ is a constant.

In most practical applications, we have used stepwise regression with backward passe after each forward step. The $\hat{Q}$-value representing level of significance for the F-distribu tion can be computed for every forward or backward step. The regression run is ter minated when $\hat{Q}$, at a forward step, crosses a predetermined level Q. Setting $Q = 0.5$ fo the present example, gave selection of 26 variables in the order shown in Table L. Unti variable 19, this run was identical to that for the previous forward selection. It is notec that acidic volcanics is part of the first six variables selected. Clearly, we could hav stopped the forward selection after about eight steps, at the point where $\hat{Q}$ crosses th $Q = 0.10$ level of significance. In most practical applications, however, we have usec $Q = 0.50$ as the cut-off level admitting considerable redundancy. The reason for doin this was that the relationship between large copper deposits and composition of geolog ical framework is exceptionally strong. In other applications, the associations may b weaker. With a cut-off of $Q = 0.5$, the $\hat{Q}$-value usually fluctuates in a rather irregula manner until $\hat{Q} \approx 0.3$ when backward passes begin to occur more frequently. These fluc tuations are also present in Table L. When the associations are weaker, a cut-off o $Q = 0.1$ or less can result in a significant loss of information.

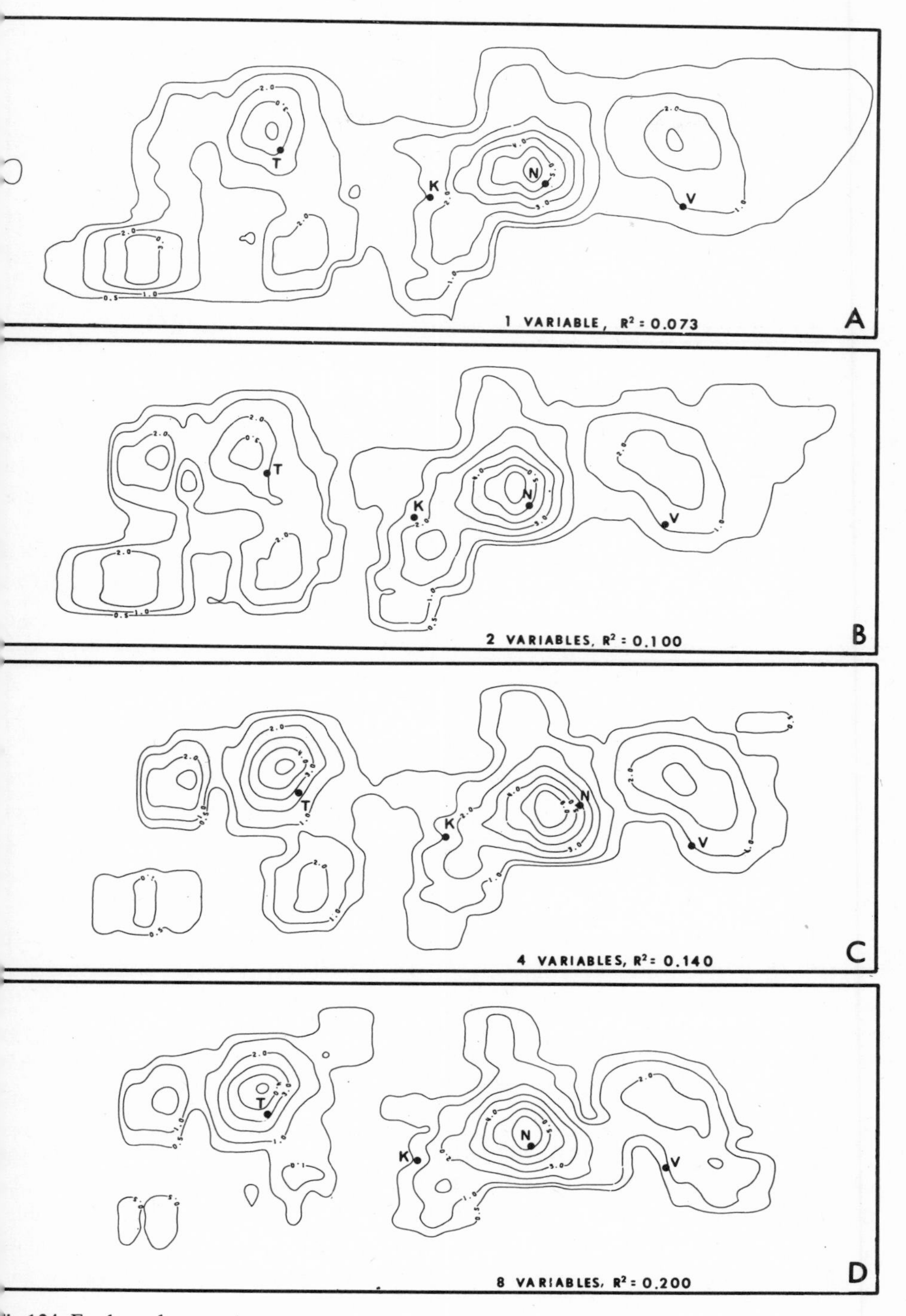

Fig. 124. For legend see next page.

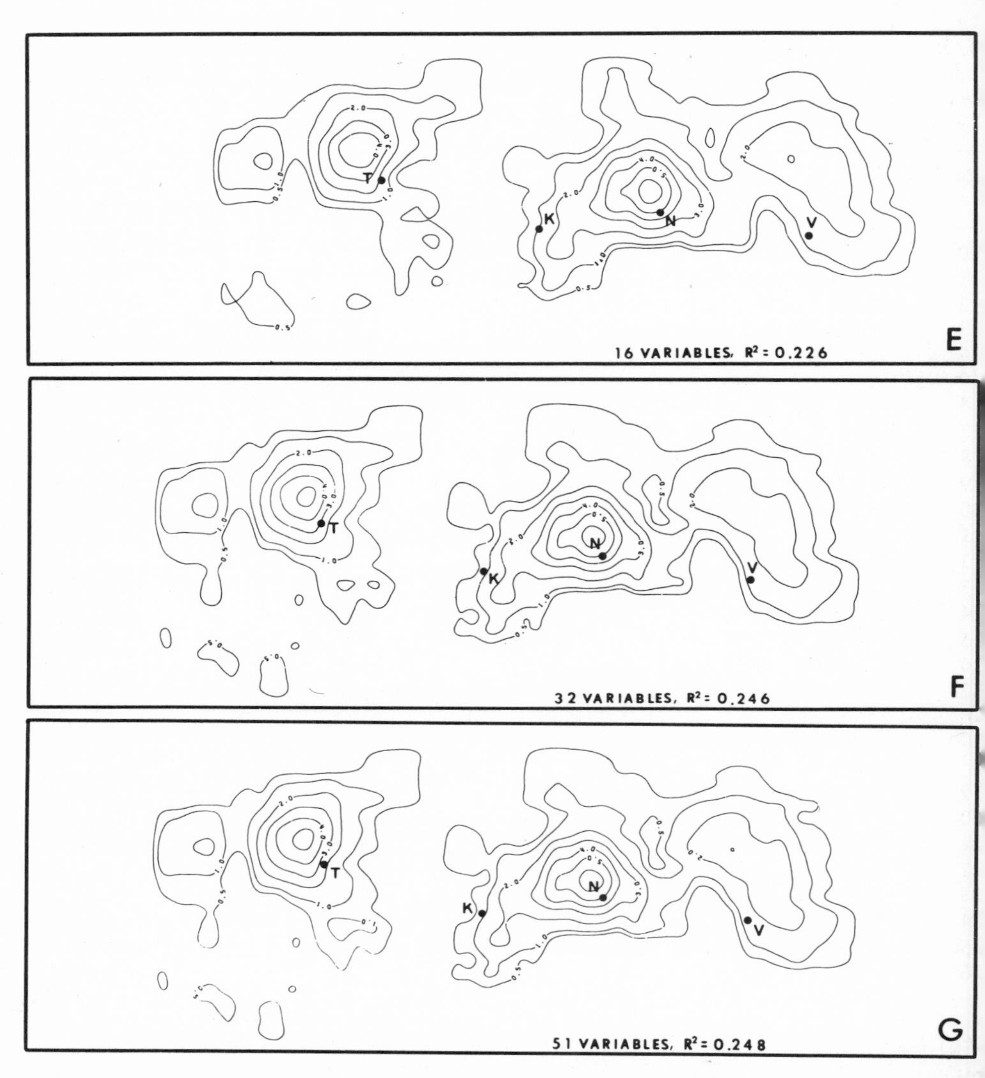

Fig.124. Seven copper probability maps representing selected steps of method of forward selection of 51 independent variables; T = Timmins; K = Kirkland Lake; N = Noranda; V = Val d'Or (cf. Fig.79). Control area consisted of 145 relatively well-explored cells instead of 50 used for Fig.123. Addition of new variables rapidly becomes redundant. (After Agterberg et al., 1972.)

It should be kept in mind that, although conventional analysis of variance is used, the results are at best approximate since the variables are far from normally distributed. The regression coefficients are also shown in Table L. Inasmuch as 26 interrelated variables were used, these coefficients have large variances (see section 8.8). The calculated values $\hat{Y}$ all are less than the observed values Y in control cells and they are larger than zero in

TABLE L

Variables selected by stepwise regression ($Q = 0.5$) of logarithmically transformed tonnage copper per cell on 55 geological and geophysical variables. Fig. 123A was based on this run. The rank of the first 19 variables is the same as that obtained by forward selection used for Fig. 124.

Rank	*Name of variable*	*$\hat{Q}$-value*	*Regression coefficient*
0	Constant term		−0.14863
1	Acidic volcanics × mafic volcanics	$0.36 \cdot 10^{-9}$	0.00045
2	Granitic rocks × acidic volcanics	$0.44 \cdot 10^{-4}$	0.00147
3	Acidic volcanics × sediments	$0.70 \cdot 10^{-3}$	0.00182
4	Acidic volcanics × iron formations	$0.45 \cdot 10^{-4}$	−0.00334
5	Acidic volcanics × aeromagn. anomaly	$0.63 \cdot 10^{-3}$	−0.01958
6	Mafic intrusions × acidic volcanics	$0.14 \cdot 10^{-3}$	0.00187
7	Metamorphic rocks × iron formations	$0.87 \cdot 10^{-3}$	0.00618
8	Mafic intrusions × metamorphic rocks	$0.91 \cdot 10^{-2}$	0.00129
9	Granitic rocks × sediments	$0.32 \cdot 10^{-1}$	0.00037
10	Sediments × Bouguer anomaly	0.13	0.00027
11	Mafic volcanics × sediments	0.021	0.00017
12	Acidic volcanics	0.13	0.05068
13	Sediments × aeromagnetic anomaly	0.12	0.00560
14	Sediments	0.14	−0.01726
15	Ultramafics × sediments	0.27	0.00049
16	Granitic rocks × iron formations	0.30	
17	Iron formations × Bouguer anomaly	0.10	−0.00490
18	Iron formations	0.19	−0.40777
19	Mafic volcanics × iron formations	0.09	0.00171
–	Backward pass, No. 16 deleted	0.65	
19	Iron formations × aeromagn. anomaly	0.20	0.02133
20	Ultramafics × iron formations	0.29	−0.00332
21	Sediments × metamorphic rocks	0.36	−0.00025
22	Acidic volcanics × methamorphic rocks	0.29	0.00123
23	Mafic intrusions × mafic volcanics	0.41	0.00026
24	Granitic rocks	0.38	0.00128
25	Granitic rocks × mafic volcanics	0.44	0.00003
26	Mafic intrusions × sediments	0.50	−0.00030

cells adjacent to the control cells. A plot of residuals against original values for all cells reflects the dichotomous nature of the dependent variable and exhibits strong positive correlation. Although a least-squares model is used for solution, the results should not be evaluated in terms of multiple regression in the framework of a multivariate normal distribution. The present method is more closely related to methods in use for estimating probabilities by least squares. A review of some methods of this type with additional references was given by Neter and Maynes (1970).

It is interesting to compare the patterns of Fig.123 and 124 with that of Fig.119A for original data. There are three places (in Timmins area, west part of Noranda area and in the northeastern part of the region) where the contours exceed density of known large copper deposits. These are places with indicated copper potential and they may be further evaluated in the search of new orebodies. However, there is one place (in vicinity of Va d'Or, Que.) where more deposits were found than is explained by the contours.

The negative anomaly near Val d'Or

To the east of Val d'Or, we have a cluster of four control cells and because two mor copper-rich cells occur farther north (see Fig.119A), we have two 40 × 40 km cells con taining as many as five smaller 10 × 10 km cells with significant copper mineralizatior On the other hand, the contoured probability index ranges from 1 to 2 at this place When the model of Table XLIX is accepted for evaluating the uncertainty in m, thi negative anomaly could be explained as a statistical fluctuation, e.g. if $m = 2$, the prob ability that $x \geqslant 5$ amounts to 4%, and this situation could occur at one or more place within a larger region. However, when the anomaly is compared with geological an metallogenic maps, a definite relationship is suggested with the so-called Cadillac Break a rather sharp stratigraphic transition zone with predominantly volcanic rocks to th north and, in part metamorphosed, sedimentary rocks to the south of it. This break i accompanied by shear-zones with graben-structures or synclines with a core of Archea sedimentary rocks belonging to a belt where these rocks contain stratified iron forma tions. Also, the Cadillac Break is characterized by extensive gold mineralization con centrated in its immediate vicinity.

The preceding statistical analysis was repeated after adding to the 55 variables 44 nev variables consisting of functions of the geographical coordinates of all cells in the regio Together these functions form an octic polynomial. The solution for the 99 variables wit a cut-off value of $Q = 0.5$ was divided into two parts similar to those for gold occurrence in the previous section. The polynomial part is shown in Fig.125. A contour map for bot components simultaneously is shown in Fig.126. We note that Fig.126 is similar t

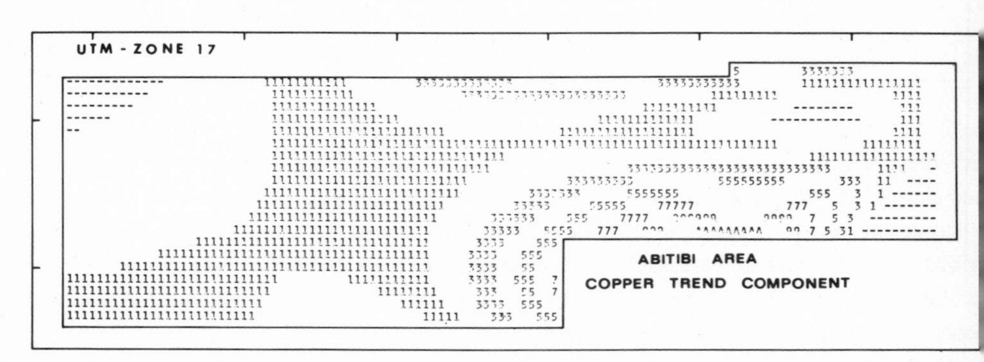

Fig.125. Octic trend component for copper; values were multiplied by 10 before printing; $A = 1$ boundaries as in Fig.119. This surface is close to zero almost everywhere; maximum occurs in th vicinity of Val d'Or and Noranda, and might be related to Cadillac Break located at the bounda between volcanics and metamorphic rocks in Fig.79A.

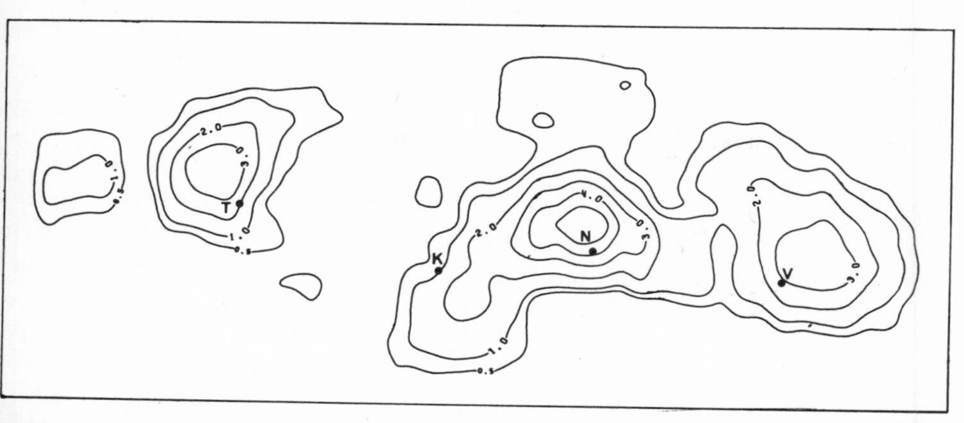

Fig.126. Copper potential map based on combined model containing geological variables and polynomial terms; effect of polynomial (shown separately in Fig.125) is higher contour values near Val d'Or than in Fig.124E–G.

Fig.123A and 124 except in the Val d'Or area. The polynomial trend surface is close to zero everywhere in the region except in the Val d'Or–Noranda area where it shows a peak. The negative anomaly tends to be eliminated when the polynomial is added.

It seems that the relatively intense copper mineralization near Val d'Or cannot be expressed adequately in terms of our 55 variables which are related to 10 × 10 km cells. Another, perhaps more regional, factor also was operative.

Dugas (1966) and Latulippe (1966) have assumed that regionally the copper and gold mineralization were in part controlled by ore-bearing solutions extending northward from the Cadillac Break into the country rock with the copper moving farther than the gold. The patterns of Fig.120 for gold and Fig.125 for copper trends would support this theory.

Single variable most strongly correlated to mineralization

The example for large copper deposits in Abitibi indicates that a useful pattern is already obtained by the careful choice of a single relevant variable. In Fig.119A, 204 of 644 cells have acidic volcanics present and 24 of 27 control cells coincide with these cells.

A somewhat better approach is to use cells with (1) acidic volcanics present and (2) Bouguer anomaly above average. Although the number of coincidences is less, the areal extent of this combination variable is more restricted. We recall that these results were based on a simplified presence–absence model. When measured amounts are used, the first variable selected is the cross-product for percentage of acidic and mafic volcanics in a cell (see Table L). A contour-map for this variable was shown in Fig.124A. The Bouguer anomaly is not part of this combination, because it assumes its largest values in parts of the area where large copper deposits have not been found. These maxima indicating places

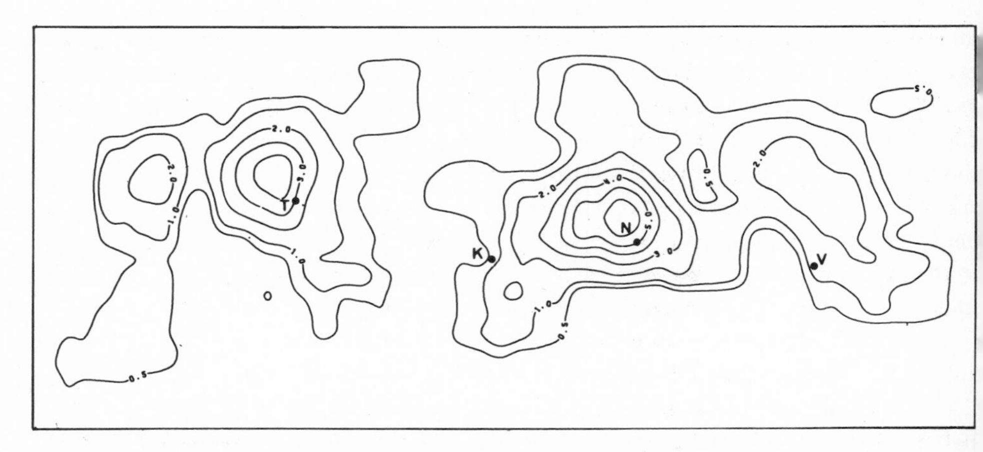

Fig.127. Map based on 9 instead of 10 basic measurement variables; mafic and acidic volcanics we lumped to "undifferentiated volcanics". Note that, although acidic volcanics singly is most importar basic variable, its elimination does not significantly change the contour pattern (cf. Fig.124E–G).

of greatest thickness for the pile of mafic volcanics, were reduced to a lower level in th yes–no model. The patterns are improved significantly if more variables are included

Unfortunately, the individual regression coefficients used for estimation rapidly be come meaningless when more than one or two variables are used, because of the stron interaction between the geological variables.

The fact that acidic volcanics singly is the most important rock type to look fo during the exploration for large sulphide deposits has been known during the past decad and led to a refined mapping of acidic volcanics.

As an experiment, we have eliminated acidic volcanics from our list of ten measure ment variables by lumping it together with mafic volcanics as "volcanics undifferen tiated". This gave 9 variables with 36 combination variables, or, in total, 45 variables fo stepwise regression. The result for a cut-off of $Q = 0.5$ is shown in Fig.127 which is nearl identical to earlier results.

It seems that, for statistical prediction, we do not need the delineation of acidi volcanics on the geological map, although singly this rock type is more strongly corre lated to occurrence of large sulphide deposits than any other feature. This does no refute the fact that most copper deposits lie on ancient volcanic centers but illustrate that the occurrence of these centers itself can be adequately expressed in terms of th other measured geological and geophysical variables.

15.8. Two-stage least-squares models

A drawback of the present approach is that a rigid network of equal-area cells superimposed on a region and used for quantifying all variables. We can hardly claim to have selected an optimum grid. If a best grid exists, the sizes and shapes of its cells probably would change from place to place in a region and in mineral-potential studies, different types of deposits might require different types of grids. The main reasons for choosing 10×10 km cells were: (1) the large extent to which geological features usually must be generalized for representation on a map at scale 1:250,000; and (2) the fact that the Universal Transverse Mercator projection method provides cells which can be readily constructed on maps of different scales everywhere. These square-shaped cells have the equal-area property and form regular patterns.

The U.T.M.-system is worldwide and consists of 60 zones, each six degrees of longitude wide, and extending from 85°S to 85°N in latitude. In practice, one can project one U.T.M.-zone into its two adjacent zones without loss of the desirable properties which remain valid by approximation. For example, the study area of Fig.118 is about 8° of longitude wide (or 590 km). It lies in U.T.M.-zone 17 and the easternmost part of the region (2° wide) was projected into this zone from the adjacent zone. The zone boundaries, however, would present a problem for continental or world-wide studies.

Some limitations of using a single grid for multiple purposes were discussed in previous sections. Ideally, we require a digitized version of the geological map, which could be three-dimensional, and may contain data for many other variables usually not shown on traditional maps. The problem of how many data should be digitized in order to preserve actual patterns was discussed by McCammon (1971). If the cells are small, one should be able to consider simultaneously not only all variables coded for a single cell but also those for the surrounding cells. To a limited extent, this requirement is met by working with models where regional variations in the variables are separated from more local multivariate relationships. In two-stage least-squares models, the values for a variable are broken into two components: one for regional trends and one for residuals. A trend value for a cell is a linear combination of the values for all other cells in the region with weights given by the familiar matrix $\mathbf{X}(\mathbf{X}'\mathbf{X})^{-1}\mathbf{X}'$. The matrix $\mathbf{X}$ in this expression now consists of values for functions of location only (e.g., the terms of a polynomial). These weights form a pattern which is different for each cell. The maximum weight usually occurs at the location of the cell considered but patterns for different cells do not have the same shape and, in this sense, the weighting scheme is nonstationary.

Mathematical model for estimating trend-free geological components

The basic model can be written in the form:

$$\mathbf{Y} = \mathbf{X}_1\boldsymbol{\beta}_1 + \mathbf{X}_2\boldsymbol{\beta}_2 + \mathbf{E} \qquad [15.13]$$

where $\mathbf{X}_1$ and $\mathbf{X}_2$ are column matrices of values for the variables in a deterministic trend function (subscript 1) and a set of geological variables (subscript 2), respectively. Without loss in generality, it can be assumed that both $\mathbf{X}_1$ and $\mathbf{X}_2$ consist of values corrected for their means. This correction also will be applied to the vector $\mathbf{Y}$ for values of the dependent variable but then it will be indicated by the subscript zero, e.g., $\mathbf{Y}_0 = \mathbf{Y} - \overline{\mathbf{Y}}$ where $\overline{\mathbf{Y}}$ consists of values all equal to $\overline{Y}$. The column vector $\mathbf{E}$ in [15.13] is for the residuals. The least-squares solution can be written as:

$$\hat{\mathbf{Y}}_0 = \hat{\mathbf{Y}}_{0T} + \hat{\mathbf{Y}}_G \qquad [15.14]$$

with:

$$\hat{\mathbf{Y}}_{0T} = \mathbf{X}_1\hat{\boldsymbol{\beta}}_1 \; ; \qquad \hat{\mathbf{Y}}_G = \mathbf{X}_2\hat{\boldsymbol{\beta}}_2 \; .$$

The calculated values of $\hat{\mathbf{Y}}_0$ are divided into a "trend" component ($\hat{\mathbf{Y}}_{0T}$) and a geological component $\hat{\mathbf{Y}}_G$. The subscript 0 is deleted from $\hat{\mathbf{Y}}_G$ to indicate that it will be used in this form which is zero on the average for all cells.

When the geological variables are deleted from the model, we retain the values $\hat{\mathbf{Y}}^*_{0T} = \mathbf{X}_1\hat{\boldsymbol{\beta}}^*_1$ with $\hat{\boldsymbol{\beta}}^*_1 = (\mathbf{X}'_1\mathbf{X}_1)^{-1}\mathbf{X}'_1\mathbf{Y}_0$ which satisfies a deterministic function. In general, this estimate of the trend in $\mathbf{Y}_0$ will be biased because $\hat{\boldsymbol{\beta}}_1 \neq \hat{\boldsymbol{\beta}}^*_1$.

Estimates of the vectors $\boldsymbol{\beta}_1$ and $\boldsymbol{\beta}_2$ are given in several textbooks (e.g., Goldberger 1964, p.194). They can be derived by using the method of partitioned inversion of section 3.8. The result can be written in the form:

$$\begin{aligned} \hat{\boldsymbol{\beta}}_1 &= \hat{\boldsymbol{\beta}}^*_1 - (\mathbf{X}'_1\mathbf{X}_1)^{-1}\mathbf{X}'_1\mathbf{X}_2\hat{\boldsymbol{\beta}}_2 \\ \hat{\boldsymbol{\beta}}_2 &= (\mathbf{X}'_2\mathbf{A}\mathbf{X}_2)^{-1}\mathbf{X}'_2\mathbf{A}\mathbf{Y}_0 \end{aligned} \qquad [15.15]$$

where $\mathbf{A} = \mathbf{I}_n - \mathbf{X}_1(\mathbf{X}'_1\mathbf{X}_1)^{-1}\mathbf{X}'_1$.

$\mathbf{I}_n$ is the n-dimensional identity matrix where n represents the number of observations. It is readily shown that the matrix $\mathbf{A}$ is symmetrical idempotent with properties $\mathbf{A} = \mathbf{A}'$ and $\mathbf{AA} = \mathbf{A}$.

Suppose that $\mathbf{X}_2$ is regressed on $\mathbf{X}_1$ and the result subtracted from $\mathbf{X}_2$. This yields a matrix of residuals $\hat{\mathbf{X}}_{2R}$ with:

$$\hat{\mathbf{X}}_{2R} = \mathbf{X}_2 - \mathbf{X}_1(\mathbf{X}'_1\mathbf{X}_1)^{-1}\mathbf{X}'_1\mathbf{X}_2 = \mathbf{A}\mathbf{X}_2 \; . \qquad [15.16]$$

When $\mathbf{Y}_0$ is regressed on $\hat{\mathbf{X}}_{2R}$, we obtain:

$$\begin{aligned} \hat{\mathbf{Y}}_{0R} &= \hat{\mathbf{X}}_{2R}(\mathbf{X}'_{2R}\hat{\mathbf{X}}_{2R})^{-1}\hat{\mathbf{X}}'_{2R}\mathbf{Y}_0 \\ &= \mathbf{A}\mathbf{X}_2(\mathbf{X}'_2\mathbf{A}\mathbf{A}\mathbf{X}_2)^{-1}\mathbf{X}'_2\mathbf{A}\mathbf{Y}_0 \; . \end{aligned} \qquad [15.17]$$

Since $\mathbf{A}$ is idempotent and by using [15.16]:

$$\hat{\mathbf{Y}}_{0R} = \mathbf{A}\mathbf{X}_2\hat{\boldsymbol{\beta}}_2 = \mathbf{X}_2\hat{\boldsymbol{\beta}}_2 - \mathbf{X}_1(\mathbf{X}'_1\mathbf{X}_1)^{-1}\mathbf{X}'_1\mathbf{X}_2\hat{\boldsymbol{\beta}}_2 \; . \qquad [15.18]$$

Then, from [15.14] and [15.15]:

$$\hat{\mathbf{Y}}_{0R} = \hat{\mathbf{Y}}_G - \hat{\mathbf{Y}}^*_{0T} \,. \qquad [15.19]$$

Suppose that, instead of $\mathbf{Y}_0$, we would regress the residuals $(\mathbf{Y}_0 - \hat{\mathbf{Y}}^*_{0T})$ on $\hat{\mathbf{X}}_{2R}$. In analogy which [15.16], we see that $\mathbf{Y}_0 - \hat{\mathbf{Y}}_{0T} = \mathbf{A}\mathbf{Y}_0$. Substitution into [15.17] gives that the calculated values for these residuals also are given by $\hat{\mathbf{Y}}_{0R}$ and satisfy [15.19]. Regression of $\hat{\mathbf{Y}}_G$ on $\mathbf{X}_1$ gives:

$$\hat{\mathbf{Y}}_{G,T} = \mathbf{X}_1(\mathbf{X}_1'\mathbf{X}_1)^{-1}\mathbf{X}_1'\mathbf{X}_2\hat{\boldsymbol{\beta}}_2 = \mathbf{X}_1(\hat{\boldsymbol{\beta}}_1^* - \hat{\boldsymbol{\beta}}_1)$$

or:

$$\hat{\mathbf{Y}}_{G,T} = \hat{\mathbf{Y}}^*_{0T} - \hat{\mathbf{Y}}_{0T} \,. \qquad [15.20]$$

At this point, we can undo the correction for the mean $\overline{\mathbf{Y}}$ and collect the following results. From [15.14] and [15.19], it follows that:

$$\hat{\mathbf{Y}} = \hat{\mathbf{Y}}_T + \hat{\mathbf{Y}}_G \qquad [15.21]$$

and $\hat{\mathbf{Y}}_R = \overline{\mathbf{Y}} + \hat{\mathbf{Y}}_G - \hat{\mathbf{Y}}^*_T$. If, by definition:

$$\hat{\mathbf{Y}}_B = \hat{\mathbf{Y}}^*_T - \hat{\mathbf{Y}}_T \qquad [15.22]$$

the second result becomes:

$$\hat{\mathbf{Y}}_R = \overline{\mathbf{Y}} - \hat{\mathbf{Y}}_B + \hat{\mathbf{Y}}_G \,. \qquad [15.23]$$

Instead of regressing $\mathbf{Y}$ on $\hat{\mathbf{X}}_{2R}$, we could have regressed $(\mathbf{Y} - \hat{\mathbf{Y}}^*_T)$ on $\hat{\mathbf{X}}_{2R}$. The result of this modified regression also is [15.23].

Finally, from [15.20] and [15.22]:

$$\hat{\mathbf{Y}}_{G,T} = \hat{\mathbf{Y}}_B \,. \qquad [15.24]$$

Discussion

The significance of these results is as follows. Suppose that a dependent variable and p independent variables all are subject to regional trends. When the information for the dependent variable ($\mathbf{Y}$) is incomplete, a regular type of bias may be obtained by eliminating the trends from all independent variables and regressing $\mathbf{Y}$ on the residuals. For convenience, we require that all trends satisfy the same type of linear function (e.g., a polynomial of the same degree). This new multiple regression leads to [15.23], a result which is valid if original values are used for $\mathbf{Y}$ but also if its residuals from the trend $\hat{\mathbf{Y}}^*_T$ are used.

If $\mathbf{Y}$ is regressed simultaneously on the deterministic trend function $\mathbf{X}_1$ and the previous set of p independent variables, the result is [15.21], where $\hat{\mathbf{Y}}_T$ represents the trend and $\hat{\mathbf{Y}}_G$ is the geological component, which also occurs in [15.23]. There is a difference between the trend functions $\hat{\mathbf{Y}}_T^*$ and $\hat{\mathbf{Y}}_T$ obtained by fitting without and with regard of the p geological variables, respectively. This difference is the bias introduced when geological differences between cells are not considered. It satisfies [15.22] but $\hat{\mathbf{Y}}_B$ also occurs in [15.23] and [15.24].

Eq. [15.24] signifies that the geological component $\hat{\mathbf{Y}}_G$ is not free of trend. In fact, its trend is exactly the bias component $\hat{\mathbf{Y}}_B$. An example of the component $\hat{\mathbf{Y}}_T$ was shown in Fig.125, and the sum $(\hat{\mathbf{Y}}_T+\hat{\mathbf{Y}}_G)$ for that situation in Fig.126. In the first part of this section, it was demonstrated that these components also can be obtained by using other procedures; notably, $\hat{\mathbf{Y}}_G$ also is the geological component obtained by regression on residuals instead of on original measurements for all geological variables. If the component $\hat{\mathbf{Y}}_T$ is due to regional variations in intensity of exploration, and if appropriate deterministic functions can be found for approximation, $\hat{\mathbf{Y}}_G$ will be biased, but this bias tends to be regular in that all coefficients in $\hat{\mathbf{Y}}_G$ may underestimate the true coefficients by a constant factor f (cf. section 15.6).

As a further example, we show in Fig.128A and B the components $\hat{\mathbf{Y}}_T$ and $\hat{\mathbf{Y}}_G$ for the example of section 8.7. A combined model leading to $\hat{\mathbf{Y}} = \hat{\mathbf{Y}}_T + \hat{\mathbf{Y}}_G$ may be regarded as the result of a trend-surface analysis on $\mathbf{Y}$ whereby, at the same time, the residuals were

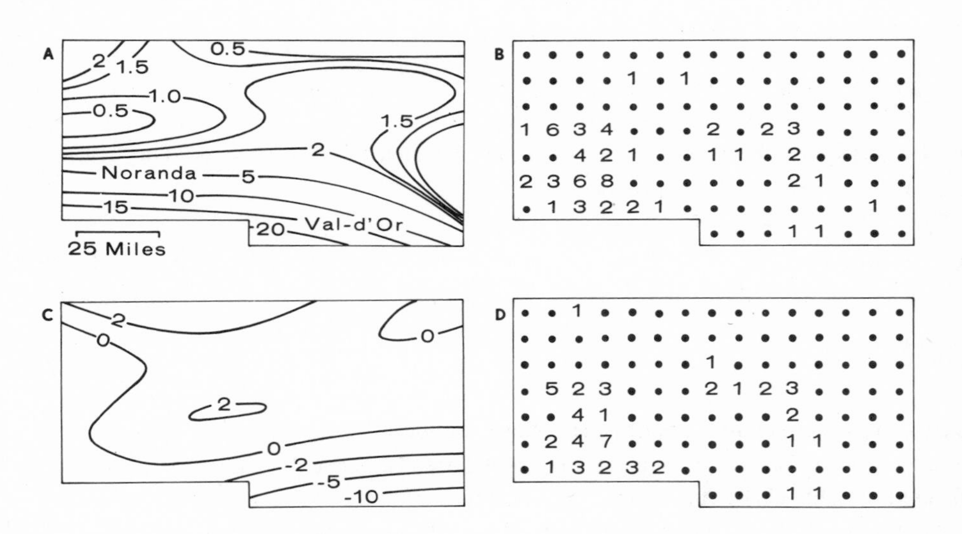

Fig.128. Two-stage least-squares model (with variable elimination) applied to number of gold occurrences per 8 × 8 mile cell in Noranda–Val d'Or region, western Quebec (from Agterberg, 1971). A and B are for cubic trend component and geological component, respectively. C. Difference between optimum linear combination of two cubic trends (for granitic rocks and mafic volcanics), and trend surface in A. D. Geological component corresponding to C, based on acidic volcanic rocks and mafic intrusions only; these two variables together account for most of B based on six variables.

expressed in terms of the geological variables. The trend of Fig.128A satisfies a cubic polynomial; it is subject to the edge effects discussed in Chapter 9.

In section 8.7, the occurrence of gold deposits in 113 cells was regressed on 6 lithological variables giving $R^2 = 0.29$ at a maximum. By using the combined model, R^2 is increased to 68% because of the importance of the trend component related to a gold belt along the southern edge of the diagram. This belt approximately coincides with the Cadillac Break discussed in the previous section.

In general, a trend component $\hat{\mathbf{Y}}_T$ may be related to geological factors of a more regional extent. It would be useful to express this type of broad-scale variation in terms of the variables that can be measured in the area. However, we must then regard the trends of these variables rather than the original data or residuals. The following model for this was originally developed by Agterberg and Cabilio (1969). The mathematical treatment given in that paper may be simplified considerably without loss in generality as indicated by Agterberg (1970b). A simplified treatment of the problem is given below.

Mathematical model to express trend components in terms of geological variables

Suppose that in the model of [15.13], the variables contained in $\mathbf{X}_2$ are regressed on those in $\mathbf{X}_1$ giving:

$$\hat{\mathbf{X}}_2 = \mathbf{X}_1\hat{\boldsymbol{\alpha}} \quad \text{with} \quad \hat{\boldsymbol{\alpha}} = (\mathbf{X}_1'\mathbf{X}_1)^{-1}\mathbf{X}_1'\mathbf{X}_2 \,. \qquad [15.25]$$

$\hat{\boldsymbol{\alpha}}$ is a matrix of regression coefficients with each of its columns defining the trend for one geological variable. Instead of regressing $\mathbf{Y}$ on $\mathbf{X}_1$ and $\mathbf{X}_2$ as before, we can regress $\mathbf{Y}$ on $\hat{\mathbf{X}}_2$ and $\mathbf{X}_2$ which gives:

$$\hat{\mathbf{Y}}_0^* = \hat{\mathbf{X}}_2\hat{\boldsymbol{\gamma}}_1 + \mathbf{X}_2\hat{\boldsymbol{\gamma}}_2 \,. \qquad [15.26]$$

The difference with [15.14] is that the trend now is a polynomial $\hat{\mathbf{X}}_2\hat{\boldsymbol{\gamma}}_1$ which is a linear combination of the individual trends for all geological variables. It will be shown that:

$$\hat{\boldsymbol{\gamma}}_2 = \hat{\boldsymbol{\beta}}_2$$

Proof. $\hat{\boldsymbol{\gamma}}_2$ satisfies [15.15] when the matrix $\mathbf{A}$ is replaced by $\mathbf{A}^*$ with:

$$\mathbf{A}^* = \mathbf{I}_n - \hat{\mathbf{X}}_2(\hat{\mathbf{X}}_2'\hat{\mathbf{X}}_2)^{-1}\hat{\mathbf{X}}_2' \qquad [15.27]$$

From [8.54], it follows that $\hat{\mathbf{X}}_2'\hat{\mathbf{X}}_2 = \mathbf{X}_2'\hat{\mathbf{X}}_2$. Hence:

$$\mathbf{X}_2'\mathbf{A}^* = \mathbf{X}_2' - \mathbf{X}_2'\hat{\mathbf{X}}_2(\mathbf{X}_2'\hat{\mathbf{X}}_2)^{-1}\hat{\mathbf{X}}_2' = \mathbf{X}_2' - \hat{\mathbf{X}}_2$$

$$= \mathbf{X}_2' - \mathbf{X}_2'\mathbf{X}_1(\mathbf{X}_1'\mathbf{X}_1)^{-1}\mathbf{X}_1' = \mathbf{X}_2'\mathbf{A} \qquad [15.28]$$

In [15.15], $\mathbf{A}$ occurs as part of the product $\mathbf{X}_2'\mathbf{A}$. Consequently, it can be replaced by $\mathbf{A}^*$, and $\hat{\boldsymbol{\beta}}_2$ by $\hat{\boldsymbol{\gamma}}_2$, which completes the proof.
Subtraction of $\hat{\mathbf{Y}}_0 = \mathbf{X}_1\hat{\boldsymbol{\beta}}_1 + \mathbf{X}_2\hat{\boldsymbol{\beta}}_2$ from $\hat{\mathbf{Y}}_0^* = \hat{\mathbf{X}}_2\hat{\boldsymbol{\gamma}}_1 + \mathbf{X}_2\hat{\boldsymbol{\beta}}_2$ gives the bias:

$$\hat{\mathbf{Y}}_0^* - \hat{\mathbf{Y}}_0 = \hat{\mathbf{X}}_2\hat{\boldsymbol{\gamma}}_1 - \mathbf{X}_1\hat{\boldsymbol{\beta}}_1$$

It is shown readily that $\mathbf{X}_2'\hat{\mathbf{X}}_2\hat{\boldsymbol{\gamma}}_1 = \mathbf{X}_2'\mathbf{X}_1\hat{\boldsymbol{\beta}}_1$. Consequently, $\mathbf{X}_2'\hat{\mathbf{Y}}_0^* = \mathbf{X}_2'\hat{\mathbf{Y}}_0$, or if $\mathbf{X}_2'\overline{\mathbf{Y}}$ is added:

$$\mathbf{X}_2'\hat{\mathbf{Y}}^* = \mathbf{X}_2'\hat{\mathbf{Y}} \qquad [15.29]$$

In general, $\hat{\mathbf{Y}}^*$ does not provide as good a fit as $\hat{\mathbf{Y}}$. The bias $\hat{\mathbf{Y}}^* - \hat{\mathbf{Y}} = \hat{\mathbf{Y}}_0^* - \hat{\mathbf{Y}}_0$ also can be estimated from the relationship:

$$\hat{\mathbf{Y}}^* - \hat{\mathbf{Y}} = \hat{\mathbf{Y}}_S - \hat{\mathbf{Y}}_T^* \qquad [15.30]$$

where $\hat{\mathbf{Y}}_T^*$ is the biased trend of $\mathbf{Y}$ (see, e.g., [15.22]) and $\hat{\mathbf{Y}}_S$ is obtained by regressing $\mathbf{Y}$ on $\hat{\mathbf{X}}_2$ only. The four sets of calculated values in [15.30] all have $\mathbf{Y}$ as a dependent variable. Hence, the sums of squares for the calculated values satisfy:

$$\hat{\mathbf{Y}}^{*\prime}\mathbf{Y} - \hat{\mathbf{Y}}'\mathbf{Y} = \hat{\mathbf{Y}}_S'\mathbf{Y} - \hat{\mathbf{Y}}_T^{*\prime}\mathbf{Y} \qquad [15.31]$$

Eq. [15.31] remains true if the subscript zero is added to all vectors.

For the example of Fig. 128, it was found that the biased estimate $\hat{\mathbf{Y}}^*$ accounts for 53% of the total variation $\mathbf{Y}_0'\mathbf{Y} = \Sigma(Y_k - \overline{Y})^2$ as compared with 68% for $\hat{\mathbf{Y}}$. If $\mathbf{Y}$ is regressed on the variables in $\hat{\mathbf{X}}_2$, we obtain a 46-% fit. The (biased) cubic trend $\hat{\mathbf{Y}}_T^*$ accounts for 61%. These four results satisfy $53 - 68 = 46 - 61 = -15\%$ which is in agreement with [15.31].

In order to illustrate these results, we have approximated the components $\hat{\mathbf{Y}}_T$ and $\hat{\mathbf{Y}}_G$ by estimates of $\hat{\mathbf{X}}_2\hat{\boldsymbol{\gamma}}_1$ and $\hat{\mathbf{Y}}_G$ which were based on a limited number of variables. First, $\mathbf{Y}$ was regressed on the six variables in $\hat{\mathbf{X}}_2$ by using the method of all possible regressions. This provided $2^6 - 1 = 63$ estimates of the biased trend $\hat{\mathbf{Y}}_S$. The single variable whose cubic trend is most strongly correlated to that for $\mathbf{Y}$ is sedimentary rocks with $R^2 = 30.3\%$. The best combination of two variables is for granitic rocks and mafic volcanics with $R^2 = 45.5\%$ which is nearly equal to the maximum value of $R^2 = 45.9\%$ obtained by using all variables. It is interesting to note that this best combination of two variables does not include the trend for the sedimentary rocks which was selected first. It illustrates that stepwise regression and forward selection do not necessarily yield the best-fitting combination of independent variables at each step.

Although we now have a good approximation of $\hat{\mathbf{Y}}_S$, this is not an estimate of $\hat{\mathbf{X}}_2\hat{\boldsymbol{\gamma}}_1$ since from previous results it follows that $\hat{\mathbf{Y}}_S = \hat{\mathbf{X}}_2\hat{\boldsymbol{\gamma}}_1 + \hat{\mathbf{Y}}_B$. In the next step of this experiment, we fitted $\mathbf{Y}$ to data for eight variables ($\hat{\mathbf{X}}_1$, $\hat{\mathbf{X}}_6$ and $\mathbf{X}_1 - \mathbf{X}_6$), again using the method of all possible regressions which gave $2^8 - 1 = 255$ solutions. The maximum R

amounted to 52.8% which was almost reached by using four variables ($\hat{\mathbf{X}}_1$,$\hat{\mathbf{X}}_6$,$\mathbf{X}_2$ and $\mathbf{X}_5$) with R^2 = 50.6%. This provided an estimate of $\hat{\mathbf{Y}}^* = \hat{\mathbf{X}}_2\hat{\gamma}_1 + \hat{\mathbf{Y}}_G$ or:

$$\hat{\mathbf{Y}}^* \cong 22.55 - 0.271\hat{\mathbf{X}}_1 - 0.259\hat{\mathbf{X}}_6 - 0.175\mathbf{X}_2 + 0.204\mathbf{X}_5 \quad [15.32]$$

where $\mathbf{X}_2$ and $\mathbf{X}_5$ represent mafic intrusions and acidic volcanics, respectively. This result was divided into the components $\hat{\mathbf{X}}_2\hat{\gamma}_1$ and $\hat{\mathbf{Y}}_G$. The approximate difference ($\hat{\mathbf{X}}_2\hat{\gamma}_1 - \hat{\mathbf{Y}}_T$) and component $\hat{\mathbf{Y}}_G$ are shown in Fig.128C and D, respectively.

The coefficients of [15.32] indicate that the gold deposits occur: (1) in areas where granitic and mafic volcanic rocks are least abundant (both $\hat{\mathbf{X}}_1$ and $\hat{\mathbf{X}}_6$ have negative coefficients); and (2) preferably, in 8 × 8 miles cells within these areas where acidic volcanics are abundant and mafic intrusions tend to be absent.

This answer is more meaningful than that given previously in section 8.7 where the estimate was $\hat{\mathbf{Y}} = 0.189 + 0.115\mathbf{X}_4 + 0.338\mathbf{X}_5$, with $\mathbf{X}_4$ and $\mathbf{X}_5$ representing sedimentary rocks and acidic volcanics, respectively. Eq.[15.32] accounts for 51% of total variation instead of 28% for the latter combination of two lithologies in 8 × 8 miles cells. In section 8.7, the problem was discussed that two types of sedimentary rocks were equally weighted; [15.32] implies, in indirect form, that regionally these two types of sedimentary rocks occur in different places.

Eq.[15.32] is not "stable" in that the result may change considerably when such factors as size and shape of study area and individual cells are altered. This instability is indicated by the fact that several other approximations are almost as good as the best solution. For example, regionally, the combination of acidic volcanics ($\hat{\mathbf{X}}_5$) and sedimentary rocks ($\hat{\mathbf{X}}_4$) accounts for 44.8% of total variation, compared to 45.5% for the best combination of granitic rocks ($\hat{\mathbf{X}}_1$) and mafic volcanics ($\hat{\mathbf{X}}_6$) used before. Likewise, the combination of variables $\hat{\mathbf{X}}_1$, $\hat{\mathbf{X}}_6$, $\mathbf{X}_5$ and $\mathbf{X}_3$ (ultramafics), accounts for 50.5% versus 50.6% for the combination of [15.32].

Concluding remarks

It is emphasized that the models discussed in this chapter are empirical. A best solution for the relationship between one variable and a set of other geological variables can be obtained by least-squares models. Except when few variables are considered, the complex interactions between all variables in the system usually prevent genetical relationships between variables emerging from the "best" multivariate solutions.

In this chapter, we have used a single dependent variable in all models. In reality, there may be a number of dependent variables which are interrelated. These interrelationships need not be considered to obtain a best solution for a set of dependent variables (Anderson, 1958, chapter 8) if they are independent of the measurement variables and geographic location in the region. For example, if copper occurs in a sulphide deposit in Abitibi there is a high probability that zinc also occurs. If this probability is the same for all sulphide deposits in the region, an optimum solution for the system of copper and zinc is obtained by treating copper and zinc separately as dependent variables. This is not usually so if the relationship between the dependent variables itself depends on the independent variables. The approach to be taken then could be similar to that used for solving certain systems of equations in econometrics (Goldberger, 1964; Malinvaux, 1966).

We have been able to account for the situation that a geological feature measured in one part of the study region has a lower probability of being indicative of specific mineral deposits than when it occurs elsewhere in the region, provided that these changes can be approximated by empirical deterministic functions. The bias introduced if not all deposits have been found in a region also was considered in detail.

Several experiments indicated that it can be advantageous to use large study areas with data coded for many cells and to quantify many features for the geological framework. For example, in Fig.122, a more complex method was required to derive results for smaller study areas whereas similar results could be derived by a simpler method when these subareas are made part of a larger study area.

The mathematical models can be made rather flexible. Situations they can cope with are: (1) the objects (mineral deposits) related to features for the geological framework need not be of the same type in that all members of a group would have to be associated with the same features; (2) the geological environment need not be of one type. In fact, we could lump environments with entirely different features together. The final prediction equation then may be divided into parts, one for each environment; and (3) a feature (e.g., a sedimentary belt) at one place may indicate something different from that same feature at another place in the study region. This flexibility is important because there is a limit to the number of clear-cut distinctions that can be made between types of deposits, broader geological environments, and individual geological features because these phenomena may be part of a continuum. Nevertheless, the choice of types of deposits, features and environments is important because resolution is lost whenever too many features are lumped together. Significant progress will be possible only when geologists who are experts in regional geology and metallogenic theory are involved both in selecting the variables and in evaluation of results.

Moreover, the features that were used in the geomathematical analysis for larger regions represent a lowest common denominator of a vast amount of information actually in existence.

BIBLIOGRAPHY

SELECTED READING

The following five lists of textbooks in geomathematics, mathematics and statistics have been compiled for those interested in further reading. References cited in the text are listed separately in the second part of the bibliography.

1. Annotated bibliography of textbooks and monographs on mathematics and statistics in geology

Blackith, R.E. and Reyment, R.A., 1971. *Multivariate Morphometrics.* Academic Press, New York, N.Y., 435 pp.
Explains methods of multivariate statistical analysis in morphometrics; applications in numerical taxonomy and biostratigraphy.

Chayes, F., 1971. *Ratio Correlation*. University of Chicago Press, Chicago, Ill., 99 pp.
Review of methods for treating ratios especially those arising from closed-number systems in petrology and geochemistry.

Davis, J.C., 1973. *Statistics and Data Analysis in Geology*. Wiley, New York, N.Y., 550 pp.
Discusses how the digital computer can be used for solving geological problems; elementary in scope; contains a number of short computer programs.

Fenner, P. (Editor), 1969. *Models of Geologic Processes; An Introduction to Mathematical Geology. A.G.I. – Short Course Lecture Notes.* Am. Geol. Inst., Washington, D.C., 359 pp.
Notes of lectures by W.C. Krumbein, M.E. Kauffman, R.B. McCammon, D.B. McIntyre and H.N. Pollack. Emphasis is on geologic process-models and multivariate analysis; computer programs include set of algorithms for petrologists in A.P.L./360 which can be used on a terminal connected to a computer installation by telephone line.

Griffiths, J.C., 1967. *Scientific Method in the Analysis of Sediments.* McGraw-Hill, New York, N.Y., 508 pp.
Thorough basic book on measurement and statistical analysis of geological features in sedimentary petrology; general discussion of scientific methods in geology and the contribution to knowledge obtained by doing quantitative measurements on geological objects.

Harbaugh, J.W. and Bonham-Carter, G. 1970. *Computer Simulation in Geology: Modelling Dynamic Systems.* Wiley, New York, N.Y., 575 pp.
Discusses computer methods for simulating time-dependent processes in stratigraphy and geology; provides background in simulation theory; many annotated references.

Harbaugh, J.W. and Merriam, D.F., 1968. *Computer Applications in Stratigraphic Analysis.* Wiley, New York, N.Y., 228 pp.
Covers large number of computer-based statistical techniques such as trend-surface analysis, harmonic analysis and factor analysis; applications to stratigraphic correlation problems.

Hazen Jr., S.W., 1967. *Some Statistical Techniques for Analyzing Mine and Mineral-Deposit Sample and Assay Data. U.S., Bur. Mines, Bull.*, 621: 223 pp.
Discusses statistics of mine sampling; extensive discussion of experimental project of sampling the Climax molybdenite deposit near Denver, Colo. by bore-holes.

Koch Jr., G.S. and Link, R.F., 1971. *Statistical Analysis of Geological Data.* Wiley, New York, N.Y. Vol. 1 (1970), 375 pp.; Vol. 2 (1971), 438 pp.
Volume 1 discusses elementary statistical techniques and methods of sampling; the authors have closely tied volume 1 to the following book on applied statistics: Li, J.C.R., 1964. *Statistical Inference.* Edwards, Ann Arbor, Mich. Vol. 1 (658 pp.); Vol. 2 (575 pp.). Volume 2 deals with spatial variability and multivariate analysis; other subjects include the theory of search in exploration for mineral deposits.

Krumbein, W.C. and Graybill, F.A., 1965. *An Introduction to Statistical Models in Geology.* McGraw-Hill, New York, N.Y., 475 pp.
Well-known basic textbook; ranges in scope from elementary to fairly advanced; excellent discussion of trend-surface analysis and other applications of the general linear model.

Laffitte, P. (Editor), 1972. *Traité d'informatique géologique.* Masson, Paris, 624 pp.
Comprehensive volume by 26 authors dealing with a great variety of techniques for electronic data processing in the geo-sciences; contains detailed descriptions of several geological data management systems, advanced statistical techniques, and graphical display methods.

Marsal, D., 1967. *Statistische Methoden für Erdwissenschaftler.* Schweizerbart, Stuttgart, 152 pp.
Discussion of elementary statistical techniques with application to geoscience data.

Matheron, G., 1962. *Traité de Géostatistique appliquée. Mém. Bur. Rech. Géol. Minières,* 14: 333 pp.
Theory and applications of G. Matheron's geostatistical theory for estimating averages and their variance in mine sampling; nomograms to represent solutions for more frequently occurring sampling schemes are based on the intrinsic model of De Wijs mainly.

Matheron, G., 1965. *Les Variables Régionalisées et leur Estimation.* Masson, Paris, 306 pp.
Basic theoretical work at an advanced level; outlines concepts of the theory of spatial variability using multi-dimensional stochastic processes theory; no practical applications.

Matheron, G., 1967. *Eléments pour une Théorie des Milieux Poreux.* Masson, Paris, 166 pp.
Develops a geomathematical theory for porous media leading to new methods to describe and treat problems of the shape of grains and textures.

Merriam, D.F. (Editor), 1969. *Computer Applications in the Earth Sciences.* Plenum, New York, N.Y. 281 pp.
First book of continuing series edited by D.F. Merriam; contains fifteen comprehensive review papers dealing with different subfields of geology and related disciplines.

Merriam, D.F. (Editor), 1970. *Geostatistics.* Plenum, New York, N.Y. 177 pp.
Proceedings of colloquium by statisticians with interests in applications in the earth sciences and geologists using statistics; twelve papers with index.

Miller, R.L. and Kahn, J.S., 1962. *Statistical Analysis in the Geological Sciences.* Wiley, New York, N.Y., 483 pp.
Structured as a textbook of mathematical statistics with many applications to earth-science data including paleontological data.

Reyment, R.A., 1971. *Introduction to Quantitative Paleoecology.* Elsevier, Amsterdam, 226 pp.
Deals with orientation analysis and spatial paleoecology; covers quantitative analysis and population dynamics of fossil assemblages.

Romanova, M.A. and Sarmanov, O.V. (Editors), 1970. *Topics in Mathematical Geology.* Consultants Bureau, New York, N.Y., 281 pp.
Translation of collection of papers originally published in Russian by Nauka, Leningrad, 1968; 24 contributions from various countries to solution of set of problems arranged by A.B. Vistelius.

Sharapov, I., 1971. *Applications of Mathematical Statistics in Geology.* Nedra, Moscow, 2nd ed. 26[illegible] pp. (In Russian).
Emphasis on probability calculus and basic concepts of mathematical statistics with examples from stratigraphy, geochemistry, and other fields.

Smith, F.G., 1966. *Geological Data Processing using FORTRAN IV.* Harper and Row, New York, N.Y., 284 pp.
Stresses that every geological scientist must, of necessity, be able to translate problems into language for machine computation; provides introduction to Boolean logic and FORTRAN-programming.

Thiergartner, H., 1968. *Grundprobleme der statistischen Behandlung geochemischer Daten.* Deutscher Verlag für Grundstoffindustrie, Leipzig, 99 pp.
Review of basic statistical techniques applied to geochemical exploration data.

Vistelius, A.B., 1967. *Studies in Mathematical Geology.* Consultants Bureau, New York, N.Y., 294 pp.
Collection of papers by A.B. Vistelius translated from the Russian; includes original work on stochastic process models for stratigraphic sections; includes applications of time-trend analysis and trend-surface theory.

2. Elementary reading

2.1. *Introduction to calculus (Chapter 2)*

Agnew, R.P., 1960. *Differential Equations.* McGraw-Hill, New York, N.Y., 485 pp.

Courant, R. and John, F., 1965. *Introduction to Calculus and Analysis.* Interscience, New York, N.Y., 661 pp.

Kaplan, W., 1952. *Advanced Calculus.* Addison-Wesley, Reading, Mass., 679 pp.

Kells, L.N., 1960. *Elementary Differential Equations.* McGraw-Hill, New York, N.Y., 318 pp.

Lowry, H.V. and Hayden, H.A., 1955. *Advanced Mathematics for Technical Students, II.* Longmans Green, London, 422 pp.

Small, L.L., 1949. *Calculus.* Appleton-Century-Crofts, New York, N.Y., 592 pp.

Sokolnikoff, I.S. and Redheffer, R.M., 1958. *Mathematics of Physics and Modern Engineering.* McGraw-Hill, New York, N.Y., 810 pp.

Thomas Jr., G.B., 1969. *Calculus.* Addison-Wesley, Cambridge, Mass., 3rd ed., 704 pp.

2.2. *Elementary matrices (Chapter 3)*

Bellman, R., 1960. *Introduction to Matrix Analysis.* McGraw-Hill, New York, N.Y., 328 pp.

Davis, P.J., 1965. *The Mathematics of Matrices.* Blaisdell, Waltham, Mass., 348 pp.

Finkbeiner, D.T., 1960. *Introduction to Matrices and Linear Transformations.* Freeman, San Francisco, Calif., 297 pp.

Frazer, R.A., Duncan, W.J. and Collar, A.R., 1960. *Elementary Matrices.* Cambridge University Press, Cambridge, 416 pp.

Gere, J.M. and Weaver Jr., W., 1965. *Matrix Algebra for Engineers.* Van Nostrand, Princeton, N.J., 168 pp.

Graybill, F.A., 1969. *An Introduction to Matrices with Application in Statistics.* Wadsworth, Belmont, Calif., 372 pp.

Hadley, G., 1961. *Linear Algebra.* Addison-Wesley, Reading, Mass., 290 pp.

.3. *Geometry (Chapter 4)*

Albert, A., 1949. *Solid Analytic Geometry.* McGraw-Hill, New York, N.Y., 162 pp.

order, H.G., 1950. *Geometry.* Hutchinson, London, 200 pp.

Maxwell, E.A., 1961. *General Homogeneous Coordinates in Space of Three Dimensions.* Cambridge University Press, Cambridge, 169 pp.

akley, C.O., 1949. *Analytic Geometry.* Barnes and Noble, New York, N.Y., 246 pp.

Phillips, H.B., 1955. *Vector Analysis.* Wiley, New York, N.Y., 236 pp.

2.4. Set theory (Section 6.2)

Halmos, P.R., 1960. *Naive Set Theory.* Van Nostrand, Princeton, N.J., 104 pp.

2.5. Elementary statistics (Chapter 6)

Dixon, W.J. and Massey Jr., F.J., 1957. *Introduction to Statistical Analysis.* McGraw-Hill, New Yor N.Y., 488 pp.
Hoel, P.G., 1960. *Elementary Statistics.* Wiley, New York, N.Y., 261 pp.
Moroney, M.J., 1956. *Facts from Figures.* Penguin Books, Baltimore, Md., 3rd ed., 472 pp.
Neville, A.M. and Kennedy, J.B., 1964. *Basic Statistical Methods for Engineers and Scientists.* Inte national Textbook, Scranton, Pa., 325 pp.
Sachs, L., 1971. *Statistische Auswertungsmethoden.* Springer, Berlin, 3rd ed., 545 pp.
Snedecor, G.W., 1956. *Statistical Methods.* Iowa State University Press, Ames, Iowa, 534 pp.
Walpole, R.E., 1968. *Introduction to Statistics.* MacMillan, New York, N.Y., 365 pp.

3. Handbooks, tables and computing

3.1. Mathematical handbooks

Abramowitz, M. and Stegun, I.A. (Editors), 1965. *Handbook of Mathematical Functions. Appl. Ma Ser., 55.* National Bureau of Standards, Washington, D.C., 1046 pp.
Burington, R.S., 1958. *Handbook of Mathematical Tables and Formulas.* Handbook Publishers, Sa dusky, Ohio, 296 pp.
Jahnke, E. and Emde, F., 1945. *Tables of Functions.* Dover, New York, N.Y., 382 pp.
Korn, G.A. and Korn, T.M., 1961. *Mathematical Handbook for Scientists and Engineers.* McGraw-Hi New York, N.Y., 943 pp.

3.2. Numerical analysis and computer algorithms

Conté, S.D., 1965. *Elementary Numerical Analysis.* McGraw-Hill, New York, N.Y., 278 pp.
Fröberg, C.E., 1969. *Introduction to Numerical Analysis.* Addison-Wesley, Reading, Mass., 2nd ed 433 pp.
Hamming, R.W., 1962. *Numerical Methods for Scientists and Engineers.* McGraw-Hill, New Yor N.Y., 411 pp.
Meyer, H.A. (Editor), 1956. *Symposium on Monte Carlo Methods.* Wiley, New York, N.Y., 382 p
Ralston, A., 1965. *A First Course in Numerical Analysis.* McGraw-Hill, New York, N.Y., 578 pp.
Ralston, A. and Wilf, H.S. (Editors), 1960. *Mathematical Methods for Digital Computers.* Wiley, Ne York, N.Y., Vol. 1, 293 pp; Vol. 2, 287 pp.

3.3. Statistical dictionaries, tables and computer programs

Delury, D.B., 1950. *Values and Integrals of the Orthogonal Polynomials up to n = 26.* University Toronto Press, Toronto, Ont., 33 pp.
Dixon, W.J. (Editor), 1967. *BMD Biomedical Computer Programs (FORTRAN IV).* University California Press, Los Angeles, Calif., 600 pp.
Fisher, R.A. and Yates, F., 1963. *Statistical Tables for Biological Agricultural and Medical Researc* Oliver and Boyd, Edinburgh, 6th ed., 146 pp.
Freund, J.E. and Williams, F.J., 1966. *Dictionary/Outline of Basic Statistics.* McGraw-Hill, New Yor N.Y., 195 pp.

Greenwood, J.A. and Hartley, H.O., 1962. *Guide to Tables in Mathematical Statistics.* Princeton University Press, Princeton, N.J., 1014 pp.

Haight, F.A., 1967. *Handbook of the Poisson Distribution.* Wiley, New York, N.Y., 168 pp.

Hald, A., 1952. *Statistical Tables and Formulas.* Wiley, New York, N.Y., 97 pp.

Hemmerle, W.J., 1967. *Statistical Computations on a Digital Computer (FORTRAN).* Blaisdell, Waltham, Mass., 230 pp.

Kendall, M.G. and Buckland, W.R., 1967. *A Dictionary of Statistical Terms.* Hafner, New York, N.Y., 2nd ed., 575 pp.

Kurtz, A.K. and Edgerton, H.A., 1967. *Statistical Dictionary of Terms and Symbols.* Hafner, New York, N.Y., 191 pp.

Lancaster, H., 1968. *Bibliography of Statistical Bibliographies.* Oliver and Boyd, Edinburgh, 103 pp.

Morice, E., 1968. *Dictionnaire de statistique.* Dunod, Paris, 196 pp.

Pearson, E.S. and Hartley, H.O., 1958. *Biometrika Tables for Statisticians, 1.* Cambridge University Press, Cambridge, 240 pp.

Sachs, L., 1970. *Statistische Methoden – Ein Soforthelfer.* Springer, Berlin, 103 pp.

4. More advanced mathematical methods

4.1. Mathematical analysis

Birkhoff, G. and MacLane, S., 1953. *A Brief Survey of Modern Algebra.* MacMillan, New York, N.Y., 276 pp.

Buck, R.C., 1956. *Advanced Calculus.* McGraw-Hill, New York, N.Y., 423 pp.

Olmsted, J.M.H., 1956. *Real Variables.* Appleton-Century-Crofts, New York, N.Y., 621 pp.

Rudin, W., 1964. *Principles of Mathematical Analysis.* McGraw-Hill, New York, N.Y., 270 pp.

4.2. Applied mathematics

Courant, R. and Hilbert, D., 1962. *Methods of Mathematical Physics.* Interscience, New York, N.Y., Vol. I, 561 pp.; Vol. II, 830pp.

Goursat, E., 1964. *Integral Equations; Calculus of Variations. Course in Mathematical Analysis, Vol. III, Part 2.* Dover, New York, N.Y., 389 pp.

Hay, G.E., 1953. *Vector and Tensor Analysis.* Dover, New York, N.Y., 193 pp.

Hildebrand, F.B., 1962. *Advanced Calculus for Applications.* Prentice-Hall, Englewood Cliffs, N.J., 646 pp.

Lovitt, W.V., 1950. *Linear Integral Equations.* Dover, New York, N.Y., 253 pp.

Smith, L.P., 1953. *Mathematical Methods for Scientists and Engineers.* Dover, New York, N.Y., 453 pp.

Sommerville, D.M.Y., 1958. *An Introduction to the Geometry of N dimensions.* Dover, New York, N.Y., 196 pp.

4.3. Theory of matrices

Schneider, H. (Editor), 1964. *Recent Advances in Matrix Theory.* University of Wisconsin Press, Madison, Wisc., 142 pp.

urnbull, H.W., 1960. *The Theory of Determinants, Matrices, and Invariants.* Dover, New York, N.Y., 374 pp.

urnbull, H.W. and Aitken, A.C., 1961. *An Introduction to the Theory of Canonical Matrices.* Dover, New York, N.Y., 200 pp.

Wedderburn, J.H.M., 1964. *Lectures on Matrices.* Dover, New York, N.Y., 200 pp.

Wilkinson, J.H., 1965. *The Algebraic Eigenvalue Problem.* Oxford University Press, Oxford, 662 pp.

4.4. Fourier series and Fourier transforms

Bochner, S. and Chandrasekharan, K., 1949. *Fourier Transforms.* Princeton University Press, Princ ton, N.J., 219 pp.
Carslaw, H.S., 1930. *Introduction to the Theory of Fourier's Series and Integrals.* Dover, New Yo N.Y., 368 pp.
Davis, H.F., 1963. *Fourier Series and Orthogonal Functions.* Allyn and Bacon, Boston, Mass., 401 p
Jackson, D., 1941. *Fourier Series and Orthogonal Polynomials. Carus Math. Monogr.*, 6. Math. Assc Am., Buffalo, N.Y., 234 pp.
Titchmarsh, E.C., 1948. *Introduction to the Theory of Fourier Integrals.* Oxford University Pre Oxford, 2nd ed., 394 pp.
Wiener, N., 1933. *The Fourier Integral and certain of its Applications.* Dover, New York, N.Y., 2 pp.

5. Statistical theory and methods

5.1. General

Anderson, R.L. and Bancroft, T.A., 1952. *Statistical Theory in Research.* McGraw-Hill, New Yo N.Y., 399 pp.
Bradley, J.V., 1968. *Distribution-free Statistical Tests.* Prentice-Hall, Englewood Cliffs, N.J., 388 p
Deming, W.E., 1948. *Statistical Adjustment of Data.* Wiley, New York, N.Y., 261 pp.
Draper, N.R. and Smith, H., 1966. *Applied Regression Analysis.* Wiley, New York, N.Y., 407 pp.
Elderton, W.P. and Johnson, N.L., 1968. *Systems of Frequency Curves.* Cambridge University Pre Cambridge, 216 pp.
Ezekiel, M. and Fox, K.A., 1959. *Methods of Correlation and Regression Analysis.* Wiley, New Yo N.Y., 548 pp.
Fraser, D.A.S., 1958. *Statistics, an Introduction.* Wiley, New York, N.Y., 398 pp.
Hald, A., 1952. *Statistical Theory with Engineering Applications.* Wiley, New York, N.Y., 783 pp.
Kullback, S., 1959. *Information Theory and Statistics.* Wiley, New York, N.Y., 395 pp.
McMillan, C. and Gonzalez, R.F., 1965. *Systems Analysis, a Computer Approach to Decision Mode* Irwin, Homewood, Ill., 336 pp.
Mood, A.M. and Graybill, F.A., 1963. *Introduction to the Theory of Statistics.* McGraw-Hill, N York, N.Y., 443 pp.
Rao, C.R., 1965. *Linear Statistical Inference and its Applications.* Wiley, New York, N.Y., 522 pp.
Scheffé, H., 1959. *The Analysis of Variance.* Wiley, New York, N.Y., 477 pp.

5.2. Sampling and design of experiments

Cochran, W.G., 1963. *Sampling Techniques.* Wiley, New York, N.Y., 2nd ed., 413 pp.
Cox, D.R., 1958. *Planning of Experiments.* Wiley, New York, N.Y., 308 pp.
Fisher, R.A., 1960. *The Design of Experiments.* Oliver and Boyd, Edinburgh, 248 pp.
Gy, P., 1967. *L'échantillonnage des minerais en vrac, 1. Mém. Bur. Rech. Géol. Minières,* 56, 186 p
Kempthorne, O., 1952. *The Design and Analysis of Experiments.* Wiley, New York, N.Y., 631 pp.
Stuart, A., 1962. *Basic Ideas of Scientific Sampling.* Griffin, London, 99 pp.

5.3. Probability and mathematical statistics

Cramér, H., 1947. *Mathematical Methods of Statistics.* Princeton University Press, Princeton, N.J., 5 pp.
Feller, W., 1968. *An Introduction to Probability Theory and its Applications.* Wiley, New York, N. Vol. I, 3rd ed., 509 pp.; Vol. II (1966), 626 pp.
Gnedenko, B.V., 1963. *The Theory of Probability.* Chelsea, New York, N.Y., 471 pp.

Kendall, M. and Moran, P.A.P., 1963. *Geometrical Probability.* Griffin, London, 125 pp.
Kendall, M.G. and Stuart, A., 1958. *The Advanced Theory of Statistics.* Hafner, New York, N.Y., Three-volume edition, Vol. 1, 433 pp.; Vol. 2, 676 pp.; Vol. 3, 552 pp.
Loève, M., 1963. *Probability Theory.* Van Nostrand, Princeton, N.J., 3rd ed., 685 pp.
Wilks, S.S., 1962. *Mathematical Statistics.* Wiley, New York, N.Y., 644 pp.

5.4. Multivariate analysis

Anderson, T.W., 1958. *Introduction to Multivariate Statistical Analysis.* Wiley, New York, N.Y., 374 pp.
Kendall, M.G., 1965. *A Course in Multivariate Statistics.* Griffin, London, 2nd ed., 185 pp.
Morrison, D.F., 1967. *Multivariate Statistical Methods.* McGraw-Hill, New York, N.Y., 338 pp.
Quenouille, M.H., 1952. *Associated Measurements.* Butterworth, London, 242 pp.
Rao, C.R., 1952. *Advanced Statistical Methods in Biometric Research.* Wiley, New York, N.Y., 390 pp.

5.5. Application of statistics and multivariate analysis in other sciences

Berry, B.J.L. and Marble, D.F. (Editors), 1968. *Spatial Analysis, a Reader in Statistical Geography.* Prentice-Hall, Englewood Cliffs, N.J., 512 pp.
Cole, J.P. and King, C.A.M., 1968. *Quantitative Geography.* Wiley, New York, N.Y., 692 pp.
Cooley, W.W. and Lohnes, P.R., 1962. *Multivariate Procedures for the Behavioral Sciences.* Wiley, New York, N.Y., 211 pp.
Dhrymes, P.J., 1970. *Econometrics; Statistical Foundations and Applications.* Harper and Row, New York, N.Y., 592 pp.
Goldberger, A.S., 1964. *Econometric Theory.* Wiley, New York, N.Y., 399 pp.
Gould, P., 1967. On the geographic interpretation of eigenvalues: An initial exploration. *Trans. Inst. Br. Geogr.* 42: 53–86.
Harman, H.H., 1960. *Modern Factor Analysis.* University of Chicago Press, Chicago, Ill., 471 pp.
King, L.J., 1969. *Statistical Analysis in Geography.* Prentice-Hall, Englewood Cliffs, N.J., 288 pp.
Malinvaux, E., 1966. *Statistical Methods of Econometrics.* Rand McNally, Chicago, Ill., 631 pp.
Panofsky, H.A. and Brier, G.W., 1965. *Some Applications of Statistics to Meteorology.* Pennsylvania State Univ., University Park, Pa., 224 pp.
Sokal, R.R. and Rohlf, F.J., 1969. *Biometry.* Freeman, San Francisco, Calif., 776 pp.
Sokal, R.R. and Sneath, P.H.A., 1963. *Principles of Numerical Taxonomy.* Freeman, San Franciso, Calif., 359 pp.
Theil, H., 1971. *Principles of Econometrics.* Wiley, New York, N.Y., 736 pp.
Wonnacott, R.J. and Wonnacott, T.H., 1970. *Econometrics.* Wiley, New York, N.Y., 445 pp.

5.6. Time series and spectral analysis

Bendat, J.S. and Piersol, A.G., 1966. *Measurement and Analysis of Random Data.* Wiley, New York, N.Y., 390 pp.
Blackman, R.B. and Tukey, J.W., 1958. *The Measurement of Power Spectra.* Dover, New York, N.Y., 190 pp.
Brown, R.G., 1963. *Smoothing, Forecasting, and Prediction of Discrete Time Series.* Prentice-Hall, Englewood Cliffs, N.J., 468 pp.
Cox, D.R. and Lewis, P.A.W., 1966. *The Statistical Analysis of Series of Events.* Methuen, London, 285 pp.
Cramér, H. and Leadbetter, M.R., 1968. *Stationary and Related Stochastic Processes.* Wiley, New York, N.Y., 348 pp.

Granger, C.W.J. and Hatanaka, M., 1964. *Spectral Analysis of Economic Time Series.* Princeton University Press, Princeton, N.J., 299 pp.
Grenander, U. and Rosenblatt, M., 1957. *Statistical Analysis of Stationary Time Series.* Wiley, New York, N.Y., 300 pp.
Jenkins, G.M. and Watts, D.G., 1968. *Spectral Analysis and its Applications.* Holden-Day, San Francisco, Calif., 525 pp.
Lee, Y.W., 1960. *Statistical Theory of Communication.* Wiley, New York, N.Y., 251 pp.
Pugachev, V.S., 1965. *Theory of Random Functions and its Application to Control Problems.* Pergamon, Oxford, 833 pp.
Whittle, P., 1963. *Prediction and Regulation.* Van Nostrand, Princeton, N.J., 147 pp.
Yaglom, A.M., 1962. *An Introduction to the Theory of Stationary Random Functions.* Prentice-Hall, Englewood Cliffs, N.J., 235 pp.

5.7. Stochastic processes

Bailey, N.T.J., 1964. *The Elements of Stochastic Processes with Applications to the Natural Sciences.* Wiley, New York, N.Y., 249 pp.
Bartlett, M.S., 1966. *An Introduction to Stochastic Processes.* Cambridge University Press, Cambridge, 2nd ed., 362 pp.
Bharucha-Reid, A.T., 1960. *Elements of the Theory of Markov Processes and their Applications.* McGraw-Hill, New York, N.Y., 457 pp.
Chiang, C.L., 1968. *Introduction to Stochastic Processes in Biostatistics.* Wiley, New York, N.Y., 312 pp.
Cox, D.R. and Miller, H.D., 1965. *The Theory of Stochastic Processes.* Wiley, New York, N.Y., 398 pp.
Kemeny, J.G. and Snell, J.L., 1960. *Finite Markov Chains.* Van Nostrand, Princeton, N.J., 210 pp.
Papoulis, A., 1965. *Probability, Random Variables and Stochastic Processes.* McGraw-Hill, New York, N.Y., 583 pp.
Parzen, E., 1962. *Stochastic Processes.* Holden-Day, San Francisco, Calif., 324 pp.
Quenouille, M.H., 1957. *The Analysis of Multiple Time-Series.* Griffin, London, 104 pp.

REFERENCES CITED

Abramowitz, M. and Stegun, I.A. (Editors), 1965. *Handbook of Mathematical Functions, Appl. Math. Ser., 55.* National Bureau of Standards, Washington, D.C., 1046 pp.
Adler, R.E. and Pilger, A., 1968. Elektronische Statistik, ein Hilfsmittel der modernen Tektonik. *Int Geol. Congr., 23rd, Prague, 1968, Proc.,* Sect. 13: 195–209.
Agterberg, F.P., 1959. On the measuring of strongly dispersed minor folds. *Geol. Mijnbouw,* 21 133–137.
Agterberg, F.P., 1961a. Tectonics of the crystalline basement of the Dolomites in North Italy. *Geol Ultraiectina,* 8: 232 pp.
Agterberg, F.P., 1961b. The skew frequency curve of some ore minerals. *Geol. Mijnbouw,* 40: 149-162.
Agterberg, F.P., 1964a. Statistical techniques for geological data. *Tectonophysics,* 1: 233–255.
Agterberg, F.P., 1964b. Methods of trend surface analysis. *Q. Colo. Sch. Mines,* 59 (4): 111–130.
Agterberg, F.P., 1964c. Statistical analysis of X-ray data for olivine. *Mineral. Mag.,* 33: 742–748.
Agterberg, F.P., 1965a. The technique of serial correlation applied to continuous series of element concentration values in homogeneous rocks. *J. Geol.,* 73: 142–154.
Agterberg, F.P., 1965b. Frequency distribution of trace elements in the Muskox layered intrusion. In J.C. Dotson and W.C. Peters (editors), *Short Course and Symposium on Computers and Compute*

Applications in Mining and Exploration. College of Mines, Univ. Arizona, Tucson, Ariz., pp. G1–G33.

Agterberg, F.P., 1966a. The use of multivariate Markov schemes in petrology. *J. Geol.*, 74: 764–785.

Agterberg, F.P., 1966b. Trend surfaces with autocorrelated residuals. *Proc. Symp. Comput. Oper. Res. Miner. Ind., Pa. State Univ.*, 1: H1–19.

Agterberg, F.P., 1967a. Computer techniques in geology. *Earth-Sci. Rev.*, 3: 47–77.

Agterberg, 1967b. Mathematical models in ore evaluation. *J. Can. Oper. Res. Soc.*, 5: 144–158.

Agterberg, F.P., 1968. Application of trend analysis in the evaluation of the Whalesback Mine, Newfoundland. *Can. Inst. Min. Metall, Spec. Vol.*, 9: 77–88.

Agterberg, F.P., 1969. Interpolation of areally distributed data. *Q. Colo. Sch. Mines*, 64 (3): 217–237.

Agterberg, F.P., 1970a. Autocorrelation functions in geology. In: D.F. Merriam (Editor), *Geostatistics*. Plenum, New York, N.Y., pp. 113–142.

Agterberg. F.P., 1970b. Multivariate prediction equations in geology. *J. Int. Assoc. Math. Geol.*, 2: 319–324.

Agterberg, F.P., 1971. A probability index for detecting favourable geological environments. *Can. Inst. Min. Metall., Spec. Vol.*, 12: 82–91.

Agterberg F.P. and Banerjee, I., 1969. Stochastic model for the deposition of varves in glacial Lake Barlow–Ojibway, Ont., Canada. *Can. J. Earth Sci.*, 6: 625–652.

Agterberg, F.P. and Briggs, G., 1963. Statistical analysis of ripple marks in Atokan and Desmoinesian rocks in the Arkoma Basin of east-central Oklahoma, *J. Sediment Petrol.*, 33: 393–410.

Agterberg, F.P. and Cabilio, P., 1969. Two-stage least-squares model for the relationship between mappable geological variables. *J. Int. Assoc. Math. Geol.*, 1: 37–153.

Agterberg, F.P. and Cameron, G.D., 1967. Computer program for the analysis of multivariate series and eigenvalue routine for asymmetrical matrices. *Geol. Surv. Can., Pap.*, 67–14. 54 pp.

Agterberg, F.P. and Fabbri, A.G., 1973. Harmonic analysis of copper and gold occurrences in the Abitibi area of the Canadian Shield. *Proc. 10th Symp. Appl. Comput. Miner. Ind., S. Afr. Inst. Min. Met.*, Johannesburg, pp. 193–201.

Agterberg, F.P. and Kelly, A.M., 1971. Geomathematical methods for use in prospecting. *Can. Min. J.*, 92 (5): 61–72.

Agterberg, F.P. and Robinson, S.C., 1972. Mathematical problems of geology. *Proc. Sess. Int. Statist. Inst., 38th, Washington, D.C., 1971, I.S.I. Bull.*, 44: 567–596.

Agterberg, F.P., Chung, C.F., Fabbri, A.G., Kelly, A.M. and Springer, J.S., 1972. Geomathematical evaluation of copper and zinc potential of the Abitibi area, Ontario and Quebec. *Geol. Surv. Can., Pap.*, 71–41: 55 pp.

Agterberg, F.P., Hills, L.V. and Trettin, H.P., 1967. Paleocurrent trend analysis of a delta in the Bjorne Formation (Lower Triassic) of northeastern Melville Island, Arctic Archipelago. *J. Sediment. Petrol.*, 37: 852–862.

Ahlberg, J.G., Nilson, E.N. and Walsh, J.L., 1967. *The Theory of Splines and their Applications*. Academic Press, New York, N.Y., 284 pp.

Ahrens, L.H., 1953. A fundamental law of geochemistry. *Nature*, 172: 1148.

Ahrens, L.H., 1964. Element distribution in igneous rocks – VII. A reconnaissance survey of the distribution of SiO_2 in granitic and basaltic rocks. *Geochim. Cosmochim. Acta*, 28: 271–290.

Aitchison, J. and Brown, J.A.C., 1957. *The Lognormal Distribution*. Cambridge Univ. Press, Cambridge, 176 pp.

Albert, A., 1949. *Solid Analytic Geometry*. McGraw-Hill, New York, N.Y., 162 pp.

Alexander-Marrack, P.D., Friend, P.F. and Yeats, A.K., 1970. Mark sensing for recording and analysis of sedimentological data. In: J.L. Cutbill. (Editor), *Data Processing in Biology and Geology*. Academic Press, London, pp. 1–16.

Allais, M., 1957. Method of appraising economic prospects of mining exploration over large territories; Algerian Sahara case study. *Manage. Sci.*, 3: 285–347.

Allegre, C., 1964. Vers une logique mathématique des séries sédimentaires. *Bull. Soc. Géol. France, Ser. 7*, 6: 214–218.

Ambrose, J.W. (Editor), 1969. Seminar on the causes and mechanics of glacier surges. *Can. J. Eart Sci.*, 6 (4): 212 pp.

Amos, D.E. and Koopmans, L.H., 1963. Tables of the distribution of the coefficient of coherence fo stationary bivariate Gaussian processes. *Sandia Corp., Mon. SCR*-483: 328 pp.

Anderson, R.Y., 1967. Sedimentary laminations in time-series study. *Kans. Geol. Surv., Compu. Contrib.*, 18: 68–72.

Anderson, R.Y. and Kirkland, D.W., 1966. Intrabasin varve correlation. *Geol. Soc. Am. Bull*, 77 241–256.

Anderson, R.Y. and Koopmans, L.H., 1965. Harmonic analysis of varve time series. *J. Geophys. Res* 68: 877–893.

Anderson, T.W., 1958. *Introduction to Multivariate Statistical Analysis.* Wiley, New York, N.Y., 37 pp.

Anderson, T.W., and Goodman, L.A., 1957. Statistical inference about Markov chains. *Ann. Math Stat.*, 28: 89–110.

Antevs, E., 1925. Retreat of the last ice-sheet in eastern Canada. *Geol. Surv. Can., Mem.*, 146: 142 pp

Arnold, K.J., 1941. *On Spherical Probability Distributions.* Unpubl. thesis, Mass. Inst. Tech., Cam bridge, Mass., 43 pp.

Baer, A.J. (Editor), 1970. *Symposium on Basins and Geosynclines of the Canadian Shield. Geol. Surv Can., Pap.*, 70–40: 265 pp.

Baragar, W.R.A., 1966. Geochemistry of the Yellowknife volcanic belt. *Can. J. Earth Sci.*, 3: 9–30.

Baragar, W.R.A. and Goodwin, A.M., 1969. Andesites and Archean volcanism of the Canadian shield In: A.R. McBirney (Editor), *Proceedings of the Andesite Conference. Oreg. Dept. Geol. Mine Ind., Bull.*, 65: 121–142.

Barry, G.S. and Freyman, A.J., 1970. Mineral endowment of the Canadian Northwest – A subjective probability assessment. *Can. Min. Metall. Bull.*, 63: 1031–1042.

Barth, T.F.W., 1962. *Theoretical Petrology.* Wiley, New York, N.Y., 2nd ed., 416 pp.

Bartlett, M.S., 1946. On the theoretical specification and sampling properties of autocorrelated time series. *J. R. Statist. Soc.* (*Suppl.*), 8 (1): 27–41.

Bartlett, M.S., 1964. The spectral analysis of two-dimensional point processes. *Biometrika*, 51: 299–311.

Bartlett, M.S., 1966. *An Introduction to Stochastic Processes*, Cambridge Univ. Press, Cambridge, 2nd ed., 362 pp.

Batschelet, E., 1965. *Statistical Methods for the Analysis of Problems in Animal Orientation and Certain Biological Rhythms.* American Institute of Biological Sciences, Washington, D.C., 57 pp.

Beale, E.M.L., 1970. A note on procedures for variable selection in multiple regression. *Technometrics*, 12: 909–914.

Beale, E.M.L., Kendall, M.G. and Mann, D.W., 1967. The discarding of variables in multivariate analysis. *Biometrika*, 54: 375–366.

Becker, R.M., 1964. Some generalized distributions with special reference to the mineral industries. 1 Sampling to *n* items per sample. *U.S., Bur. Mines, Rep. Invest.*, 6329: 53 pp.

Berner, H., Ekström, T., Lillequist, R., Stephansson, O. and Wikström, A., 1971. Data storage and processing in geological mapping. 1: Field data sheet. *Geol. För. Stockh. Förh.*, 93: 85–101; 2 Data file. *Ibidem*, 93: 693–705.

Bhattacharyya, B.K., 1965. Two-dimensional harmonic analysis as a tool for magnetic interpretation *Geophysics*, 30: 829–857.

Bhattacharyya, B.K. and Holroyd, M.T., 1971. Numerical treatment and automatic mapping of two dimensional data in digital form. *Can. Inst. Min. Metall., Spec. Vol.*, 12: 148–158.

Billings, M.P., 1972. *Structural Geology.* Prentice-Hall, Englewood Cliffs, N.J., 3rd ed. 606 pp.

Blackman, R.B. and Tukey, J.W., 1958. *The Measurement of Power Spectra.* Dover, New York, N.Y 190 pp.

Blais, R.A. and Carlier, P.A., 1968. Application of geostatistics in ore valuation. *Can. Inst. Min. Metall. Spec. Vol.*, 9: 41–68.

Bliss, C.I. and Fisher, R.A., 1953. Fitting the negative binomial distribution to biological data and note on the efficient fitting of the negative binomial. *Biometrics,* 9: 176–200.

Blondel, F. and Ventura, E., 1954. Estimation de la valeur de la production minière mondiale en 1950. *Ann. Mines, Paris,* 10: 25–81.

Boole, G., 1854. *Laws of Thought.* Dover, New York, N.Y., 424 pp. (reprint).

Bonyun, D. and Stevens, G., 1971. A general purpose computer program to produce geological stereo net diagrams. In: J.L. Cutbill (Editor), *Data Processing in Biology and Geology.* Academic Press, London, pp. 165–188.

Bostick, N.H., 1970. Electronic data processing applied to uranium resource prediction and exploration. *Trans. Soc. Min. Eng. – A.I.M.E.,* 247: 4–10.

Botbol, J.M., 1971. Characteristic analysis as applied to mineral exploration. *Can. Inst. Min. Metall., Spec. Vol,,* 10: 92–99.

Boyle, R.W., 1968. The source of metals and gangue elements in epigenetic deposits. *Miner. Deposita,* 3: 174–177.

Brier, G.W., 1968. Long-range prediction of the zonal westerlies and some problems in data analysis. *Rev. Geophys.* 6: 525–551.

Briggs, L.I. and Briggs, D.Z., 1969. Stratigraphic analysis. In: D.F. Merriam (Editor), *Computer Applications in the Earth Sciences.* Plenum, New York, N.Y., pp. 13–39.

Briggs, L.I. and Pollack, H.N., 1967. Digital model of evaporite sedimentation. *Science,* 155: 453–456.

Brinck, J., 1971. MIMIC, The prediction of mineral resources and long-term price trends in the non-ferrous metal mining industry is no longer Utopian. *Eurospectra,* 10: 46–56.

Burk, Jr., C.F., 1968. Data in the earth sciences. In: E.R.W. Neale (Editor), *The Earth Sciences in Canada: A Centennial Appraisal and Forecast. R. Soc. Can., Spec. Publ.,* 11: 75–81.

Burr, I.W., 1942. Cumulative frequency functions. *Ann. Math. Statist.,* 13: 215–232.

Busk, H.G., 1929. *Earth Flexures.* Cambridge Univ. Press, Cambridge, 106 pp.

Cameron, E.M., 1968. A geochemical profile of the Swan Hills Reef. *Can. J. Earth Sci.,* 5: 287–310.

Cameron, E.M., Siddeley, G. and Durham, C.C., 1971. Distribution of ore elements in rocks for evaluating ore potential: Nickel, copper, cobalt, and sulfur in ultramafic rocks of the Canadian shield. *Can. Min. Metall., Spec. Vol.,* 11: 298–313.

Carr, D.D., Horowitz, A., Hrabar, S.V., Ridge, K.F., Rooney, R., Straw, W.T., Webb, W. and Potter, P.E., 1966. Stratigraphic sections, bedding sequences, and random processes. *Science,* 154: 1162–1164.

Carrs, B.W. and Neidell, N.S., 1966. A geological cyclicity detected by means of polarity coincidence correlation. *Nature,* 218 (5058): 136–137.

Cattell, R.B., 1965. Factor analysis: an introduction to essentials. *Biometrics,* 21: 190–215, 405–435.

Chamberlin, T.C., 1897 and 1931. The method of multiple working hypotheses. *J. Geol.,* 5: 837–848 and 39: 155–165.

Chamberlin, T.C., 1899. Lord Kelvin's address on the age of the earth as an abode fitted for life. *Science,* 9: 889–901; and 10: 11–18.

Chayes, F., 1956. *Petrographic Modal Analysis.* Wiley, New York, N.Y., 113 pp.

Chayes, F., 1960. On correlation between variables of constant sum. *J. Geophys, Res.,* 65: 4185–4193.

Chayes, F., 1964. A petrographic distinction between Cenozoic volcanics in and around the open oceans. *J. Geophys. Res.,* 69: 1573–1588.

Chayes, F., 1968. On locating field boundaries in simple phase diagrams by means of discriminant functions. *Am. Mineral.,* 53: 359–371.

Chayes, F., 1970. Effect of a single non-zero open covariance on the simple closure test. In: D.F. Merriam (Editor), *Geostatistics,* Plenum, New York, N.Y., pp. 11–22.

Chayes, F., 1971. *Ratio Correlation.* Univ. Chicago Press, Chicago, Ill., 99 pp.

Chayes F., 1972. *Rock Information System (Information and User's Guide).* Geophys. Lab., Carnegie Inst., Washington, D.C., Version III, 18+XVIII pp.

Chayes, F. and Kruskal, W., 1966. An approximate statistical test for correlations between proportions. *J. Geol.*, 74: 692–702.

Cochran, W.G., Mosteller, F. and Tukey, J.W., 1954. Principles of sampling. *J. Am. Statist. Assoc.*, 49: 13–35.

Cole, J.P. and King, C.A.M., 1968. *Quantitative Geography*. Wiley, London, 692 pp.

Collyer, P.L. and Merriam, D.F., in preparation. An application of cluster analysis in mineral exploration.

Conté, S.D., 1965. *Elementary Numerical Analysis*. McGraw-Hill, New York, N.Y., 278 pp.

Cooley, J.W., 1966. Applications of the Fast Fourier Transform method. *Proc. I.B.M. Sci. Comput. Symp., Thomas J. Watson Res. Center, Yorktown Heights, N.Y., 1966*, 32 pp.

Cooley, J.W. and Lohnes, P.R., 1962. *Multivariate Procedures for the Behaviorial Sciences*. Wiley, New York, N.Y., 211 pp.

Cooley, J.W. and Tukey, J.W., 1965. An algorithm for the machine calculation of complex Fourier series. *Math. Comput.*, 19: 297–307.

Courant, R. and Hilbert, D., 1962. *Methods of Mathematical Physics*. Interscience, New York, N.Y., Vol. I, 561 pp.; Vol. II, 830 pp.

Cox, D.R. and Miller, H.D., 1965. *The Theory of Stochastic Processes*. Wiley, New York, N.Y., 398 pp.

Cramér, H., 1947. *Mathematical Methods of Statistics*. Princeton Univ. Press, Princeton, N.J., 575 pp.

Cramér, H. and Leadbetter, M.R., 1968. *Stationary and Related Stochastic Processes*. Wiley, New York, N.Y., 348 pp.

Cruden, D.M. and Charlesworth, H.A.K., 1966. The Mississippian-Jurassic unconformity near Nordegg, Alberta. *Bull. Can. Pet. Geol.*, 14: 266–272.

Dalkey, N.C. and Rourke, D.L., 1971. Experimental assessment of Delphi procedures with group value judgement. *RAND Corp. Rep. R612ARPA*, Santa Monica, Calif., 49 pp.

Daly, R.A., 1968. *Igneous Rocks and the Depths of the Earth* (originally published 1933). Hafner, New York, N.Y., 598 pp.

D'Amico, C., 1964. Petrography and tectonics in the Agordo–Cereda region (Crystalline of the southern Alps). *Geol. Mijnbouw*, 43: 236–244.

David, M., 1969. The notion of "extension variance" and its application to the grade estimation of stratiform deposits. In: A. Weiss (Editor), *A Decade of Digital Computing in the Mineral Industry*. Am. Inst. Min. Metall., Petrol. Eng., New York, N.Y., pp. 63–81.

Davis, J.C., 1971. Optical processing of microporous fabrics. In: J.L. Cutbill (Editor), *Data Processing in Biology and Geology*. Academic Press, London, pp. 69–88.

Davis, J.C., 1973. *Statistics and Data Analysis in Geology*. Wiley, New York, N.Y., 550 pp.

De Geer, G., 1940. Geochronologia Suecica principles. *K. Sven. Vetenskapsakad. Handl., Ser. 3*, 18: 1–367.

DeGeoffroy, J.G. and Wignall, T.K., 1971. A probabilistic appraisal of mineral resources in a portion of the Grenville Province of the Canadian shield. *Econ. Geol.*, 66: 466–479.

DeGeoffroy, J.G. and Wu, S.M., 1970. A statistical study of ore occurrences in the greenstone belts of the Canadian shield. *Econ. Geol.*, 65: 496–509.

Deming, W.E., 1943. *Statistical Adjustment of Data*. Wiley, New York, N.Y., 261 pp.

Derry, D.R., 1968. Exploration. *Can. Min. Metall. Bull.*, 61: 200–205.

De Sitter, L.U., 1964. *Structural Geology*. McGraw-Hill, New York, N.Y., 2nd ed., 551 pp.

De Wijs, H.J., 1951. Statistics of ore distribution, I. *Geol. Mijnbouw*, 30: 365–375; Part II (1953), *Ibid.*, 32: 12–24.

Dieterich, J.H., 1970. Computer experiments on mechanics of finite amplitude folds. *Can. J. Earth Sci.*, 7: 467–476.

Dixon, C.J., 1970. Machine languages for representation of geological information. In: J.L. Cutbill (Editor), *Data Processing in Biology and Geology*. Academic Press, London, pp. 123–134.

Doeglas, D.J., 1946. Interpretation of the results of mechanical analysis. *J. Sediment Petrol.*, 16: 19–40.

Douglas, R.J.W. (Editor), 1970. *Geology and Economic Minerals of Canada. Econ. Geol. Rep. 1.* Geol. Surv. Can., Ottawa, 799 pp.

Dowds, J.P., 1969. Statistical geometry of petroleum reservoirs in exploration and exploitation. *J. Pet. Techn.*, 1969 (7): 841–852.

Draper, N.R. and Smith, H., 1966. *Applied Regression Analysis.* Wiley, New York, N.Y., 407 pp.

Duff, P.McL.D., Hallam, A. and Walton, E.K., 1967. *Cyclic Sedimentation.* Elsevier, Amsterdam, 280 pp.

Dugas, J., 1966. The relationship of mineralization to Precambrian stratigraphy in the Rouyn–Noranda area, Quebec. *Geol. Assoc. Can., Spec. Pap.*, 3: 43–55.

Durbin, J. and Watson, G.S., 1950. Testing for serial correlation in least-squares regression. *Biometrika*, 37: 409–428; Part II, *Ibidem*, 38 (1951): 159–178.

Efroymson, M.A., 1960. Multiple regression analysis. In: A. Ralston and H.S. Wilf (Editors), *Mathematical Methods for Digital Computers.* Wiley, New York, N.Y., pp. 191–203.

Eisenhart, C., 1963. The background and evolution of the method of least squares. *Proc. Sess. Int. Statist. Inst., 34th, Ottawa* (Preprint).

Elderton, W.P. and Johnson, N.L., 1969. *Systems of Frequency Curves.* Cambridge Univ. Press, Cambridge, 216 pp.

Elliott, D., 1965. The quantitative mapping of directional minor structures. *J. Geol.*, 73: 865–880.

Engel, J.H., 1957. Use of clustering in mineralogical and other surveys. *Proc. 1st Int. Conf. Oper. Res.*, Oper. Res. Soc. Am., Baltimore, pp. 176–192.

Fahrig, W.F., Gaucher, E.H. and Larochelle, A., 1965. Paleomagnetism of diabase dykes of the Canadian shield. *Can. J. Earth Sci.*, 2:278–298.

Fallot, P., 1955. Les dilemmes tectoniques des Alpes Orientales. *Ann. Soc. Géol. Belg.*, 78: 147–170.

Feller, W., 1968. *An Introduction to Probability Theory and its Applications.* Wiley, New York, N.Y., Vol. I, 3rd ed., 509 pp.; Vol. II (1966), 626 pp.

Findlay, D.C. and Smith, C.H., 1965. The Muskox Drilling Project. *Geol. Surv. Can., Pap.*, 64–44: 170 pp.

Finney, P.J., 1941. On the distribution of a variate whose logarithm is normally distributed. *J.R. Statist. Soc., Suppl.* 7: 151–161.

Fisher, R.A., 1922. On the mathematical foundations of theoretical statistics. *Philos. Trans. R. Soc. (Lond.), Ser. A.*, 222: 209–368.

Fisher, R.A., 1936. The use of multiple measurements in taxonomic problems. *Ann. Eugenics,* 7: 179–188. (Also *ibidem*, 8: 376–386.)

Fisher, R.A., 1953. Dispersion on a sphere. *Proc. R. Soc. Lond. Ser. A.*, 217: 295–305.

Fisher, R.A., 1954. Expansion of statistics. *Am. Sci.*, 42: 275–282.

Fisher, R.A., 1960. *The Design of Experiments.* Oliver and Boyd, Edinburgh, 248 pp.

Forgotson Jr., J.M., 1963. How computers help find oil. *Oil Gas J.*, 61: 100–109.

Fox, W.T., 1967. FORTRAN IV program for vector trend analyses of directional data. *Kans, Geol. Surv., Comput. Contrib.*, 11, 36 pp.

Francis, J.G.F., 1961. The *QR*-transformation. *Comput. J.*, 4: 265–271; Part II, *Ibidem*, 4 (1962): 332–345.

Frazer, R.A., Duncan, W.J. and Collar, A.R., 1960. *Elementary Matrices.* Cambridge University Press, Cambridge, 416 pp.

Fröberg, C.E., 1969. *Introduction to Numerical Analysis.* Addison-Wesley, Reading, Mass., 2nd ed., 433 pp.

Garrett, R.G. and Nichol, I., 1969. Factor analysis as an aid in the interpretation of regional stream sediment data. *Q. Colo. Sch. Mines,* 64 (1): 245–264.

Gentleman, W.M. and Sande, G., 1966. Fast Fourier Transform – for fun and profit. *Am. Fed. Inform. Soc., Conf. Proc.*, 29: 563–578.

GEOCOM Bulletin, 1968–70. International abstracts and current information on mathematical geoscience and computer technology. Geosystems, London.

Gibb, R.A., Van Boeckel, J.J.G.M. and Hornal, R.W., 1969. A preliminary analysis of the gravity anomaly field in the Timmins–Senneterre mining area. *Gravity Map Ser. No. 58.* Dept. Energy, Mines Res., Dom. Obs., Ottawa, 25 pp. (with map).

Giger, H., 1968. Zufallsmoirés. *Opt. Acta,* 15: 511–519.

Giger, H. and Erkan, Y., 1968. Zur Bestimmung der fundamentalen Dichten in einem Festkörpergefüge. *Z. Wiss. Mikrosk. Mikrosk. Tech.,* 69: 36–48.

Gingerich, P.D., 1969. Markov analysis of cyclic alluvial sediments. *J. Sediment. Petrol.,* 39: 330–332.

Gnedenko, B.V., 1963. *The Theory of Probability.* Chelsea, New York, N.Y., 471 pp.

Goldberger, A.S., 1964. *Econometric Theory.* Wiley, New York, N.Y., 399 pp.

Good, D.I., 1964. FORTRAN II trend-surface program for the IBM 1620. *Kans. Geol. Surv. Spec. Dist. Publ.,* 14: 54 pp.

Goodman, N.R., 1957. On the joint estimation of the spectra, cospectrum and quadrature spectrum of a two-dimensional stationary Gaussian process. *O.T.S.-U.S. Dep. of Commerce Publ.,* 162–074: 168 pp.

Goodwin, A.M. and Ridler, R.H., 1970. The Abitibi orogenic belt. *Geol. Surv. Can., Pap.,* 70–40: 1–30.

Gower, J.C., 1970. Classification and geology. *Rev. Int. Stat. Inst.,* 38: 35–41.

Grad, J. and Brebner, M.A., 1968. Eigenvalues and eigenvectors of a real general matrix (Algorithm 343). *Communications, Assoc. Computing Machinery,* 11 (12): 820–826.

Granger, C.W.J. and Hatanaka, M., 1964. *Spectral Analysis of Economic Time Series.* Princeton Univ. Press, Princeton, N.J., 299 pp.

Gray, H.L. and Schucany, W.R., 1972. *The Generalized Jackknife Statistic.* Dekker, New York, N.Y., 308 pp.

Grenander, U. and Rosenblatt, M., 1957. *Statistical Analysis of Stationary Time Series.* Wiley, New York, N.Y., 300 pp.

Griffiths, J.C., 1960. Frequency distributions in accessory mineral analysis. *J. Geol.,* 68: 353–365.

Griffiths, J.C., 1962. Statistical methods in sedimentary petrography. In: H.B. Milner (Editor), *Sedimentary Petrography, 1.* MacMillan, New York, N.Y., 4th ed., pp. 565–617.

Griffiths, J.C., 1966a. Future trends in geomathematics. *Bull. Miner. Ind. Exp. Stn., Pa. State Univ.,* 35 (5): 1–8.

Griffiths, J.C., 1966b. Exploration for natural resources. *Oper. Res.,* 14: 189–209.

Griffiths, J.C., 1966c. Genetic model for interpretive petrology. *J. Geol.,* 74: 655–672.

Griffiths, J.C., 1967. Mathematical exploration strategy and decision-making. *Proc. 7th World Pet. Congr., Mexico.* 2: 599–604.

Griffiths, J.C., 1970. Current trends in geomathematics. *Earth Sci. Rev.,* 6: 121–140.

Griffiths, J.C. and Drew, L.J., 1966. Grid spacing and success ratios in exploration for natural resources. *Miner. Ind. Exp. Stn, Pa. State Univ., Spec. Publ.,* 2–65: R1–R48.

Griffiths, J.C. and Singer, D.A., 1971a. Unit regional value of nonrenewable natural resources as a measure of potential for development of large regions. *Geol. Soc. Aust., Spec. Publ.,* 3: 227–238.

Griffiths, J.C. and Singer, D.A., 1971b. A first generation simulation model for selecting amongst exploration programs with special application to the search for uranium ore bodies. GEOCOM Programs 2. *GEOCOM Bull.,* 4 (11–12): 1–42.

Griffiths, J.C. and Singer, D.A., 1973. The Engel simulator and the search for uranium. *Proc. 10th Int. Symp. Appl. Comput. Miner. Ind.,* S. Afr. Inst. Min. Metall., Johannesburg, pp. 9–16.

Gross, G.A., 1965. Geology of iron deposits of Canada. *Geol. Surv. Can. Econ. Geol. Rep.,* 22 (1), 181 pp.

Grover, J.E. and Orville, P.M., 1969. The partitioning of cations between coexisting single- and multisite phases with application to the assemblages: orthopyroxene–clinopyroxene and ortho-pyroxene–olivine. *Geochim. Cosmochim. Acta,* 33: 205–226.

Gujarati, D., 1970. Use of dummy variables in testing for equality between sets of coefficients in linear regressions: A generalization. *Am. Stat.,* 24 (5): 18–21.

Gumbel, E.J., Greenwood, J.A. and Durand, D., 1953. The circular normal distribution: Theory and Tables. *J. Am. Stat. Assoc.*, 48: 131–152.

Hald, A., 1949. Maximum likelihood estimation of the parameters of a normal distribution which is truncated at a known point. *Skand. Aktuarietidskr.*, 1949: 119–134.

Hald, A., 1952. *Statistical Theory with Engineering Applications.* Wiley, New York, N.Y., 783 pp.

Hamaker, H.C., 1962. On multiple regression analysis. *Stat. Neerl.* 16: 31–56.

Harbaugh, J.W. (Chairman), 1965. Mathematics recommendations for undergraduate geology students. *J. Geol. Educ.*, 13: 91–92.

Harbaugh, J.W. and Bonham-Carter, G., 1970. *Computer Simulation in Geology: Modelling Dynamic Systems.* Wiley, New York, N.Y., 575 pp.

Harbaugh, J.W. and Merriam, D.F., 1968. *Computer Applications in Stratigraphic Analysis.* Wiley, New York, N.Y., 228 pp.

Harbaugh, J.W. and Preston, F.W., 1965. Fourier series analysis in geology. In: J.C. Dotson and W.C.J. Peters (Editors), *Short Course and Symposium on Computers and Computer Application in Mining and Exploration.* College of Mines, Univ. Arizona, Tucson, Ariz., pp. R1–R46.

Harman, H.H., 1960. *Modern Factor Analysis.* Univ. Chicago Press, Chicago, Ill., 471 pp.

Harris, D.P., 1965. *An Application of Multivariate Statistical Analysis to Mineral Exploration.* Thesis, Pennsylvania State Univ., University Park, Pa., 261 pp.

Harris, D.P., 1969. Alaska's base and precious metals resources: A probabilistic regional appraisal. *Q. Colo. Sch. Mines,* 64 (3): 295–327.

Harris, D.P. and Euresty, D.A., 1973. The impact of transportation network upon the potential supply of base and precious metals. *Proc. 10th Int. Symp. Appl. Comput. Methods Miner. Ind.*, S. Afr. Inst. Min. Metall., Johannesburg, pp. 99–108.

Harris, D.P., Freyman, A.J. and Barry, G.S., 1971. A mineral resource appraisal of the Canadian Northwest using subjective probabilities and geological opinion. *Can. Inst. Min. Metall., Spec. Vol.*, 12: 100–116.

Harrison, J.M., 1963. Nature and significance of geological maps. In: C.C. Albritton, Jr. (Editor). *The Fabric of Geology.* Addison-Wesley, Cambridge, Mass., pp. 225–232.

Haugh, I., Brisbin, W.C. and Turek, A., 1967. A computer-oriented field sheet for structural data. *Can. J. Earth Sci.*, 4: 657–662.

Hazen Jr., S.W., 1967. Some statistical techniques for analyzing mine and mineral-deposit sample and assay data. *U.S. Bur. Mines., Bull.*, 621: 223 pp.

Hazen Jr., S.W. and Berkenkotter, R.D., 1962. Experimental mine-sampling project designed for statistical analysis. *U.S. Bur. Mines, Rep. Invest.*, 5962: 111 pp.

Hazen Jr., S.W. and Meyer, W.L., 1966. Using probability models as a basis for making decisions during mineral deposit exploration. *U.S. Bur. Mines, Rep. Invest.*, 6778: 83 pp.

Heien, D.M., 1968. A note on log-linear regression. *J. Am. Stat. Assoc.*, 63: 1034–1038.

Heiskanen, W.A. and Vening Meinesz, F.A., 1958. *The Earth and its Gravity Field.* McGraw-Hill, New York, N.Y., 470 pp.

Hempkins. W.B., 1970. A FORTRAN IV program for two-dimensional autocorrelation analysis of geologic and remotely-sensed data. *Northwestern Univ. Rep.*, 21 (NGR–14–007–027): 54 pp.

Hess, H.H., 1952. Orthopyroxenes of the Bushveld type, ion substitutions and changes in unit cell dimensions. *Am. J. Sci. (Bowen Vol.)*, pp. 173–187.

Hoerl, A.E. and Kennard, R.W., 1970. Ridge regression: Biased estimation for nonorthogonal problems. *Technometrics,* 12: 55–67; Applications, *ibidem,* pp. 69–82.

Hollin, J.T., 1969. Ice-sheet surges and the geological record. *Can. J. Earth Sci.*, 6: 903–910.

Hopwood, T., 1968. Derivation of a coefficient of degree of preferred orientation from contoured fabric diagrams. *Geol. Soc. Am. Bull.*, 79: 1651–1654.

Horton, C.W., Hempkins, W.B. and Hoffman, A.A.J., 1964. A statistical analysis of some aeromagnetic maps from the northwestern Canadian Shield. *Geophysics,* 29: 584–601.

Hotz, P.E., 1953. Petrology of granophyre in diabase near Dillsburg, Pennsylvania. *Geol. Soc. Am. Bull.*, 64: 675–704.

Howard, J.C. (Editor), 1968. *Bibliography of Statistical Applications in Geology. Counc. Educ. Geol. Sci. Publ.*, 2, Am. Geol. Inst., Washington, D.C., 24 pp.

Hruška, J. and Burk Jr., C.F., 1971. Computer-based storage and retrieval of geoscience information: Bibliography 1946–69. *Geol. Surv. Can., Pap.*, 71–40: 52 pp. (336 references).

Hubaux, A., 1972. A new geological tool – the data. *Int. Union Geol. Sci., COGEODATA, Doc.*, 36: 39 pp.

Hubbert, M.K., 1937. Theory of scale models applied to the study of geologic structures. *Bull.Geol. Soc. Am.* 48: 1459–1520.

Hughes, O.L., 1965. Surficial geology of part of the Cochrane district, Ontario, Canada. In: H.E. Wright and D.G. Frey (Editors), *International Studies on the Quaternary. Geol. Soc. Am., Spec. Pap.*, 84: 71–102.

Huijbregts, C. and Matheron, G., 1971. Universal kriging (An optimal method for estimating and contouring in trend surface analysis). *Can. Inst. Min. Metall., Spec. Vol.*, 12: 159–169.

Hutchinson, R.W., Ridler, R.H. and Suffel, G.G., 1971. Metallogenic relationships in the Abitibi Belt, Canada: A model for Archean metallogeny. *Can. Min. Metall. Bull.*, 64: 48–57.

Imbrie, J., 1963. Factor and vector analysis program for analyzing geologic data. *Off. Nav. Res., Geogr. Branch, Tech. Rep.*, 6: 83 pp.

Imbrie, J. and Purdy, E.G., 1962. Classification of modern Bahamian carbonate sediments. *Am. Assoc. Pet. Geol., Mem.*, 1: 253–272.

Imbrie, J. and Van Andel, T.H., 1964. Vector analysis of heavy mineral data. *Geol. Soc. Am. Bull.*, 75: 1131–1156.

Irvine, T.N., 1965, 1967. Chromian spinel as a petrogenetic indicator. *Can. J. Earth Sci.*, Part 1, 2: 648–672; Part 2, 4: 71–103.

Irvine, T.N., 1970. Heat transfer during solidification of layered intrusions. *Can. J. Earth Sci.*, 7:1031–1061.

Irving, E., 1964. *Paleomagnetism and its Applications to Geological and Geophysical Problems*, Wiley, New York, N.Y., 399 pp.

Isnard, P., Mallet, J.L., Cazes, P. and Sattran, V., 1972. Corrélations géologiques. Méthodes statistiques de traitement des données. In: P. Laffitte (Editor), *Traité d'informatique géologique*. Masson, Paris, pp. 379–536.

Jacod, J. and Joathon, P., 1971. Use of random-genetic models in the study of sedimentary processes. *J. Int. Assoc. Math. Geol.*, 3: 265–279.

Jambor, J.L. and Smith, C.H., 1964. Olivine composition determination with small-diameter X-ray powder cameras. *Mineral. Mag.*, 33: 730–741.

James, W.R., 1966. The Fourier series model in map analysis. *Northwestern Univ., Dep. Geol. Tech. Rep. 1*, 37 pp.

James, W.R., 1967. Nonlinear models for trend analysis in geology. *Kans. Geol. Surv., Comput. Contrib.*, 12: 26–30.

Jenkins, G.M., 1961. General considerations in the analysis of spectra. *Technometrics*, 3: 133–166.

Jenkins, G.M. and Watts, D.G., 1968. *Spectral Analysis and its Applications.* Holden-Day, San Francisco, Calif., 525 pp.

Jizba, Z.V., 1959. Frequency distribution of elements in rocks. *Geochim. Cosmochim. Acta*, 16: 79–82.

Johnson, N.L., 1949. Systems of frequency curves generated by methods of translations. *Biometrika*, 36: 149–176.

Jones, T.A., 1968. Statistical analysis of orientation data. *J. Sediment. Petrol.*, 38: 61–67.

Jones, T.A. and James, W.R., 1969. Analysis of bimodal orientation data. *J. Int. Assoc. Math. Geol.*, 1: 129–136.

Jose, P.D., 1965. Sun's motion and sunspots. *Astron. J.*, 70: 193–200.

Jowett, G.H., 1955. Least-squares regression analysis for trend reduced time series. *J. R. Stat. Soc., Ser. B*, 17: 91–104.

Kaiser, H.F., 1958. The varimax criterion for analytic rotation in factor analysis. *Psychometrika*, 23: 187–200.

Kapteyn, J.C., 1903. *Skew Frequency Curves in Biology and Statistics.* Noordhoff, Groningen, 133 pp.

Kaufman, G.M., 1963. *Statistical Decision and Related Techniques in Oil and Gas Exploration.* Prentice-Hall, Englewood Cliffs, N.J., 307 pp.

Kelly, A.M. and Sheriff, W.J., 1969. A statistical examination of the metallic mineral resources of British Columbia. *Proc. Symp. Decision-making Miner. Ind., Univ. Br. Columbia, Vancouver,* pp. 221–243.

Kendall, M.G., 1965. *A Course in Multivariate Statistics.* Griffin, London, 2nd ed., 185 pp.

Kendall, M.G. and Stuart, A., 1958. *The Advanced Theory of Statistics.* Hafner, New York, N.Y., three-volume ed., Vol. 1, 433 pp.; Vol. 2, 676 pp.; Vol. 3, 552 pp.

Klovan, J.E. and Billings, G.K., 1967. Classification of geological samples by discriminant-function analysis. *Bull. Can. Pet. Geol.,* 3: 313–330.

Koch Jr., G.S. and Link, R.F., 1971. *Statistical Analysis of Geological Data.* Wiley, New York, N.Y., Vol. 1 (1970), 375 pp.; Vol. 2 (1971), 438 pp.

Kolmogorov, A., 1931. Ueber die analytischen Methoden in der Wahrscheinlichkeitsrechnung. *Math. Ann.,* 104: 415–458.

Kolmogorov, A.N., 1941. On the logarithmic normal distribution law for the sizes of splitting particles. *Dokl. Akad. Nauk S.S.S.R.,* 31: 99–100.

Kolmogorov, A.N., 1951. Solution of a problem in probability theory connected with the problem of the mechanism of stratification. *Am. Math. Soc. Trans.,* 53: 8 pp.

Koopmans, L.H., 1967. A comparison of coherence and correlation as measures of association for time or spacially indexed data. *Kans. Geol. Surv., Comput. Contrib.,* 18: 1–4.

Kossinna, E., 1933. Die Erdoberfläche. In: B. Gutenberg (Editor), *Handbuch der Geophysik, 2.* Borntraeger, Berlin, pp. 869–954.

Köster, E., 1964. *Granulometrische und Morphometrische Messmethoden an Mineralkörnern, Steinen und sonstigen Stoffen.* Enke, Stuttgart, 336 pp.

Kretz, R., 1961. Some applications of thermodynamics to coexisting minerals of variable composition. Examples; orthopyroxene-clinopyroxene and orthopyroxene-garnet. *J. Geol.,* 69: 361–387.

Kretz. R., 1963. Distribution of magnesium and iron between orthopyroxene and calcic pyroxene in natural mineral assemblages. *J. Geol.,* 71: 773–785.

Kretz. R., 1969. On the spatial distribution of crystals in rocks. *Lithos,* 2: 39–65.

Kretz, R., 1970. Variation in the composition of muscovite and albite in a pegmatite dike near Yellowknife. *Can. J. Earth Sci.,* 7: 1219–1235.

Krige, D.G., 1951. A statistical approach to some basic valuation problems on the Witwatersrand. *J.S. Afr. Inst. Min. Metall.,* 52: 119–139.

Krige, D.G., 1960. On the departure of ore value distributions from the lognormal model in South African gold mines. *J.S. Afr. Inst. Min. Metall.,* 61: 231–244.

Krige, D.G., 1962. Economic aspects of stoping through unpayable ore. *J.S. Afr. Inst. Min. Metall.,* 63: 364–374.

Krige, D.G., 1966a. Two-dimensional weighted moving average trend surfaces for ore valuation. *Proc. Symp. Math. Statist. Comput. Appl. Ore Valuation.* S. Afr. Inst. Min. Metall., Johannesburg, pp. 13–38.

Krige, D.G., 1966b. A study of gold and uranium distribution patterns in the Klerkdorp goldfield. *Geoexploration,* 4: 43–53.

Krige, D.G. and Ueckermann, H.J., 1963. Value contours and improved regression techniques for ore reserve valuations. *J. S. Afr. Inst. Min. Metall.,* 63: 429–452.

Krige, D.G., Watson, M.I., Oberholzer, W.J. and Du Toit, S.R., 1969. The use of contour surfaces as predictive models for ore values. *Proc. Symp. Comput. Appl. Oper. Res. Miner. Industry, 8th, Salt Lake City.* Am. Inst. Min. Metall., pp. 127–161.

Krumbein, W.C., 1936. Application of logarithmic moments to size frequency distributions of sediments. *J. Sediment. Petrol.*, 6: 35–47.

Krumbein, W.C., 1939. Preferred orientation of pebbles in sedimentary deposits. *J. Geol.*, 47: 673–706.

Krumbein, W.C., 1955. Experimental design in the earth sciences. *Trans. Am. Geophys. Union*, 36: 1–11.

Krumbein, W.C., 1963. Confidence intervals on low-order polynomial trend surfaces. *J. Geophys. Res.*, 68: 5869–5878.

Krumbein, W.C., 1966. A comparison of polynomial and Fourier models in map analysis. *Northwest. Univ., Dep. Geol., Tech. Rep. 2*: 45 pp.

Krumbein, W.C., 1967. FORTRAN IV computer programs for Markov chain experiments in geology. *Kans. Geol. Surv., Comp. Contrib.*, 13: 38 pp.

Krumbein, W.C., 1969. The computer in geological perspective. In: D.F. Merriam (Editor), *Computer Applications in the Earth Sciences.* Plenum, New York, N.Y., pp. 251–275.

Krumbein, W.C., 1970. Geological models in transition. In: D.F. Merriam (Editor), *Geostatistics.* Plenum, New York, N.Y., pp. 143–161.

Krumbein, W.C. and Dacey, M.F., 1969. Markov chains and embedded Markov chains in geology. *J. Int. Assoc. Math. Geol.*, 1: 79–96.

Krumbein, W.C. and Graybill, F.A., 1965. *An Introduction to Statistical Models in Geology.* McGraw-Hill, New York, N.Y., 475 pp.

Krumbein, W.C. and Jones, T.A., 1970. The influence of areal trends on correlations between sedimentary properties. *J. Sediment. Petrol.*, 40: 656–665.

Krumbein, W.C., Benson, B.T. and Hempkins, W.B., 1964. Whirlpool, a computer program for "sorting out" independent variables by sequential multiple linear regression. *Off. Nav. Res., Geogr. Branch, Tech. Rep. 14*: 49 pp.

Kuenen, P.H., 1951. Mechanics of varve formation and the action of turbidity currents. *Geol. Fören. Stockh. Förh.*, 73: 69–84.

Kullback, S., 1959. *Information Theory and Statistics.* Wiley, New York, N.Y., 395 pp.

Kuno, H., Yamasaki, K., Iida, C. and Nagashima, K., 1957. Differentiation of Hawaiian magmas. *Jap. J. Geol. Geogr.*, 28: 179–218.

Kutina, J., 1968. On the application of the principle of equidistances in the search for ore veins. *Proc. 23rd Int. Geol. Congr.*, 7: 99–110.

Kutina, J. and Fabbri, A.G., 1972. Relationship of structural lineaments and mineral occurrences in the Abitibi area of the Canadian shield. *Geol. Surv. Can., Pap.*, 71–9: 36 pp.

Lafeber, D., 1966. Soil structural concepts. *Eng. Geol.*, 1: 261–290.

Laffitte, P., 1969. La codification sémantique et informatique géologique. *Ann. Mines*, 1969: 75–83.

Lang, A.H., 1970. Base metals, Canadian shield. In: R.H.W. Douglas (Editor), *Geology and Economic Minerals of Canada. Geol. Surv. Can. Econ. Geol. Rep.*, 1: 185–200.

Latulippe, M., 1966. The relationship of mineralization to Precambrian stratigraphy in the Matagami and Val d'Or districts of Quebec. *Geol. Assoc. Can., Spec. Pap.*, 3: 21–42.

Lee, P.J., 1969. The theory and application of canonical trend surfaces. *J. Geol.*, 77: 303–318.

Lee, Y.W., 1960. *Statistical Theory of Communication.* Wiley, New York, N.Y., 251 pp.

Leet, L.D. and Judson, S., 1954. *Physical Geology.* Prentice-Hall, New York, N.Y., 466 pp.

Le Maitre, R.W., 1968. Chemical variation within and between volcanic rock series – a statistical approach. *J. Petrol.*, 9: 220–252.

Leonardi, P., 1967. *Le Dolomiti: Geologia dei Monti tra Isarco e Piave.* Cons. Nazion. Ric. e Giunta Prov., Trento, 1019 pp.

Leopold, L.B. and Langbein, W.B., 1966. River meanders. *Sci. Am.* 214 (6): 60–70.

Leymarie, P., 1970. Contribution aux méthodes d'acquisition, de représentation et de traitement de l'information en géologie. *Sci. Terre, Mém.*, 18: 170 pp.

Link, R.F., Koch Jr., G.S. and Gladfelder, G.W., 1964. Computer methods of fitting surfaces to assay and other data by regression analysis. *U.S., Bur. Mines Rep. Invest.*, 6508: 69 pp.

Link, W.K., 1954. Robot geology. *Am. Assoc. Pet. Geol., Bull.*, 38: 2411.
Lliboutry, L.A., 1969. Contribution à la théorie des ondes glaciaires. *Can. J. Earth Sci.*, 6: 943–954.
Loève, M., 1963. *Probability Theory.* Van Nostrand, Princeton, N.J., 3rd ed., 685 pp.
Loudon, T.V., 1964. Computer analysis of orientation data in structural geology. *Off. Nav. Res., Geogr. Branch, Tech. Rep.*, 13: 130 pp.
Loudon, T.V., 1971. Some geological data structures: arrays, networks, trees and forests. In: J.L. Cutbill (Editor), *Data Processing in Biology and Geology.* Academic Press, London, pp. 135–146.
Lowdon, J.A., Stockwell, C.H., Tipper, H.W. and Wanless, R.K., 1963. Age determinations and geological studies. *Geol. Surv. Can., Pap.*, 62–17: 140 pp.
Lyell, C., 1833. *Principles of Geology, 3.* Murray, London, 398 pp. (+ 109 pp. supplements).
MacGregor. I.D. and Smith, C.H., 1963. The use of chrome spinels in petrographic studies of ultramafic intrusions. *Can. Mineral.* 7: 403–412.
Mahalanobis, P.C., 1960. A method of fractile graphical analysis. *Econometrica,* 28: 325–351.
Malinvaux, E., 1966. *Statistical Methods of Econometrics.* Rand McNally, Chicago, Ill., 631 pp.
Matalas, N.C., 1963. Autocorrelation of rainfall and streamflow minimums. *U.S., Geol. Surv., Prof. Pap.*, 434–B (and D): 10 pp.
Matern, B., 1960. Spatial variation. *Medd. Skogsforskningsinst.*, 49: 144 pp.
Matheron, G., 1962. Traité de géostatistique appliquée. *Mém. Bur. Rech. Géol. Minières,* 14: 333 pp.
Matheron, G., 1963. Principles of geostatistics. *Econ. Geol.*, 58: 1246–1266.
Matheron, G., 1965. *Les Variables Régionalisées et leur Estimation.* Masson, Paris, 306 pp.
Matheron, G., 1967. Kriging or polynomial interpolation procedures? *Trans. Can. Inst. Min. Metall.*, 70: 240–244.
Matheron, G., 1969. Cours de géostatistique. *Cah. Cent. Morphol. Math.*, 4: 82 pp.
Matheron, G., 1970a. Random functions and their applications in geology. In: D.F. Merriam (Editor), *Geostatistics.* Plenum, New York, N.Y., pp. 79–88.
Matheron, G., 1970b. Structures aléatoires et géologie mathématique. *Rev. Int. Statist. Inst.*, 38: 1–11.
Matheron, G., 1971a. The theory of regionalized variables and its applications. *Cah. Cent. Morphol. Math.*, 5: 211 pp.
Matheron, G., 1971b. Random sets theory and its applications to stereology. *J. Microsc.*, 95: 15–23.
McCammon, R.B., 1969. Multivariate methods in geology. In: *Models of Geologic Processes.* Short course, Am. Geol. Inst., Washington, D.C., pp. RMA1–RMF6.
McCammon, R.B., 1970. Component estimation under uncertainty. In: D.F. Merriam (Editor), *Geostatistics.* Plenum, New York, N.Y., pp. 45–61.
McCammon, R.B., 1971. Environmental pattern reconstruction from sample data 1. Mississippi Delta region. *Dep. Geol. Sci., Univ. Ill, Chic. Circle, Tech. Rep.*, 71 (1): 102 pp.
McCrossan, R.G., 1969. An analysis of size frequency distribution of oil and gas reserves of Western Canada. *Can. J. Earth Sci.*, 6: 201–212.
McGee, B.A., 1969. The Canadian Index to geoscience data. In: *Report of Activities. Geol. Surv. Can., Pap.*, 69–1 (Part B), p. 49.
McGlynn, J.C., 1970. Superior Province. In: R.H.W. Douglas (Editor), *Geology and Economic Minerals of Canada. Geol. Surv. Can., Econ. Geol. Rep.*, 1: 54–84.
McIntyre, D.B., 1969. Supplementary notes on APL/360. In: *Models of geologic processes.* Short Course, Am. Geol. Inst., Washington, D.C., pp. DBM–B1–C4.
McKelvey, V.E., 1972. Mineral resource estimates and public policy. *Am. Sci.*, 60: 32–40.
Merriam, D.F., 1963. The geologic history of Kansas. *Kans. Geol. Surv., Bull.*, 162, 317 pp.
Merriam, D.F. (Editor), 1964. Symposium on cyclic sedimentation. *Geol. Surv. Kans. Bull.*, 169 (volumes I and II): 633 pp.
Merriam, D.F., 1966. Geologic use of the computer. *Wyo. Geol. Assoc. Ann. Conf., 20th, Proc.*, pp. 109–112.
Merriam, D.F. (Editor), 1969. *Computer Applications in the Earth Sciences.* Plenum, New York, N.Y., 281 pp.

Merriam, D.F., 1970. Comparison of British and American Carboniferous cyclic rock sequences. *J. Int. Assoc. Math. Geol.*, 2: 241–264.

Merriam, D.F., 1971. Computer applications in stratigraphic problem solving. *Can. Inst. Min. Metall., Spec. Vol.*, 12: 139–147.

Middleton, G.V., 1965. The Tukey chi-square test. *J. Geol.*, 73: 547–549.

Miesch, A.T., 1969. The constant sum problem in geochemistry. In: D.F. Merriam (Editor), *Computer Applications in the Earth Sciences.* Plenum, New York, N.Y., pp. 161–176.

Miesch, A.T. and Connor, J.J., 1968. Stepwise regression and nonpolynomial models in trend analysis. *Kans. Geol. Surv., Comput. Contrib.*, 27: 40 pp.

Miller, R.L. and Kahn, J.S., 1962. *Statistical Analysis in the Geological Sciences.* Wiley, New York, N.Y., 483 pp.

Moore, R.C., 1936. Stratigraphic classification of the Pennsylvanian rocks of Kansas. *Kans. Geol. Surv. Bull.*, 22: 256 pp.

Moorhouse, W.W., 1965. Stratigraphic position of sulphides in the Archean. *Trans. Can. Inst. Min. Metall.*, 58: 947–950.

Mosteller, F., 1971. The Jackknife. *Rev. Int. Stat. Inst.*, 39: 363–368.

Mueller, R.F., 1961. Analysis of relations of Mg, Fe and Mn in certain metamorphic minerals. *Geochim. Cosmochim. Acta*, 25: 267–296.

Murata, W.J. and Richter, D.H., 1966. Chemistry of the lavas of the 1959–1960 eruption of Kilauea Volcano, Hawaii. *U.S., Geol. Surv., Prof. Pap.*, 537–A (also, B and D): 26 pp.

Nafziger, R.H. and Muan, A., 1967. Equilibrium phase compositions and thermodynamic properties of olivines and pyroxenes in the system MgO–"FeO"–SiO_2. *Am. Mineral.*, 52: 1364–1385.

Neter, J. and Maynes, E.S., 1970. Correlation coefficient with 0,1 dependent variable. *J. Am. Stat. Assoc.*, 65: 501–509.

Nieuwenkamp, W., 1968. Natuurfilosofie en de geologie van Leopold von Buch. *K. Ned. Akad. Wet. Proc., Ser. B*, 71 (4): 267–278.

Nockolds, S.R. and Allen, R., 1953. The geochemistry of some igneous rock series. *Geochim. Cosmochim. Acta*, 4: 105–142.

Nolan, T.B., 1950. The search for new mining districts. *Econ. Geol.*, 45: 601–608.

Ondrick, C.W. and Griffiths, J.C., 1969. FORTRAN IV computer program for fitting observed count data to discrete distribution models of binomial, Poisson and negative binomial. *Kans. Geol. Surv., Comput. Contrib.*, 35: 20 pp.

Ostwald, W., 1910. *Grosse Männer.* Adakemische Verlagsgesellschaft, Leipzig., 424 pp.

Papoulis, A., 1968. *Systems and Transforms with Applications in Optics.* McGraw-Hill, New York, N.Y., 474 pp.

Parks, J.M., 1972. Cluster and factor analysis in classification and correlation of paleo-environmental data. *Proc. Int. Stat. Inst., 38th, Washington, D.C., 1971, I.S.I. Bull.*, 44: 539–551.

Parzen, E., 1962. *Stochastic Processes.* Holden-Day, San Francisco, Calif., 324 pp.

Parzen, E., 1967. Time series analysis for models of signal plus white noise. In: B. Harris (Editor), *Spectral Analysis of Time Series.* Wiley, New York, N.Y., pp. 233–257.

Pearson, E.S. and Hartley, H.O., 1958. *Biometrika Tables for Statisticians, 1.* Cambridge University Press. Cambridge, 240 pp.

Pearson, K., 1895. Contributions to the mathematical theory of evolution. *Philos. Trans. R. Soc. Lond, Ser. A.*, 186: 343–414.

Pelletier, B.R., 1958. Pocono paleocurrents in Pennsylvania and Maryland. *Bull. Geol. Soc. Am.*, 6[illegible]: 1033–1064.

Pettijohn, F.J., 1970. The Canadian shield – a status report, 1970. *Geol. Surv. Can., Pap.*, 70–4[illegible]: 239–255.

Phillips, F.C., 1972. *The Use of Stereographic Projection in Structural Geology.* Arnold, London, 3rd ed., 90 pp.

Pincus, H.J., 1956. Some vector and arithmetic operations on two-dimensional orientation variates with applications to geological data. *J. Geol.*, 64: 533–557.

Pincus, H.J. and Dobrin, M.B., 1966. Geological applications of optical data processing. *J. Geophys. Res.*, 71: 4861–4869.

Piper, D.J.W., Harland, W.B. and Cutbill, J.L., 1970. Recording of geological data in the field using forms for input to the IBM handwriting reader. In: J.L. Cutbill (Editor), *Data Processing in Biology and Geology*. Academic Press, London, pp. 17–38.

Poldervaart, A., 1964. Chemical definition of alkali basalts and tholeiites. *Geol. Soc. Am. Bull.*, 75: 229–232.

Pope, P.T., 1969. *On the Stepwise Construction of a Prediction Equation.* Thesis, Southern Methodist Univ., Dallas, Texas, 49 pp.

Potter, P.E. and Blakely, R.F., 1968. Random processes and lithologic transitions. *J. Geol.*, 76: 154–170.

Potter, P.E. and Olson, J.S., 1954. Variance components of cross-bedding direction in some basal Pennsylvanian sandstones of the eastern Interior basin: Geological application. *J. Geol.*, 62: 50–73.

Potter, P.E. and Pettijohn, F.J., 1963. *Paleocurrents and Basin Analysis.* Academic Press, New York, N.Y., 296 pp.

Preston, F.W. and Esler, J.E., 1967. A FORTRAN program to compute two-dimensional power spectra. *Geol. Surv. Kans., Comput. Contrib.*, 16: 23 pp.

Pretorius, D.A., 1968. Mineral exploration in Southern Africa: Problems and prognosis for the next twenty years. *Econ. Geol. Res. Unit, Witwatersrand Univ., Inf. Circ.*, 50: 15 pp.

Pugachev, V.S., 1965. *Theory of Random Functions and its Application to Control Problems.* Pergamon, Oxford, 833 pp.

Quenouille, M.H., 1952. *Associated Measurements.* Butterworth, London, 242 pp.

Quenouille, M.H., 1957. *The Analysis of Multiple Time-Series.* Griffin, London, 104 pp.

Ralston, A., 1965. *A First Course in Numerical Analysis.* McGraw-Hill, New York, N.Y., 578 pp.

Ralston, A. and Wilf, H.S. (Editors), 1960. *Mathematical Methods for Digital Computers.* Wiley, New York, N.Y., Vol. 1, 293 pp.; Vol. 2, 287 pp.

Ramberg, H., 1952. *The Origin of Metamorphic and Metasomatic Rocks.* Univ. Chicago Press, Chicago, Ill., 317 pp.

Ramberg, H., 1967. *Gravity, Deformation and the Earth's Crust.* Academic Press, New York, N.Y., 214 pp.

Ramberg, H. and DeVore, G., 1951. The distribution of Fe^{2+} and Mg^{2+} in coexisting olivines and pyroxenes. *J. Geol.*, 59: 193–210.

Ramsay, J.G., 1960. The deformation of early linear structures in areas of repeated folding. *J. Geol.*, 68: 75–93.

Ramsay, J.G., 1967. *Folding and Fracturing of Rocks.* McGraw-Hill, New York, N.Y., 568 pp.

Rao, C.R., 1952. *Advanced Statistical Methods in Biometric Research.* Wiley, New York, N.Y., 390 pp.

Rao, J.S. and Sengupta, S., 1970. An optimum hierarchical sampling procedure for cross-bedding data. *J. Geol.*, 78: 533–544.

Read, W.A., 1969. Analysis and simulation of Namurian sediments in Central Scotland using a Markov-process model. *J. Int. Assoc. Math. Geol.*, 1: 190–220.

Reiche, P., 1938. An analysis of cross-lamination of the Coconino sandstone. *J. Geol.*, 46: 905–932.

Richardson, W.A. and Sneesby, G., 1922. The frequency distribution of igneous rocks. *Mineral. Mag.*, 19: 303–313.

Roberts, M.C. and Mark, D.M., 1970. The use of trend surfaces in till fabric analysis. *Can. J. Earth Sci.*, 7: 1179–1183.

Robinson, J.E., 1970. Spatial filtering of geological data. *Rev. Int. Stat. Inst.*, 38: 21–34.

Robinson, J.E. and Merriam, D.F., 1972. Enhancement of patterns in geologic data by spatial filtering. *J. Geol.*, 80: 333–345.

Robinson, S.C., 1969. International aspects of geological data storage and retrieval. In: A. Weiss (Editor), *A Decade of Digital Computing in the Mineral Industry.* Am. Inst. Min. Metall., Pet. Eng., New York, N.Y., pp. 319–329.

Robinson, S.C., 1970. A review of data processing in the earth sciences in Canada. *J. Int. Assoc. Math. Geol.,* 2: 377–398.

Roddick, J.A. and Hutchison, W.W., 1972. A computer-based system for geological field data on the Coast Mountains project, British Columbia, Canada. *Int. Geol. Congr. 24th,* Sect. 16: 36–59.

Rogers, J.J.W. and Adams, J.A.S., 1963. Lognormality of thorium concentrations in the Conway granite. *Geochim. Cosmochim. Acta,* 27: 775–783.

Roscoe, S.M., 1965. Geochemical and isotopic studies, Noranda and Mattagami areas. *Trans. Can. Inst. Min. Metall.,* 68: 279–285.

Rosenblatt, M. (Editor), 1963. *Time Series Analysis.* Wiley, New York, N.Y., 497 pp.

Runcorn, S.K., 1965. Changes in the convection pattern in the earth's mantle and continental drift: evidence for a cold origin of the earth. *Philos. Trans. R. Soc. Lond., Ser. A,* 258: 228–251.

Sander, B., 1948. *Einführung in die Gefügekunde der geologischen Körper, 1.* Springer, Vienna, 215 pp.

Scheidegger, A.E., 1963. *Principles of Geodynamics.* Academic Press, New York, N.Y., 2nd ed., 362 pp.

Scheidegger, A.E., 1964. The tectonic stress and tectonic motion direction in Europe and western Asia as calculated from earthquake fault plane solutions. *Seismol. Soc. Am. Bull.,* 54: 1519–1528.

Scheidegger, A.E., 1965. On the statistics of the orientation of bedding planes, grain axes and similar sedimentological data. *U.S., Geol. Surv. Prof. Pap.,* 525–C: C164–C167.

Scheidegger, A.E. and Langbein, W.B., 1966. Probability concepts in geomorphology. *U.S., Geol. Surv., Prof. Pap.,* 500–C: 14 pp.

Scherer, W., 1968. *Applications of Markov Chains to Cyclical Sedimentation in the Oficina Formation, Eastern Venezuela.* Unpublished thesis, Northwestern Univ., Evanston, Ill., 93 pp.

Schove, D.J., 1971. Varve-teleconnection across the Baltic. *Geogr. Ann., Ser. A.,* 53: 214–234.

Schove, D.J., 1972. A varve teleconnection project. *Proc. VIIIe Congr. INQUA, Paris, 1969:* 928–935

Schwarzacher, W., 1967. Some experiments to simulate the Pennsylvanian rock sequence of Kansas. *Kans. Geol. Surv., Comput. Contrib.,* 18: 5–14.

Schwarzacher, W., 1969. The use of Markov chains in the study of sedimentary cycles. *J. Int. Assoc. Math. Geol.,* 1: 17–40.

Schwarzacher, W., 1972. The semi-Markov process as a general sedimentation model. In: D.F. Merriam (Editor), *Mathematical Models of Sedimentary Processes.* Plenum, New York, N.Y., pp. 247–268.

Sengupta, S. and Rao, J.S., 1966. Statistical analysis of cross-bedding azimuths from the Kamthi formation around Bheemaram Pranhita–Godavari Valley. *Indian. J. Stat., Ser. B.,* 28: 165–174.

Serra, J., 1966. Remarques sur une lame mince de minerai lorrain. *Bull. B.R.G.M.,* 6: 1–36.

Serra, J., 1971. Stereology and structural elements. *J. Microsc.,* 95: 93–103.

Shannon, C.E., 1948. A mathematical theory of communications. *Bell System Tech. J.,* 27: 379–423 and 623–656.

Shaw, D.M. and Bankier, J.D., 1954. Statistical methods applied to geochemistry. *Geochim. Cosmochim. Acta,* 5: 111–123.

Sichel, H.S., 1952. New methods in the statistical evaluation of mine sampling data. *Trans. Inst. Min. Metall.,* 61: 261–288.

Sichel, H.S., 1961. On the departure of ore value distributions from the lognormal model in South African gold mines. Contribution to discussion of paper by D.G. Krige (1960). *J.S. Afr. Inst. Min. Metall.,* 62: 333–338.

Sichel, H.S., 1966. The estimation of means and associated confidence limits for small samples from lognormal populations. *Proc. Symp. Math. Stat. Comput. Appl. Ore Valuation.* S. Afr. Inst. Min Metall., Johannesburg, pp. 106–122.

Sichel, H.S., 1973. Statistical valuation of diamondiferous deposits. *Proc. 10th Int. Symp. Appl. Comput. Miner. Ind.,* S. Afr. Inst. Min. Metall., Johannesburg, pp. 17–26.

Simpson, T., 1755. On the advantage of taking the mean of a number of observations in practical astronomy. *Philos. Trans. R. Soc. Lond.*, 46: 82 (p. 579 in Vol. 10, 1809 Edition, Baldwin, London).

Sinclair, A.J. and Woodsworth, G.L., 1970. Multiple regression as a method of estimating exploration potential in an area near Terrace, B.C. *Econ. Geol.*, 65: 998–1003.

Singer, D.A., 1972. *Multivariate Statistical Analysis of the Unit Regional Value of Mineral Resources.* Thesis, Penn. State Univ., Final Report U.S. Bur. Mines, 211 pp.

Slichter, L.B., 1960. The need of a new philosophy of prospecting. *Min. Eng.*, 12: 570–576.

Smith, C.H., 1962. Notes on the Muskox intrusion, Coppermine River area, District of Mackenzie. *Geol. Surv. Can., Pap.*, 61–25: 16 pp.

Smith, C.H. and Kapp, H.E., 1963. The Muskox intrusion, a recently discovered layered intrusion in the Coppermine area, Northwest Territories, Canada. *Mineral. Soc. Am., Spec. Pap.*, 1: 30–35.

Smith, C.H., Irvine, T.N. and Findlay, D.C., 1963. Legend of revised *Geol. Surv. Can., Prel. Map*, 36–1961.

Smith, F.G., 1966. *Geological Data Processing using FORTRAN IV.* Harper and Row, New York, N.Y., 284 pp.

Smith, F.G., 1972. A computer program to plot random or real proportions of three components of a mixture. *J. Int. Assoc. Math. Geol.*, 4: 263–268.

Sokal, R.R. and Sneath, P.H.A., 1963. *Principles of Numerical Taxonomy.* Freeman, San Francisco, Calif., 359 pp.

Sommerfeld, A., 1948. *Vorlesungen über theoretischen Physik, 1.* Dieterich, Wiesbaden, 276 pp.

Spector, A. and Grant, F.S., 1970. Statistical models for interpreting aeromagnetic data. *Geophysics*, 35: 293–302.

Spencer, D., 1966. Factors affecting element distributions in a Silurian graptolite band. *Chem. Geol.*, 1: 221–249.

Stauffer, M.R., 1966. An empirical-statistical study of three-dimensional fabric diagrams as used in structural analysis. *Can. J. Earth Sci.*, 3: 473–498.

Steinmetz, R., 1962. Analysis of vectorial data. *J. Sediment. Petrol.*, 32: 801–812.

Stephens, M.A., 1962. *The Statistics of Directions – the Fisher and Von Mises Distributions.* Unpubl. thesis, Univ. of Toronto, Toronto, Ont., 211 pp.

Stockwell, C.H., 1964. Fourth report on structural provinces, orogenic and time-classification of rocks of the Canadian Precambrian shield. *Geol. Surv. Can., Pap.*, 64–17: 1–21.

Sullivan, J., 1970. Relative discovery potential of the principal economic metals. *Bull. Can. Inst. Min. Metall.*, 63: 773–783.

Sutterlin, P.G., 1971. The design of computer-processible information systems. *Can. Inst. Min. Metall., Spec. Vol.*, 12: 399–403.

Switzer, P., 1965. A random set process in the plane with a Markovian property. *Ann. Math. Stat.*, 36: 1859–1863.

Switzer, P., 1970. Numerical classification. In: D.F. Merriam (Editor), *Geostatistics.* Plenum, New York, N.Y., pp. 31–44.

Szameitat, K. and Schäffer, K.A., 1963. Imperfect frames in statistics and the consequences for their use in sampling. *Bull., 34th Sess., Int. Stat. Inst., The Hague*, 1: 517–538.

Thöni, H., 1969. A table for estimating the mean of a lognormal distribution. *J. Am. Stat. Assoc.*, 64: 632–636.

Thornton, C.P. and Tuttle, O.F., 1960. Chemistry of igneous rocks, 1. Differentiation index. *Am. J. Sci.*, 258: 664–684.

Tomlinson, R.F. (Editor), 1972. *Geographical Data Handling (2 volumes).* Int. Geogr. Union, Commission on Geographical Data Sensing and Processing, Ottawa, approx. 1200 pp.

Trettin, H.P., 1971. Geology of Lower Paleozoic formations, Hazen Plateau and southern Grant Land Mountains, Ellesmere Island, Arctic Archipelago. *Geol. Surv. Can., Bull.*, 203: 134 pp.

Trettin, H.P. and Hills, L.V., 1966. Lower Triassic tar sands of northwestern Melville Island, Arctic Archipelago. *Geol. Surv. Can., Pap.*, 66–34: 122 pp.

Tribus, M., 1962. The use of the maximum entropy estimate in the estimation of reliability. In: R. Machol and P. Gray (Editors), *Recent Developments in Information and Decision Processes*. MacMillan, New York, N.Y., pp. 102–140.

Trooster, G., 1950. Fundamentele beschouwing van profielconstructies. *K. Ned. Akad. Wet., Proc., Ser. B.,* 53: 913–918.

Tukey, J.W., 1962. The future of data analysis. *Ann. Math. Stat.,* 33: 1–67.

Tukey, J.W., 1967. An introduction to the calculations of numerical spectrum analysis. In: B. Harris (Editor), *Spectral Analysis of Time Series.* Wiley, New York, N.Y., pp. 25–46.

Tukey. J.W., 1970. Some further inputs. In: D.F. Merriam (Editor), *Geostatistics.* Plenum, New York, N.Y., pp. 163–174.

Turner, F.J. and Weiss, L.E., 1963. *Structural Analysis of Metamorphic Tectonites.* McGraw-Hill, New York, N.Y., 545 pp.

Twenhofel, W.H., 1950. *Principles of Sedimentation,* McGraw-Hill, New York, N.Y., 2nd ed., 673 pp.

Uhler, R.S. and Bradley, P.G., 1970. A stochastic model for petroleum exploration over large regions. *J. Am. Stat. Assoc.,* 65: 623–630.

Umbgrove, J.H., 1947. *The Pulse of the Earth.* Nijhoff, The Hague, 2nd ed., 358 pp.

Van Bemmelen, R.W., 1954. *Mountain Building.* Nijhoff, The Hague, 177 pp.

Van Bemmelen, R.W., 1960. Zur Mechanik der ostalpinen Deckenbildung. *Geol. Rundsch.,* 50: 474–499.

Van Bemmelen, R.W., 1961. The scientific character of geology. *J. Geol.,* 69: 453–463.

Van Bemmelen, R.W., 1964. Phénomènes géodynamiques. *Mém. Soc. Belg. Géol. Paléontol. Hydrol.,* 8: 127 pp.

Van Bemmelen, R.W., 1966. The structural evolution of the Southern Alps. *Geol. Mijnbouw,* 45: 405–444.

Van Bemmelen, R.W., 1972. *Geodynamic models.* Elsevier, Amsterdam, 267 pp.

Van Landewijk, J.E.J.M., 1954. Construction of geological sections by graphic integration. *Geol. Mijnbouw,* 16: 321–325.

Vistelius, A.B., 1949. On the question of the mechanism of formation of strata. *Dokl. Akad. Nauk S.S.S.R.,* 65: 191–194.

Vistelius, A.B., 1960. The skew frequency distributions and the fundamental law of the geochemical processes. *J. Geol.,* 68: 1–22.

Vistelius, A.B., 1961. Sedimentation time trend functions and their application for correlation of sedimentary deposits. *J. Geol.,* 69: 703–728.

Vistelius, A.B., 1966a. Genesis of the Mount Belaya granite (an experiment in stochastic modelling). *Dokl. Akad. Nauk S.S.S.R.,* 167: 48–50.

Vistelius, A.B., 1966b. *Structural Diagrams.* Pergamon, Oxford, 178 pp. (translated from the Russian).

Vistelius, A.B., 1968. On the checking of the theoretical backgrounds of the stochastical models in concrete geological investigations. *Int. Geol. Congr., 23rd, Prague, 1968, Proc.,* 13: 153–161.

Vistelius, A.B. (Editor), 1969. *Mathematical Geology. Systematic Index of Reference.* Lab. Math. Geology, Leningrad, 246 pp.

Vistelius, A.B. and Janovskaya, T.B., 1967. The programming of geological and geochemical problems for all-purpose electronic computers. In: A.B. Vistelius, *Studies in Mathematical Geology,* Consultants Bureau, New York, N.Y., pp. 29–45 (translated from the Russian).

Von Buch, L., 1842. Ueber Granit und Gneiss, vorzüglich in Hinsicht der äusseren Form, mit welcher diese Gebirgsarten auf der Erdflache erscheinen. Read 15 December, 1842. Abh. K. Akad. Berlin, IV/2 (1884): 717–738.

Vyshemirskiy, V.S., Dmitriev, A.N. and Trofimnk, A.A., 1971. Criteria for prediction of giant oil pools. *World Petrol. Congr., Moscow, 1971, Spec. Pap.,* 8: 15 pp.

Wager, L.R., 1960. The major element variation of the layered series of the Skaergaard intrusion and a re-estimation of the average composition of the hidden layered series and of the successive residual magmas. *J. Petrol.,* 1: 364–398.

Wald, A., 1947. *Sequential Analysis.* Wiley, New York, N.Y., 212 pp.

Walker, A.M., 1960. Some consequences of superimposed error in time series analysis. *Biometrika,* 47: 33–43.

Walters, F.F., 1969. Contouring by machine: A user's guide. *Bull. Am. Assoc. Petrol.,* 53: 2324–2340.

Wampler, R.H., 1970. On the accuracy of least-squares computer programs. *J. Am. Stat. Assoc.,* 65: 549–565.

Wanless, R.K., Stevens, R.D., Lachance, G.R. and Rimsaite, J.Y.H., 1965. Age determinations and geological studies. 1, Isotopic ages (Rep. 5). *Geol. Surv. Can., Pap.,* 64–17: 126 pp.

Washington, H.S., 1917. Chemical analyses of igneous rocks. *U.S., Geol. Surv., Prof. Pap.,* 99: 1201 pp.

Watson, G.S., 1960. More significance tests on the sphere. *Biometrika,* 47: 87–91.

Watson, G.S., 1965. Equatorial distributions on a sphere. *Biometrika,* 52: 193–201.

Watson, G.S., 1966. The statistics of orientation data. *J. Geol.,* 74: 786–797.

Watson, G.S., 1967. Linear least-squares regression. *Ann. Math. Stat.,* 38: 1679–1699.

Watson, G.S., 1970. Orientation statistics in the earth sciences. *Acta Univ. Uppsaliensis, Upps.,* 2: 73 pp.

Watson, G.S., 1971. Trend surface analysis. *J. Int. Assoc. Math. Geol.,* 3: 215–226.

Watson, G.S. and Irving, E., 1957. Statistical methods in rock-magnetism. *Mon. Not. R. Astron. Soc., Geophys. Suppl.,* 7 (6): 289–300.

Watson, G.S. and Williams, E.J., 1956. On the construction of significance tests on the circle and sphere. *Biometrika,* 43: 344–352.

Watson, M.I., 1968. Methods and models in ore evaluation. *S. Afr. Chamber Mines, Res. Rep.* 55: 28 pp.

Weertman, J., 1969. Water lubrication mechanism of glacier surges. *Can. J. Earth Sci.,* 6: 929–942.

Wegmann, E., 1958. Das Erbe Werner's und Hutton's. *Geologie,* 7: 531–559.

Weller, J.M., 1964. Development of the concept and interpretation of cyclic sedimentation. *Kans. Geol. Surv., Bull.,* 169: 607–621.

Westergård, A.H., Johansson, S. and Sundius, N., 1943. Beskrivning till kartbladet Lidköping. *Sver. Geol. Unders., Ser. Aa,* No. 182: 197 pp.

Whitten, E.H.T., 1966a. *Structural Geology of Folded Rocks.* Rand McNally, Chicago, Ill., 663 pp.

Whitten, E.H.T., 1966b. The general linear equation in prediction of gold content in Witwatersrand rocks. *Proc. Symp. Math. Stat. Comput. Ore Valuation, Johannesburg, 1966,* pp. 124–156.

Whitten, E.H.T., 1970. Orthogonal polynomial trend surfaces for irregularly spaced data. *J. Int. Assoc. Math. Geol.,* 2: 141–152.

Whittle, P., 1954a. On stationary processes in the plane. *Biometrika,* 41: 434–449.

Whittle, P., 1954b. A statistical investigation of sunspot observations with special reference to H. Alfvén's sunspot model. *Astrophys. J.,* 119: 251–260.

Whittle, P., 1962. Topographic correlation, power-law covariance functions and diffusion. *Biometrika,* 49: 305–314.

Whittle, P., 1963a. *Prediction and Regulation.* Van Nostrand, Princeton, N.J., 147 pp.

Whittle, P., 1963b. Stochastic processes in several dimensions. *Bull. 34th Sess. Int. Stat. Inst., The Hague,* pp. 974–985.

Wilkinson, J.H., 1965. *The Algebraic Eigenvalue Problem.* Oxford University Press, Oxford, 662 pp.

Wilks, S.S., 1962. *Mathematical Statistics.* Wiley, New York, N.Y., 644 pp.

Wilson, A.T., 1969. The climatic effects of large-scale surges of ice sheets. *Can. J. Earth Sci.,* 6: 911–918.

Wilson, H.D.B., 1967. Volcanism and ore deposits in the Canadian Archaean. *Proc. Geol. Assoc. Can.,* 18: 11–31.

Wood, W.H. and Wood, R.M., 1966. Arithmetic means of circular data. *J. Sediment. Petrol.,* 36: 50–56.

Working, H. and Hotelling, H., 1929. The application of the theory of error to the interpretation of trends. *J. Am. Stat. Assoc.,* 24: 73–89.

Wright, C.W., 1958. Order and disorder in nature. *Geol. Assoc. Can., Proc.,* 69: 77–82.

Wright, H.E., 1957. Stone orientation in Wadena drumlin field, Minnesota, *Geogr. Ann. Ser. A.*, 39: 19–31.

Wynne-Edwards, H.R. and Hason, Z.U., 1970. Intersecting orogenic belts across the North Atlantic. *Am. J. Sci.*, 268: 289–308.

Wynne-Edwards, H.R., Laurin, A.F., Sharma, K.N.M., Nandi, A., Kehlenbeck, M.M. and Franconi, A., 1970. Computerized geological mapping in the Grenville Province, Quebec. *Can. J. Earth Sci.*, 7: 1357–1373.

Yaglom, A.M., 1962. *An Introduction to the Theory of Stationary Functions.* Prentice-Hall, Englewood Cliffs, N.J., 235 pp.

Yoder, H.S. and Sahama, T.G., 1957. Olivine X-ray determinative curve. *Am. Mineral.*, 42: 475–491.

York, D., 1966. Least-squares fitting of a straight line. *Can. J. Phys.*, 44: 1079–1086.

Yule, G.U., 1927. On the method of investigating periodicities in disturbed series, with special reference to Wolfer's sunspot numbers. *Philos. Trans. R. Soc. Lond. Ser. A.*, 226: 267–298.

Zwart, H.J., 1969. Metamorphic facies series in the European orogenic belts. *Geol. Assoc. Can., Spec. Pap.*, 5: 7–16.

INDEX